LIGAND FIELD
ENERGY DIAGRAMS

LIGAND FIELD
ENERGY DIAGRAMS

E. König and S. Kremer

University of Erlangen-Nürnberg
Erlangen, West Germany

SPRINGER SCIENCE+BUSINESS MEDIA, LLC

Library of Congress Cataloging in Publication Data

König, E   1929-
   Ligand field energy diagrams.

   Bibliography: p.
   Includes index.
   1. Ligand field theory—Charts, diagrams, etc. I. Kremer, S., 1942-
joint author. II. Title.
QD475.K58   541'.2242   76-45670
ISBN 978-1-4757-1531-6   ISBN 978-1-4757-1529-3 (eBook)
DOI 10.1007/978-1-4757-1529-3

PREFACE

Twenty years ago Tanabe and Sugano published the first ligand field energy diagrams which are applicable to d^N electronic configurations. These diagrams are limited in scope in that they can be used only for octahedral symmetry and for a limited number of terms. The present volume is an attempt to fill the gap by providing a reasonable number of complete and accurate ligand field energy diagrams for d^N configurations in the most commonly encountered symmetries.

Despite their limited nature, the diagrams of Tanabe and Sugano were extensively used in the past in order to rationalize optical and luminescence spectra and to discuss various electronic properties of transition metal ions, their coordination compounds and solids. Moreover, Tanabe–Sugano diagrams have an established place in the theory of transition metal compounds and are included in most textbooks of inorganic and coordination chemistry. It is expected that the present diagrams will be found useful for a similar purpose.

In the last few years, ligand field theory and related models have somewhat depreciated in importance as a consequence of increasing efforts to apply general computer programs like IBMOL or POLYATOM. However, in the area of complex inorganic systems, self-consistent field (SCF), and *ab initio* methods capable of doing the same job as semiempirical theories are not yet in sight. As it slowly becomes apparent that it would be extremely difficult to classify the wealth of experimental data on the basis of *ab initio* calculations, some disillusionment takes place and the interest in the traditional methods of theoretical inorganic chemistry increases again.

The present situation is characterized by the large number of compounds of distorted or unusual geometries and of compounds in new or uncommon electronic states. In addition, x-ray structural, optical and UV spectroscopic, magnetic, and other physical data on these compounds have been collected over the last years by the combined efforts of preparative inorganic chemists and chemical physicists. In order to rationalize this large number of data, a new and complete set of ligand field energy diagrams is urgently needed which would provide the same type of information on ground and excited states of these compounds as Tanabe–Sugano diagrams did for the much more limited data which were available at the time of their first presentation in literature.

It is with these intentions that we set out to compute ligand field energy diagrams for the more important geometrical arrangements of d^N compounds. It soon became clear that calculating matrix elements by the established methods of ligand field theoretical textbooks and plotting diagrams by hand is a road to nowhere. That is to say, these methods are too cumbersome and costly to produce a definite result in a reasonable amount of time. Irreducible tensor operator methods were therefore chosen for the computation of matrix elements, and suitable plotter routines were developed in order to generate the diagrams from the computer output by using a digital plotter. Many chemists as well as other potential users of this volume will not be familiar with these methods. Therefore, in the first part of the volume, we provide the theoretical basis required for the underlying calculations. We do not intend to give a rigorous treatment of the theory but rather to provide a short but readable introduction to the specific methods used here. For further details, the reader is encouraged to turn to the references given in the text. The detailed expressions for all the quantities which directly enter the calculations like the various matrix elements, coupling coefficients, symmetry adaptation coefficients, coefficients of fractional parentage, etc., are also provided for convenience of the user.

The reader who is not interested in following the steps of the theoretician may turn directly to Chapter 8, where he will find a description of the diagrams and a guide to their use. For conversion of the parameters of this volume into ligand field parameters employed by other authors, the reader may consult the relations listed in Section 3.3. For application of the diagrams for studies in the angular overlap model, in several molecular orbital treatments, or in crystal field theory, relevant details are given in Chapter 9. It is important to realize that for the application of the diagrams, the assumption of a ligand field is not necessarily required. The observed interactions are equally well understood in terms of the corresponding λ antibonding effects.

We are most grateful to the Computing Center of this University, in particular to its director, Dr. F. Wolf, for assistance in the development of plotter routines and for special assignment of computer and plotter time. The interest of one of the authors (E.K.) in applications of ligand field theory was stimulated by the late H. L. Schläfer, and this volume may be understood as a tribute to his memory. Thanks are due to R. Schnakig for discussions and helpful comments as well as to H. A. Goodwin and E. Hoefs for careful reading of the manuscript. Finally, the authors owe a debt of gratitude to their wives, Gerti and Karin, for their patience during preparation of the manuscript.

E. König
S. Kremer

Erlangen, Germany

CONTENTS

PART I

PART I
THEORY AND METHODS

1

INTRODUCTION

There have been various attempts to extend the diagrams of Tanabe and Sugano[59] to cover additional geometries of some interest, particularly those of tetragonally and trigonally distorted octahedra. In general, these treatments are not suitable for our purpose, since only a specific problem was covered or incomplete diagrams were produced. A notable exception is the work of Perumareddi[43] on the d^2 and d^8 configurations. Unfortunately, the larger part of these diagrams is available only as a microfilm edition.

The calculations performed in the present work make extensive use of the tensor operator formalism. The method of irreducible tensor operators was introduced by Racah[46-49] in order to calculate efficiently matrix elements of the perturbation Hamiltonian for free atoms and ions. This procedure may be easily extended to the study of ions in ligand fields by adding the corresponding potential to the free ion Hamiltonian. The resulting Hamiltonian defines the weak-field coupling case which, in its traditional as well as in tensor algebraic form,[24,62] has been extensively applied in the past. The case of intermediate-field coupling is characterized by the sequence of perturbation terms according to Coulomb interaction, ligand field, and spin–orbit coupling. This coupling scheme is realistic in many crystals and complexes of iron series metals, although, particularly in its tensor algebraic formulation, it seems to have escaped the attention of most workers in the field.[30] The case of strong-field coupling is defined by the following succession of terms: ligand field, Coulomb interaction, and spin–orbit coupling. The tensor operator form of this coupling was studied by Griffith[11] in some detail. Explicit calculations remained scarce due to the complex algebra and the fact that the method is incompletely

formulated, thus being unsuitable for direct computation. Consequently, the present program made it necessary to develop the tensor operator algebra for the coupling cases of the intermediate and the strong field in a computer compatible form. In the strong-field coupling case, this was achieved, e.g., by employing the method of fractional parentage coefficients which was extended to non-simply reducible (non-SR) groups and to mixed configurations of arbitrary symmetry. It should be noted that complete calculations in all three coupling cases produce identical results if no additional approximations are introduced.

The plan of the volume is roughly as follows. In Chapter 2, irreducible tensor operators, their Kronecker and scalar products, the Wigner–Eckart theorem, and the coefficients of fractional parentage are introduced in spherical symmetry. The algebra of these quantities is well established since it is widely used in calculations on gaseous transition metal and rare earth atoms. In Chapter 3, irreducible tensor operators and their algebra are generalized to point groups of arbitrary symmetry. The concept of the ligand field is then introduced, and the coupling schemes of the weak field, intermediate field, and strong field are defined. The parameters employed in the diagrams are defined here in terms of matrix elements of the ligand field operator. Thus Dq is used in cubic (O_h, T_d); Dq, Ds, Dt in tetragonal (D_{4h}, D_4, C_{4v}); Dq, $D\sigma$, $D\tau$ in trigonal (D_{3d}, D_3, C_{3v}); and $D\mu$, $D\nu$ in cylindrical ($D_{\infty h}$, D_∞, $C_{\infty v}$) symmetry. The relationships between these quantities and the parameters introduced by other authors are given for easy conversion of data from the literature. No interpretation of the parameters is given in the present context, the reader being referred, in this respect, to several recent reviews.[4,10] Subsequently, in Chap-

1

ters 4, 5, and 6, the expressions required for the calculation of matrix elements of Coulomb repulsion, ligand field, and spin–orbit interactions are given in detail for the three coupling cases. It should be noted that the relevant algebra in the intermediate-field and strong-field coupling cases is presented here for the first time and in a consistent and computer compatible form. In particular, expressions are derived (in Chapter 6) on the basis of the symmetric group, for the calculation of fractional parentage coefficients in arbitrary non-SR groups and in mixed configurations. Chapter 7 follows with a description of the programs and the plotter routines developed for the present purpose. A description of the diagrams and an introduction to their application have been included as Chapter 8. Another section of some importance covers the relation between the ligand field approach and other theoretical methods. Thus it is shown, in Chapter 9, how the diagrams may be applied to crystal fields, to the angular overlap model, and to various molecular orbital treatments. All relevant conversion relations are listed in detail.

The second and major part of the volume consists of about 400 ligand field energy diagrams. The diagrams cover all d^N electron configurations of transition metal ions—except the trivial cases d^1 and d^9—in cubic (O_h, T_d), tetragonal (D_{4h}, D_4, C_{4v}), trigonal (D_{3d}, D_3, C_{3v}), and cylindrical ($D_{\infty h}$, D_∞, $C_{\infty v}$) symmetry. The energy levels are displayed both as functions of the cubic field parameter Dq and as functions of the axial field parameter Dt (in tetragonal symmetry) or $D\tau$ (in trigonal symmetry), for a representative choice of the ratios $K = Ds/Dt$ or $K = D\sigma/D\tau$. In cylindrical symmetry, the dependence on Dv is displayed for several values of $K = D\mu/Dv$. The size of the diagrams is sufficient to enable an accurate comparison with experimental data. It should be realized that due to the relation between the parameters of the ligand field and the angular overlap model, as discussed in Section 9.2 below, ligand field interactions may be understood in terms of λ antibonding effects and *vice versa*. The presentation of the results in terms of ligand field parameters has been chosen for convenience rather than necessity, and may be abandoned whenever possible by reference to the relations listed in Chapter 9.

2

TENSOR OPERATOR ALGEBRA IN FREE ATOMS AND IONS

2.1. THE DEFINITION OF IRREDUCIBLE TENSOR OPERATORS

When considering atomic problems it is usually convenient to operate in spherical polar coordinates rather than in Cartesian coordinates. The spherical harmonics $Y_{jm} = |jm\rangle$ then arise in a natural way as the simultaneous eigenfunctions of the Hamiltonian and the operators $\mathbf{J}^2$ and J_z of the square of the total angular momentum and of its z component. This fact is intimately related to the transformation properties of the $|jm\rangle$ eigenvectors under rotations. The rotation through the infinitesimal angle ε about the z axis of the physical system may be described by the operator[6,8]

$$R_z(\varepsilon) = 1 - i\varepsilon J_z \qquad (2.1)$$

The obvious generalization to a finite rotation about an arbitrary axis then follows as

$$R(\alpha\beta\gamma) = e^{-i\gamma J_{z''}} e^{-i\beta J_{y'}} e^{-i\alpha J_z} \qquad (2.2)$$

where α, β, and γ are the Eulerian angles. According to Eq. (2.2), the general rotation is thus decomposed into three successive rotations, the first being by the angle α about the original z axis, followed by a rotation by the angle β about the intermediate y' axis, and finally by a rotation by the angle γ about the new z'' axis. The rotations are positive if they would advance a right-handed screw in the positive direction along the axis. The angular momentum eigenvectors $|jm\rangle$ transform under a rotation according to

$$R(\alpha\beta\gamma)|jm\rangle = \sum_{m'=-j}^{j} |jm'\rangle D^{(j)}_{m'm}(\alpha\beta\gamma) \qquad (2.3)$$

Since the set of matrices $\mathbf{D}^{(j)}(\alpha\beta\gamma)$ with elements

$$D^{(j)}_{m'm}(\alpha\beta\gamma) = \langle jm'|R(\alpha\beta\gamma)|jm\rangle \qquad (2.4)$$

forms the irreducible representation $\mathbf{D}^{(j)}$ of the three-dimensional rotation group $SO(3) \equiv R_3$, the corresponding basis functions are just the $2j+1$ kets $|jm\rangle$.

A spherical *tensor operator* $\mathbf{T}^{(k)}$ of rank k may now be defined by the set of its $2k+1$ components $\mathbf{T}^{(k)}_q$ ($q = -k, \ldots, k$), which transform under rotations of the system in the same way as the spherical harmonics of order k:[39]

$$R\mathbf{T}^{(k)}_q R^{-1} = \sum_p \mathbf{T}^{(k)}_p D^{(k)}_{pq}(\alpha\beta\gamma) \qquad (2.5)$$

Here, it should be recalled that if the basis functions are changed on transformation by R according to $\psi \to R\psi$, an operator T is changed according to $T \to RTR^{-1}$. The tensor operator $\mathbf{T}^{(k)}$ is called irreducible, if representation $\mathbf{D}^{(k)}$ is irreducible.

An alternate definition is based on the commutation relations with respect to the angular momentum operators

$$[(J_x \pm iJ_y), \mathbf{T}^{(k)}_q] = [(k \mp q)(k \pm q + 1)]^{1/2}\mathbf{T}^{(k)}_{q\pm1}$$
$$[J_z, \mathbf{T}^{(k)}_q] = q\mathbf{T}^{(k)}_q \qquad (2.6)$$

This is in fact the definition given by Racah.[47] It may be easily shown that the definitions of Eqs. (2.5) and (2.6) are equivalent.

If $k = 0$ the tensor operator $\mathbf{T}^{(k)}_q$ is simply a scalar

$$\mathbf{T}^{(0)}_0 = T \qquad (2.7)$$

whereas for $k = 1$ we obtain

$$\mathbf{T}_{\pm 1}^{(1)} = \mp \frac{1}{\sqrt{2}} (\mathbf{T}_x \pm i \mathbf{T}_y) \qquad \mathbf{T}_0^{(1)} = \mathbf{T}_z \qquad (2.8)$$

It follows that $\mathbf{J}$ is a tensor operator of rank 1 with components $\pm (1/\sqrt{2}) J_{\mp}$ and J_z. Also spherical harmonics may be considered as tensor operators since they transform in the same way as the $\mathbf{T}_q^{(k)}$. Especially useful for this purpose is their rationalized form introduced by Racah:[47]

$$\mathbf{C}_q^{(k)} = \left(\frac{4\pi}{2k+1} \right)^{1/2} Y_{kq} \qquad (2.9)$$

With the quantities $\mathbf{C}_q^{(k)}$ the spherical harmonic addition theorem[7] assumes the particularly simple form

$$P_k(\cos \omega_{ij}) = \frac{4\pi}{2k+1} \sum_q Y_{kq}^*(\theta_i, \varphi_i) Y_{kq}(\theta_j, \varphi_j)$$

$$= \sum_q (-1)^q (\mathbf{C}_{-q}^{(k)})_i (\mathbf{C}_q^{(k)})_j$$

$$= (\mathbf{C}_i^{(k)} \cdot \mathbf{C}_j^{(k)}) \qquad (2.10)$$

where the definition for the scalar product of tensor operators, i.e., Eq. (2.45), has been employed.

We now consider matrix elements of $\mathbf{T}_q^{(k)}$ in the system of angular momentum eigenvectors $|jm\rangle$

$$\langle \alpha j m | \mathbf{T}_q^{(k)} | \alpha' j' m' \rangle$$

Here α refers to any additional quantum numbers which may be required to specify the states. The dependence of the matrix element on m, m', and q is then determined by the *Wigner–Eckart theorem* as[8]

$$\langle \alpha j m | \mathbf{T}_q^{(k)} | \alpha' j' m' \rangle$$

$$= (-1)^{j-m} \langle \alpha j \| \mathbf{T}^{(k)} \| \alpha' j' \rangle \begin{pmatrix} j & k & j' \\ -m & q & m' \end{pmatrix} \qquad (2.11)$$

This is the basic relation within the algebra of tensor operators. The quantity written with double rules around the tensor operator is independent of m, m', and q and is called a *reduced matrix element*. The last factor on the right-hand side of Eq. (2.11) is a 3-j symbol which is related to the Wigner coefficients according to[6,8]

$$\begin{pmatrix} j_1 & j_2 & j_3 \\ m_1 & m_2 & m_3 \end{pmatrix}$$

$$= (-1)^{j_1 - j_2 - m_3} [j_3]^{-1/2} \langle j_3 - m_3 | j_1 m_1 j_2 m_2 \rangle \qquad (2.12)$$

Here, and in what follows, we use $[j]$ to denote $2j + 1$. The 3-j symbols have important symmetry properties facilitating their use. Thus they are left invariant under an even permutation of the columns, while an odd permutation multiplies it by $(-1)^{j_1 + j_2 + j_3}$ as does a

reversal of signs of the indices m_1, m_2, and m_3. The 3-j symbol is zero unless $m_1 + m_2 + m_3 = 0$ and the triangular conditions

$$j_1 + j_2 - j_3 \geq 0 \qquad j_1 - j_2 + j_3 \geq 0$$

$$-j_1 + j_2 + j_3 \geq 0 \qquad (2.13)$$

are satisfied. From these conditions it follows that the matrix elements of Eq. (2.11) are nonzero only if

$$j + j' \geq k \geq |j - j'| \qquad \text{and} \qquad m - m' = q \qquad (2.14)$$

For the reduced matrix elements of the adjoint tensor operator $\mathbf{T}^{(k)\dagger}$ it may be shown that

$$\langle \alpha j \| \mathbf{T}^{(k)\dagger} \| \alpha' j' \rangle = (-1)^{j-j'} \langle \alpha' j' \| \mathbf{T}^{(k)} \| \alpha j \rangle^* \qquad (2.15)$$

If $\mathbf{T}^{(k)}$ is self-adjoint (Hermitian), the relation between the components

$$\mathbf{T}_q^{(k)\dagger} = (-1)^q \mathbf{T}_{-q}^{(k)} \qquad (2.16)$$

follows the conventions of the usual definition of spherical harmonics. In this case Eq. (2.15) assumes the form

$$\langle \alpha j \| \mathbf{T}^{(k)} \| \alpha' j' \rangle = (-1)^{j-j'} \langle \alpha' j' \| \mathbf{T}^{(k)} \| \alpha j \rangle^* \qquad (2.17)$$

and thus the reduced matrix is not Hermitian.

Sometimes the value of a specific reduced matrix element is required. In this case the following procedure is usually adopted: First, the simplest (from the point of view of calculation) matrix element $\langle \alpha j m | \mathbf{T}_q^{(k)} | \alpha' j' m' \rangle$ is selected, and subsequently this is compared with the general expression Eq. (2.11). Let us consider, for example, the reduced matrix element of the angular momentum, $\langle \alpha j \| \mathbf{J}^{(1)} \| \alpha' j' \rangle$. Noting that the eigenvalue of $\mathbf{J}_z = \mathbf{J}_0^{(1)}$ equals m, we obtain

$$\langle \alpha j m | \mathbf{J}_z | \alpha' j' m' \rangle = m \delta(\alpha, \alpha') \delta(j, j') \delta(m, m') \qquad (2.18)$$

Introducing Eq. (2.18) into (2.11) for $m = m' = \frac{1}{2}$ and putting $q = 0$ gives

$$\langle \alpha j \tfrac{1}{2} | \mathbf{J}_0^{(1)} | \alpha' j' \tfrac{1}{2} \rangle = \tfrac{1}{2} \delta(\alpha, a') \delta(j, j')$$

$$= (-1)^{j-1/2} \begin{pmatrix} j & 1 & j \\ -\tfrac{1}{2} & 0 & \tfrac{1}{2} \end{pmatrix} \langle \alpha j \| \mathbf{J}^{(1)} \| \alpha' j' \rangle \qquad (2.19)$$

Since[52]

$$\begin{pmatrix} j & 1 & j \\ -m & 0 & m \end{pmatrix} = (-1)^{j-m} m [j(j+1)(2j+1)]^{-1/2} \qquad (2.20)$$

we finally obtain the result

$$\langle \alpha j \| \mathbf{J}^{(1)} \| \alpha' j' \rangle$$

$$= [j(j+1)(2j+1)]^{1/2} \delta(\alpha, \alpha') \delta(j, j') \qquad (2.21)$$

In the particular cases of orbital and spin angular momentum, Eq. (2.21) assumes the forms

$$\langle l\|\mathbf{l}\|l'\rangle = [l(l+1)(2l+1)]^{1/2}\delta(l,l') \qquad (2.22)$$

and

$$\langle s\|\mathbf{s}\|s'\rangle = [s(s+1)(2s+1)]^{1/2}\delta(s,s')$$
$$= (\tfrac{3}{2})^{1/2}\delta(s,s') \qquad (2.23)$$

The important reduced matrix element of the spherical harmonic tensor operator $\langle l\|\mathbf{C}^{(k)}\|l'\rangle$ may be obtained by applying Eq. (2.11) to $Y_q^{(k)}$ thus:

$$\langle lm|Y_q^{(k)}|l'm'\rangle$$
$$= (-1)^{l-m}\langle l\|\mathbf{Y}^{(k)}\|l'\rangle \begin{pmatrix} l & k & l' \\ -m & q & m' \end{pmatrix} \qquad (2.24)$$

On the other hand, the triple integral over spherical harmonics is

$$\int Y_{lm}^* Y_{kq} Y_{l'm'} \sin\theta\, d\theta\, d\varphi$$
$$= (-1)^m \int Y_{l-m} Y_{kq} Y_{l'm'} \sin\theta\, d\theta\, d\varphi$$
$$= (-1)^m \left(\frac{(2l+1)(2k+1)(2l'+1)}{4\pi}\right)^{1/2}$$
$$\times \begin{pmatrix} l & k & l' \\ 0 & 0 & 0 \end{pmatrix}\begin{pmatrix} l & k & l' \\ -m & q & m' \end{pmatrix} \qquad (2.25)$$

Comparing Eqs. (2.24) and (2.25) gives for $l+k+l' = 2g$, where g an integer,

$$\langle l\|\mathbf{Y}^{(k)}\|l'\rangle$$
$$= (-1)^l \left(\frac{(2l+1)(2k+1)(2l'+1)}{4\pi}\right)^{1/2}$$
$$\times \begin{pmatrix} l & k & l' \\ 0 & 0 & 0 \end{pmatrix} \qquad (2.26)$$

and, consequently,

$$\langle l\|\mathbf{C}^{(k)}\|l'\rangle$$
$$= (-1)^l [(2l+1)(2l'+1)]^{1/2}\begin{pmatrix} l & k & l' \\ 0 & 0 & 0 \end{pmatrix} \qquad (2.27)$$

If $l+k+l' \neq 2g$, it is $\langle l\|\mathbf{C}^{(k)}\|l'\rangle = 0$. The numerical values for various l, k, and l' have been listed, e.g., in Table 16 of Sobelman.[56]

Of particular interest are the *unit tensor operators* $\mathbf{u}^{(k)}$ of Racah,[47] which play an important role in the treatment of the N-electron problem. The $\mathbf{u}^{(k)}$ are defined by the condition

$$\langle \alpha j\|\mathbf{u}^{(k)}\|\alpha'j'\rangle = \delta(\alpha,\alpha')\delta(j,j') \qquad (2.28)$$

i.e., their reduced matrix elements are unity between states of the same α and j, and zero otherwise. The matrix elements therefore result as

$$\langle \alpha jm|u_q^{(k)}|\alpha jm'\rangle = (-1)^{j-m}\begin{pmatrix} j & k & j \\ -m & q & m' \end{pmatrix} \qquad (2.29)$$

From the Wigner–Eckart theorem it then follows that the matrix elements of a tensor operator $\mathbf{T}^{(k)}$ between the same states are proportional to those of $\mathbf{u}^{(k)}$, the constant of proportionality being the reduced matrix element of $\mathbf{T}^{(k)}$

$$\langle \alpha jm|T_q^{(k)}|\alpha jm'\rangle$$
$$= \langle \alpha j\|\mathbf{T}^{(k)}\|\alpha j\rangle\langle \alpha jm|u_q^{(k)}|\alpha jm'\rangle \qquad (2.30)$$

It should be noted that, for a given basis characterized by a certain value of j, the operators $u_q^{(k)}$ for $k>2j$ are identically zero, since the corresponding coupling coefficients vanish.

It may be shown that the operators $u_q^{(k)}$ are expressible in terms of the angular momentum operators J_+, J_-, J_z, and $\mathbf{J}^2$. On the other hand, if Eq. (2.30) is applied to the tensor operators $Y_q^{(k)}$, the reduced matrix elements are given by Eq. (2.26). However, the $Y_q^{(k)}$ have nonzero matrix elements between states with different j values satisfying the triangular condition, while the $u_q^{(k)}$ do not, by definition. The relation

$$\mathbf{Y}^{(k)} = \mathbf{u}^{(k)}\langle j\|\mathbf{Y}^{(k)}\|j\rangle \qquad (2.31)$$

is thus valid only within subspaces of fixed j values. In addition, assuming integral j values, the $\langle j\|\mathbf{Y}^{(k)}\|j\rangle$ vanish for k odd and thus the $u_q^{(k)}$ cannot always be expressed in terms of the $\mathbf{Y}^{(k)}$.

Finally, the $\mathbf{U}^{(k)}$ are N-electron tensor operators corresponding to the unit tensor operators $\mathbf{u}^{(k)}$ introduced above:

$$\mathbf{U}^{(k)} = \sum_i^N \mathbf{u}^{(k)}(i) \qquad (2.32)$$

the matrix elements of the $\mathbf{U}^{(k)}$ being given by

$$\langle \alpha JM|U_q^{(k)}|\alpha'J'M'\rangle$$
$$= (-1)^{J-M}\langle \alpha J\|\mathbf{U}^{(k)}\|\alpha'J'\rangle\begin{pmatrix} J & k & J' \\ -M & q & M' \end{pmatrix} \qquad (2.33)$$

The detailed calculation of the reduced matrix elements $\langle \alpha J\|\mathbf{U}^{(k)}\|\alpha'J'\rangle$ requires coefficients of fractional parentage and will be therefore treated in Section 2.4 below.

2.2. KRONECKER AND SCALAR PRODUCTS OF TENSOR OPERATORS

We consider the properties of a tensor operator $\mathbf{X}_Q^{(K)}$ which satisfies the commutation relations of Eq.

(2.6) and is itself the Kronecker (direct, tensor) product of the tensor operators $\mathbf{T}^{(k_1)}$ and $\mathbf{U}^{(k_2)}$ such that

$$\mathbf{X}_Q^{(K)} = \{\mathbf{T}^{(k_1)} \otimes \mathbf{U}^{(k_2)}\}_Q^{(K)} \quad (2.34)$$

Here, $K = k_1 + k_2, \ldots, k_1 - k_2$, $\mathbf{X}^{(K)}$ is irreducible, and, by complete analogy with the coupling of angular momenta, the components $\mathbf{X}_Q^{(K)}$ are determined by

$$\mathbf{X}_Q^{(K)} = \sum_{q_1,q_2} \mathbf{T}_{q_1}^{(k_1)} \mathbf{U}_{q_2}^{(k_2)} \langle k_1 q_1 k_2 q_2 | KQ \rangle \quad (2.35)$$

If the two operators $\mathbf{T}^{(k_1)}$ and $\mathbf{U}^{(k_2)}$ act on the same coordinates, the expression for the matrix elements of $\mathbf{X}_Q^{(K)}$ obtains as[24,64]

$$\langle \alpha j m | \mathbf{X}_Q^{(K)} | \alpha' j' m' \rangle$$

$$= \sum_{\substack{q_1 q_2 \\ \alpha'' j'' m''}} \langle \alpha j m | \mathbf{T}_{q_1}^{(k_1)} | \alpha'' j'' m'' \rangle \langle \alpha'' j'' m'' | \mathbf{U}_{q_2}^{(k_2)} | \alpha' j' m' \rangle$$

$$\times \langle k_1 q_1 k_2 q_2 | KQ \rangle \quad (2.36)$$

Application of Eq. (2.11) then gives for the reduced matrix element of the Kronecker product

$$\langle \alpha j \| \mathbf{X}^{(K)} \| \alpha' j' \rangle$$

$$= (-1)^{j+K+j'} [K]^{1/2} \sum_{\alpha'' j''} \langle \alpha j \| \mathbf{T}^{(k_1)} \| \alpha'' j'' \rangle$$

$$\times \langle \alpha'' j'' \| \mathbf{U}^{(k_2)} \| \alpha' j' \rangle \begin{Bmatrix} j & K & j' \\ k_2 & j'' & k_1 \end{Bmatrix} \quad (2.37)$$

Here, the sum over three Wigner coefficients has been replaced by a 6-j symbol.

Besides the 3-j symbols introduced in Eq. (2.12), 6-j and 9-j symbols are frequently encountered in tensor algebra. The 6-j symbol is associated with the coupling of three angular momenta. Thus the coupling of j_1, j_2, and j_3 to produce states with a definite value of the total angular momentum J may be carried out in two different ways. One may first couple j_1 and j_2 to states characterized by an intermediate j_{12} and then couple these to j_3 to give final states with definite total J. Alternatively, one may couple j_1 to an intermediate j_{23} resulting from the coupling of j_2 and j_3. The two sets of functions obtained according to these two schemes are linearly related according to

$$|j_1, (j_2 j_3) j_{23}; J\rangle$$

$$= \sum_{j_{12}} \langle (j_1 j_2) j_{12}, j_3; J | j_1, (j_2 j_3) j_{23}; J \rangle$$

$$\times |(j_1 j_2) j_{12}, j_3; J\rangle \quad (2.38)$$

The 6-j symbol is then defined by[8,52]

$$\begin{Bmatrix} j_1 & j_2 & j_{12} \\ j_3 & J & j_{23} \end{Bmatrix} = (-1)^{j_1+j_2+j_3+J} ([j_{12}][j_{23}])^{-1/2}$$

$$\times \langle (j_1 j_2) j_{12}, j_3; J | j_1, (j_2 j_3) j_{23}; J \rangle \quad (2.39)$$

The 6-j symbol is invariant with respect to any permutation of columns or the interchange of the upper and lower arguments of any two columns. It is nonzero only for those sets of values of the arguments for which the four triads $(j_1 j_{23} J)$, $(j_{12} j_3 J)$, $(j_1 j_2 j_{12})$, and $(j_2 j_3 j_{23})$ satisfy the triangular conditions of Eq. (2.13).

The 9-j symbol is related to the transformation coefficients arising in the coupling of four angular momenta and is defined by[8,52]

$$\begin{Bmatrix} j_1 & j_2 & j_{12} \\ j_3 & j_4 & j_{34} \\ j_{13} & j_{24} & J \end{Bmatrix} = ([j_{12}][j_{34}][j_{13}][j_{24}])^{-1/2}$$

$$\times \langle (j_1 j_2) j_{12}, (j_3 j_4) j_{34}; J | (j_1 j_3) j_{13}, (j_2 j_4) j_{24}; J \rangle$$

$$= \sum_t (-1)^{2t} [t] \begin{Bmatrix} j_1 & j_2 & j_{12} \\ j_{34} & J & t \end{Bmatrix}$$

$$\times \begin{Bmatrix} j_3 & j_4 & j_{34} \\ j_2 & t & j_{24} \end{Bmatrix} \begin{Bmatrix} j_{13} & j_{24} & J \\ t & j_1 & j_3 \end{Bmatrix} \quad (2.40)$$

Obviously, the 9-j symbol may be expressed as a summation over products of three 6-j symbols or as a summation over products of six 3-j symbols. The 9-j symbol is invariant with respect to an even permutation of rows or columns, whereas an odd permutation introduces a factor $(-1)^r$, where r is the sum of its nine arguments.

We now examine the properties of the Kronecker product Eq. (2.35) when the tensor operators $\mathbf{T}^{(k_1)}$ and $\mathbf{U}^{(k_2)}$ operate on different parts of the system. Thus $\mathbf{T}^{(k_1)}$ may act on the coordinates of particle 1 and $\mathbf{U}^{(k_2)}$ on the coordinates of particle 2, or $\mathbf{T}^{(k_1)}$ may operate alone on the spin coordinates and $\mathbf{U}^{(k_2)}$ on the orbital coordinates of one particle. If j_1, j_2, and jm refer to the parts 1, 2, and the total system, respectively, we obtain in this case

$$\langle \alpha_1 j_1 \alpha_2 j_2 j m | \mathbf{X}_Q^{(K)} | \alpha_1' j_1' \alpha_2' j_2' j' m' \rangle$$

$$= \sum_{\substack{q_1 q_2 \\ m_1 m_2 m_1' m_2'}} \langle k_1 q_1 k_2 q_2 | KQ \rangle \langle j m | j_1 m_1 j_2 m_2 \rangle$$

$$\times \langle j_1' m_1' j_2' m_2' | j' m' \rangle \langle \alpha_1 j_1 m_1 | \mathbf{T}_{q_1}^{(k_1)} | \alpha_1' j_1' m_1' \rangle$$

$$\times \langle \alpha_2 j_2 m_2 | \mathbf{U}_{q_2}^{(k_2)} | \alpha_2' j_2' m_2' \rangle \quad (2.41)$$

By applying Eq. (2.11) the reduced matrix element of the Kronecker product may be written in terms of the reduced matrix elements of $\mathbf{T}^{(k_1)}$ and $\mathbf{U}^{(k_2)}$ as[64]

$$\langle \alpha_1 j_1 \alpha_2 j_2 j \| \mathbf{X}^{(K)} \| \alpha_1' j_1' \alpha_2' j_2' j' \rangle$$

$$= ([K][j][j'])^{1/2} \langle \alpha_1 j_1 \| \mathbf{T}^{(k_1)} \| \alpha_1' j_1' \rangle \langle \alpha_2 j_2 \| \mathbf{U}^{(k_2)} \| \alpha_2' j_2' \rangle$$

$$\times \begin{Bmatrix} j_1 & j_1' & k_1 \\ j_2 & j_2' & k_2 \\ j & j' & K \end{Bmatrix} \quad (2.42)$$

From Eq. (2.42) we may obtain the reduced matrix element of an operator $\mathbf{T}^{(k)}$ acting only on part 1 in the $\alpha_1 j_1 j_2 j$ scheme. To this end we put $\mathbf{U}^{(k_2)} = 1$, $k_2 = 0$, and $k_1 = k$ and find

$$\langle \alpha_1 j_1 j_2 j \| \mathbf{T}^{(k)} \| \alpha'_1 j'_1 j'_2 j' \rangle$$

$$= (-1)^{j_1 + j_2 + j' + k} ([j][j'])^{1/2} \delta(j_2, j'_2)$$

$$\times \langle \alpha_1 j_1 \| \mathbf{T}^{(k)} \| \alpha'_1 j'_1 \rangle \begin{Bmatrix} j & j' & k \\ j'_1 & j_1 & j_2 \end{Bmatrix} \quad (2.43)$$

Similarly, for the operator $\mathbf{U}^{(k)}$ acting only on part 2, we may show

$$\langle j_1 \alpha_2 j_2 j \| \mathbf{U}^{(k)} \| j'_1 \alpha'_2 j'_2 j' \rangle$$

$$= (-1)^{j_1 + j'_2 + j + k} ([j][j'])^{1/2} \delta(j_1, j'_1) \langle \alpha_2 j_2 \| \mathbf{U}^{(k)} \| \alpha'_2 j'_2 \rangle$$

$$\times \begin{Bmatrix} j & j' & k \\ j'_2 & j_2 & j_1 \end{Bmatrix} \quad (2.44)$$

The scalar product of the tensor operators $\mathbf{T}^{(k)}$ and $\mathbf{U}^{(k)}$ of the same rank is traditionally defined by

$$(\mathbf{T}^{(k)} \cdot \mathbf{U}^{(k)}) = \sum_q (-1)^q \mathbf{T}_q^{(k)} \mathbf{U}_{-q}^{(k)} \quad (2.45)$$

where k is integral. This expression may be compared with Eq. (2.35), where the rank of the product $K = 0$ only if $k_1 = k_2 = k$, thus producing

$$\{\mathbf{T}^{(k)} \otimes \mathbf{U}^{(k)}\}_0^{(0)} = \sum_q \mathbf{T}_q^{(k)} \mathbf{U}_{-q}^{(k)} \langle kqk - q | 0 0 \rangle \quad (2.46)$$

Consequently, we obtain

$$(\mathbf{T}^{(k)} \cdot \mathbf{U}^{(k)}) = (-1)^k [k]^{1/2} \{\mathbf{T}^{(k)} \otimes \mathbf{U}^{(k)}\}_0^{(0)} \quad (2.47)$$

Apart from the difference arising in Eq. (2.47), the definition Eq. (2.45) has the advantage that for $k = 1$ it coincides with the usual definition of the scalar product of two vectors.

The matrix elements of the scalar product in the αjm scheme may be found from Eqs. (2.36) and (2.37) by putting $K = 0$, $k_1 = k_2 = k$, and using Eq. (2.47) to give

$$\langle \alpha jm | (\mathbf{T}^{(k)} \cdot \mathbf{U}^{(k)}) | \alpha' j' m' \rangle$$

$$= \delta(j, j') \delta(m, m') [j]^{-1/2} \sum_{\alpha'' j''} (-1)^{j - j''} \langle \alpha j \| \mathbf{T}^{(k)} \| \alpha'' j'' \rangle$$

$$\times \langle \alpha'' j'' \| \mathbf{U}^{(k)} \| \alpha' j' \rangle \quad (2.48)$$

Similarly, the matrix elements of the scalar product in the $\alpha_1 j_1 \alpha_2 j_2 jm$ scheme may be obtained from Eqs. (2.41) and (2.42) as[64]

$$\langle \alpha_1 j_1 \alpha_2 j_2 jm | (\mathbf{T}^{(k)} \cdot \mathbf{U}^{(k)}) | \alpha'_1 j'_1 \alpha'_2 j'_2 j' m' \rangle$$

$$= (-1)^{j_1 + j'_2 + j} \delta(j, j') \delta(m, m') \begin{Bmatrix} j_1 & j'_1 & k \\ j'_2 & j_2 & j \end{Bmatrix}$$

$$\times \langle \alpha_1 j_1 \| \mathbf{T}^{(k)} \| \alpha'_1 j'_1 \rangle \langle \alpha_2 j_2 \| \mathbf{U}^{(k)} \| \alpha'_2 j'_2 \rangle \quad (2.49)$$

where the 6-j symbol results from simplification of a 9-j symbol of Eq. (2.42).

2.3. FRACTIONAL PARENTAGE COEFFICIENTS AND THE MATRIX ELEMENTS OF N-ELECTRON SYSTEMS

The eigenfunction for a system of N electrons must be written such that it is antisymmetric with respect to the exchange of coordinates of any two electrons. For $N \geq 3$ *equivalent* electrons, this may be achieved by making use of the concept of *fractional parentage*.[24,48] Although, in general, it is not possible to write the eigenfunction as the simple product of an antisymmetric function for j^{N-1} times a function for the Nth electron, it is possible to express it as a sum of such products:

$$|j^N \alpha JM\rangle = \sum_{\bar{\alpha}\bar{J}} |j^{N-1}(\bar{\alpha}\bar{J})j; JM\rangle$$

$$\times \langle j^{N-1}(\bar{\alpha}\bar{J}), j | \} j^N \alpha J \rangle \quad (2.50)$$

Here, the summation extends over all states $|j^{N-1}(\bar{\alpha}\bar{J})\rangle$ of the complete set for the j^{N-1} configuration. The barred states $|\bar{\alpha}\bar{J}\rangle$ of the j^{N-1} configuration are orthonormal and fully antisymmetric in the first $N-1$ electrons and are known as the *parents* of the state αJ of the j^N configuration. The coefficients of the expansion (2.50), $\langle j^{N-1}(\bar{\alpha}\bar{J}), j | \} j^N \alpha J \rangle$, sometimes abbreviated to $\langle \bar{\psi} | \} \psi \rangle$, are known as the *coefficients of fractional parentage* (cfp). The cfp are normalized and follow the relation

$$\sum_{\bar{\alpha}\bar{J}} \langle j^N \alpha J \{ | j^{N-1}(\bar{\alpha}\bar{J}), j \rangle \langle j^{N-1}(\bar{\alpha}\bar{J}), j | \} j^N \alpha' J' \rangle$$

$$= \delta(\alpha, \alpha') \delta(J, J') \quad (2.51)$$

Since the cfp are chosen in such a way as to produce the properly antisymmetrized eigenfunctions for j^N, they vanish for all forbidden states of the configuration. Accordingly, the cfp satisfy the system of equations

$$\sum_{\alpha' J'} \langle J'' j^2 (J'''); J | J'' j(J')j; J \rangle \langle j^{N-2}(\alpha'' J''), j | \} j^{N-1} \alpha' J' \rangle$$

$$\times \langle j^{N-1}(\alpha' J'), j | \} j^N \alpha J \rangle$$

$$= 0 \quad \text{for } (J''' - 2j) \text{ even} \quad (2.52)$$

Provided the cfp of the configuration j^{N-1} are known, Eq. (2.52) may be employed to calculate the cfp for j^N. In the case of Russell–Saunders coupling, Nielson and Koster[40] have calculated and tabulated the cfp for all p^N, d^N, and f^N configurations up to the half-filled shell. The cfp for the more than half-filled shell are rarely needed, since the matrix elements of the l^{4l+2-N} configuration may always be obtained from those of l^N. An alternative way for the calculation of cfp will be discussed in Section 6.1.

The system of equations (2.52) does not fix the phases of the eigenfunctions of different terms. For the f^N configuration in Russell–Saunders coupling, the cfp may be factorized into three parts characterized by the quantum numbers W, U, and vSL.[24,62] The partitions of N into three and two integers according to

$$W \equiv (w_1, w_2, w_3) \qquad 2 \geq w_1 \geq w_2 \geq w_3 \geq 0$$
$$U \equiv (u_1, u_2) \qquad 2 \geq u_1 \geq u_2 \geq 0 \qquad (2.53)$$

correspond to irreducible representations of the groups R_7 and G_2, respectively. Here, $R_7 \equiv SO(7)$ is the seven-dimensional rotation group in the space of the basis functions for f-electrons, and G_2 the exceptional group, the relation being that of the group chain

$$R_7 \supset G_2 \supset R_3 \qquad (2.54)$$

Also SL has the usual significance with respect to the group R_3, whereas some doubly occurring SL states are separated by the additional label v. Although the phases are partly fixed by this classification, some arbitrary choice has to be made (cf. Table 7-2 of Judd[24]).

In the d^N configuration, terms possessing identical SL labels may be separated according to irreducible representations of the group R_5 of rotations in the space of basis functions for d electrons. The characterization employs partitions of N into two integers according to

$$W \equiv (w_1, w_2) \qquad 2 \geq w_1 \geq w_2 \geq 0 \qquad (2.55)$$

An alternative scheme which is consistent with that of Eq. (2.55) uses the *seniority number* v.[48,49] In this case the states are classified according to eigenvalues of the operator

$$Q = \sum_{i<j}^{N} q_{ij} \qquad (2.56)$$

where q_{ij} is the scalar operator defined by

$$\langle l^2 LM | q_{ij} | l^2 LM \rangle = (2l+1)\delta(L, 0) \qquad (2.57)$$

It may be shown[48] that the eigenvalues of operator Q for the configuration l^N are related to those for l^{N-2} according to

$$Q(l^N \alpha SL) = Q(l^{N-2} \alpha SL) + 2l + 3 - N \qquad (2.58)$$

Accordingly, for every term of l^N with a nonvanishing Q, a state of the same symmetry type arises in l^{N-2}, and this series may be continued until, for the configuration l^v, $Q(l^v \alpha SL) = 0$. It follows that to each state

a seniority number v may be assigned corresponding to that value of N for which the particular term first appears. Then Eq. (2.58) may be written as a function of N and v only:

$$Q(N, v) = \tfrac{1}{4}(N-v)(4l+4-N-v) \qquad (2.59)$$

If the seniority number v is introduced into Eq. (2.52) in place of α, this system of equations needs to be solved alone for the configurations l^v. The cfp for the remaining configurations then follow from recursion formulas listed by Racah.[48] It should be noted that the phases of the cfp are fixed by these relations, except for a general phase factor.

In many of the systems of interest there are more than two equivalent electrons. In these cases, methods based on the concept of fractional parentage may be used to advantage. Two simple types of operators will be considered. The one-electron operator (type **F** operator) is defined by

$$\mathbf{F} = \sum_i^N \mathbf{f}_i \qquad (2.60)$$

where $\mathbf{f}_i$ is a single-electron operator which acts only on the coordinates of the ith electron. The two-electron operator (type **G** operator) is defined by

$$\mathbf{G} = \sum_{i>j}^N \mathbf{g}_{ij} \qquad (2.61)$$

where $\mathbf{g}_{ij} = \mathbf{g}_{ji}$ operates only on the coordinates of electrons i and j, and $i > j$ denotes a sum over all pairs of electrons.

If we restrict our attention to configurations of equivalent electrons, the matrix element of operator **F** between states $|\psi\rangle$ and $|\psi'\rangle$ of the l^N configuration will be

$$\langle l^N \psi | \mathbf{F} | l^N \psi' \rangle = N \langle \psi | \mathbf{f}_N | \psi' \rangle \qquad (2.62)$$

where $\mathbf{f}_N$ operates on the coordinates of the Nth electron only. Let us now assume that **F** is a tensor of rank k, $\mathbf{F}_k$. Application of Eqs. (2.44) and (2.50) gives for the reduced matrix element of $\mathbf{F}_k$

$$\langle j^N vJ \| \mathbf{F}_k \| j^N v'J' \rangle = N([J][J'])^{1/2} \langle j \| \mathbf{f}_k(N) \| j \rangle$$

$$\times \sum_{\bar{v}\bar{J}} (-1)^{J+j+\bar{J}+k} \begin{Bmatrix} J & J' & k \\ j & j & \bar{J} \end{Bmatrix} \langle j^N vJ \{| j^{N-1}(\bar{v}\bar{J}), j \rangle$$

$$\times \langle j^{N-1}(\bar{v}\bar{J}), j |\} j^N v'J' \rangle \qquad (2.63)$$

The calculation of reduced matrix elements of the type $\langle j \| \mathbf{f}_k(N) \| j \rangle$ has been discussed in Section 2.1. Obviously, the matrix element of $\mathbf{F}_k$ will vanish, if the two states differ in the coordinates of more than one electron.

For the ratio of the reduced matrix elements of two F-type operators $\mathbf{F}_k$ and $\mathbf{T}_k$, we obtain

$$\langle j^N vJ \| \mathbf{F}_k \| j^N v'J' \rangle$$

$$= \frac{\langle j \| \mathbf{f}_k(N) \| j \rangle}{\langle j \| \mathbf{t}_k(N) \| j \rangle} \langle j^N vJ \| \mathbf{T}_k \| j^N v'J' \rangle \qquad (2.64)$$

The matrix elements of operator $\mathbf{G}$ may be calculated using two-particle coefficients of fractional parentage.[48] In practice, a more convenient method is usually employed. Thus of the $\frac{1}{2} N(N-1)$ components $\mathbf{g}_{ij}$ $(i > j)$ comprising $\mathbf{G}$, $\frac{1}{2}(N-1)(N-2)$ do not involve the Nth electron. Consequently,

$$\langle \psi | \mathbf{G} | \psi' \rangle = \frac{N}{N-2} \langle \psi | \sum_{j < i \neq N} \mathbf{g}_{ij} | \psi' \rangle \qquad (2.65)$$

The expansion (2.50) then gives for the matrix element of $\mathbf{G}$

$$\langle j^N vJ | \mathbf{G} | j^N v'J' \rangle$$

$$= \delta(J, J') \frac{N}{N-2} \sum_{\bar{v}\bar{J}\bar{v}'} \langle j^N vJ \{| j^{N-1}(\bar{v}\bar{J}), j \rangle$$

$$\times \langle j^{N-1}\bar{v}\bar{J} | \sum_{j < i \neq N} \mathbf{g}_{ij} | j^{N-1}\bar{v}'\bar{J} \rangle \langle j^{N-1}(\bar{v}'\bar{J}), j |\} j^N v'J \rangle \qquad (2.66)$$

Equation (2.66) relates the matrix elements of $\mathbf{G}$ for j^N to those for j^{N-1}. The matrix elements for j^{N-1} may then be expressed similarly in terms of those for j^{N-2}. This recursion may be continued until the matrix element for two electrons $\langle j^2 vJ | \sum \mathbf{g}_{ij} | j^2 v'J \rangle$ is obtained. The latter will be discussed in more detail in Section 2.4 below. Obviously, matrix elements of $\mathbf{G}$ between two states are zero if the states differ in the coordinates of more than two electrons.

2.4. ENERGY LEVELS OF GASEOUS TRANSITION METAL IONS

In this section, we are primarily concerned with the calculation of energy levels which arise from the principal electronic configurations of gaseous transition metal ions. The transition metals are characterized by the progressive filling of the $3d$, $4d$, or $5d$ shells of their electronic configurations. The neutral transition metal atoms of the iron series possess the common feature of an argon core ($1s^2 2s^2 2p^6 3s^2 3p^6$), a partly filled $3d$ shell, and one or two outer electrons ($4s^1$ or $4s^2$). In the neutral transition metal atoms of the $4d$ or $5d$ series, the closed inner shells consist of the krypton ($1s^2 2s^2 2p^6 3s^2 3p^6 3d^{10} 4s^2 4p^6$) or the xenon ($1s^2 2s^2 2p^6 3s^2 3p^6 3d^{10} 4s^2 4p^6 4d^{10} 5s^2 5p^6$)

core, respectively. The partly filled shell is then $4d$ or $5d$, the outer electrons being characterized as $5s$ or $6s$.

On ionization, transition metals first lose one and then, if available, a second $4s$, $5s$, or $6s$ electron. Consequently, ionized transition metals normally involve substantially simpler configurations, particularly if ionized sufficiently so as to leave an incompletely filled d^N shell outside the inert gas core. In view of our interest in transition metal ions in crystals and complex ions, we will limit our treatment in what follows to the d^N configuration outside of closed shells.

The Hamiltonian for an N-electron atom with nuclear charge Ze and a nucleus fixed in space may be approximated by

$$\mathcal{H} = -\frac{\hbar^2}{2m} \sum_{i=1}^{N} \nabla_i^2 - \sum_{i=1}^{N} \frac{Ze^2}{r_i} + \sum_{i>j}^{N} \frac{e^2}{r_{ij}}$$

$$+ \sum_{i=1}^{N} \xi(r_i)(\mathbf{s}_i \cdot \mathbf{l}_i) \qquad (2.67)$$

Here, the first and the second term are the operators of the kinetic energy and the potential energy in the field of the nucleus, respectively, of all the electrons; the third term is the operator of Coulomb interaction between pairs of electrons, and the last term is the operator of spin–orbit interaction. It is well known that exact solutions of the Schrödinger equation cannot be obtained for systems with more than one electron. Therefore, the Hamiltonian equation (2.67) is usually dealt with in terms of the *central field approximation*.[7,55] In this case each electron is assumed to move independently in a field formed by the effect of the nucleus and the spherically averaged potentials of the other electrons. Each electron may then be said to move in a *screening potential* $-U(r_i)/e$. Approximate solutions to the problem of Eq. (2.67) may be obtained by first solving the eigenvalue equation of the Hamiltonian

$$\mathcal{H}_0 = \sum_{i=1}^{N} \left(-\frac{\hbar^2}{2m} \nabla_i^2 + U(r_i) \right) \qquad (2.68)$$

and by treating the difference $\mathcal{H} - \mathcal{H}_0 = V$ as a perturbation:

$$V = \sum_{i=1}^{N} \left(-\frac{Ze^2}{r_i} - U(r_i) \right)$$

$$+ \sum_{i>j}^{N} \frac{e^2}{r_{ij}} + \sum_{i=1}^{N} \xi(r_i)(\mathbf{s}_i \cdot \mathbf{l}_i) \qquad (2.69)$$

The Schrödinger equation for the Hamiltonian of Eq. (2.68) can be separated into N single-electron eigenvalue equations, each of which differs from the

Schrödinger equation for the hydrogen atom alone in the form of the potential energy operator, $U(r)$ in place of $-e^2/r$. Hence the angular part of the equation may be solved exactly, the solutions being the spherical harmonics $Y_{lm}(\theta, \varphi)$. On the other hand, only an approximate solution may be obtained, in general, for the radial equation, the resulting radial functions being dependent on the form of the potential $U(r)$. The different eigenvalues of the Schrödinger equation for $\mathcal{H}_0$ then correspond to the energies of the various electronic configurations which are defined in terms of the quantum numbers $(n_i l_i)$.

The eigenfunctions resulting from solution of the central field problem may be subsequently employed in the calculation of matrix elements of the perturbation operator V [Eq. (2.69)]. It should be noted that the results will be always obtained in terms of radial integrals times their angular parts. Whereas the latter may be calculated exactly, e.g., by tensor operator methods, the radial integrals are often left as parameters which are determined by experiment. Turning our attention to the detailed form of the perturbation potential of Eq. (2.69), we note that the first term is purely radial and thus will contribute only energy shifts. These are the same for all the levels arising from a given configuration and do not affect the energy level structure of the configuration. The calculations for gaseous transition metal ions, particularly for those of the iron series, are usually performed in *Russell–Saunders coupling*, i.e., assuming in Eq. (2.69)

$$\sum_{i>j}^{N} \frac{e^2}{r_{ij}} > \sum_{i=1}^{N} \xi(r_i)(\mathbf{s}_i \cdot \mathbf{l}_i) \qquad (2.70)$$

The Coulomb interaction $\sum e^2/r_{ij}$ is usually different for different states of the same configuration, and thus configurations are split into SL terms. It may be shown that, apart from a constant shift of energy, the term energies are independent of closed shells. Hence the summation in Eq. (2.69) may be restricted to electrons in incomplete shells. The required matrix elements are of the form[7]

$$\left\langle \alpha SL \left| \sum_{i>j} \frac{e^2}{r_{ij}} \right| \alpha' SL \right\rangle$$

$$= \sum_k e^2 \left\langle \alpha SL \left| \sum_{i>j} \frac{r_<^k}{r_>^{k+1}} P_k(\cos \omega_{ij}) \right| \alpha' SL \right\rangle$$

$$= \sum_k e^2 \left\langle \alpha SL \left| \sum_{i>j} \frac{r_<^k}{r_>^{k+1}} (\mathbf{C}_i^{(k)} \cdot \mathbf{C}_j^{(k)}) \right| \alpha' SL \right\rangle \qquad (2.71)$$

and are thus diagonal in S and L and independent of J and M. The operator e^2/r_{ij} in Eq. (2.71) has been expanded in terms of Legendre polynomials of the cosine of angle ω_{ij} between the vectors from the nucleus to the electrons i and j. Also $r_<$ is the smaller and $r_>$ the larger of the two distances between the nucleus and the electrons. Application of the spherical harmonic addition theorem of Eq. (2.10) gives the last expression of Eq. (2.71).

The matrix element of Coulomb interaction for a two-electron configuration may then be written, by application of Eqs. (2.71) and (2.49),

$$\left\langle (n_a l_a, n_b l_b); SL \left| \frac{e^2}{r_{12}} \right| (n_a l_a, n_b l_b); SL \right\rangle$$

$$= \sum_k [f_k(l_a, l_b) F^k(n_a l_a, n_b l_b) \\ + g_k(l_a, l_b) G^k(n_a l_a, n_b l_b)] \qquad (2.72)$$

Here

$$f_k(l_a, l_b) = (-1)^{l_a + l_b + L} \langle l_a \| \mathbf{C}_1^{(k)} \| l_a \rangle \langle l_b \| \mathbf{C}_2^{(k)} \| l_b \rangle \\ \times \begin{Bmatrix} l_a & l_a & k \\ l_b & l_b & L \end{Bmatrix} \qquad (2.73)$$

and

$$g_k(l_a, l_b) = (-1)^S \langle l_a \| \mathbf{C}^{(k)} \| l_b \rangle^2 \begin{Bmatrix} l_a & l_b & k \\ l_a & l_b & L \end{Bmatrix} \qquad (2.74)$$

In addition, the radial integrals F^k (direct integrals) and G^k (exchange integrals) of Eq. (2.72) are known as *Slater–Condon parameters*. Both quantities are positive, and F^k as well as $G^k/[k]$ are decreasing functions of k. Often it is more convenient to use the *reduced* radial integrals F_k and G_k, where

$$F_k = \frac{F^k}{D_k} \qquad G_k = \frac{G^k}{D_k} \qquad (2.75)$$

and the D_k are the denominators given in the Tables 1^6 and 2^6 of Condon and Shortley.[7] For d electrons, in particular, Eqs. (2.75) assume the form

$$F_0 = F^0 \qquad F_2 = \tfrac{1}{49} F^2 \qquad F_4 = \tfrac{1}{441} F^4 \qquad (2.76)$$

with similar expressions for G_k, $k = 0, 2, 4$. Again for d electrons, the quantities A, B, and C known as *Racah parameters* and defined by[47]

$$A = F_0 - 49 F_4 \qquad B = F_2 - 5 F_4 \qquad C = 35 F_4 \qquad (2.77)$$

are also frequently used.

If all the electrons are equivalent, Eq. (2.72) assumes the form

$$\left\langle (nl)^2; SL \left| \frac{e^2}{r_{12}} \right| (nl)^2; SL \right\rangle = \sum_k f_k(l, l) F^k(nl, nl) \qquad (2.78)$$

where

$$f_k(l, l) = (-1)^L \langle l \| \mathbf{C}^{(k)} \| l \rangle^2 \begin{Bmatrix} l & l & k \\ l & l & L \end{Bmatrix} \quad (2.79)$$

For other forms of the two-electron Coulomb interaction one may consult, e.g., Condon and Shortley.[7]

We have seen above that the effect of Coulomb interaction is to remove partly the degeneracy of the configurations; it thus leads to a number of terms, each being characterized by the quantum numbers S and L. The degeneracy may be further removed by the relativistic effect of spin–orbit interaction approximately represented by the terms [cf. Eq. (2.69)][7,55]

$$\mathcal{H}_{so} = \sum_{i=1}^{N} \xi(r_i)(\mathbf{s}_i \cdot \mathbf{l}_i) \quad (2.80)$$

where

$$\xi(r_i) = \frac{\hbar^2}{2m^2c^2 r_i} \frac{dU(r_i)}{dr_i} \quad (2.81)$$

The spin–orbit interaction is diagonal in l, and interaction between configurations of different n will be needed rarely. The matrix elements of $\mathcal{H}_{so}$ are diagonal in J and independent of M. However, since they are not diagonal in L and S, they may couple terms differing in S and L by not more than one unit. The matrix elements of spin–orbit interaction in the configuration l^N may then be obtained by application of Eq. (2.49) as

$$\left\langle l^N \alpha SLJM \left| \zeta_{nl} \sum_{i=1}^{N} (\mathbf{s}_i \cdot \mathbf{l}_i) \right| l^N \alpha' S'L'JM \right\rangle$$

$$= \zeta_{nl}(-1)^{J+L+S'}[l(l+1)(2l+1)]^{1/2} \begin{Bmatrix} L & L' & 1 \\ S' & S & J \end{Bmatrix}$$

$$\times \langle l^N \alpha SL \| \mathbf{V}^{(11)} \| l^N \alpha' S'L' \rangle \quad (2.82)$$

Here, ζ_{nl} is the spin–orbit coupling constant for the nl configuration[7]

$$\zeta_{nl} = \int_0^\infty [R_{nl}(r)]^2 \xi(r) r^2 \, dr \quad (2.83)$$

In Russell–Saunders coupling, frequently the spin–orbit coupling constant for a given SL state λ is employed,

$$\lambda = \pm \frac{\zeta_{nl}}{2S} \quad (2.84)$$

where the positive sign is taken for $N \leq 2l + 1$ and the negative sign for $N > 2l + 1$. The reduced matrix element of the double tensor $\mathbf{V}^{(11)}$ appearing in Eq. (2.82) will be considered below.

The Coulomb and spin–orbit interactions account by far for the greatest part of the energy of the states in gaseous transition metal ions. However, there are several additional interactions which may be needed in precise treatments or under special circumstances. These include the effects of spin–spin, orbit–orbit, and spin–other-orbit interactions. In addition, Coulomb and spin–orbit interactions in mixed configurations like $l^N l'$ may be of interest. Finally, the effects of external electric and magnetic fields, the Stark and Zeeman effects, cannot be overlooked. The treatment of these interactions by tensor operator methods has been discussed in the volumes by Judd[24] and Wybourne,[62] where details may be found.

Often matrix elements of the more complicated interactions may be expressed in terms of reduced matrix elements of the tensor operators $\mathbf{U}^{(k)}$ and $\mathbf{V}^{(1k)}$. The N-electron unit tensor operator $\mathbf{U}^{(k)}$ has been defined in Eq. (2.32), and the corresponding reduced matrix elements may be obtained by application of Eqs. (2.63) and (2.44) as

$$\langle \psi_1 \| \mathbf{U}^{(k)} \| \psi_1' \rangle$$

$$= N([L_1][L_1'])^{1/2} \sum_{\bar{\psi}} \langle \psi_1 \{ | \bar{\psi} \rangle \langle \bar{\psi} | \} \psi_1' \rangle$$

$$\times (-1)^{\bar{L}+L_1+l+k} \begin{Bmatrix} L_1 & L_1' & k \\ l & l & \bar{L} \end{Bmatrix} \quad (2.85)$$

Here, the coefficients of fractional parentage have been abbreviated to $\langle \psi_1 \{ | \bar{\psi} \rangle$ and $\langle \bar{\psi} | \} \psi_1' \rangle$. The double tensors $\mathbf{V}^{(1k)}$ are defined by[47]

$$\mathbf{V}^{(1k)} = \sum_i (\mathbf{s} \cdot \mathbf{u}^{(k)})_i \quad (2.86)$$

where the $\mathbf{u}^{(k)}$ are the unit tensor operators introduced in Eq. (2.28). The reduced matrix elements of $\mathbf{V}^{(1k)}$ follow from the application of Eq. (2.63) as

$$\langle l^N \psi_1 \| \mathbf{V}^{(1k)} \| l^N \psi_1' \rangle$$

$$= N[s(s+1)(2s+1)]([S_1][L_1][S_1'][L_1'])^{1/2}$$

$$\times \sum_{\bar{\psi}} \langle \psi_1 \{ | \bar{\psi} \rangle \langle \bar{\psi} | \} \psi_1' \rangle (-1)^{\bar{S}+\bar{L}+S_1+L_1+l+s+k+1}$$

$$\times \begin{Bmatrix} S_1 & S_1' & 1 \\ s & s & \bar{S} \end{Bmatrix} \begin{Bmatrix} L_1 & L_1' & k \\ l & l & \bar{L} \end{Bmatrix} \quad (2.87)$$

The rank k in $\mathbf{U}^{(k)}$ and $\mathbf{V}^{(1k)}$ may assume both odd and even integral values. For even k the matrix elements of $\mathbf{U}^{(k)}$ will change sign under conjugation, whereas those of $\mathbf{V}^{(1k)}$ will remain unchanged. For odd k the opposite will hold. The complete reduced matrix elements of the unit tensor operators $\mathbf{U}^{(k)}$, $k = 2$ to 6, and $\mathbf{V}^{(1k)}$ have been computed and tabulated by Nielson and Koster[40] for the configurations p^N, d^N, and f^N, and are thus available.

3

TENSOR OPERATOR ALGEBRA IN CRYSTALS AND COMPLEX IONS

3.1. IRREDUCIBLE TENSOR OPERATORS IN POINT GROUPS

The concepts and results of Sections 2.1 and 2.2 may be easily generalized to groups of symmetry other than spherical. We shall consider, in this section, the application to point groups which is relevant in the treatment of transition metal ions imbedded in crystals or transition metal ions forming the central part of complex ions.

In spherical symmetry, the eigenvectors $|jm\rangle$ transform according to irreducible representation $\mathbf{D}^{(j)}$, component m, of the three-dimensional rotation group $SO(3) \equiv R_3$. In point group G, the rotation operators $R(\alpha\beta\gamma)$ have to be replaced by the symmetry operators $A, B, \ldots, R, \ldots$. Also, the functions $|\Gamma\gamma\rangle$ are used in place of the $|jm\rangle$, Γ indicating the irreducible representation, and γ a particular component of Γ. Additional labels, abbreviated here by α, may be required, since there may be several sets of functions $|\Gamma\gamma\rangle$ having the same transformation properties. Under the effect of an operator R of G, the functions $|\alpha\Gamma\gamma\rangle$ then transform according to

$$R|\alpha\Gamma\gamma\rangle = \sum_{\lambda=1}^{d_\Gamma} |\alpha\Gamma\lambda\rangle\langle\Gamma\lambda|R|\Gamma\gamma\rangle \qquad (3.1)$$

The group of matrices $\mathbf{D}^{(\Gamma)}(R)$ with elements

$$D_{\gamma\lambda}^{(\Gamma)}(R) = \langle\Gamma\gamma|R|\Gamma\lambda\rangle \qquad (3.2)$$

forms the irreducible representation Γ of G, the d_Γ functions $|\alpha\Gamma\gamma\rangle$ being the corresponding basis functions. Here, d_Γ is the dimension of irreducible representation Γ. We will also assume that, in general, the matrices of the irreducible representations are chosen to be unitary such that

$$[\mathbf{D}^{(\Gamma)}(R)]^{-1} = [\mathbf{D}^{(\Gamma)}(R)]^\dagger \qquad (3.3)$$

or

$$\langle\Gamma\gamma|R^{-1}|\Gamma\lambda\rangle = \langle\Gamma\lambda|R|\Gamma\gamma\rangle^* = \langle\overline{\Gamma\lambda}|R|\overline{\Gamma\gamma}\rangle \qquad (3.4)$$

where the latter notation has been introduced for future convenience. The orthogonality relations of the matrix elements then take the form

$$\sum_R \langle\Gamma\gamma|R|\Gamma\lambda\rangle^*\langle\Gamma'\gamma'|R|\Gamma'\lambda'\rangle$$
$$= \frac{g}{d_\Gamma}\delta(\Gamma,\Gamma')\delta(\gamma,\gamma')\delta(\lambda,\lambda') \qquad (3.5)$$

where g is the order of group G.

We now consider the expansion of an arbitrary function $|f\rangle$ according to

$$|f\rangle = \sum_{\Gamma,\lambda} |f\Gamma\lambda\rangle \qquad (3.6)$$

The *projection operator* $P_{\lambda\lambda}^\Gamma$ and the *shift operator* $P_{\lambda\mu}^\Gamma$ may then be defined by the equations

$$\begin{aligned} P_{\lambda\lambda}^\Gamma|f\rangle &= |f\Gamma\lambda\rangle \\ P_{\lambda\mu}^\Gamma|f\rangle &= |f\Gamma\lambda\rangle \end{aligned} \qquad (3.7)$$

It may be easily shown that the projection operators are determined in terms of symmetry operators R and their matrix elements according to [28,38]

$$P_{\lambda\mu}^\Gamma = \frac{d_\Gamma}{g}\sum_R \langle\Gamma\lambda|R|\Gamma\mu\rangle^* R \qquad (3.8)$$

The operators $P_{\lambda\mu}^\Gamma$ operate in the same form on each other:

$$P_{\lambda\mu}^{\Gamma}P_{\mu'\nu}^{\Gamma'} = P_{\lambda\nu}^{\Gamma}\delta(\Gamma, \Gamma')\delta(\mu, \mu') \qquad (3.9)$$

and are real.

Projection operators are frequently employed to generate *symmetry functions*, i.e., sets of functions transforming according to the different rows of the individual irreducible representations of a given point group G. Since, for the symmetry groups of interest, G is a subgroup of $SO(3)$, the angular momentum eigenvectors $|jm\rangle$ may be taken as the starting point:

$$|j\Gamma\gamma a\rangle = \sum_m |jm\rangle\langle jm|j\Gamma\gamma a\rangle \qquad (3.10)$$

The index a in Eq. (3.10) is required whenever representation Γ occurs more than once in the decomposition of $\mathbf{D}^{(j)}$. The determination of the symmetry adaptation coefficients $\langle jm|j\Gamma\gamma a\rangle$ may be based on application of Eq. (3.7), thus

$$P_{\gamma\gamma}^{\Gamma}|jm\rangle = \frac{d_\Gamma}{g}\sum_R \langle\Gamma\gamma|R|\Gamma\gamma\rangle^* R|jm\rangle = |j\Gamma\gamma a\rangle$$

$$(3.11)$$

The actual computation of the coefficients employs the derived system of equations[28]

$$\sum_a \langle jm'|j\Gamma\gamma'a\rangle\langle j\Gamma\gamma a|jm\rangle$$

$$= \frac{d_\Gamma}{g}\sum_R \langle\Gamma\gamma'|R|\Gamma\gamma\rangle^*\langle jm'|R|jm\rangle \quad (3.12)$$

It should be noted that Eq. (3.10) defines a unitary transformation. Consequently, the orthogonality relations of the coefficients assume the form

$$\sum_m \langle j\Gamma\gamma a|jm\rangle^*\langle j\Gamma'\gamma'a'|jm\rangle$$

$$= \delta(\Gamma, \Gamma')\delta(\gamma, \gamma')\delta(a, a')$$

$$\sum_{\Gamma\gamma a} \langle jm|j\Gamma\gamma a\rangle^*\langle j'm'|j'\Gamma\gamma a\rangle$$

$$= \delta(j, j')\delta(m, m') \qquad (3.13)$$

By complete analogy with Eq. (2.5), we now consider a set of operators $\mathbf{T}_\mu^{(\Gamma)}$ which under the operations R of the group transform into linear combinations of each other[38]:

$$R\mathbf{T}_\mu^{(\Gamma)}R^{-1} = \sum_\lambda \mathbf{T}_\lambda^{(\Gamma)}\langle\Gamma\lambda|R|\Gamma\mu\rangle \qquad (3.14)$$

In Eq. (3.14), λ assumes d_Γ different values, and the matrix elements $\langle\Gamma\lambda|R|\Gamma\mu\rangle$ have been introduced by Eq. (3.2). If representation Γ is irreducible, the set of d_Γ operators is said to constitute an *irreducible tensor operator* $\mathbf{T}^{(\Gamma)}$ belonging to the irreducible representa-

tion Γ of group G. The individual operators $\mathbf{T}_\mu^{(\Gamma)}$ are considered as the components of $\mathbf{T}^{(\Gamma)}$ and, in particular, $\mathbf{T}_\lambda^{(\Gamma)}$ is said to belong to the row λ of representation Γ. We also assume that, under the group operations, the tensor components $\mathbf{T}_\gamma^{(\Gamma)}$ and the basis functions $|\Gamma\gamma\rangle$ transform according to the same representation matrices [cf. Eqs. (3.14) and (3.1)].

By analogy with the definition of projection operators, we may now define a *projection transformation* as

$$\mathbf{O}_{\lambda\mu}^{\Gamma} = \frac{d_\Gamma}{g}\sum_R \langle\Gamma\lambda|R|\Gamma\mu\rangle^* R| \;\; |R^{-1} \qquad (3.15)$$

where $R| \;\; |R^{-1}$ should indicate that $\mathbf{O}_{\lambda\mu}^{\Gamma}$ acts on an operator $\mathbf{T}$ by means of a similarity transformation RTR^{-1}.

The transformation properties of the adjoint of a tensor operator $\mathbf{T}^{(\Gamma)}$ follow from Eq. (3.14):

$$R\mathbf{T}_\mu^{(\Gamma)\dagger}R^{-1} = \sum_\lambda \mathbf{T}_\lambda^{(\Gamma)\dagger}\langle\Gamma\lambda|R|\Gamma\mu\rangle^* \qquad (3.16)$$

Since the R are unitary, $\mathbf{T}^{(\Gamma)\dagger}$ belongs to the complex conjugate representation $\Gamma^* = \bar{\Gamma}$. For most of the groups of interest, the representations provided by the matrices $\mathbf{D}^{(\Gamma)}(R)$ and $\mathbf{D}^{(\Gamma)}(R)^*$ are equivalent and thus $\mathbf{T}^{(\Gamma)}$ is self-adjoint.

The matrix elements of the tensor operator $\mathbf{T}_\gamma^{(\Gamma)}$ in an arbitrary finite or compact group G are determined by the *Wigner–Eckart theorem* which, in this case, may be written as[32]

$$\langle\alpha_1\Gamma_1\gamma_1|\mathbf{T}_\gamma^{(\Gamma)}|\alpha_2\Gamma_2\gamma_2\rangle$$

$$= \sum_b (d_{\Gamma_1})^{-1/2}\langle\Gamma_1\gamma_1 b|\Gamma_2\gamma_2\Gamma\gamma\rangle\langle\alpha_1\Gamma_1\|\mathbf{T}^{(\Gamma)}\|\alpha_2\Gamma_2\rangle_b$$

$$(3.17)$$

In Eq. (3.17), d_{Γ_1} is the dimension of representation Γ_1 and the index b is required if Γ_1 occurs more than once in the direct product of representations $\Gamma_2 \otimes \Gamma$. The reader should note that there are several forms of the Wigner–Eckart theorem in use. These forms differ in the definition of the tensor components, the coupling coefficients, the reduced matrix elements, and/or the phase factors. Failure to observe this variance in using the algebra and/or the numerical data from different sources may lead to erroneous results.

In the three-dimensional rotation group $SO(3)$, the Wigner coefficient is usually replaced by a 3-j symbol, and a factor $(-1)^{2k}$ is dropped. Then the form of Eq. (2.11) follows directly from Eq. (3.17).

The quantity $\langle\Gamma_1\gamma_1 b|\Gamma_2\gamma_2\Gamma\gamma\rangle$ in Eq. (3.17) is a symmetry coupling coefficient of the type as defined by[28,58]

$$|(\Gamma_1\Gamma_2)\Gamma\gamma b\rangle = \sum_{\gamma_1\gamma_2} |\Gamma_1\gamma_1\rangle|\Gamma_2\gamma_2\rangle\langle\Gamma_1\gamma_1\Gamma_2\gamma_2|\Gamma\gamma b\rangle$$

(3.18)

In Eq. (3.18), basis functions $|\Gamma_1\gamma_1\rangle$ and $|\Gamma_2\gamma_2\rangle$ transforming, respectively, as the irreducible representations Γ_1 and Γ_2 of G are coupled to yield a function $|(\Gamma_1\Gamma_2)\Gamma\gamma b\rangle$, where the irreducible representation Γ occurs in the direct product

$$\Gamma_1\otimes\Gamma_2 = \sum_i n_{\Gamma_i}\Gamma_i$$

(3.19)

Here, function $|(\Gamma_1\Gamma_2)\Gamma\gamma b\rangle$ is often denoted as $|\Gamma\gamma b\rangle$ for brevity. The coupling coefficients of Eq. (3.18) form a unitary matrix

$$\langle\Gamma_1\gamma_1\Gamma_2\gamma_2|\Gamma\gamma b\rangle = \langle\Gamma\gamma b|\Gamma_1\gamma_1\Gamma_2\gamma_2\rangle^*$$

(3.20)

and conform to the orthogonality relations

$$\sum_{\gamma_1\gamma_2}\langle\Gamma\gamma b|\Gamma_1\gamma_1\Gamma_2\gamma_2\rangle\langle\Gamma_1\gamma_1\Gamma_2\gamma_2|\Gamma'\gamma'b'\rangle$$
$$= \delta(\Gamma,\Gamma')\delta(\gamma,\gamma')\delta(b,b')$$

(3.21)

$$\sum_{\Gamma\gamma b}\langle\Gamma_1\gamma_1\Gamma_2\gamma_2|\Gamma\gamma b\rangle\langle\Gamma\gamma b|\Gamma_1\gamma_1'\Gamma_2\gamma_2'\rangle$$
$$= \delta(\gamma_1,\gamma_1')\delta(\gamma_2,\gamma_2')$$

(3.22)

Again, symmetry coupling coefficients may be computed in numerical form on the basis of relations following from the application of projection operators.

In order to establish definite properties with respect to the interchange of arguments in these coefficients, representations Γ, Γ_1, and Γ_2 are associated with the quantum numbers j, j_1, and j_2, serially, including the index b, such that

$$\langle\Gamma\gamma b|\Gamma_1\gamma_1\Gamma_2\gamma_2\rangle = (-1)^{(j_1+j_2-j)_b}\langle\Gamma\gamma b|\Gamma_2\gamma_2\Gamma_1\gamma_1\rangle$$

(3.23)

This relation follows from application of a lemma by Racah (cf. Section 3.2). Another consequence is the definition of a symmetrized coupling coefficient, usually denoted as a 3-Γ symbol,[29]

$$\begin{pmatrix}\Gamma_1 & \Gamma_2 & \Gamma\\ \gamma_1 & \gamma_2 & \gamma\end{pmatrix}_b = (-1)^{(j_1-j_2+j)_b}[d_\Gamma]^{-1/2}\begin{pmatrix}j\\ \overline{\Gamma\gamma a} & \overline{\overline{\Gamma\gamma a}}\end{pmatrix}$$
$$\times\langle\overline{\Gamma\gamma}b|\Gamma_1\gamma_1\Gamma_2\gamma_2\rangle$$

(3.24)

The quantity $\begin{pmatrix}j\\ \overline{\Gamma\gamma a} & \overline{\overline{\Gamma\gamma a}}\end{pmatrix}$ in Eq. (3.24) is a metric

tensor which will be discussed below [cf. Eq. (3.44)]. The properties of the 3-Γ symbol with respect to the interchange of columns are then equivalent to those of the 3-j symbol in $SO(3)$:

$$\begin{pmatrix}\Gamma_1 & \Gamma_2 & \Gamma\\ \gamma_1 & \gamma_2 & \gamma\end{pmatrix}_b = \begin{pmatrix}\Gamma_2 & \Gamma & \Gamma_1\\ \gamma_2 & \gamma & \gamma_1\end{pmatrix}_b$$
$$= (-1)^{(j_1+j_2+j)_b}\begin{pmatrix}\Gamma_2 & \Gamma_1 & \Gamma\\ \gamma_2 & \gamma_1 & \gamma\end{pmatrix}_b$$

(3.25)

Some useful relations concerning the 3-Γ symbols may be found in Appendix II.

Similar to $SO(3)$, the combination of four symmetry coupling coefficients within point group G may be employed to introduce a W coefficient:

$$W(\Gamma_1\Gamma_2\Gamma_3\Gamma_4;\Gamma_5\Gamma_6)_{b_1b_2b_3b_4}$$
$$= (d_{\Gamma_5})^{-1/2}(d_{\Gamma_6})^{-1/2}$$
$$\times\langle\Gamma_1\Gamma_2(\Gamma_5b_3)\Gamma_4;\Gamma_3b_1|\Gamma_1(\Gamma_2\Gamma_4)\Gamma_6b_2;\Gamma_3b_4\rangle$$
$$= (d_{\Gamma_5})^{-1/2}(d_{\Gamma_6})^{-1/2}\sum_{\gamma_1\gamma_2\gamma_4\gamma_5\gamma_6}\langle\Gamma_5\gamma_5\Gamma_4\gamma_4|\Gamma_3\gamma_3b_1\rangle^*$$
$$\times\langle\Gamma_2\gamma_2\Gamma_4\gamma_4|\Gamma_6\gamma_6b_2\rangle\langle\Gamma_1\gamma_1\Gamma_2\gamma_2|\Gamma_5\gamma_5b_3\rangle^*$$
$$\times\langle\Gamma_1\gamma_1\Gamma_6\gamma_6|\Gamma_3\gamma_3b_4\rangle$$

(3.26)

This definition may be then rewritten, in combination with Eq. (3.24), such as to introduce a 6-Γ symbol:[29]

$$\begin{Bmatrix}\Gamma_1 & \Gamma_2 & \Gamma_5\\ \Gamma_4 & \Gamma_3 & \Gamma_6\end{Bmatrix}_{b_1b_2b_4b_3}$$
$$= (-1)^{(j_3+j_4-j_5)_{b_1}+(j_1+j_2+j_5)_{b_3}}$$
$$\times W(\bar\Gamma_1\bar\Gamma_2\bar\Gamma_3\bar\Gamma_4;\Gamma_5\bar\Gamma_6)_{b_1b_2b_3b_4}$$

(3.27)

It should be observed that, in an arbitrary finite or compact group, both W coefficients and 6-Γ symbols follow, on interchange of arguments, somewhat more involved relations than those of the corresponding quantities in $SO(3)$.[29] A number of these relations which are often useful may be found in Appendix III.

Turning our attention to products of tensor operators, we consider the properties of a tensor operator $\mathbf{X}_\gamma^{(\Gamma b)}$, which is the Kronecker product of two irreducible tensor operators $\mathbf{T}^{(\Gamma_1)}$ and $\mathbf{U}^{(\Gamma_2)}$. By complete analogy with Eq. (2.34) it is

$$\mathbf{X}_\gamma^{(\Gamma b)} = \{\mathbf{T}^{(\Gamma_1)}\otimes\mathbf{U}^{(\Gamma_2)}\}_\gamma^{(\Gamma b)}$$

(3.28)

where Γ is contained in the direct product $\Gamma_1\otimes\Gamma_2$ [cf. Eq. (3.19)], $\mathbf{X}^{(\Gamma b)}$ is irreducible, and the components of this product are given by

$$\mathbf{X}_\gamma^{(\Gamma b)} = \sum_{\gamma_1\gamma_2}\mathbf{T}_{\gamma_1}^{(\Gamma_1)}\mathbf{U}_{\gamma_2}^{(\Gamma_2)}\langle\Gamma_1\gamma_1\Gamma_2\gamma_2|\Gamma\gamma b\rangle$$

(3.29)

This coupling thus proceeds analogously to that in Eq. (3.18).

We next consider the Kronecker product of Eq. (3.28) if the tensor operators $\mathbf{T}^{(\Gamma_1)}$ and $\mathbf{U}^{(\Gamma_2)}$ operate on two different parts of the system. The function $|(\alpha_1\Gamma_1, \alpha_2\Gamma_2)\Gamma\gamma b\rangle$ describing the combined system is then set up from the basis functions $|\alpha_1\Gamma_1\gamma_1\rangle$ and $|\alpha_2\Gamma_2\gamma_2\rangle$ of parts 1 and 2, respectively, according to Eq. (3.18). Although a general expression for the reduced matrix element may be conceived as an extension of Eq. (2.42), we consider here the situation where the operator $\mathbf{T}^{(\Gamma_1)}$ acts only on part 1 of the system. Similar to Eq. (2.43) we find, for an arbitrary point group G,

$$\langle\alpha_1\Gamma_1\Gamma_2, \Gamma_3 b_3\|\mathbf{T}^{(\Gamma)}\|\alpha_1'\Gamma_1'\Gamma_2', \Gamma_3'b_3'\rangle_b$$

$$= (d_{\Gamma_3})^{1/2}(d_{\Gamma_3'})^{1/2}\delta(\Gamma_2, \Gamma_2')$$

$$\times \sum_{b_1} (-1)^{(j_1'+j_3'+j_2)b_3'+(j_1-j_1'+j)b_1}$$

$$\times \langle\alpha_1\Gamma_1\|\mathbf{T}^{(\Gamma)}\|\alpha_1'\Gamma_1'\rangle_{b_1}\begin{Bmatrix}\Gamma_1 & \bar{\Gamma}_1' & \bar{\Gamma} \\ \Gamma_3' & \Gamma_3 & \Gamma_2\end{Bmatrix}_{bb_3'b_3b_1} \quad (3.30)$$

Here, we have set in Eq. (3.29) $\mathbf{U}^{(\Gamma_2)} = 1$, $\Gamma_2 = A_1$, and $\Gamma_1 = \Gamma$. If the operator $\mathbf{U}^{(\Gamma_2)}$ acts only on part 2 of the system, we may show, in complete analogy to Eq. (2.44),

$$\langle\alpha_2\Gamma_1\Gamma_2, \Gamma_3 b_3\|\mathbf{U}^{(\Gamma)}\|\alpha_2'\Gamma_1'\Gamma_2', \Gamma_3'b_3'\rangle_b$$

$$= (d_{\Gamma_3})^{1/2}(d_{\Gamma_3'})^{1/2}\delta(\Gamma_1, \Gamma_1')$$

$$\times \sum_{b_2} (-1)^{(j_1-j_2+j_3)b_3+(j_2+j_2'+j)b_2}$$

$$\times \langle\alpha_2\Gamma_2\|\mathbf{U}^{(\Gamma)}\|\alpha_2'\Gamma_2'\rangle_{b_2}\begin{Bmatrix}\Gamma_2 & \bar{\Gamma}_2' & \bar{\Gamma} \\ \Gamma_3' & \Gamma_3 & \Gamma_1\end{Bmatrix}_{bb_3'b_3b_2} \quad (3.31)$$

The scalar product of two irreducible tensor operators $\mathbf{T}^{(\Gamma)}$ and $\mathbf{U}^{(\Gamma)}$ may be conveniently defined according to [cf. Eq. (2.45)]

$$(\mathbf{T}^{(\Gamma)} \cdot \mathbf{U}^{(\Gamma)})$$

$$= (-1)^j(d_\Gamma)^{1/2}\delta(\bar{\Gamma}, \Gamma')\{\mathbf{T}^{(\Gamma)}\otimes\mathbf{U}^{(\bar{\Gamma})}\}_{a_1}^{(A_1)}$$

$$= (-1)^j(d_\Gamma)^{1/2}\delta(\bar{\Gamma}, \Gamma')\sum_\gamma \mathbf{T}_\gamma^{(\Gamma)}\mathbf{U}_{\bar{\gamma}}^{(\bar{\Gamma})}\langle\Gamma\gamma\bar{\Gamma}\bar{\gamma}|A_1a_1\rangle$$

$$(3.32)$$

Here, the phase factor $(-1)^j$ is defined in the same way as in Eq. (3.24).

The most important products of tensor operators are those transforming according to the totally symmetric representation A_1. Scalar products of the form as in Eq. (3.32) are therefore of special interest. In

order to investigate the calculation of matrix elements of the operator $(\mathbf{T}^{(\Gamma)} \cdot \mathbf{U}^{(\Gamma)})$, we assume that the basis functions have been constructed by coupling, e.g., the two functions $|\alpha_1\Gamma_1\gamma_1\rangle$ and $|\alpha_2\Gamma_2\gamma_2\rangle$ to the coupled functions $|(\alpha_1\Gamma_1, \alpha_2\Gamma_2)\Gamma_3\gamma_3 b\rangle$. It then follows that

$$\langle(\alpha_1\Gamma_1, \alpha_2\Gamma_2)\Gamma_3\gamma_3 b_3|(\mathbf{T}^{(\Gamma)} \cdot \mathbf{U}^{(\Gamma)})|(\alpha_1'\Gamma_1', \alpha_2'\Gamma_2')\Gamma_3'\gamma_3'b_3'\rangle$$

$$= (-1)^j(d_\Gamma)^{1/2}\sum_\gamma \sum_{\gamma_1\gamma_2\gamma_1'\gamma_2'} \langle\Gamma_1\gamma_1\Gamma_2\gamma_2|\Gamma_3\gamma_3 b_3\rangle^*$$

$$\times \langle\Gamma_1'\gamma_1'\Gamma_2'\gamma_2'|\Gamma_3'\gamma_3'b_3'\rangle\langle\Gamma\gamma\bar{\Gamma}\bar{\gamma}|A_1a_1\rangle$$

$$\times \langle\alpha_1\Gamma_1\gamma_1|\mathbf{T}_\gamma^{(\Gamma)}|\alpha_1'\Gamma_1'\gamma_1'\rangle\langle\alpha_2\Gamma_2\gamma_2|\mathbf{U}_{\bar{\gamma}}^{(\bar{\Gamma})}|\alpha_2'\Gamma_2'\gamma_2'\rangle$$

$$(3.33)$$

If the coupling coefficients are combined into a 6-Γ symbol, Eq. (3.33) may be expressed by analogy to Eq. (2.49) as

$$\langle(\alpha_1\Gamma_1, \alpha_2\Gamma_2)\Gamma_3\gamma_3 b_3|(\mathbf{T}^{(\Gamma)} \cdot \mathbf{U}^{(\Gamma)})|(\alpha_1'\Gamma_1', \alpha_2'\Gamma_2')\Gamma_3'\gamma_3'b_3'\rangle$$

$$= \delta(\Gamma_3, \Gamma_3')\delta(\gamma_3, \gamma_3')\sum_{b_1b_2} (-1)^{(j_1-j_3-j_2)b_3+(j_1+j_1'+j)b_1+j}$$

$$\times \begin{Bmatrix}\Gamma_1 & \bar{\Gamma}_1' & \bar{\Gamma} \\ \bar{\Gamma}_2' & \bar{\Gamma}_2 & \bar{\Gamma}_3\end{Bmatrix}_{b_2b_3'b_3b_1} \langle\alpha_1\Gamma_1\|\mathbf{T}^{(\Gamma)}\|\alpha_1'\Gamma_1'\rangle_{b_1}$$

$$\times \langle\alpha_2\Gamma_2\|\mathbf{U}^{(\Gamma)}\|\alpha_2'\Gamma_2'\rangle_{b_2} \quad (3.34)$$

The matrix elements of the individual tensor operators in Eq. (3.33) or Eq. (3.34) may be subsequently calculated on the basis of Eq. (3.17).

3.2. IRREDUCIBLE TENSOR OPERATORS IN SUBGROUPS OF $SO(3)$

In most problems of physical and chemical interest the symmetry point group G is a subgroup of the three-dimensional rotation group, $G \subset SO(3)$. In this case, representation $D^{(j)}$ of the basis kets $|jm\rangle$ subduces a representation $D^{(j)}(G)$, which, in general, may be decomposed according to

$$D^{(j)}(G) = \sum_i a_i^{(j)}\Gamma_i \quad (3.35)$$

It follows that basis functions of any irreducible representation of G may be obtained as given by Eq. (3.10). In complete analogy, an irreducible tensor operator within $G \subset SO(3)$ may be expressed in terms of irreducible spherical tensors $\mathbf{T}_q^{(k)}$ thus[30]

$$\mathbf{T}_\gamma^{(k\Gamma a)} = \sum_q \mathbf{T}_q^{(k)}\langle kq|k\Gamma\gamma a\rangle \quad (3.36)$$

It may be shown that, in this case, the *Wigner–Eckart*

theorem assumes the particular form[30]

$$\langle \alpha_1 j_1 \Gamma_1 \gamma_1 a_1 | \mathbf{T}_\gamma^{(k\Gamma a)} | \alpha_2 j_2 \Gamma_2 \gamma_2 a_2 \rangle$$

$$= (-1)^{2k} [j_1]^{-1/2} \langle j_1 \Gamma_1 \gamma_1 a_1 | j_2 \Gamma_2 \gamma_2 a_2, k\Gamma\gamma a \rangle$$

$$\times \langle \alpha_1 j_1 \| \mathbf{T}^{(k)} \| \alpha_2 j_2 \rangle \qquad (3.37)$$

The reduced matrix element on the right-hand side of Eq. (3.37) is defined within $SO(3)$ and thus may be calculated using the methods of Chapter 2. The coupling coefficient in $G \subset SO(3)$ which arises in Eq. (3.37) is defined by

$$\langle j_1 \Gamma_1 \gamma_1 a_1 | j_2 \Gamma_2 \gamma_2 a_2, k\Gamma\gamma a \rangle$$

$$= \sum_{m_1 m_2 q} \langle j_1 m_1 | j_1 \Gamma_1 \gamma_1 a_1 \rangle^* \langle j_2 m_2 | j_2 \Gamma_2 \gamma_2 a_2 \rangle$$

$$\times \langle kq | k\Gamma\gamma a \rangle \langle j_1 m_1 | j_2 m_2 kq \rangle \qquad (3.38)$$

To illustrate the nature of this coefficient let us consider the basis functions $|j\Gamma\gamma a\rangle$ resulting from the decomposition of $D^{(j)}(G)$ according to Eq. (3.35). The collection of bases for all representations Γ of G which are consistent with Eq. (3.35) forms a possible basis for the representation $D^{(j)}$ of $SO(3)$. The coupling of two basis kets $|j_1 \Gamma_1 \gamma_1 a_1\rangle$ and $|j_2 \Gamma_2 \gamma_2 a_2\rangle$ transforming according to irreducible representations Γ_1 and Γ_2 of $G \subset SO(3)$ then follows as[28]

$$|(j_1 j_2) j\Gamma\gamma a\rangle = \sum_{\substack{\Gamma_1 \gamma_1 a_1 \\ \Gamma_2 \gamma_2 a_2}} |j_1 \Gamma_1 \gamma_1 a_1\rangle |j_2 \Gamma_2 \gamma_2 a_2\rangle$$

$$\times \langle j_1 \Gamma_1 \gamma_1 a_1, j_2 \Gamma_2 \gamma_2 a_2 | j\Gamma\gamma a \rangle \qquad (3.39)$$

Here, $|(j_1 j_2) j\Gamma\gamma a\rangle$ is often abbreviated as $|j\Gamma\gamma a\rangle$, and the coupling coefficient has been defined in Eq. (3.38). It should be noted that the product ket in Eq. (3.39) is reduced only with respect to the covering group $SO(3)$. A symmetric form similar to the 3-j symbol has been introduced for both the coupling coefficients of Eq. (3.18) [cf. Eq. (3.24)] and those of Eq. (3.38),[26,29] and this may be usefully employed in certain applications.

The basis functions of Eq. (3.39) may be equally expressed in terms of the basis $|(j_1 \Gamma_1 a_1, j_2 \Gamma_2 a_2)\Gamma\gamma b\rangle$, cf. Eq. (3.18), which is also reduced with respect to the subgroup $G \subset SO(3)$

$$|(j_1 j_2) j\Gamma\gamma a\rangle = \sum_{\Gamma_1 a_1 \Gamma_2 a_2 b} |(j_1 \Gamma_1 a_1, j_2 \Gamma_2 a_2)\Gamma\gamma b\rangle$$

$$\times (j_1 \Gamma_1 a_1, j_2 \Gamma_2 a_2 | j\Gamma ab) \qquad (3.40)$$

Here, the coefficient in parentheses, the *isoscalar factor*,[63] is independent of the components of the representations involved. From Eq. (3.40) then follows, in conjunction with Eqs. (3.18) and (3.39),[28,49]

$$\langle j_1 \Gamma_1 \gamma_1 a_1, j_2 \Gamma_2 \gamma_2 a_2 | j\Gamma\gamma a \rangle$$

$$= \sum_b (j_1 \Gamma_1 a_1, j_2 \Gamma_2 a_2 | j\Gamma ab)\langle \Gamma_1 \gamma_1 \Gamma_2 \gamma_2 | \Gamma\gamma b \rangle \qquad (3.41)$$

This relation, the *lemma of Racah*, evidently provides a connection between the coupling coefficients of Eq. (3.18) and those of Eq. (3.39). It should be noted that if, in Eq. (3.19), $n_\Gamma > 1$, the different linearly independent and orthogonal sets of coefficients are distinguished by the index b.

Equation (3.41) may be employed to derive a simple relation without the summation over b. To this end, we identify b formally with one of the quantum numbers j, j_1, or j_2, i.e., we use the orthogonality of the functions $|(j_1 j_2) j\Gamma\gamma a\rangle$ with different j, j_1, or j_2 in order to obtain the orthogonal sets of coupling coefficients. With a suitable choice of the j, j_1, and j_2 we obtain, after normalization according to Eq. (3.21), for each different value of the index b one set of n_Γ linearly independent coupling coefficients:[28]

$$\langle \Gamma_1 \gamma_1 \Gamma_2 \gamma_2 | \Gamma\gamma b \rangle$$

$$= \frac{\langle j_1 \Gamma_1 \gamma_1 a_1, j_2 \Gamma_2 \gamma_2 a_2 | j\Gamma\gamma a \rangle}{(\sum_{\gamma_1' \gamma_2'} |\langle j_1 \Gamma_1 \gamma_1' a_1, j_2 \Gamma_2 \gamma_2' a_2 | j\Gamma\gamma' a \rangle|^2)^{1/2}} \qquad (3.42)$$

The discussed procedure is capable of providing, on the basis of Eq. (3.42), coupling coefficients with the required properties even for the rare case $\Gamma = \Gamma_1 = \Gamma_2$ if $n_\Gamma > 1$. With respect to the standardization of phase it is usually agreed that the smallest possible values should be assumed for the quantum numbers j_i of Eq. (3.10).

In this way, quantum numbers j, j_1, and j_2, in conjunction with the index b, have been assigned to the representation symbols Γ, Γ_1, and Γ_2, serially. It should be noted that the behavior of the coupling coefficients on interchange of arguments [cf. Eq. (3.23)] then arises as a consequence of symmetry properties of the coefficient in Eq. (3.39).

We now define, in analogy to the metric tensor within $SO(3)$ as introduced by Wigner[61]

$$\begin{pmatrix} j \\ m & m' \end{pmatrix} = (-1)^{j+m}\delta(m, -m') \qquad (3.43)$$

a similar quantity in the subgroup $G \subset SO(3)$ by (cf. Appendix I)[29]

$$\begin{pmatrix} j \\ \Gamma\gamma a & \overline{\Gamma\gamma a} \end{pmatrix} = \sum_{mm'} \langle j\Gamma\gamma a | jm \rangle^* \begin{pmatrix} j \\ m & m' \end{pmatrix} \langle jm' | j\overline{\Gamma\gamma a} \rangle \qquad (3.44)$$

It is now possible to derive[29] from the coupling coefficient of Eq. (3.42), if the definition Eq. (3.44) is employed, the 3-Γ symbol [cf. Eq. (3.24)]. This quantity shows the same behavior on interchange of arguments as the 3-j symbol in $SO(3)$ [cf. Eq. (3.25)].

We next examine the properties of a tensor operator $\mathbf{X}^{(k\Gamma a)}$ which is the Kronecker product of the tensor operators $\mathbf{T}^{(k_1\Gamma_1a_1)}$ and $\mathbf{U}^{(k_2\Gamma_2a_2)}$. In analogy to the coupling of basis kets [cf. Eq. (3.39)], the components of this product are determined by[30]

$$\mathbf{X}_\gamma^{(k\Gamma a)} = \sum_{\substack{\Gamma_1\gamma_1a_1 \\ \Gamma_2\gamma_2a_2}} \mathbf{T}_{\gamma_1}^{(k_1\Gamma_1a_1)}\mathbf{U}_{\gamma_2}^{(k_2\Gamma_2a_2)}$$
$$\times \langle k_1\Gamma_1\gamma_1a_1, k_2\Gamma_2\gamma_2a_2|k\Gamma\gamma a\rangle \qquad (3.45)$$

Here $\mathbf{X}_\gamma^{(k\Gamma a)}$, which is used as condensed notation according to

$$\mathbf{X}_\gamma^{(k\Gamma a)} = \{\mathbf{T}^{(k_1)} \otimes \mathbf{U}^{(k_2)}\}_\gamma^{(k\Gamma a)} \qquad (3.46)$$

is reduced with respect to $SO(3)$ only. If it is assumed, however, that the commuting tensor operators $\mathbf{T}^{(k_1\Gamma_1a_1)}$ and $\mathbf{U}^{(k_2\Gamma_2a_2)}$ act on different parts of a system within $G \subset SO(3)$, the reduction of $\mathbf{X}^{(\Gamma b)}$ with respect to the group G should be considered. In this case, it is

$$\{\mathbf{T}^{(k_1\Gamma_1a_1)} \otimes \mathbf{U}^{(k_2\Gamma_2a_2)}\}_\gamma^{(\Gamma b)}$$
$$= \sum_{\gamma_1\gamma_2} \mathbf{T}_{\gamma_1}^{(k_1\Gamma_1a_1)}\mathbf{U}_{\gamma_2}^{(k_2\Gamma_2a_2)}\langle \Gamma_1\gamma_1\Gamma_2\gamma_2|\Gamma\gamma b\rangle \qquad (3.47)$$

Similar to the case in Eq. (3.40), a relation exists between the tensor operators in Eqs. (3.45) and (3.47), which results in

$$\{\mathbf{T}^{(k_1)} \otimes \mathbf{U}^{(k_2)}\}_\gamma^{(k\Gamma a)}$$
$$\sum_{\substack{\Gamma_1a_1 \\ \Gamma_2a_2b}} \{\mathbf{T}^{(k_1\Gamma_1a_1)}\otimes\mathbf{U}^{(k_2\Gamma_2a_2)}\}_\gamma^{(\Gamma b)}$$
$$\times (k_1\Gamma_1a_1, k_2\Gamma_2a_2|k\Gamma ab) \quad\cdots\cdots \quad (3.48)$$

The totally symmetric scalar product of the tensor operators $\mathbf{T}^{(k\Gamma a)}$ and $\mathbf{U}^{(k\Gamma a')}$ has been defined for fixed Γ and Γ' in Eq. (3.32). The quantities of Eqs. (3.32) and (3.47) are related according to

$$(\mathbf{T}^{(k\Gamma a)} \cdot \mathbf{U}^{(k\overline{\Gamma a})}) = (-1)^j(d_\Gamma)^{1/2}\{\mathbf{T}^{(k\Gamma a)} \otimes \mathbf{U}^{(k\overline{\Gamma a})}\}_{a_1}^{(A_1)} \qquad (3.49)$$

This equation is analogous to Eq. (2.47) with integral j. If, instead of Eq. (3.45), the definition of the scalar product of Eq. (2.47) is taken as the starting point, one obtains using Eq. (3.48) and noting that

$$(k\Gamma a, k\overline{\Gamma a}|0A_1) = \left(\frac{d_\Gamma}{2k+1}\right)^{1/2}\begin{pmatrix} j \\ \Gamma\gamma a \quad \overline{\Gamma\gamma a}\end{pmatrix}$$
$$\times \begin{pmatrix} k \\ \Gamma\gamma a \quad \overline{\Gamma\gamma a}\end{pmatrix}^* \qquad (3.50)$$

the relation

$$(\mathbf{T}^{(k)} \cdot \mathbf{U}^{(k)})$$
$$= \sum_{\Gamma a} (-1)^{k+j}\begin{pmatrix} j \\ \Gamma\gamma a \quad \overline{\Gamma\gamma a}\end{pmatrix}$$
$$\times \begin{pmatrix} k \\ \Gamma\gamma a \quad \overline{\Gamma\gamma a}\end{pmatrix}^* (\mathbf{T}^{(k\Gamma a)} \cdot \mathbf{U}^{(k\overline{\Gamma a})}) \qquad (3.51)$$

Thus Eq. (3.51) provides a connection between the scalar products in $SO(3)$ and in the subgroup G. If Eq. (3.32) is introduced into Eq. (3.51) and the relation (cf. Appendix II)

$$\langle \Gamma\gamma\overline{\Gamma\gamma}|A_1a_1\rangle = (d_\Gamma)^{-1/2}\begin{pmatrix} j \\ \Gamma\gamma a \quad \overline{\Gamma\gamma a}\end{pmatrix}^* \qquad (3.52)$$

is observed, the result is

$$(\mathbf{T}^{(k)} \cdot \mathbf{U}^{(k)}) = (-1)^k \sum_{\Gamma\gamma a}\begin{pmatrix} k \\ \Gamma\gamma a \quad \overline{\Gamma\gamma a}\end{pmatrix}^* \mathbf{T}_\gamma^{(k\Gamma a)}\mathbf{U}_{\overline{\gamma}}^{(k\overline{\Gamma a})} \qquad (3.53)$$

This relation provides the decomposition of the scalar product of two irreducible tensor operators in $SO(3)$ into tensor operators which are irreducible within $G \subset SO(3)$.

3.3. CONCEPT OF THE LIGAND FIELD

We shall now consider a transition metal atom or ion in the environment formed by a number of ligands in a complex ion, complex molecule, or crystal.

The Hamiltonian for an N-electron system of this type may be written as

$$\mathcal{H} = \mathcal{H}_0 + \mathcal{H}_G \qquad (3.54)$$

where $\mathcal{H}_0$ is the Hamiltonian of the free ion, given by Eq. (2.67), and

$$[\mathcal{H}_0, R] = 0 \qquad R \in [SO(3)]^N \qquad (3.55)$$

In Eq. (3.55), it is $R \in G$, and $[SO(3)]^N$ is the Nth-rank inner direct product (cf. Section 3.4) of the three-dimensional rotation group $SO(3)$. Irreducible representations $D^{(j)}$ of $[SO(3)]^N$ are characterized by integral or half-integral values of J which are quantum numbers for $\mathcal{H}_0$. The corresponding levels are $(2J+1)$-fold degenerate. Hamiltonian $\mathcal{H}_G$ represents the additional effect of the ligands and

$$[\mathcal{H}_G, R] = 0 \qquad R \in [G]^N \qquad (3.56)$$

Here $[G]^N$ is the Nth-rank inner direct product (cf. Section 3.4) of point group G which characterizes the symmetry of the complex. Irreducible representations $\Gamma(G)$ of $[G]^N$ are quantum numbers for $\mathscr{H}$. The connection between $D^{(j)}$ and $\Gamma(G)$ is then established by the subduction relation, Eq. (3.35). This relation determines how each level characterized by J is split by the effect of the descending symmetry of its surroundings.

Hamiltonian $\mathscr{H}_G$ may be considered as a sum of N one-electron ligand field potentials

$$\mathscr{H}_G = V_{LF} = \sum_{i=1}^{N} V_{LF}(\mathbf{r}_i) \qquad (3.57)$$

In order to account for the spatial distribution of ligand electrons, particularly toward the metal ion, the density of ligand electrons at the position of metal electron i, $\rho(\mathbf{r}_i)$, is introduced. The ligand field potential $V_{LF}(\mathbf{r}_i)$ then results as solution of the Poisson equation

$$\nabla^2 V_{LF}(\mathbf{r}_i) = 4\pi e \rho(\mathbf{r}_i) \qquad (3.58)$$

It has been shown[20] that this solution is of the form

$$V_{LF}(\mathbf{r}_i) = \int \frac{e\rho(\mathbf{R}_a)}{|\mathbf{R}_a - \mathbf{r}_i|} d\mathbf{R}_a \qquad (3.59)$$

Here, $\mathbf{r}_i$ is the position vector of the electron i with coordinates r_i, θ_i, and φ_i; and $\mathbf{R}_a$ characterizes a general point of the environment, the variables of the integral being R_a, Θ_a, and Φ_a. If we use the spherical harmonic addition theorem Eq. (2.10), the quantity $1/|\mathbf{R}_a - \mathbf{r}_i|$ may be completely factorized to give

$$\frac{1}{|\mathbf{R}_a - \mathbf{r}_i|} = \sum_{k,q} \frac{r_<^k}{r_>^{k+1}} \mathbf{C}_q^{(k)}(\Theta_a, \Phi_a)^* \mathbf{C}_q^{(k)}(\theta_i, \varphi_i) \qquad (3.60)$$

With this result, the solution to Eq. (3.58) may be written in terms of tensor operators as

$$V_{LF}(r_i, \theta_i, \varphi_i) = \sum_{kq} B_q^k(r_i) \mathbf{C}_q^{(k)}(\theta_i, \varphi_i) \qquad (3.61)$$

where

$$B_q^k(r_i) = \int e\rho(\mathbf{R}_a) \frac{r_<^k}{r_>^{k+1}} \mathbf{C}_q^{(k)}(\Theta_a, \Phi_a)^* d\mathbf{R}_a \qquad (3.62)$$

Obviously, the operators $\mathbf{C}_q^{(k)}(\theta_i, \varphi_i)$ with $q = k$, $k-1, \ldots, -k$ are the basis for an irreducible representation of $SO(3)$. In more simple terms, Eq. (3.61) may be regarded as an expansion of perturbation operator $V_{LF}(r_i, \theta_i, \varphi_i)$ of Eq. (3.57) in terms of operators transforming as a complete set of eigenvectors of the unperturbed problem, i.e., of $\mathscr{H}_0$.

An alternate treatment consists of the expansion of potential $V_{LF}(r_i, \theta_i, \varphi_i)$ in terms of tensor operators

$\mathbf{S}_\gamma^{(k\Gamma a)}$ which transform according to an irreducible representation of G,[30]

$$V_{LF}(r_i, \theta_i, \varphi_i) = \sum_{k\Gamma\gamma a} B_\gamma^{k\Gamma a}(r_i) \mathbf{S}_\gamma^{(k\Gamma a)}(\theta_i, \varphi_i) \qquad (3.63)$$

The operators $\mathbf{S}_\gamma^{(k\Gamma a)}$ are linear combinations

$$\mathbf{S}_\gamma^{(k\Gamma a)}(\theta_i, \varphi_i) = \sum_q \mathbf{C}_q^{(k)}(\theta_i, \varphi_i)\langle kq | k\Gamma\gamma a\rangle \qquad (3.64)$$

where the symmetry adaptation coefficients have been defined in Eq. (3.10). The invariance requirement against symmetry transformations of group G then restricts Γ in $\mathbf{S}_\gamma^{(k\Gamma a)}$ to the totally symmetric representation A_1, and thus Eq. (3.63) may be replaced by

$$V_{LF}(r_i, \theta_i, \varphi_i) = \sum_{ka} B^{ka}(r_i) \mathbf{S}^{(ka)}(\theta_i, \varphi_i) \qquad (3.65)$$

where the $\mathbf{S}^{(ka)}$ are determined by Eq. (3.64) for $\Gamma = A_1$, $\gamma = a_1$. The $B^{ka}(r_i)$ follow from Eqs. (3.61) and (3.65) according to

$$B_q^k(r_i) = \sum_a B^{ka}(r_i)\langle kq | kA_1 a_1 a\rangle \qquad (3.66)$$

If primarily the charge distribution external to the metal ion is considered, i.e., $\mathbf{r}_i < \mathbf{R}_a$ is assumed, Eq. (3.61) may be rewritten as

$$V_{CF}(r_i, \theta_i, \varphi_i) = \sum_{kq} A_q^k r_i^k \mathbf{C}_q^{(k)}(\theta_i, \varphi_i) \qquad (3.67)$$

where

$$A_q^k = \int \frac{e\rho(\mathbf{R}_a)}{R_a^{k+1}} \mathbf{C}_q^{(k)}(\Theta_a, \Phi_a)^* d\mathbf{R}_a \qquad (3.68)$$

Evidently, the A_q^k are simply numerical constants characterizing the environment. The potential of Eq. (3.67) then is a solution of the Laplace equation

$$\nabla^2 V_{CF}(\mathbf{r}_i) = 0 \qquad (3.69)$$

and thus corresponds to the point-charge model of simple crystal field theory as it was conceived by Bethe.[5] Since this electrostatic model is physically not well founded, it will be considered in more detail only in Chapter 9 below.

The ligand field potential [Eq. (3.61)] acting on the ith electron may be expressed, for the symmetries which are of interest in this volume, in terms of the tensor operators $\mathbf{C}_q^{(k)}$ as follows:[58]

(a) Cubic (O_h, T_d) symmetry:

$$V_{\text{cub}}(r_i, \theta_i, \varphi_i) = \tfrac{7}{2} B_c^4(r_i)[\mathbf{C}_0^{(4)} + (\tfrac{5}{14})^{1/2}(\mathbf{C}_4^{(4)} + \mathbf{C}_{-4}^{(4)})] \qquad (3.70)$$

(b) Tetragonal (D_{4h}, D_4, C_{4v}) symmetry:

$$V_{D_4}(r_i, \theta_i, \varphi_i) = B_0^2(r_i)\mathbf{C}_0^{(2)} + B_0^4(r_i)\mathbf{C}_0^{(4)}$$

$$+ \frac{1}{\sqrt{2}}B_4^4(r_i)(\mathbf{C}_4^{(4)} + \mathbf{C}_{-4}^{(4)}) \qquad (3.71)$$

(c) Trigonal (D_{3d}, D_3, C_{3v}) symmetry:

$$V_{D_3}(r_i, \theta_i, \varphi_i) = B_0^{2'}(r_i)\mathbf{C}_0^{(2)} + B_0^{4'}(r_i)\mathbf{C}_0^{(4)}$$

$$+ \frac{1}{\sqrt{2}}B_3^{4'}(r_i)(\mathbf{C}_3^{(4)} - \mathbf{C}_{-3}^{(4)}) \qquad (3.72)$$

(d) Cylindrical ($D_{\infty h}$, D_∞, $C_{\infty v}$) symmetry:

$$V_{D_\infty}(r_i, \theta_i, \varphi_i) = B_0^{2''}(r_i)\mathbf{C}_0^{(2)} + B_0^{4''}(r_i)\mathbf{C}_0^{(4)}$$
$$\qquad (3.73)$$

In these potentials, the term in $\mathbf{C}_0^{(0)}$ has been omitted since this term contributes only a uniform shift to all energies. The potential of Eq. (3.72) applies for a choice of axes where the z axis is taken as the trigonal axis and the y axis as the C_2 axis. The potential of Eq. (3.73) is applicable to all D_n symmetries with $n > 4$ and is thus denoted by D_∞.

The alternate formulation of the ligand field potential is in terms of the tensor operators $\mathbf{S}^{(ka)}$ [cf. Eq. (3.65)], which are defined as follows:

(a) Cubic (O_h, T_d) symmetry:

$$\mathbf{S}^{(4)} = \frac{1}{2}(\frac{7}{3})^{1/2}[\mathbf{C}_0^{(4)} + (\frac{5}{14})^{1/2}(\mathbf{C}_4^{(4)} + \mathbf{C}_{-4}^{(4)})] \quad (3.74)$$

(b) Tetragonal (D_{4h}, D_4, C_{4v}) *symmetry*:

$$\mathbf{S}^{(2)} = \mathbf{C}_0^{(2)} \qquad \mathbf{S}^{(41)} = \mathbf{C}_0^{(4)}$$

$$\mathbf{S}^{(42)} = \frac{1}{\sqrt{2}}(\mathbf{C}_4^{(4)} + \mathbf{C}_{-4}^{(4)}) \qquad (3.75)$$

(c) Trigonal (D_{3d}, D_3, C_{3v}) symmetry:

$$\mathbf{S}^{(2)} = \mathbf{C}_0^{(2)} \qquad \mathbf{S}^{(41)} = \mathbf{C}_0^{(4)}$$
$$\qquad (3.76)$$
$$\mathbf{S}^{(42)} = \frac{1}{\sqrt{2}}(\mathbf{C}_3^{(4)} - \mathbf{C}_{-3}^{(4)})$$

(d) Cylindrical ($D_{\infty h}$, D_∞, $C_{\infty v}$) symmetry:

$$\mathbf{S}^{(2)} = \mathbf{C}_0^{(2)} \qquad \mathbf{S}^{(4)} = \mathbf{C}_0^{(4)} \qquad (3.77)$$

The matrix elements of the ligand field potential Eq. (3.61) between the single-electron functions $|lm\rangle$

may be expressed as

$$\langle lm|V_{LF}(r_i, \theta_i, \varphi_i)|lm'\rangle$$

$$= \sum_k \langle B_{m-m'}^k(r_i)\rangle(-1)^m(2l+1)\begin{pmatrix} l & k & l \\ 0 & 0 & 0 \end{pmatrix}$$

$$\times \begin{pmatrix} l & k & l \\ m' & m-m' & -m \end{pmatrix} \qquad (3.78)$$

The quantities $\langle B_q^k(r_i)\rangle$ of Eq. (3.78) may serve directly as empirical parameters to be evaluated from experiment. However, the expansion of Eq. (3.61) often produces more than one term in the matrix element Eq. (3.78). The expansion of $V_{LF}(r_i, \theta_i, \varphi_i)$ in terms of tensor operators of the appropriate symmetry group [cf. Eq. (3.65)] is then more convenient since it produces the most diagonal matrix. Here, the quantities $\langle B^{ka}(r_i)\rangle$ may be again considered as effective parameters of the ligand field.

The parameters employed in the ligand field diagrams of the present volume are the well-known cubic ligand field parameter Dq and, for the remaining symmetries, the parameters introduced by Ballhausen[2] and Liehr.[33] These parameters are defined best in terms of matrix elements of the type $\langle lm|V_{LF}(r_i, \theta_i, \varphi_i)|lm'\rangle$.

In cubic symmetry, one obtains, if complex wave functions $|lm\rangle$ quantized along the fourfold axis are used,

$$\langle 2 \pm 2|V_{cub}|2 \pm 2\rangle = Dq$$
$$\langle 2 \pm 2|V_{cub}|2 \mp 2\rangle = 5Dq$$
$$\langle 2 \pm 1|V_{cub}|2 \pm 1\rangle = -4Dq \qquad (3.79)$$
$$\langle 2 \ 0|V_{cub}|2 \ 0\rangle = 6Dq$$

whereas in terms of real d orbitals the result is

$$\langle d_{x^2-y^2}|V_{cub}|d_{x^2-y^2}\rangle = \langle d_{z^2}|V_{cub}|d_{z^2}\rangle = 6Dq$$
$$\langle d_{xy}|V_{cub}|d_{xy}\rangle = \langle d_{xz}|V_{cub}|d_{xz}\rangle \qquad (3.80)$$
$$= \langle d_{yz}|V_{cub}|d_{yz}\rangle = -4Dq$$

In tetragonal symmetry, the parameters Dq, Ds, and Dt are required. If the definition of Dq from cubic symmetry is retained, the corresponding expressions for complex wave functions read

$$\langle 2 \pm 2|V'_{D_4}|2 \pm 2\rangle = 2Ds - Dt$$
$$\langle 2 \pm 1|V'_{D_4}|2 \pm 1\rangle = -Ds + 4Dt \qquad (3.81)$$
$$\langle 2 \ 0|V'_{D_4}|2 \ 0\rangle = -2Ds - 6Dt$$

Here only matrix elements of the difference potential $V'_{D_4} = V_{D_4} - V_{cub}$ have been listed. To these must be added the matrix element of V_{cub} from Eq. (3.79), which gives

$$\langle 2 \pm 2|V_{D_4}|2 \pm 2\rangle = Dq + 2Ds - Dt \quad \text{etc.} \tag{3.82}$$

Similarly, using real wave functions, one obtains

$$\langle d_{x^2-y^2}|V'_{D_4}|d_{x^2-y^2}\rangle = 2Ds - Dt$$

$$\langle d_{z^2}|V'_{D_4}|d_{z^2}\rangle = -2Ds - 6Dt \tag{3.83}$$

$$\langle d_{xy}|V'_{D_4}|d_{xy}\rangle = 2Ds - Dt$$

$$\langle d_{xz}|V'_{D_4}|d_{xz}\rangle = \langle d_{yz}|V'_{D_4}|d_{yz}\rangle = -Ds + 4Dt$$

The various ligand field parameters introduced above for cubic and tetragonal symmetries are related according to

$$\langle B^4(r_i)\rangle = \sqrt{21}\langle B_c^4(r_i)\rangle = 6\sqrt{21}Dq$$

$$\langle B^2(r_i)\rangle = \langle B_0^2(r_i)\rangle = -7Ds$$

$$\langle B^{41}(r_i)\rangle = \langle B_0^4(r_i)\rangle = 21(Dq - Dt) \tag{3.84}$$

$$\langle B^{42}(r_i)\rangle = \langle B_4^4(r_i)\rangle = 3\sqrt{35}Dq$$

If the trigonal axis is chosen as the axis of quantization, the nonzero matrix elements of V_{cub} result as

$$\langle 2 \pm 2|V_{cub}|2 \pm 2\rangle = -\tfrac{2}{3}Dq$$

$$\langle 2 \pm 1|V_{cub}|2 \pm 1\rangle = \tfrac{8}{3}Dq$$

$$\langle 2\ 0|V_{cub}|2\ 0\rangle = -4Dq \tag{3.85}$$

$$\langle 2 \pm 2|V_{cub}|2 \mp 1\rangle = \langle 2 \mp 1|V_{cub}|2 \pm 2\rangle$$

$$= \pm\tfrac{10}{3}\sqrt{2}Dq$$

In trigonal symmetry, it is usually most convenient to use wave functions quantized along the threefold axis. Similar to Ds and Dt in tetragonal symmetry, the trigonal splitting parameters $D\sigma$ and $D\tau$ may be introduced. The nonvanishing matrix elements in the case of weak-field coupling are then obtained as

$$\langle 2 \pm 2|V'_{D_3}|2 \pm 2\rangle = 2D\sigma - D\tau$$

$$\langle 2 \pm 1|V'_{D_3}|2 \pm 1\rangle = -D\sigma + 4D\tau \tag{3.86}$$

$$\langle 2\ 0|V'_{D_3}|2\ 0\rangle = -2D\sigma - 6D\tau$$

Here again only matrix elements of the difference potential $V'_{D_3} = V_{D_3} - V_{cub}$ are listed. These have to be implemented by the matrix elements of V_{cub}, given by Eq. (3.85), thus producing

$$\langle 2 \pm 2|V_{D_3}|2 \pm 2\rangle = -\tfrac{2}{3}Dq + 2D\sigma - D\tau \quad \text{etc.} \tag{3.87}$$

In the strong-field coupling case, the matrix elements

of the trigonal potential result as

$$\langle e_\pm(t_{2g})|V'_{D_3}|e_\pm(t_{2g})\rangle = D\sigma + \tfrac{2}{3}D\tau$$

$$\langle a_1(t_{2g})|V'_{D_3}|a_1(t_{2g})\rangle = -2D\sigma - 6D\tau$$

$$\langle e_\pm(e_g)|V'_{D_3}|e_\pm(e_g)\rangle = \tfrac{7}{3}D\tau \tag{3.88}$$

$$\langle e_\pm(t_{2g})|V'_{D_3}|e_\pm(e_g)\rangle = -\frac{\sqrt{2}}{3}(3D\sigma - 5D\tau)$$

The different ligand field parameters for trigonal symmetry introduced above are related to each other by

$$\langle B^2(r_i)\rangle = \langle B_0^{2'}(r_i)\rangle = -7D\sigma$$

$$\langle B^{41}(r_i)\rangle = \langle B_0^{4'}(r_i)\rangle = -14(Dq + \tfrac{3}{2}D\tau) \tag{3.89}$$

$$\langle B^{42}(r_i)\rangle = \langle B_3^{4'}(r_i)\rangle = -4\sqrt{35}Dq$$

Another set of trigonal ligand field parameters, K and K', has been introduced by Sugano and Tanabe,[57] and this set is equivalent to the parameters v and v' of Pryce and Runciman.[45] These parameters are defined by the following relations:

$$\langle e_\pm(t_{2g})|V'_{D_3}|e_\pm(t_{2g})\rangle = -\tfrac{1}{3}v = K$$

$$\langle a_1(t_{2g})|V'_{D_3}|a_1(t_{2g})\rangle = \tfrac{2}{3}v = -2K \tag{3.90}$$

$$\langle e_\pm(t_{2g})|V'_{D_3}|e_\pm(e_g)\rangle = v' = K'$$

It is characteristic for the definitions of Eq. (3.90) that the center of gravity is maintained for the trigonally split t_{2g} orbitals. This is possible since the "cubic" parameter, which is analogous to Dq, has been defined here in a different way. If this parameter, consistent with Eq. (3.90), is denoted by Dq', the following relations between the two sets of parameters arise:[44]

$$Dq + \tfrac{7}{18}D\tau = Dq'$$

$$\tfrac{1}{3}(9D\sigma + 20D\tau) = -v = 3K \tag{3.91}$$

$$\tfrac{1}{3}\sqrt{2}(3D\sigma - 5D\tau) = -v' = -K'$$

It should be noted that only by proper regard of Eq. (3.91), including the first equation in the set, it is possible to transform the trigonal matrices of any d^N configuration from one scheme into the other.

It should be also mentioned that another author, Perumareddi,[42] although employing the parameters Ds, Dt for D_4 symmetry and $D\sigma$, $D\tau$ for D_3 symmetry, uses these parameters with a sign opposite to the definition by Ballhausen[2,33] which is applied in this volume.

Several less frequently employed definitions of ligand field (LF) parameters from the literature are listed below.

(a) LF parameters of Otsuka for D_4 symmetry:[41]

$$X = 10(Dq - \tfrac{7}{12}Dt)$$

$$Y = 2Ds + \tfrac{5}{2}Dt \tag{3.92}$$

$$Z = -2Ds + \tfrac{10}{3}Dt$$

(b) LF parameters of Hempel et al.[16] for D_4 symmetry:

$$DQ = 6\sqrt{21}Dq - \tfrac{7}{2}\sqrt{21}Dt$$

$$DS = -7Ds \tag{3.93}$$

$$DT = \tfrac{7}{2}\sqrt{15}Dt$$

(c) LF parameters of Kammer[25] for D_3 symmetry:

$$D = 10Dq$$

$$E = -\tfrac{5}{3}D\tau \tag{3.94}$$

$$F = -D\sigma$$

(d) LF parameters of Varga et al.[60] for D_3 symmetry:

$$\begin{aligned} Ds &= -D\sigma \\ Dt &= -D\tau \end{aligned} \tag{3.95}$$

In cylindrical symmetry D_∞, the parameters $D\mu$ and $D\nu$ may be employed. The matrix elements for complex wave functions in weak-field coupling obtain from Eq. (3.81) if Ds and Dt are replaced by $D\mu$ and $D\nu$, respectively,

$$\langle 2 \pm 2 | V_{D_\infty} | 2 \pm 2 \rangle = 2D\mu - D\nu$$

$$\langle 2 \pm 1 | V_{D_\infty} | 2 \pm 1 \rangle = -D\mu + 4D\nu \tag{3.96}$$

$$\langle 2\ 0 | V_{D_\infty} | 2\ 0 \rangle = -2D\mu - 6D\nu$$

The resulting matrix elements apply to all point groups D_n, where $n > 4$. Similarly, for real wave functions in strong-field coupling, Eq. (3.83) obtains if the above replacement is effected. Finally, the ligand field parameters resulting from Eqs. (3.73) and (3.77) are related to $D\mu$ and $D\nu$ according to

$$\begin{aligned} \langle B^2(r_i) \rangle &= \langle B_0^{2''}(r_i) \rangle = -7D\mu \\ \langle B^4(r_i) \rangle &= \langle B_0^{4''}(r_i) \rangle = -21D\nu \end{aligned} \tag{3.97}$$

Obviously, there are numerous advantages and disadvantages as regards the various definitions of ligand field parameters. No evidence is available with respect to the dependence of these parameters on particular ligands or ligands in a specific direction of space. It follows that no relation exists between ligand field parameters pertaining to different molecular symmetries. In particular, the definition of Dq in D_4

symmetry is not identical to that in D_3 symmetry. However, in Chapter 9, this relation will be elaborated upon on the basis of crystal field theory.

3.4. THE THREE COUPLING CASES: WEAK FIELD, INTERMEDIATE FIELD, AND STRONG FIELD

We consider a transition metal ion which is imbedded in a crystal or surrounded, within a complex, by a certain number of negatively charged ions or neutral molecules. The Hamiltonian of the system may then be approximated, on the basis of Eqs. (2.68), (2.69), and (3.57), by

$$\begin{aligned} \mathcal{H} = \sum_{i=1}^{N} & \left\{ -\frac{\hbar^2}{2m} \nabla_i^2 + U(r_i) \right\} + \sum_{i>j} \frac{e^2}{r_{ij}} \\ & + \sum_i \kappa\xi(r_i)(\mathbf{s}_i \cdot \mathbf{l}_i) + \sum_i V_{LF}(r_i, \theta_i, \varphi_i) \\ & + \sum_i \frac{\mu_B}{\hbar}(\kappa\mathbf{l}_i + g_e\mathbf{s}_i)\mathbf{H} \end{aligned} \tag{3.98}$$

The individual terms in this expansion have been discussed above except the last term, the Zeeman operator, which is nonzero only if an external magnetic field is present. This term has been included in Eq. (3.98) to remind us of the fact that additional small terms exist which may become important in certain applications. In what follows, these additional terms will be neglected.

In Éq. (3.98), the term in braces, i.e., $\mathcal{H}_0$ of Eq. (2.68), is always the largest one, whereas the following three terms may vary over a considerable range of magnitude. Consequently, three different coupling schemes may be distinguished which differ, at least formally, by the sequence of magnitudes assumed for the three central terms of Eq. (3.98). The coupling schemes may be characterized as follows.

(i) Weak-field coupling:

$$\sum_{i>j} \frac{e^2}{r_{ij}} > \sum_i \kappa\xi(r_i)(\mathbf{s}_i \cdot \mathbf{l}_i) > \sum_i V_{LF}(r_i, \theta_i, \varphi_i) \tag{3.99}$$

(ii) Intermediate-field coupling:

$$\sum_{i>j} \frac{e^2}{r_{ij}} > \sum_i V_{LF}(r_i, \theta_i, \varphi_i) > \sum_i \kappa\xi(r_i)(\mathbf{s}_i \cdot \mathbf{l}_i) \tag{3.100}$$

(iii) Strong-field coupling:

$$\sum_i V_{LF}(r_i, \theta_i, \varphi_i) > \sum_{i>j} \frac{e^2}{r_{ij}} > \sum_i \kappa\xi(r_i)(\mathbf{s}_i \cdot \mathbf{l}_i) \tag{3.101}$$

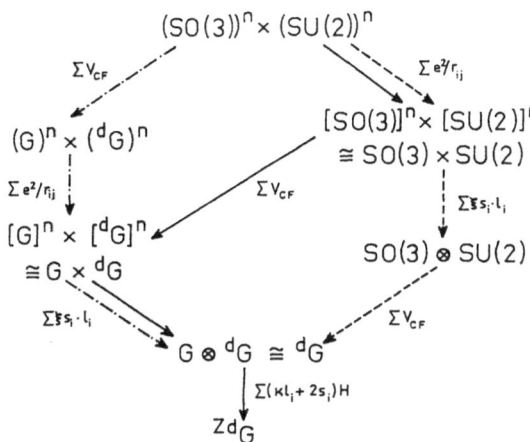

FIGURE 1. Group lattice for configuration d^n in a ligand field of symmetry G according to weak-field coupling ($\text{-}\text{-}\text{-}\rightarrow$), intermediate-field coupling ($\longrightarrow$), and strong-field coupling ($\text{---}\cdot\rightarrow$).

Thus, for example, the case of intermediate-field coupling is specified by the requirement that the ligand field interaction should be larger than the effect of spin–orbit coupling, although smaller than the Coulomb interaction between the electrons. In this way, a sequence of Hamiltonians of increasing accuracy is established. To clarify further this situation, the theory of group chains[36] may be applied.

Let us recall to this end the definition of the direct product of groups H and K, usually denoted by $H \times K$ and defined by the elements

$$H \times K = \{H_a \times K_b, a = 1 \text{ to } h; b = 1 \text{ to } k\} \tag{3.102}$$

Here h and k are the orders of groups H and K, respectively. The inner direct product of H and K, will be denoted by $H \otimes K$ and is defined accordingly by

$$H \otimes K = \{H_{ja} \times K_a, j = 1 \text{ to } h/k; a = 1 \text{ to } k\} \tag{3.103}$$

if the homomorphism $H \to K$ holds. It should be observed that it is

$$H \times K \supset H \otimes K \cong H \tag{3.104}$$

Also, the n-fold inner direct product of H with itself

is denoted here by $[H]^n$, whereby

$$[H]^n = H \otimes H \otimes \cdots \otimes H$$
$$= \{H_a \times H_a \times \cdots \times H_a, a = 1 \text{ to } h\} \tag{3.105}$$

and similarly to Eq. (3.104),

$$(H)^n \supset [H]^n \cong H \tag{3.106}$$

With this in mind, the different coupling schemes may be conveniently characterized by the corresponding chain of symmetry groups. Starting with $\mathcal{H}_0$ we thus obtain in intermediate-field coupling:

$$(SO(3))^N \times (SU(2))^N \supset [SO(3)]^N \times [SU(2)]^N$$
$$\supset [G]^N \times [{}^dG]^N \supset G \otimes {}^dG \cong {}^dG \tag{3.107}$$

The mutual relationships between the three coupling schemes are indicated by the group lattice shown in Fig. 1. Here we have denoted the crystallographic (or molecular) point group determined by the potential $V_{LF}(r_i, \theta_i, \varphi_i)$ by G, the corresponding double group by ${}^dG$, and the Zeeman group by ${}^{Zd}G$. Of course, the characterization of the individual groups of Fig. 1 may be complemented by S_n, the group of permutations of n particles, i.e., the N electrons of the incompletely filled d shell.

In the following three chapters, the coupling cases of the weak field, the intermediate field, and the strong field will be discussed individually in more detail.

4

WEAK-FIELD COUPLING

4.1. STATES AND WAVE FUNCTIONS

The levels resulting from Russell–Saunders terms of the free ion may be conventionally denoted by $^{2S+1}_v L_J$, the corresponding wave functions being $|vSLJM\rangle$, where $M = J, J-1, \ldots, -J$ (cf. Section 2.4). The effect of a ligand field of particular symmetry consists, in general, in splitting the free ion levels labeled by J into ligand field states. The decomposition is equivalent to the reduction

$$D^{(J)}(G) = \sum_i a_i^{(J)} \Gamma_i \qquad (4.1)$$

where $D^{(J)}(G)$ is subduced from representation $D^{(J)}$ of $SO(3)$ and where G denotes the symmetry group appropriate to the problem. The basis functions conforming to the Hamiltonian terms in the order of Eq. (3.99) and transforming according to irreducible representation Γ, component γ, of G are then given by

$$|vSLJ\Gamma\gamma a\rangle = \sum_M |vSLJM\rangle\langle JM|J\Gamma\gamma a\rangle \qquad (4.2)$$

[cf. Eq. (3.10)]. In the d^2 configuration and if the symmetry is cubic, we always have $a_i^{(J)} \equiv 1$ in Eq. (4.1), and thus the branching multiplicity index a in Eq. (4.2) is not required. In all higher configurations or if the symmetry is tetragonal or trigonal, $a_i^{(J)} > 1$ may arise for a particular value of i in Eq. (4.1). In this case, Γ_i occurs more than once, the equivalent states then being distinguished by the index a. The symmetry adaptation coefficients $\langle JM|J\Gamma\gamma a\rangle$ of Eq. (4.2) may be determined on the basis of Eq. (3.12).

The various ligand field states arising from the configurations d^N, $N = 2, \ldots, 8$ in cubic, tetragonal, trigonal, and cylindrical symmetry are listed in Table I in conjunction with the dimensions of the corresponding energy matrices. In accord with the diagrams for d^N, $N = 3, \ldots, 7$, in Part II of the volume, the effect of spin–orbit interaction has been disregarded in this listing. It should be noted that for d^2 and d^8 in cubic symmetry *with* spin–orbit coupling, the irreducible representations Γ_1, Γ_2, Γ_3, Γ_4, and Γ_5 occur, the energy matrices having here the dimensions 4, 1, 5, 4, and 6, respectively. If the symmetry is tetragonal, the levels Γ_{t1}, Γ_{t2}, Γ_{t3}, Γ_{t4}, and Γ_{t5} arise, the corresponding degeneracies being 9, 4, 6, 6, and 10. If the symmetry is trigonal, the resulting irreducible representations are Γ_1^T, Γ_2^T, and Γ_3^T, the dimensions of the energy matrices then being 10, 5, and 15, respectively. The notation for irreducible representations follows the usage introduced by Low.[34]

4.2. COULOMB INTERACTION

Matrix elements of the Coulomb interelectronic repulsion operator between the functions of Eq. (4.2) are determined by

$$\left\langle vSLJ\Gamma\gamma a \left| \sum_{i>j} \frac{e^2}{r_{ij}} \right| v'S'L'J'\Gamma'\gamma'a' \right\rangle$$

$$= \left\langle vSL \left| \sum_{i>j} \frac{e^2}{r_{ij}} \right| v'S'L' \right\rangle$$

$$\times \delta(J, J')\delta(\Gamma, \Gamma')\delta(\gamma, \gamma')\delta(a, a') \qquad (4.3)$$

Obviously, the nonzero matrix elements on the right-hand side of Eq. (4.3) apply to the d^N configuration of the free transition metal ion. These matrix elements are identical for all components M_S and M_L and they may be reduced to matrix elements for two electrons by the procedure discussed in Section 2.3. The two-electron matrix elements of the operator $\sum e^2/r_{ij}$ for

TABLE I

The representations of energy terms in the configurations d^N, $N = 2, 3, \ldots, 8$ of O_h, D_4, D_3, and D_∞ symmetry and the dimensions of the corresponding energy matrices

O_h	d_Γ	D_4	d_Γ	D_3	d_Γ	D_∞	d_Γ
Configurations d^2, d^8							
Γ_1	4	Γ_{t1}	9	Γ_1^T	10	Γ_1^C	7
Γ_2	1	Γ_{t2}	4	Γ_2^T	5	Γ_2^C	2
Γ_3	5	Γ_{t3}	6	Γ_3^T	15	Γ_3^C	7
Γ_4	4	Γ_{t4}	6			Γ_4^C	6
Γ_5	6	Γ_{t5}	10			Γ_5^C	3
						Γ_6^C	2
Configurations d^3, d^7							
$^4A_2$	1	$^4A_2$	2	$^4A_1$	1	$^4A_2$	2
$^4T_1$	2	$^4B_1$	1	$^4A_2$	3	$^4E_1$	2
$^4T_2$	1	$^4B_2$	1	^4E	3	$^4E_2$	1
$^2A_1$	1	^4E	3	$^2A_1$	6	$^4E_3$	1
$^2A_2$	1	$^2A_1$	5	$^2A_2$	6	$^2A_1$	3
^2E	4	$^2A_2$	5	^2E	14	$^2A_2$	3
$^2T_1$	5	$^2B_1$	5			$^2E_1$	6
$^2T_2$	5	$^2B_2$	5			$^2E_2$	5
		^2E	10			$^2E_3$	3
						$^2E_4$	2
						$^2E_5$	1
Configurations d^4, d^6							
^5E	1	$^5A_1$	1	$^5A_1$	1	$^5A_1$	1
$^5T_2$	1	$^5B_1$	1	^5E	2	$^5E_1$	1
$^3A_1$	1	$^5B_2$	1	$^3A_1$	6	$^5E_2$	1
$^3A_2$	2	^5E	1	$^3A_2$	9	$^3A_1$	2
^3E	3	$^3A_1$	4	^3E	15	$^3A_2$	5
$^3T_1$	7	$^3A_2$	7	$^1A_1$	12	$^3E_1$	7
$^3T_2$	5	$^3B_1$	5	$^1A_2$	6	$^3E_2$	5
$^1A_1$	5	$^3B_2$	5	^1E	16	$^3E_3$	4
$^1A_2$	2	^3E	12			$^3E_4$	2
^1E	5	$^1A_1$	10			$^3E_5$	1
$^1T_1$	4	$^1A_2$	4			$^1A_1$	7
$^1T_2$	7	$^1B_1$	7			$^1A_2$	1
		$^1B_2$	7			$^1E_1$	6
		^1E	11			$^1E_2$	6
						$^1E_3$	4
						$^1E_4$	3
						$^1E_5$	1
						$^1E_6$	1
Configuration d^5							
$^6A_1$	1	$^6A_1$	1	$^6A_1$	1	$^6A_1$	1
$^4A_1$	1	$^4A_1$	3	$^4A_1$	4	$^4A_1$	2
$^4A_2$	1	$^4A_2$	3	$^4A_2$	4	$^4A_2$	2
^4E	2	$^4B_1$	3	^4E	8	$^4E_1$	4
$^4T_1$	3	$^4B_2$	3	$^2A_1$	14	$^4E_2$	3
$^4T_2$	3	^4E	6	$^2A_2$	11	$^4E_3$	2
$^2A_1$	4	$^2A_1$	11	^2E	25	$^4E_4$	1
$^2A_2$	3	$^2A_2$	8			$^2A_1$	7
^2E	7	$^2B_1$	10			$^2A_2$	4
$^2T_1$	8	$^2B_2$	10			$^2E_1$	10
$^2T_2$	10	^2E	18			$^2E_2$	9
						$^2E_3$	6
						$^2E_4$	4
						$^2E_5$	2
						$^2E_6$	1

equivalent electrons are then determined by Eqs. (2.78) and (2.79). Consequently, the computation of the matrix elements, Eq. (4.3), is converted to that of the reduced matrix elements $\langle l \| \mathbf{C}^{(k)} \| l' \rangle$ [cf. Eq. (2.27)], and the 6-j symbols [cf. Eq. (2.39)]. The quantities F^k ($k = 0, 2, 4$) or the related Racah parameters A, B, C, given by Eq. (2.77), are usually considered as adjustable parameters, and this practice is followed in the diagrams that follow. Since Racah parameter A provides a uniform shift of all energies only, this quantity need not be considered further.

4.3. SPIN–ORBIT COUPLING

Matrix elements of the spin–orbit coupling operator between the functions of Eq. (4.2) may be expressed as

$$\langle vSLJ\Gamma\gamma a | \sum_i \kappa \xi(r_i)(\mathbf{s}_i \cdot \mathbf{l}_i) | v'S'L'J'\Gamma'\gamma'a' \rangle$$

$$= (-1)^{J+L+S'}[l(l+1)(2l+1)]^{1/2} \begin{Bmatrix} L & L' & 1 \\ S' & S & J \end{Bmatrix}$$

$$\times \langle vSL \| \mathbf{V}^{(11)} \| v'S'L' \rangle \delta(J, J')$$

$$\times \delta(\Gamma, \Gamma')\delta(\gamma, \gamma')\delta(a, a')\zeta_{nl} \tag{4.4}$$

Here, ζ_{nl} is the spin–orbit coupling *parameter* which is usually taken as being reduced in magnitude by the amount κ as compared to ζ_{nl}^{free} of the free ion. The quantity κ is denoted the *orbital reduction factor*.[9] In addition, $\langle vSL \| \mathbf{V}^{(11)} \| v'S'L' \rangle$ is a reduced matrix element of the double tensor operator $\mathbf{V}^{(11)}$ [cf. Eq. (2.87)]. The computation of the matrix elements [Eq. (4.4)] thus requires coefficients of fractional parentage and 6-j symbols, the quantity ζ_{nl} or κ being determined on a semiempirical basis. It should be noted that, for convenience of presentation, the major part of the diagrams in the second part of the volume is displayed for zero spin–orbit forces. A notable exception are the configurations d^2 and d^8 in cubic, tetragonal, trigonal, and cylindrical symmetry.

4.4. LIGAND FIELD INTERACTION

Matrix elements of a ligand field potential of cubic (O_h, T_d) symmetry [cf. Eq. (3.70)] between the functions of Eq. (4.2) may be written as

$$\langle vSLJ\Gamma\gamma a | \sum_i V_{\text{cub}}(r_i, \theta_i, \varphi_i) | v'S'L'J'\Gamma'\gamma'a' \rangle$$

$$= (-1)^{S+L'} 3\sqrt{70}[(2J+1)(2J'+1)]^{1/2} \begin{Bmatrix} J & J' & 4 \\ L' & L & S \end{Bmatrix}$$

$$\times \langle vSL \| \mathbf{U}^{(4)} \| v'S'L' \rangle \sum_{MM'} (-1)^M \langle J\Gamma\gamma a | JM \rangle$$

$$\times \left\{ \begin{pmatrix} J & 4 & J' \\ -M & 0 & M' \end{pmatrix} + \left(\frac{5}{14} \right)^{1/2} \right.$$

$$\left. \times \left[\begin{pmatrix} J & 4 & J' \\ -M & 4 & M' \end{pmatrix} + \begin{pmatrix} J & 4 & J' \\ -M & -4 & M' \end{pmatrix} \right] \right\}$$

$$\times \langle J'M' | J'\Gamma'\gamma'a' \rangle \delta(\Gamma, \Gamma') \delta(\gamma, \gamma') \delta(S, S') Dq \quad (4.5)$$

Here, the symmetry adaptation coefficients $\langle JM | J\Gamma\gamma a \rangle$ have been defined in Eq. (4.2) above, $\langle vSL \| \mathbf{U}^{(4)} \| v'S'L' \rangle$ is a reduced matrix element of the unit tensor operator $\mathbf{U}^{(4)}$ [cf. Eq. (2.85)], and Dq is the cubic ligand field parameter [cf. Eq. (3.79)]. Matrix elements of a ligand field potential of tetragonal (D_{4h}, D_4, C_{4v}) symmetry [cf. Eq. (3.71)] may be expressed similarly as

$$\langle vSLJ\Gamma\gamma a | \sum_i V_{D_4}(r_i, \theta_i, \varphi_i) | v'S'L'J'\Gamma'\gamma'a' \rangle$$

$$= (-1)^{S+L} [(2J+1)(2J'+1)]^{1/2}$$

$$\times \left\{ \left(\sqrt{70} \begin{Bmatrix} J & J' & 2 \\ L' & L & S \end{Bmatrix} \right. \right.$$

$$\times \langle vSL \| \mathbf{U}^{(2)} \| v'S'L' \rangle \sum_M (-1)^M \langle J\Gamma\gamma a | JM \rangle$$

$$\left. \times \begin{pmatrix} J & 2 & J' \\ -M & 0 & M \end{pmatrix} \langle J'M | J'\Gamma'\gamma'a' \rangle Ds \right)$$

$$+ 3\sqrt{70} \begin{Bmatrix} J & J' & 4 \\ L' & L & S \end{Bmatrix} \langle vSL \| \mathbf{U}^{(4)} \| v'S'L' \rangle$$

$$\times \left[\left(\sum_M (-1)^M \langle J\Gamma\gamma a | JM \rangle \begin{pmatrix} J & 4 & J' \\ -M & 0 & M \end{pmatrix} \right. \right.$$

$$\left. \times \langle J'M | J'\Gamma'\gamma'a' \rangle (Dq - Dt) \right)$$

$$+ \left(\sqrt{\tfrac{5}{14}} \sum_{MM'} (-1)^M \langle J\Gamma\gamma a | JM \rangle \right.$$

$$\times \left[\begin{pmatrix} J & 4 & J' \\ -M & 4 & M' \end{pmatrix} + \begin{pmatrix} J & 4 & J' \\ -M & -4 & M' \end{pmatrix} \right]$$

$$\left. \left. \left. \times \langle J'M' | J'\Gamma'\gamma'a' \rangle Dq \right) \right] \right\} \delta(\Gamma, \Gamma') \delta(\gamma, \gamma') \delta(S, S')$$
$$(4.6)$$

Also, matrix elements of a ligand field potential of trigonal (D_{3d}, D_3, C_{3v}) symmetry [cf. Eq. (3.72)] may be obtained according to

$$\langle vSLJ\Gamma\gamma a | \sum_i V_{D_3}(r_i, \theta_i, \varphi_i) | v'S'L'J'\Gamma'\gamma'a' \rangle$$

$$= (-1)^{S+L} [(2J+1)(2J'+1)]^{1/2}$$

$$\times \left\{ \left(\sqrt{70} \begin{Bmatrix} J & J' & 2 \\ L' & L & S \end{Bmatrix} \right. \right.$$

$$\times \langle vSL \| \mathbf{U}^{(2)} \| v'S'L' \rangle \sum_M (-1)^M \langle J\Gamma\gamma a | JM \rangle$$

$$\times \begin{pmatrix} J & 2 & J' \\ -M & 0 & M \end{pmatrix} \langle J'M | J'\Gamma'\gamma'a' \rangle D\sigma \right)$$

$$+ 2\sqrt{70} \begin{Bmatrix} J & J' & 4 \\ L' & L & S \end{Bmatrix} \langle vSL \| \mathbf{U}^{(4)} \| v'S'L' \rangle$$

$$\times \left[\left(\sum_M (-1)^M \langle J\Gamma\gamma a | Jm \rangle \begin{pmatrix} J & 4 & J' \\ -M & 0 & M \end{pmatrix} \right. \right.$$

$$\left. \times \langle J'M | J'\Gamma'\gamma'a' \rangle (-Dq + \tfrac{3}{2} D\tau) \right)$$

$$- \left(\sqrt{\tfrac{10}{7}} \sum_{MM'} (-1)^M \langle J\Gamma\gamma a | JM \rangle \right.$$

$$\times \left[\begin{pmatrix} J & 4 & J' \\ -M & 3 & M' \end{pmatrix} - \begin{pmatrix} J & 4 & J' \\ -M & -3 & M' \end{pmatrix} \right]$$

$$\left. \left. \left. \times \langle J'M' | J'\Gamma'\gamma'a' \rangle Dq \right) \right] \right\} \delta(\Gamma, \Gamma') \delta(\gamma, \gamma') \delta(S, S')$$
$$(4.7)$$

Here, in addition to the quantities defined above, $\langle vSL \| \mathbf{U}^{(2)} \| v'S'L' \rangle$ is a reduced matrix element of the unit tensor operator $\mathbf{U}^{(2)}$ [cf. Eq. (2.85)]. In Eq. (4.6), Dq, Ds, and Dt are the tetragonal ligand field parameters [cf. Eq. (3.81)]. Similarly, Dq, $D\sigma$, and $D\tau$ in Eq. (4.7) are the parameters of the trigonal ligand field [cf. Eqs. (3.85) and (3.86)]. Matrix elements of a ligand field potential of cylindrical ($D_{\infty h}$, D_∞, $C_{\infty v}$) symmetry [cf. Eq. (3.73)] simply result from Eq. (4.6) if $Dq = 0$. Evidently, the computation of matrix elements Eqs. (4.5)–(4.7) requires only coefficients of fractional parentage, 3-j and 6-j symbols, the quantities Dq, Ds, Dt, etc. being considered as adjustable parameters. The diagrams in the second part of the volume are therefore presented, for tetragonal symmetry, e.g., as function of Dq, Dt, and Ds or rather $K = Ds/Dt$.

4.5. ADDITIONAL INTERACTIONS

The energy matrices resulting on the basis of the matrix elements of Sections 4.2–4.4 are partly in diagonal form. The dimensions of the individual submatrices follow from Table I. The diagonalization of these matrices then produces eigenvalues of energy E_n which are displayed in the second part of the volume. In addition, a linear combination of the p_Γ basis functions $|vSLJ\Gamma\gamma a \rangle$ of representation Γ, component γ, results as

$$| n\Gamma\gamma \rangle = \sum_{k=1}^{p_\Gamma} | vSLJ\Gamma\gamma a \rangle C_k \quad (4.8)$$

Here, p_Γ is the dimension of the energy matrix representation Γ and the summation index $k = vSLJa$. The coefficients C_k squared determine the composition of the individual levels. The functions Eq. (4.8) may be employed to determine the effect of additional interactions. Of the latter, the energy of interaction with an external magnetic field (Zeeman effect) is of considerable interest, since it enables the calculation of magnetic susceptibility. This topic is dealt with elsewhere.[31]

The inclusion of the normal Coulomb, spin–orbit, and ligand field interactions accounts by far for the greatest part of the energy of transition metal ions in crystals or complex molecules. In an accurate treatment of these systems, additional interactions need to be considered. These remaining interactions comprise the spin–spin, orbit–orbit, and spin–other-orbit interactions as well as the effects of magnetic hyperfine structure, etc. The corresponding energy contributions are usually small and, therefore, a perturbation approach within the lowest levels of Eq. (4.8) is very often adequate. For rare earth atoms, the remaining interactions have been considered in irreducible tensor formalism by Wybourne.[62] The relevant algebra may be easily applied to transition metal ions in crystals and complexes if the procedure discussed above is followed.

5

INTERMEDIATE-FIELD COUPLING

5.1. STATES AND WAVE FUNCTIONS

The Russell–Saunders terms of the free ion are usually denoted $^{2S+1}_vL$, the corresponding wave functions being $|vSM_S\rangle|vLM_L\rangle$, where $M_S = S, S-1, \ldots, -S$ and $M_L = L, L-1, \ldots, -L$ (cf. Section 2.4). If a ligand field characterized by symmetry group G is applied, the $^{2S+1}_vL$ terms are split, in general, according to the reduction

$$D^{(L)}(G) = \sum_i a_i^{(L)} \Gamma_{Li}$$
$$D^{(S)}(G) = \sum_j a_j^{(S)} \Gamma_{Sj} \tag{5.1}$$

Here, the representations $D^{(L)}(G)$ and $D^{(S)}(G)$ of orbital and spin angular momentum are subduced from representations $D^{(L)}$ and $D^{(S)}$ of $SO(3)$, respectively. The basis functions conforming to the Hamiltonian terms in the order of Eq. (3.100) and transforming according to irreducible representations Γ_L and Γ_S, components γ_L and γ_S, of G are then determined by[30]

$$|vL\Gamma_L\gamma_La_L\rangle = \sum_{M_L} |vLM_L\rangle\langle LM_L|L\Gamma_L\gamma_La_L\rangle$$
$$|vS\Gamma_S\gamma_Sa_S\rangle = \sum_{M_S} |vSM_S\rangle\langle SM_S|S\Gamma_S\gamma_Sa_S\rangle \tag{5.2}$$

[cf. Eq. (3.10)]. In the d^2 configuration and if the symmetry is cubic, it is in Eq. (5.1) always $a_i^{(L)} \equiv 1$ and $a_j^{(S)} \equiv 1$, and thus the branching multiplicity indices a_L and a_S in Eq. (5.2) are not required. In all higher configurations or if the symmetry is tetragonal or trigonal, $a_i^{(L)} > 1$ or $a_j^{(S)} > 1$ may arise for a particular

value of i or j in Eq. (5.1). In this case, Γ_{Li} or Γ_{Sj} occurs more than once, the equivalent states then being distinguished by the index a_L or a_S. The symmetry adaptation coefficients $\langle LM_L|L\Gamma_L\gamma_La_L\rangle$ and $\langle SM_S|S\Gamma_S\gamma_Sa_S\rangle$ of Eq. (5.2) may be determined again on the basis of Eq. (3.12).

If spin–orbit coupling is introduced next [cf. Eq. (3.100)], basis functions are required which transform according to representation Γ_i following from the Kronecker product

$$\Gamma_S \otimes \Gamma_L = \sum_i b_i \Gamma_i \tag{5.3}$$

The coupled basis functions may then be written as

$$|(vS\Gamma_Sa_SL\Gamma_La_L)\Gamma\gamma b\rangle$$
$$= \sum_{\gamma_L\gamma_S} |vS\Gamma_S\gamma_Sa_S\rangle|vL\Gamma_L\gamma_La_L\rangle\langle\Gamma_S\gamma_S\Gamma_L\gamma_L|\Gamma\gamma b\rangle \tag{5.4}$$

Here, the Kronecker multiplicity index b is required if, in Eq. (5.3), $b_i > 1$, i.e., if Γ of Eq. (5.4) occurs more than once in the reduction. This situation is encountered in the odd electron configurations d^3, d^5, and d^7 of cubic symmetry if spin–orbit coupling is included.

For a listing of the various ligand field states arising from the d^N configurations in cubic, tetragonal, trigonal, and cylindrical symmetry, consult Table I.

5.2. COULOMB INTERACTION

Matrix elements of the Coulomb interelectronic repulsion operator between the functions of Eq. (5.4) may be written, similar to Eq. (4.3), as[30]

$$\left\langle (vS\Gamma_S a_S L\Gamma_L a_L)\Gamma\gamma b \left| \sum_{i>j} \frac{e^2}{r_{ij}} \right| (v'S'\Gamma'_S a'_S L'\Gamma'_L a'_L)\Gamma'\gamma'b' \right\rangle$$

$$= \left\langle vSL \left| \sum_{i>j} \frac{e^2}{r_{ij}} \right| v'S'L' \right\rangle \delta(\Gamma_S, \Gamma'_S)\delta(a_S, a'_S)$$

$$\times \delta(\Gamma_L, \Gamma'_L)\delta(a_L, a'_L)\delta(\Gamma, \Gamma')\delta(\gamma, \gamma')\delta(b, b') \quad (5.5)$$

Again, matrix elements applying to the free transition metal ion appear on the right-hand side of Eq. (5.5). The subsequent reduction of these matrix elements follows the outline in Section 4.2. Equivalent to the case of weak-field coupling, the Coulomb interaction may be finally expressed in terms of the Slater–Condon parameters F^k ($k = 0, 2, 4$) or the Racah parameters A, B, C.

5.3. LIGAND FIELD INTERACTION

In order to calculate matrix elements of the ligand field operator, it is convenient to introduce the general tensor operators $(\mathbf{S}^{(ka)})^{A_1}$ of A_1 symmetry [cf. Eq. (3.65)]. If a cubic (O_h, T_d) ligand field potential is considered [cf. Eqs. (3.70) and (3.74)], matrix elements between the functions of Eq. (5.4) may be expressed as[30]

$$\langle (vS\Gamma_S a_S L\Gamma_L a_L)\Gamma\gamma b | \sum_i V_{\text{cub}}(r_i, \theta_i, \varphi_i) | (v'S'\Gamma'_S a'_S L'\Gamma'_L a'_L)\Gamma'\gamma'b' \rangle$$

$$= 6\sqrt{30}[2L+1]^{-1/2}\langle vSL\|\mathbf{U}^{(4)}\|v'S'L'\rangle\langle L\Gamma_L\gamma_L a_L | L'\Gamma'_L\gamma'_L a'_L, 4A_1 a_1 \rangle$$

$$\times \delta(S, S')\delta(\Gamma_S, \Gamma'_S)\delta(a_S, a'_S)\delta(\Gamma_L, \Gamma'_L)\delta(\gamma_L, \gamma'_L)\delta(\Gamma, \Gamma')\delta(\gamma, \gamma')\delta(b, b')Dq \quad (5.6)$$

Here, the long coupling coefficient is defined by [cf. Eq. (3.38)]

$$\langle L\Gamma_L\gamma_L a_L | L'\Gamma'_L\gamma'_L a'_L, 4A_1 a_1 \rangle = \sum_{M_L M_{Lq}} \langle L'M'_L 4q | LM_L \rangle \langle L'M'_L | L'\Gamma'_L\gamma'_L a'_L \rangle \langle 4q | 4A_1 a_1 \rangle \langle LM_L | L\Gamma_L\gamma_L a_L \rangle^*$$

$$(5.7)$$

Also, $\langle vSL\|\mathbf{U}^{(4)}\|v'S'L'\rangle$ is a reduced matrix element of the unit tensor operator $\mathbf{U}^{(4)}$ [cf. Eq. (2.85)], and Dq is the cubic ligand field parameter [cf. Eq. (3.79)]. Matrix elements of a ligand field potential of tetragonal (D_{4h}, D_4, C_{4v}) symmetry [cf. Eq. (3.71)] may be written similarly as

$$\langle (vS\Gamma_S a_S L\Gamma_L a_L)\Gamma\gamma b | \sum_i V_{D_4}(r_i, \theta_i, \varphi_i) | (v'S'\Gamma'_S a'_S L'\Gamma'_L a'_L)\Gamma'\gamma'b' \rangle$$

$$= [2L+1]^{-1/2}\{\sqrt{70}\langle vSL\|\mathbf{U}^{(2)}\|v'S'L'\rangle\langle L\Gamma_L\gamma_L a_L | L'\Gamma'_L\gamma'_L a'_L, 2A_1 a_1 \rangle Ds + 3\sqrt{70}\langle vSL\|\mathbf{U}^{(4)}\|v'S'L'\rangle$$

$$\times [\langle L\Gamma_L\gamma_L a_L | L'\Gamma'_L\gamma'_L a'_L, 4A_1 a_1 1 \rangle(Dq - Dt) + \sqrt{\tfrac{5}{7}}\langle L\Gamma_L\gamma_L a_L | L'\Gamma'_L\gamma'_L a'_L, 4A_1 a_1 2 \rangle Dq]\}$$

$$\times \delta(S, S')\delta(\Gamma_S, \Gamma'_S)\delta(a_S, a'_S)\delta(\Gamma_L, \Gamma'_L)\delta(\gamma_L, \gamma'_L)\delta(\Gamma, \Gamma')\delta(\gamma, \gamma')\delta(b, b') \quad (5.8)$$

Also, matrix elements of a ligand field potential of trigonal (D_{3d}, D_3, C_{3v}) symmetry [cf. Eq. (3.72)] may be obtained according to

$$\langle (vS\Gamma_S a_S L\Gamma_L a_L)\Gamma\gamma b | \sum_i V_{D_3}(r_i, \theta_i, \varphi_i) | (v'S'\Gamma'_S a'_S L'\Gamma'_L a'_L)\Gamma'\gamma'b' \rangle$$

$$= [2L+1]^{-1/2}\{\sqrt{70}\langle vSL\|\mathbf{U}^{(2)}\|v'S'L'\rangle\langle L\Gamma_L\gamma_L a_L | L'\Gamma'_L\gamma'_L a'_L, 2A_1 a_1 \rangle D\sigma + 2\sqrt{70}\langle vSL\|\mathbf{U}^{(4)}\|v'S'L'\rangle$$

$$\times [\langle L\Gamma_L\gamma_L a_L | L'\Gamma'_L\gamma'_L a'_L, 4A_1 a_1 1 \rangle(\tfrac{3}{2}D\tau - Dq) - 2\sqrt{\tfrac{5}{7}}\langle L\Gamma_L\gamma_L a_L | L'\Gamma'_L\gamma'_L a'_L, 4A_1 a_1 2 \rangle Dq]\}$$

$$\times \delta(S, S')\delta(\Gamma_S, \Gamma'_S)\delta(a_S, a'_S)\delta(\Gamma_L, \Gamma'_L)\delta(\gamma_L, \gamma'_L)\delta(\Gamma, \Gamma')\delta(\gamma, \gamma')\delta(b, b') \quad (5.9)$$

Here, in addition to the quantities defined above, $\langle vSL\|\mathbf{U}^{(2)}\|v'S'L'\rangle$ is a reduced matrix element of the unit tensor operator $\mathbf{U}^{(2)}$ [cf. Eq. (2.85)]. In Eq. (5.8), Dq, Ds, and Dt are the tetragonal ligand field parameters [cf. Eq. (3.81)]. Similarly, Dq, $D\sigma$, and $D\tau$ in Eq. (5.9) are the parameters of the trigonal ligand field [cf. Eqs. (3.85) and

(3.86)]. Matrix elements of a ligand field potential of cylindrical ($D_{\infty h}$, D_∞, $C_{\infty v}$) symmetry [cf. Eq. (3.73)] result from Eq. (5.8) if $Dq = 0$. It follows, similar to Section 4.4 above, that the computation of matrix elements Eqs. (5.6), (5.8), and (5.9) requires, besides expansion coefficients Eq. (3.10), coefficients of fractional parentage, 3-j and 6-j symbols. The quantities Dq, Ds, Dt, etc. are again considered as adjustable parameters to be determined from a comparison with experimental data.

5.4. SPIN–ORBIT COUPLING

The calculation of spin–orbit interaction proceeds along the general lines discussed in Section 3.2 above. First the scalar product $(\mathbf{s}_i \cdot \mathbf{l}_i)$ of the two spherical tensor operators is decomposed according to Eq. (3.53), thus producing

$$(\mathbf{s}^{(1)} \cdot \mathbf{l}^{(1)}) = -\sum_{\Gamma_1 \gamma_1} \left(\begin{matrix} & 1 & \\ \Gamma_1 \gamma_1 & & \overline{\Gamma_1 \gamma_1} \end{matrix}\right)^* \mathbf{s}_{\gamma_1}^{(1\Gamma_1)} \mathbf{l}_{\bar{\gamma}_1}^{(1\bar{\Gamma}_1)} \tag{5.10}$$

Here, the metric tensor in group $G \subset SO(3)$ which has been defined in Eq. (3.44) is, in fact, real and it is always $k = 1$. The matrix element of spin–orbit coupling between the functions of Eq. (5.4) then follows by application of Eqs. (3.18) and (3.37) as[30]

$$\langle (vS\Gamma_S a_S L\Gamma_L a_L)\Gamma\gamma b| \sum_i \kappa \xi(r_i)(\mathbf{s}_i \cdot \mathbf{l}_i)|(v'S'\Gamma_S' a_S' L'\Gamma_L' a_L')\Gamma'\gamma'b'\rangle$$

$$= -\sqrt{30}[(2L+1)(2S+1)]^{-1/2}\langle vSL\|\mathbf{V}^{(11)}\|v'S'L'\rangle \sum_{\substack{\gamma_S \gamma_L \\ \gamma_{\dot{S}} \gamma_{\dot{L}}}} \langle \Gamma\gamma b|\Gamma_S\gamma_S \Gamma_L\gamma_L\rangle\langle\Gamma_S'\gamma_S'\Gamma_L'\gamma_L'|\Gamma'\gamma'b'\rangle$$

$$\times \sum_{\Gamma_1 \gamma_1} \left(\begin{matrix} & 1 & \\ \Gamma_1 \gamma_1 & & \overline{\Gamma_1 \gamma_1} \end{matrix}\right) \langle S\Gamma_S\gamma_S a_S|S'\Gamma_S'\gamma_S'a_S', 1\Gamma_1\gamma_1\rangle\langle L\Gamma_L\gamma_L a_L|L'\Gamma_L'\gamma_L'a_L', 1\overline{\Gamma_1\gamma_1}\rangle\delta(\Gamma,\Gamma')\delta(\gamma,\gamma')\zeta_{nl} \tag{5.11}$$

In order to obtain Eq. (5.11), Eq. (3.37) has been applied here to the matrix elements of the individual tensor operators $\mathbf{s}_{i\gamma_1}^{(1\Gamma_1)}$ and $\mathbf{l}_{i\bar{\gamma}_1}^{(1\bar{\Gamma}_1)}$ and the double tensor operator $\mathbf{V}^{(11)}$ has been introduced according to

$$\langle vSL\|\sum_i \mathbf{s}_i^{(1)}\|v'S'L'\rangle\langle vSL\|\sum_i \mathbf{l}_i^{(1)}\|v'S'L'\rangle = \langle vSL\|\sum_i \mathbf{s}_i^{(1)}\mathbf{l}_i^{(1)}\|v'S'L'\rangle = \sqrt{30}\langle vSL\|\mathbf{V}^{(11)}\|v'S'L'\rangle \tag{5.12}$$

In cubic symmetry, the one-electron vector operators $\mathbf{s}_i$ and $\mathbf{l}_i$ transform according to irreducible representation T_1. The totally symmetric product operator then assumes the form

$$(\mathbf{s}^{(1T_1)} \cdot \mathbf{l}^{(1T_1)}) = (-1)\sum_{\gamma_1} \left(\begin{matrix} & 1 & \\ T_1\gamma_1 & & T_1\gamma_1 \end{matrix}\right) \mathbf{s}_{\gamma_1}^{(1T_1)} \mathbf{l}_{\bar{\gamma}_1}^{(1T_1)}$$

$$= -\left[\left(\begin{matrix} & 1 & \\ T_11 & & T_1-1 \end{matrix}\right)\mathbf{s}_1^{(1T_1)}\mathbf{l}_{-1}^{(1T_1)} + \left(\begin{matrix} & 1 & \\ T_10 & & T_10 \end{matrix}\right)\mathbf{s}_0^{(1T_1)}\mathbf{l}_0^{(1T_1)} + \left(\begin{matrix} & 1 & \\ T_1-1 & & T_11 \end{matrix}\right)\mathbf{s}_{-1}^{(1T_1)}\mathbf{l}_1^{(1T_1)}\right] \tag{5.13}$$

Since $\mathbf{j}_\pm$ ($\mathbf{j} = \mathbf{s}, \mathbf{l}$) and $\mathbf{j}_z$ transform in the same way as $\mathbf{j}_{\pm1}^{(1T_1)}$ and $\mathbf{j}_0^{(1T_1)}$, respectively, Eq. (5.13) may be written in the equivalent form

$$(\mathbf{s}^{(1T_1)} \cdot \mathbf{l}^{(1T_1)}) = \tfrac{1}{2}(\mathbf{s}_+^{(1)}\mathbf{l}_-^{(1)}) + \tfrac{1}{2}(\mathbf{s}_-^{(1)}\mathbf{l}_+^{(1)}) + (\mathbf{s}_z^{(1)}\mathbf{l}_z^{(1)}) \tag{5.14}$$

If the symmetry is D_4, D_3, or D_∞, the transformation properties of the operators may be described by representation A_2 or E of the appropriate symmetry group. In Eqs. (5.10) and (5.11), Γ_1 then has to be substituted accordingly. In Eq. (5.11), ζ_{nl} is the spin–orbit coupling *parameter* which is, in general, reduced in magnitude by the amount κ as compared to ζ_{nl}^{free} of the free ion. The quantity κ is the *orbital reduction factor*,[9] its anisotropy being neglected by the assumption $\kappa_\parallel = \kappa_\perp = \kappa$. In addition, $\langle vSL\|\mathbf{V}^{(11)}\|v'S'L'\rangle$ is a reduced matrix element of the double tensor operator $\mathbf{V}^{(11)}$ [cf. Eq. (2.87)]. The symmetry coupling coefficients $\langle \Gamma\gamma b|\Gamma_S\gamma_S\Gamma_L\gamma_L\rangle$ have been defined in Eq. (3.18), the long coupling coefficients in Eq. (3.38). A completely general formulation for the matrix elements of $(\mathbf{s}_i \cdot \mathbf{l}_i)$ has been given elsewhere [cf. Reference 30, Eq. (43)]. It follows, similar to Section

4.3 above, that the computation of matrix elements of spin–orbit coupling requires, besides expansion coefficients Eq. (3.10), coefficients of fractional parentage, 3-j and 6-j symbols. The quantities ζ_{nl} and κ are again determined on a semiempirical basis.

5.5. ADDITIONAL INTERACTIONS

The energy matrices obtained by combination of the matrix elements which were calculated in Sections 5.2–5.4 are partly in diagonal form. The dimensions of the individual submatrices are evident from Table I. The diagonalization of the matrices produces eigenvalues of energy E_n which are identical to those obtained in case of weak-field coupling. These energies are displayed in Part II of the volume as function of suitable parameters. In addition, to each eigenvalue E_n, the corresponding eigenfunction is obtained as linear combination of the p_Γ basis functions $|(vS\Gamma_S a_S L\Gamma_L a_L)\Gamma\gamma b\rangle$, namely,

$$|n'\Gamma\gamma\rangle = \sum_{k'=1}^{p_\Gamma} |(vS\Gamma_S a_S L\Gamma_L a_L)\Gamma\gamma b\rangle C'_{k'} \quad (5.15)$$

Here, the dimension p_Γ is dependent on the symmetry of the problem and the summation index $k' \equiv vS\Gamma_S a_S L\Gamma_L a_L$. The coefficients $C'_{k'}$ squared determine the composition of the individual levels in the intermediate-field coupling scheme. As in Section 4.5, the functions Eq. (5.15) may be employed to determine the effect of additional interactions. Again the interaction energy with an external magnetic field (Zeeman effect) is of primary interest as the starting point for the calculation of magnetic susceptibility. With respect to this aspect of the theory and its applications we refer to a separate treatment.[31]

It has been mentioned above that the greatest part of the energy of transition metal ions in crystals or complex molecules is accounted for by the normal Coulomb, ligand field, and spin–orbit interactions. However, additional though small interactions exist. These are usually treated by a perturbation approach on the lowest levels of Eq. (5.15). Wybourne[62] has discussed these remaining interactions for free atoms in terms of irreducible tensor operators. In as much as the algebra for intermediate-field coupling has been developed in detail,[30] it may be easily extended to cover any of the smaller interactions that may be of interest.

6

STRONG-FIELD COUPLING

6.1. THE SYMMETRIC GROUP AND THE METHOD OF FRACTIONAL PARENTAGE COEFFICIENTS

There are two different methods for the formation of antisymmetric wave functions for an N-electron system. According to the more common approach, the various permutations of the N-electron function are arranged into a Slater determinant. In this case, the advantages of vector coupling procedures cannot be utilized. An alternate method consists of application of the coefficients of fractional parentage (cfp) originally introduced by Racah[48] for the simple situation of N equivalent electrons in $SO(3)$. The antisymmetric N-electron wave function is then generated as a sum of products of the antisymmetric $(N-1)$-electron function times a function for the Nth electron. The coefficients of this expansion are the cfp (cf. Section 2.3). The method has been extended by Horie[17] and others[15,21] and may be readily generalized to the problem at hand, i.e., to an arbitrary configuration of inequivalent electron states in an arbitrary point group G.

Let us characterize a basis function, subject to its permutation properties, by the appropriate irreducible representation associated with partition $[\lambda]$ of the symmetric group S_N.[51] The total N-electron wave function, if antisymmetric against the interchange of electrons, will then transform according to the antisymmetric representation $[1^N]$ of S_N. Moreover, if the wave function is the product of a spin and an orbital function, the two functions will transform according to the adjoint representations $[\lambda]$ and $[\tilde{\lambda}]$ and we may write

$$|[\lambda][\tilde{\lambda}],[1^N]\rangle = \sum_{(r)} |[\lambda](r)\rangle|[\tilde{\lambda}](\tilde{r})\rangle\langle[\lambda](r),[\tilde{\lambda}](\tilde{r})|[1^N]\rangle$$

(6.1)

In Eq. (6.1), the summation extends over all d_λ possible Yamanouchi symbols (r) associated with representation $[\lambda]$, where d_λ denotes the dimension of $[\lambda]$. It should be noted that, for the Yamanouchi standard orthogonal representations which will be used exclusively in what follows, the coupling coefficient in Eq. (6.1) assumes the value $1/\sqrt{d_\lambda}$. The antisymmetric N-electron wave function of Eq. (6.1) may then be written more specifically as

$$|(\alpha\beta)SM_S\Gamma\gamma[\lambda]\rangle$$
$$= \frac{1}{\sqrt{d_\lambda}}\sum_{(r)}|\alpha SM_S[\tilde{\lambda}](\tilde{r})\rangle|\beta\Gamma\gamma[\lambda](r)\rangle$$

(6.2)

Here, α and β denote additional quantum numbers that may be required, and

$$|(l^N\alpha\beta)SM_S\Gamma\gamma,[\tilde{\lambda}][\lambda],[1^N]\rangle$$

has been abbreviated by $|(\alpha\beta)SM_S\Gamma\gamma[\lambda]\rangle$.

The N-electron orbital function of Eq. (6.2) may then be expressed as a linear combination of products of the function for $(N-1)$ electrons and a function for the Nth electron. If the N-electron function transforms according to representation $[\lambda]$ and the $(N-1)$-electron function according to representation $[\lambda_1]$, one obtains

$$|l^N\beta\Gamma\gamma[\lambda](r)\rangle$$
$$= \sum_{\beta_1\Gamma_1\Gamma_2a_2b}|l^{N-1}\beta_1\Gamma_1[\lambda_1](r_1),\Gamma_2a_2;l^N\Gamma\gamma b)$$
$$\times\langle l^{N-1}\beta_1\Gamma_1[\lambda_1],\Gamma_2a_2,b|\}l^N\beta\Gamma[\lambda]\rangle$$

(6.3)

Here, the function enclosed within $|$ $)$ denotes the above-mentioned linear combination of product functions

$$|l^{N-1}\beta_1\Gamma_1[\lambda_1](r_1), \Gamma_2a_2; l^N\Gamma\gamma b)$$

$$= \sum_{\gamma_1\gamma_2} |l^{N-1}\beta_1\Gamma_1\gamma_1[\lambda_1](r_1))|\Gamma_2\gamma_2a_2\rangle$$

$$\times \langle\Gamma_1\gamma_1\Gamma_2\gamma_2|\Gamma\gamma b\rangle \qquad (6.4)$$

where $(r_1) = (r_{N-1}r_{N-2}\ldots 1)$. The expansion, in Eq. (6.3), thus defines the orbital cfp, the brace $|\}$ indicating that the cfp do not form a quadratic matrix. A summation over $[\lambda_1](r_1)$ is not required since this representation and its component are fixed by $[\lambda](r)$. Also the cfp conform to the orthogonality relations

$$\sum_{\beta_1\Gamma_1\Gamma_2a_2b} \langle l^N\beta\Gamma[\lambda]\{|l^{N-1}\beta_1\Gamma_1[\lambda_1], \Gamma_2a_2, b\rangle$$

$$\times\langle l^{N-1}\beta_1\Gamma_1[\lambda_1], \Gamma_2a_2, b|\}l^N\beta'\Gamma'[\lambda']\rangle$$

$$= \delta(\beta, \beta')\delta(\Gamma, \Gamma')\delta(\lambda, \lambda')$$

$$\sum_{\lambda} \langle l^{N-1}\beta_1\Gamma_1[\lambda_1], \Gamma_2a_2, b|\}l^N\beta\Gamma[\lambda]\rangle$$

$$\times\langle l^N\beta\Gamma[\lambda]\{|l^{N-1}\beta_1'\Gamma_1'[\lambda_1'], \Gamma_2'a_2', b'\rangle$$

$$= \delta(\beta_1\Gamma_1\lambda_1, \beta_1'\Gamma_1'\lambda_1')\delta(\Gamma_2a_2, \Gamma_2'a_2')\delta(b, b') \qquad (6.5)$$

The representations $[\bar{\lambda}]$ of the N-electron spin function in Eq. (6.2) are restricted in that the Young tableaux of $[\bar{\lambda}]$ consist of two rows only. Each representation $D^{(S)}(SU(2))$ has a corresponding partition $[\lambda_1, \lambda_2]$ of the symmetric group S_N according to[13,24]

$$[\bar{\lambda}] = [\lambda_1, \lambda_2] = \left[\frac{N+2S}{2}, \frac{N-2S}{2}\right]$$

$$S = \tfrac{1}{2}(\lambda_1 - \lambda_2) \qquad (6.6)$$

In particular, the totally symmetric representation of one row $[N]$ is connected to the maximum spin S of that state. In the N-electron spin function of Eq. (6.2) the representation $[\bar{\lambda}]$ is related to a specific value of S and *vice versa*. As a consequence of orthogonality and normalization of the spin functions, it follows that

$$|\langle(N-1)\alpha_1S_1[\bar{\lambda}_1], \tfrac{1}{2}|\}N\alpha S[\bar{\lambda}]\rangle|^2 = 1 \qquad (6.7)$$

and the spin cfp assume the value $+1$ provided $S_1 + \tfrac{1}{2} \geq S$ and $|S_1 - \tfrac{1}{2}| \leq S$.

If Eq. (6.3) and the corresponding expression for the N-electron spin function are introduced into Eq. (6.2), and if Eq. (6.7) is observed, the antisymmetric total N-electron wave function is obtained as

$$|l^N\alpha\beta SM_S\Gamma\gamma[\lambda])$$

$$= \frac{1}{d_\lambda} \sum_{(r)} \sum_{\beta_1\Gamma_1\Gamma_2a_2b} |l^{N-1}\beta_1\Gamma_1[\lambda_1](r_1), \Gamma_2a_2; l^N\Gamma\gamma b)$$

$$\times |(N-1)\alpha_1S_1[\bar{\lambda}_1](\tilde{r}_1), \tfrac{1}{2}; NSM_S)$$

$$\times\langle l^{N-1}\beta_1\Gamma_1[\lambda_1], \Gamma_2a_2, b|\}l^N\beta\Gamma[\lambda]\rangle \qquad (6.8)$$

The sum over all (r) of $[\lambda]$ in Eq. (6.8) may be replaced by a sum over all $[\lambda_1]$ and (r_1), whereby the symbol (r) may be dropped. The sum over (r_1) of the vector-coupled functions in Eq. (6.8) may then be rewritten similar to Eq. (6.2) to give

$$\sum_{(r_1)} |l^{N-1}\beta_1\Gamma_1[\lambda_1](r_1), \Gamma_2a_2; l^N\Gamma\gamma b)$$

$$\times |(N-1)\alpha_1S_1[\bar{\lambda}_1](\tilde{r}_1), \tfrac{1}{2}; NSM_S)$$

$$= \sqrt{d_{\lambda_1}}|l^{N-1}\alpha_1\beta_1S_1\Gamma_1[\lambda_1], \tfrac{1}{2}\Gamma_2a_2; l^NSM_S\Gamma\gamma b) \qquad (6.9)$$

If Eq. (6.9) is introduced into Eq. (6.8), simultaneously replacing the sum over (r) by a sum over $[\lambda_1]$, it becomes

$$|l^N\alpha\beta SM_S\Gamma\gamma[\lambda]) = \sum_{[\lambda_1]\beta_1\Gamma_1\Gamma_2a_2} \sum_b |l^{N-1}\alpha_1\beta_1S_1\Gamma_1[\lambda_1], \tfrac{1}{2}\Gamma_2a_2; l^NSM_S\Gamma\gamma b)$$

$$\times\langle l^{N-1}\alpha_1\beta_1S_1\Gamma_1[\lambda_1], \tfrac{1}{2}\Gamma_2a_2, b|\}l^N\alpha\beta S\Gamma[\lambda]\rangle \qquad (6.10)$$

This is the final expression for the total antisymmetric N-electron wave function. The fractional parentage coefficient in Eq. (6.10) is defined by

$$\langle l^{N-1}\alpha_1\beta_1S_1\Gamma_1[\lambda_1], \tfrac{1}{2}\Gamma_2a_2, b|\}l^N\alpha\beta S\Gamma[\lambda]\rangle = (d_{\lambda_1}/d_\lambda)^{1/2}\langle l^{N-1}\beta_1\Gamma_1[\lambda_1], \Gamma_2a_2, b|\}l^N\beta\Gamma[\lambda]\rangle \qquad (6.11)$$

and the function on the right-hand side of Eq. (6.10) may be expressed in detail by

$$|l^{N-1}\alpha_1\beta_1S_1\Gamma_1[\lambda_1], \tfrac{1}{2}\Gamma_2a_2; l^NSM_S\Gamma\gamma b) = \sum_{\gamma_1\gamma_2M_{S_1}m_{s_2}} |l^{N-1}\alpha_1\beta_1S_1M_{S_1}\Gamma_1\gamma_1[\lambda_1])|\tfrac{1}{2}m_{s_2}\Gamma_2\gamma_2a_2\rangle$$

$$\times\langle\Gamma_1\gamma_1\Gamma_2\gamma_2|\Gamma\gamma b\rangle\langle S_1M_{S_1}\tfrac{1}{2}m_{s_2}|SM_S\rangle \qquad (6.12)$$

Evidently, in order to calculate matrix elements of irreducible tensor operators, the relations derived above may be employed. In addition, the fractional parentage coefficients of Eq. (6.10) will be required.

An expression for the direct calculation of the orbital cfp of mixed configurations within the three-dimensional rotation group $SO(3)$ has been derived by Horie.[17] This expression may be easily generalized to arbitrary and non-simply-reducible (non-SR) groups. To this end, two specific properties have to be accounted for: firstly, in the direct product of two irreducible representations Γ_1 and Γ_2 [cf. Eq. (3.19)] an index b_j may be required if the resulting representation Γ_j occurs more than once. Secondly, in the reduction of subduced representation $D^{(l)}(G)$ [cf. Eq. (3.35)] an additional label a_i may be needed in order to distinguish the individual single-electron functions. Thus, e.g., an orbital function for two electrons may be characterized as $|(\Gamma_1 a_1 \Gamma_2 a_2, b)\Gamma\gamma[\lambda]\rangle$, the corresponding one-electron functions being denoted as $|\Gamma_1\gamma_1 a_1\rangle$ and $|\Gamma_2\gamma_2 a_2\rangle$. The final expression for the cfp which is applicable to arbitrary non-SR-groups as well as to mixed configurations results as

$$\langle l^{N-1}\beta_1'\Gamma_1'[\lambda_1'], \Gamma_2' a_2', b'|\}(l^{N-1}\beta_1\Gamma_1[\lambda_1], \Gamma_2 a_2, b)\Gamma[\lambda]\rangle$$

$$= N(l^{N-1}\beta_1\Gamma_1[\lambda_1], \Gamma_2 a_2, b; \Gamma[\lambda])\frac{d_\lambda}{Nd_{\lambda_1}}\Big\{\delta(\beta_1'\Gamma_1'[\lambda_1'], \beta_1\Gamma_1[\lambda_1])\delta(\Gamma_2' a_2', \Gamma_2 a_2)\delta(b', b)$$

$$+ \sum_{[\lambda_1'']}\frac{(N-1)d_{\lambda_1''}}{d_{\lambda_1}}\langle[\lambda_1''][\lambda_1'], [\lambda]|P_{N-1N}|[\lambda_1''][\lambda_1], [\lambda]\rangle \sum_{\beta_1''\Gamma_1''b_1b_1}\langle\Gamma_1'[\lambda_1']\{(l^{N-1}\beta_1')|\beta_1''\Gamma_1''[\lambda_1''], \Gamma_2 a_2, b_1'\rangle$$

$$\times\langle\Gamma_1''\Gamma_2(\Gamma_1'b_1')\Gamma_2', \Gamma b'|\Gamma_1''\Gamma_2'(\Gamma_1 b_1)\Gamma_2, \Gamma b\rangle\langle\beta_1''\Gamma_1''[\lambda_1''], \Gamma_2' a_2', b_1|\}(l^{N-1}\beta_1)\Gamma_1[\lambda_1]\rangle\Big\} \qquad (6.13)$$

Here, $\Gamma_1[\lambda_1]$ evidently specifies the representations of the antisymmetric $(N-1)$-electron function, β_1 denotes all the additional symbols that may be required, whereas $\Gamma_2 a_2$ determines the one-electron function of the Nth electron. The corresponding primed quantities specify other functions of the same kind as the unprimed ones. The normalization constant in Eq. (6.13) is defined by

$$N(l^{N-1}\beta_1\Gamma_1[\lambda_1], \Gamma_2 a_2, b; \Gamma[\lambda]) = |\langle l^{N-1}\beta_1\Gamma_1[\lambda_1], \Gamma_2 a_2, b|\}(l^{N-1}\beta_1\Gamma_1[\lambda_1], \Gamma_2 a_2, b)\Gamma[\lambda]\rangle|^{-1} \qquad (6.14)$$

and is real and positive, thus fixing the phases of all coefficients in Eq. (6.13). The expression in parentheses within the cfp, i.e., $(l^{N-1}\beta_1\Gamma_1[\lambda_1], \Gamma_2 a_2, b)$, indicates which coefficient of Eq. (6.3) was taken as real and positive. The symbols in parentheses thus specify the so-called "principal parent"[50] of the term considered. The coupling coefficient in Eq. (6.13) may be expressed in terms of a W coefficient by

$$\langle\Gamma_1''\Gamma_2(\Gamma_1'b_1')\Gamma_2', \Gamma b'|\Gamma_1''\Gamma_2'(\Gamma_1 b_1)\Gamma_2, \Gamma b\rangle = (-1)^{(j_1+j_2-j)b_i+(j_1+j_2-j)b}(d_{\Gamma_1}\cdot d_{\Gamma_i})^{1/2}W(\Gamma_2\Gamma_1''\Gamma\Gamma_2'; \Gamma_1'\Gamma_1)_{b'b_1b_ib} \qquad (6.15)$$

The matrix elements of the permutation operator P_{N-1N} between the representation symbols indicated, i.e., $\langle[\lambda_1''][\lambda_1'], [\lambda]|P_{N-1N}|[\lambda_1''][\lambda_1], [\lambda]\rangle$, may be easily obtained and are tabulated. In the reduction considered here, i.e., $S_N \rightarrow S_{N-1} \rightarrow S_{N-2}$, etc., only one particle is being separated at one time. Thus if the Young tableaux $[\lambda_1'']$, $[\lambda_1]$, and $[\lambda]$ in S_{N-2}, S_{N-1}, and S_N are defined, the corresponding Yamanouchi symbols are likewise fixed. Let us consider, e.g., the representations in S_5 given below:

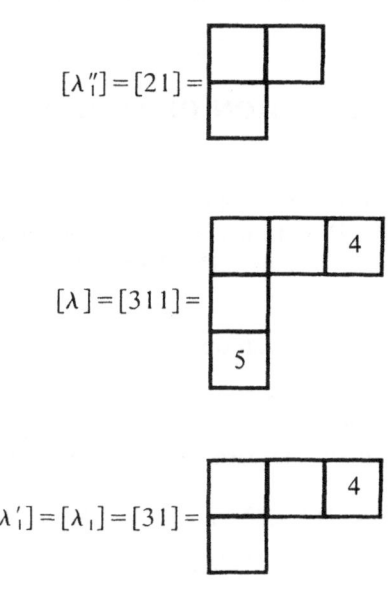

$$[\lambda_1''] = [21] =$$

$$[\lambda] = [311] =$$

$$[\lambda_1'] = [\lambda_1] = [31] =$$

In S_5, the Yamanouchi symbol $(31r_3r_21)$ corresponds to the sequence of symbols in $(\lambda_1'', \lambda_1, \lambda)$. The matrix element to be determined thus is

$$\langle(31r_3r_21)|P_{45}|(31r_3r_21)\rangle$$

If the phases are fixed as usual,[22] we obtain for our example

$$\langle[21],[31];[311]|P_{45}|[21],[31];[311]\rangle = -\tfrac{1}{4}$$

For the symmetric groups S_2 to S_5 the numerical factors

$$a(\lambda_1'', \lambda_1', \lambda_1, \lambda) = \frac{(N-1)d_{\lambda_1''}}{d_{\lambda_i}}$$

$$\times \langle[\lambda_1''][\lambda_1'], [\lambda]|P_{N-1\,N}|[\lambda_1''][\lambda_1], [\lambda]\rangle \quad (6.16)$$

which directly occur in the expression Eq. (6.13) for the cfp have been calculated and the results are listed in Table II.

In what follows, the total antisymmetric N-electron wave function will be written, on the basis of Eq. (6.10) and employing the cfp of Eq. (6.13) as

$$|(l^N\alpha_1\beta_1S_1\Gamma_1, \tfrac{1}{2}\Gamma_2a_2, b)SM_S\Gamma\gamma\rangle$$

$$= \sum_{\substack{\alpha_1'\beta_1'S_1'\Gamma_1' \\ \Gamma_2'a_2'b'}} |l^{N-1}\alpha_1'\beta_1'S_1'\Gamma_1', \tfrac{1}{2}\Gamma_2'a_2'; l^NSM_S\Gamma\gamma b')$$

$$\times \langle l^{N-1}\alpha_1'\beta_1'S_1'\Gamma_1', \tfrac{1}{2}\Gamma_2'a_2', b'|\}(l^{N-1}\alpha_1\beta_1S_1\Gamma_1, \tfrac{1}{2}\Gamma_2a_2, b)S\Gamma\rangle \quad (6.17)$$

Here, as a consequence, the representations $[\lambda]$ are not explicitly specified. This expression will be used, in particular, in the calculation of matrix elements (cf. Sections 6.2–6.6).

6.2. APPLICATION TO THE CALCULATION OF MATRIX ELEMENTS OF IRREDUCIBLE TENSOR OPERATORS

Matrix elements of irreducible tensor operators between antisymmetric N-electron wave functions may be easily expressed, on the basis of Eq. (6.17), in terms of matrix elements between simple product functions. The resulting matrix elements are subsequently calculated by the tensor operator method which we have developed in previous sections.

Let us consider first the matrix element of an operator $\mathbf{F} = \sum_{i=1}^{N} \mathbf{f}_i$, which is itself the sum of single-electron operators $\mathbf{f}_i$. Due to the equivalence of electrons, the matrix elements of all the operators $\mathbf{f}_i$ are identical [cf. Eq. (2.62)]. If we concentrate, in particular, on an operator $\mathbf{T}^{(\Gamma'')} = \sum_i \mathbf{t}^{(\Gamma'')}(i)$ acting on the orbital part of a many-electron function only, we obtain, on the basis of Eqs. (3.31) and (6.17)

$$\langle(l^{N-1}\alpha_1\beta_1S_1\Gamma_1, \tfrac{1}{2}\Gamma_2a_2, b)S\Gamma\|\mathbf{T}^{(\Gamma'')}\|(l^{N-1}\alpha_1'\beta_1'S_1'\Gamma_1', \tfrac{1}{2}\Gamma_2'a_2', b')S'\Gamma'\rangle_{b_4}$$

$$= \delta(S, S')N(d_\Gamma \cdot d_{\Gamma'})^{1/2} \sum \langle S\Gamma(l^{N-1}\alpha_1\beta_1S_1\Gamma_1, \tfrac{1}{2}\Gamma_2a_2, b)\{|l^{N-1}\alpha_4\beta_4S_4\Gamma_4, \tfrac{1}{2}\Gamma_3a_3, b_1\rangle$$

TABLE II

The coefficients $a(\lambda_1'', \lambda_1', \lambda_1, \lambda)$ of Eq. (6.16) for the symmetric groups S_2 to S_5. The coefficients follow the relations $a(\lambda_1'', \lambda_1, \lambda_1, \lambda) = -a(\tilde{\lambda}_1'', \tilde{\lambda}_1, \tilde{\lambda}_1, \tilde{\lambda})$, $a(\lambda_1'', \lambda_1', \lambda_1, \lambda) = (d_{\lambda_1}/d_{\lambda_1'})\,a(\lambda_1'', \lambda_1, \lambda_1', \lambda)$ and $a(\lambda_1'', \lambda_1', \lambda_1, \lambda) = a(\tilde{\lambda}_1'', \tilde{\lambda}_1', \tilde{\lambda}_1, \tilde{\lambda})$ for $\lambda_1' \neq \lambda_1$.

	λ_1''	λ_1'	λ_1	λ	a
S_2		[1]	[1]	[2]	1
		[1]	[1]	[11]	−1
S_3	[1]	[2]	[2]	[3]	2
	[1]	[2]	[2]	[21]	−1
	[1]	[2]	[11]	[21]	$\sqrt{3}$
S_4	[2]	[21]	[21]	[22]	3/2
	[2]	[21]	[21]	[211]	−3/2
	[11]	[21]	[21]	[211]	−1/2
	[11]	[21]	[111]	[211]	$\sqrt{2}$
	[11]	[111]	[111]	[211]	1
	[11]	[111]	[111]	[1111]	−3
S_5	[3]	[31]	[31]	[311]	−4/3
	[21]	[31]	[31]	[311]	−2/3
	[21]	[31]	[211]	[311]	$\sqrt{20/3}$
	[21]	[22]	[22]	[221]	−2
	[21]	[22]	[211]	[221]	$\sqrt{12}$
	[21]	[211]	[211]	[221]	4/3
	[21]	[211]	[211]	[2111]	−8/3
	[111]	[211]	[211]	[221]	−4/3
	[111]	[211]	[211]	[2111]	−1/3
	[111]	[211]	[1111]	[2111]	$\sqrt{5/3}$
	[111]	[1111]	[1111]	[2111]	1
	[111]	[1111]	[1111]	[11111]	−4

$$\times \langle l^{N-1}\alpha_4\beta_4 S_4\Gamma_4, \tfrac{1}{2}\Gamma_3'a_3', b_2|\}(l^{N-1}\alpha_1'\beta_1'S_1'\Gamma_1', \tfrac{1}{2}\Gamma_2'a_2', b')S\Gamma'\rangle \sum_{b_3} (-1)^{(j_4-j_3+j)_{b_1}+(j_3+j_5'+j'')_{b_3}}$$

$$\times \begin{Bmatrix} \Gamma_3 & \bar{\Gamma}_3' & \bar{\Gamma}'' \\ \Gamma' & \Gamma & \Gamma_4 \end{Bmatrix}_{b_4 b_2 b_1 b_3} \langle a_3\Gamma_3\|\mathbf{t}^{(\Gamma'')}\|a_3'\Gamma_3'\rangle_{b_3} \tag{6.18}$$

This expression simplifies to a form similar to Eq. (2.44) whenever the indices b_i are not required. In this case, the sums over b_1 and b_2 disappear and the phase factor becomes $(-1)^{j_4+j+j_3'+j''}$.

The next operator of interest is the scalar two-electron operator $\mathbf{G} = \sum_{i>j}^{N} \mathbf{g}_{ij}$, which is the sum of operators $\mathbf{g}_{ij}$ acting on the coordinates of electrons i and j only. Again assuming that $\mathbf{G}$ operates on the orbital part of a many-electron function only, we obtain on the basis of Eq. (6.17) [cf. also Eq. (2.66)]

$$\langle (l^{N-1}\alpha_1\beta_1 S_1\Gamma_1, \tfrac{1}{2}\Gamma_2 a_2, b)S\Gamma|\mathbf{G}^{(N)}|(l^{N-1}\alpha_1'\beta_1'S_1'\Gamma_1', \tfrac{1}{2}\Gamma_2'a_2', b')S'\Gamma'\rangle$$

$$= \delta(S, S')\delta(\Gamma, \Gamma')\frac{N}{N-2} \sum_{\substack{\alpha_4 S_4\Gamma_4\beta_4 \\ \beta_4'b_1\Gamma_3 a_3}} \langle S\Gamma(l^{N-1}\alpha_1\beta_1 S_1\Gamma_1, \tfrac{1}{2}\Gamma_2 a_2, b)\{|l^{N-1}\alpha_4\beta_4 S_4\Gamma_4, \tfrac{1}{2}\Gamma_3 a_3, b_1\rangle$$

$$\times \langle l^{N-1}\alpha_4\beta_4'S_4\Gamma_4, \tfrac{1}{2}\Gamma_3 a_3, b_1|\}(l^{N-1}\alpha_1'\beta_1'S_1'\Gamma_1', \tfrac{1}{2}\Gamma_2'a_2', b')S\Gamma\rangle\langle l^{N-1}\alpha_4\beta_4 S_4\Gamma_4|\mathbf{G}^{(N-1)}|l^{N-1}\alpha_4\beta_4'S_4\Gamma_4\rangle \tag{6.19}$$

Equation (6.19) thus relates the matrix elements of a $\mathbf{G}$-type operator for a N-electron system to the matrix elements for a $(N-1)$-electron system. In case the index b is not required, Eq. (6.19) may be simplified by deletion of the sum over b_1.

It is evident that all previously derived relations for the matrix elements of irreducible tensor operators over product wave functions may be similarly applied to the calculation of matrix elements over antisymmetric N-electron functions.

6.3. STATES AND WAVE FUNCTIONS

In strong-field coupling, the starting point is the single-electron functions $|lm_l\rangle$ of the free ion. If a ligand field characterized by symmetry group G is applied, the corresponding states are split, in general, according to the reduction

$$D^{(l)}(G) = \sum_i a_i^{(l)}\Gamma_{li} \tag{6.20}$$

Here, representation $D^{(l)}(G)$ of orbital angular momentum is subduced from representation $D^{(l)}$ of $SO(3)$. The basis functions conforming to the Hamiltonian inclusive of the first term in Eq. (3.101) and transforming according to irreducible representation Γ_l, component γ_l, of G are then determined by

$$|\alpha l\Gamma_l\gamma_l a_l\rangle = \sum_{m_l} |\alpha lm_l\rangle\langle lm_l|l\Gamma_l\gamma_l a_l\rangle \tag{6.21}$$

In Eq. (6.21), a_l is a branching multiplicity index which is introduced in order to distinguish equivalent states if, in Eq. (6.20), $a_i^{(l)} > 1$. The expansion coefficients $\langle lm_l|l\Gamma_l\gamma_l a_l\rangle$ in Eq. (6.21) may be determined again on the basis of Eq. (3.12).

If Coulomb interaction is introduced according to Eq. (3.101), the single-electron orbital states of the N-electron system are coupled according to the Kronecker product of the irreducible representations $\Gamma_i^{(k)}$ of the N electrons involved $(k = 1, \ldots, N)$:

$$\Gamma_i^{(1)} \otimes \Gamma_j^{(2)} \otimes \cdots \otimes \Gamma_m^{(N)} = \sum_r b_r\Gamma_r \tag{6.22}$$

Similarly, the single-electron spin functions are coupled in correspondence to the Kronecker product of the irreducible representations $D^{(1/2)}(SU(2))$ of the N electrons:

$$D^{(1/2)}(SU(2)) \otimes D^{(1/2)}(SU(2)) \otimes \cdots \otimes D^{(1/2)}(SU(2)) = \sum_{S}^{N/2} D^{(S)}(SU(2)) \tag{6.23}$$

In a two-electron system, the coupled basis functions may be written as

$$|(\Gamma_1 a_1 \Gamma_2 a_2) S M_S \Gamma_L \gamma_L b_L\rangle = \sum_{\substack{m_{s_1} m_{s_2} \\ \gamma_1 \gamma_2}} |m_{s_1} \Gamma_1 \gamma_1 a_1\rangle |m_{s_2} \Gamma_2 \gamma_2 a_2\rangle \langle \tfrac{1}{2} m_{s_1} \tfrac{1}{2} m_{s_2} | S M_S\rangle \langle \Gamma_1 \gamma_1 \Gamma_2 \gamma_2 | \Gamma_L \gamma_L b_L\rangle \qquad (6.24)$$

The functions Eq. (6.24) may be subsequently coupled to the function of a third electron,

$$|m_{s_3} \Gamma_3 \gamma_3 a_3\rangle = |\tfrac{1}{2} m_{s_3}\rangle |\Gamma_3 \gamma_3 a_3\rangle$$

etc. An antisymmetric N-electron wave function subject to a ligand field of certain symmetry and to Coulomb interaction may then be represented by

$$|\{\Gamma_1 a_1 \Gamma_2 a_2 (S_3 \Gamma_3 b_3) \ldots\} S M_S \Gamma_L \gamma_L\rangle$$

Here, the symbols enclosed in curled brackets characterize the complete term and thus correspond to the "parent term" introduced in Section 6.1.

If spin–orbit coupling is introduced next [cf. Eq. (3.101)], basis functions Γ_i transforming according to the Kronecker product $\Gamma_S \otimes \Gamma_L$ are required. To this end, representation $D^{(S)}(G)$ of total spin S which is subduced from representation $D^{(S)}$ of $SU(2)$ [cf. Eq. (6.23)] is reduced according to

$$D^{(S)}(G) = \sum_j a_j^{(S)} \Gamma_{Sj} \qquad (6.25)$$

Subsequently, the Kronecker product

$$\Gamma_S \otimes \Gamma_L = \sum_i b_i \Gamma_i \qquad (6.26)$$

is formed. If the basis functions corresponding to the decomposition of Eq. (6.25) are written as

$$|(\text{parent}) S \Gamma_S \gamma_S a_S \Gamma_L \gamma_L\rangle = \sum_{M_S} |(\text{parent}) S M_S \Gamma_L \gamma_L\rangle \langle S M_S | S \Gamma_S \gamma_S a_S\rangle \qquad (6.27)$$

the coupled basis functions finally result as

$$|(\text{parent}) S \Gamma_S a_S \Gamma_L; \Gamma \gamma b\rangle = \sum_{\gamma_S \gamma_L} |(\text{parent}) S \Gamma_S \gamma_S a_S \Gamma_L \gamma_L\rangle \langle \Gamma_S \gamma_S \Gamma_L \gamma_L | \Gamma \gamma b\rangle \qquad (6.28)$$

As before, the Kronecker multiplicity index b is required if, in Eq. (6.26), $b_i > 1$, i.e., if Γ of Eq. (6.28) occurs more than once in the reduction.

The number of the various ligand field states which arise from the d^N configurations in cubic, tetragonal, trigonal, and cylindrical symmetry is again given in Table I.

6.4. LIGAND FIELD INTERACTION

It has been demonstrated in Section 3.3 that the ligand field potential may be always expressed as a linear combination of irreducible tensor operators $\mathbf{S}_\gamma^{(k\Gamma a)}(\theta_i, \varphi_i)$

$$V_{LF} = \sum_i V_{LF}(r_i, \theta_i, \varphi_i) = \sum_i \sum_{k\Gamma\gamma a} B_\gamma^{k\Gamma a}(r_i) \mathbf{S}_\gamma^{(k\Gamma a)}(\theta_i, \varphi_i) \qquad (6.29)$$

In Eq. (6.29), the tensor operators $\mathbf{S}_\gamma^{(k\Gamma a)}(\theta_i, \varphi_i)$ represent that particular component of the spherical tensor operator $\mathbf{C}_q^{(k)}(\theta_i, \varphi_i)$ which transforms according to irreducible representation Γ,

$$\mathbf{S}_\gamma^{(k\Gamma a)}(\theta_i, \varphi_i) = \sum_q \mathbf{C}_q^{(k)}(\theta_i, \varphi_i) \langle kq | k\Gamma \gamma a\rangle \qquad (6.30)$$

Since the potential should be invariant against symmetry transformations, it is always in Eq. (6.29) $\Gamma = A_1$ [cf. also Eq. (3.65)].

Matrix elements of the ligand field operator of arbitrary symmetry between the functions of Eq. (6.28) may be expressed as

$$\langle (\text{parent}) S \Gamma_S a_S \Gamma_L; \Gamma \gamma b | V_{LF} | (\text{parent})' S' \Gamma'_S a'_S \Gamma'_L; \Gamma' \gamma' b' \rangle$$

$$= \delta(S, S') \delta(\Gamma_S, \Gamma'_S) \delta(a_S, a'_S) \delta(\Gamma_L, \Gamma'_L) \delta(\Gamma, \Gamma') \delta(\gamma, \gamma') \delta(b, b')$$

$$\times (1/\sqrt{d_{\Gamma_L}}) \langle (\text{parent}) S \Gamma_L \| V_{LF} \| (\text{parent})' S \Gamma_L \rangle \tag{6.31}$$

This expression is obtained simply on the basis of Eq. (3.17). If it is observed that

$$\begin{Bmatrix} \Gamma_3 & \Gamma_3 & A_1 \\ \Gamma_L & \Gamma_L & \Gamma_4 \end{Bmatrix} = (-1)^{(j_3 + j_L + j_4)} (d_{\Gamma_3} d_{\Gamma_L})^{-1/2} \tag{6.32}$$

and Eq. (3.31) is used, the reduced matrix element in Eq. (6.31) results as

$$\langle (l^{N-1} \alpha_1 S_1 \Gamma_1, \Gamma_2 a_2, b) S \Gamma_L \| V_{LF} \| (l^{N-1} \alpha'_1 S'_1 \Gamma'_1, \Gamma'_2 a'_2, b') S \Gamma_L \rangle$$

$$= N (d_{\Gamma_L})^{1/2} \sum_{\substack{\alpha_4 S_4 \Gamma_4 \Gamma_3 a_3 \\ a'_3 b_1}} \langle l^N S \Gamma_L (l^{N-1} \alpha_1 S_1 \Gamma_1, \Gamma_2 a_2, b) \{ | l^{N-1} \alpha_4 S_4 \Gamma_4, \Gamma_3 a_3, b_1 \rangle$$

$$\times \langle l^{N-1} \alpha_4 S_4 \Gamma_4, \Gamma_3 a'_3, b_1 | \} (l^{N-1} \alpha'_1 S'_1 \Gamma'_1, \Gamma'_2 a'_2, b') S \Gamma_L \rangle (d_{\Gamma_3})^{-1/2} \langle a_3 \Gamma_3 \| V_{LF}(r_i, \theta_i, \varphi_i) \| a'_3 \Gamma_3 \rangle \tag{6.33}$$

The reduced matrix element between single-electron functions may be further decomposed by application of Eq. (6.29) such that

$$\langle a_3 \Gamma_3 \| V_{LF}(r_i, \theta_i, \varphi_i) \| a'_3 \Gamma_3 \rangle = \sum_{ka} (d_{\Gamma_3})^{1/2} \langle a_3 \Gamma_3 \gamma_3 | \mathbf{S}^{(ka)} | a'_3 \Gamma_3 \gamma_3 \rangle \langle B^{ka}(r_i) \rangle \tag{6.34}$$

and

$$\langle a_3 \Gamma_3 \gamma_3 | \mathbf{S}^{(ka)} | a'_3 \Gamma_3 \gamma_3 \rangle = (-1)^{2k} (2l+1)^{-1/2} \langle l \| \mathbf{C}^{(k)} \| l \rangle \langle l \Gamma_3 \gamma_3 a_3 | l \Gamma_3 \gamma_3 a'_3, k A_1 a_1 a \rangle \tag{6.35}$$

The reduced matrix elements of the tensor operators $\mathbf{C}^{(k)}$ are determined by Eq. (2.27) or, more specifically, by

$$\langle l \| \mathbf{C}^{(k)} \| l \rangle = (-1)^l (2l+1) \begin{pmatrix} l & k & l \\ 0 & 0 & 0 \end{pmatrix} \tag{6.36}$$

The quantities $\langle B^{ka}(r_i) \rangle$ may be expressed in terms of any other type of ligand field parameters on the basis of relations listed in Section 3.3.

Alternatively, the product of the dimension factor and the single-electron reduced matrix element in Eq. (6.33),

$$(d_{\Gamma_3})^{-1/2} \langle a_3 \Gamma_3 \| V_{LF}(r_i, \theta_i, \varphi_i) \| a'_3 \Gamma_3 \rangle \tag{6.37}$$

may be expressed in terms of the parameters explicitly employed in the diagrams of Part II of this volume. Thus if a ligand field potential of cubic (O_h, T_d) symmetry is considered [cf. Eq. (3.70)], expression (6.37) may be directly identified with Eq. (3.80). Similarly, if the potential is of tetragonal symmetry [cf. Eq. (3.71)], expression (6.37) may be identified with Eqs. (3.80) and (3.83). Finally, if a trigonal potential is considered [cf. Eq. (3.72)], expression (6.37) may be identified with Eqs. (3.80) and (3.88), the correspondence always depending on the function chosen for $|a_3 \Gamma_3 \rangle$.

By either way, the computation of matrix elements Eq. (6.31) requires only coefficients of fractional parentage, given by Eq. (6.13), and the long coupling coefficient in Eq. (6.35), which is defined by Eq. (3.38).

6.5. COULOMB INTERACTION

The Coulomb interelectronic repulsion operator is a scalar operator and transforms as such according to representation A_1 of the symmetry group in question. The corresponding matrix elements between the functions

Eq. (6.28) are then obtained according to

$$\left\langle (\text{parent})S\Gamma_S a_S \Gamma_L ; \Gamma\gamma b \left| \sum_{i>j} \frac{e^2}{r_{ij}} \right| (\text{parent})'S'\Gamma'_S a'_S \Gamma'_L ; \Gamma'\gamma' b' \right\rangle$$

$$= \delta(S, S')\delta(\Gamma_S, \Gamma'_S)\delta(a_S, a'_S)\delta(\Gamma_L, \Gamma'_L)\delta(\Gamma, \Gamma')\delta(\gamma, \gamma')\delta(b, b') \left\langle (\text{parent})S\Gamma_L \left| \sum_{i>j} \frac{e^2}{r_{ij}} \right| (\text{parent})'S\Gamma_L \right\rangle \tag{6.38}$$

The matrix elements of a system of N electrons may be calculated, if $N > 2$, from those of a $(N-1)$-electron system [cf. Eq. (6.19)],

$$\left\langle (\text{parent})S\Gamma_L \left| \sum_{i>j} \frac{e^2}{r_{ij}} \right| (\text{parent})'S\Gamma_L \right\rangle$$

$$= \frac{N}{N-2} \sum_{\substack{\alpha_4 S_4 \Gamma_4 \alpha'_4 \\ b_1 \Gamma_3 a_3}} \langle (\text{parent})S\Gamma_L \{ | l^{N-1}\alpha_4 S_4\Gamma_4, \tfrac{1}{2}\Gamma_3 a_3, b_1 \rangle \langle \alpha'_4 S_4\Gamma_4, \tfrac{1}{2}\Gamma_3 a_3, b_1 | \} (\text{parent})'S\Gamma_L \rangle$$

$$\times \left\langle (l^{N-1}\alpha_4)S_4\Gamma_4 \left| \sum_{i>j} \frac{e^2}{r_{ij}} \right| (l^{N-1}\alpha'_4)S_4\Gamma_4 \right\rangle \tag{6.39}$$

In addition, the matrix element between two-electron functions ($N = 2$) is required. For the most general case we obtain

$$\left\langle (\Gamma_3 a_3 \Gamma_4 a_4, b)S\Gamma \left| \frac{e^2}{r_{12}} \right| (\Gamma'_3 a'_3 \Gamma'_4 a'_4, b')S\Gamma \right\rangle$$

$$= \sum_{\substack{\Gamma_1 a_1 \Gamma_2 a_2 \\ \Gamma'_1 a'_1 \Gamma'_2 a'_2}} \langle S\Gamma(\Gamma_3 a_3 \Gamma_4 a_4, b)\{ | \Gamma_1 a_1 \Gamma_2 a_2, b \rangle \langle \Gamma'_1 a'_1 \Gamma'_2 a'_2, b' | \}(\Gamma'_3 a'_3 \Gamma'_4 a'_4, b')S\Gamma \rangle$$

$$\times \sum_{\gamma_1 \gamma_2 \gamma'_1 \gamma'_2} \langle \Gamma_1 \gamma_1 \Gamma_2 \gamma_2 | \Gamma\gamma b \rangle^* \langle \Gamma'_1 \gamma'_1 \Gamma'_2 \gamma'_2 | \Gamma\gamma b' \rangle \sum_{m_1 m_2 m'_1 m'_2} \langle lm'_1 | l\Gamma'_1 \gamma'_1 a'_1 \rangle \langle lm'_2 | l\Gamma'_2 \gamma'_2 a'_2 \rangle$$

$$\times \langle lm_1 | l\Gamma_1 \gamma_1 a_1 \rangle^* \langle lm_2 | l\Gamma_2 \gamma_2 a_2 \rangle^* \left\langle lm_1 lm_2 \left| \frac{e^2}{r_{12}} \right| lm'_1 lm'_2 \right\rangle \tag{6.40}$$

Here

$$\left\langle lm_1 lm_2 \left| \frac{e^2}{r_{12}} \right| lm'_1 lm'_2 \right\rangle$$

$$= \delta(m_1 + m_2, m'_1 + m'_2)(-1)^{m_1+m_2}(2l+1)^2 \sum_k F^k(nl, nl) \begin{pmatrix} l & l & k \\ 0 & 0 & 0 \end{pmatrix}^2 \begin{pmatrix} l & l & k \\ m_1 & -m'_1 & m'_1 - m_1 \end{pmatrix}$$

$$\times \begin{pmatrix} l & l & k \\ m_2 & -m'_2 & m'_2 - m_2 \end{pmatrix} \tag{6.41}$$

It should be noted that the cfp in

$$|(\Gamma_1 a_1 \Gamma_2 a_2, b)S\Gamma_L \rangle = \sum_{\Gamma'_1 a'_1 \Gamma'_2 a'_2} |(\Gamma'_1 a'_1 \Gamma'_2 a'_2, b)S\Gamma_L \rangle \langle \tfrac{1}{2}\Gamma'_1 a'_1, \tfrac{1}{2}\Gamma'_2 a'_2, b | \}(\Gamma_1 a_1 \Gamma_2 a_2, b)S\Gamma_L \rangle \tag{6.42}$$

equals 1 only for identical electron states and, in this case, it may be expressed by the vector-coupled functions alone. Also, the index b remains unchanged in a two-electron function since the interchange $\Gamma_1 \leftrightarrow \Gamma_2$ does not produce a different value of b. Therefore no b' arises in Eq. (6.42).

Evidently, the computation of the matrix elements Eq. (6.38) is reduced lastly to the reduced matrix elements of Eq. (6.40) and to coefficients of fractional parentage. The calculation according to Eq. (6.40) then requires cfp, symmetry coupling coefficients $\langle \Gamma_1 \gamma_1 \Gamma_2 \gamma_2 | \Gamma\gamma b \rangle$ of Eq. (3.18), symmetry adaptation coefficients $\langle lm | l\Gamma\gamma a \rangle$ of Eq. (3.10), and 3-j symbols. The quantities F^k or the related Racah parameters A, B, C [cf. Eq. (2.77)] are again considered as adjustable parameters.

6.6. SPIN-ORBIT COUPLING AND ADDITIONAL INTERACTIONS

The operator of spin–orbit coupling is a scalar single-electron operator and may thus be written, according to Eq. (3.53), in terms of irreducible tensor operators as

$$\sum_i \kappa\xi(r_i)(\mathbf{s}_i^{(1)}\cdot\mathbf{l}_i^{(1)}) = -\sum_{\Gamma_3\gamma_3 a_3}\left(\begin{array}{cc}1\\\Gamma_3\gamma_3 a_3 & \overline{\Gamma_3\gamma_3 a_3}\end{array}\right)^* \sum_i \kappa^{\Gamma_3 a_3}\xi^{\Gamma_3 a_3}(r_i)(\mathbf{s}_{i\gamma_3}^{(1\Gamma_3 a_3)}\cdot\mathbf{l}_{i\overline{\gamma_3}}^{(1\overline{\Gamma_3 a_3})}) \tag{6.43}$$

Here, the metric tensor has been defined in Eq. (3.44). The matrix element of the spin–orbit coupling operator between the functions Eq. (6.28) then follows by application of Eqs. (3.37) and (6.43) as

$$\langle(\text{parent})S\Gamma_s a_s\Gamma_L;\Gamma\gamma b_1|\sum_i \kappa\xi(r_i)(\mathbf{s}_i\cdot\mathbf{l}_i)|(\text{parent})'S'\Gamma_s' a_s'\Gamma_L';\Gamma'\gamma' b_2\rangle$$

$$= \delta(\Gamma,\Gamma')\delta(\gamma,\gamma')\sum_{\gamma_L\gamma_S\gamma_L'\gamma_S'}\langle\Gamma_S\gamma_S\Gamma_L\gamma_L|\Gamma\gamma b_1\rangle^*\langle\Gamma_S'\gamma_S'\Gamma_L'\gamma_L'|\Gamma\gamma b_2\rangle(-1)\sum_{\Gamma_3\gamma_3 a_3}\left(\begin{array}{cc}1\\\Gamma_3\gamma_3 a_3 & \overline{\Gamma_3\gamma_3 a_3}\end{array}\right)^*$$

$$\times\langle(\text{parent})S\Gamma_S\gamma_S a_s|\sum_i \mathbf{s}_{i\gamma_3}^{(1\Gamma_3 a_3)}|(\text{parent})'S'\Gamma_S'\gamma_S'a_S'\rangle\langle(\text{parent})\Gamma_L\gamma_L|\sum_i \mathbf{l}_{i\overline{\gamma_3}}^{(1\overline{\Gamma_3 a_3})}|(\text{parent})'\Gamma_L'\gamma_L'\rangle\zeta_{nl}^{\Gamma_3 a_3} \tag{6.44}$$

The matrix element of $\sum_i \mathbf{s}_{i\gamma_3}^{(1\Gamma_3 a_3)}$ may be further reduced by application of Eq. (3.37):

$$\langle(\text{parent})S\Gamma_S\gamma_S a_S|\sum_i \mathbf{s}_{i\gamma_3}^{(1\Gamma_3 a_3)}|(\text{parent})'S'\Gamma_S'\gamma_S'a_S'\rangle$$

$$= [S]^{-1/2}\langle S\Gamma_S\gamma_S a_S|S'\Gamma_S'\gamma_S'a_S', 1\Gamma_3\gamma_3 a_3\rangle$$

$$\times\langle(\text{parent})S\|\sum_i \mathbf{s}_i^{(1)}\|(\text{parent})'S'\rangle \tag{6.45}$$

For the product of the two matrix elements in Eq. (6.44) then follows, on the basis of Eqs. (6.18), (2.44), and (3.17),

$$\langle(\text{parent})S\Gamma_S\gamma_S a_S|\sum_i \mathbf{s}_{i\gamma_3}^{(1\Gamma_3 a_3)}|(\text{parent})'S'\Gamma_S'\gamma_S'a_S'\rangle$$

$$\times\langle(\text{parent})\Gamma_L\gamma_L|\sum_i \mathbf{l}_{i\overline{\gamma_3}}^{(1\overline{\Gamma_3 a_3})}|(\text{parent})'\Gamma_L'\gamma_L'\rangle$$

$$= [S']^{1/2}\langle S\Gamma_S\gamma_S a_S|S'\Gamma_S'\gamma_S'a_S', 1\Gamma_3\gamma_3 a_3\rangle(d_{\Gamma_L'})^{1/2}$$

$$\times\sum_{b_4}\langle\Gamma_L\gamma_L b_4|\Gamma_L'\gamma_L'\overline{\Gamma_3\gamma_3}\rangle$$

$$\times\sum_{\substack{\alpha_1 S_1\Gamma_1\Gamma_2 a_2\\\Gamma_2 a_2 bb'}}\langle S\Gamma_L(\text{parent})\{|\alpha_1 S_1\Gamma_1,\Gamma_2 a_2, b\rangle$$

$$\times\langle\alpha_1 S_1\Gamma_1,\Gamma_2' a_2', b'|\}(\text{parent})'S'\Gamma_L'\rangle N(-1)^{S_1+S+3/2}$$

$$\times\left\{\begin{array}{ccc}\frac{1}{2} & \frac{1}{2} & 1\\S & S' & S_1\end{array}\right\}\langle\tfrac{1}{2}\|\mathbf{s}^{(1)}\|\tfrac{1}{2}\rangle\sum_{b_5}(-1)^{(j_1-j_2+j_L)_b+(j_2+j_2'+j_3)_{bs}}$$

$$\times\left\{\begin{array}{ccc}\Gamma_2 & \overline{\Gamma_2'} & \Gamma_3\\\Gamma_L' & \Gamma_L & \Gamma_1\end{array}\right\}_{b_4 b' bb_5}\langle l\Gamma_2 a_2\|\mathbf{l}^{(1\overline{\Gamma_3 a_3})}\|l\Gamma_2'a_2'\rangle_{bs} \tag{6.46}$$

The single-electron reduced matrix elements of Eq. (6.46) may be expressed as

$$\langle\tfrac{1}{2}\|\mathbf{s}^{(1)}\|\tfrac{1}{2}\rangle = \sqrt{3/2}$$

$$\langle l\Gamma_2 a_2\|\mathbf{l}^{(1\overline{\Gamma_3 a_3})}\|l\Gamma_2'a_2'\rangle_{bs}$$

$$= [d_{\Gamma_2}/(2l+1)]^{1/2}\langle l\Gamma_2 a_2 b_5|l\Gamma_2'a_2', 1\overline{\Gamma_3 a_3}\rangle\langle l\|\mathbf{l}^{(1)}\|l\rangle$$

$$\langle l\|\mathbf{l}^{(1)}\|l\rangle = [l(l+1)(2l+1)]^{1/2} \tag{6.47}$$

In Eq. (6.44), $\zeta_{nl}^{\Gamma_3 a_3}$ is the spin–orbit coupling parameter which is, in general, reduced in magnitude by the amount κ as compared to ζ_{nl}^{free} of the free ion. The quantity $\kappa^{\Gamma_3 a_3}$ is the orbital reduction factor. It should be noted that the anisotropy of spin–orbit coupling is rarely taken into account. Consequently, $\zeta_{nl}^{\Gamma_3 a_3}$ may be replaced, in most cases, by ζ_{nl} and $\kappa^{\Gamma_3 a_3}$ may be substituted by κ. Of the quantities required for computation, the symmetry coupling coefficients $\langle\Gamma_S\gamma_S\Gamma_L\gamma_L|\Gamma\gamma b\rangle$ have been defined in Eq. (3.18), and the long coupling coefficients are given by Eq. (3.38). In addition, coefficients of fractional parentage [cf. Eq. (6.13)], the symmetry adaptation coefficients of Eq. (3.10) as well as 6-Γ symbols of Eq. (3.27) are needed. The quantities ζ_{nl} and κ are again determined on a semiempirical basis.

The energy matrices which are obtained by a combination of the matrix elements of Sections 6.4–6.6 are partly in diagonal form. The dimensions of the individual submatrices may be taken from Table I. The diagonalization of the matrices produces eigenvalues of energy E_n which are identical to those resulting in weak-field and intermediate-field coupling. These energies are displayed in Part II of this volume. In addition, to each eigenvalue E_n, the corresponding eigenfunction obtains as a linear combination of the p_Γ basis functions,

$$|n''\Gamma\gamma\rangle = \sum_{k''=1}^{p_\Gamma}|(l^{N-1}\alpha_1\beta_1 S_1\Gamma_1, \tfrac{1}{2}\Gamma_2 a_2, b)SM_S\Gamma\gamma\rangle C_{k''}'' \tag{6.48}$$

Here, p_Γ depends on the symmetry of the problem and $k'' \equiv l^{N-1}\alpha_1\beta_1 S_1\Gamma_1, \frac{1}{2}\Gamma_2 a_2, b$. The coefficients $C_{k''}''$ squared determine the composition of the individual levels in the strong-field coupling scheme, this composition being, in general, different from that in weak-field or intermediate-field coupling. Similar to Sections 4.5 and 5.5, the functions Eq. (6.48) may then be used to determine the effect of additional interactions. For more details these previous sections should be consulted.

7

DESCRIPTION OF PROGRAMS

7.1. GENERAL OUTLINE

The programs to be described below have been written in order to generate by computer the diagrams presented in Part II of this volume. The diagrams display the relative term energies of d^N transition metal complex ions ($N = 2, \ldots, 8$) within the approximation of ligand field theory. The energies are plotted as functions of certain parameters, e.g., Dq, Ds, Dt in tetragonal (D_{4h}, D_4, C_{4v}) symmetry.

The calculations performed in the construction of the diagrams require expressions for matrix elements of the ligand field, Coulomb repulsion, and spin–orbit interaction operators within the coupling schemes of the weak, intermediate, and strong fields. For this purpose, a number of various coupling and recoupling coefficients as well as the coefficients of fractional parentage are needed. Detailed algebraic expressions for these quantities have been listed in earlier chapters. Once the matrices are calculated, diagonalization yields the relevant eigenvalues which may subsequently be plotted as functions of a selected parameter.

Essentially three sets of programs may be distinguished which serve different purposes:

(i) Calculation of matrix elements of the given Hamiltonian;

(ii) Diagonalization of the matrix and reordering of eigenvalues;

(iii) Plotting of the diagrams.

The three sets of programs are self-contained. They are linked by tape units and disk files. The programs were written for a CDC 3300 in Fortran IV and require a high-speed store of not more than 25K words.

7.2. COMPUTATION OF MATRIX ELEMENTS

A. Strong-Field Version

The main program MAIN 1 controls the run of the complete program and calls the various subroutines. First subroutine SYM computes all symmetry coupling coefficients of a given point group on the basis of Eq. (3.42). The Yamanouchi symbols and some other quantities of the symmetric group S_N are calculated by subroutine YAMAN. Subroutine CFP then computes the coefficients of fractional parentage for the N-electron system according to Eq. (6.13). In this process, the supposedly available cfp for the ($N-1$)-electron system are read from the disk file, the cfp for all terms of d^N are computed, and the resulting functions are orthogonalized by a Schmidt procedure.

If, e.g., spin–orbit interaction is to be included, the term designations

$$|(\gamma_1 \gamma_2 S_1 \Gamma_1, \gamma_3 \ldots S\Gamma_s a_s \Gamma_L)\Gamma b\rangle$$

are set up in subroutine MTERM and subsequently transferred into MAIN 1. The program MAIN 1 initiates the computation of matrix elements between these terms. For this purpose MAIN 1 calls subroutine FIELD for the calculation of matrix elements of the ligand field operator according to Eq. (6.31) ff, subroutine COULOMB for the calculation of matrix elements of Coulomb interaction according to Eq. (6.38) ff, and subroutine SPINORBT for the calculation of matrix elements of spin–orbit interaction employing Eq. (6.44) ff. The various coefficients which are required in these computations are provided by separate subroutines. Thus FLATO computes the coefficients $\langle j\Gamma\gamma a|j_1\Gamma_1\gamma_1 a_1, j_2\Gamma_2\gamma_2 a_2\rangle$, WIGNER the well-known quantities $\begin{pmatrix} j_1 & j_2 & j \\ m_1 & m_2 & m \end{pmatrix}$, GAMMA the coefficients $\begin{Bmatrix} \Gamma_1 & \Gamma_2 & \Gamma_3 \\ \Gamma_4 & \Gamma_5 & \Gamma_6 \end{Bmatrix}$, and CLEB the quantities $\begin{pmatrix} \Gamma_1 & \Gamma_2 & \Gamma \\ \gamma_1 & \gamma_2 & \gamma \end{pmatrix}$. Finally, subroutine CFP1 draws the coefficients $\langle\ldots S_1\Gamma_1, \gamma_N|\}(\text{parent})S\Gamma\rangle$ from a matrix. If all terms have been passed, the term designations and matrix elements are written onto a disk file. The essential steps of the program are indicated in the flow-chart of Fig. 2.

B. Weak-Field Version

The control is effected by program MAIN 2. For convenience, the cfp for electron configurations d^N ($N = 2, \ldots, 5$) as well as the almost diagonal matrices

of Coulomb interaction are taken from tables by Racah and stored by routine BLOCKDATA in COMMON blocks.

By calling subroutine PROSYM, all basis functions $|j\Gamma\gamma a\rangle$ for $j = 0, 1/2, 1, \ldots, 13/2$ and for representations Γ of the given symmetry group are calculated by the method of projection operators [cf. Eq. (3.11)]. Subroutine MTERM then determines the quantum numbers of the terms without or with inclusion of spin–orbit coupling, i.e., $|vSL\Gamma a\rangle$ or $|vSLJ\Gamma a\rangle$, respectively, and transfers these to MAIN 2. The main program calls various subroutines to compute the required matrix elements between each two of these terms. Thus COULOMB calculates matrix elements of Coulomb interaction according to Eq. (4.3). SPINORBT similarly computes matrix elements of spin–orbit interaction using Eq. (4.4). Finally, FIELD calculates matrix elements of ligand

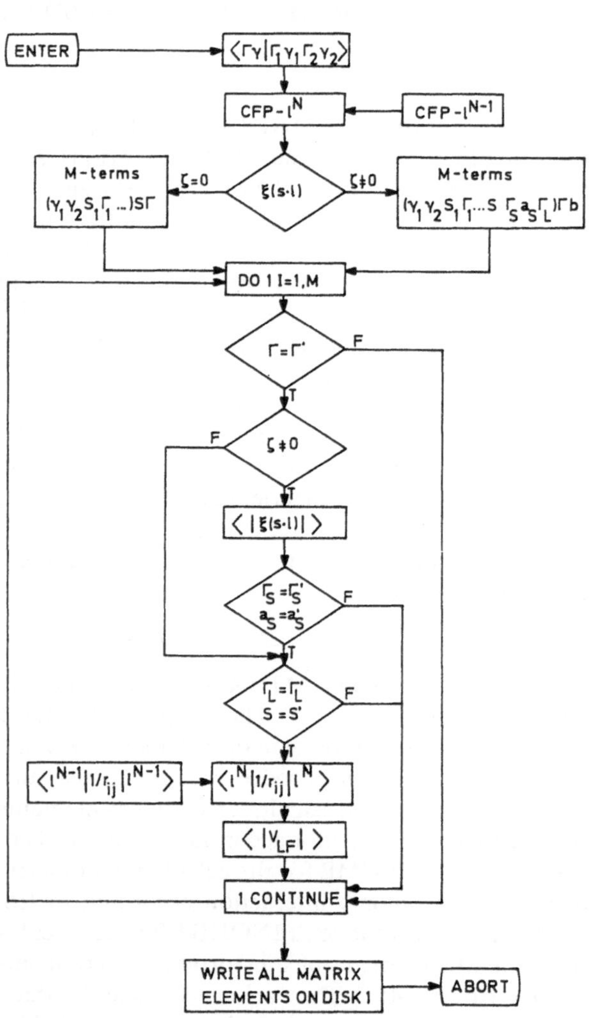

FIGURE 2. Flow-chart of the computer program for calculation of matrix elements in the strong-field version.

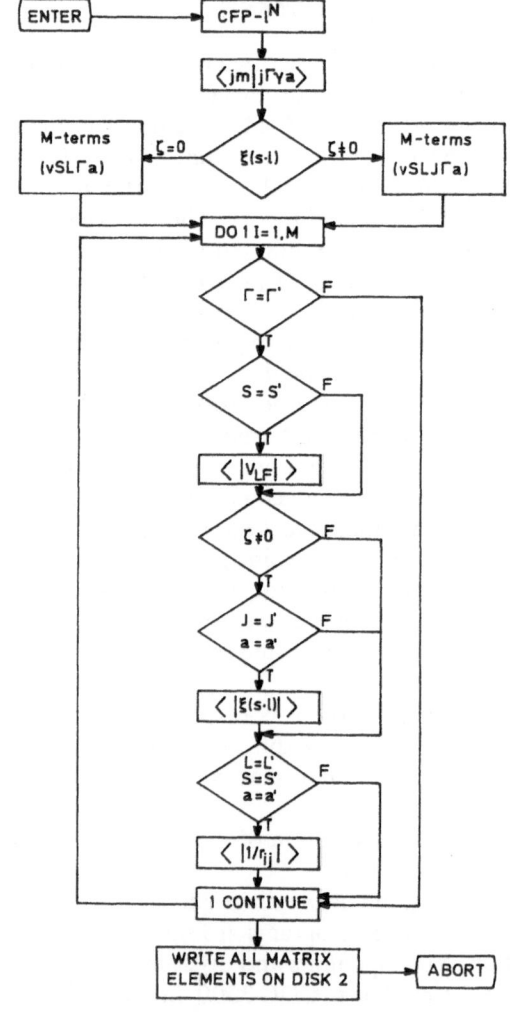

FIGURE 3. Flow-chart of the computer program for calculation of matrix elements in the weak-field version.

field interaction on the basis of Eqs. (4.5), (4.6), or (4.7). Again, all required coefficients are calculated or provided by separate subroutines. Thus the subroutines FLATO, WIGNER, CLEB, and CFP were mentioned in the strong-field version above. In addition, subroutine RACAH calculates the 6-j symbols $\begin{Bmatrix} j_1 & j_2 & j_3 \\ j_4 & j_5 & j_6 \end{Bmatrix}$. If matrix elements between all terms are calculated, these are written, in conjunction with the appropriate term designations, onto a disk file. The structure of the program is illustrated by the flow-chart of Fig. 3.

7.3. DIAGONALIZATION OF THE ENERGY MATRIX

The complete program is controlled here by MAIN 3. First, all those parameters are read which have been given fixed values, including another parameter which specifies whether the diagonalization should be performed within the strong-field or the weak-field version. In the following text, only these two versions are covered, since if spin–orbit interaction is neglected, the coupling schemes of the weak field and the intermediate field are identical. As soon as the first value of the ligand field parameter to be varied, PX, is read, subroutine SFDIAG or subroutine WFDIAG is called in order to compute the strong-field or the weak-field eigenvalues respectively. These subroutines read from the disk file the matrix elements calculated in part 1 of the total program (cf. Section 7.2), multiply these with the given values of the parameters, and effect diagonalization of the matrices. In case that, compared to the previous diagonalization, the lowest energy state has changed, the subroutine independently determines the accurate position of the cross-point by continuous variation of the parameter involved. The eigenvalues are then ordered according to their magnitude. MAIN 3 investigates whether these energy values are situated on or outside of the assumed diagram borders. If this is the case, the appropriate term designations for the left or the right border are written into the blocks DESIGL or DESIGR, respectively. Subroutine ORDER inserts the computed energies according to increasing values of the varied parameter and from left to right into the correct column of a $n \times m$ matrix. Here n determines the number of eigenvalues, whereas m gives the number of those values of the varied parameter PX for which calculations have already been performed. If the next value of PX is read, the procedure is repeated, until after the last value of PX all the

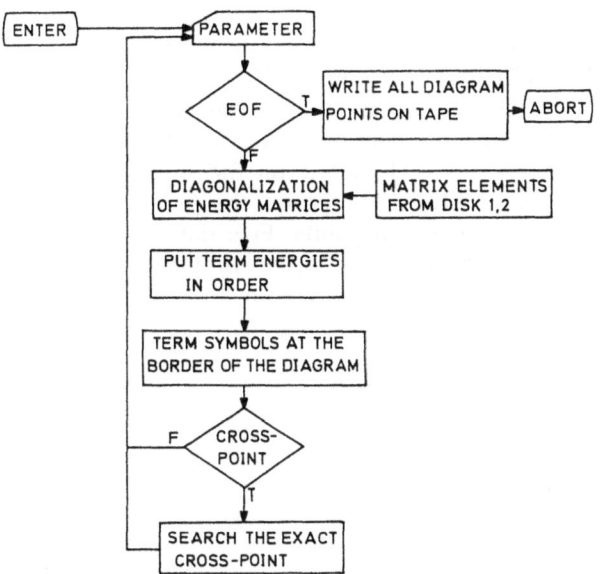

FIGURE 4. Flow-chart of the computer program for calculation of specific points in the ligand field energy diagrams.

information is stored on tape by calling WRITETAP. In order to extend the number of supporting points, the procedure may be started at some later instant. In this case, at first all stored data are read from tape by calling READTAPE, whereupon the program proceeds as described above. A flow-chart for this part of the program is shown in Fig. 4.

7.4. PLOTTING THE ENERGY DIAGRAMS

The complete procedure is controlled here by the main program PLOTTER. First of all, the data for a diagram computed in part 2 of the total program are read from tape, and the various parameters are given fixed values in order to determine the scale on the abscissa and the ordinate, the upper limit for the energies to be plotted, and the left-hand smallest as well as the right-hand largest value of the ligand field parameter employed. Subsequently, captions for the diagram are read and subroutine COORD is called to draw the axes including all required lettering into an area of format A 3. Subroutine DESIGL then writes all designations for terms, as applicable at the smallest value of the abscissa parameter used, on the left margin of the diagram from bottom to top. Weak-field and/or strong-field notation may be chosen. No designations are written onto energy curves which do not start at the left border of the diagram. Next the individual energy curves are drawn according to increasing energies, the designations on the right or upper margin of the diagram being added. To this end, each individual row of the $n \times m$ matrix as computed in part 2 of the program, i.e., an energy curve with m supporting points, is transferred from PLOTTER to subroutine GRAPH, one at a time. INTERPOL then calculates, from three supporting points and from the slope of the first point, the curve between points 1 and 2 on the basis of a cubic equation. GRAPH draws the connecting line between these two points. Five differently marked lines are available in order to distinguish terms belonging to different representations. On the right margin, the same program writes the designation of the term, whereby weak-field or strong-field notation may be selected. Altogether GRAPH is called n times in the course of production of the diagram. After all of the calculations have been performed, the data are copied into the plotter file and the diagram proper is produced by a plotter of type BENSON 122.

8

DESCRIPTION OF THE DIAGRAMS AND THEIR APPLICATION

The energy diagrams collected in Part II of this volume have been obtained on the basis of the tensor operator algebra described briefly in Chapters 2–5 and by application of the programs outlined above. The diagrams present the relative term energies in cm^{-1} (on the ordinate) as functions of a specific parameter of the ligand field likewise in cm^{-1} (on the abscissa), all the remaining parameters being assigned fixed values. The electron configurations d^N, where $N = 2, 3, \ldots, 8$ in cubic (O_h, T_d), tetragonal (D_{4h}, D_4, C_{4v}), trigonal (D_{3d}, D_3, C_{3v}), and cylindrical ($D_{\infty h}$, D_∞, $C_{\infty v}$) symmetry of the ligand field are considered. No diagrams have been included for configurations d^1 and d^9 since, in this case, the energies may be easily obtained by inserting the given parameter values into the corresponding expressions for the one-electron matrix elements. The upper limit for the energies included in the diagrams has been chosen as $90,000\ cm^{-1}$, energies above $30,000\ cm^{-1}$ being drawn on a compressed scale. More specific details concerning construction and use of the diagrams will be given in the following sections.

First of all, suitable fixed values have to be selected for all those parameters which are not explicitly varied in the diagrams. Thus the Racah parameter B has always been given a mean value approximately corresponding to that of a hexaquo-ion. In addition, $C = 4B$ has been assumed. It should be observed that this is no serious restriction on the use of the diagrams, as will be shown below. For clarity of presentation, spin–orbit interaction has

been included for configurations d^2 and d^8 only. In this case, the spin–orbit coupling parameter has been chosen such as to approximate that of important free transition metal ions, i.e., $\zeta = 210\ cm^{-1}$ in configuration d^2, $\zeta = 650\ cm^{-1}$ in configuration d^8.

For each electron configuration, two types of diagrams have been constructed. In one series of diagrams, Dq has been varied on the abscissa from 0 to $3000\ cm^{-1}$, all other parameters being fixed. Separate diagrams are presented for values of the axial ligand field parameters Dt or $D\tau$ of 0, ± 500 and $\pm 1000\ cm^{-1}$ as well as for values of $K = Ds/Dt$ or $K = D\sigma/D\tau$ of ± 1 and ± 5. The range of values chosen for the parameters characterizing the axial ligand field is determined by the analysis of optical spectra and other physical properties of transition metal ions in solids and complex compounds. The above type of diagrams is presented for all symmetries treated except for cylindrical symmetry where $Dq = 0$. It follows that for each configuration in the symmetries D_4 and D_3, 16 diagrams are required. Diagrams characterized by $Ds = Dt = 0$ or $D\sigma = D\tau = 0$ correspond to octahedral symmetry and are identical to the well-known diagrams of Tanabe and Sugano.

In a second series of diagrams, one parameter of the axial ligand field, i.e., Dt in D_4 symmetry, $D\tau$ in D_3 symmetry, and Dv in D_∞ symmetry has been varied on the abscissa. The range of values for these parameters has been chosen as -3000 to $+3000\ cm^{-1}$, the case of cubic symmetry thus arising in the center of the diagram, i.e., for $Dt = 0$, $D\tau = 0$, and $D\nu = 0$. Separate diagrams again were drawn for values of $K = \pm 1$ and ± 5. In the symmetries D_4 and D_3, a mean Dq value has been chosen approximately corresponding to that of hexaquo ions. It follows that, in general, four diagrams of this type are required. In certain electron configurations where the strength of the ligand field is considerably different for two- and three-valent transition metal ions, diagrams for two different Dq values were constructed, thus raising the number of these diagrams to eight. The configurations that are involved here are d^3(V^{2+}, Cr^{3+}), d^4(Cr^{2+}, Mn^{3+}), d^5(Mn^{2+}, Fe^{3+}), and d^6(Fe^{2+}, Co^{3+}).

Notational details in the diagrams will be considered next. Designations for the individual energy terms which are displayed in the diagrams have been added at the left and right border of the diagrams. The designation always gives only the term providing the largest contribution to the wave function. Of course, this specification applies alone for that combination of parameter values characterizing the particular point where the designation has been copied into the diagram.

In the first place, the designation lists the appropriate representation within the symmetry group, thus specifying the transformation behavior of the underlying wave function. If spin–orbit interaction has been included (in configurations d^2 and d^8 only), modified representation symbols according to Bethe are employed. Thus, e.g., Γ_{T5} is written for the two-dimensional representation E of symmetry D_4, Γ_3^T for representation E of D_3 symmetry, and Γ_3^C for representation E_1 of D_∞ symmetry. If spin–orbit interaction has not been included (all configurations except d^2 and d^8), Mulliken representation symbols are used. In order to facilitate identification of the various terms, differently marked lines are employed for the different representations.

In those diagrams which show the variation with Dq, weak-field term notations are given on the left border in addition to the representation symbol. On the right border of the diagrams, strong-field term notations are listed for the configurations d^2 and d^8. Diagrams for all the remaining configurations carry on the right border both weak-field and strong-field term notations. Due to limitations of space, strong-field notations are given in terms of running numbers (except for configurations d^2 and d^8) which are explained on the cover page for the series of diagrams concerned. Diagrams showing the variation with Dt, $D\tau$, or $D\nu$ carry weak-field as well as strong-field notation on both left and right borders except for d^2 and d^8 configurations, where only the strong-field

TABLE III
Complementary subshell configurations for d^N and d^{10-N}

Symmetry	d^N	d^{10-N}
O_h	$t_2^{N_1} e^{N_2}$	$t_2^{6-N_1} e^{4-N_2}$
D_4	$a_1^{N_1} b_1^{N_2} b_2^{N_3} e^{N_4}$	$a_1^{2-N_1} b_1^{2-N_2} b_2^{2-N_3} e^{4-N_4}$
D_3	$a_1^{N_1} e_e^{N_2} e_t^{N_3}$	$a_1^{2-N_1} e_e^{4-N_2} e_t^{4-N_3}$
D_∞	$a_1^{N_1} e_1^{N_2} e_2^{N_3}$	$a_1^{2-N_1} e_1^{4-N_2} e_2^{4-N_3}$

notation is listed. It should be observed that capital letters are always employed for the designation of one-electron states instead of the more conventional lower case letters, i.e., Γ_{T3}(^3E, B_2E) instead of $\Gamma_{T3}[^3E(b_2e)]$. In the strong-field notation for D_3 symmetry, the one-electron states are always reduced to those in octahedral symmetry. Here, the symbol E_T is used for the usual symbol $e(t_2)$ and the symbol E_E for $e(e)$. It is well known that energy matrices for configuration d^{10-N} where $N = 2$, 3, 4 correspond to the matrices for d^N with opposite sign of the ligand field parameters. Therefore, the same strong-field-based notations have been employed in the configurations d^N and d^{10-N}. It follows that in order to obtain the complete strong-field notation for energy terms arising from configuration d^{10-N}, the complementary subshell configurations have to be formed according to the scheme in Table III.

If, as a consequence of the variation of one of the ligand field parameters, the characterization of the ground state in terms of representation symbol and/or spin multiplicity is changed, the resulting crosspoint is indicated, in the diagram, by a vertical line. The symbols required for identification of the ground states involved are indicated by spin multiplicity and representation symbol. Thus a crossover indicated by $^3\Gamma_{T5} \to {}^1\Gamma_{T3}$ in configuration d^2 corresponds to the crosspoint $\Gamma_{T5}[^3X\ldots] \to \Gamma_{T3}[^1Y\ldots]$, the notation being self-evident.

A few additional remarks may be useful. Each diagram is specified in the caption in terms of number of d electrons, symmetry, and the specific values of the parameters chosen. Each set of diagrams is preceded by a cover page giving all necessary information including the variation of the parameters performed and a list of the numbers abbreviating the strong-field designations (see above for more details). For technical reasons, only capital letters have been used in this text, e.g., DQ instead of Dq, etc. In addition, the axial ligand field parameters in fields of D_3 and D_∞ symmetry have been denoted by the same symbols as in D_4 symmetry, i.e., DS and DT, instead of the conventional notation $D\sigma$, $D\tau$ and $D\mu$, $D\nu$, respectively. Finally, the same notation as in lower symmetry has been retained for the energy terms in diagrams of cubic symmetry (e.g., $Dt = Ds = 0$). This particular choice of notation should be convenient if a comparison with lower symmetry diagrams is attempted.

We add some specific comments concerning the construction of the many-electron wave functions and the implications for term designations following therefrom. The coupling of states in the configurations d^2 and d^3 is obvious. In the electron configuration d^4 of D_4 and D_3 symmetry, the strong-field states were obtained by the coupling of two-electron states as indicated by the employed term symbols

$$^{2S+1}\Gamma(\gamma_1\gamma_2(^{2S_1+1}\Gamma_1)\gamma_3\gamma_4(^{2S_2+1}\Gamma_2))$$

In cylindrical symmetry, d^4-state functions were constructed from those of the d^3 configuration according to

$$^{2S+1}\Gamma(^{2S_1'+1}\Gamma_1'(^{2S_2'+1}\Gamma_2'(\gamma_1\gamma_2)\gamma_3)\gamma_4)$$

In configuration d^5, three- and two-electron states were coupled in D_4 and D_3 symmetry according to

$$^{2S+1}\Gamma(\gamma_1\gamma_2\gamma_3(^{2S_1+1}\Gamma_1)\gamma_4\gamma_5(^{2S_2+1}\Gamma_2))$$

whereas in D_∞ symmetry, d^5 electron states were obtained from states of d^4 according to

$$^{2S+1}\Gamma(^{2S_1'+1}\Gamma_1'(^{2S_2'+1}\Gamma_2'(^{2S_3'+1}\Gamma_3'(\gamma_1\gamma_2)\gamma_3)\gamma_4)\gamma_5)$$

An important field of application of the present energy diagrams is in the assignment of electronic transitions in the optical spectra of transition metal ions incorporated in solids and complex compounds. As a preliminary step to accurate calculations and to assignments resulting therefrom, an as complete set of diagrams as possible, i.e., for all possible parameter values, is required. On the other hand, the available number of diagrams will be always limited, depending on the magnitude of the steps chosen for variation of the "fixed" parameter values. It is therefore important that the best possible use is made of a limited set of available energy diagrams. This may be achieved in a number of ways.

The variation of a particular energy level between the diagrams presented here may be easily deduced, at least in a qualitative way, by carefully comparing diagrams which are characterized by the enclosing parameter values. Thus, e.g., the variation

of term energies between diagrams specified by

$$Dt = -1000 \quad \rightarrow \quad Dt = -500 \quad \rightarrow$$
$$Dt = +500 \quad \rightarrow \quad Dt = +1000 \text{ cm}^{-1}$$

for a fixed value of K may be followed, and the actual position of energy levels may be obtained by numerical interpolation. Difficulties arise only in the immediate neighborhood of crosspoints.

If diagrams are required for other values of the Racah parameter B than those employed here, all energies may be simply enlarged or reduced in the same proportion. Thus if $E = f(B)$ for a certain diagram D, it easily follows that for a diagram D' where $B' = xB$, it is $E' = xE = f(xB)$.

In fact, even a much more general relation of this kind holds for all diagrams included in this volume. Let us assume that, for a given diagram D, the relation holds that

$$E = f(B, Dq, Dt, K) \qquad (8.1)$$

Let us further assume that parameters B, Dq, and Dt of diagram D are related to parameters B', Dq', and Dt' of another diagram D' by

$$B' = xB \qquad Dq' = xDq \qquad Dt' = xDt \quad (8.2)$$

It then may be easily shown that the term energies E and E' of diagrams D and D', respectively, are related according to

$$E' = xE = f(xB, xDq, xDt, K) \qquad (8.3)$$

This relation enables the user to obtain, from the diagrams of this volume, term energies for any other values of Dq and Dt. Naturally, the same relation still holds if Dt in D_4 symmetry is replaced by $D\tau$ in D_3 symmetry or by $D\nu$ in D_∞ symmetry. The application of Eq. (8.3) simply corresponds to a change of scale in the diagrams.

A special case is that of configurations d^2 and d^8. Here, in addition to the parameters explicitly involved above, the spin–orbit coupling parameter ζ has to be considered. It is easily demonstrated that ζ is affected by a change of scale in the same way as, say, the parameters B, Dq, and Dt. Consequently, Eq. (8.3) has to be modified according to

$$E' = xE = f(xB, xDq, xDt, x\zeta, K) \qquad (8.4)$$

However, in most cases the effect of spin–orbit interaction is rather small and changes in this parameter need rarely to be explicitly considered.

9

RELATIONS BETWEEN LIGAND FIELD THEORY AND OTHER METHODS

9.1. LIGAND FIELDS AND CRYSTAL FIELD THEORY

In the purely electrostatic crystal field model, the potential produced by a number of point charges $Z_a e$ with polar coordinates R_a, Θ_a, Φ_a may be written as [cf. Eq. (3.67)][12,58]

$$V_{CF}(r_i, \theta_i, \varphi_i)$$

$$= \sum_a Z_a e^2 \sum_{k=0}^{\infty} \frac{r_i^k}{R_a^{k+1}} P_k(\cos \omega_{ai}) \qquad (9.1)$$

where

$$P_k(\cos \omega_{ai})$$

$$= \sum_q (-1)^q \mathbf{C}_q^{(k)}(\theta_i, \varphi_i) \cdot \mathbf{C}_{-q}^{(k)}(\Theta_a, \Phi_a) \qquad (9.2)$$

Here, r_i, θ_i, φ_i are the polar coordinates of electron i and ω_{ai} is the angle between the position vectors in the directions (θ_i, φ_i) and (Θ_a, Φ_a), i.e., $\mathbf{r}_i$ and $\mathbf{R}_a$. In general, summation of Eq. (9.1) over the electrons i will be required in order to obtain the total crystal field potential. The ligands are approximated here by simple point charges, as should be evident from the treatment in Section 3.3 above, the electron density of ligand electrons being completely neglected. It should also be noted that, in contrast to ligand field theory, the contribution of individual ligands is an additive property.

The matrix elements of operator $V_{CF}(r_i, \theta_i, \varphi_i)$ between the one-electron functions $|lm\rangle$ may be expressed here according to

$$\langle lm | V_{CF}(r_i, \theta_i, \varphi_i) | lm' \rangle$$

$$= \sum_a \sum_k I_k(a)(-1)^{m'}(2l+1) \begin{pmatrix} l & k & l \\ 0 & 0 & 0 \end{pmatrix}$$

$$\times \begin{pmatrix} l & k & l \\ m' & m-m' & -m \end{pmatrix} \mathbf{C}_{m'-m}^{(k)}(\Theta_a, \Phi_a) \qquad (9.3)$$

If the ligand coordinates Θ_a, Φ_a are specified, the tensor operators $\mathbf{C}_q^{(k)}(\Theta_a, \Phi_a)$ in Eq. (9.3) may be converted into simple numerical factors. Also, the quantity $I_k(a)$ may be considered as a crystal field parameter defined by

$$I_k(a) = \frac{Z_a e^2}{R_a^{k+1}} \langle r_i^k \rangle \qquad (9.4)$$

In Eq. (9.4), the expectation value $\langle r_i^k \rangle$ is determined by

$$\langle r_i^k \rangle = \int_0^{\infty} [R_{nl}(r_i)]^2 r_i^{k+2} \, dr_i \qquad (9.5)$$

$R_{nl}(r)$ being the radial part of the atomic orbital $\psi = R_{nl}(r) Y_{lm}(\theta, \varphi)$. The parameters $I_k(a)$ are related to the integrals $G^k(a)$ of early crystal field theory,[19]

$$I_k(a) = Z_a e^2 G^k(a) \qquad (9.6)$$

The integrals $G^k(a)$ as well as their first and second derivatives with respect to R_a have been tabulated for $k = 0, 2, 4$.[3] We will now consider matrix elements Eq. (9.3) for specific ligand arrangements and the relations between the $I_k(a)$ and ligand field parameters resulting therefrom.

Let us consider first a single ligand situated on the positive z axis (symmetry $C_{\infty v}$). The one-electron matrix elements result on the basis of Eq. (9.3) as

$$\langle 2 \pm 2|V_\infty|2 \pm 2\rangle = -\tfrac{2}{7}I_2(z) + \tfrac{1}{21}I_4(z)$$

$$\langle 2 \pm 1|V_\infty|2 \pm 1\rangle = \tfrac{1}{7}I_2(z) - \tfrac{4}{21}I_4(z) \qquad (9.7)$$

$$\langle 2\ 0|V_\infty|2\ 0\rangle = \tfrac{2}{7}I_2(z) + \tfrac{2}{7}I_4(z)$$

Comparing these expressions with Eq. (3.81) as modified for $C_{\infty v}$ symmetry, one obtains a relation between the $I_k(a)$ and parameters of the ligand field:

$$D\mu = -\tfrac{1}{7}I_2(z)$$
$$D\nu = -\tfrac{1}{21}I_4(z) \qquad (9.8)$$

Here and in what follows, the quantities $I_0(z)$ which arise equally in all diagonal elements are neglected.

Similarly one obtains, if two identical ligands on the positive and the negative z axis are considered (symmetry $D_{\infty h}$),

$$D\mu = -\tfrac{2}{7}I_2(z)$$
$$D\nu = -\tfrac{2}{21}I_4(z) \qquad (9.9)$$

If n identical ligands are situated in the xy plane (symmetry D_{nh}), application of Eq. (9.3) produces the one-electron matrix elements

$$\langle 2 \pm 2|V_n(xy)|2 \pm 2\rangle = \tfrac{1}{7}I_2(xy)n + \tfrac{1}{56}I_4(xy)n$$

$$\langle 2 \pm 1|V_n(xy)|2 \pm 1\rangle = -\tfrac{1}{14}I_2(xy)n - \tfrac{1}{14}I_4(xy)n$$

$$\langle 2\ 0|V_n(xy)|2\ 0\rangle = -\tfrac{1}{7}I_2(xy)n + \tfrac{3}{28}I_4(xy)n$$

$$(9.10)$$

In addition to Eq. (9.10) there is a nonzero off-diagonal matrix element if $n \geq 3$*

$$\langle 2 \pm 2|V_n(xy)|2 \mp 2\rangle = \sum_\nu^n \tfrac{5}{24} e^{\mp 4i\varphi_\nu} I_4(\nu) \quad (9.11)$$

where

$$e^{\mp 4i\varphi_\nu} = \cos\left(\frac{8\pi\nu}{n}\right) \mp i \sin\left(\frac{8\pi\nu}{n}\right) \qquad (9.12)$$

Consequently, matrix element Eq. (9.11) has to be considered only for $n = 4$, where

$$\langle 2 \pm 2|V_{D_4}|2 \mp 2\rangle = \tfrac{5}{6}I_4(xy) \qquad (9.13)$$

Combination of the expressions derived above produces one-electron matrix elements for various ligand arrangements. From these matrix elements, relations between the parameters of the crystal and

ligand fields may be deduced. The particular relations given below are of interest in the context of the present volume.

For an octahedral arrangement of six ligands (symmetry O_h) we obtain

$$Dq = \tfrac{1}{6}I_4(xyz) \qquad (9.14)$$

whereas six ligands in tetragonal (D_{4h}, D_4, C_{4v}) symmetry give

$$Dq = \tfrac{1}{6}I_4(xy)$$
$$Ds = \tfrac{2}{7}[I_2(xy) - I_2(z)] \qquad (9.15)$$
$$Dt = \tfrac{2}{21}[I_4(xy) - I_4(z)]$$

If the arrangement of five ligands in trigonal symmetry (trigonal bipyramid, C_{3v}) is considered, one obtains

$$D\sigma = \tfrac{3}{14}I_2(xy) - \tfrac{2}{7}I_2(z)$$
$$D\tau = -\tfrac{2}{21}[\tfrac{9}{16}I_4(xy) + I_4(z)] \qquad (9.16)$$

If, on the other hand, a trigonally distorted octahedron of six ligands (symmetry D_{3d}, D_3) is the situation of interest, the result is

$$Dq' = \frac{9}{2\sqrt{2}}\sin^3\theta\cos\theta\, Dq = \frac{3}{4\sqrt{2}}\sin^3\theta\cos\theta\, I_4(xy)$$

$$D\sigma = -\tfrac{3}{7}(3\cos^2\theta - 1)I_2(xy)$$

$$D\tau = -[\tfrac{1}{28}(35\cos^4\theta - 30\cos^2\theta + 3)$$
$$+ \tfrac{1}{4}\sqrt{2}\sin^3\theta\cos\theta]I_4(xy)$$

$$(9.17)$$

In Eqs. (9.17), θ is the angle between the axis C_3^z and the metal-ligand vector. Also, Dq' corresponds to the quantity Dq introduced for ligand fields of trigonal symmetry [cf. Eq. (3.85)], whereas Dq in Eq. (9.17) refers to the usual definition [cf. Eq. (9.15)]. In the octahedron, θ is 54° 44′ and, consequently, $Dq' = Dq$ in this situation.

The relations relevant to two ligands in cylindrical symmetry have been given in Eq. (9.9) above. Corresponding relations for ligand arrangements of symmetry D_n with $n \geq 5$ may be obtained using Eq. (9.10) in conjunction with Eq. (3.81) as modified for cylindrical symmetry. Thus, for symmetry D_5 with five ligands, one obtains

$$D\mu = \tfrac{5}{14}I_2(xy)$$
$$D\nu = -\tfrac{5}{56}I_4(xy) \qquad (9.18)$$

whereas six ligands in D_6 symmetry yield

$$D\mu = \tfrac{3}{7}I_2(xy)$$
$$D\nu = -\tfrac{3}{28}I_4(xy) \qquad (9.19)$$

*The case $n = 2$, where nonzero off-diagonal matrix elements also arise, will not be considered here.

The energy matrices for symmetries D_n, $n = 5, \ldots, \infty$, may thus be given in terms of the same ligand field parameters, i.e., $D\mu$, $D\nu$, whereas the corresponding crystal field expressions are different.

Finally, it should be noted that the parameter $C_p(a)$ introduced by Gerloch and Slade[10] is defined by

$$C_p(a) = \tfrac{2}{7} I_2(a) \qquad (9.20)$$

The diagrams presented in Part II of the present volume may then be employed to determine crystal field rather than ligand field parameters. For this purpose, the relations between the two types of parameters listed above will be useful.

9.2. LIGAND FIELDS AND THE ANGULAR OVERLAP MODEL

Considering the perturbing effect of an isolated ligand on the positive z axis as represented by the operator V_{AOM}, the principal assumption of the angular overlap model (AOM)[23,53,54] concerns the diagonal form of the matrix of single-electron d orbitals on the metal. Consequently, the bond between the metal ion and the ligand is treated as axially symmetric, the orbitals transforming according to irreducible representations of group $C_{\infty v}$:

$$\langle l\Gamma\gamma | V_{AOM}(r_i, 0, 0) | l\Gamma'\gamma' \rangle = \delta(\Gamma\gamma, \Gamma'\gamma') e_\Gamma \qquad (9.21)$$

Here, representations Γ of $C_{\infty v}$ arise from the decomposition of subduced representation $D^{(l)}(C_{\infty v})$ according to

$$D^{(l)}(C_{\infty v}) = \sum_i a_i^{(l)} \Gamma_i \qquad (9.22)$$

The quantities e_Γ in Eq. (9.21) are parameters of the AOM which are usually determined by comparison with experiment. The e_Γ are closely related to the reduced matrix elements of the AOM operator according to

$$e_\Gamma = (d_\Gamma)^{-1/2} \langle l\Gamma \| V_{AOM}(r_i, 0, 0) \| l\Gamma \rangle \qquad (9.23)$$

Here, the allocation of the radial and angular contributions comprised in the quantities of Eq. (9.23) is open to discussion.

The e_Γ are usually denoted, in correspondence to the irreducible representations of group $C_{\infty v}$, by e_σ if $\Gamma = A_1 = \Sigma$, e_π if $\Gamma = E_1 = \Pi$, and e_δ if $\Gamma = E_2 = \Delta$. It should be observed that, in agreement with the assumption of axial symmetry of the metal–ligand bond, the contributions of individual ligands are treated as additive properties. Each individual metal–ligand bond may then be characterized by $l+1$

parameters, e_Γ. Only l parameters are required if energy differences are considered.

Let us now examine ligand i in a general position characterized by the coordinates r_i, θ_i, φ_i. The matrix elements of the AOM operator are determined in this case by[53]

$$\langle lt | V_{AOM}(r_i, \theta_i, \varphi_i) | lt' \rangle$$

$$= \sum_j \sum_{t''} D_{tt''}^{(l)}(\varphi_j, \theta_j, \psi_j) D_{t't''}^{(l)}(\varphi_j, \theta_j, \psi_j) e_{t''}(j) \qquad (9.24)$$

Here $D^{(l)}$ denotes real representations of $SO(3)$ and $|lt\rangle$ are basis functions within $C_{\infty v}$ ($t = \lambda\varsigma$, $\lambda = \sigma, \pi, \delta$, $\varsigma = c, s$),

$$|l\sigma\rangle_r = |l0\rangle_c$$

$$|l\lambda c\rangle_r = \frac{1}{\sqrt{2}}[(-1)^\lambda |l\lambda\rangle_c + |l-\lambda\rangle_c] \qquad (9.25)$$

$$|l\lambda s\rangle_r = -\frac{i}{\sqrt{2}}[(-1)^\lambda |l\lambda\rangle_c - |l-\lambda\rangle_c]$$

In Eqs. (9.25) the additional abbreviations $c = \cos$, $s = \sin$, $r = $ real and $c = $ complex have been employed.

The additivity of ligand contributions shows the close relation between the AOM and the crystal field theory of Section 9.1. Evidently it is immaterial for the mathematical structure of the theory which of the premises is employed in order to account for the effect of the ligands. In both models the matrix elements between one-electron orbitals are considered as semiempirical parameters, although the interpretation of the latter involves completely different approaches to the nature of the metal–ligand interaction. In crystal field theory, the separation of one-electron energies is caused by the Coulomb potential of the ligands which in turn are represented by point-charges. In the AOM, the different form of weak covalent interactions between metal and ligands is responsible for the observed energy splitting.

In order to obtain a relation between the parameters of the AOM and those of the crystal field theory, the crystal field potential of Eq. (9.1) is expressed in terms of real spherical harmonics $Z_u^{(k)}(\theta_i, \varphi_i)$ as

$$V_{CF}(r_i, \theta_i, \varphi_i)$$

$$= \sum_a Z_a e^2 \sum_{k,u} \frac{r_i^k}{R_a^{k+1}} Z_u^k(\theta_i, \varphi_i) Z_u^k(\Theta_a, \Phi_a) \qquad (9.26)$$

The matrix elements of $V_{CF}(r_i, \theta_i, \varphi_i)$ between the basis functions of Eq. (9.25) then result as

$$\langle lt|V_{CF}(r_i, \theta_i, \varphi_i)|lt'\rangle$$

$$=\sum_a \sum_{k,u} I_k(a)\begin{pmatrix} l & k & l \\ t & u & t' \end{pmatrix}$$

$$\times (2l+1)\begin{pmatrix} l & k & l \\ 0 & 0 & 0 \end{pmatrix} Z_u^{(k)}(\Theta_a, \Phi_a) \quad (9.27)$$

where $t, u = \sigma, \lambda c, \lambda s$ and where the real 3-l symbols of Schäffer[14] are used. If now the matrix element of Eq. (9.27) is equated to that of Eq. (9.24) and, without loss of generality, ligand a is placed on the positive z axis, one obtains[27]

$$\sum_k I_k(a)(2l+1)(-1)^\lambda \begin{pmatrix} l & k & l \\ \lambda & 0 & -\lambda \end{pmatrix}\begin{pmatrix} l & k & l \\ 0 & 0 & 0 \end{pmatrix} = e_\lambda(a)$$

$$(9.28)$$

Here, $\lambda = 0, \pm 1, \pm 2$; $t = \lambda c, \lambda s$; and, in addition, the 3-l symbols have been replaced by the more familiar 3-j symbols. If, for convenience,

$$b_{\lambda k}(l) = (-1)^\lambda (2l+1)\begin{pmatrix} l & k & l \\ \lambda & 0 & -\lambda \end{pmatrix}\begin{pmatrix} l & k & l \\ 0 & 0 & 0 \end{pmatrix} \quad (9.29)$$

is introduced, Eq. (9.28) may be written with $e_{\lambda c} = e_{\lambda s} = e_\lambda$:

$$e_\lambda(a) = \sum_k b_{\lambda k}(l) I_k(a) \quad (9.30)$$

The coefficients $b_{\lambda k}(l)$ are listed for d electrons ($l = 2$) in Table IV. The reversed relation to Eq. (9.30) is ($\lambda = \sigma, \pi c, \pi s$)

$$I_k(a) = \sum_\lambda a_{k\lambda}(l) e_\lambda(a) \quad (9.31)$$

where

$$a_{k\lambda}(l) = (-1)^\lambda \frac{2k+1}{2l+1}\begin{pmatrix} l & k & l \\ \lambda & 0 & -\lambda \end{pmatrix}\begin{pmatrix} l & k & l \\ 0 & 0 & 0 \end{pmatrix}^{-1}$$

$$(9.32)$$

The coefficients $a_{k\lambda}(l)$ are listed for d electrons ($l = 2$) in Table V. Consequently, Eqs. (9.30) and (9.31) may be generally employed to relate the parameters of the AOM and the crystal field. It should be noted that, if

TABLE IV

Values of the coefficient $b_{\lambda k}(l)$ for $l = 2$

$b_{\lambda k}(2)$	0	2	4
σ	1	$\frac{2}{7}$	$\frac{2}{7}$
π	1	$\frac{1}{7}$	$-\frac{4}{21}$
δ	1	$-\frac{2}{7}$	$\frac{1}{21}$

TABLE V

Values of the coefficient $a_{k\lambda}(l)$ for $l = 2$

$a_{k\lambda}(2)$	σ	π	δ
0	$\frac{1}{5}$	$\frac{1}{5}$	$\frac{1}{5}$
2	1	$-\frac{1}{2}$	-1
4	$\frac{9}{5}$	$-\frac{6}{5}$	$\frac{3}{10}$

not only energy differences are considered, the quantity $I_0(a)$ will arise as an additive term in Eq. (9.30).

In addition, if the relations of Section 9.1 are employed, conversion of AOM parameters into ligand field parameters and *vice versa* becomes possible. We list below the relations obtained for the various ligand arrangements discussed previously.

For an octahedron of six ligands it holds that

$$10Dq = 3e_\sigma - 4e_\pi + e_\delta \quad (9.33)$$

whereas for six ligands in tetragonal symmetry (D_{4h}, D_4, C_{4v}) the result is

$$Dq = \frac{3}{10}e_\sigma(xy) - \frac{4}{10}e_\pi(xy) + \frac{1}{10}e_\delta(xy)$$

$$Ds = \frac{2}{7}[e_\sigma(xy) - e_\sigma(z)] + \frac{2}{7}[e_\pi(xy) - e_\pi(z)]$$

$$\quad - \frac{4}{7}[e_\delta(xy) - e_\delta(z)] \quad (9.34)$$

$$Dt = \frac{6}{35}[e_\sigma(xy) - e_\sigma(z)] - \frac{8}{35}[e_\pi(xy) - e_\pi(z)]$$

$$\quad + \frac{2}{35}[e_\delta(xy) - e_\delta(z)]$$

In the trigonal symmetry formed by five ligands (C_{3v}) one obtains

$$D\sigma = \frac{1}{14}[3e_\sigma(xy) - 4e_\sigma(z) + 3e_\pi(xy) - 4e_\pi(z)$$

$$\quad - 6e_\delta(xy) + 8e_\delta(z)]$$

$$D\tau = -\frac{2}{7}[\frac{27}{80}e_\sigma(xy) + \frac{3}{5}e_\sigma(z) - \frac{9}{20}e_\pi(xy) - \frac{4}{5}e_\pi(z)$$

$$\quad + \frac{9}{80}e_\delta(xy) + \frac{1}{5}e_\delta(z)] \quad (9.35)$$

Six ligands in a trigonally distorted octahedral arrangement (D_{3d}, D_3) give

$$Dq' = \frac{3\sqrt{2}}{8}\sin^3\theta\cos\theta[\frac{9}{5}e_\sigma(xy) - \frac{12}{5}e_\pi(xy) + \frac{3}{5}e_\delta(xy)]$$

$$D\sigma = -\frac{3}{7}(3\cos^2\theta - 1)[e_\sigma(xy) + e_\pi(xy) - 2e_\delta(xy)]$$

$$D\tau = -[\frac{1}{28}(35\cos^4\theta - 30\cos^2\theta + 3) \quad (9.36)$$

$$\quad + \frac{1}{4}\sqrt{2}\sin^3\theta\cos\theta]$$

$$\quad \times [\frac{9}{5}e_\sigma(xy) - \frac{12}{5}e_\pi(xy) + \frac{3}{5}e_\delta(xy)]$$

In the cylindrical symmetry $(D_{\infty h}, D_\infty, C_{\infty v})$ provided by two ligands one obtains

$$D\mu = -\frac{2}{7}[e_\sigma(z) + e_\pi(z) - 2e_\delta(z)]$$

$$D\nu = -\frac{2}{7}[\frac{3}{5}e_\sigma(z) - \frac{4}{5}e_\pi(z) + \frac{1}{5}e_\delta(z)] \quad (9.37)$$

Analogous relations may be obtained for symmetries D_{nh}, with $n \geq 5$, where the energy matrices are identical with those in $D_{\infty h}$.

The relations discussed above clearly demonstrate that ligand field parameters; crystal field parameters of order k, $I_k(a)$; and the λ-antibonding parameters of the AOM, $e_\lambda(a)$; are intimately related. In fact each contribution of the AOM can be obtained from a particular crystal field potential. It is important to realize that a completely different interpretation of the parameter values resulting from the diagrams in Part II of the volume is now available. In particular, ligand field parameters may be understood in terms of λ-antibonding effects, and the assumption of a ligand field is not necessarily required.

9.3. LIGAND FIELDS AND MOLECULAR ORBITAL THEORY

In ligand field theory, matrix elements of the type $\langle lm_l | V_{LF}(r_i, \theta_i, \varphi_i) | lm_l' \rangle$ between the complex one-electron d functions $|lm_l\rangle$ are often considered as semiempirical parameters [cf. Eq. (3.79) ff]. Diagonalization of the resulting 5×5 matrix V_{LF} produces eigenfunctions $C_i' = |a\Gamma_i\gamma_i\rangle$ and eigenvalues E_i' of the problem in the one-electron d orbital basis according to

$$\mathbf{V}_{LF}\mathbf{C}' = \mathbf{C}'\mathbf{E}' \qquad (9.38)$$

If a many-electron calculation should be performed, the evaluation of matrix elements of the type $\langle LM_L | \sum_i V_{LF}(r_i, \theta_i, \varphi_i) | L'M_L' \rangle$ again requires the matrix elements $\langle lm_l | V_{LF}(r_i, \theta_i, \varphi_i) | lm_l' \rangle$.

The results of one-electron molecular orbital (MO) calculations are obtained in terms of absolute MO energies and the corresponding eigenfunctions

$$|MO i, \Gamma\gamma\rangle = \sum_j A_{ij} |\psi_M, \Gamma_j\gamma_j\rangle + \sum_j \alpha_{ij} |\varphi_{\text{Lig}}, \Gamma_j\gamma_j\rangle \quad (9.39)$$

which are composed of s, p, and d orbitals on the metal, $|\psi_M, \Gamma_j\gamma_j\rangle$, and a suitable combination of orbitals on the ligands, $|\varphi_{\text{Lig}}, \Gamma_j\gamma_j\rangle$, usually given in LCAO form as $\varphi_{\text{Lig}} = \sum_n \lambda_n \chi_n$. The energies of the molecular orbitals Eq. (9.39) form the diagonal matrix $\mathbf{E}$. In what follows, we investigate a method[18] of constructing, from matrix $\mathbf{E}$, a new matrix $\mathbf{V}_{\text{eff}}$, where

$$\mathbf{V}_{\text{eff}}\mathbf{C} = \mathbf{C}\mathbf{E} \qquad (9.40)$$

Here, the elements of $\mathbf{V}_{\text{eff}}$ are $\langle a\Gamma\gamma | V_{\text{eff}} | a'\Gamma\gamma \rangle$ and the eigenvectors of matrix $\mathbf{C}$ should most nearly resemble the MO eigenfunctions, Eq. (9.39).

A suitable procedure to this effect consists of truncating the orbitals Eq. (9.39) such as to eliminate all but the d orbital contribution. On normalization, the matrix $\mathbf{A}$ of coefficients A_{ij} from Eq. (9.39) yields matrix $\mathbf{B}$, which is, in general, nonorthogonal. The required orthonormal matrix $\mathbf{C}$ may then be constructed, employing the orthogonalization of Löwdin,[35] according to

$$\mathbf{C} = \mathbf{B} \cdot \mathbf{S}^{-1/2} \qquad (9.41)$$

Here, $\tilde{\mathbf{B}} \cdot \mathbf{B} = \mathbf{S}$ is the overlap matrix and matrix $\mathbf{S}^{-1/2}$ is obtained as follows. First, matrix $\mathbf{S}$ is diagonalized such that $\bar{\mathbf{U}}\mathbf{S}\mathbf{U} = \mathbf{D}$, where $\mathbf{D}$ is a diagonal matrix. From $\mathbf{D}$, matrix $\mathbf{D}^{-1/2}$ is easily constructed in as much $D_i^{-1/2} = (D_i)^{-1/2}$ and consequently

$$\mathbf{S}^{-1/2} = \mathbf{U} \cdot \mathbf{D}^{-1/2} \cdot \bar{\mathbf{U}} \qquad (9.42)$$

It has been shown[1] that the eigenvectors of matrix $\mathbf{C}$ thus obtained most nearly resemble the original MO eigenfunctions of Eq. (9.39). From matrix $\mathbf{C}$ and the energies E_j resulting as above, $\mathbf{V}_{\text{eff}}$ may be calculated on the basis of Eq. (9.40). It should be observed that, in general, matrix $\mathbf{V}_{\text{eff}}$ is a real matrix $\mathbf{V}_{\text{eff}}^{\text{real}}$. If a strong-field calculation is to be performed, the matrix elements of $\mathbf{V}_{\text{eff}}^{\text{real}}$ may be introduced directly as parameters. If a many-electron weak-field calculation is appended, a transformation of the real basis $\mathbf{C}^{\text{real}}$ into a complex d orbital basis $\mathbf{C}^{\text{complex}}$ is required,

$$\mathbf{C}^{\text{complex}} = \mathbf{C}^{\text{real}} \cdot \mathbf{T} \qquad (9.43)$$

From this it follows that

$$\mathbf{V}_{\text{eff}}^{\text{complex}} = \mathbf{T}^{-1} \cdot \mathbf{V}_{\text{eff}}^{\text{real}} \cdot \mathbf{T} \qquad (9.44)$$

It should be also noted that if each of the five d orbitals belongs to a different irreducible representation, i.e., if in

$$D^{(2)}(G) = \sum_i a_i^{(2)}\Gamma_i \qquad (9.45)$$

$a_i = 1$, Eq. (9.40) becomes an identity, $\mathbf{V}_{\text{eff}}^{\text{real}} = \mathbf{E}$. In the corresponding point groups (group A according to Horrocks[18]), MO energy differences may thus be directly equated to the relative energies in the ligand field method. For some other point groups (group B according to Horrocks[18]), $a_i > 1$ in Eq. (9.45), i.e., two or more d orbitals belong to the same irreducible representation. It is in these point groups that the above procedure is particularly useful since the required off-diagonal matrix elements are easily determined.

Thus an alternative parametrization scheme is introduced involving one-electron orbital energies which, in a way, are physically more meaningful quantities than the ligand field parameters introduced above. With reference to the diagrams in Part II of the volume, the point groups of cubic (O_h, T_d), tetragonal (D_{4h}, D_4, C_{4v}), and cylindrical ($D_{\infty h}$, D_∞, $C_{\infty v}$) symmetry belong to group A, whereas those of trigonal symmetry (D_{3d}, D_3, C_{3v}) belong to group B. In particular, relations between ligand field and MO parameters are obtained as given below.

In cubic (O_h, T_d) symmetry, it is simply

$$Dq = \tfrac{1}{6}\langle e_g|V_{\text{eff}}|e_g\rangle = -\tfrac{1}{4}\langle t_{2g}|V_{\text{eff}}|t_{2g}\rangle \quad (9.46)$$

whereas in tetragonal (D_{4h}, D_4, C_{4v}) symmetry, one obtains

$$
\begin{aligned}
6Dq + 2Ds - Dt &= \langle b_{1g}|V_{\text{eff}}|b_{1g}\rangle \\
6Dq - 2Ds - 6Dt &= \langle a_{1g}|V_{\text{eff}}|a_{1g}\rangle \\
-4Dq - Ds + 4Dt &= \langle e_g|V_{\text{eff}}|e_g\rangle \\
-4Dq + 2Ds - Dt &= \langle b_{2g}|V_{\text{eff}}|b_{2g}\rangle
\end{aligned}
\quad (9.47)
$$

and similar relations follow in cylindrical ($D_{\infty h}$, D_∞, $C_{\infty v}$) symmetry

$$
\begin{aligned}
-2D\mu - 6D\nu &= \langle a_{1g}|V_{\text{eff}}|a_{1g}\rangle \\
-D\mu + 4D\nu &= \langle e_{1g}|V_{\text{eff}}|e_{1g}\rangle \\
2D\mu - D\nu &= \langle e_{2g}|V_{\text{eff}}|e_{2g}\rangle
\end{aligned}
\quad (9.48)
$$

Finally, in the more complicated case of trigonal (D_{3d}, D_3, C_{3v}) symmetry the result is

$$
\begin{aligned}
-4Dq - 2D\sigma - 6D\tau &= \langle a_1|V_{\text{eff}}|a_1\rangle \\
-4Dq + D\sigma + \tfrac{2}{3}D\tau &= \langle e|V_{\text{eff}}|e\rangle \\
6Dq + \tfrac{7}{3}D\tau &= \langle e'|V_{\text{eff}}|e'\rangle
\end{aligned}
\quad (9.49)
$$

$$-\frac{\sqrt{2}}{3}(3D\sigma - 5D\tau) = \langle e|V_{\text{eff}}|e'\rangle$$

9.4. LIGAND FIELDS AND THE σ- AND π-ANTIBONDING MODEL OF McCLURE

An empirical model based on the concept of molecular orbitals has been proposed by McClure[37] in order to rationalize the various band splittings encountered in substituted octahedral complexes. The model considers the separations of e_g and t_{2g} orbitals to arise from the σ- and π-antibonding effects of different ligands.

Assuming that the e_g orbitals are only σ-bonding, their splitting is determined by the parameter $d\sigma = \sigma_z - \sigma_{xy}$, where σ_z and σ_{xy} are characteristic of the bonding contribution of the ligands along the z axis and in the xy plane, respectively. Similarly, the splitting of presumably only π-bonding t_{2g} orbitals is determined by the parameter $d\pi = \pi_z - \pi_{xy}$. The parameters $d\sigma$ and $d\pi$ thus indicate the difference in antibonding capacity of the ligands in the two directions of space. For substituted octahedral complexes, the relative orbital energy changes have been listed.[37] A conversion of the parameters introduced by McClure into the usual ligand field parameters may be easily performed. Thus, for an MA_4B_2 complex of D_4 symmetry, one obtains

$$
\begin{aligned}
Ds &= -\tfrac{8}{21}d\sigma - \tfrac{2}{7}d\pi \\
Dt &= -\tfrac{8}{35}(d\sigma - d\pi)
\end{aligned}
\quad (9.50)
$$

The corresponding result for an MA_5B complex of D_4 symmetry is

$$
\begin{aligned}
Ds &= -\tfrac{4}{21}d\sigma - \tfrac{1}{7}d\pi \\
Dt &= -\tfrac{4}{35}(d\sigma - d\pi)
\end{aligned}
\quad (9.51)
$$

Appendix I

THE METRIC TENSOR

Definition in $SO(3)$ according to Wigner:

$$\begin{pmatrix} j \\ m \quad m' \end{pmatrix} = (-1)^{j+m}\delta(m, -m') \quad \text{(A-I.1)}$$

Definition in point group $G \subset SO(3)$:

$$\begin{pmatrix} j \\ \Gamma\gamma a \quad \overline{\Gamma\gamma a} \end{pmatrix} = \sum_{m,m'} \langle j\Gamma\gamma a | jm \rangle^*$$

$$\times \begin{pmatrix} j \\ m \quad m' \end{pmatrix} \langle jm' | \overline{j\Gamma\gamma a} \rangle \quad \text{(A-I.2)}$$

Orthogonality:

$$\begin{pmatrix} j \\ \overline{\Gamma\gamma a} \quad \Gamma\gamma a \end{pmatrix}^* \begin{pmatrix} j \\ \overline{\Gamma\gamma a} \quad \Gamma'\gamma'a' \end{pmatrix}$$

$$= \delta(\Gamma, \Gamma')\delta(\gamma, \gamma')\delta(a, a') \quad \text{(A-I.3)}$$

Symmetry property:

$$\begin{pmatrix} j \\ \Gamma\gamma a \quad \overline{\Gamma\gamma a} \end{pmatrix} = (-1)^{2j}\begin{pmatrix} j \\ \overline{\Gamma\gamma a} \quad \Gamma\gamma a \end{pmatrix} \quad \text{(A-I.4)}$$

Appendix II

SYMMETRY COUPLING COEFFICIENTS AND 3-Γ SYMBOLS

The symmetry coupling coefficients are defined according to the transformation

$$|(\Gamma_1\Gamma_2)\Gamma\gamma b\rangle$$

$$= \sum_{\gamma_1\gamma_2} |\Gamma_1\gamma_1\rangle|\Gamma_2\gamma_2\rangle\langle\Gamma_1\gamma_1\Gamma_2\gamma_2|\Gamma\gamma b\rangle \qquad \text{(A-II.1)}$$

The coefficients are zero if the direct product $\Gamma_1 \otimes \Gamma_2$ does not contain representation Γ. The 3-Γ symbol may be defined according to the relation:

$$\begin{pmatrix} \Gamma_1 & \Gamma_2 & \Gamma \\ \gamma_1 & \gamma_2 & \gamma \end{pmatrix}_b = (-1)^{(j_1-j_2+j)_b}$$

$$\times \frac{1}{\sqrt{d_\Gamma}}\begin{pmatrix} & j & \\ \Gamma\gamma a & & \overline{\Gamma\gamma a} \end{pmatrix}\langle\overline{\Gamma\gamma b}|\Gamma_1\gamma_1\Gamma_2\gamma_2\rangle$$

$$\text{(A-II.2)}$$

Orthogonality relations:

$$\sum_{\gamma_1\gamma_2}\begin{pmatrix} \Gamma_1 & \Gamma_2 & \Gamma \\ \gamma_1 & \gamma_2 & \gamma \end{pmatrix}_b\begin{pmatrix} \Gamma_1 & \Gamma_2 & \Gamma' \\ \gamma_1 & \gamma_2 & \gamma' \end{pmatrix}_{b'}^* d_\Gamma$$

$$= \delta(\Gamma,\Gamma')\delta(\gamma,\gamma')\delta(b,b')$$

$$\sum_{\Gamma\gamma b}\begin{pmatrix} \Gamma_1 & \Gamma_2 & \Gamma \\ \gamma_1 & \gamma_2 & \gamma \end{pmatrix}_b\begin{pmatrix} \Gamma_1 & \Gamma_2 & \Gamma \\ \gamma_1' & \gamma_2' & \gamma \end{pmatrix}_b^* d_\Gamma$$

$$= \delta(\gamma_1,\gamma_1')\delta(\gamma_2,\gamma_2') \qquad \text{(A-II.3)}$$

Symmetry properties:

$$\begin{pmatrix} \Gamma_1 & \Gamma_2 & \Gamma \\ \gamma_1 & \gamma_2 & \gamma \end{pmatrix}_b = \begin{pmatrix} \Gamma_2 & \Gamma & \Gamma_1 \\ \gamma_2 & \gamma & \gamma_1 \end{pmatrix}_b$$

$$= (-1)^{(j_1+j_2+j)_b}\begin{pmatrix} \Gamma_2 & \Gamma_1 & \Gamma \\ \gamma_2 & \gamma_1 & \gamma \end{pmatrix}_b \quad \text{etc.} \qquad \text{(A-II.4)}$$

$$\langle\Gamma\gamma b|\Gamma_1\gamma_1\Gamma_2\gamma_2\rangle = (-1)^{(j_1+j_2-j)_b}\langle\Gamma\gamma b|\Gamma_2\gamma_2\Gamma_1\gamma_1\rangle$$

$$= \sqrt{\frac{d_\Gamma}{d_{\Gamma_1}}}\begin{pmatrix} & j & \\ \Gamma\gamma a & & \overline{\Gamma\gamma a} \end{pmatrix}^*\begin{pmatrix} & j_1 & \\ \Gamma_1\gamma_1 a_1 & & \overline{\Gamma_1\gamma_1 a_1} \end{pmatrix}$$

$$\times (-1)^{(j_1+j_2-j)_b}\langle\overline{\Gamma_1\gamma_1}b|\overline{\Gamma\gamma}\Gamma_2\gamma_2\rangle$$

$$= \sqrt{\frac{d_\Gamma}{d_{\Gamma_2}}}\begin{pmatrix} & j & \\ \Gamma\gamma a & & \overline{\Gamma\gamma a} \end{pmatrix}^*\begin{pmatrix} & j_2 & \\ \Gamma_2\gamma_2 a_2 & & \overline{\Gamma_2\gamma_2 a_2} \end{pmatrix}$$

$$\times (-1)^{(j_1+j-j_2)_b}\langle\overline{\Gamma_2\gamma_2}b|\Gamma_1\gamma_1\overline{\Gamma\gamma}\rangle \qquad \text{(A-II.5)}$$

$$\begin{pmatrix} \Gamma_1 & \Gamma_2 & \Gamma \\ \gamma_1 & \gamma_2 & \gamma \end{pmatrix}_b^* = \begin{pmatrix} & j_1 & \\ \Gamma_1\gamma_1 a_1 & & \overline{\Gamma_1\gamma_1 a_1} \end{pmatrix}^*\begin{pmatrix} & j_2 & \\ \Gamma_2\gamma_2 a_2 & & \overline{\Gamma_2\gamma_2 a_2} \end{pmatrix}^*$$

$$\times \begin{pmatrix} & j & \\ \Gamma\gamma a & & \overline{\Gamma\gamma a} \end{pmatrix}^*\begin{pmatrix} \bar\Gamma_1 & \bar\Gamma_2 & \bar\Gamma \\ \bar\gamma_1 & \bar\gamma_2 & \bar\gamma \end{pmatrix}_b \qquad \text{(A-II.6)}$$

Special forms:

$$\langle\Gamma\gamma|\Gamma_1\gamma_1 A_1 a_1\rangle = \delta(\Gamma_1,\Gamma)\delta(\gamma_1,\gamma) \quad \text{(A-II.7)}$$

$$\begin{pmatrix} \Gamma_1 & A_1 & \Gamma \\ \gamma_1 & a_1 & \gamma \end{pmatrix} = \frac{1}{\sqrt{d_\Gamma}}\begin{pmatrix} & j & \\ \overline{\Gamma\gamma a} & & \Gamma\gamma a \end{pmatrix}\delta(\bar\Gamma,\Gamma_1)\delta(\bar\gamma,\gamma_1)$$

$$\text{(A-II.8)}$$

$$\langle A_1 a_1|\Gamma_1\gamma_1\Gamma_2\gamma_2\rangle = (-1)^{2j_1}\begin{pmatrix} & j_1 & \\ \Gamma_1\gamma_1 a_1 & & \overline{\Gamma_1\gamma_1 a_1} \end{pmatrix}$$

$$\times \frac{1}{\sqrt{d_{\Gamma_1}}}\delta(\bar\Gamma_1,\Gamma_2)\delta(\bar\gamma_1,\gamma_2) \qquad \text{(A-II.9)}$$

$$\begin{pmatrix} \Gamma_1 & \Gamma_2 & A_1 \\ \gamma_1 & \gamma_2 & a_1 \end{pmatrix} = \langle A_1 a_1|\Gamma_1\gamma_1\Gamma_2\gamma_2\rangle \qquad \text{(A-II.10)}$$

Appendix III

RACAH W COEFFICIENTS AND 6-Γ SYMBOLS

Definition of the Racah W coefficient:

$$W(\Gamma_1\Gamma_2\Gamma_3\Gamma_4; \Gamma_5\Gamma_6)_{b_1b_2b_3b_4}$$

$$= (d_{\Gamma_5})^{-1/2}(d_{\Gamma_6})^{-1/2}$$

$$\times \langle\Gamma_1\Gamma_2(\Gamma_5b_3)\Gamma_4; \Gamma_3b_1|\Gamma_1(\Gamma_2\Gamma_4)\Gamma_6b_2; \Gamma_3b_4\rangle$$

$$= (d_{\Gamma_5})^{-1/2}(d_{\Gamma_6})^{-1/2}\sum_{\gamma_1\gamma_2\gamma_4\gamma_5\gamma_6}\langle\Gamma_5\gamma_5\Gamma_4\gamma_4|\Gamma_3\gamma_3b_1\rangle^*$$

$$\times\langle\Gamma_2\gamma_2\Gamma_4\gamma_4|\Gamma_6\gamma_6b_2\rangle\langle\Gamma_1\gamma_1\Gamma_2\gamma_2|\Gamma_5\gamma_5b_3\rangle^*$$

$$\times\langle\Gamma_1\gamma_1\Gamma_6\gamma_6|\Gamma_3\gamma_3b_4\rangle \qquad \text{(A-III.1)}$$

Definition of the 6-Γ symbol:

$$\begin{Bmatrix}\Gamma_1 & \Gamma_2 & \Gamma_5 \\ \Gamma_4 & \Gamma_3 & \Gamma_6\end{Bmatrix}_{b_1b_2b_4b_3} = (-1)^{(j_3+j_4-j_5)_{b_1}+(j_1+j_2+j_5)_{b_3}}$$

$$\times W(\bar\Gamma_1\bar\Gamma_2\bar\Gamma_3\bar\Gamma_4; \Gamma_5\bar\Gamma_6)_{b_1b_2b_3b_4} \qquad \text{(A-III.2)}$$

Symmetry properties:

$$\begin{Bmatrix}\Gamma_1 & \Gamma_2 & \Gamma_3 \\ \Gamma_4 & \Gamma_5 & \Gamma_6\end{Bmatrix}_{b_1b_2b_3b_4} = \begin{Bmatrix}\bar\Gamma_1 & \bar\Gamma_2 & \bar\Gamma_3 \\ \bar\Gamma_4 & \bar\Gamma_5 & \bar\Gamma_6\end{Bmatrix}^*_{b_1b_2b_3b_4}$$

$$= \begin{Bmatrix}\Gamma_2 & \Gamma_3 & \Gamma_1 \\ \Gamma_5 & \Gamma_6 & \Gamma_4\end{Bmatrix}_{b_3b_1b_2b_4}$$

$$= \begin{Bmatrix}\bar\Gamma_1 & \Gamma_5 & \bar\Gamma_6 \\ \bar\Gamma_4 & \Gamma_2 & \bar\Gamma_3\end{Bmatrix}_{b_2b_1b_4b_3}$$

$$= \begin{Bmatrix}\Gamma_4 & \bar\Gamma_5 & \bar\Gamma_3 \\ \Gamma_1 & \bar\Gamma_2 & \bar\Gamma_6\end{Bmatrix}_{b_4b_3b_2b_1}$$

$$= (-1)^{(j_3-j_4+j_5)_{b_1}+(j_2+j_4-j_6)_{b_2}+(j_1-j_5+j_6)_{b_3}}$$

$$\times(-1)^{(j_1+j_2+j_3)_{b_4}}\begin{Bmatrix}\Gamma_2 & \Gamma_1 & \Gamma_3 \\ \bar\Gamma_5 & \bar\Gamma_4 & \bar\Gamma_6\end{Bmatrix}_{b_1b_3b_2b_4}$$

$$= (-1)^{(j_3-j_4+j_5)_{b_1}+(j_2+j_4-j_6)_{b_2}+(j_1-j_5+j_6)_{b_3}}$$

$$\times(-1)^{(j_1+j_2+j_3)_{b_4}}\begin{Bmatrix}\Gamma_1 & \Gamma_3 & \Gamma_2 \\ \bar\Gamma_4 & \bar\Gamma_6 & \bar\Gamma_5\end{Bmatrix}_{b_2b_1b_3b_4} \qquad \text{(A-III.3)}$$

Triangular condition: The Racah W coefficient is zero if the direct products of the three representations given below do not contain representation A_1:

$$(\bar\Gamma_3\Gamma_4\Gamma_5) \qquad (\Gamma_2\Gamma_4\bar\Gamma_6) \qquad (\Gamma_1\Gamma_2\bar\Gamma_5) \qquad (\Gamma_1\Gamma_6\bar\Gamma_3)$$

Special forms:

$$W(\Gamma_1\Gamma_2\Gamma_3\Gamma_4; A_1\Gamma_6)_{b_1b_2b_3b_4}$$

$$= (-1)^{j_1+j_3-j_6}(d_{\Gamma_1}\cdot d_{\Gamma_3})^{-1/2}$$

$$\times\delta(\Gamma_3,\Gamma_4)\delta(\Gamma_1,\bar\Gamma_2)\delta(b_2,b_3)\delta(b_1,b_4) \qquad \text{(A-III.4)}$$

$$\begin{Bmatrix}\Gamma_1 & \Gamma_2 & A_1 \\ \Gamma_4 & \Gamma_3 & \Gamma_6\end{Bmatrix}_{b_1b_2b_3b_4} = (-1)^{j_1+j_3+j_6}(d_{\Gamma_1}\cdot d_{\Gamma_3})^{-1/2}$$

$$\times\delta(\Gamma_3,\Gamma_4)\delta(\Gamma_1,\bar\Gamma_2)\delta(b_1,b_4)\delta(b_2,b_3) \qquad \text{(A-III.5)}$$

REFERENCES

1. Aiken, J. G., Jonassen, H. B., and Aldrick, H. S., *J. Chem. Phys.* **62**, 2745 (1975).
2. Ballhausen, C. J., *Introduction to Ligand Field Theory*, McGraw-Hill Book Co., New York (1962).
3. Ballhausen, C. J., and Ancmon, E. M., *Mat. Fys. Medd. Dan. Vid. Selsk.* **31**, No. 9 (1958).
4. Berthier, G., *Adv. Quantum Chem.* **8**, 183 (1974).
5. Bethe, H. A., *Ann. Physik* **3**, 133 (1929).
6. Brink, D. M., and Satchler, G. R., *Angular Momentum*, Oxford University Press (1968).
7. Condon, E. U., and Shortley, G. H., *The Theory of Atomic Spectra*, Cambridge University Press (1935).
8. Edmonds, A. R., *Angular Momentum in Quantum Mechanics*, Princeton University Press (1960).
9. Gerloch, M., and Miller, J. R., *Prog. Inorg. Chem.* **10**, 1 (1968).
10. Gerloch, M., and Slade, R. C., *Ligand-Field Parameters*, Cambridge University Press (1973).
11. Griffith, J. S., *The Irreducible Tensor Method for Molecular Symmetry Groups*, Prentice-Hall, Englewood Cliffs, N.J. (1962).
12. Griffith, J. S., *The Theory of Transition Metal Ions*, Cambridge University Press (1964).
13. Hamermesh, M., *Group Theory*, Addison-Wesley Publishing Co., Reading, Mass. (1962).
14. Harnung, S. E., and Schäffer, C. E., *Struct. Bonding* **12**, 257 (1972).
15. Hassitt, A., *Proc. R. Soc. Lond.* A **229**, 110 (1955).
16. Hempel, J. C., Donini, J. C., Hollebone, B. R., and Lever, A. B. P., *J. Am. Chem. Soc.* **96**, 1693 (1974).
17. Horie, H., *J. Phys. Soc. Jap.* **19**, 1783 (1964).
18. Horrocks, W. DeW., *J. Am. Chem. Soc.* **94**, 656 (1972).
19. Ilse, F. E., and Hartmann, H., *Z. Phys. Chem.* **197**, 239 (1951).
20. Jackson, J. D., *Classical Electrodynamics*, John Wiley, New York (1962).
21. Jahn, H. A., *Philos. Trans. R. Soc. Lond.* A **253**, 27 (1960).
22. Jahn, H. A., and Van Wieringen, H., *Proc. R. Soc. Lond.* A **209**, 502 (1951).
23. Jørgensen, C. K., *Modern Aspects of Ligand Field Theory*, North-Holland Publishing Co., Amsterdam (1971).
24. Judd, B. R., *Operator Techniques in Atomic Spectroscopy*, McGraw-Hill Book Co., New York (1963).
25. Kammer, H., *Acta Chim. Acad. Sci. Hung.* **66**, 189 (1970); **66**, 203 (1970).
26. Kibler, M., *J. Mol. Spectrosc.* **26**, 111 (1968).
27. Kibler, M. R., *J. Chem. Phys.* **61**, 3859 (1974).
28. König, E., and Kremer, S., *Theor. Chim. Acta* **32**, 27 (1973).
29. König, E., and Kremer, S., *Z. Naturforsch.* **29A**, 1179 (1974).
30. König, E., and Kremer, S., *Int. J. Quantum Chem.* **8**, 347 (1974).
31. König, E., and Kremer, S., unpublished.
32. Koster, G. F., *Phys. Rev.* **109**, 227 (1958).
33. Liehr, A. D., *J. Phys. Chem.* **64**, 43 (1960).
34. Low, W., *Paramagnetic Resonance in Solids, in:* Solid State Physics, F. Seitz and D. Turnbull, eds., Suppl. 2, Academic Press, New York (1960).
35. Löwdin, P. O., *J. Chem. Phys.* **18**, 365 (1950).
36. Matsen, F. A., and Plummer, O. R., *in: Group Theory and its Applications*, E. M. Loebl, ed., Academic Press, New York (1968), Vol. 1, p. 221.
37. McClure, D. S., *Advances in the Chemistry of the Coordination Compounds*, S. Kirschner, ed., Macmillan, New York (1961), p. 498.
38. McWeeny, R., *Symmetry, an Introduction to Group Theory and its Applications*, Pergamon Press, London (1963).
39. Messiah, A., *Quantum Mechanics*, North-Holland Publishing Co., Amsterdam (1966).
40. Nielson, C. W., and Koster, G. F., *Spectroscopic Coefficients for p^n, d^n, and f^n Configurations*, MIT Press, Cambridge, Mass. (1964).
41. Otsuka, J., *J. Phys. Soc. Jap.* **21**, 596 (1966).
42. Perumareddi, J. R., *J. Phys. Chem.* **71**, 3144 (1967).
43. Perumareddi, J. R., *Z. Naturforsch.* **27A**, 1820 (1972).
44. Perumareddi, J. R., *Phys. Status Solidi* **55B**, K 97 (1973).
45. Pryce, M., and Runciman, W. A., *Disc. Faraday Soc.* **26**, 34 (1958).
46. Racah, G., *Phys. Rev.* **61**, 186 (1942).
47. Racah, G., *Phys. Rev.* **62**, 438 (1942).
48. Racah, G., *Phys. Rev.* **63**, 367 (1943).
49. Racah, G., *Phys. Rev.* **76**, 1352 (1949).
50. Redmond, P. J., *Proc. R. Soc. Lond.* A **222**, 84 (1954).
51. Robinson, G. De B., *Representation Theory of the Symmetric Group*, Edinburgh University Press (1961).
52. Rotenberg, M., Bivins, R., Metropolis, N., and Wooten, J. K., *The 3j and 6j Symbols*, MIT Press, Cambridge, Mass. (1960).
53. Schäffer, C. E., *Pure Appl. Chem.* **24**, 361 (1970).

54. Schäffer, C. E., *Struct. Bonding* **5**, 68 (1968).

55. Slater, J. C., *The Quantum Theory of Atomic Structure* McGraw-Hill Book Co., New York (1960).

56. Sobelman, I. I., *An Introduction to the Theory of Atomic Spectra*, Pergamon Press, Oxford (1972).

57. Sugano, S., and Tanabe, Y., *J. Phys. Soc. Jap.* **13**, 880 (1958).

58. Sugano, S., Tanabe, Y., and Kamimura, H., *Multiplets of Transition-Metal Ions in Crystals*, Academic Press, New York (1970).

59. Tanabe, Y., and Sugano, S., *J. Phys. Soc. Jap.* **9**, 753 (1954); **9**, 766 (1954).

60. Varga, J. A., and Becker, C. A. L., *J. Phys. Chem.* **76**, 2907 (1972).

61. Wigner, E. P., On the Matrices which Reduce the Kronecker Product of Representations of S. R. Groups *in*: *Quantum Theory of Angular Momentum*, L. C. Biedenharn and H. Van Dam, eds., Academic Press, New York (1965).

62. Wybourne, B. G., *Spectroscopic Properties of Rare Earths*, Interscience Publishers, New York (1965).

63. Wybourne, B. G., *Symmetry Principles and Atomic Spectroscopy*, Wiley–Interscience, New York (1970).

64. Yutsis, A. P., Levinson, I. B., and Vanagas, V. V., *Mathematical Apparatus of the Theory of Angular Momentum*, Israel Program for Scientific Translations, Jerusalem (1962).

PART II
DIAGRAMS

PART II

DIAGRAMS

2 D ELECTRONS, TETRAGONAL SYMMETRY

B=630   C=4×B   ZETA=210

ENERGY AS FUNCTION OF DQ
DT=0, 500, -500, 1000, -1000
K=1, -1, 5, -5

ENERGY AS FUNCTION OF DT
DQ=1840
K=1, -1, 5, -5

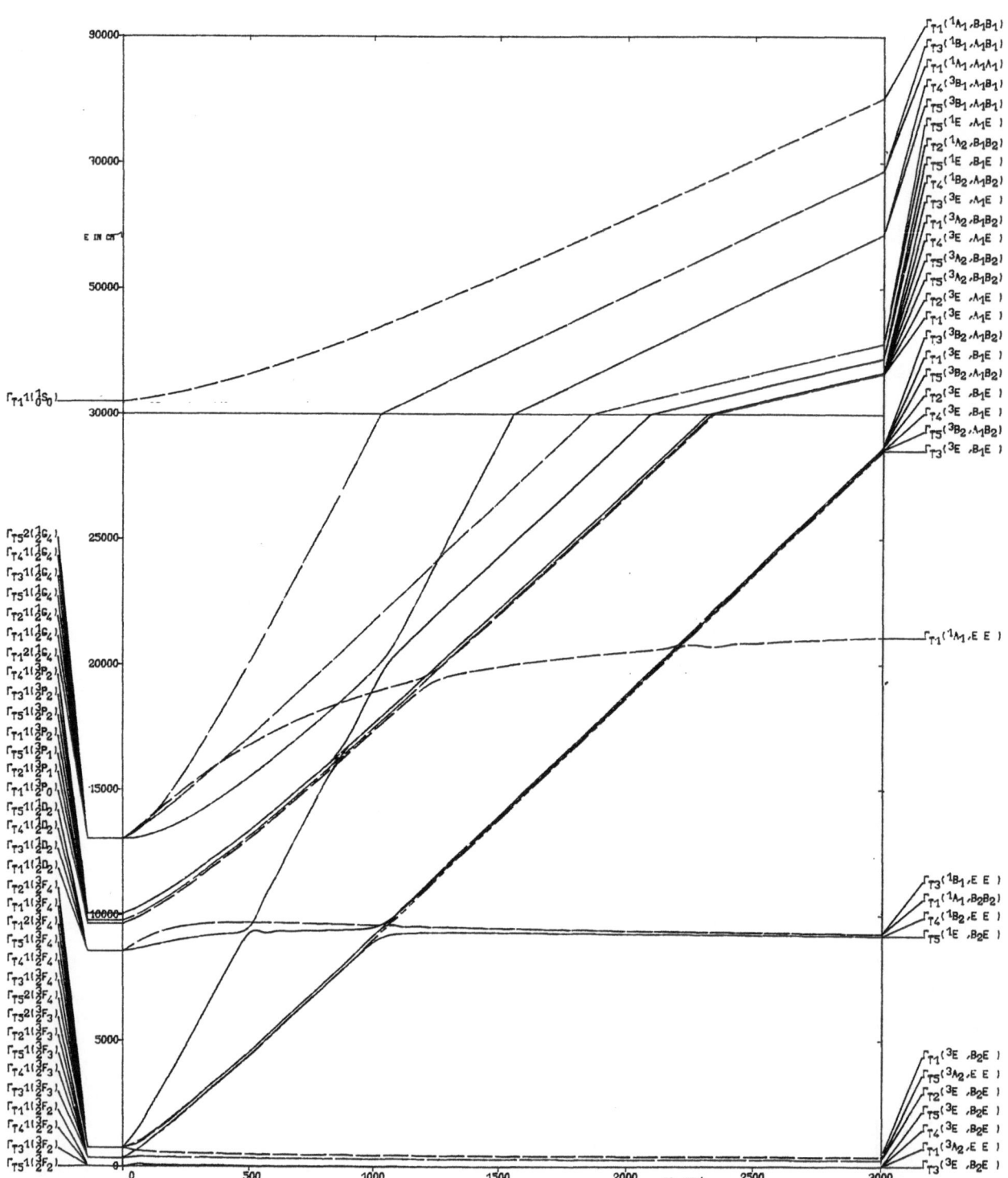

2 D ELECTRONS, TETRAGONAL SYMMETRY   B=630 C=4xB DT=0 K=0   DS=KxDT ZETA=210

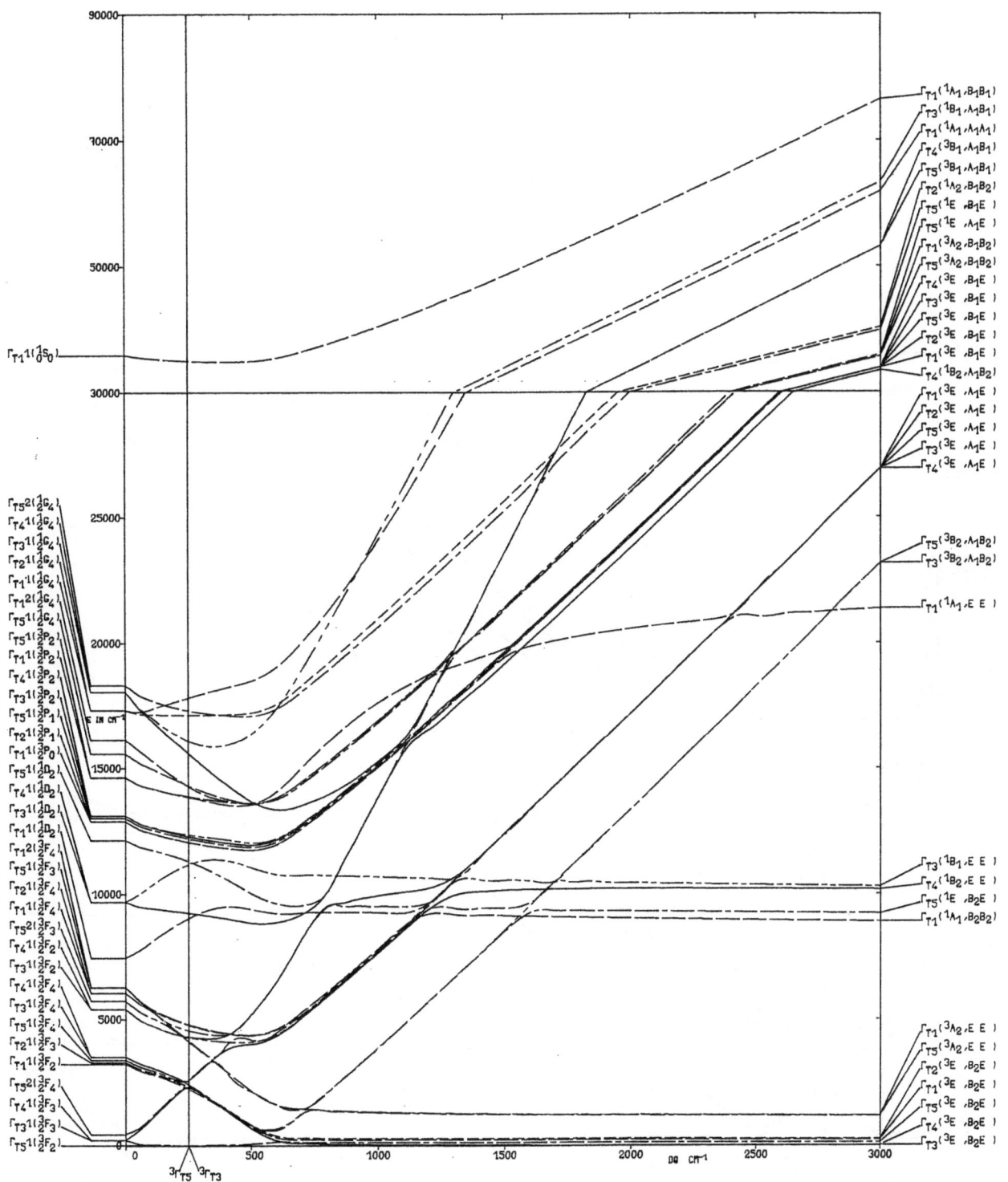

2 D ELECTRONS, TETRAGONAL SYMMETRY

B=630 C=4×B DT=500 K=1 DS=K×DT ZETA=210

69

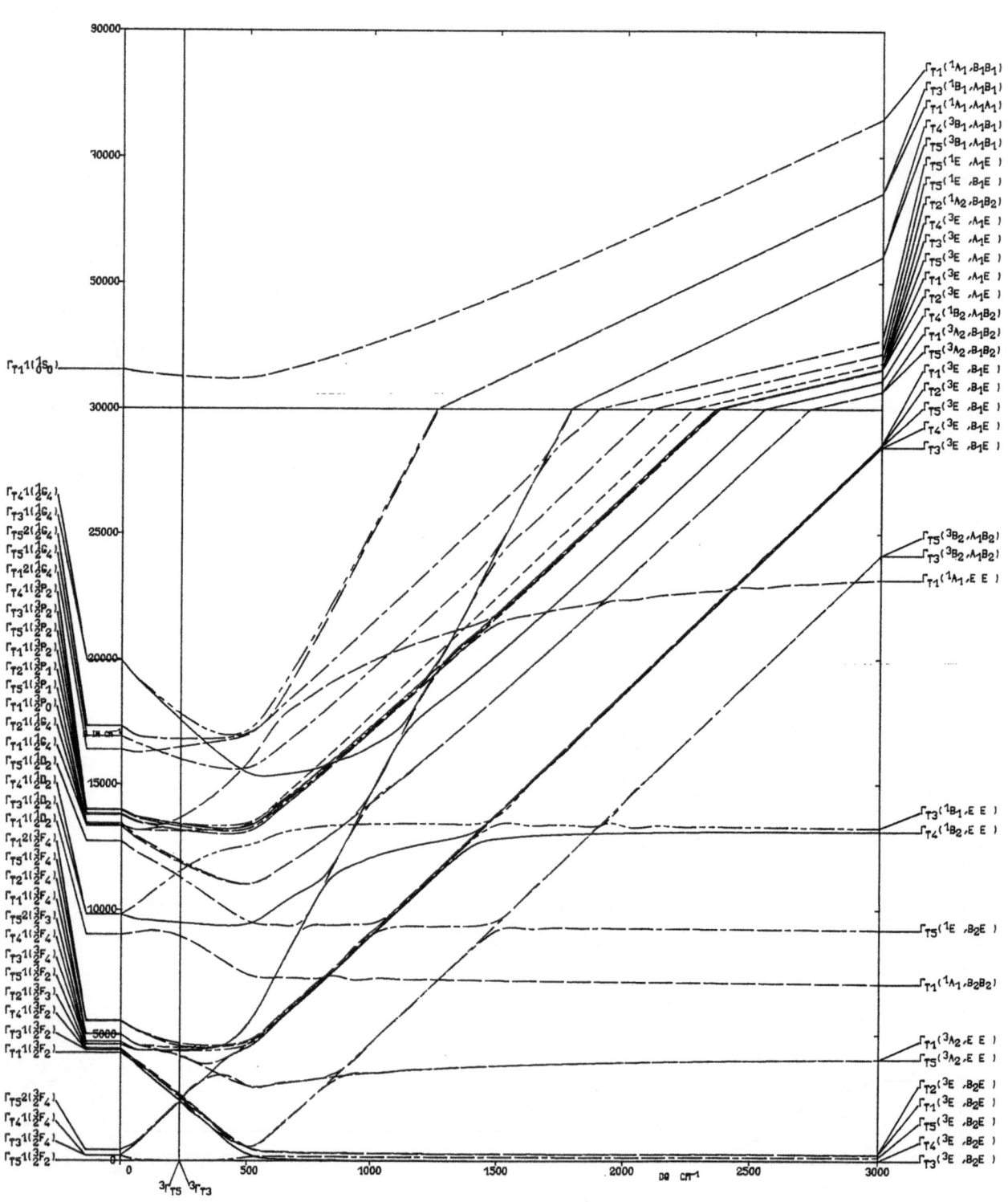

$\Gamma_{T1}1({}^1_0S_0)$

$\Gamma_{T4}1({}^1_2G_4)$
$\Gamma_{T3}1({}^1_2G_4)$
$\Gamma_{T5}2({}^1_2G_4)$
$\Gamma_{T5}1({}^1_2G_4)$
$\Gamma_{T1}2({}^1_2G_4)$
$\Gamma_{T4}1({}^3_2P_2)$
$\Gamma_{T3}1({}^3_2P_2)$
$\Gamma_{T5}1({}^3_2P_2)$
$\Gamma_{T1}1({}^3_2P_2)$
$\Gamma_{T2}1({}^3_2P_1)$
$\Gamma_{T5}1({}^3_2P_1)$
$\Gamma_{T1}1({}^3_2P_0)$
$\Gamma_{T2}1({}^1_2G_4)$
$\Gamma_{T1}1({}^1_2G_4)$
$\Gamma_{T5}1({}^1_2D_2)$
$\Gamma_{T4}1({}^1_2D_2)$
$\Gamma_{T3}1({}^1_2D_2)$
$\Gamma_{T1}1({}^1_2D_2)$
$\Gamma_{T1}2({}^3_2F_4)$
$\Gamma_{T5}1({}^3_2F_4)$
$\Gamma_{T4}1({}^3_2F_4)$
$\Gamma_{T2}1({}^3_2F_4)$
$\Gamma_{T1}1({}^3_2F_4)$
$\Gamma_{T5}2({}^3_2F_3)$
$\Gamma_{T4}1({}^3_2F_4)$
$\Gamma_{T3}1({}^3_2F_2)$
$\Gamma_{T2}1({}^3_2F_3)$
$\Gamma_{T4}1({}^3_2F_4)$
$\Gamma_{T3}1({}^3_2F_2)$
$\Gamma_{T1}1({}^3_2F_2)$

$\Gamma_{T5}2({}^3_2F_4)$
$\Gamma_{T4}1({}^3_2F_4)$
$\Gamma_{T3}1({}^3_2F_4)$
$\Gamma_{T5}1({}^3_2F_2)$

$\Gamma_{T1}({}^1A_1,B_1B_1)$
$\Gamma_{T3}({}^1B_1,A_1B_1)$
$\Gamma_{T1}({}^1A_1,A_1A_1)$
$\Gamma_{T4}({}^3B_1,A_1B_1)$
$\Gamma_{T5}({}^3B_1,A_1B_1)$
$\Gamma_{T5}({}^1E,A_1E)$
$\Gamma_{T5}({}^1E,B_1E)$
$\Gamma_{T2}({}^1A_2,B_1B_2)$
$\Gamma_{T4}({}^3E,A_1E)$
$\Gamma_{T3}({}^3E,A_1E)$
$\Gamma_{T5}({}^3E,A_1E)$
$\Gamma_{T1}({}^3E,A_1E)$
$\Gamma_{T2}({}^3E,A_1E)$
$\Gamma_{T4}({}^3B_2,A_1B_2)$
$\Gamma_{T3}({}^3A_2,B_1B_2)$
$\Gamma_{T5}({}^3A_2,B_1B_2)$
$\Gamma_{T2}({}^3E,B_1E)$
$\Gamma_{T1}({}^3E,B_1E)$
$\Gamma_{T4}({}^3E,B_1E)$
$\Gamma_{T3}({}^3E,B_1E)$

$\Gamma_{T5}({}^3B_2,A_1B_2)$
$\Gamma_{T3}({}^3B_2,A_1B_2)$
$\Gamma_{T1}({}^1A_1,EE)$

$\Gamma_{T3}({}^1B_1,EE)$
$\Gamma_{T4}({}^1B_2,EE)$

$\Gamma_{T5}({}^1E,B_2E)$

$\Gamma_{T1}({}^1A_1,B_2B_2)$

$\Gamma_{T1}({}^3A_2,EE)$
$\Gamma_{T5}({}^3A_2,EE)$

$\Gamma_{T2}({}^3E,B_2E)$
$\Gamma_{T1}({}^3E,B_2E)$
$\Gamma_{T5}({}^3E,B_2E)$
$\Gamma_{T4}({}^3E,B_2E)$
$\Gamma_{T3}({}^3E,B_2E)$

${}^3\Gamma_{T5}$ ${}^3\Gamma_{T3}$

2 D ELECTRONS, TETRAGONAL SYMMETRY

B=630 C=4×B DT=500 K=-1 DS=K×DT ZETA=210

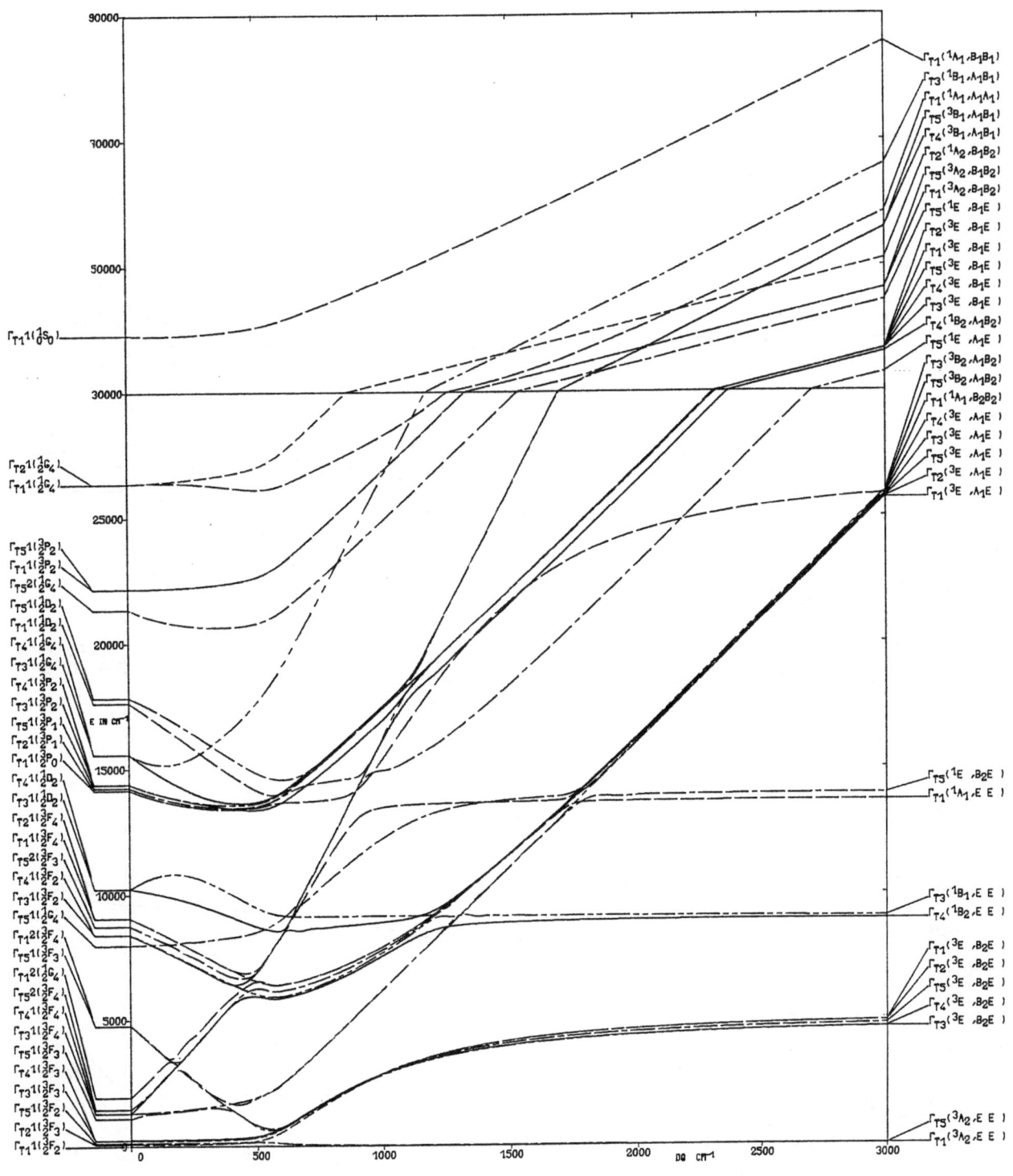

2 D ELECTRONS, TETRAGONAL SYMMETRY   B=630 C=4×B DT=500 K=5   DS=K×DT ZETA=210

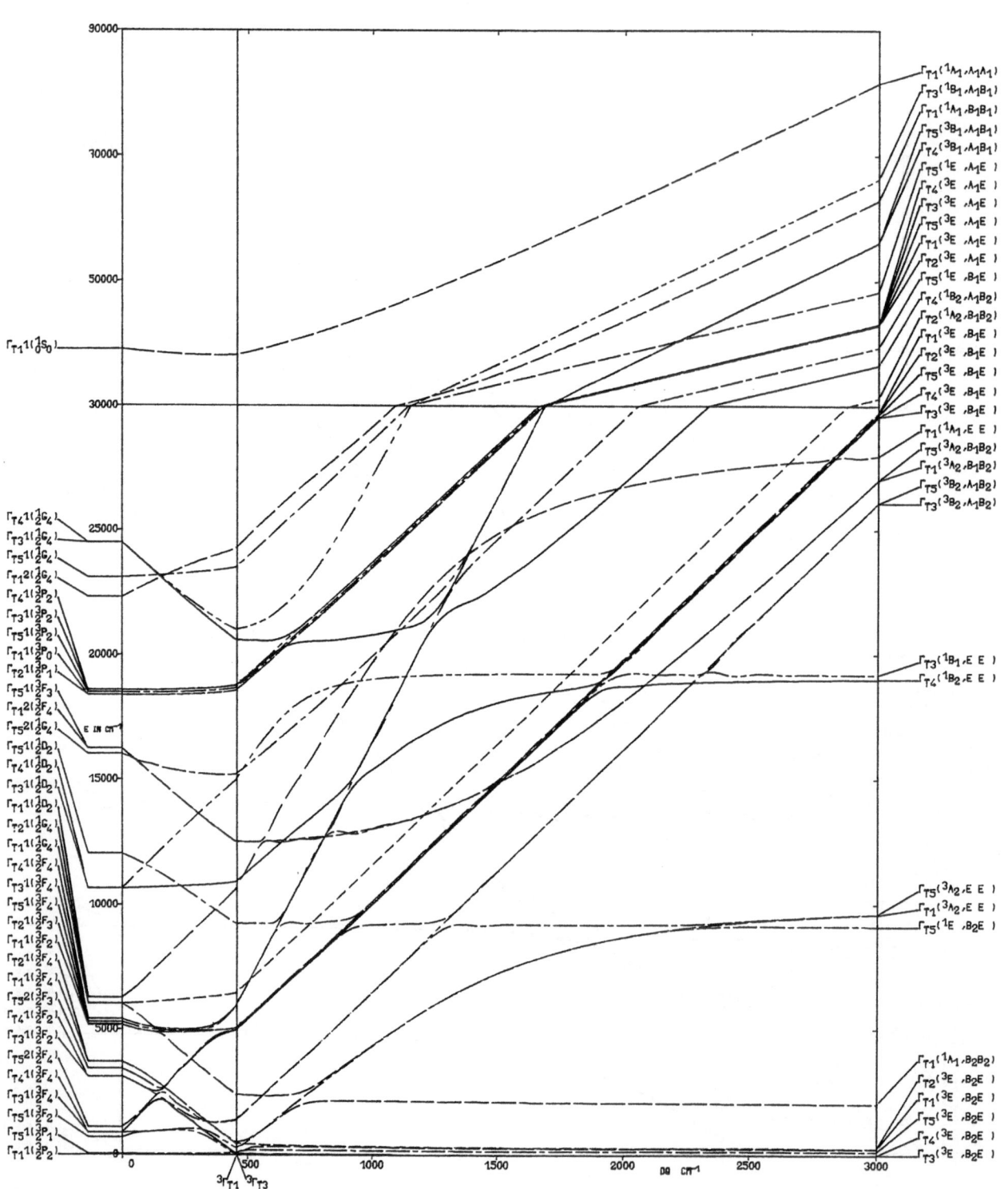

$\Gamma_{T1}1(^1_0S_0)$

$\Gamma_{T4}1(^1_2G_4)$
$\Gamma_{T3}1(^1_2G_4)$
$\Gamma_{T5}1(^1_2G_4)$
$\Gamma_{T1}2(^1_2G_4)$
$\Gamma_{T4}1(^3_2P_2)$
$\Gamma_{T3}1(^3_2P_2)$
$\Gamma_{T5}1(^3_2P_2)$
$\Gamma_{T1}1(^3_2P_0)$
$\Gamma_{T2}1(^3_2P_1)$
$\Gamma_{T5}1(^3_2F_3)$
$\Gamma_{T1}2(^1_2G_4)$
$\Gamma_{T5}2(^1_2G_4)$
$\Gamma_{T5}1(^1_2D_2)$
$\Gamma_{T4}1(^1_2D_2)$
$\Gamma_{T3}1(^1_2D_2)$
$\Gamma_{T1}1(^1_2D_2)$
$\Gamma_{T2}1(^1_2G_4)$
$\Gamma_{T1}1(^1_2G_4)$
$\Gamma_{T4}1(^3_2F_4)$
$\Gamma_{T3}1(^3_2F_4)$
$\Gamma_{T2}1(^3_2F_3)$
$\Gamma_{T5}1(^3_2F_2)$
$\Gamma_{T2}1(^3_2F_4)$
$\Gamma_{T1}1(^3_2F_4)$
$\Gamma_{T5}2(^3_2F_3)$
$\Gamma_{T4}1(^3_2F_2)$
$\Gamma_{T3}1(^3_2F_2)$
$\Gamma_{T5}2(^3_2F_4)$
$\Gamma_{T4}1(^3_2F_4)$
$\Gamma_{T3}1(^3_2F_4)$
$\Gamma_{T5}1(^3_2F_2)$
$\Gamma_{T2}1(^3_2P_1)$
$\Gamma_{T5}1(^3_2P_1)$
$\Gamma_{T1}1(^3_2P_2)$

$\Gamma_{T1}(^1A_1,A_1A_1)$
$\Gamma_{T3}(^1B_1,A_1B_1)$
$\Gamma_{T1}(^1A_1,B_1B_1)$
$\Gamma_{T5}(^3B_1,A_1B_1)$
$\Gamma_{T4}(^3B_1,A_1B_1)$
$\Gamma_{T5}(^3E,A_1E)$
$\Gamma_{T4}(^3E,A_1E)$
$\Gamma_{T3}(^3E,A_1E)$
$\Gamma_{T1}(^3E,A_1E)$
$\Gamma_{T2}(^3E,A_1E)$
$\Gamma_{T5}(^1E,B_1E)$
$\Gamma_{T4}(^1B_2,A_1B_2)$
$\Gamma_{T2}(^1A_2,B_1B_2)$
$\Gamma_{T1}(^3E,B_1E)$
$\Gamma_{T2}(^3E,B_1E)$
$\Gamma_{T5}(^3E,B_1E)$
$\Gamma_{T4}(^3E,B_1E)$
$\Gamma_{T3}(^3E,B_1E)$
$\Gamma_{T1}(^1A_1,E E)$
$\Gamma_{T5}(^3A_2,B_1B_2)$
$\Gamma_{T1}(^3A_2,B_1B_2)$
$\Gamma_{T5}(^3B_2,A_1B_2)$
$\Gamma_{T3}(^3B_2,A_1B_2)$

$\Gamma_{T3}(^1B_1,E E)$
$\Gamma_{T4}(^1B_2,E E)$

$\Gamma_{T5}(^3A_2,E E)$
$\Gamma_{T1}(^3A_2,E E)$
$\Gamma_{T5}(^1E,B_2E)$

$\Gamma_{T1}(^1A_1,B_2B_2)$
$\Gamma_{T2}(^3E,B_2E)$
$\Gamma_{T1}(^3E,B_2E)$
$\Gamma_{T5}(^3E,B_2E)$
$\Gamma_{T4}(^3E,B_2E)$
$\Gamma_{T3}(^3E,B_2E)$

E IN CM^{-1}

$^3\Gamma_{T1}$ $^3\Gamma_{T3}$

DQ CM^{-1}

2 D ELECTRONS, TETRAGONAL SYMMETRY

B=630 C=4×B DT=500 K=-5 DS=K×DT ZETA=210

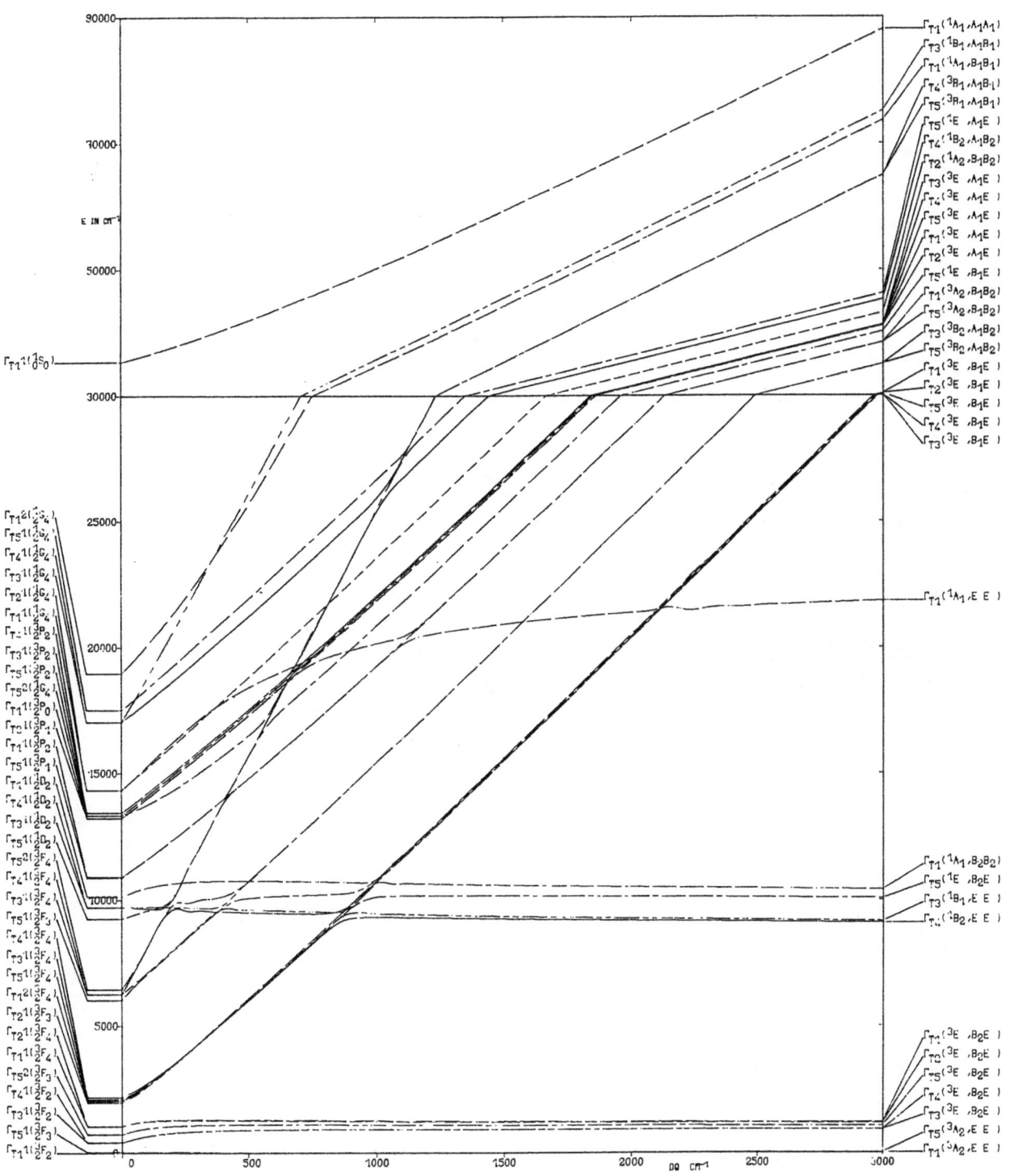

2 D ELECTRONS, TETRAGONAL SYMMETRY   B=630 C=4×B   DT=-500 K=1   DS=K×DT   ZETA=210

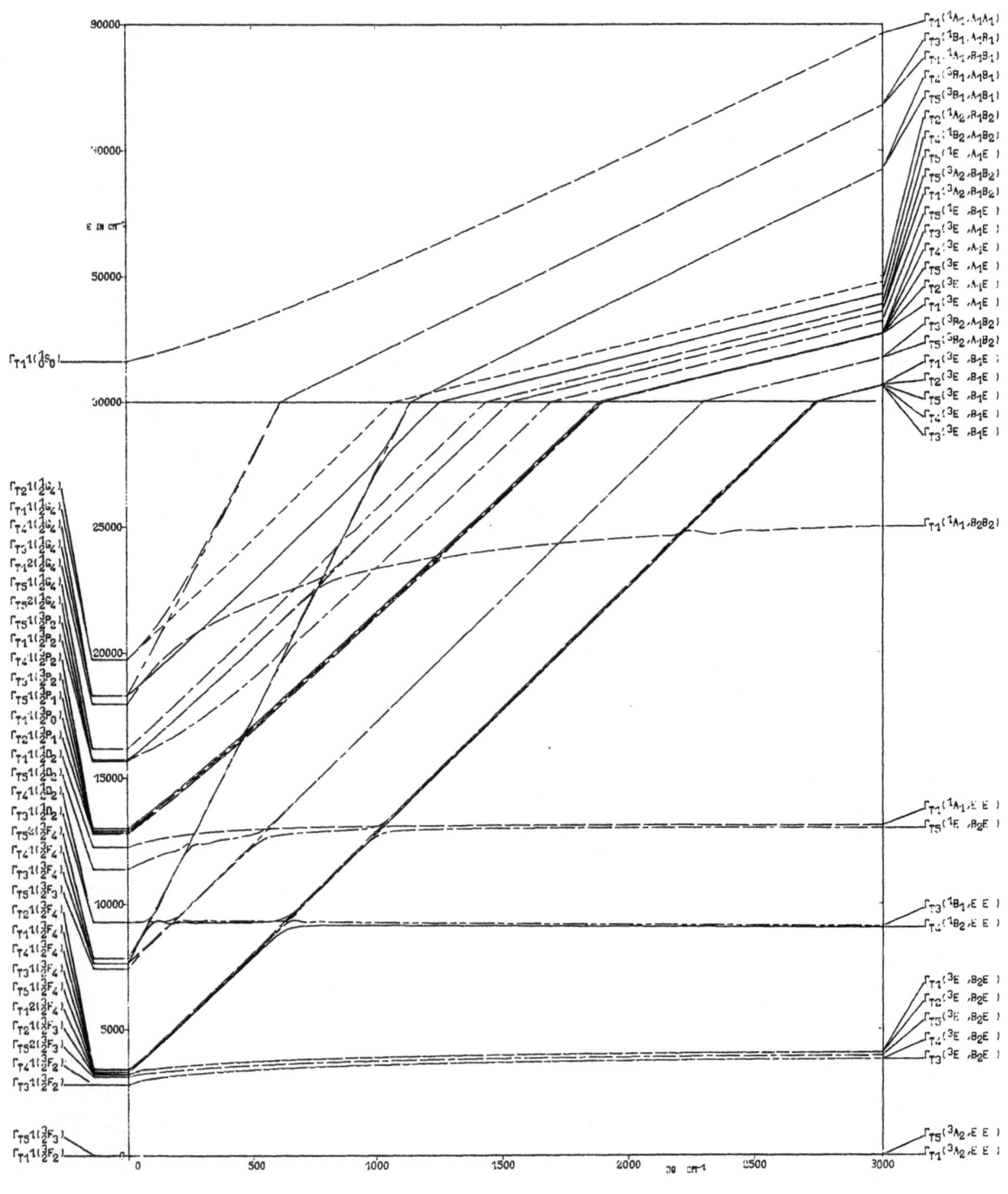

2 D ELECTRONS, TETRAGONAL SYMMETRY   B=630 C=4*B DT=-500 K=-1 DS=K*DT ZETA=210

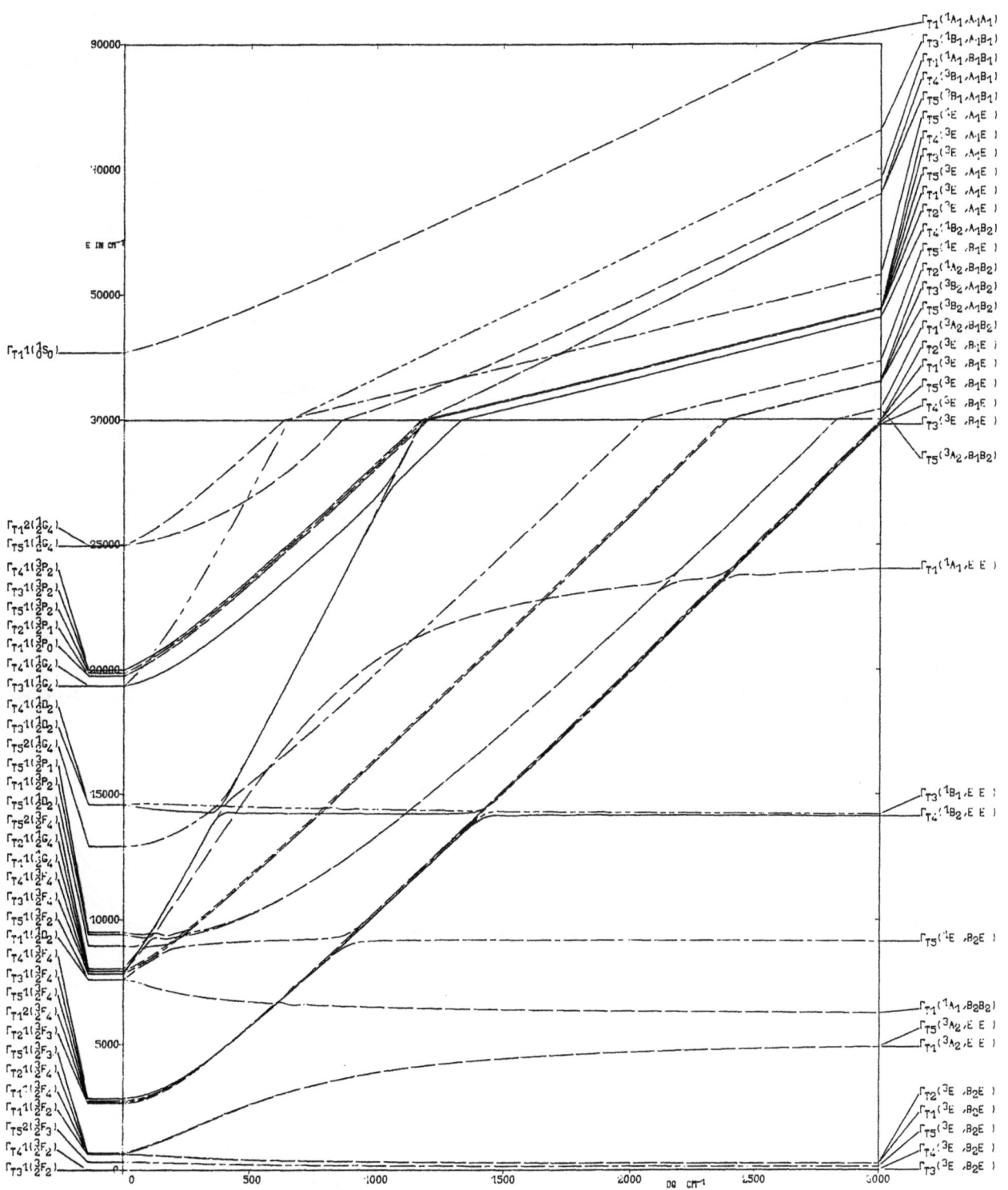

2 D ELECTRONS, TETRAGONAL SYMMETRY   R=630 C=4×B DT=-500 K=5   DS=K×DT ZETA=210

75

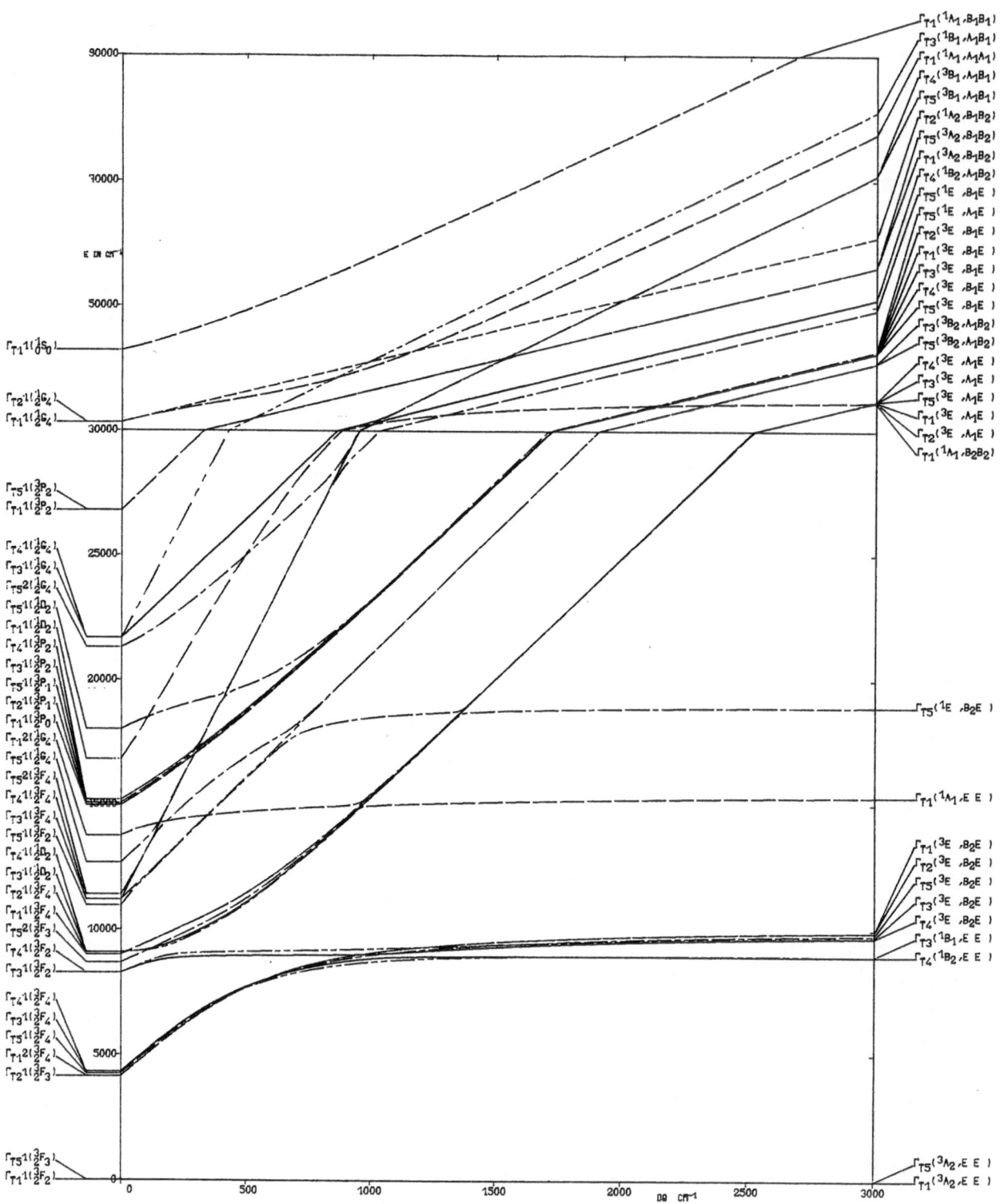

90000

70000

E IN CM⁻¹

$\Gamma_{T1}1(^1S_0)$

50000

$\Gamma_{T2}1(^1G_4)$
$\Gamma_{T1}1(^1G_4)$

30000

$\Gamma_{T5}1(^3P_2)$
$\Gamma_{T1}1(^3P_2)$

$\Gamma_{T4}1(^1G_4)$
$\Gamma_{T3}1(^1G_4)$
$\Gamma_{T5}2(^1G_4)$
$\Gamma_{T5}1(^1D_2)$
$\Gamma_{T1}1(^1D_2)$
$\Gamma_{T4}1(^3P_2)$
$\Gamma_{T3}1(^3P_2)$
$\Gamma_{T5}1(^3P_1)$
$\Gamma_{T2}1(^3P_1)$
$\Gamma_{T1}1(^3P_0)$
$\Gamma_{T1}2(^1G_4)$
$\Gamma_{T5}1(^1G_4)$
$\Gamma_{T5}2(^3F_4)$
$\Gamma_{T4}1(^3F_4)$
$\Gamma_{T3}1(^3F_4)$
$\Gamma_{T5}1(^3F_2)$
$\Gamma_{T4}1(^1D_2)$
$\Gamma_{T3}1(^1D_2)$
$\Gamma_{T2}1(^3F_4)$
$\Gamma_{T1}1(^3F_4)$
$\Gamma_{T5}2(^3F_3)$
$\Gamma_{T4}1(^3F_2)$
$\Gamma_{T3}1(^3F_2)$

25000

20000

15000

10000

$\Gamma_{T4}1(^3F_4)$
$\Gamma_{T3}1(^3F_4)$
$\Gamma_{T5}1(^3F_4)$
$\Gamma_{T1}2(^3F_4)$
$\Gamma_{T2}1(^3F_3)$

5000

$\Gamma_{T5}1(^3F_3)$
$\Gamma_{T1}1(^3F_2)$

0

Right side labels:
$\Gamma_{T1}(^1A_1, B_1B_1)$
$\Gamma_{T3}(^1B_1, A_1B_1)$
$\Gamma_{T1}(^1A_1, A_1A_1)$
$\Gamma_{T4}(^3B_1, A_1B_1)$
$\Gamma_{T5}(^3B_1, A_1B_1)$
$\Gamma_{T2}(^1A_2, B_1B_2)$
$\Gamma_{T5}(^3A_2, B_1B_2)$
$\Gamma_{T1}(^3A_2, B_1B_2)$
$\Gamma_{T4}(^1B_2, A_1B_2)$
$\Gamma_{T5}(^1E, B_1E)$
$\Gamma_{T5}(^1E, A_1E)$
$\Gamma_{T2}(^3E, B_1E)$
$\Gamma_{T1}(^3E, B_1E)$
$\Gamma_{T3}(^3E, B_1E)$
$\Gamma_{T4}(^3E, B_1E)$
$\Gamma_{T3}(^3B_2, A_1B_2)$
$\Gamma_{T5}(^3B_2, A_1B_2)$
$\Gamma_{T4}(^3E, A_1E)$
$\Gamma_{T3}(^3E, A_1E)$
$\Gamma_{T5}(^3E, A_1E)$
$\Gamma_{T1}(^3E, A_1E)$
$\Gamma_{T2}(^3E, A_1E)$
$\Gamma_{T1}(^1A_1, B_2B_2)$

$\Gamma_{T5}(^1E, B_2E)$

$\Gamma_{T1}(^1A_1, E E)$

$\Gamma_{T1}(^3E, B_2E)$
$\Gamma_{T2}(^3E, B_2E)$
$\Gamma_{T5}(^3E, B_2E)$
$\Gamma_{T3}(^3E, B_2E)$
$\Gamma_{T4}(^3E, B_2E)$
$\Gamma_{T3}(^1B_1, E E)$
$\Gamma_{T4}(^1B_2, E E)$

$\Gamma_{T5}(^3A_2, E E)$
$\Gamma_{T1}(^3A_2, E E)$

0   500   1000   1500   2000   2500   3000

DQ CM⁻¹

76

2 D ELECTRONS, TETRAGONAL SYMMETRY

B=630   C=4×B   DT=-500   K=-5   DS=K×DT   ZETA=210

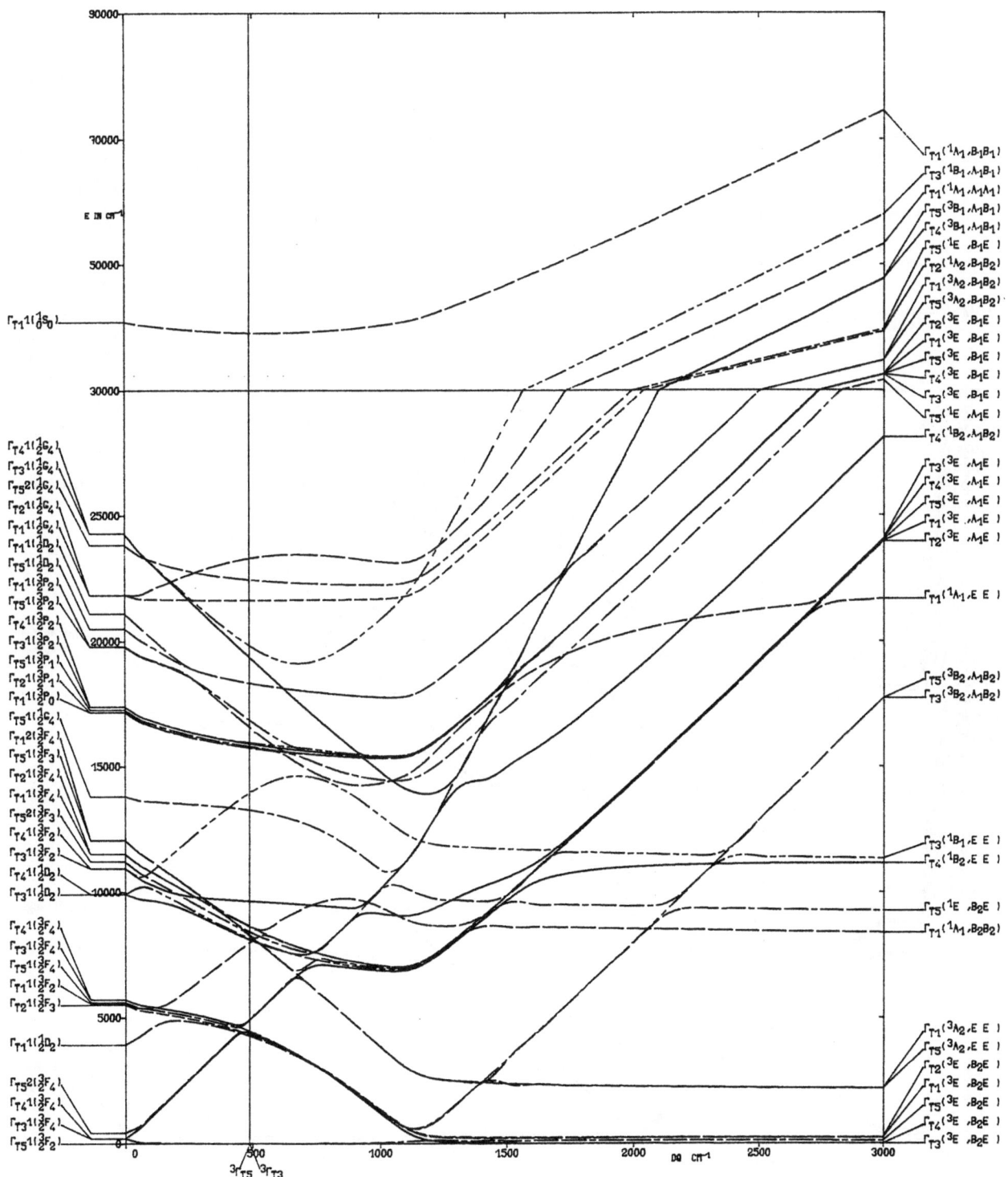

2 D ELECTRONS, TETRAGONAL SYMMETRY   B=630 C=4×B DT=1000 K=1   DS=K×DT ZETA=210

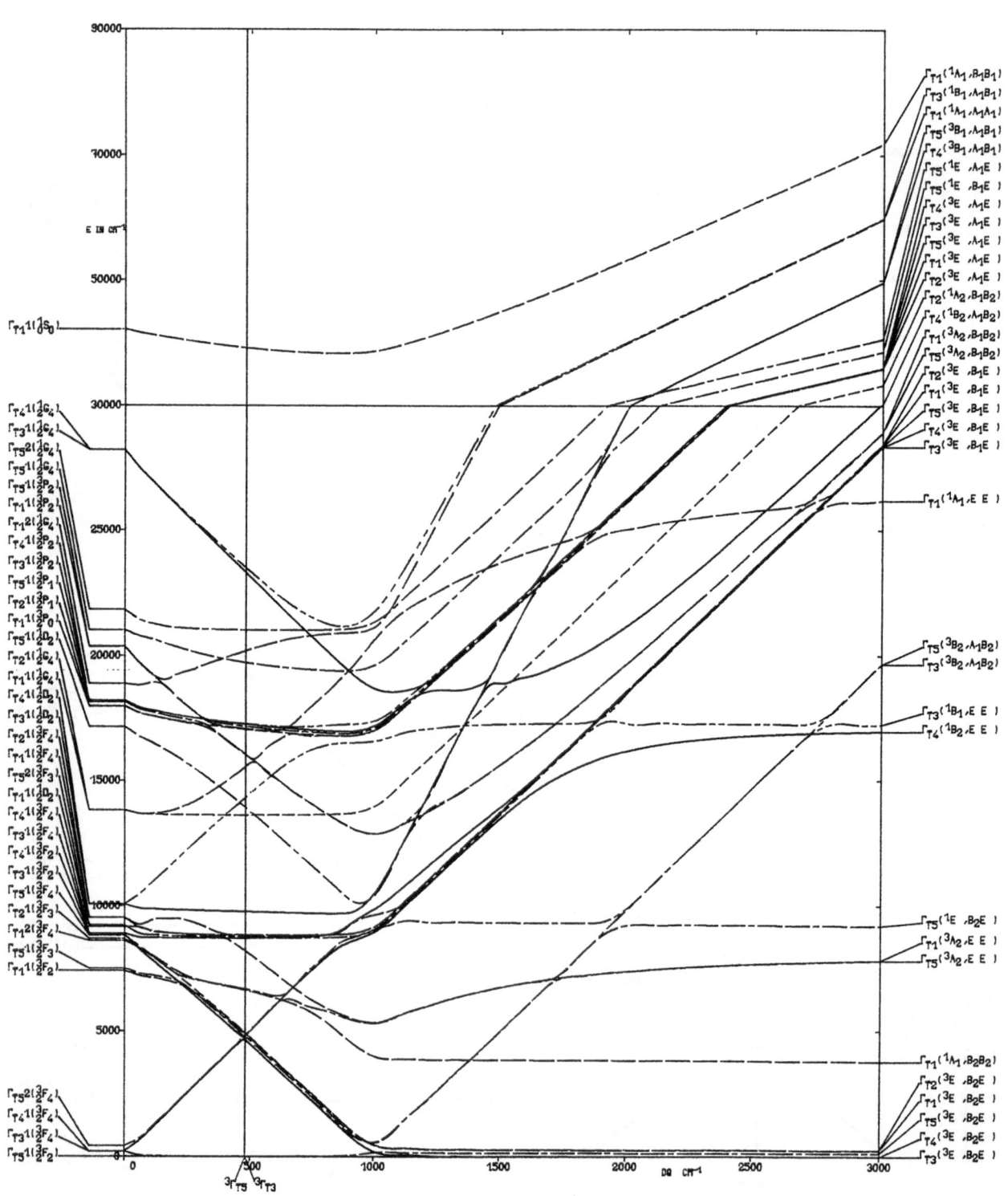

2 D ELECTRONS, TETRAGONAL SYMMETRY   B=630 C=4×B DT=1000 K=-1 DS=K×DT ZETA=210

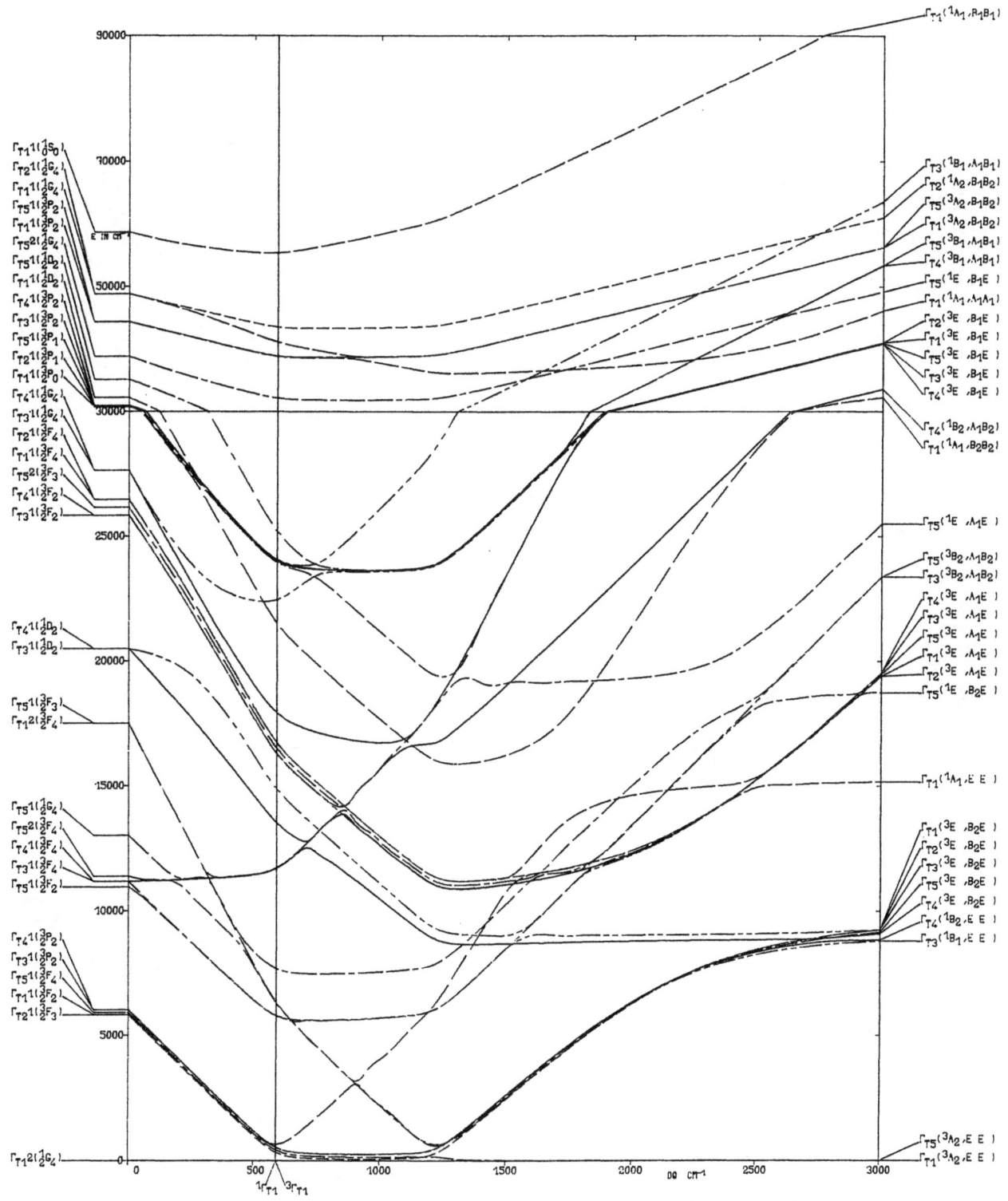

2 D ELECTRONS, TETRAGONAL SYMMETRY

B=630 C=4*B DT=1000 K=5 DS=K*DT ZETA=210

79

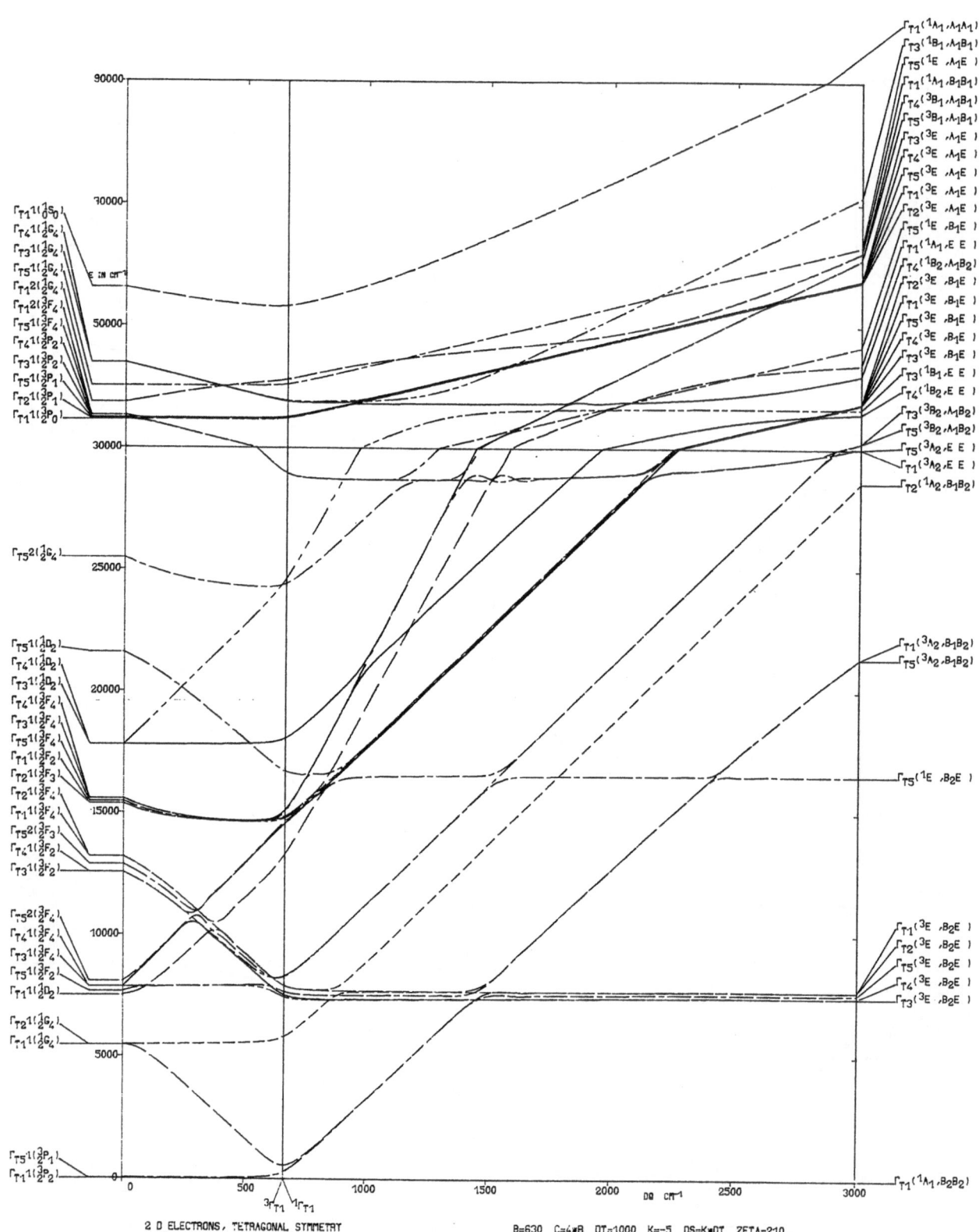

90000

70000

$\Gamma_{T1}{}^{1}({}^{1}S_0)$
$\Gamma_{T4}{}^{1}({}^{1}G_4)$
$\Gamma_{T3}{}^{1}({}^{1}G_4)$
$\Gamma_{T5}{}^{1}({}^{1}G_4)$
$\Gamma_{T1}{}^{2}({}^{1}G_4)$
$\Gamma_{T2}{}^{1}({}^{3}F_4)$
$\Gamma_{T5}{}^{1}({}^{3}F_4)$

50000

$\Gamma_{T4}{}^{1}({}^{3}P_2)$
$\Gamma_{T3}{}^{1}({}^{3}P_2)$
$\Gamma_{T5}{}^{1}({}^{3}P_1)$
$\Gamma_{T2}{}^{1}({}^{3}P_1)$
$\Gamma_{T1}{}^{1}({}^{3}P_0)$

30000

$\Gamma_{T5}{}^{2}({}^{1}G_4)$

25000

$\Gamma_{T5}{}^{1}({}^{1}D_2)$
$\Gamma_{T4}{}^{1}({}^{1}D_2)$
$\Gamma_{T3}{}^{1}({}^{1}D_2)$
$\Gamma_{T1}{}^{1}({}^{1}D_2)$

20000

$\Gamma_{T4}{}^{1}({}^{3}F_4)$
$\Gamma_{T5}{}^{1}({}^{3}F_4)$
$\Gamma_{T1}{}^{1}({}^{3}F_4)$
$\Gamma_{T2}{}^{1}({}^{3}F_3)$
$\Gamma_{T1}{}^{1}({}^{3}F_4)$
$\Gamma_{T5}{}^{2}({}^{3}F_3)$

15000

$\Gamma_{T4}{}^{1}({}^{3}F_2)$
$\Gamma_{T3}{}^{1}({}^{3}F_2)$

$\Gamma_{T5}{}^{2}({}^{3}F_4)$
$\Gamma_{T4}{}^{1}({}^{3}F_4)$
$\Gamma_{T3}{}^{1}({}^{3}F_4)$
$\Gamma_{T5}{}^{1}({}^{3}F_2)$
$\Gamma_{T1}{}^{1}({}^{1}D_2)$

10000

$\Gamma_{T2}{}^{1}({}^{1}G_4)$
$\Gamma_{T1}{}^{1}({}^{1}G_4)$

5000

$\Gamma_{T5}{}^{1}({}^{3}P_1)$
$\Gamma_{T1}{}^{1}({}^{3}P_2)$

0

$\Gamma_{T1}({}^{1}A_1,A_1A_1)$
$\Gamma_{T3}({}^{1}B_1,A_1B_1)$
$\Gamma_{T5}({}^{1}E,A_1E)$
$\Gamma_{T1}({}^{1}A_1,B_1B_1)$
$\Gamma_{T4}({}^{3}B_1,A_1B_1)$
$\Gamma_{T5}({}^{3}B_1,A_1B_1)$
$\Gamma_{T3}({}^{3}E,A_1E)$
$\Gamma_{T4}({}^{3}E,A_1E)$
$\Gamma_{T5}({}^{3}E,A_1E)$
$\Gamma_{T1}({}^{3}E,A_1E)$
$\Gamma_{T2}({}^{3}E,A_1E)$
$\Gamma_{T5}({}^{1}E,B_1E)$
$\Gamma_{T1}({}^{1}A_1,EE)$
$\Gamma_{T4}({}^{3}B_2,A_1B_2)$
$\Gamma_{T2}({}^{3}E,B_1E)$
$\Gamma_{T3}({}^{3}E,B_1E)$
$\Gamma_{T5}({}^{3}E,B_1E)$
$\Gamma_{T4}({}^{3}E,B_1E)$
$\Gamma_{T3}({}^{1}B_1,EE)$
$\Gamma_{T4}({}^{1}B_2,EE)$
$\Gamma_{T3}({}^{3}B_2,A_1B_2)$
$\Gamma_{T5}({}^{3}B_2,A_1B_2)$
$\Gamma_{T5}({}^{3}A_2,EE)$
$\Gamma_{T1}({}^{3}A_2,EE)$
$\Gamma_{T2}({}^{1}A_2,B_1B_2)$

$\Gamma_{T1}({}^{3}A_2,B_1B_2)$
$\Gamma_{T5}({}^{3}A_2,B_1B_2)$

$\Gamma_{T5}({}^{1}E,B_2E)$

$\Gamma_{T1}({}^{3}E,B_2E)$
$\Gamma_{T2}({}^{3}E,B_2E)$
$\Gamma_{T5}({}^{3}E,B_2E)$
$\Gamma_{T4}({}^{3}E,B_2E)$
$\Gamma_{T3}({}^{3}E,B_2E)$

$\Gamma_{T1}({}^{1}A_1,B_2B_2)$

${}^{3}\Gamma_{T1}$ ${}^{1}\Gamma_{T1}$

0   500   1000   1500   2000   2500   3000

DQ CM^{-1}

80

2 D ELECTRONS, TETRAGONAL SYMMETRY

B=630 C=4×B DT=1000 K=-5 DS=K×DT ZETA=210

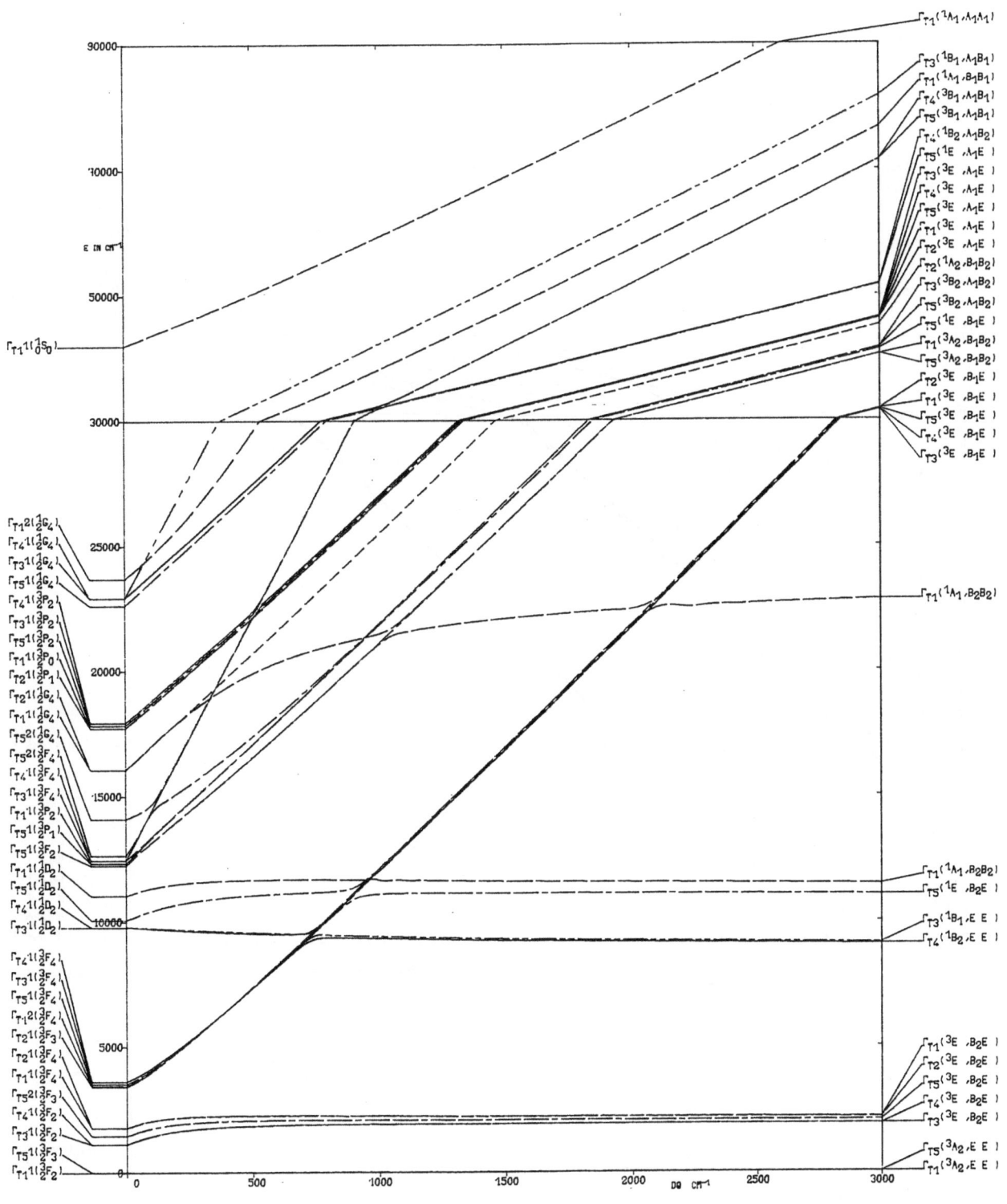

2 D ELECTRONS, TETRAGONAL SYMMETRY   B=630 C=4xB DT=-1000 K=1   DS=KxDT ZETA=210

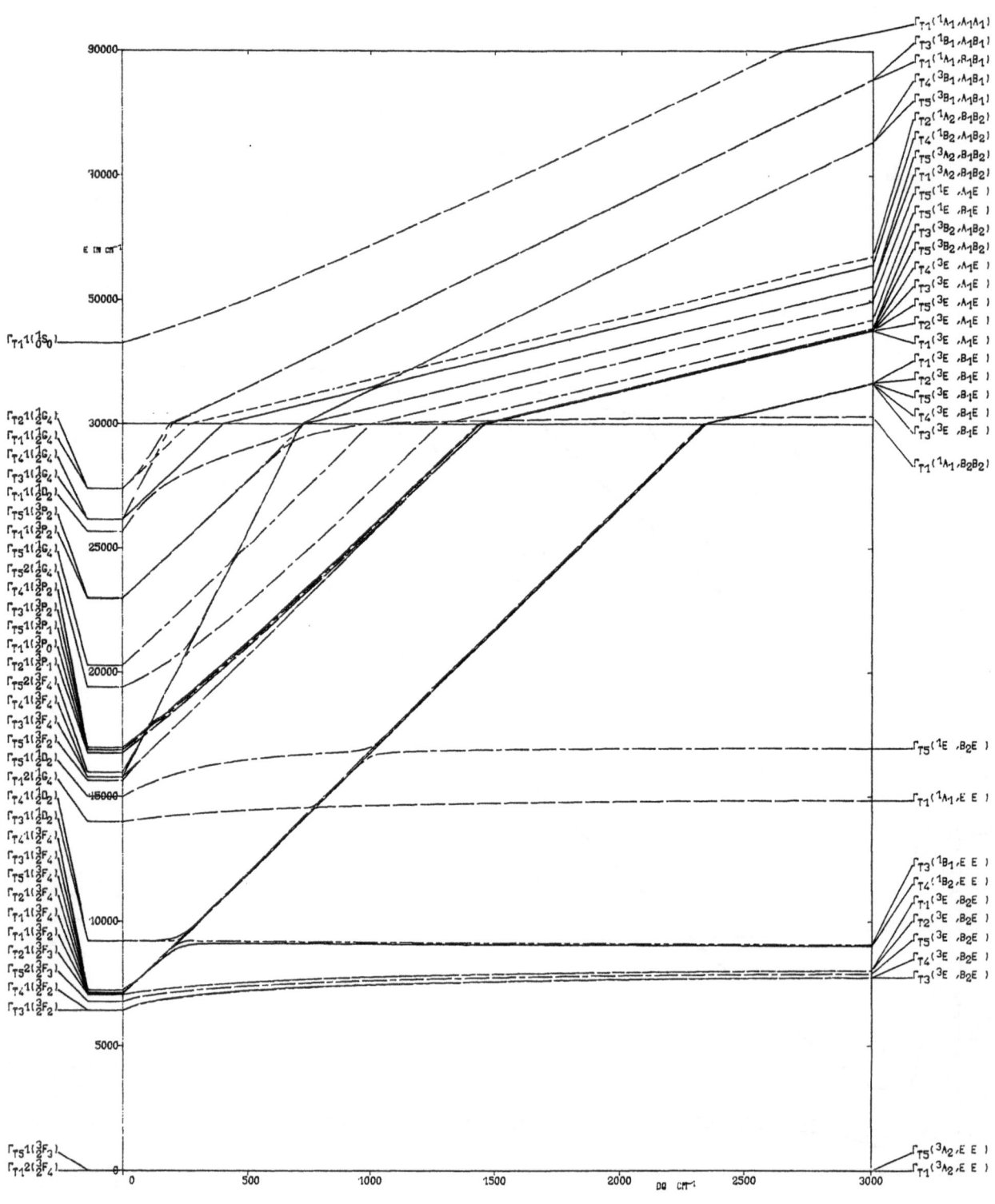

$\Gamma_{T1}{}^1(^1A_1,A_1A_1)$
$\Gamma_{T3}{}^1(^1B_1,A_1B_1)$
$\Gamma_{T1}{}^1(^1A_1,A_1B_1)$
$\Gamma_{T4}{}^1(^3B_1,A_1B_1)$
$\Gamma_{T5}{}^1(^3B_1,A_1B_1)$
$\Gamma_{T2}{}^1(^1A_2,B_1B_2)$
$\Gamma_{T4}{}^1(^1B_2,A_1B_2)$
$\Gamma_{T5}{}^1(^3A_2,B_1B_2)$
$\Gamma_{T1}{}^1(^3A_2,A_1B_2)$
$\Gamma_{T5}{}^1(^1E,A_1E)$
$\Gamma_{T5}{}^1(^1E,B_1E)$
$\Gamma_{T5}{}^1(^3B_2,A_1B_2)$
$\Gamma_{T5}{}^1(^3B_2,A_1B_2)$
$\Gamma_{T4}{}^1(^3E,A_1E)$
$\Gamma_{T3}{}^1(^3E,A_1E)$
$\Gamma_{T5}{}^1(^3E,A_1E)$
$\Gamma_{T2}{}^1(^3E,A_1E)$
$\Gamma_{T1}{}^1(^3E,A_1E)$
$\Gamma_{T2}{}^1(^3E,B_1E)$
$\Gamma_{T5}{}^1(^3E,B_1E)$
$\Gamma_{T4}{}^1(^3E,B_1E)$
$\Gamma_{T3}{}^1(^3E,B_1E)$

$\Gamma_{T1}{}^1(^1A_1,B_2B_2)$

$\Gamma_{T1}{}^1(^1S_0)$

$\Gamma_{T2}{}^1(^1G_4)$
$\Gamma_{T1}{}^1(^1G_4)$
$\Gamma_{T4}{}^1(^1G_4)$
$\Gamma_{T3}{}^1(^1G_4)$
$\Gamma_{T1}{}^1(^1D_2)$
$\Gamma_{T5}{}^1(^3P_2)$
$\Gamma_{T1}{}^1(^3P_2)$
$\Gamma_{T5}{}^1(^1G_4)$
$\Gamma_{T5}{}^2(^1G_4)$
$\Gamma_{T4}{}^1(^3P_2)$
$\Gamma_{T3}{}^1(^3P_2)$
$\Gamma_{T5}{}^1(^3P_1)$
$\Gamma_{T1}{}^1(^3P_0)$
$\Gamma_{T2}{}^1(^3P_1)$
$\Gamma_{T5}{}^2(^3F_4)$
$\Gamma_{T4}{}^1(^3F_4)$
$\Gamma_{T3}{}^1(^3F_4)$
$\Gamma_{T5}{}^1(^3F_2)$
$\Gamma_{T5}{}^1(^1D_2)$
$\Gamma_{T1}{}^2(^1G_4)$
$\Gamma_{T4}{}^1(^1D_2)$
$\Gamma_{T3}{}^1(^1D_2)$
$\Gamma_{T4}{}^1(^3F_4)$
$\Gamma_{T1}{}^1(^3F_4)$
$\Gamma_{T2}{}^1(^3F_4)$
$\Gamma_{T1}{}^1(^3F_4)$
$\Gamma_{T1}{}^1(^3F_2)$
$\Gamma_{T2}{}^1(^3F_3)$
$\Gamma_{T5}{}^2(^3F_3)$
$\Gamma_{T4}{}^1(^3F_2)$
$\Gamma_{T3}{}^1(^3F_2)$

$\Gamma_{T5}{}^1(^1E,B_2E)$

$\Gamma_{T1}{}^1(^1A_1,E E)$

$\Gamma_{T3}{}^1(^1B_1,E E)$
$\Gamma_{T4}{}^1(^1B_2,E E)$
$\Gamma_{T1}{}^1(^3E,B_2E)$
$\Gamma_{T2}{}^1(^3E,B_2E)$
$\Gamma_{T5}{}^1(^3E,B_2E)$
$\Gamma_{T4}{}^1(^3E,B_2E)$
$\Gamma_{T3}{}^1(^3E,B_2E)$

$\Gamma_{T5}{}^1(^3F_3)$
$\Gamma_{T1}{}^2(^3F_4)$

$\Gamma_{T5}{}^1(^3A_2,E E)$
$\Gamma_{T1}{}^1(^3A_2,E E)$

E IN cm⁻¹

DQ cm⁻¹

2 D ELECTRONS, TETRAGONAL SYMMETRY   B=630 C=4*B DT=-1000 K=-1 DS=K*DT ZETA=210

82

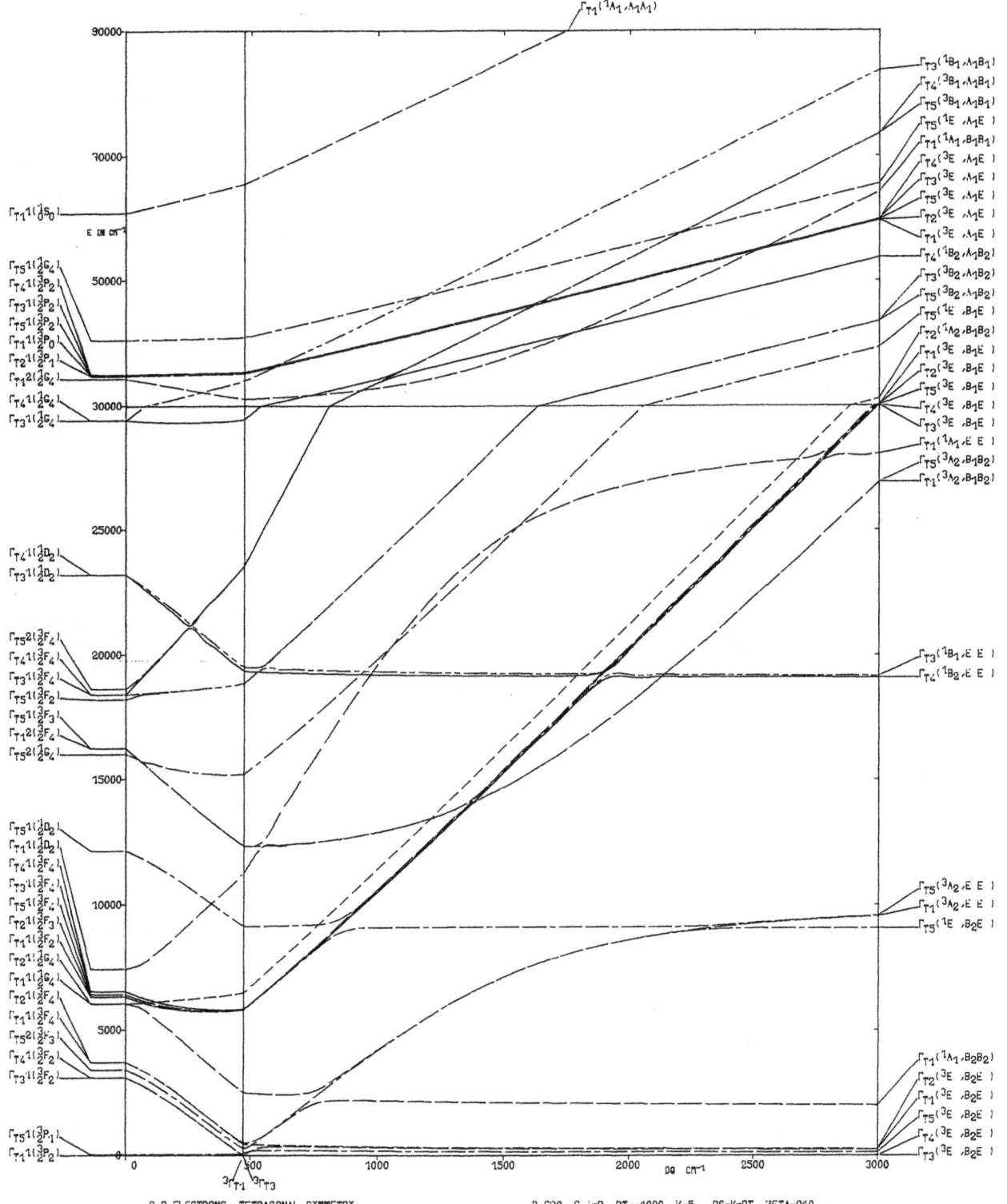

2 D ELECTRONS, TETRAGONAL SYMMETRY   B=630 C=4×B DT=-1000 K=5 DS=K×DT ZETA=210

83

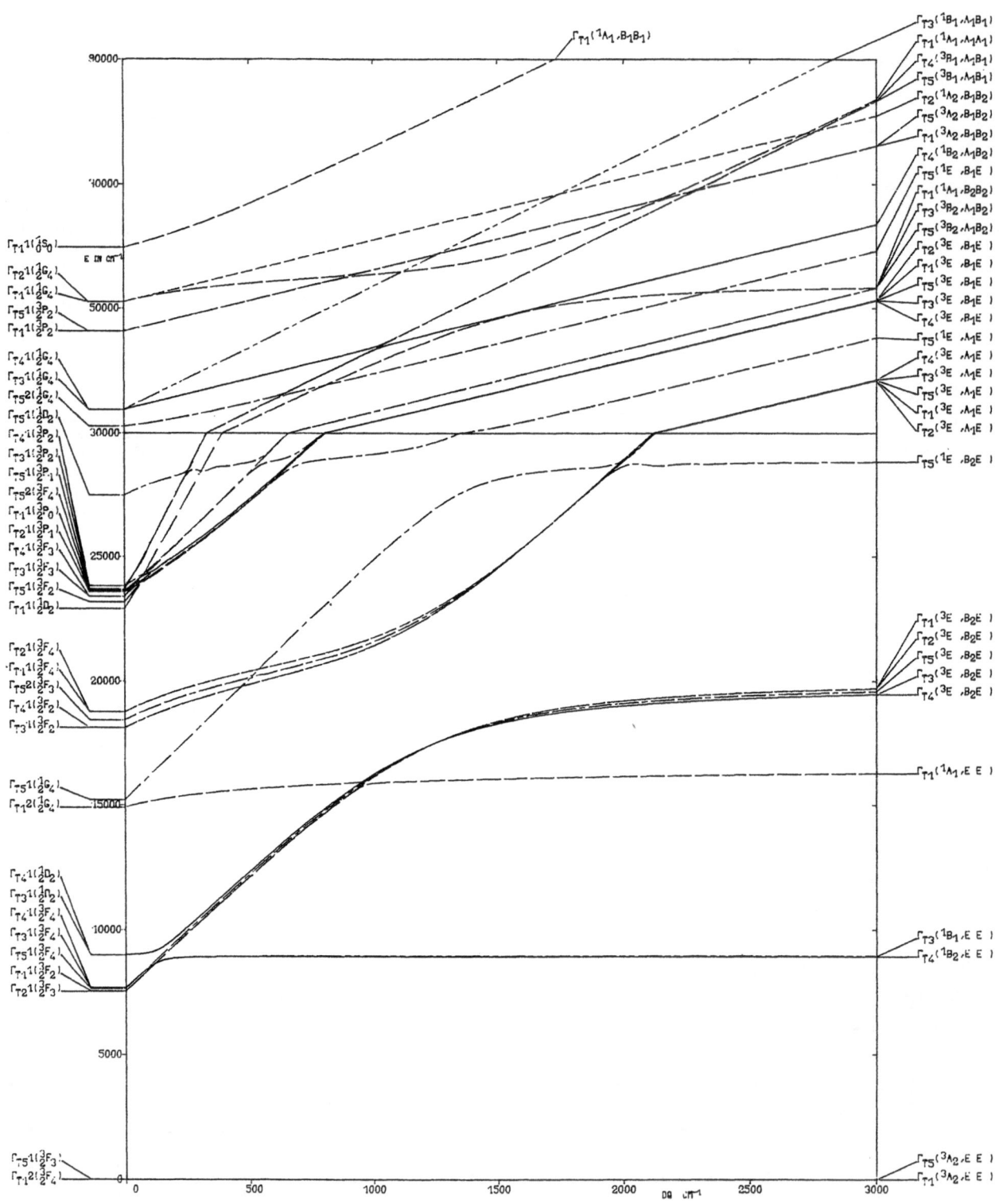

2 D ELECTRONS, TETRAGONAL SYMMETRY   B=630 C=4*B DT=-1000 K=-5 DS=K*DT ZETA=210

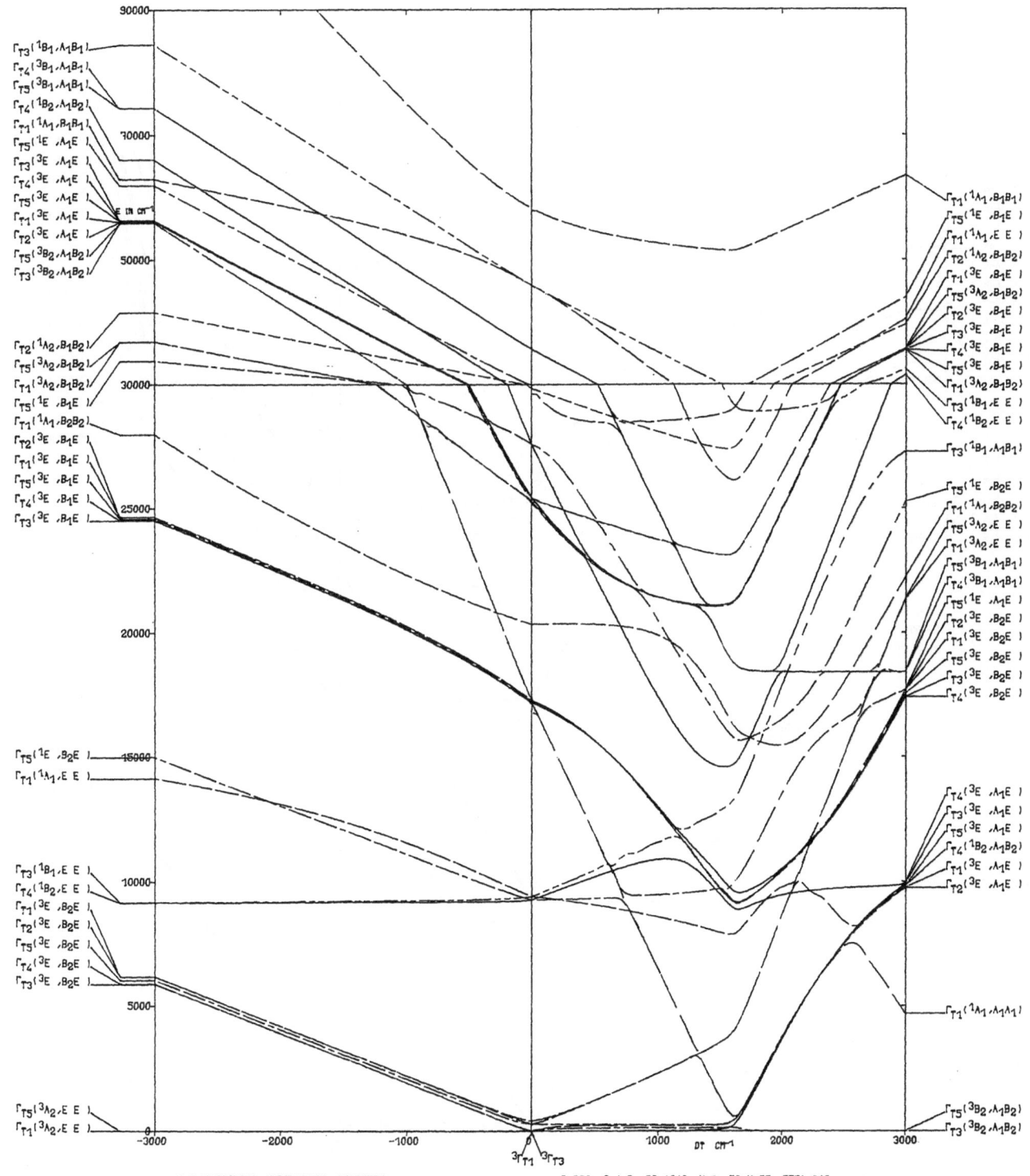

$\Gamma_{T3}(^1B_1, A_1B_1)$
$\Gamma_{T4}(^3B_1, A_1B_1)$
$\Gamma_{T5}(^3B_1, A_1B_1)$
$\Gamma_{T4}(^1B_2, A_1B_2)$
$\Gamma_{T1}(^1A_1, B_1B_1)$
$\Gamma_{T5}(^1E, A_1E)$
$\Gamma_{T3}(^3E, A_1E)$
$\Gamma_{T4}(^3E, A_1E)$
$\Gamma_{T5}(^3E, A_1E)$
$\Gamma_{T1}(^3E, A_1E)$
$\Gamma_{T2}(^3E, A_1E)$
$\Gamma_{T5}(^3B_2, A_1B_2)$
$\Gamma_{T3}(^3B_2, A_1B_2)$

$\Gamma_{T2}(^1A_2, B_1B_2)$
$\Gamma_{T3}(^3A_2, B_1B_2)$
$\Gamma_{T5}(^3A_2, B_1B_2)$
$\Gamma_{T1}(^1E, B_1E)$
$\Gamma_{T1}(^1A_1, B_2B_2)$
$\Gamma_{T2}(^3E, B_1E)$
$\Gamma_{T1}(^3E, B_1E)$
$\Gamma_{T5}(^3E, B_1E)$
$\Gamma_{T4}(^3E, B_1E)$
$\Gamma_{T3}(^3E, B_1E)$

$\Gamma_{T5}(^1E, B_2E)$
$\Gamma_{T1}(^1A_1, E E)$

$\Gamma_{T3}(^1B_1, E E)$
$\Gamma_{T4}(^1B_2, E E)$
$\Gamma_{T1}(^3E, B_2E)$
$\Gamma_{T2}(^3E, B_2E)$
$\Gamma_{T5}(^3E, B_2E)$
$\Gamma_{T4}(^3E, B_2E)$
$\Gamma_{T3}(^3E, B_2E)$

$\Gamma_{T5}(^3A_2, E E)$
$\Gamma_{T1}(^3A_2, E E)$

$\Gamma_{T1}(^1A_1, B_1B_1)$
$\Gamma_{T5}(^1E, B_1E)$
$\Gamma_{T1}(^1A_2, E E)$
$\Gamma_{T2}(^1A_2, B_1E)$
$\Gamma_{T1}(^3E, B_1E)$
$\Gamma_{T5}(^3A_2, B_1B_2)$
$\Gamma_{T2}(^3E, B_1E)$
$\Gamma_{T3}(^3E, B_1E)$
$\Gamma_{T4}(^3E, B_1E)$
$\Gamma_{T5}(^3E, B_1E)$
$\Gamma_{T1}(^3A_2, B_1B_2)$
$\Gamma_{T3}(^1B_1, E E)$
$\Gamma_{T4}(^1B_2, E E)$

$\Gamma_{T3}(^1B_1, A_1B_1)$

$\Gamma_{T5}(^1E, B_2E)$
$\Gamma_{T1}(^1A_1, B_2B_2)$
$\Gamma_{T5}(^3A_2, E E)$
$\Gamma_{T1}(^3A_2, E E)$
$\Gamma_{T5}(^3B_1, A_1B_1)$
$\Gamma_{T4}(^3B_1, A_1B_1)$
$\Gamma_{T5}(^1E, A_1E)$
$\Gamma_{T2}(^3E, B_2E)$
$\Gamma_{T1}(^3E, B_2E)$
$\Gamma_{T5}(^3E, B_2E)$
$\Gamma_{T3}(^3E, B_2E)$
$\Gamma_{T4}(^3E, B_2E)$

$\Gamma_{T4}(^3E, A_1E)$
$\Gamma_{T3}(^3E, A_1E)$
$\Gamma_{T5}(^3E, A_1E)$
$\Gamma_{T4}(^1B_2, A_1B_2)$
$\Gamma_{T1}(^3E, A_1E)$
$\Gamma_{T2}(^3E, A_1E)$

$\Gamma_{T1}(^1A_1, A_1A_1)$

$\Gamma_{T5}(^3B_2, A_1B_2)$
$\Gamma_{T3}(^3B_2, A_1B_2)$

$\Gamma_{T5}(^3A_2, E E)$
$\Gamma_{T1}(^3A_2, E E)$

$^3\Gamma_{T1}$ $^3\Gamma_{T3}$

-3000   -2000   -1000   1000   DT CM⁻¹   2000   3000

90000
80000
70000
E IN CM⁻¹
50000
30000
25000
20000
15000
10000
5000
0

2 D ELECTRONS, TETRAGONAL SYMMETRY   B=630 C=4*B DQ=1840 K=1 DS=K*DT ZETA=210

85

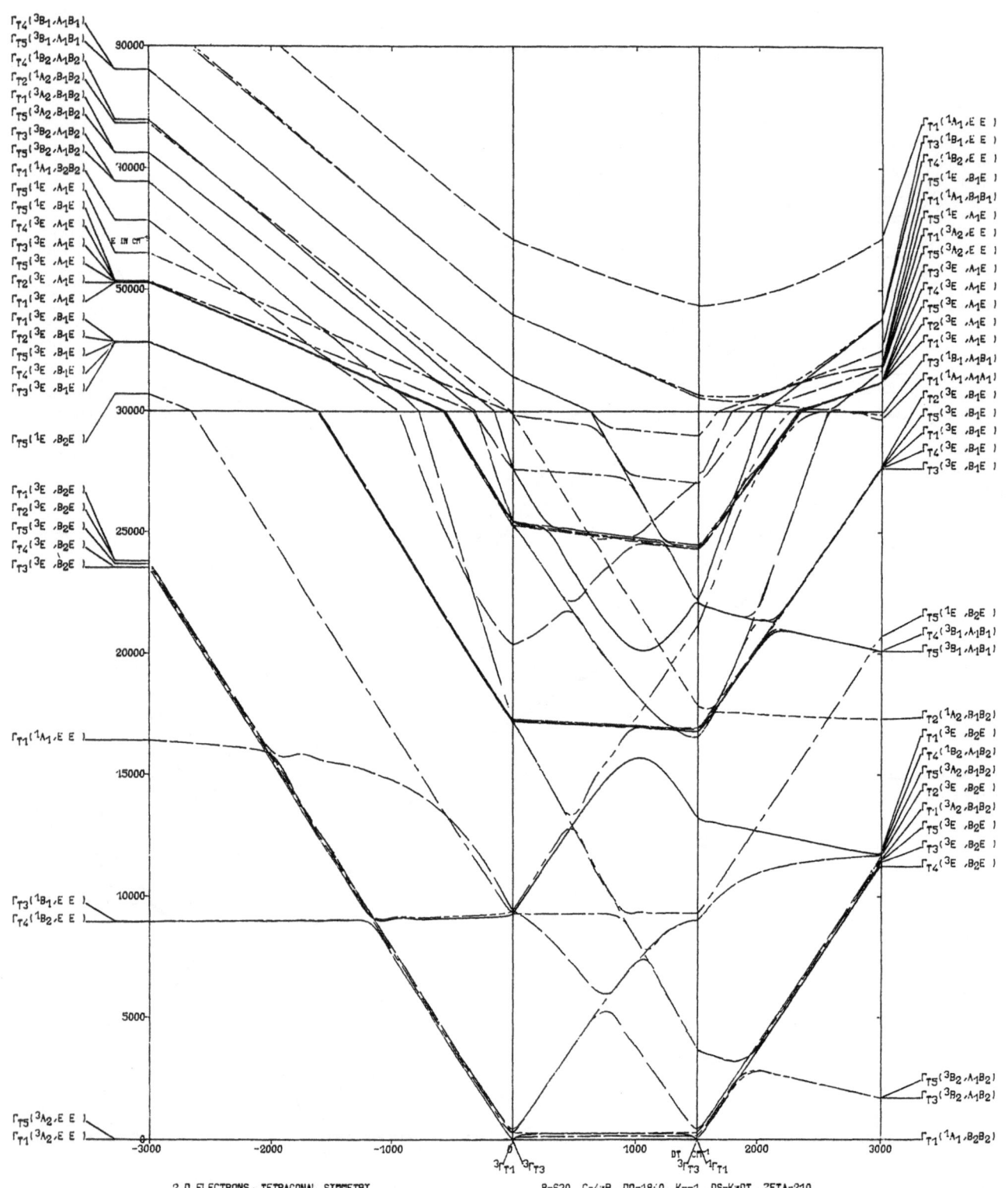

2 D ELECTRONS, TETRAGONAL SYMMETRY   B=630 C=4*B DQ=1840 K=-1 DS=K*DT ZETA=210

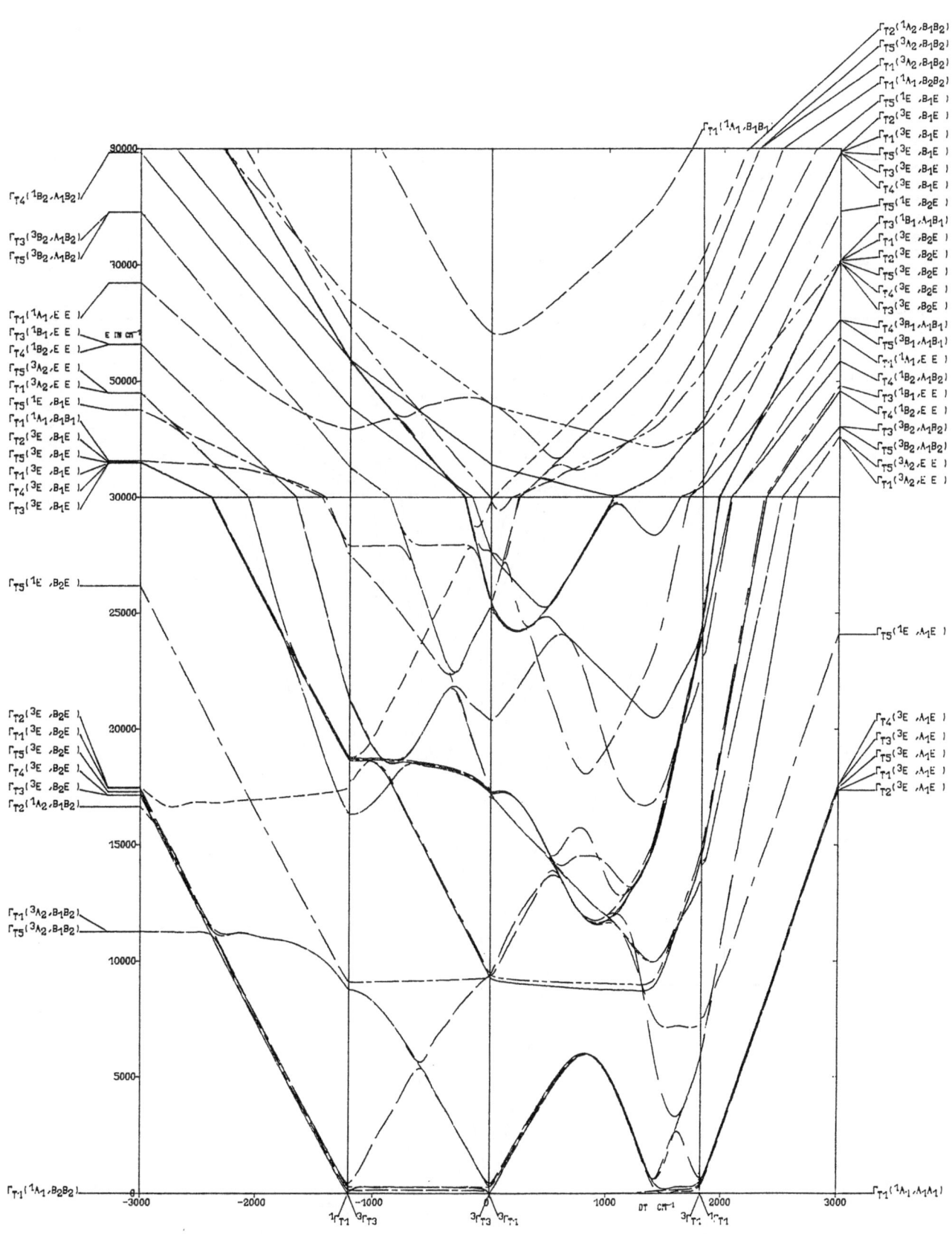

$\Gamma_{T4}(^1B_2,A_1B_2)$
$\Gamma_{T3}(^3B_2,A_1B_2)$
$\Gamma_{T5}(^3B_2,A_1B_2)$
$\Gamma_{T1}(^1A_1,E\ E\ )$
$\Gamma_{T3}(^1B_1,E\ E\ )$
$\Gamma_{T4}(^1B_2,E\ E\ )$
$\Gamma_{T5}(^3A_2,E\ E\ )$
$\Gamma_{T1}(^3A_2,E\ E\ )$
$\Gamma_{T5}(^1E\ ,B_2\ )$
$\Gamma_{T1}(^1A_1,B_1B_1)$
$\Gamma_{T2}(^3E\ ,B_1E\ )$
$\Gamma_{T5}(^3E\ ,B_1E\ )$
$\Gamma_{T1}(^3E\ ,B_1E\ )$
$\Gamma_{T4}(^3E\ ,B_1E\ )$
$\Gamma_{T3}(^3E\ ,B_1E\ )$

$\Gamma_{T5}(^1E\ ,B_2E\ )$

$\Gamma_{T2}(^3E\ ,B_2E\ )$
$\Gamma_{T1}(^3E\ ,B_2E\ )$
$\Gamma_{T5}(^3E\ ,B_2E\ )$
$\Gamma_{T4}(^3E\ ,B_2E\ )$
$\Gamma_{T3}(^3E\ ,B_2E\ )$
$\Gamma_{T2}(^1A_2,B_1B_2)$

$\Gamma_{T1}(^3A_2,B_1B_2)$
$\Gamma_{T5}(^3A_2,B_1B_2)$

$\Gamma_{T1}(^1A_1,B_2B_2)$

$\Gamma_{T2}(^1A_2,B_1B_2)$
$\Gamma_{T5}(^3A_2,B_1B_2)$
$\Gamma_{T1}(^3A_2,B_1B_2)$
$\Gamma_{T1}(^1A_1,B_2B_2)$
$\Gamma_{T5}(^1E\ ,B_1E\ )$
$\Gamma_{T2}(^3E\ ,B_1E\ )$
$\Gamma_{T1}(^3E\ ,B_1E\ )$
$\Gamma_{T5}(^3E\ ,B_1E\ )$
$\Gamma_{T3}(^3E\ ,B_1E\ )$
$\Gamma_{T4}(^3E\ ,B_1E\ )$
$\Gamma_{T5}(^1E\ ,B_2E\ )$
$\Gamma_{T3}(^1B_1,A_1B_1)$
$\Gamma_{T1}(^3E\ ,B_2E\ )$
$\Gamma_{T2}(^3E\ ,B_2E\ )$
$\Gamma_{T5}(^3E\ ,B_2E\ )$
$\Gamma_{T4}(^3E\ ,B_2E\ )$
$\Gamma_{T3}(^3E\ ,B_2E\ )$
$\Gamma_{T4}(^3B_1,A_1B_1)$
$\Gamma_{T5}(^3B_1,A_1B_1)$
$\Gamma_{T1}(^1A_1,E\ E\ )$
$\Gamma_{T4}(^1B_2,A_1B_2)$
$\Gamma_{T3}(^1B_1,E\ E\ )$
$\Gamma_{T4}(^1B_2,E\ E\ )$
$\Gamma_{T3}(^3B_2,A_1B_2)$
$\Gamma_{T5}(^3B_2,A_1B_2)$
$\Gamma_{T5}(^3A_2,E\ E\ )$
$\Gamma_{T1}(^3A_2,E\ E\ )$

$\Gamma_{T1}(^1A_1,B_1B_1)$

$\Gamma_{T2}(^1A_2,B_1B_2)$
$\Gamma_{T5}(^3A_2,B_1B_2)$
$\Gamma_{T1}(^3A_2,B_1B_2)$
$\Gamma_{T1}(^1A_1,B_2B_2)$
$\Gamma_{T5}(^1E\ ,B_1E\ )$
$\Gamma_{T2}(^3E\ ,B_1E\ )$
$\Gamma_{T1}(^3E\ ,B_1E\ )$
$\Gamma_{T5}(^3E\ ,B_1E\ )$
$\Gamma_{T3}(^3E\ ,B_1E\ )$
$\Gamma_{T4}(^3E\ ,B_1E\ )$

$\Gamma_{T5}(^1E\ ,A_1E\ )$

$\Gamma_{T4}(^3E\ ,A_1E\ )$
$\Gamma_{T3}(^3E\ ,A_1E\ )$
$\Gamma_{T5}(^3E\ ,A_1E\ )$
$\Gamma_{T1}(^3E\ ,A_1E\ )$
$\Gamma_{T2}(^3E\ ,A_1E\ )$

$\Gamma_{T1}(^1A_1,A_1A_1)$

E IN CM^{-1}

90000
70000
50000
30000
25000
20000
15000
10000
5000
0

-3000   -2000   -1000   0   1000   2000   3000

DT CM^{-1}

$^1\Gamma_{T1}$ $^3\Gamma_{T3}$ $^3\Gamma_{T3}$ $^3\Gamma_{T1}$ $^3\Gamma_{T1}$ $^1\Gamma_{T1}$

2 D ELECTRONS, TETRAGONAL SYMMETRY   B=630   C=4*B   DQ=1840   K=5   DS=K*DT   ZETA=210

87

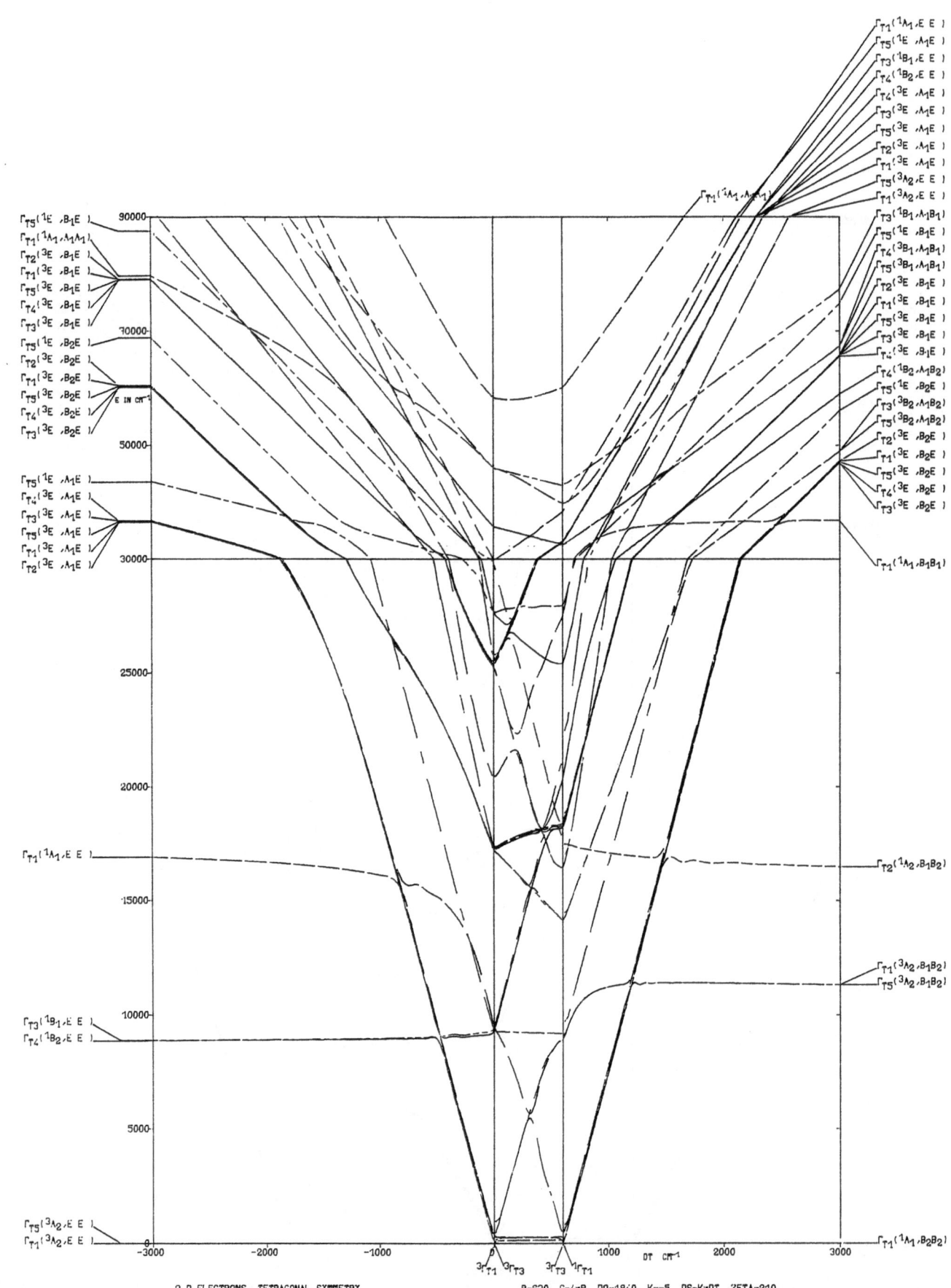

2 D ELECTRONS, TETRAGONAL SYMMETRY   B=630 C=4×B DQ=1840 K=-5 DS=K×DT ZETA=210

2 D ELECTRONS, TRIGONAL SYMMETRY

B=630   C=4×B   ZETA=210

ENERGY AS FUNCTION OF DQ
DT=0, 500, -500, 1000, -1000
K=1, -1, 5, -5

ENERGY AS FUNCTION OF DT
DQ=1840
K=1, -1, 5, -5

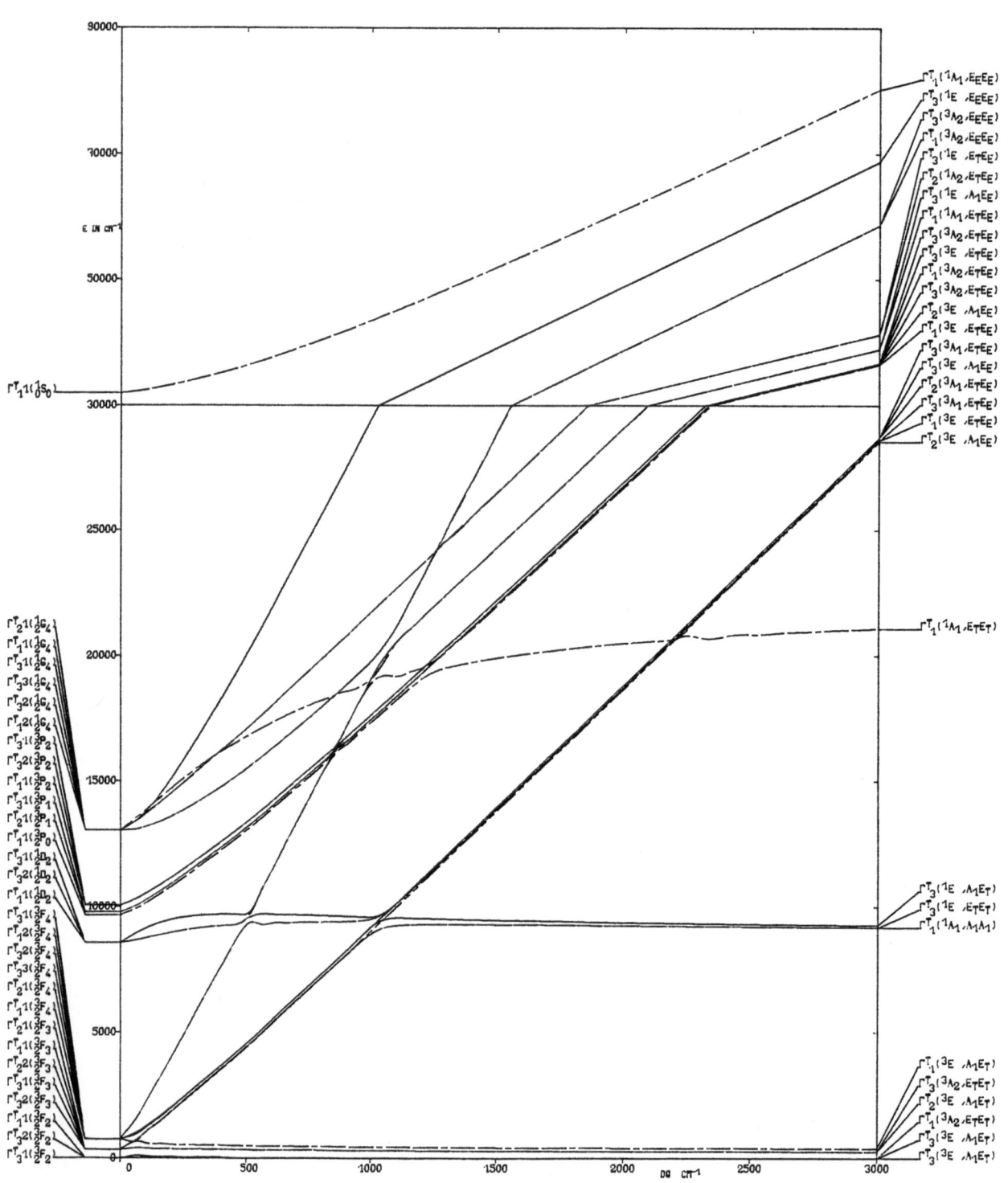

90

2 D ELECTRONS, TRIGONAL SYMMETRY

B=630 C=4×B DT=0 K=0 DS=K×DT ZETA=210

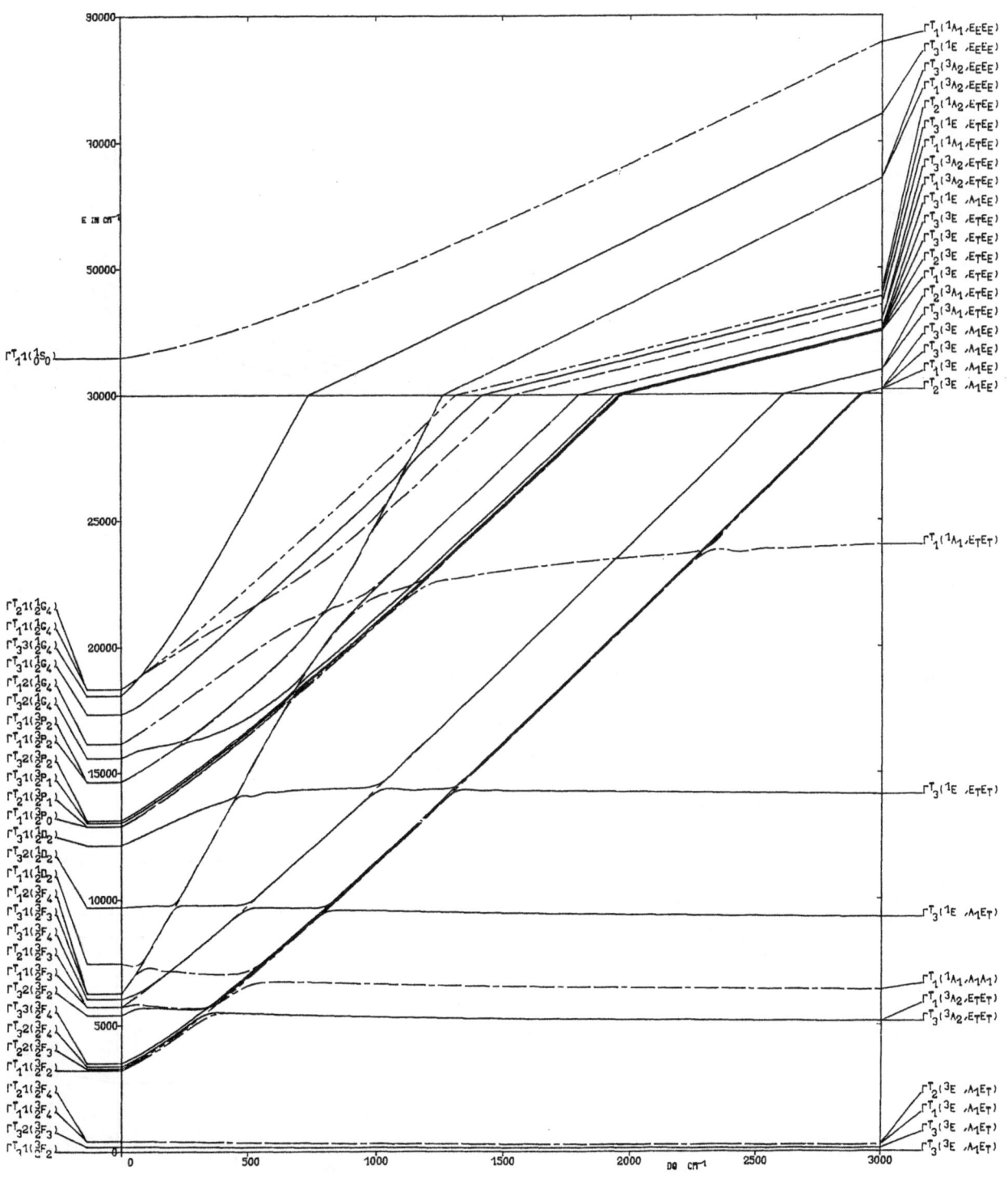

2 D ELECTRONS, TRIGONAL SYMMETRY   B=630 C=4×B DT=500 K=1 DS=K×DT ZETA=210

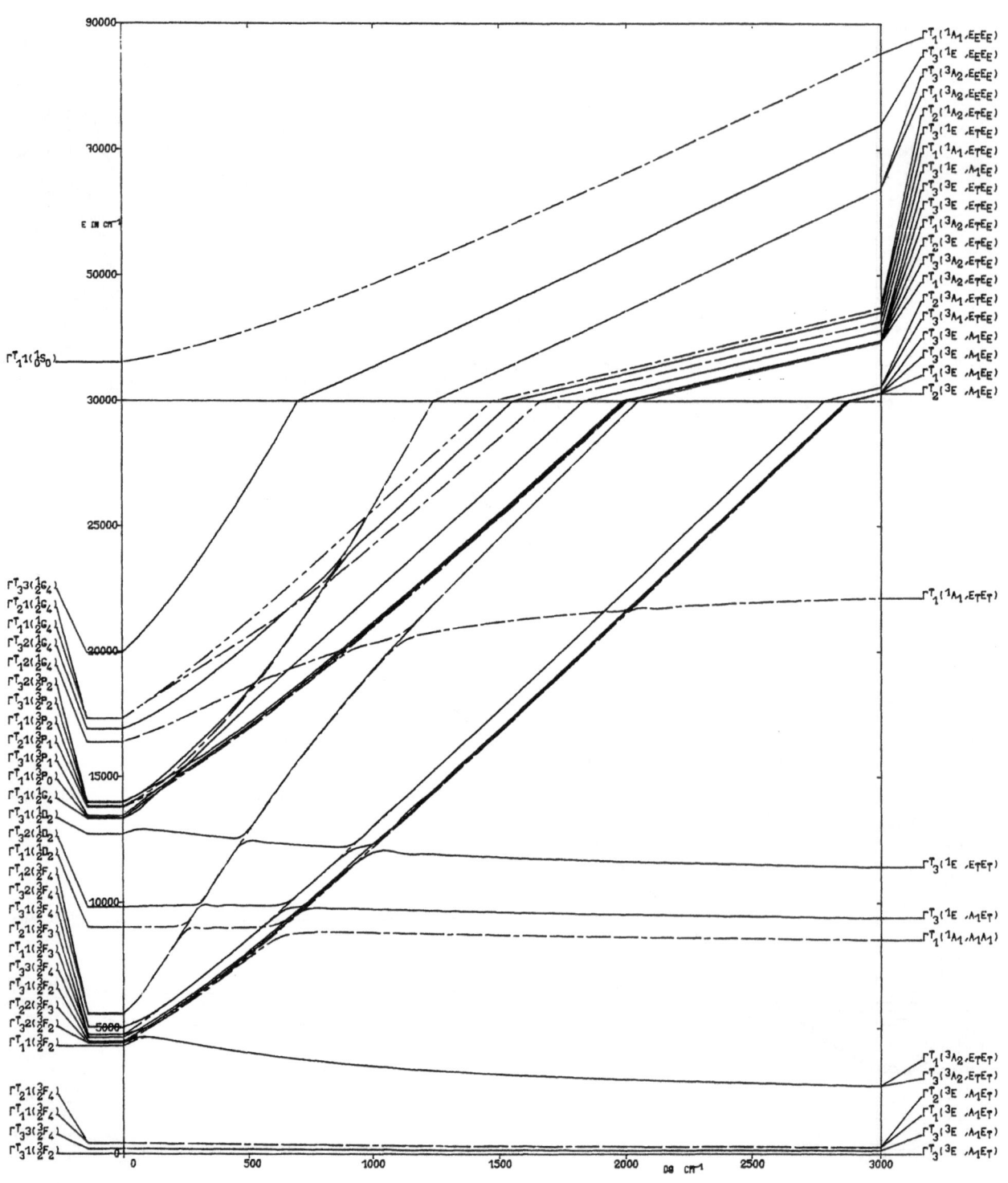

2 D ELECTRONS, TRIGONAL SYMMETRY

B=630   C=4*B   DT=500   K=-1   DS=K*DT   ZETA=210

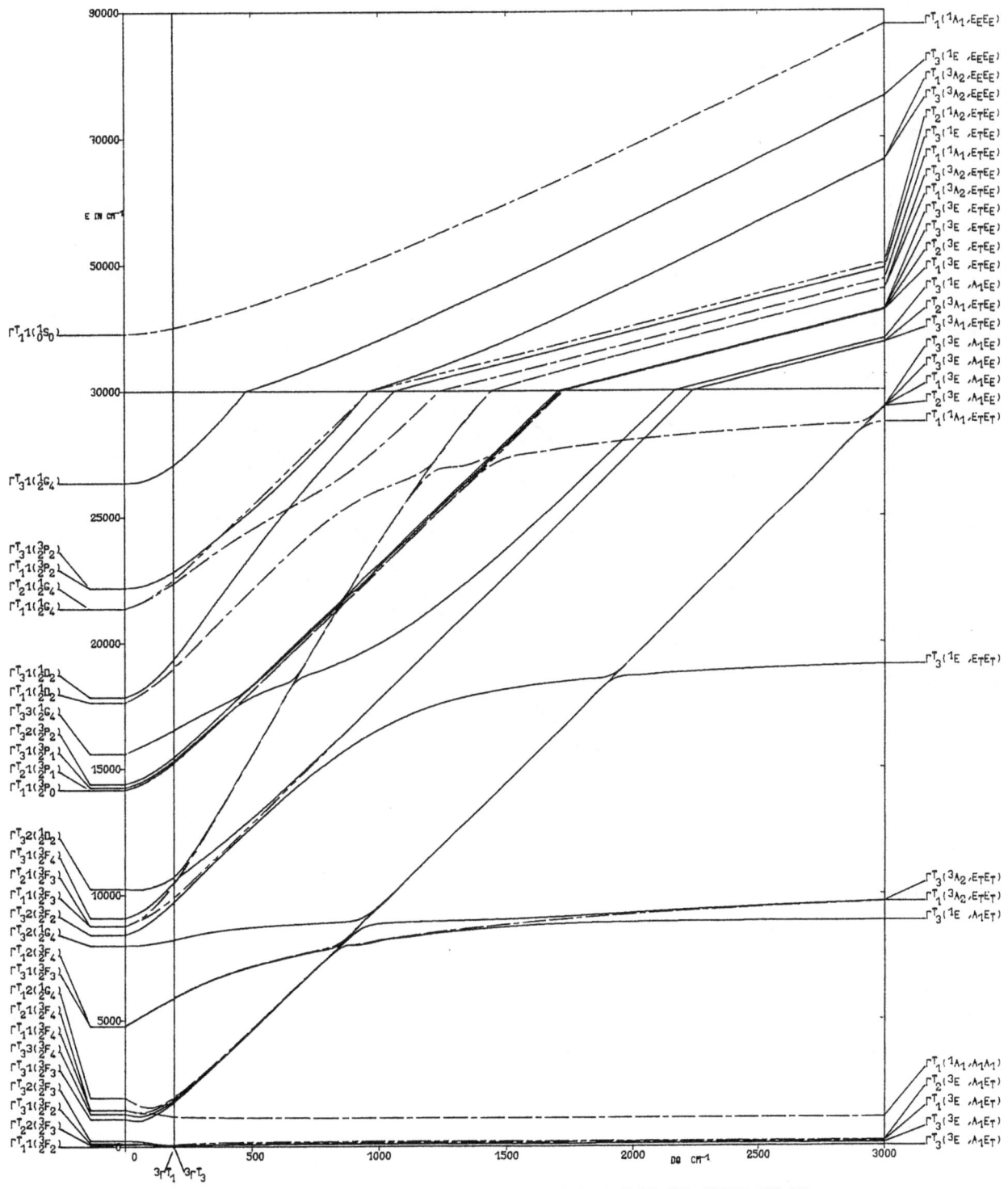

2 D ELECTRONS, TRIGONAL SYMMETRY   B=630 C=4*B DT=500 K=5 DS=K*DT ZETA=210

93

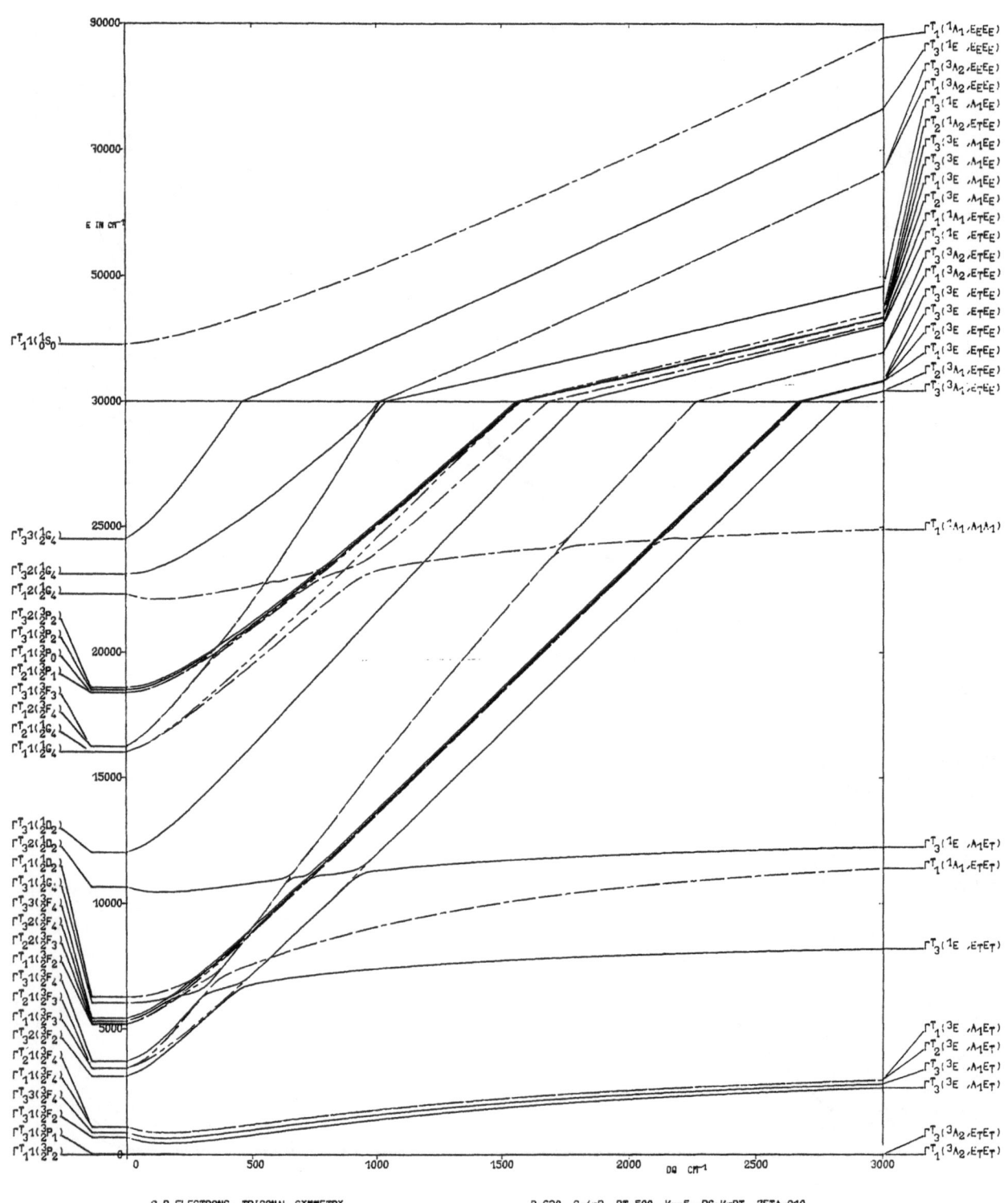

2 D ELECTRONS, TRIGONAL SYMMETRY   B=630 C=4×B DT=500 K=-5 DS=K×DT ZETA=210

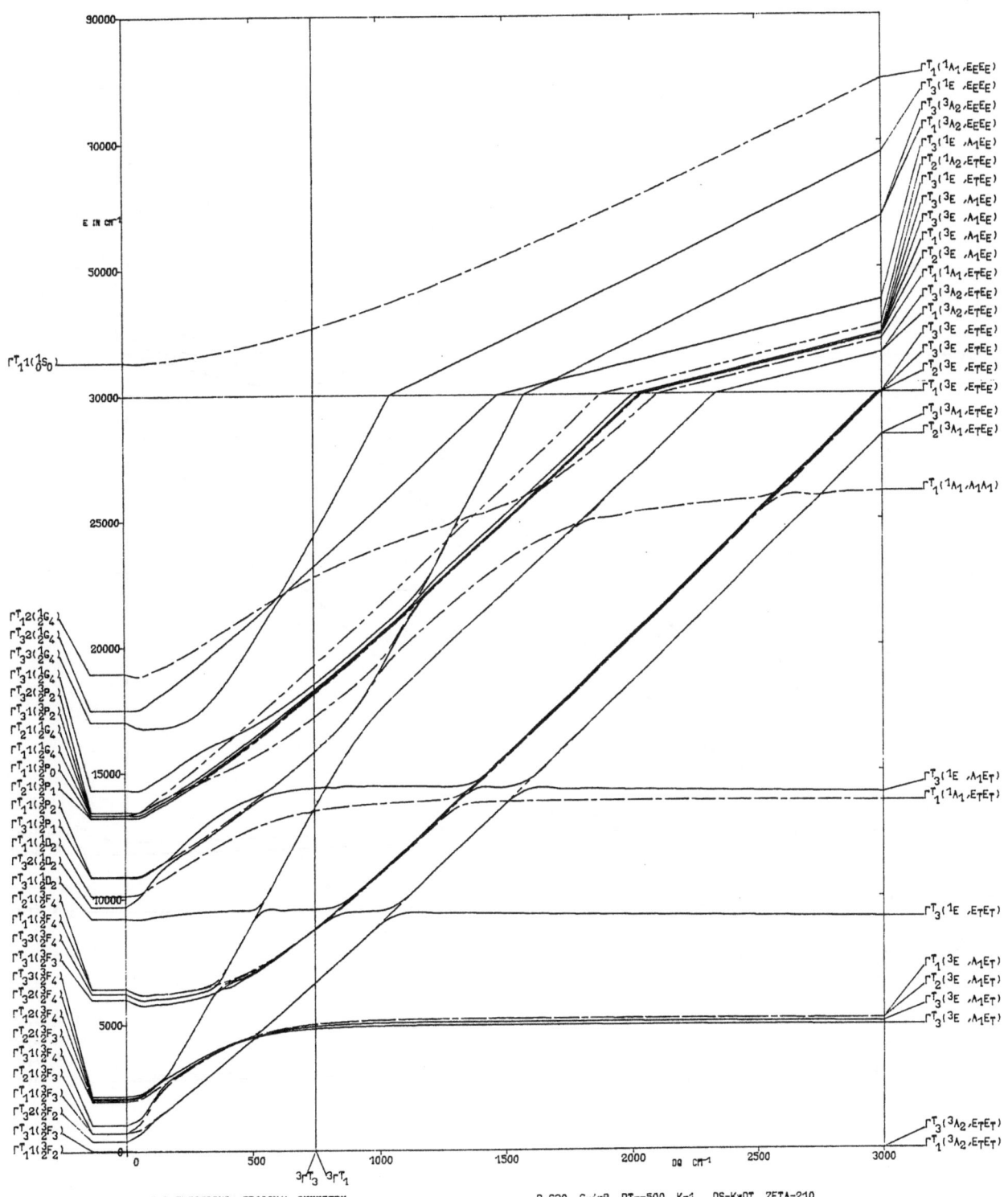

2 D ELECTRONS, TRIGONAL SYMMETRY   B=630 C=4×B DT=-500 K=1   DS=K×DT ZETA=210

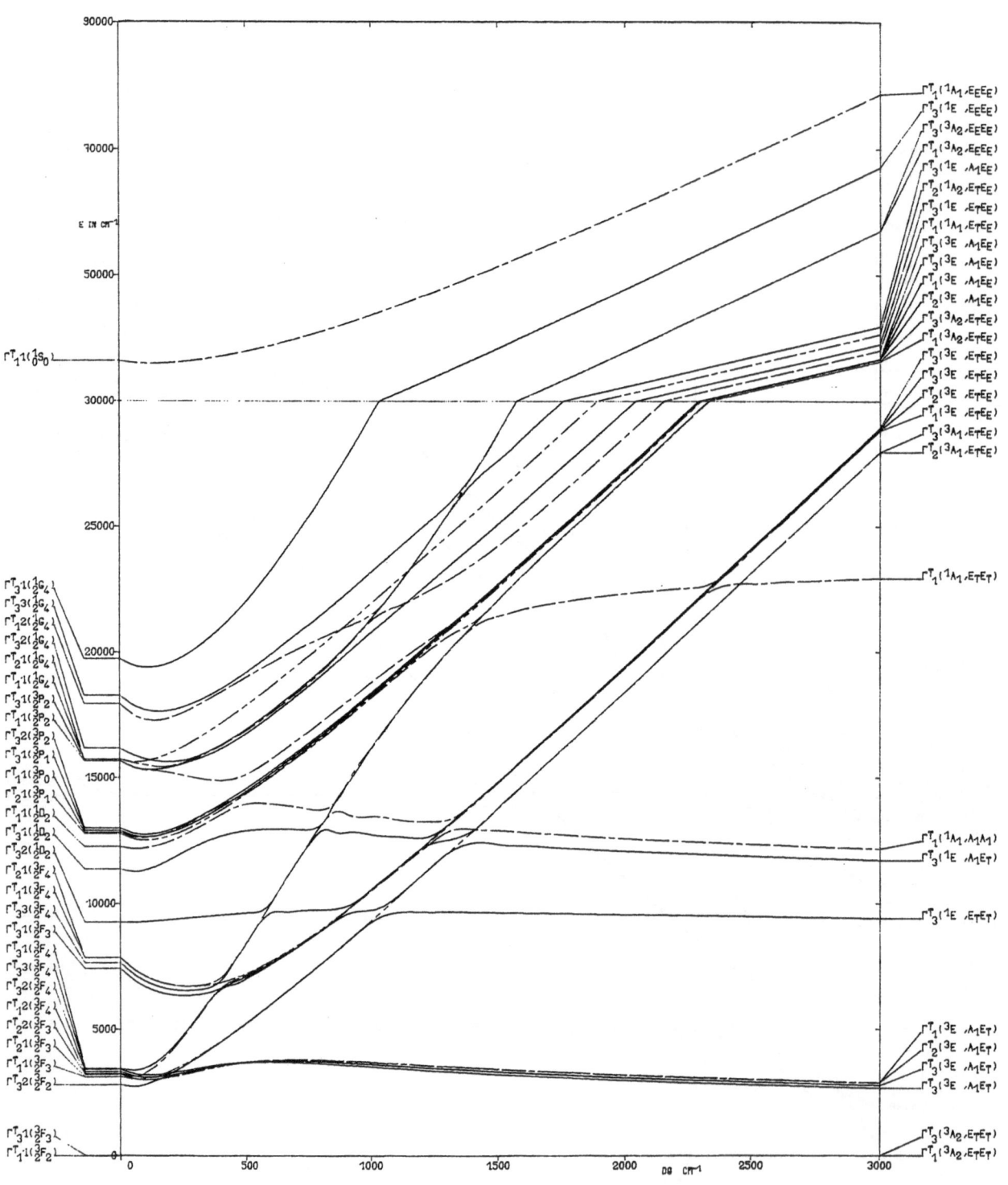

96

2 D ELECTRONS, TRIGONAL SYMMETRY

B=630   C=4∗B   DT=-500   K=-1   DS=K∗DT   2ETA=210

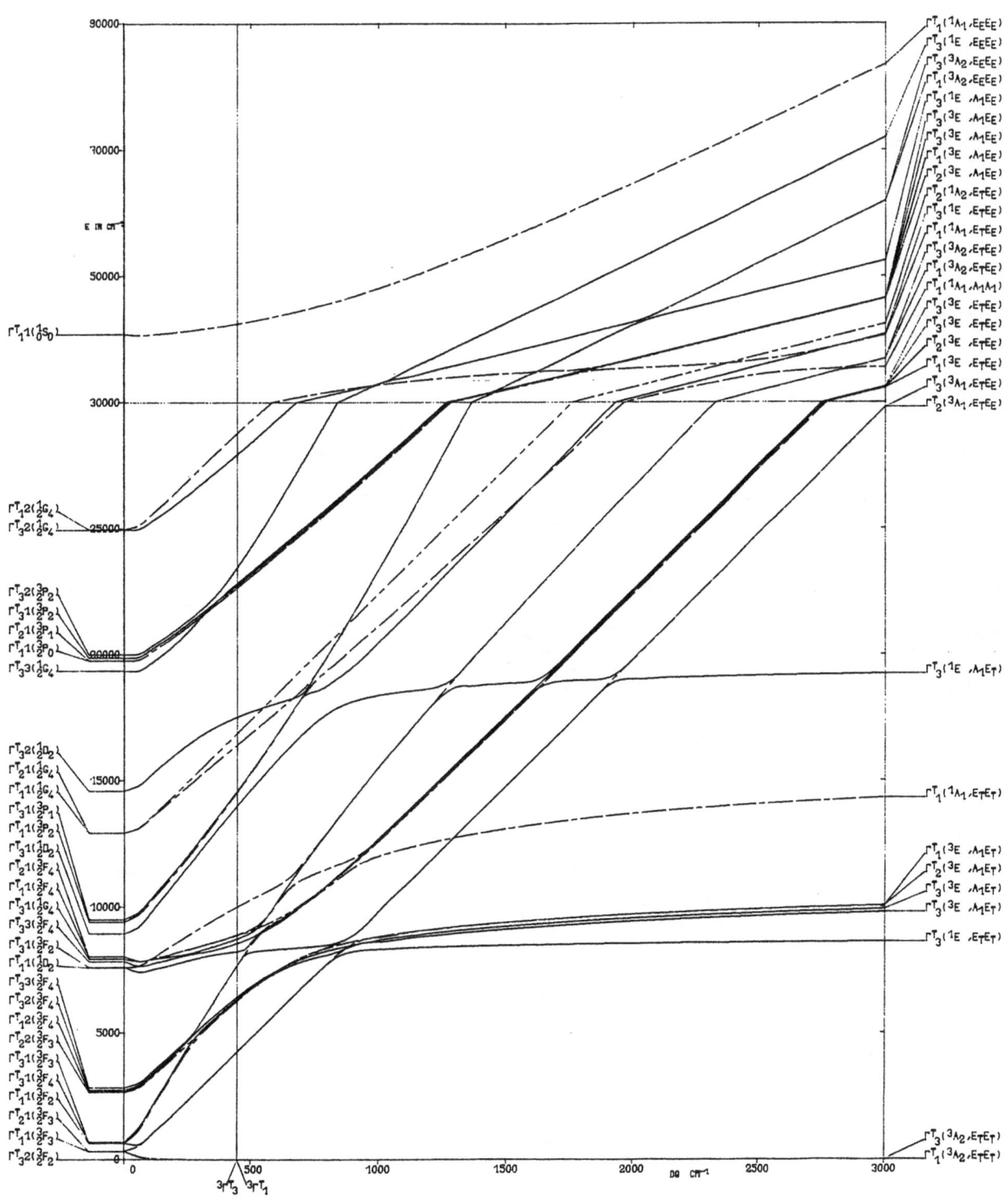

2 D ELECTRONS, TRIGONAL SYMMETRY   B=630 C=4×B DT=-500 K=5 DS=K×DT ZETA=210

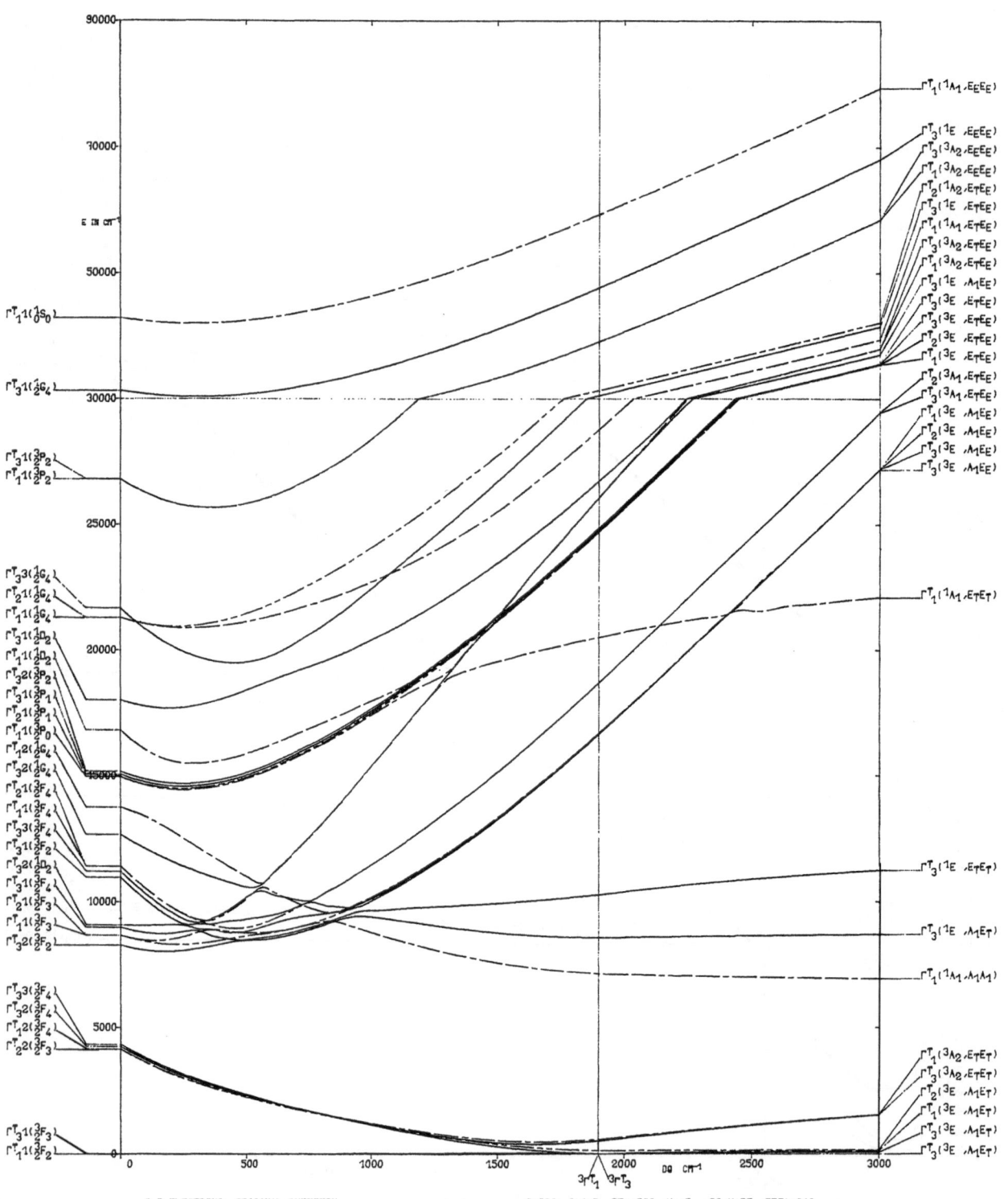

2 D ELECTRONS, TRIGONAL SYMMETRY   B=630   C=4×B   DT=-500   K=-5   DS=K×DT   ZETA=210

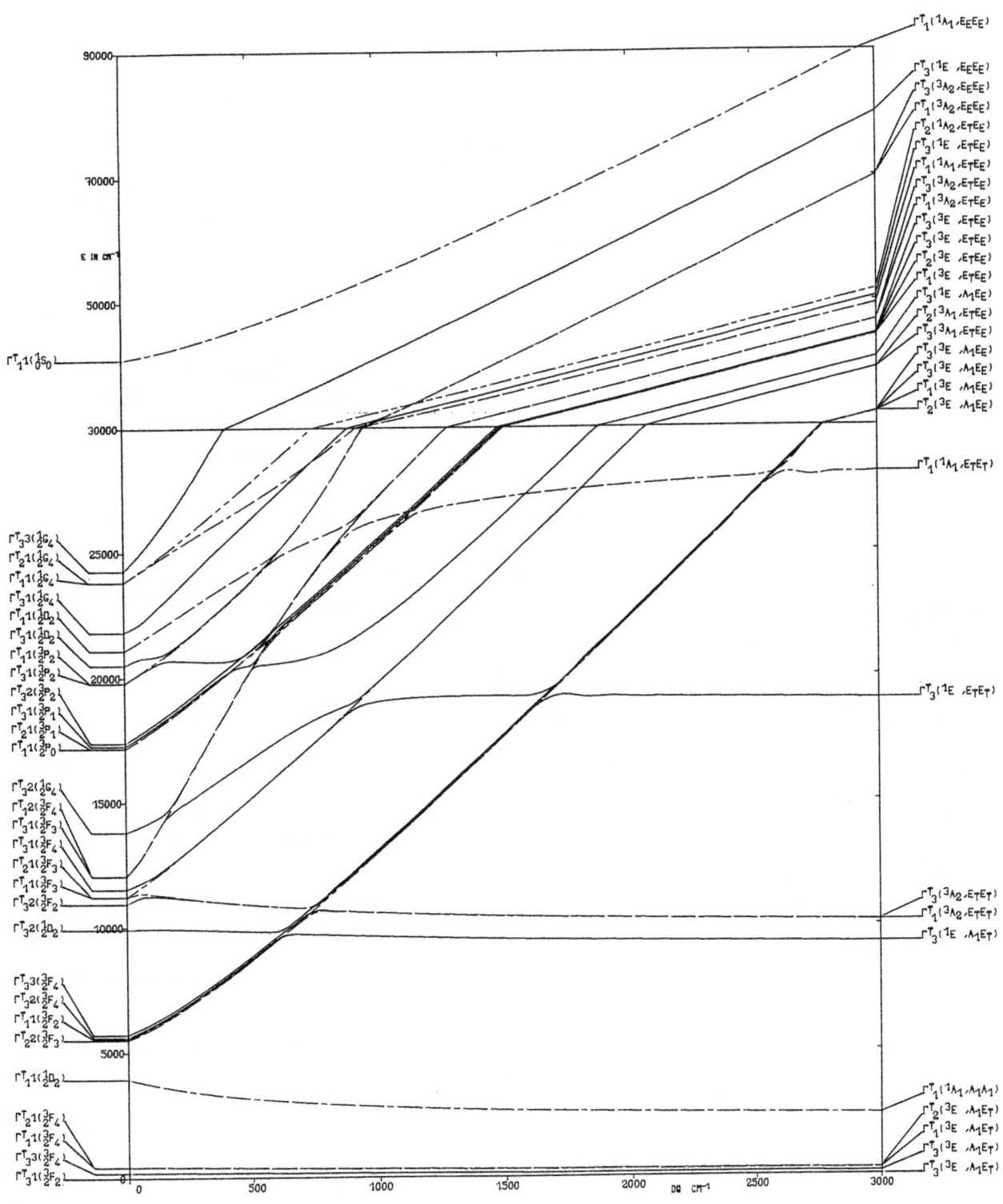

2 D ELECTRONS, TRIGONAL SYMMETRY   B=630 C=4×B DT=1000 K=1   DS=K×DT ZETA=210

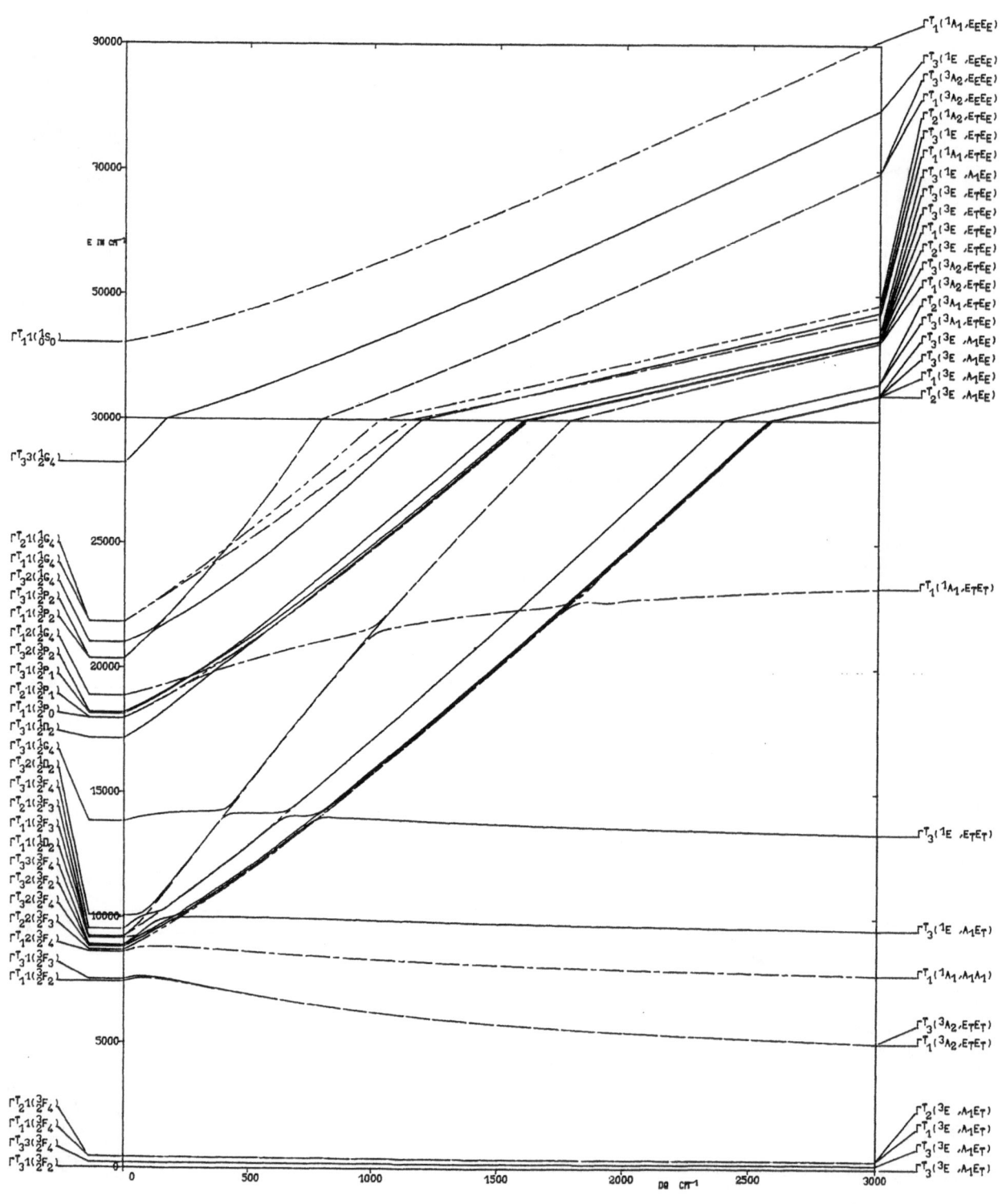

100

2 D ELECTRONS, TRIGONAL SYMMETRY

B=630   C=4×B   DT=1000   K=-1   DS=K×DT   ZETA=210

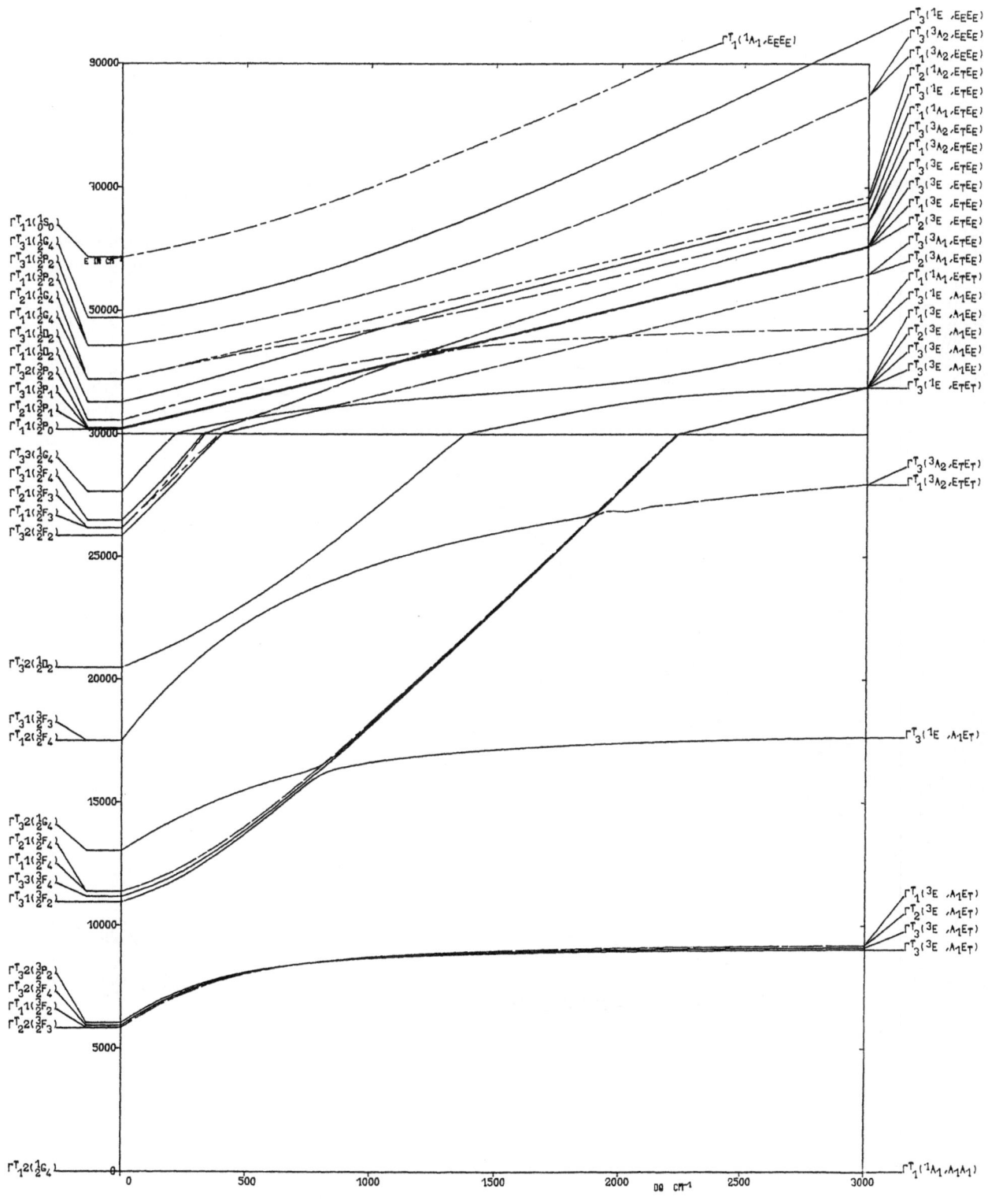

2 D ELECTRONS, TRIGONAL SYMMETRY   B=630 C=4*B DT=1000 K=5 DS=K*DT ZETA=210

101

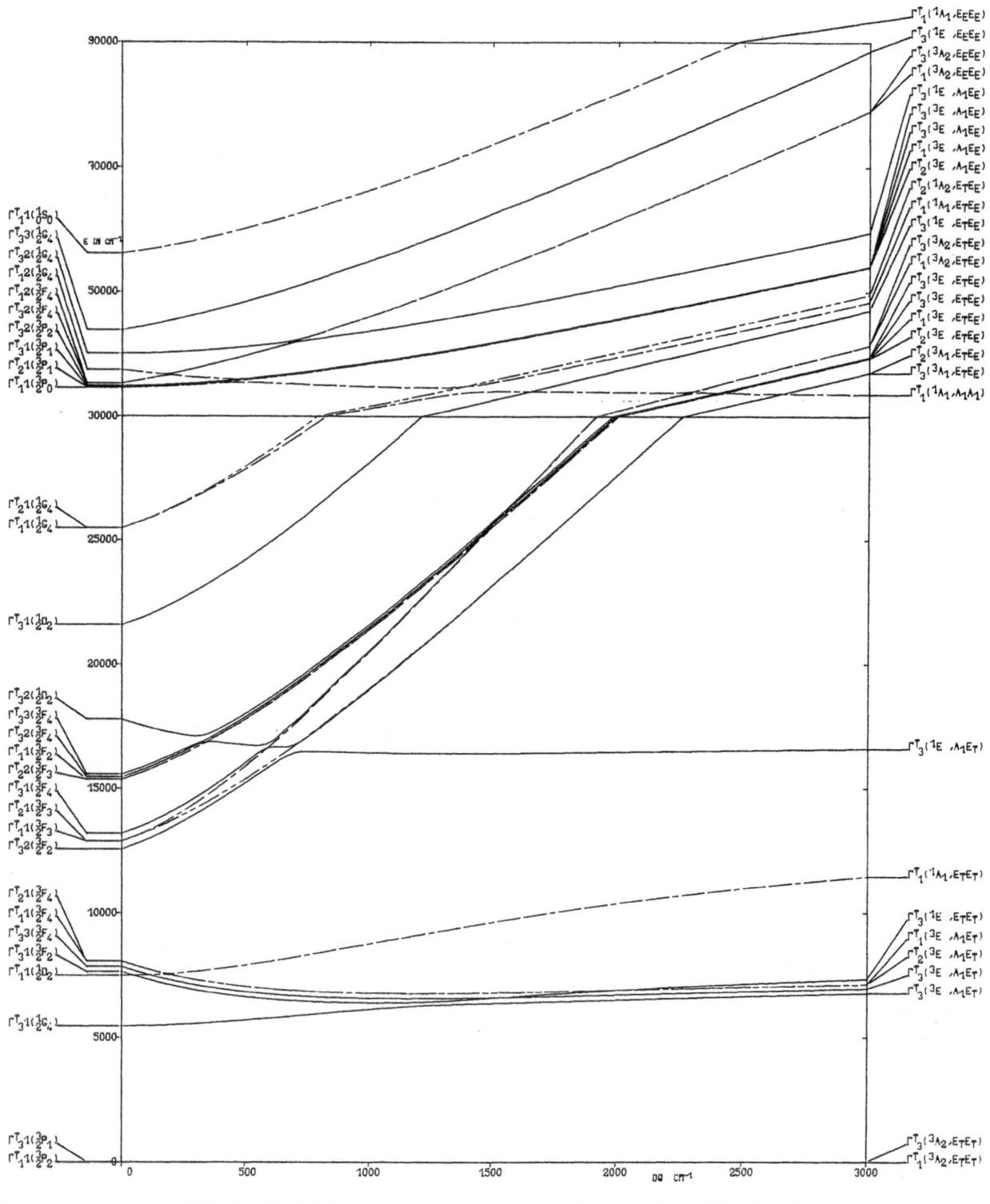

2 D ELECTRONS, TRIGONAL SYMMETRY   B=630 C=4×B DT=1000 K=·5 DS=K×DT ZETA=210

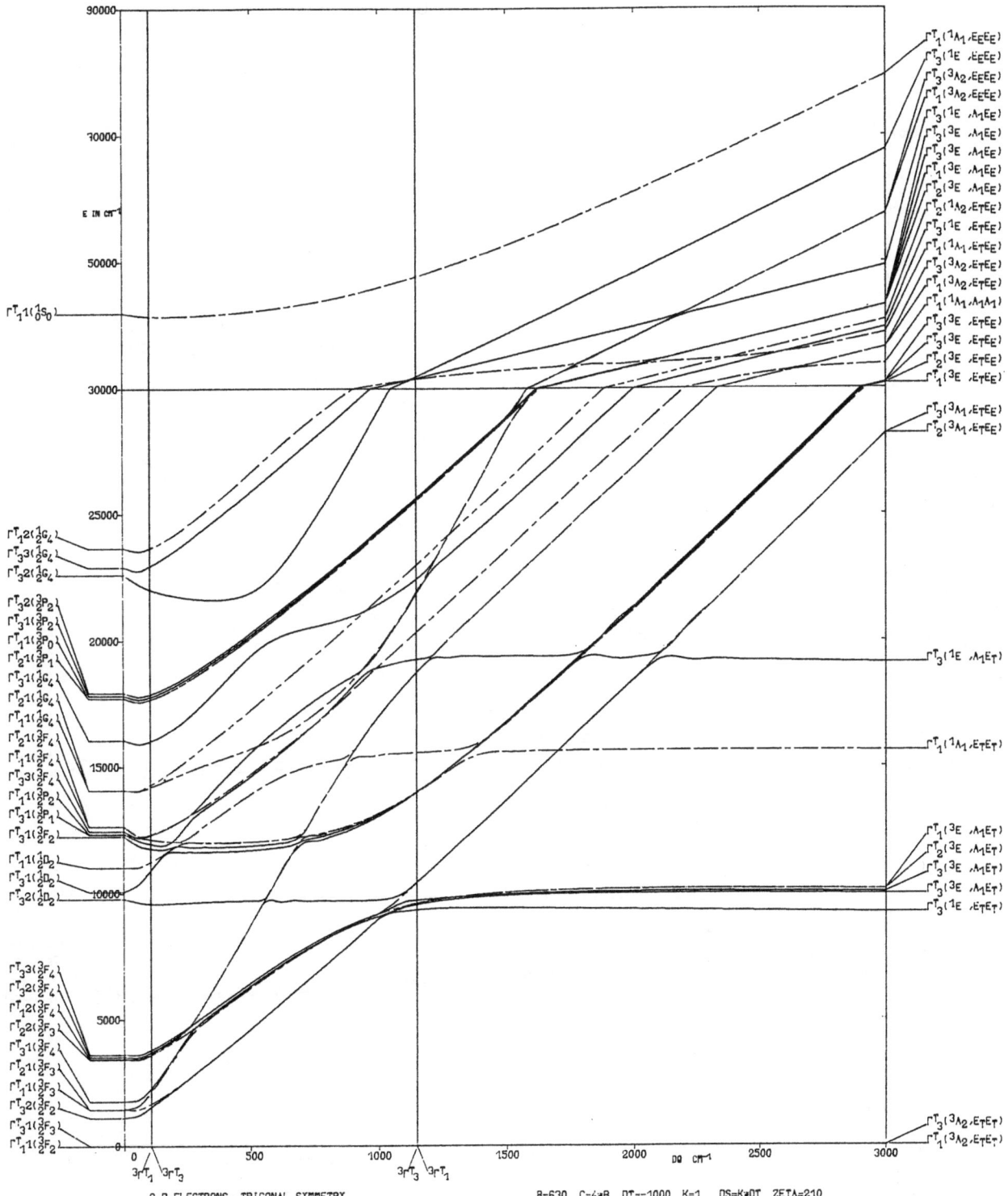

2 D ELECTRONS, TRIGONAL SYMMETRY   B=630 C=4xB DT=-1000 K=1   DS=KxDT ZETA=210

103

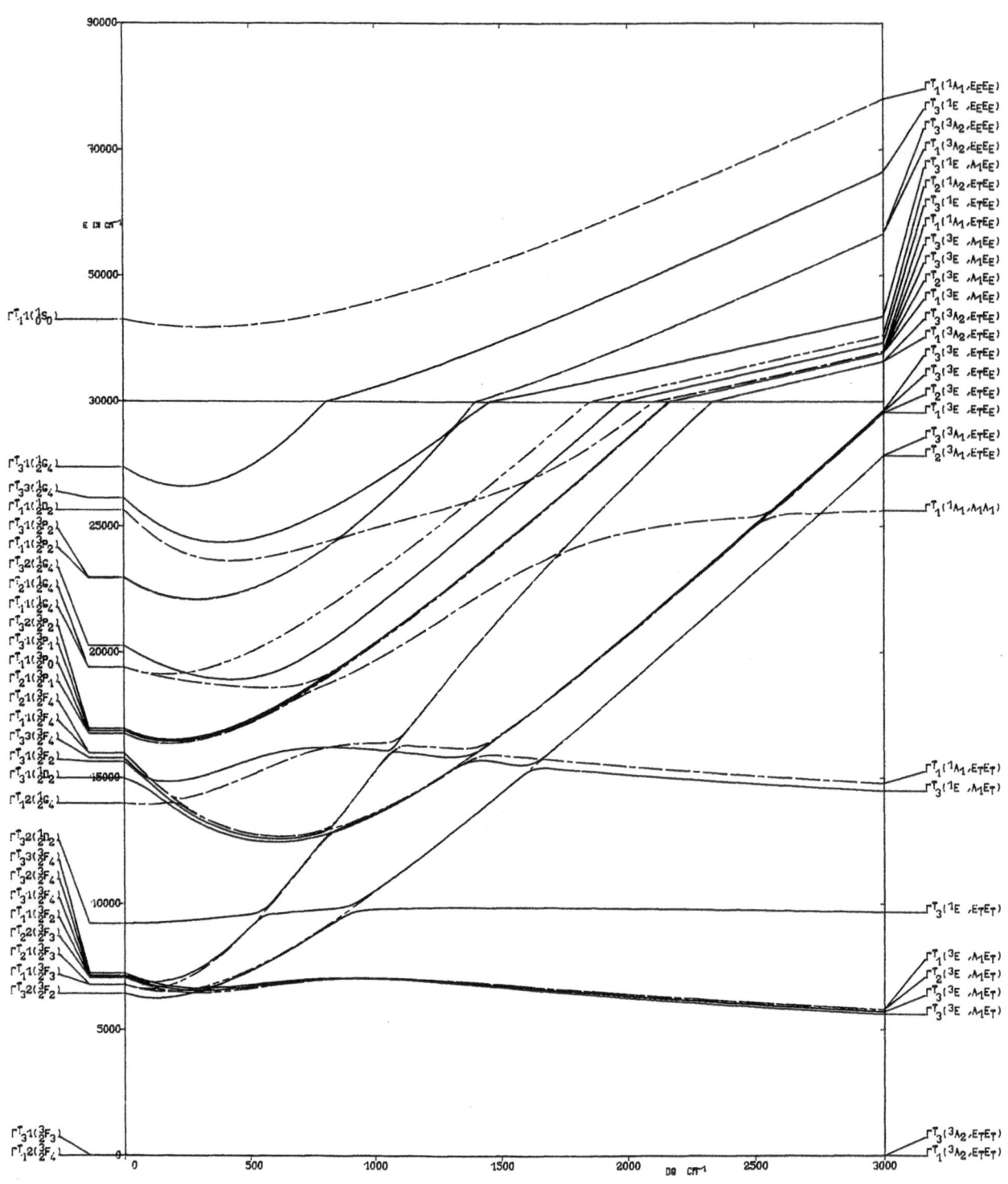

104

2 D ELECTRONS, TRIGONAL SYMMETRY

B=630 C=4×B DT=-1000 K=-1 DS=K×DT 2ETA=210

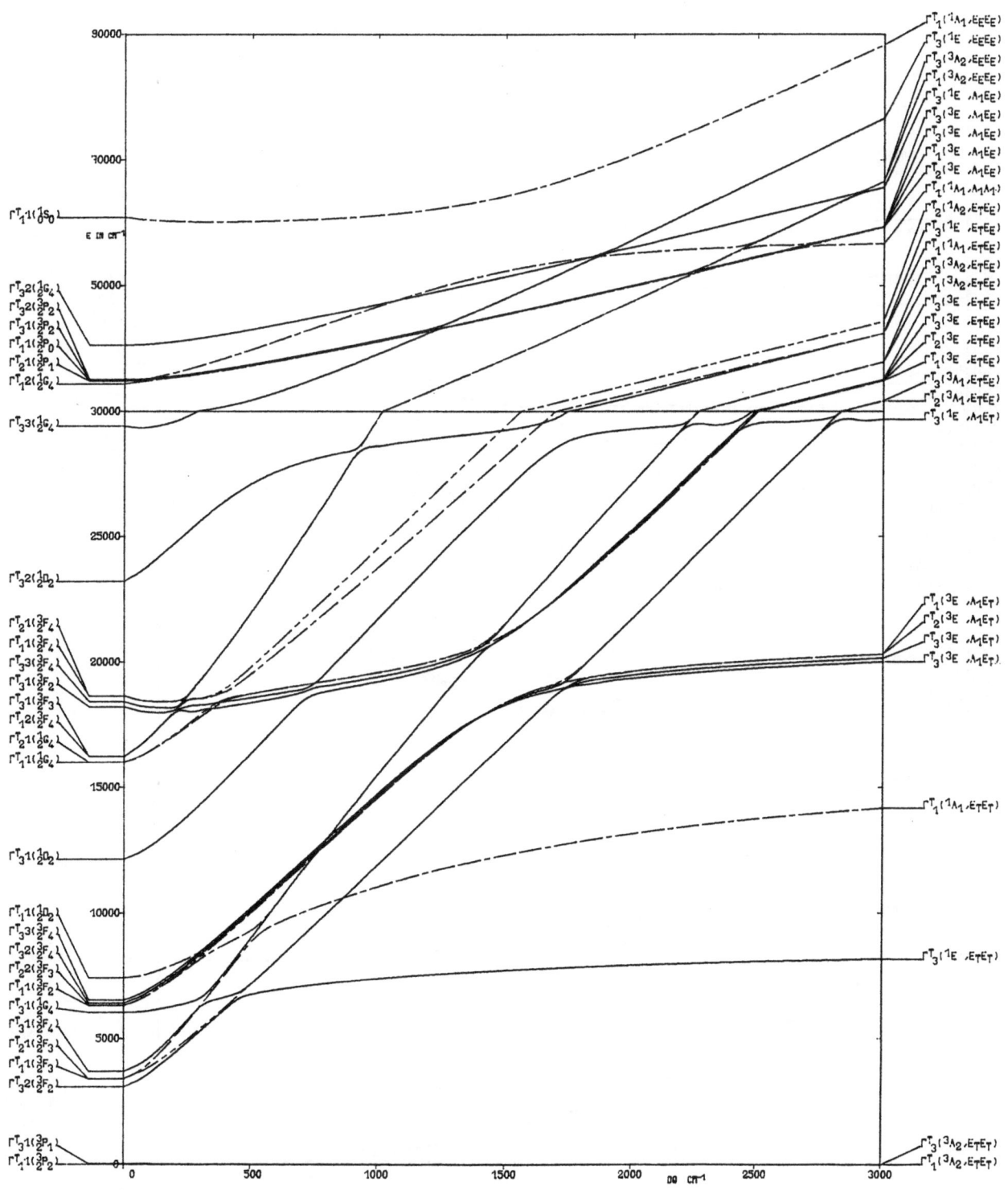

2 D ELECTRONS, TRIGONAL SYMMETRY   B=630 C=4*B DT=-1000 K=5   DS=K*DT 2ETA=210

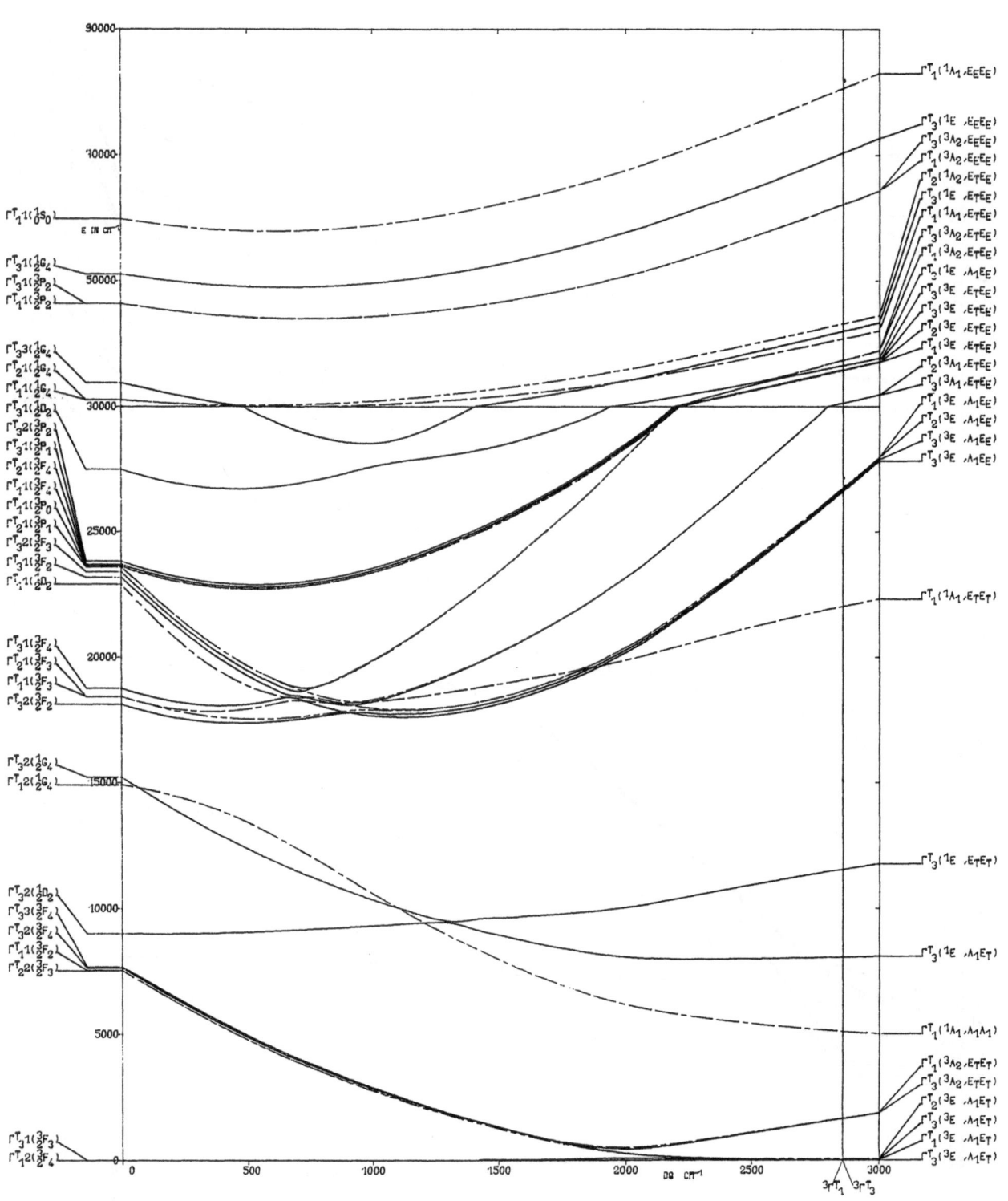

106

2 D ELECTRONS, TRIGONAL SYMMETRY

B=630 C=4*B DT=-1000 K=-5 DS=K*DT ZETA=210

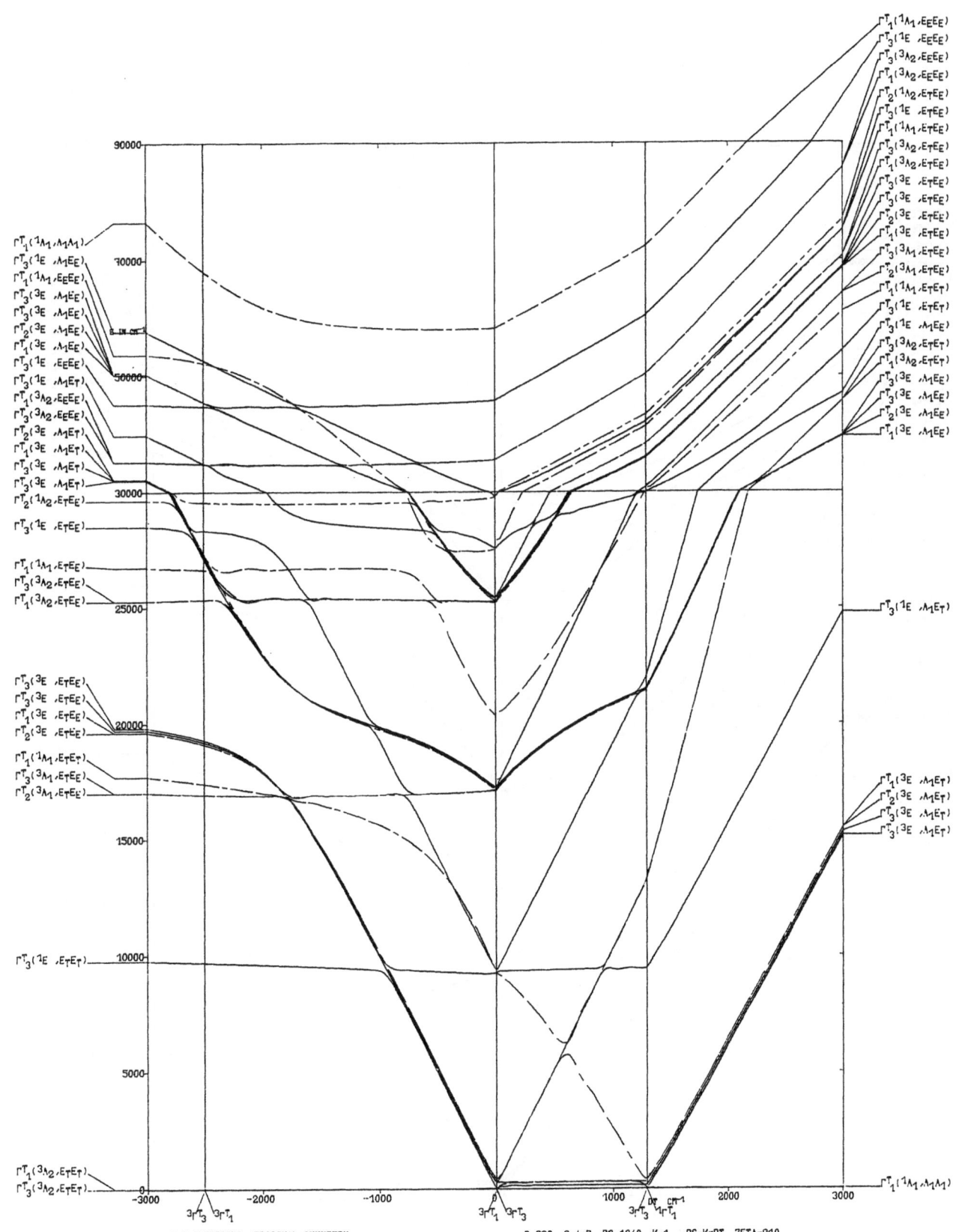

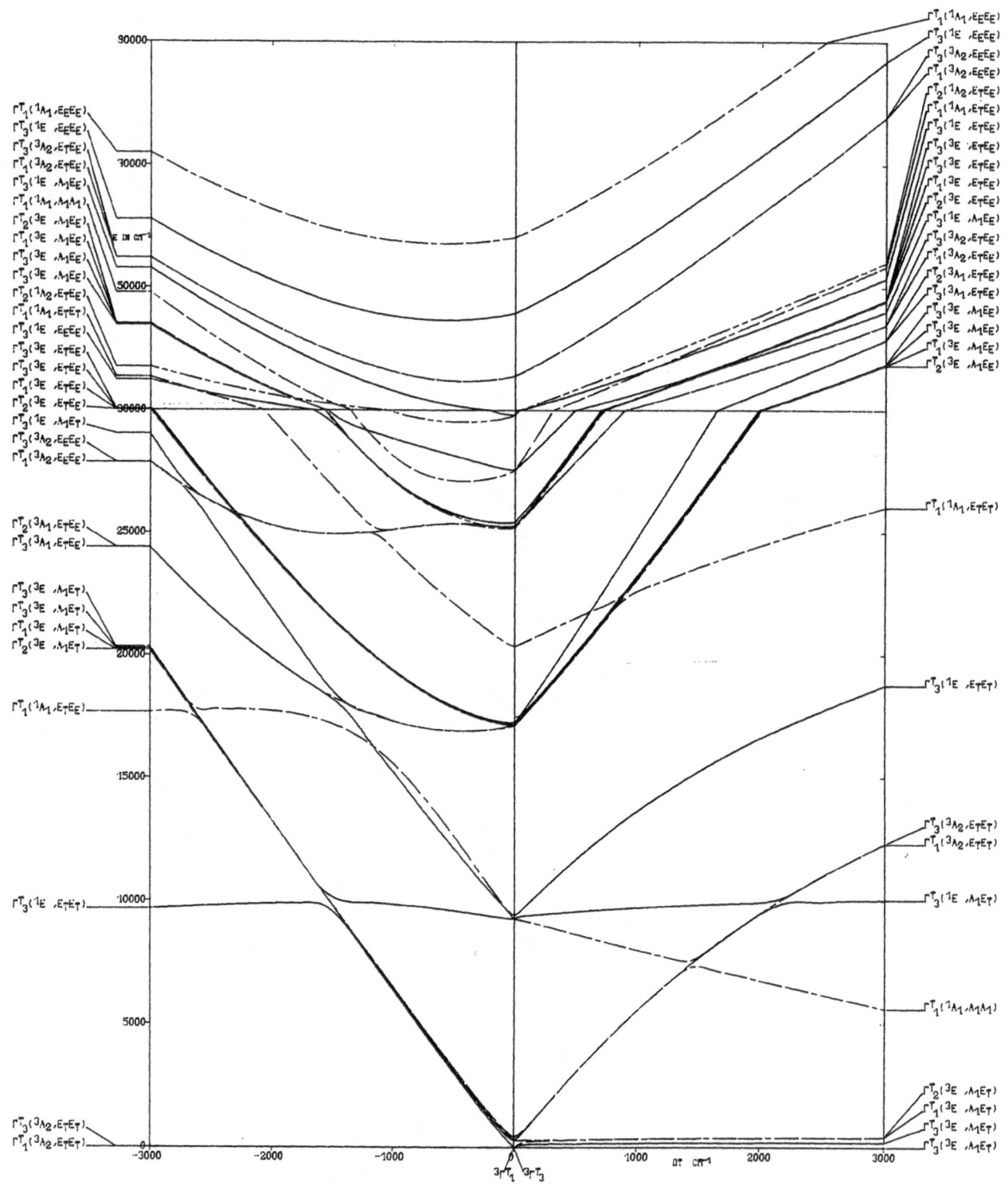

2 D ELECTRONS, TRIGONAL SYMMETRY

B=630 C=4*B DQ=1840 K=-1 DS=K*DT ZETA=210

108

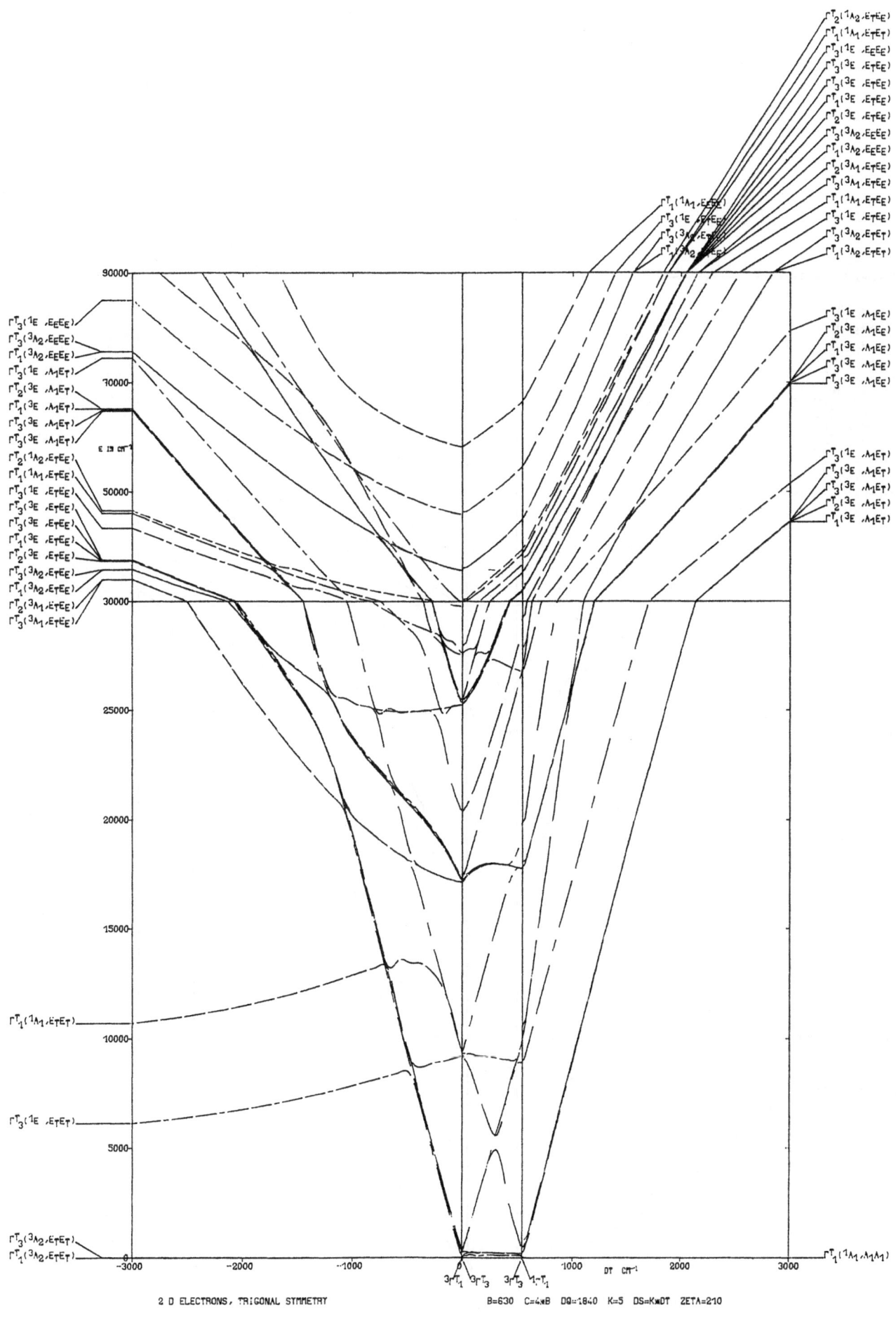

2 D ELECTRONS, TRIGONAL SYMMETRY   B=630 C=4×B DQ=1840 K=5 DS=K×DT ZETA=210

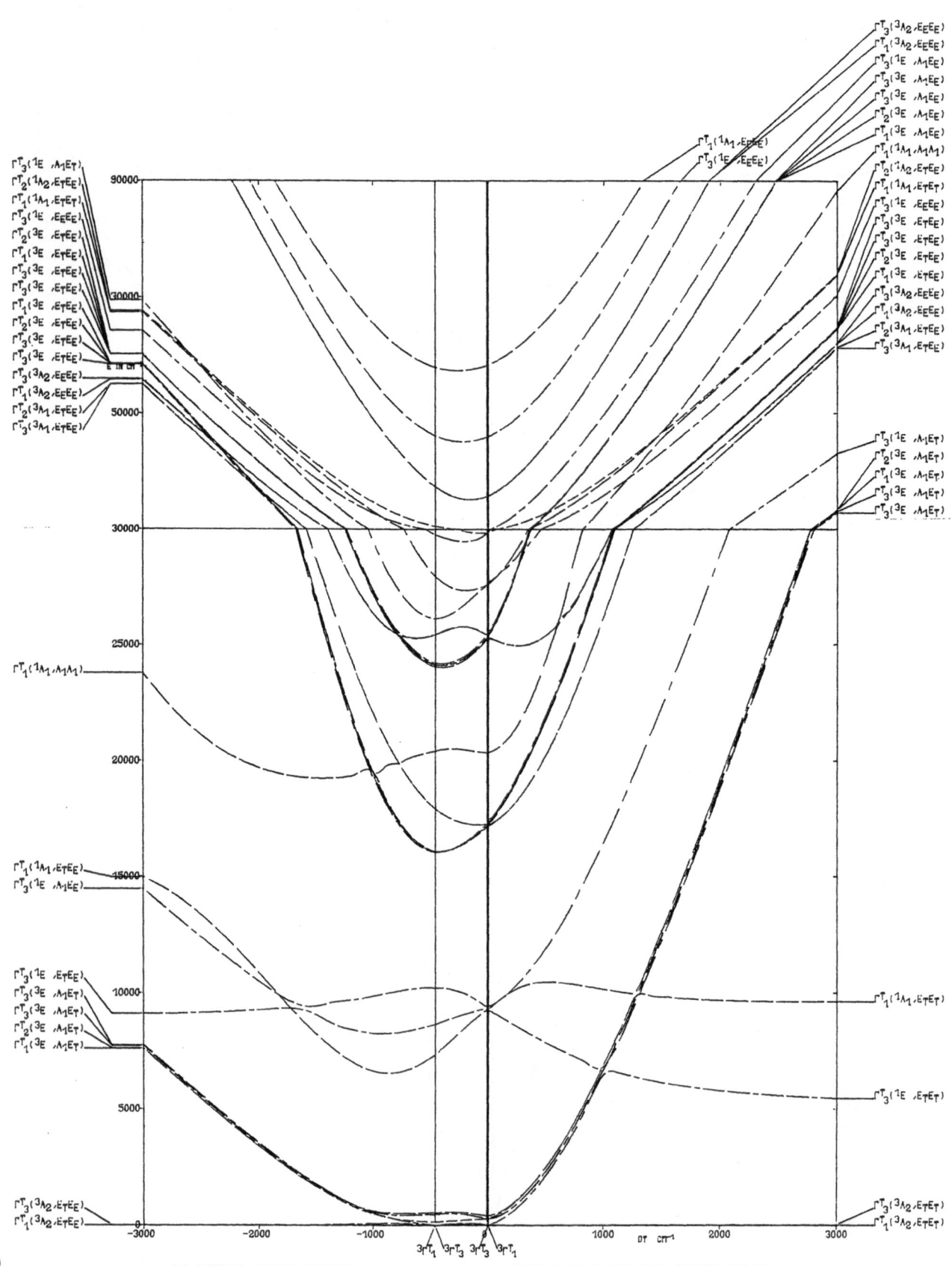

2 D ELECTRONS, TRIGONAL SYMMETRY

R=630   C=4×B   DQ=1840   K=-5   DS=K×DT   2ETA=210

2 D ELECTRONS, CYLINDRICAL SYMMETRY

B=630   C=4*B   ZETA=210

ENERGY AS FUNCTION OF DT
K=1, -1, 5, -5

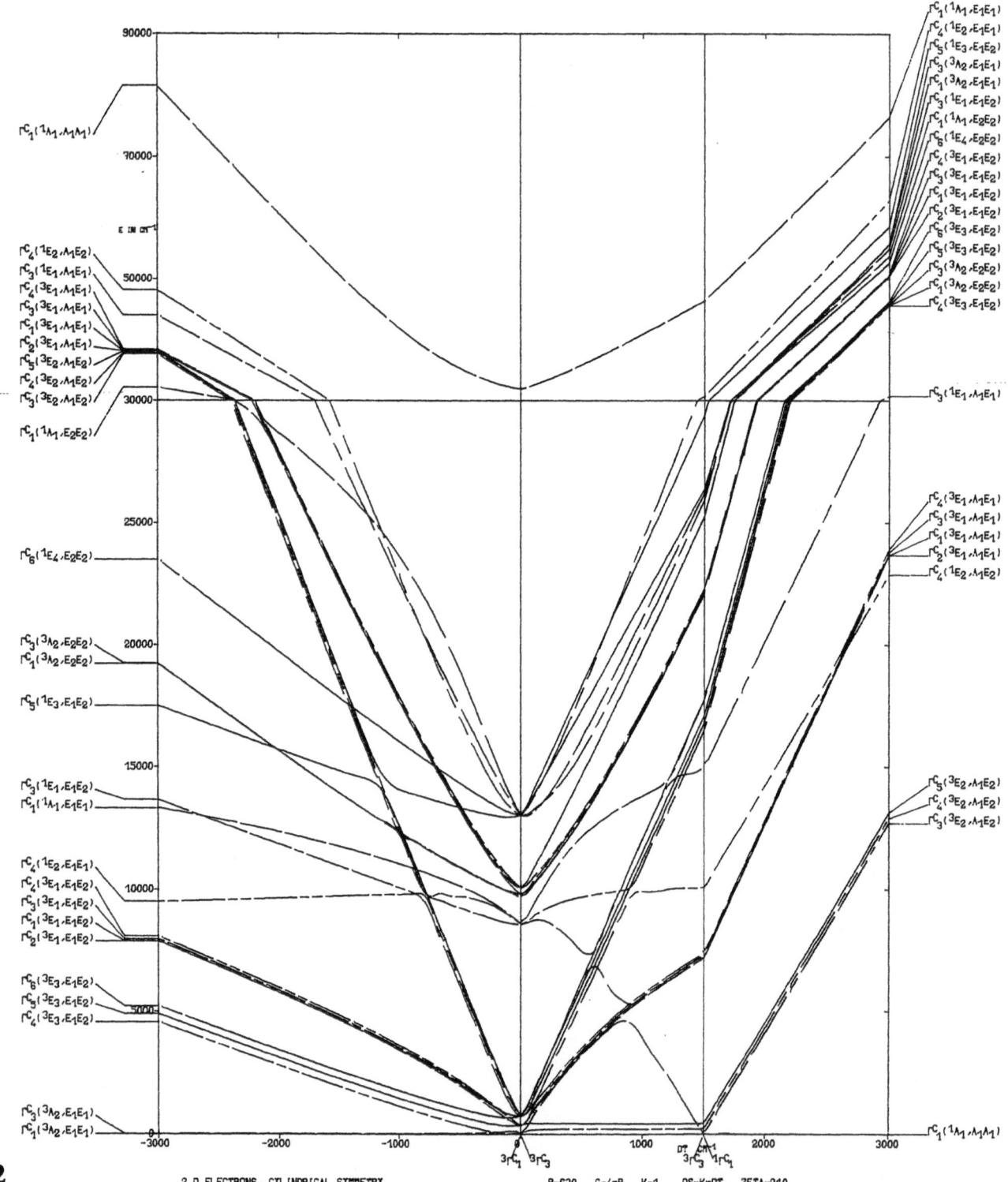

2 D ELECTRONS, CYLINDRICAL SYMMETRY   B=630   C=4×B   K=1   DS=K×DT   ZETA=210

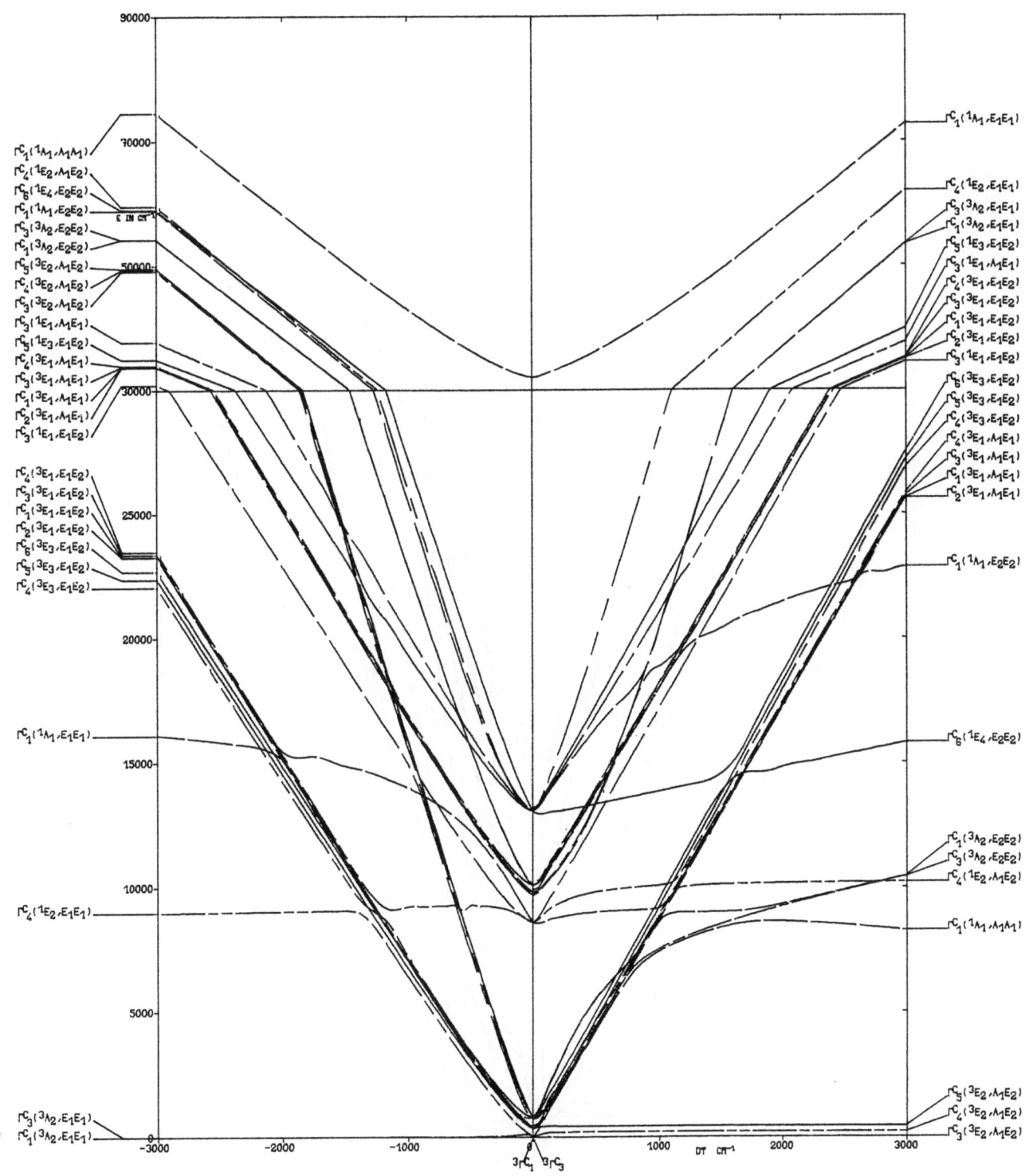

$\Gamma C_1(^1A_1, A_1A_1)$

$\Gamma C_4(^1E_2, A_1E_2)$
$\Gamma C_6(^1E_4, E_2E_2)$
$\Gamma C_1(^1A_1, E_2E_2)$
$\Gamma C_3(^3A_2, E_2E_2)$
$\Gamma C_1(^3A_2, E_2E_2)$
$\Gamma C_5(^3E_2, A_1E_2)$
$\Gamma C_4(^3E_2, A_1E_2)$
$\Gamma C_3(^3E_2, A_1E_2)$
$\Gamma C_3(^1E_1, A_1E_1)$
$\Gamma C_5(^1E_3, E_1E_2)$
$\Gamma C_4(^3E_1, A_1E_1)$
$\Gamma C_3(^3E_1, A_1E_1)$
$\Gamma C_1(^3E_1, A_1E_1)$
$\Gamma C_2(^3E_1, A_1E_1)$
$\Gamma C_3(^1E_1, E_1E_2)$

$\Gamma C_4(^3E_1, E_1E_2)$
$\Gamma C_3(^3E_1, E_1E_2)$
$\Gamma C_1(^3E_1, E_1E_2)$
$\Gamma C_2(^3E_1, E_1E_2)$
$\Gamma C_6(^3E_3, E_1E_2)$
$\Gamma C_5(^3E_3, E_1E_2)$
$\Gamma C_4(^3E_3, E_1E_2)$

$\Gamma C_1(^1A_1, E_1E_1)$

$\Gamma C_4(^1E_2, E_1E_1)$

$\Gamma C_3(^3A_2, E_1E_1)$
$\Gamma C_1(^3A_2, E_1E_1)$

$\Gamma C_1(^1A_1, E_1E_1)$

$\Gamma C_4(^1E_2, E_1E_1)$
$\Gamma C_3(^3A_2, E_1E_1)$
$\Gamma C_3(^3A_2, E_1E_1)$
$\Gamma C_5(^1E_3, E_1E_2)$
$\Gamma C_1(^3E_1, A_1E_1)$
$\Gamma C_4(^3E_1, E_1E_2)$
$\Gamma C_3(^3E_1, E_1E_2)$
$\Gamma C_2(^3E_1, E_1E_2)$
$\Gamma C_1(^1E_1, E_1E_2)$
$\Gamma C_6(^3E_3, E_1E_2)$
$\Gamma C_5(^3E_3, E_1E_2)$
$\Gamma C_4(^3E_3, E_1E_2)$
$\Gamma C_4(^3E_3, A_1E_2)$
$\Gamma C_3(^3E_1, A_1E_1)$
$\Gamma C_1(^3E_1, A_1E_1)$
$\Gamma C_2(^3E_1, A_1E_1)$

$\Gamma C_1(^1A_1, E_2E_2)$

$\Gamma C_6(^1E_4, E_2E_2)$

$\Gamma C_1(^3A_2, E_2E_2)$
$\Gamma C_3(^3A_2, E_2E_2)$
$\Gamma C_4(^1E_2, A_1E_2)$

$\Gamma C_1(^1A_1, A_1A_1)$

$\Gamma C_5(^3E_2, A_1E_2)$
$\Gamma C_4(^3E_2, A_1E_2)$
$\Gamma C_3(^3E_2, A_1E_2)$

$^3\Gamma C_1$ $^3\Gamma C_3$

$\Gamma C_3(^3A_2, E_1E_1)$
$\Gamma C_1(^3A_2, E_1E_1)$

2 D ELECTRONS, CYLINDRICAL SYMMETRY   B=630   C=4×B   K=-1   DS=K×DT   ZETA=210

113

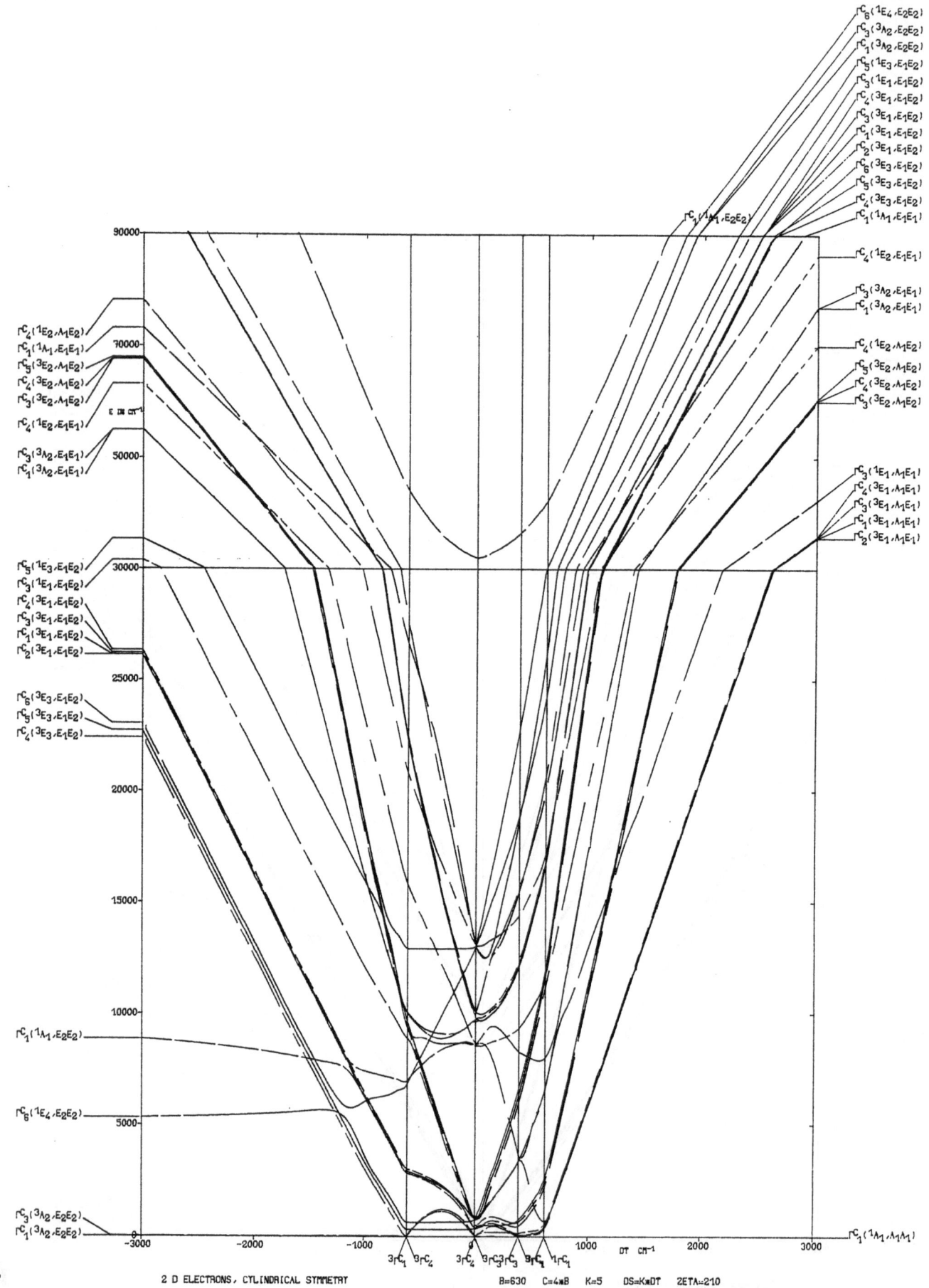

90000

^ΓC₄(¹E₂,Λ₁E₂)
^ΓC₁(¹A₁,E₁E₁) 70000
^ΓC₅(³E₂,Λ₁E₂)
^ΓC₄(³E₂,Λ₁E₂)
^ΓC₃(³E₂,Λ₁E₂)

^ΓC₄(¹E₂,E₁E₁)

^ΓC₃(³A₂,E₁E₁) 50000
^ΓC₃(³A₂,E₁E₁)

^ΓC₅(¹E₃,E₁E₂) 30000
^ΓC₃(¹E₁,E₁E₂)
^ΓC₄(³E₁,E₁E₂)
^ΓC₃(³E₁,E₁E₂)
^ΓC₁(³E₁,E₁E₂)
^ΓC₂(³E₁,E₁E₂)

25000

^ΓC₆(³E₃,E₁E₂)
^ΓC₅(³E₃,E₁E₂)
^ΓC₄(³E₃,E₁E₂)

20000

15000

^ΓC₁(¹A₁,E₂E₂) 10000

^ΓC₆(¹E₄,E₂E₂) 5000

^ΓC₃(³A₂,E₂E₂)
^ΓC₁(³A₂,E₂E₂) 0

E IN CM⁻¹

-3000   -2000   -1000   1000   2000   3000

DT CM⁻¹

3Γ C₁ 9Γ C₄   3Γ C₄ 9Γ C₃ 9Γ C₁ 1Γ C₁

^ΓC₈(¹E₄,E₂E₂)
^ΓC₃(³A₂,E₂E₂)
^ΓC₁(³A₂,E₂E₂)
^ΓC₅(¹E₃,E₁E₂)
^ΓC₃(³E₁,E₁E₂)
^ΓC₄(³E₁,E₁E₂)
^ΓC₃(³E₁,E₁E₂)
^ΓC₁(³E₁,E₁E₂)
^ΓC₂(³E₁,E₁E₂)
^ΓC₆(³E₃,E₁E₂)
^ΓC₅(³E₃,E₁E₂)
^ΓC₄(³E₃,E₁E₂)
^ΓC₁(¹A₁,E₁E₁)

^ΓC₁(¹A₁,E₂E₂)

^ΓC₄(¹E₂,E₁E₁)

^ΓC₃(³A₂,E₁E₁)
^ΓC₁(³A₂,E₁E₁)

^ΓC₄(¹E₂,Λ₁E₂)
^ΓC₅(³E₂,Λ₁E₂)
^ΓC₄(³E₂,Λ₁E₂)
^ΓC₃(³E₂,Λ₁E₂)

^ΓC₃(³E₁,Λ₁E₁)
^ΓC₄(³E₁,Λ₁E₁)
^ΓC₃(³E₁,Λ₁E₁)
^ΓC₁(³E₁,Λ₁E₁)
^ΓC₂(³E₁,Λ₁E₁)

^ΓC₁(¹A₁,Λ₁A₁)

114

2 D ELECTRONS, CYLINDRICAL SYMMETRY

B=630   C=4×B   K=5   DS=K×DT   ZETA=210

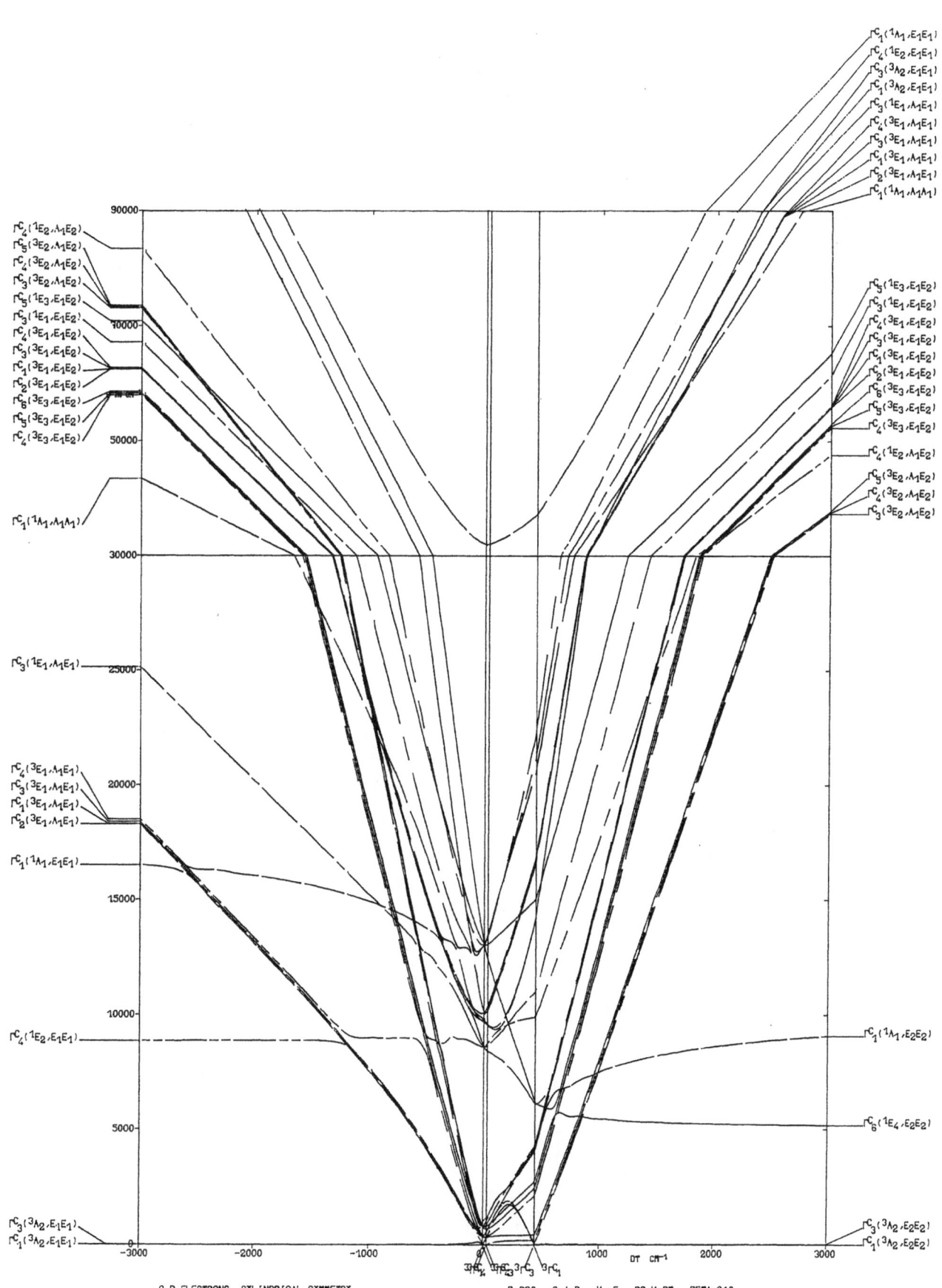

2 D ELECTRONS, CYLINDRICAL SYMMETRY   B=630 C=4*B K=-5 DS=K*DT ZETA=210

115

3 D ELECTRONS, TETRAGONAL SYMMETRY

B=700   C=4×B   ZETA=0

ENERGY AS FUNCTION OF DQ
DT=0, 500, -500, 1000, -1000
K=1, -1, 5, -5

ENERGY AS FUNCTION OF DT
DQ=1230, 1740
K=1, -1, 5, -5

(37) ^4E $(B_1B_2(^3A_2)E$)
(36) ^4E $(A_1B_2(^3B_2)E$)
(35) ^4E $(A_1B_1(^3B_1)E$)
(34) $^4B_2(B_1E$ $(^3E$ $)E$)
(33) $^4B_1(B_2E$ $(^3E$ $)E$)
(32) $^4A_2(A_1E$ $(^3E$ $)E$)
(31) $^4A_2(A_1B_1(^3B_1)B_2)$
(30) ^2E $(E$ E $(^1A_1)E$)
(29) ^2E $(B_2B_2(^1A_1)E$)
(28) ^2E $(B_1B_2(^3A_2)E$)
(27) ^2E $(B_1B_2(^1A_2)E$)
(26) ^2E $(B_1B_1(^1A_1)E$)
(25) ^2E $(A_1B_2(^3B_2)E$)
(24) ^2E $(A_1B_2(^1B_2)E$)
(23) ^2E $(A_1B_1(^3B_1)E$)
(22) ^2E $(A_1B_1(^1B_1)E$)
(21) ^2E $(A_1A_1(^1A_1)E$)
(20) $^2B_2(B_2E$ $(^1E$ $)E$)
(19) $^2B_2(B_1E$ $(^1E$ $)E$)
(18) $^2B_2(B_1B_1(^1A_1)B_2)$
(17) $^2B_2(A_1E$ $(^1E$ $)E$)
(16) $^2B_2(A_1A_1(^1A_1)B_2)$
(15) $^2B_1(B_2E$ $(^1E$ $)E$)
(14) $^2B_1(B_1E$ $(^1E$ $)E$)
(13) $^2B_1(B_1B_2(^1A_2)B_2)$
(12) $^2B_1(A_1E$ $(^1E$ $)E$)
(11) $^2B_1(A_1A_1(^1A_1)B_1)$
(10) $^2A_2(B_2E$ $(^1E$ $)E$)
(9) $^2A_2(B_1E$ $(^1E$ $)E$)
(8) $^2A_2(A_1E$ $(^1E$ $)E$)
(7) $^2A_2(A_1B_1(^3B_1)B_2)$
(6) $^2A_2(A_1B_1(^1B_1)B_2)$
(5) $^2A_1(B_2E$ $(^1E$ $)E$)
(4) $^2A_1(B_1E$ $(^1E$ $)E$)
(3) $^2A_1(A_1E$ $(^1E$ $)E$)
(2) $^2A_1(A_1B_2(^1B_2)B_2)$
(1) $^2A_1(A_1B_1(^1B_1)B_1)$

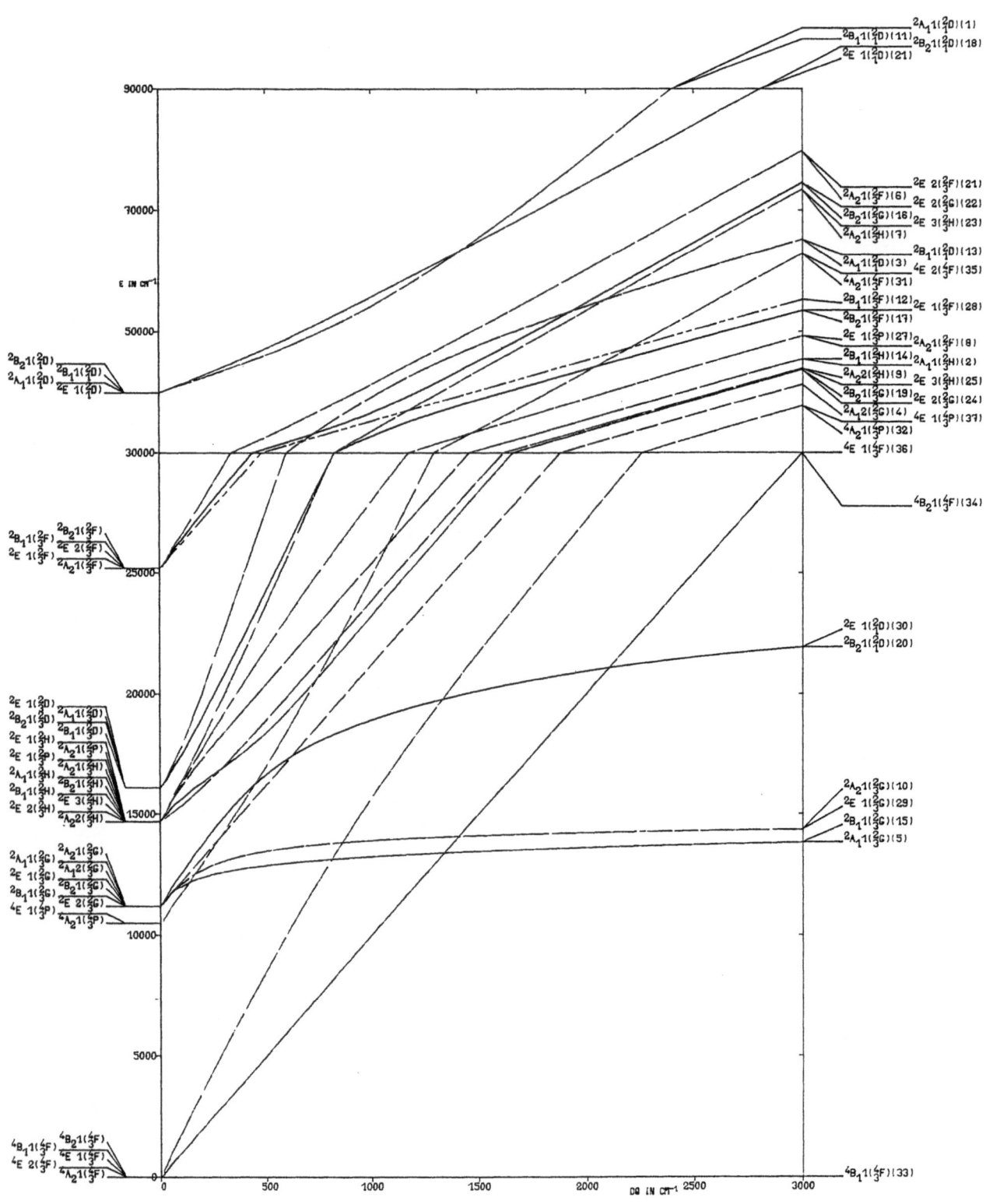

3 D ELECTRONS, TETRAGONAL SYMMETRY   B=700 C=4*B DT=0 K=0 DS=K*DT ZETA=0

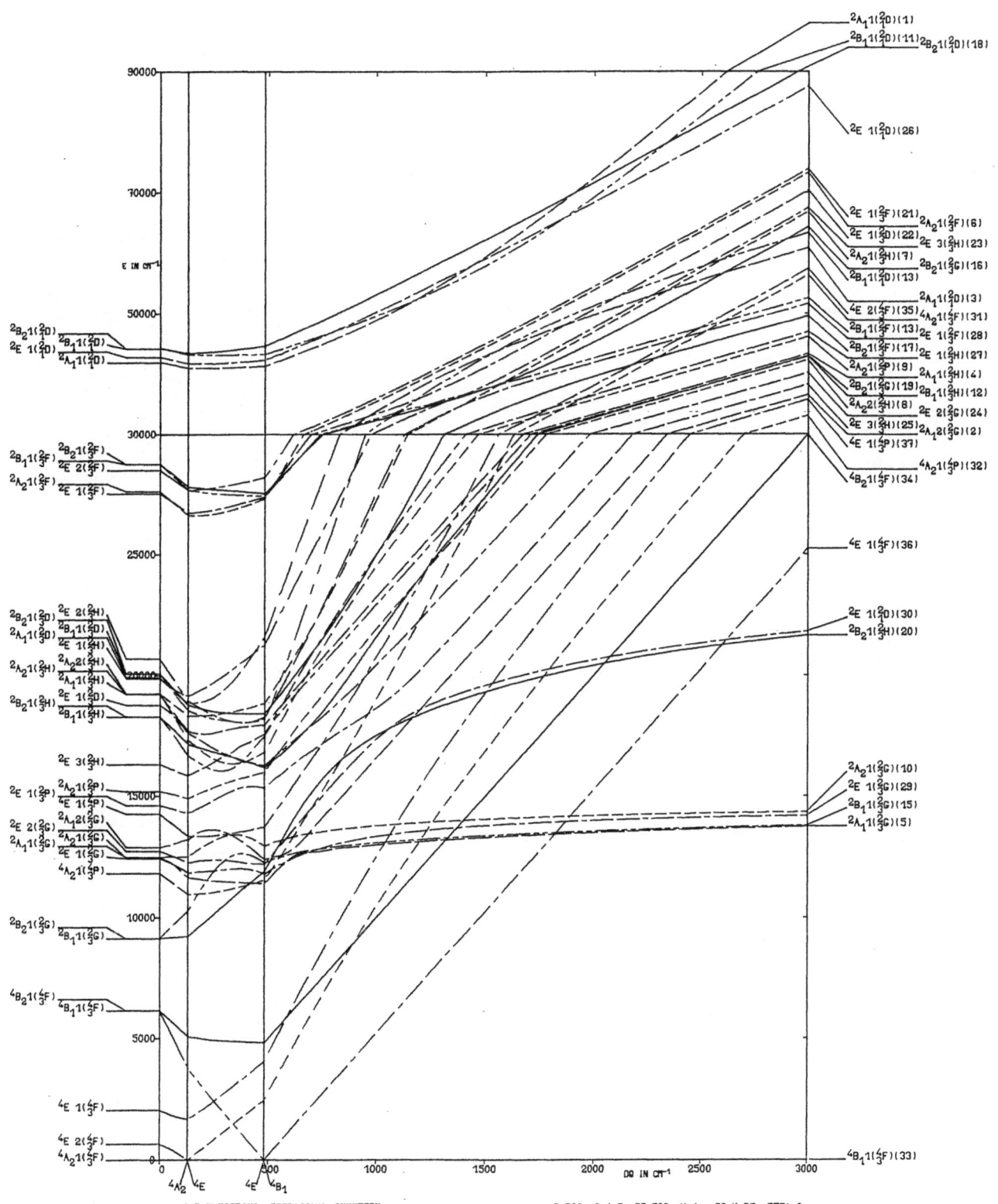

3 D ELECTRONS, TETRAGONAL SYMMETRY   B=700 C=4×B DT=500 K=1 DS=K×DT ZETA=0

119

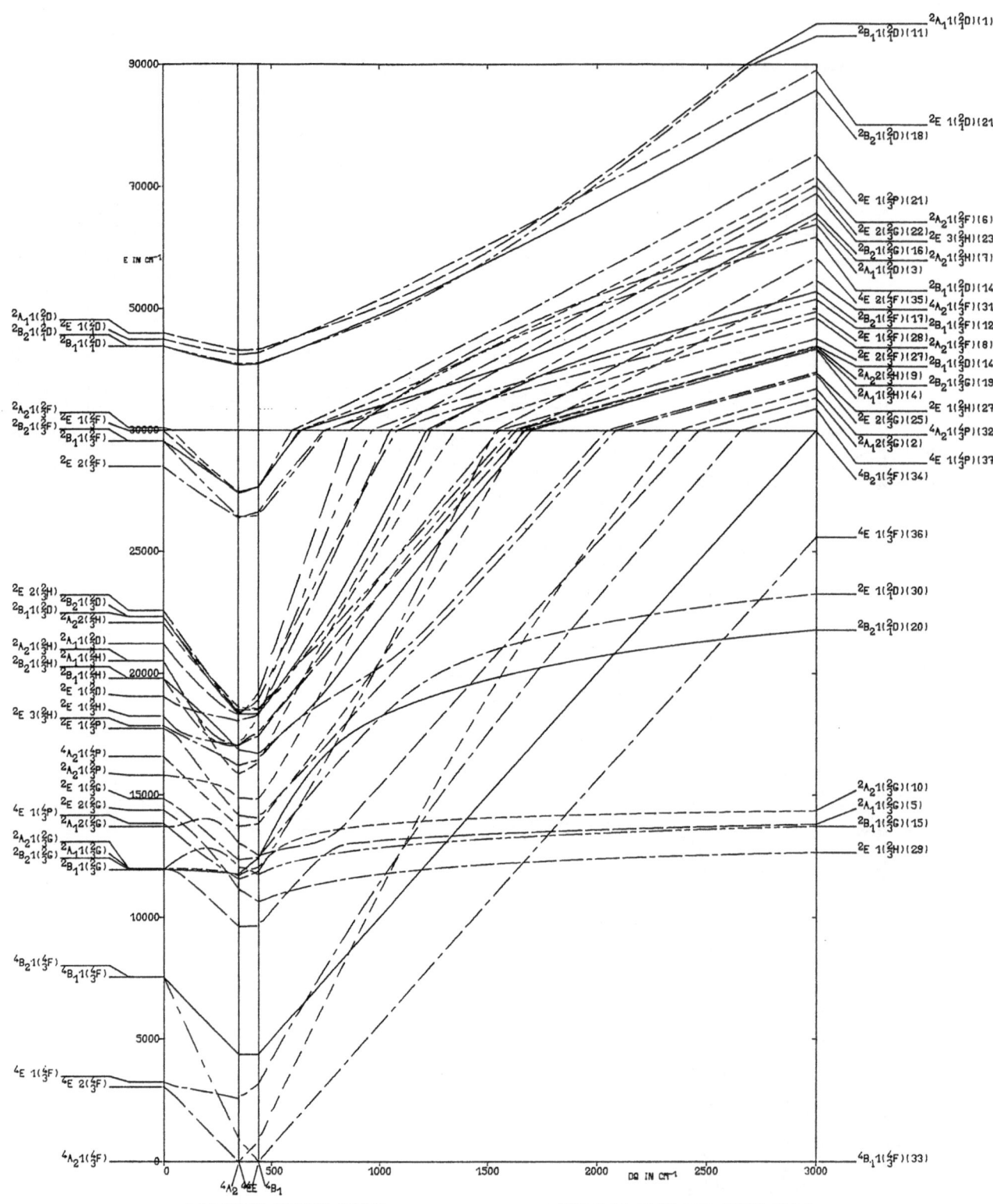

120

3 D ELECTRONS, TETRAGONAL SYMMETRY   B=700 C=4xB DT=500 K=-1 DS=KxDT ZETA=0

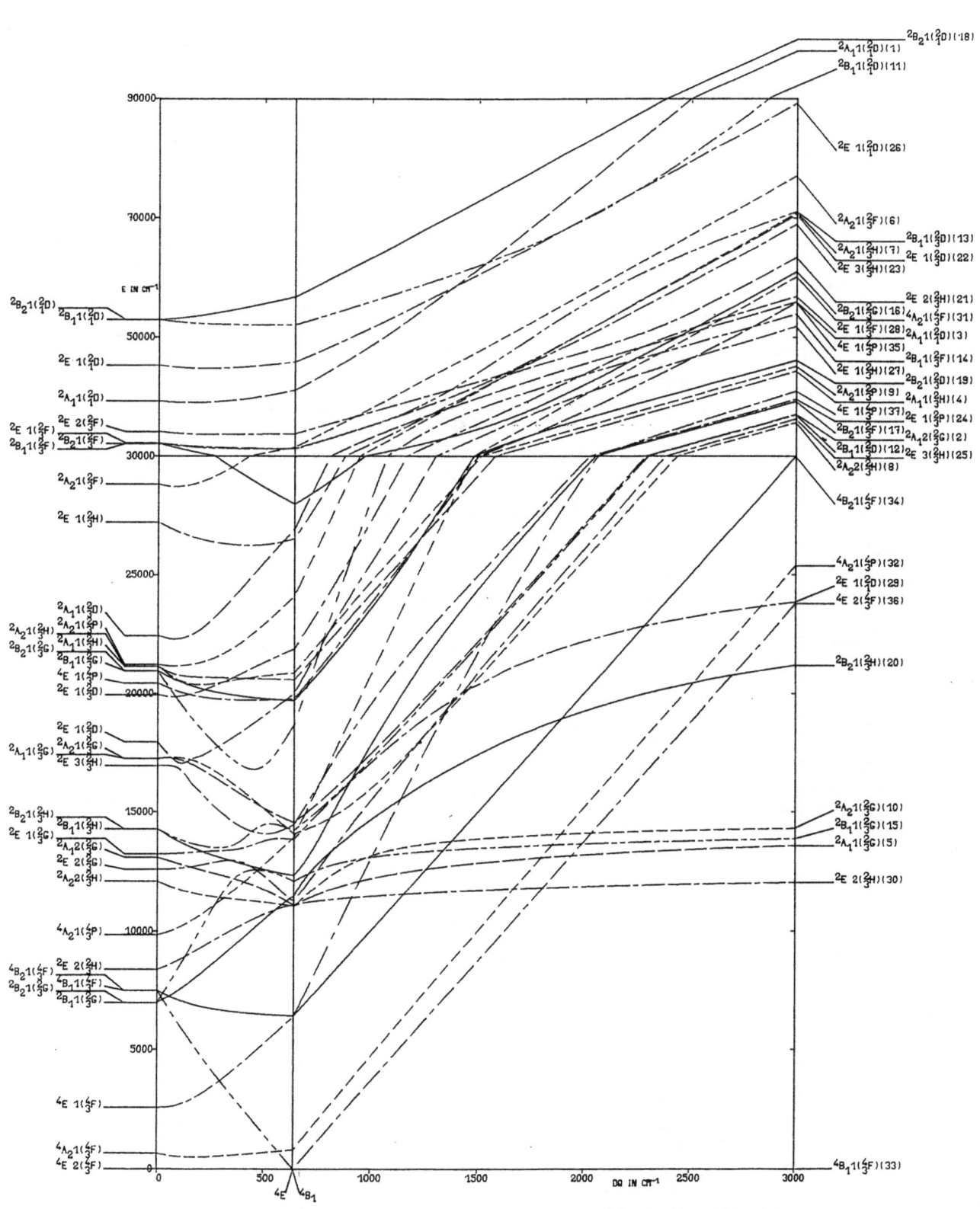

3 D ELECTRONS, TETRAGONAL SYMMETRY   B=700 C=4×B DT=500 K=5   DS=K×DT ZETA=0

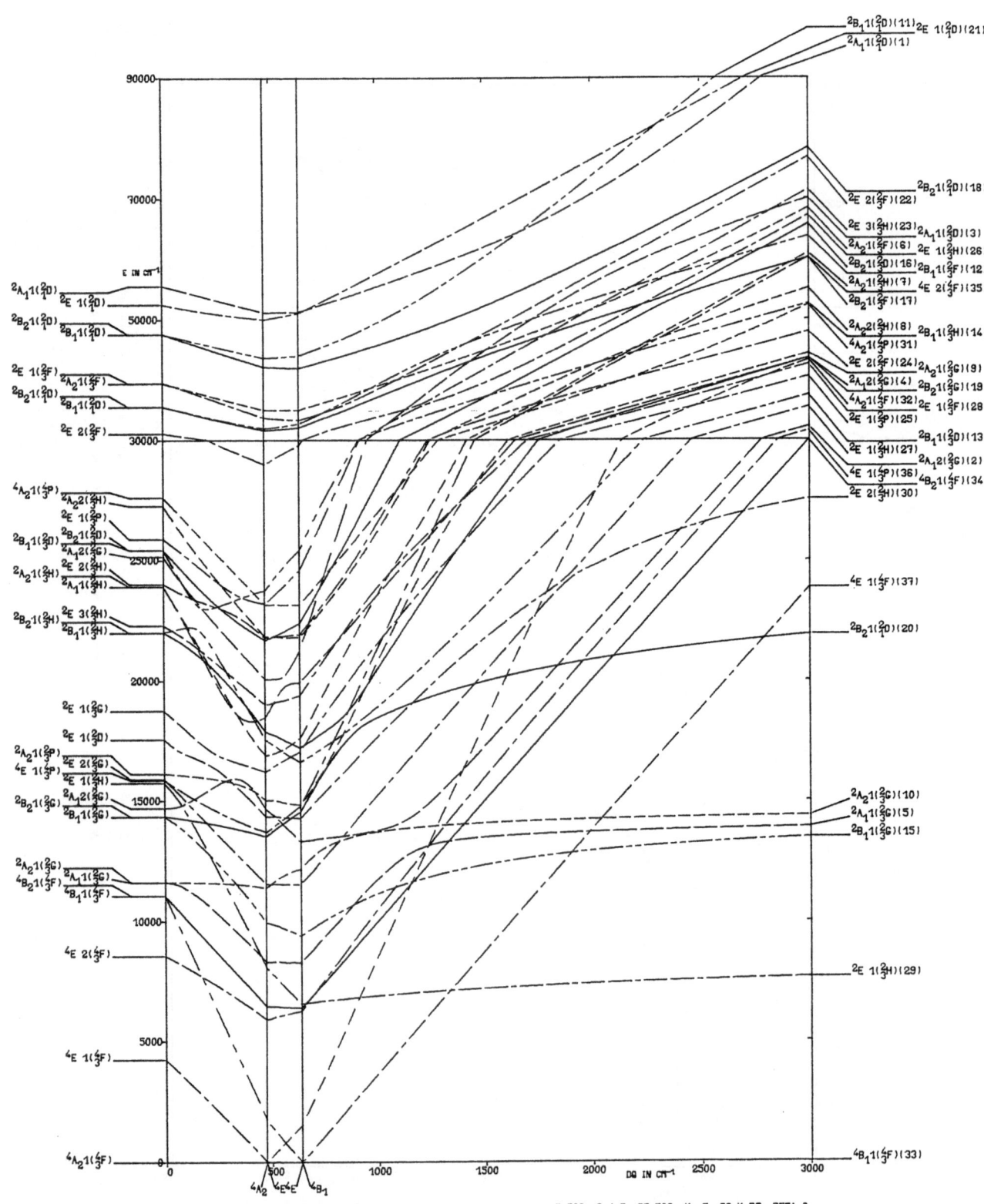

3 D ELECTRONS, TETRAGONAL SYMMETRY

B=700 C=4×B DT=500 K=-5 DS=K×DT ZETA=0

122

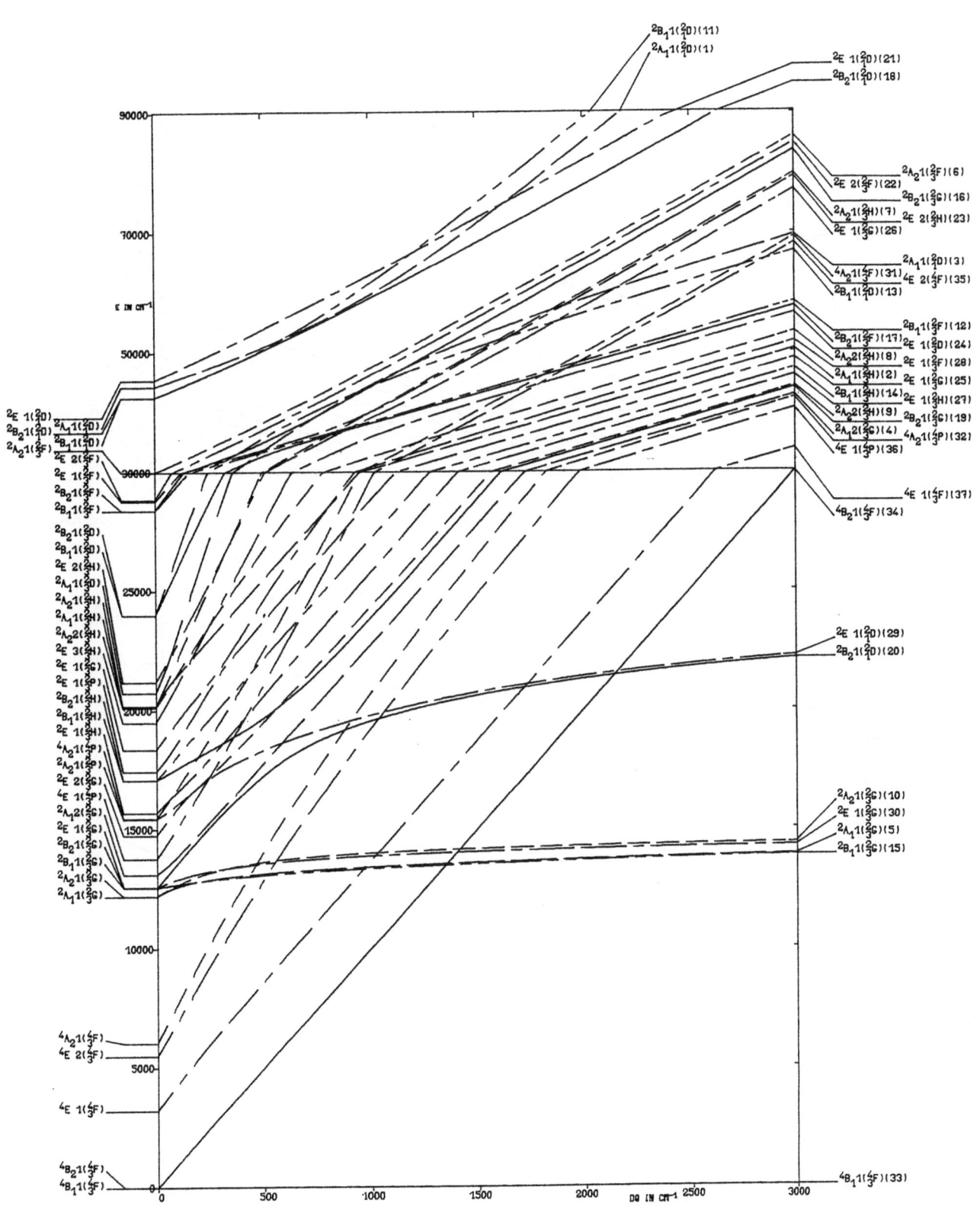

3 D ELECTRONS, TETRAGONAL SYMMETRY   B=700 C=4×B DT=-500 K=1 DS=K×DT ZETA=0

123

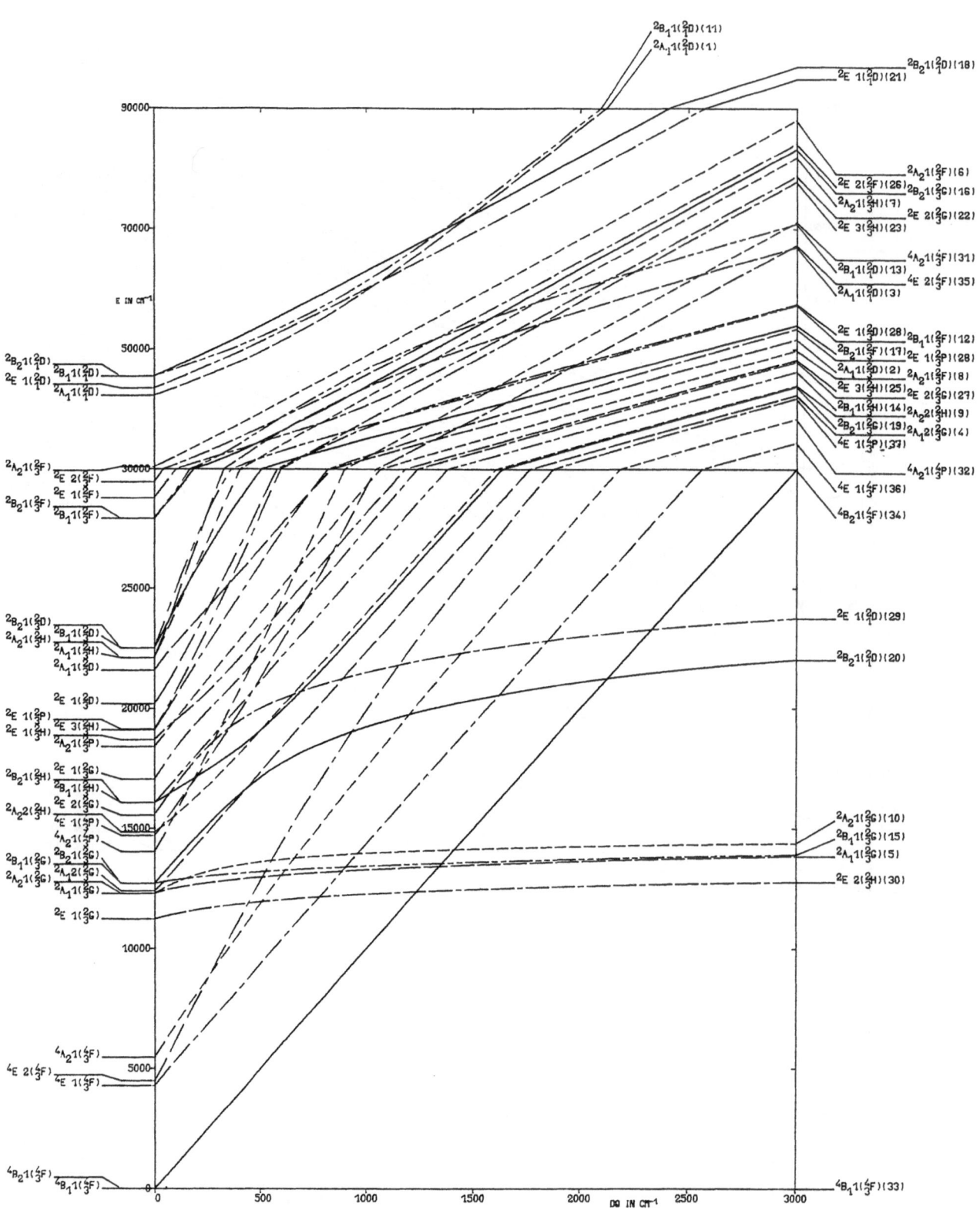

3 D ELECTRONS, TETRAGONAL SYMMETRY   B=700 C=4×B DT=-500 K=-1 DS=K×DT ZETA=0

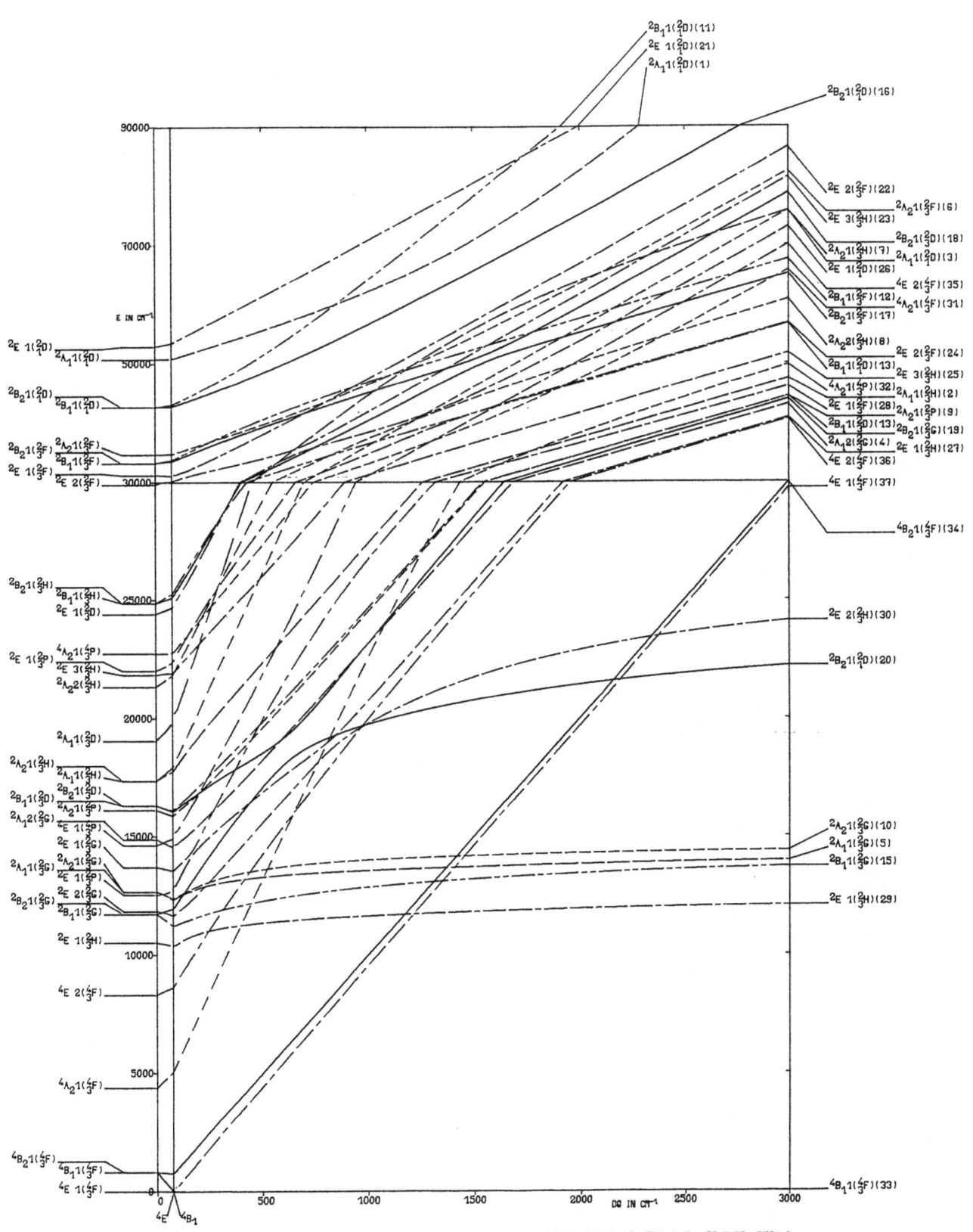

3 D ELECTRONS, TETRAGONAL SYMMETRY   B=700 C=4×B DT=-500 K=5 DS=K×DT ZETA=0

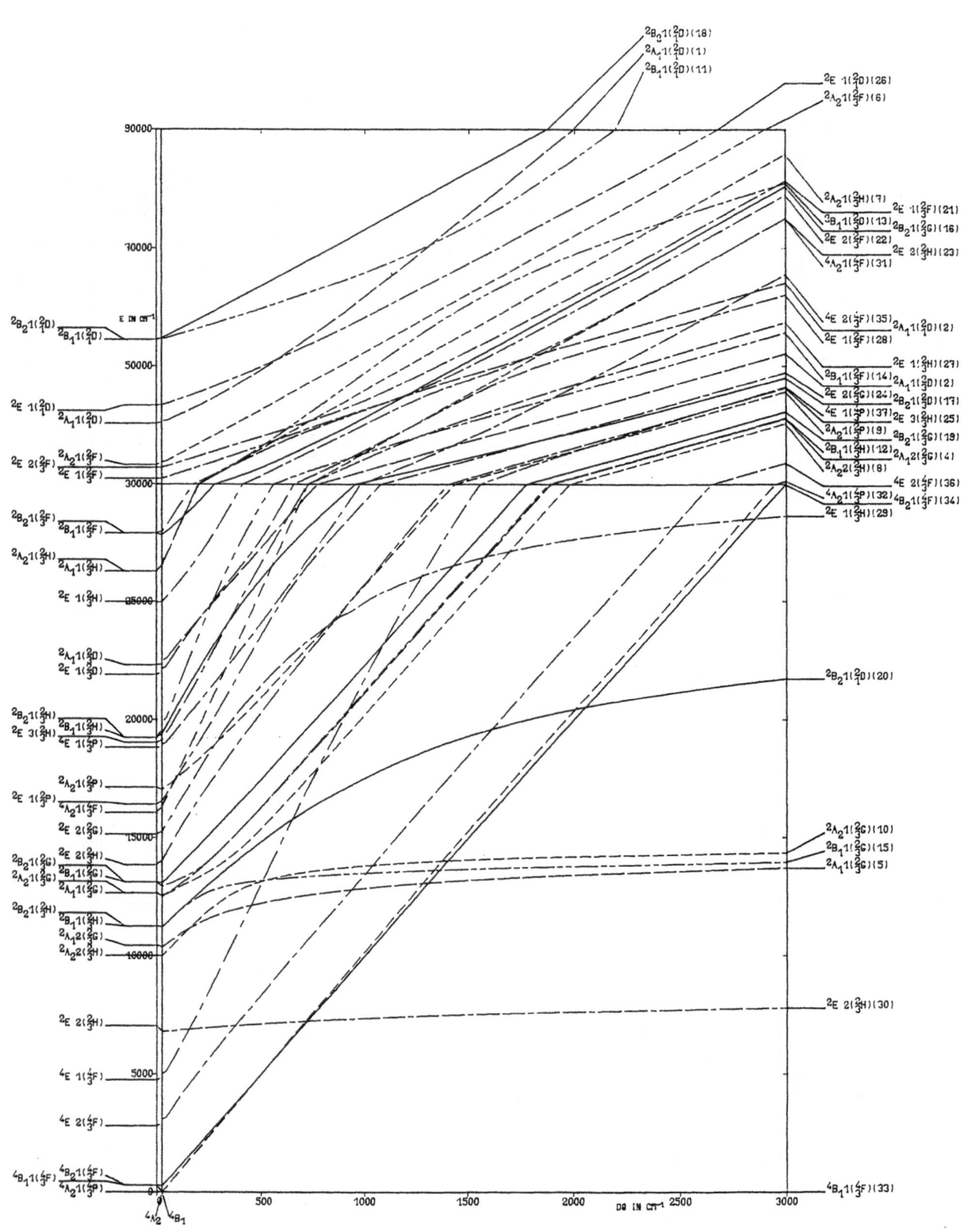

3 D ELECTRONS, TETRAGONAL SYMMETRY   B=700 C=4×B DT=-500 K=-5 DS=K×DT ZETA=0

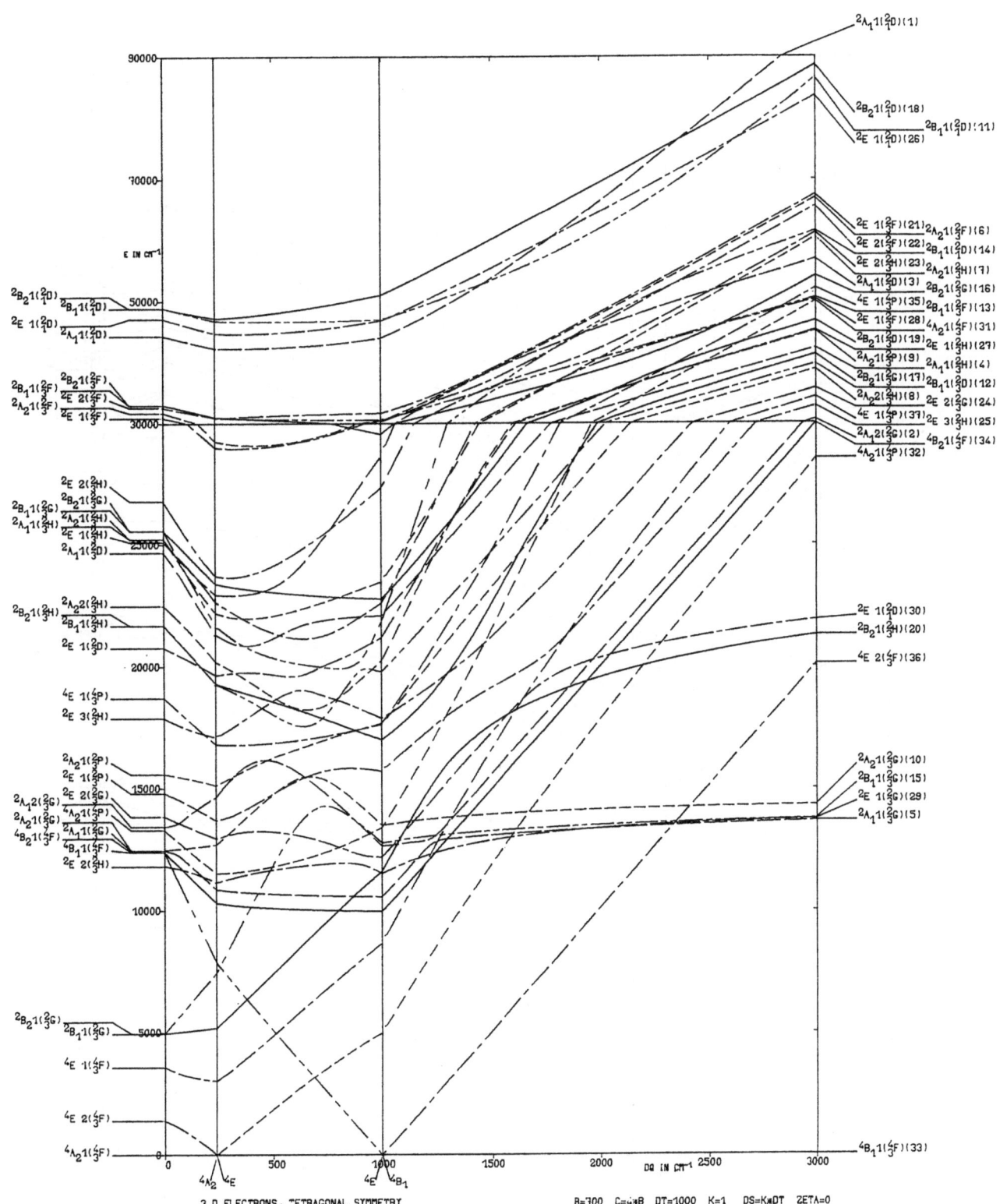

3 D ELECTRONS, TETRAGONAL SYMMETRY   B=700 C=4×B DT=1000 K=1 DS=K×DT ZETA=0

127

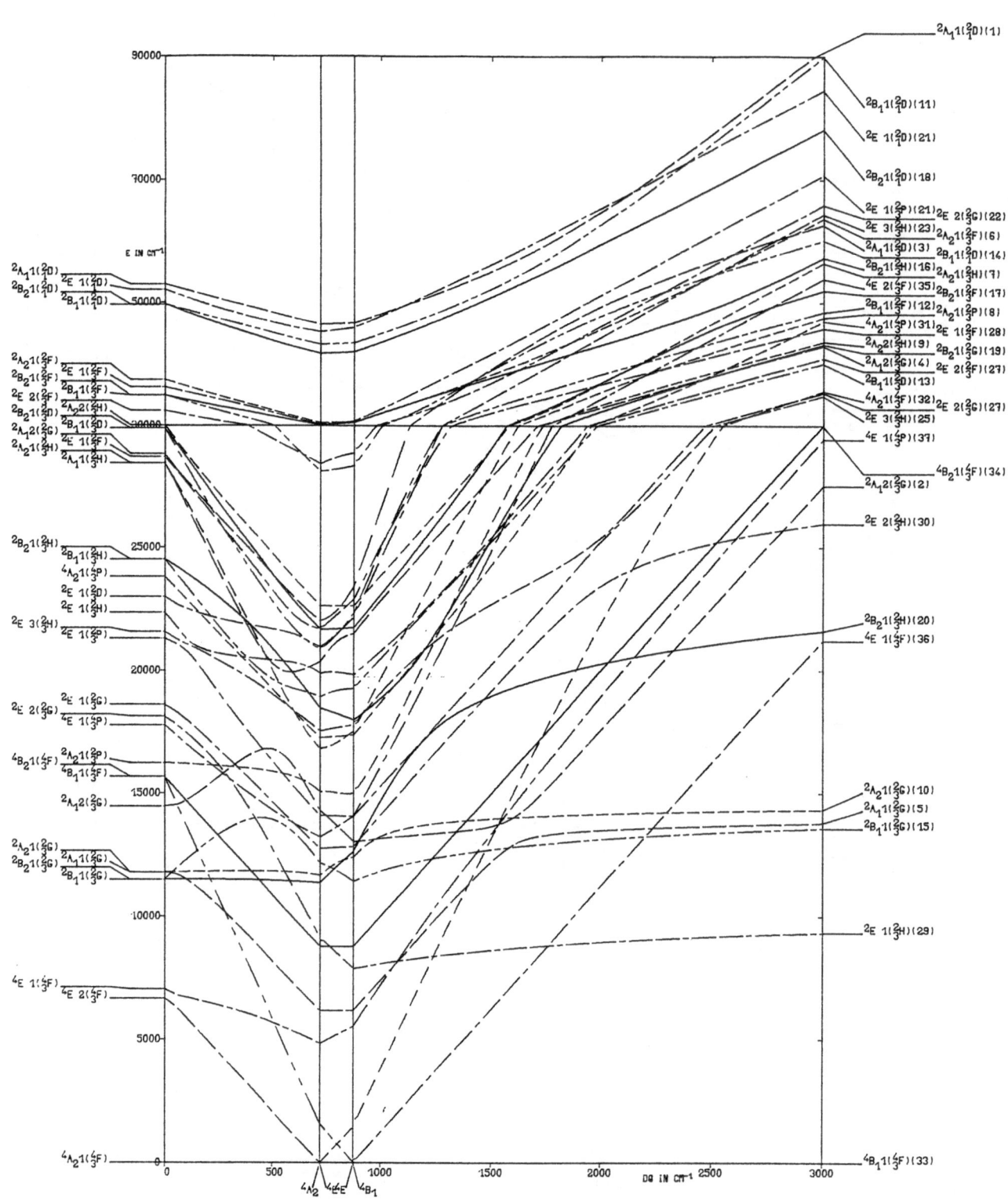

128

3 D ELECTRONS, TETRAGONAL SYMMETRY   B=700 C=4×B DT=1000 K=−1 DS=K×DT ZETA=0

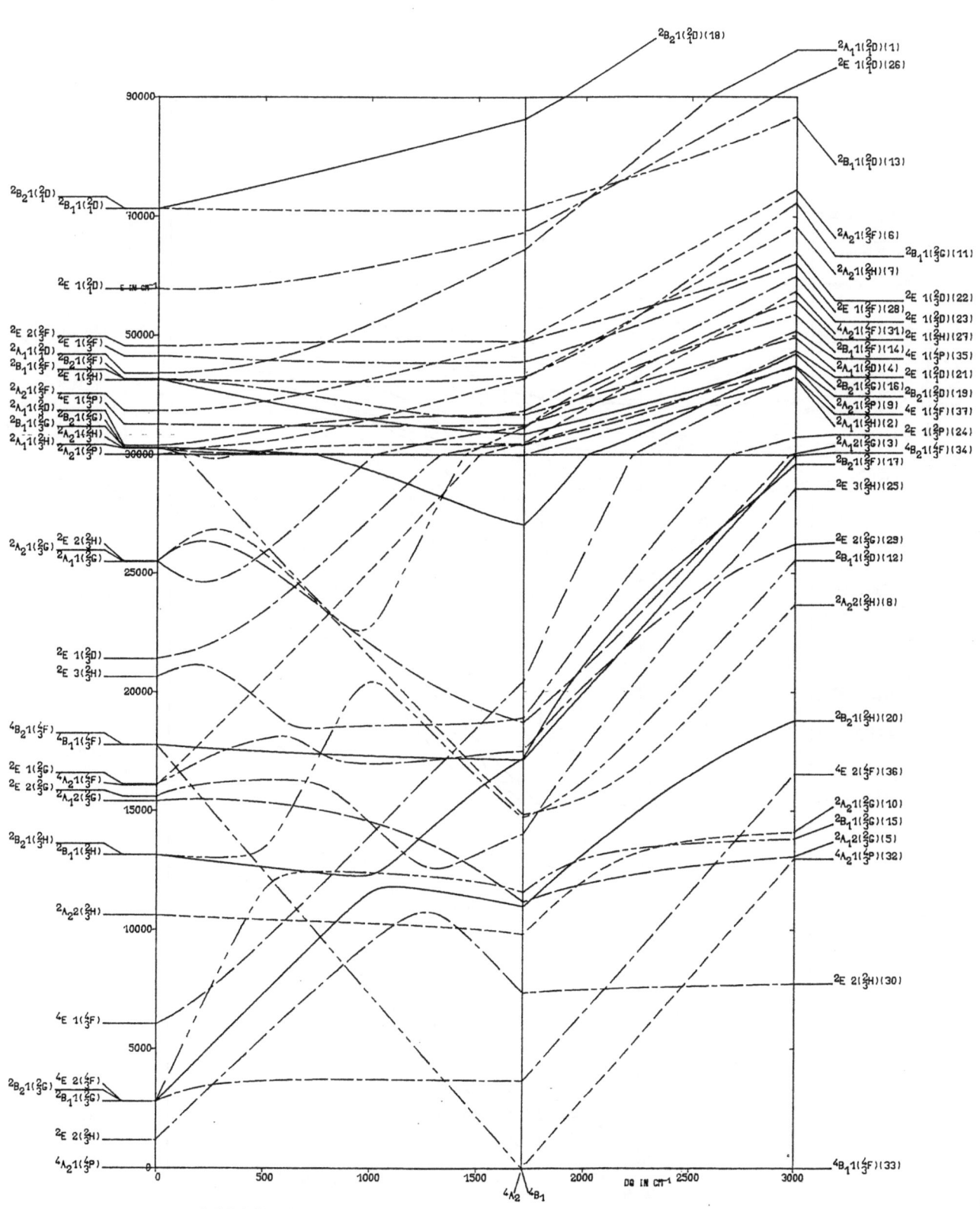

3 D ELECTRONS, TETRAGONAL SYMMETRY

B=700   C=4×B   DT=1000   K=5   DS=K×DT   ZETA=0

129

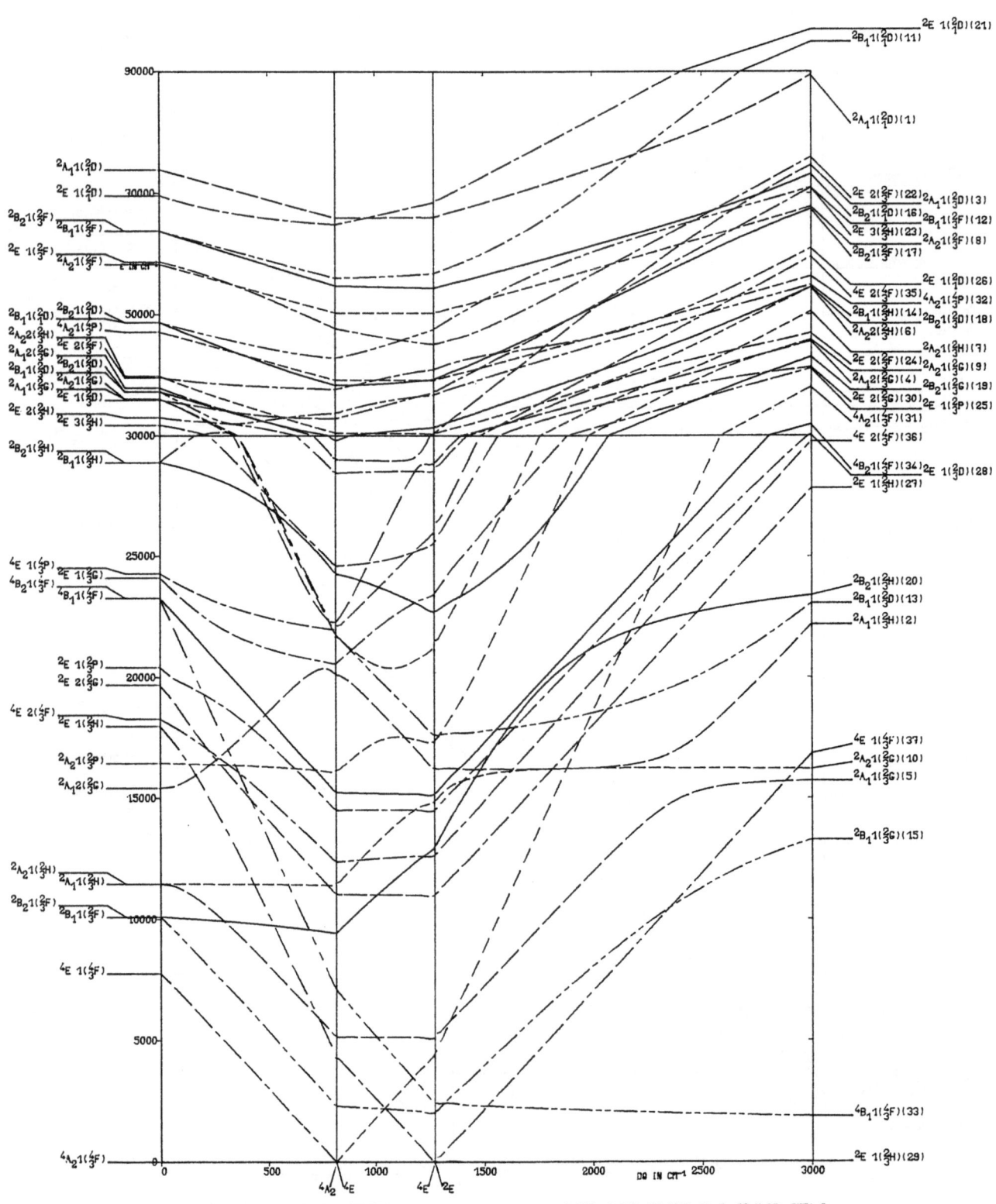

130

3 D ELECTRONS, TETRAGONAL SYMMETRY   B=700 C=4мB DT=1000 K=-5 DS=KмDT ZETA=0

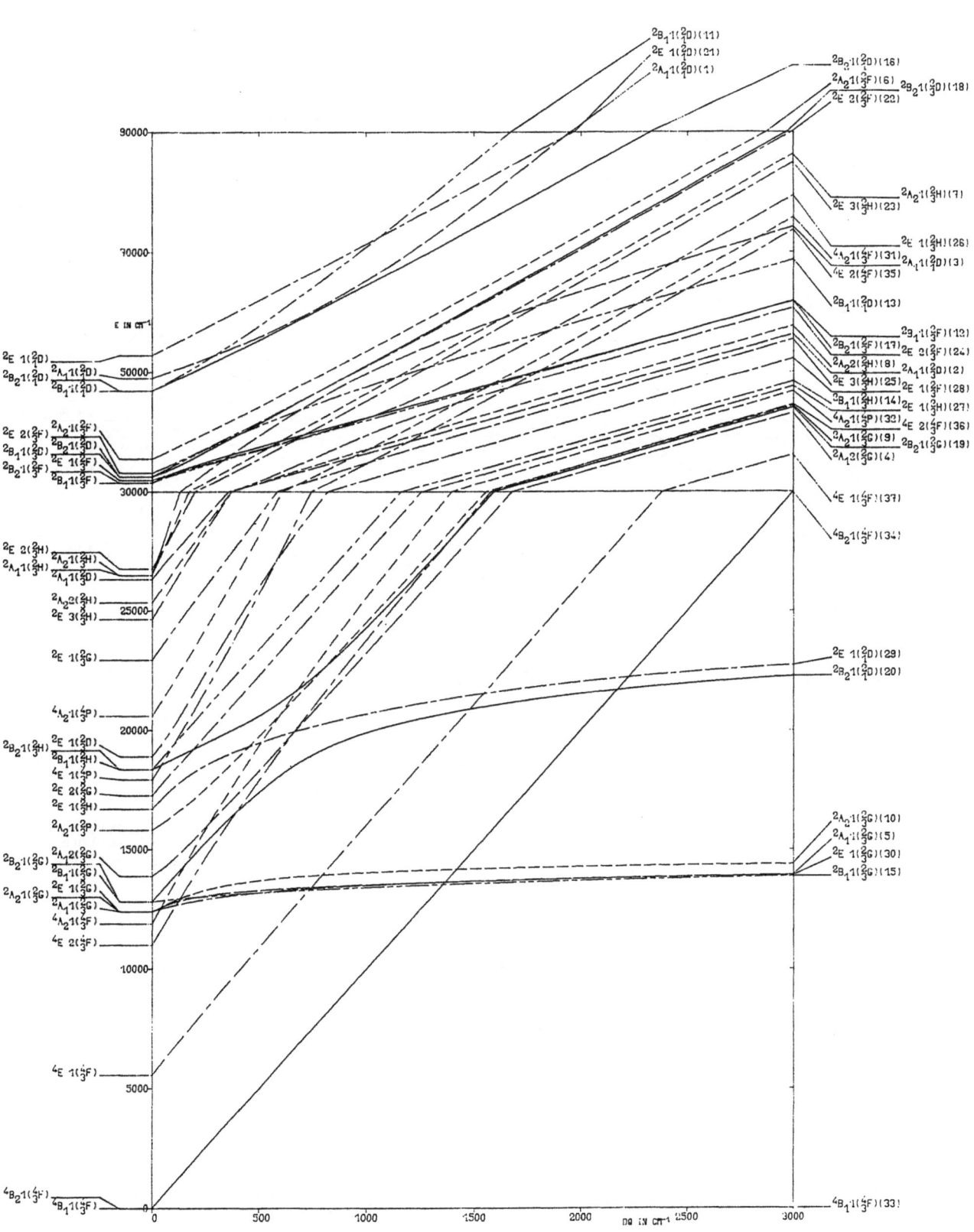

3 D ELECTRONS, TETRAGONAL SYMMETRY   B=700 C=4*B DT=-1000 K=1   DS=K*DT ZETA=0

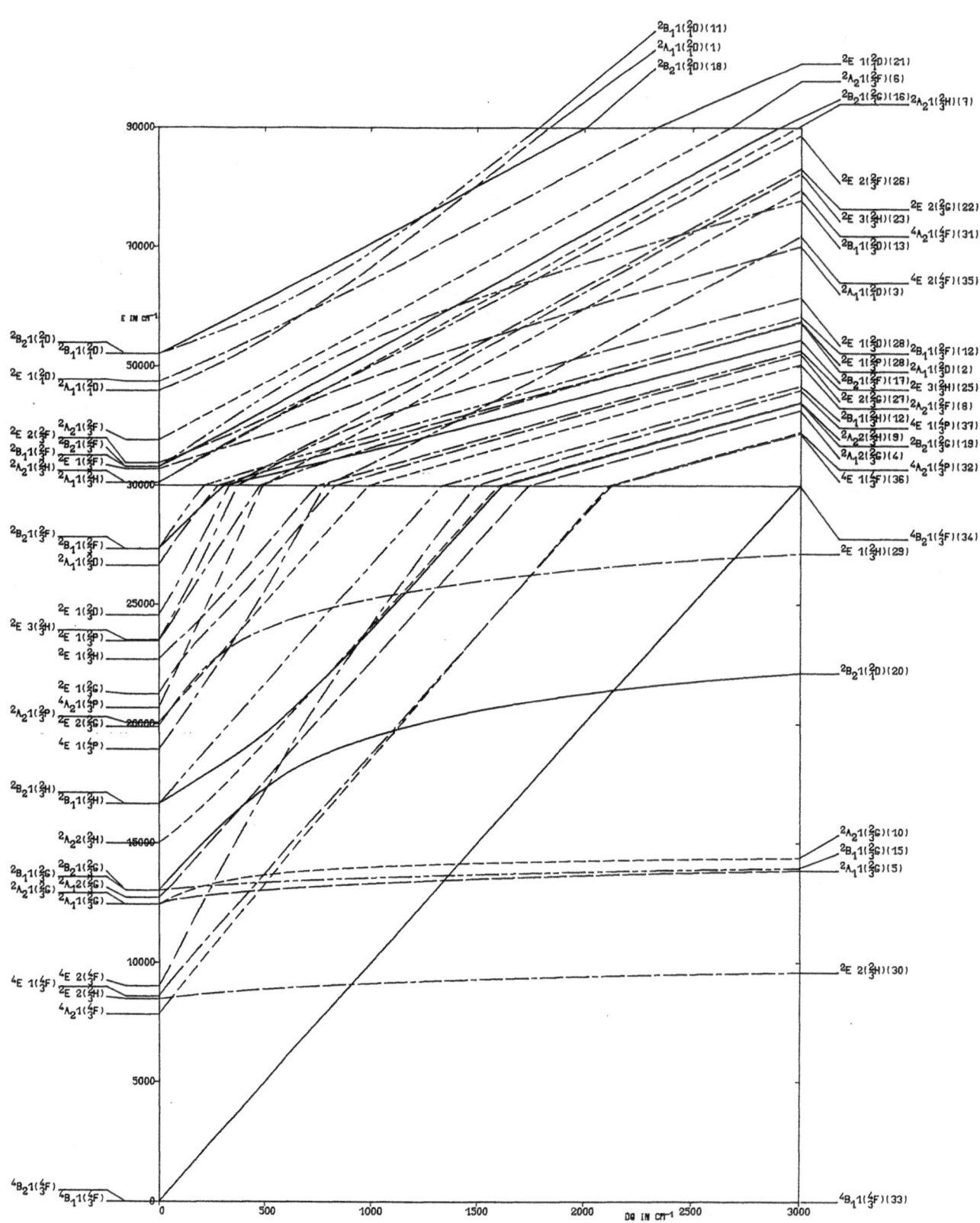

3 D ELECTRONS, TETRAGONAL SYMMETRY

B=700 C=4πB DT=-1000 K=-1 DS=KπDT ZETA=0

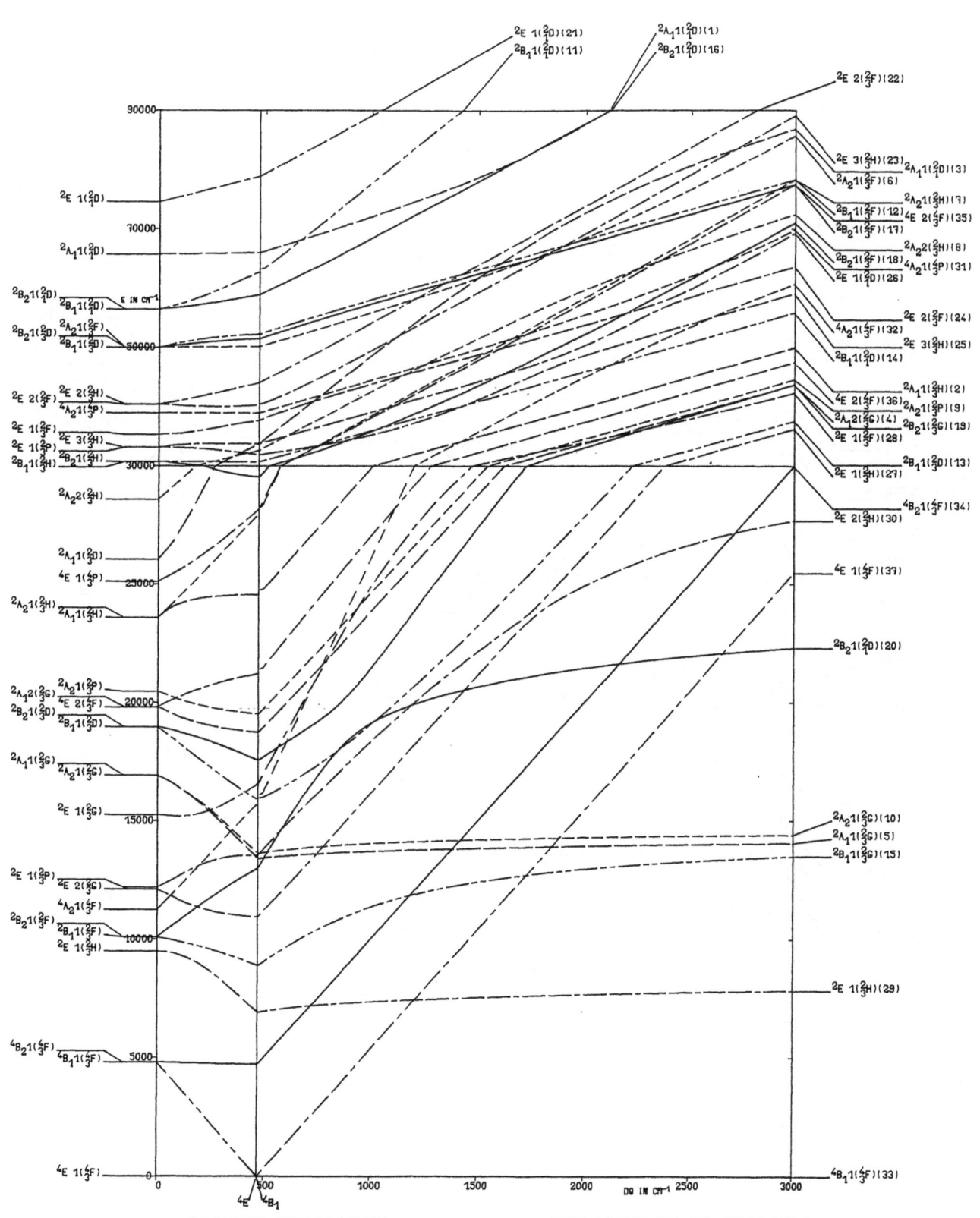

3 D ELECTRONS, TETRAGONAL SYMMETRY   B=700   C=4×B   DT=-1000   K=5   DS=K×DT   ZETA=0

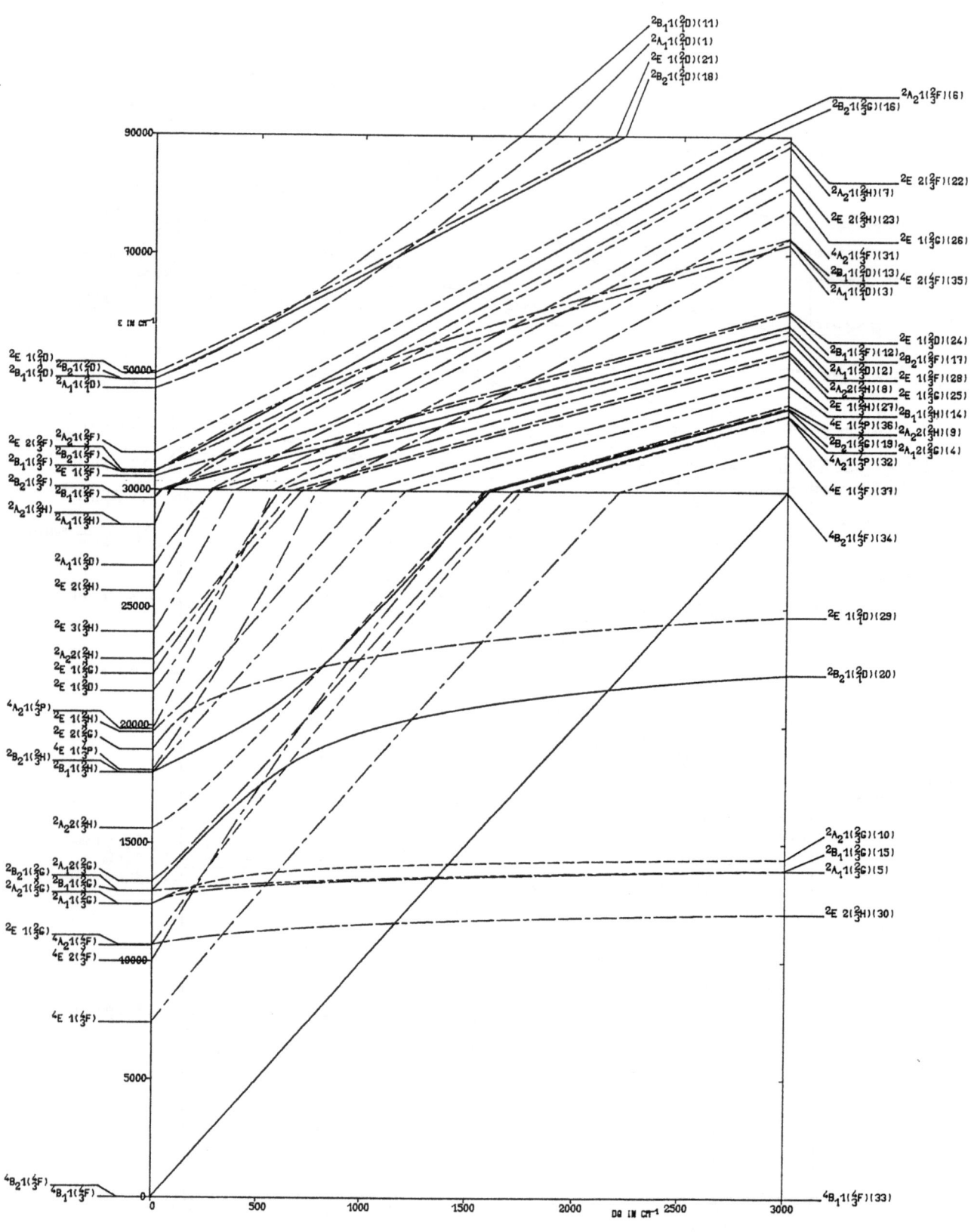

134

3 D ELECTRONS, TETRAGONAL SYMMETRY   B=700 C=4×B DT=-1000 K=-5 DS=K×DT ZETA=0

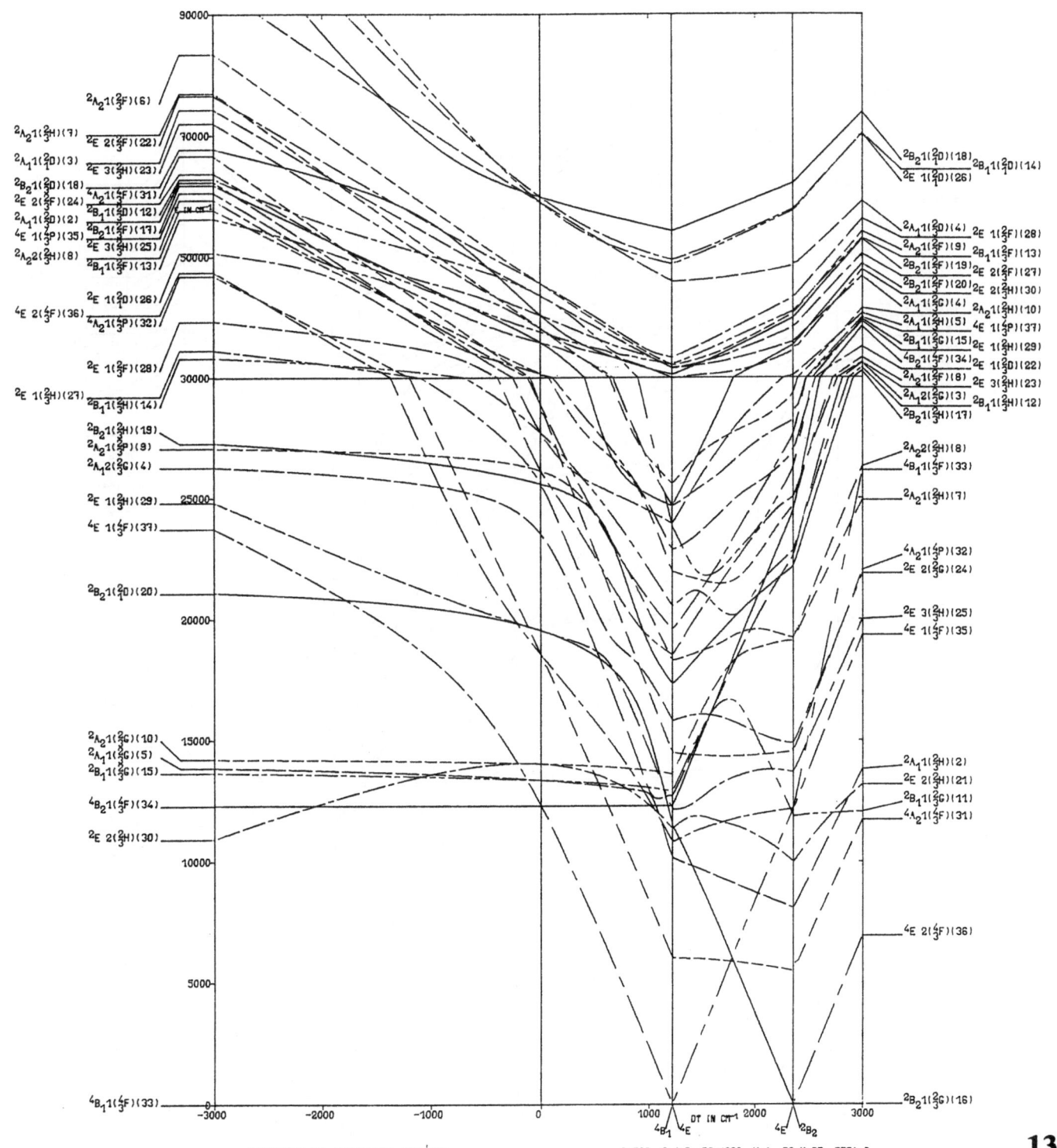

3 D ELECTRONS, TETRAGONAL SYMMETRY   B=700 C=4×B DQ=1230 K=1 DS=K×DT ZETA=0

135

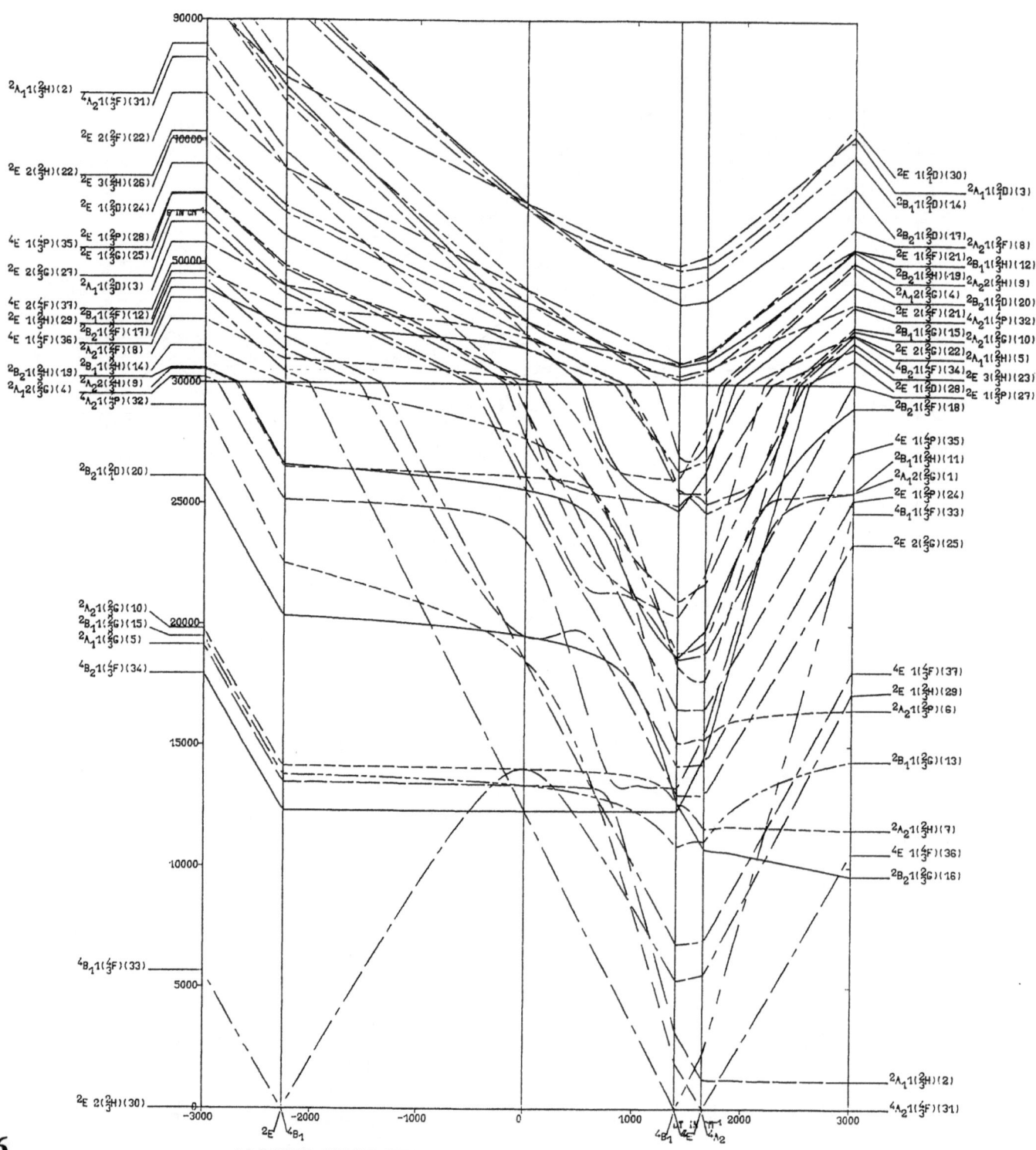

3 D ELECTRONS, TETRAGONAL SYMMETRY

B=700 C=4×B DQ=1230 K=-1 DS=K×DT 2ETA=0

136

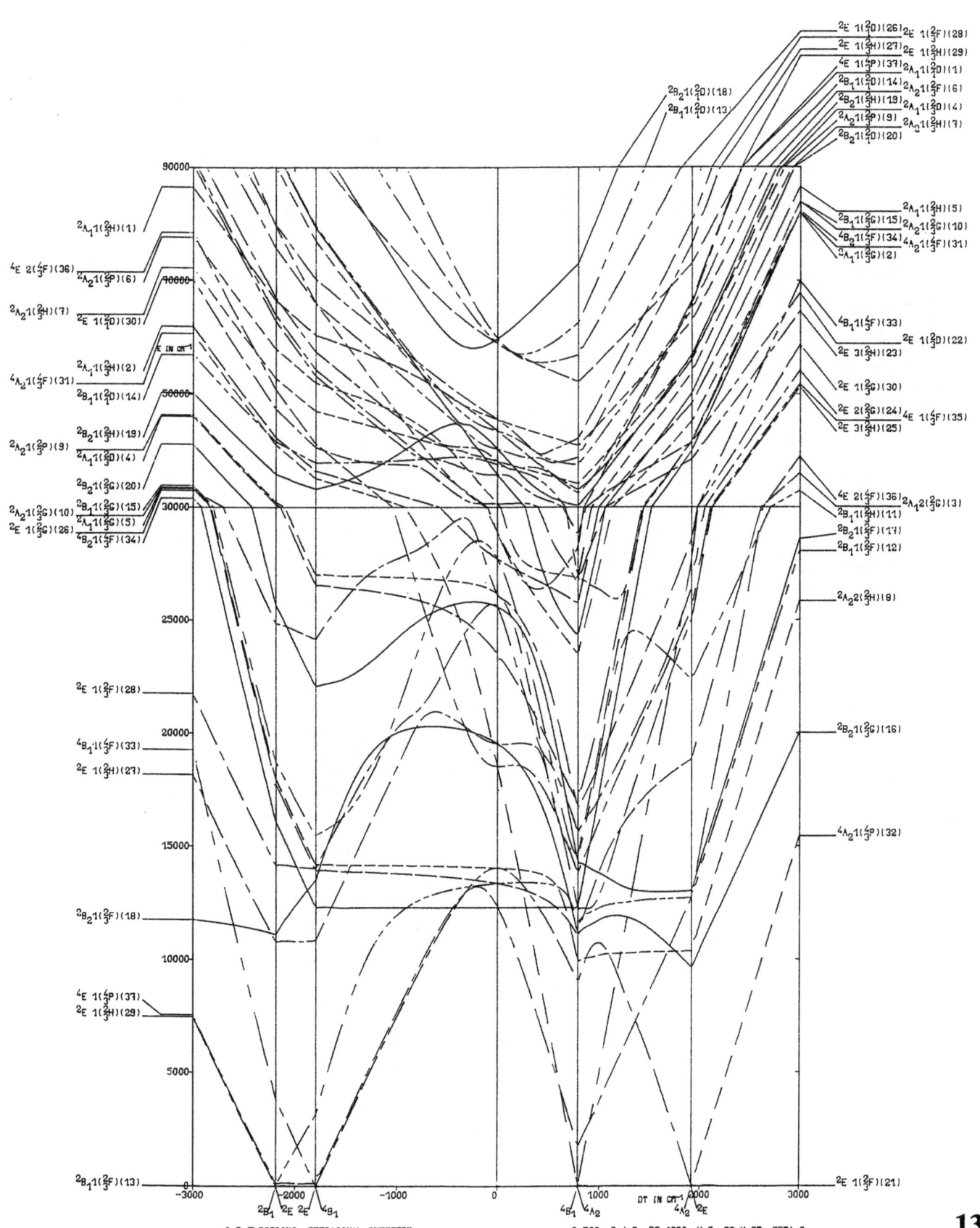

$^2A_1 1(^2H)(1)$

$^4E\ 2(^4F)(36)$ $^4A_2 1(^4P)(6)$

$^2A_2 1(^2H)(7)$ $^2E\ 1(^2D)(30)$

$^4A_2 1(^4F)(31)$ $^2A_1 1(^2H)(2)$ $^2B_1 1(^2D)(14)$

$^2A_2 1(^2P)(9)$ $^2B_1 1(^2H)(19)$ $^2A_1 1(^2D)(4)$

$^2A_2 1(^2G)(10)$ $^2B_2 1(^2G)(20)$
$^2A_2 1(^2G)(10)$ $^2B_1 1(^2G)(15)$
$^2E\ 1(^2G)(26)$ $^2A_1 1(^2G)(5)$
  $^4B_2 1(^4F)(34)$

$^2E\ 1(^2F)(28)$

$^4B_1 1(^4F)(33)$

$^2E\ 1(^2H)(27)$

$^4E\ 1(^4P)(37)$
$^2E\ 1(^2H)(29)$

$^2B_1 1(^2F)(13)$

$^2B_2 1(^2D)(18)$
$^2B_1 1(^2D)(13)$

$^2E\ 1(^2D)(26)$ $^2E\ 1(^2F)(28)$
$^2E\ 1(^2H)(27)$ $^2E\ 1(^2H)(29)$
$^4E\ 1(^4P)(37)$ $^2A_1 1(^2D)(1)$
$^2B_1 1(^2D)(14)$ $^2A_2 1(^2F)(6)$
$^2B_2 1(^2H)(19)$ $^2A_1 1(^2D)(4)$
$^2A_2 1(^2P)(9)$ $^2A_2 1(^2H)(7)$
$^2B_2 1(^2D)(20)$

$^2A_1 1(^2H)(5)$
$^2B_1 1(^2G)(15)$ $^2A_2 1(^2G)(10)$
$^4B_2 1(^4F)(34)$ $^4A_2 1(^4F)(31)$
$^2A_1 1(^2G)(2)$

$^4B_1 1(^4F)(33)$
$^2E\ 1(^2D)(22)$
$^2E\ 3(^2H)(23)$

$^2E\ 1(^2G)(30)$

$^2E\ 2(^2G)(24)$ $^4E\ 1(^4F)(35)$
$^2E\ 3(^2H)(25)$

$^4E\ 2(^4F)(36)$ $^2A_1 2(^2G)(3)$
$^2B_1 1(^2G)(11)$
$^2B_1 1(^2F)(17)$
$^2B_1 1(^2F)(12)$

$^2A_2 2(^2H)(8)$

$^2B_2 1(^2G)(16)$

$^4A_2 1(^4P)(32)$

$^2E\ 1(^2F)(21)$

-3000   -2000   -1000   0   1000 $DT\ IN\ CM^{-1}$ 2000   3000

$^2B_1$ ^2E ^2E $^4B_1$ $^4B_1$ $^4A_2$ $^4A_2$ ^2E

3 D ELECTRONS, TETRAGONAL SYMMETRY   B=700 C=4×B DQ=1230 K=5 DS=K×DT 2ETA=0

E IN CM⁻¹

90000

80000

70000

60000

50000

40000

30000

25000

20000

15000

10000

5000

0

137

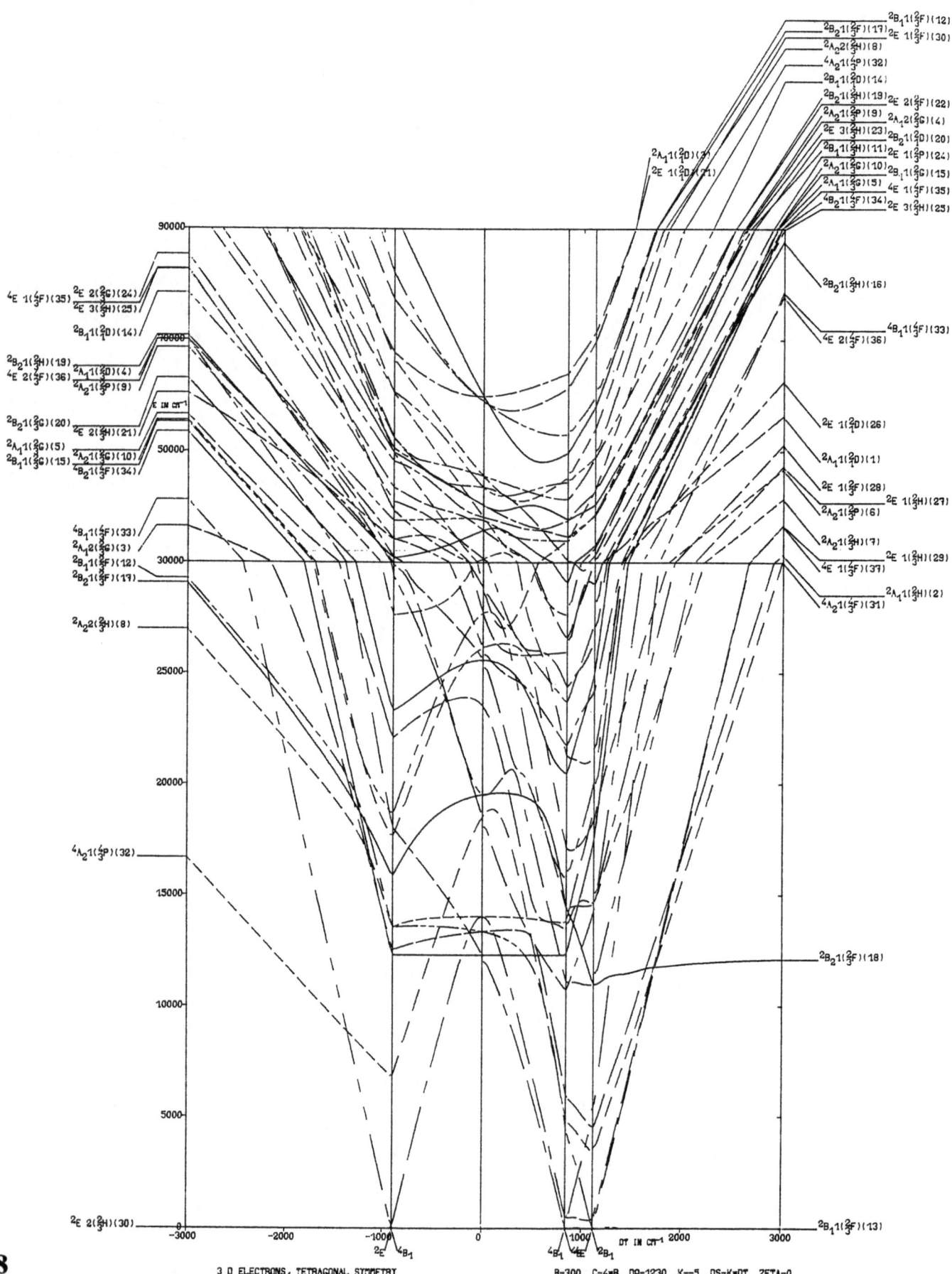

3 D ELECTRONS, TETRAGONAL SYMMETRY

B=700 C=4×B DQ=1230 K=-5 DS=K×DT ZETA=0

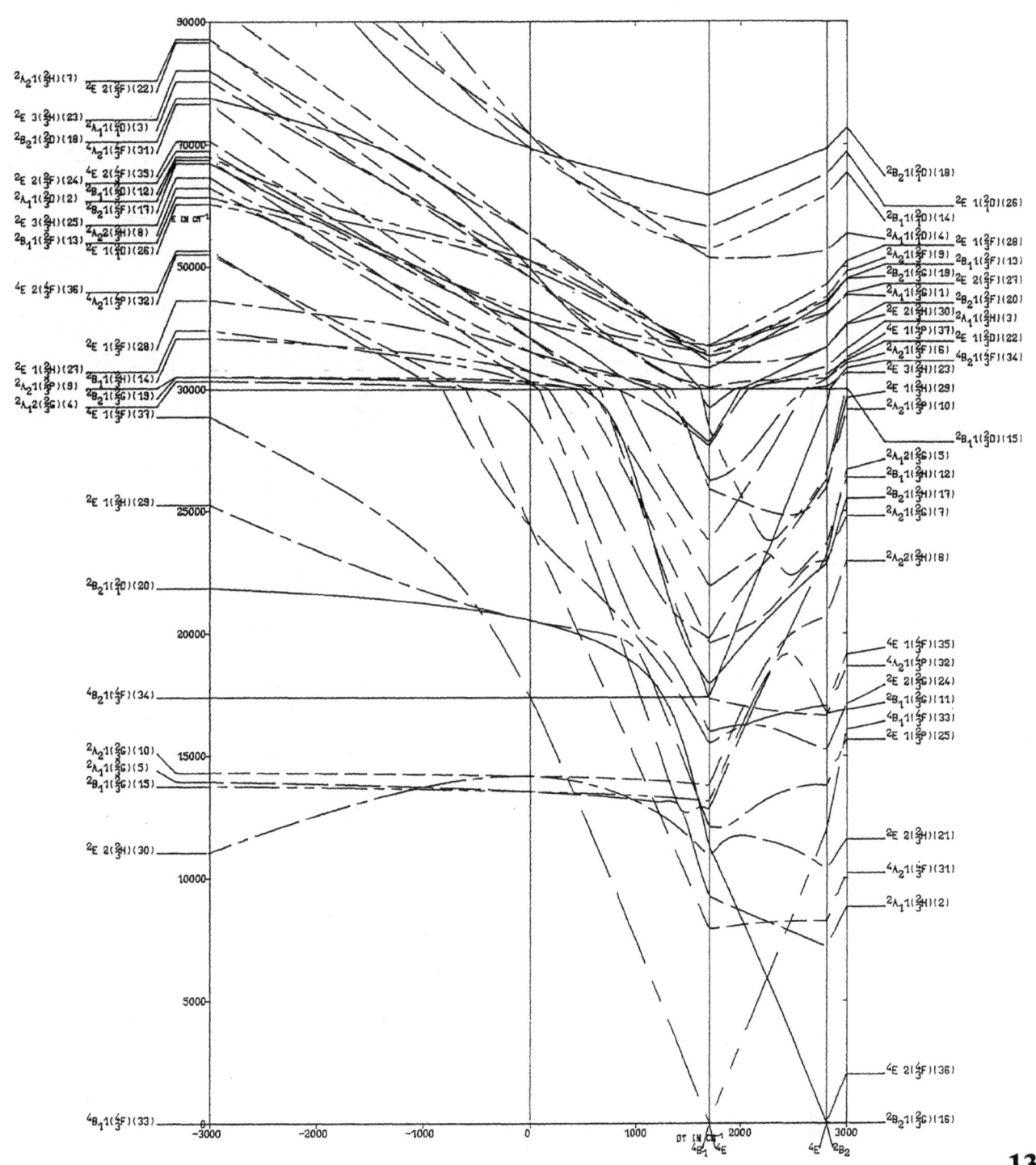

3 D ELECTRONS, TETRAGONAL SYMMETRY   B=700 C=4*B DQ=1740 K=1 DS=K*DT ZETA=0

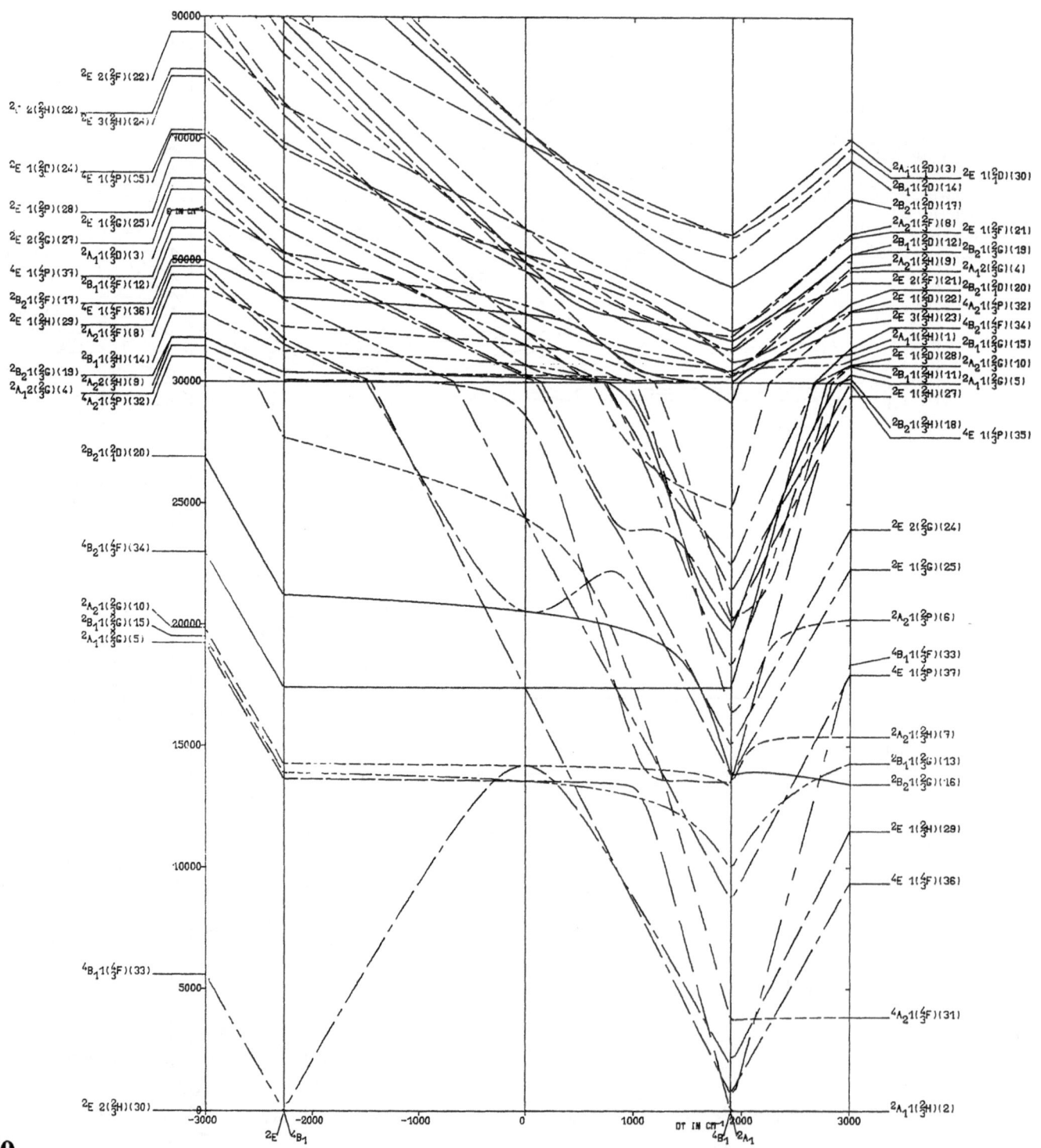

3 D ELECTRONS, TETRAGONAL SYMMETRY

B=700 C=4×B DQ=1940 K=-1 DS=K×DT ZETA=0

140

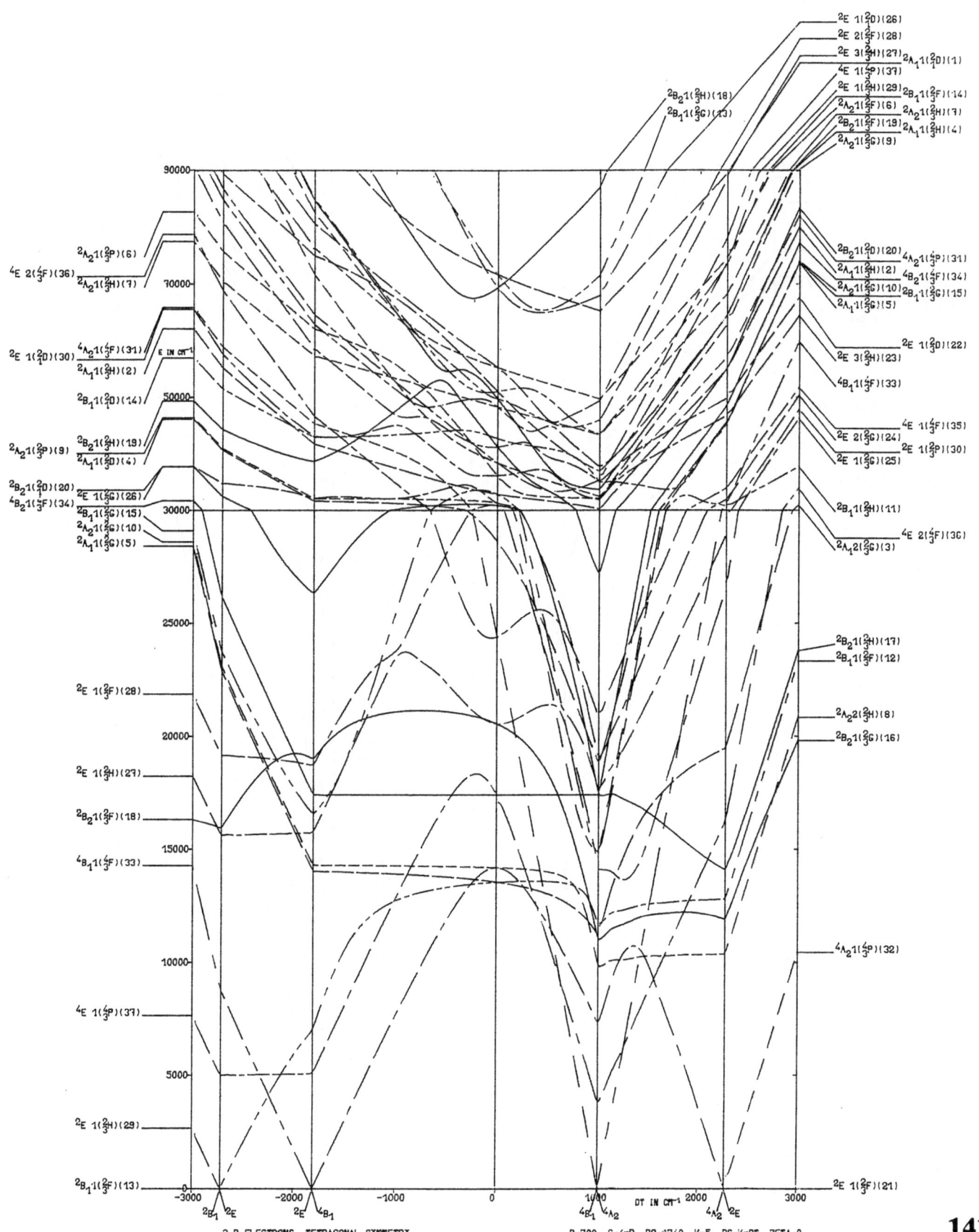

$^2E\ 1(^2_2D)(26)$
$^2E\ 2(^2_2F)(28)$
$^2E\ 3(^2_3H)(27)$ $_{2A_11(^2_2D)(1)}$
$^4E\ 1(^4_3P)(37)$
$^2E\ 1(^2_3H)(29)$ $_{2B_11(^2_3F)(14)}$
$^2B_21(^2_3H)(18)$ $^2A_21(^2_3F)(6)$ $_{2A_21(^2_3H)(7)}$
$^2B_11(^2_3G)(13)$ $^2B_21(^2_3F)(19)$ $_{2A_11(^2_3H)(4)}$
$_{2A_21(^2_3G)(9)}$

$^2A_21(^2_2P)(6)$
$^4E\ 2(^4_3F)(36)$ $_{2A_21(^2_3H)(7)}$
$^2B_21(^2_2D)(20)$
$^2A_11(^2_3H)(2)$ $^4A_21(^4_3P)(31)$
$^2A_11(^2_3G)(10)$ $^4B_21(^4_3F)(34)$
$^2A_11(^2_3G)(5)$ $^2B_11(^2_3G)(15)$

$^2E\ 1(^2_2D)(30)$ $^4A_21(^4_3F)(31)$
$^2A_11(^2_3H)(2)$
$^2E\ 1(^2_2D)(22)$

$^2B_11(^2_2D)(14)$
$^2E\ 3(^2_3H)(23)$

$^4B_11(^4_3F)(33)$

$^2A_21(^2_2P)(9)$ $^2B_21(^2_3H)(19)$ $^4E\ 1(^4_3F)(35)$
$^2A_11(^2_2D)(4)$ $^2E\ 2(^2_3G)(24)$
$^2E\ 1(^2_3P)(30)$
$^2E\ 1(^2_3G)(25)$

$^2B_21(^2_2D)(20)$
$^4B_21(^4_3F)(34)$ $^2E\ 1(^2_3G)(26)$ $^2B_11(^2_3H)(11)$
$^2B_11(^2_3G)(15)$
$^2A_21(^2_3G)(10)$
$^2A_11(^2_3G)(5)$ $^4E\ 2(^4_3F)(36)$
$^2A_12(^2_3G)(3)$

$^2B_21(^2_3H)(17)$
$^2E\ 1(^2_3F)(28)$ $^2B_11(^2_3F)(12)$

$^2E\ 1(^2_3H)(27)$

$^2A_22(^2_3H)(8)$
$^2B_21(^2_3F)(18)$ $^2B_21(^2_3G)(16)$

$^4B_11(^4_3F)(33)$

$^4A_21(^4_3P)(32)$

$^4E\ 1(^4_3P)(37)$

$^2E\ 1(^2_3H)(29)$

$^2B_11(^2_3F)(13)$ $^2E\ 1(^2_3F)(21)$

-3000   -2000   -1000   0   1000   DT IN CM^{-1} 2000   3000

$^2B_1$ ^2E ^2E $^4B_1$ $^4B_1$ $^4A_2$ $^4A_2$ ^2E

3 D ELECTRONS, TETRAGONAL SYMMETRY   B=700 C=4*B DQ=1740 K=5 DS=K*DT ZETA=0

141

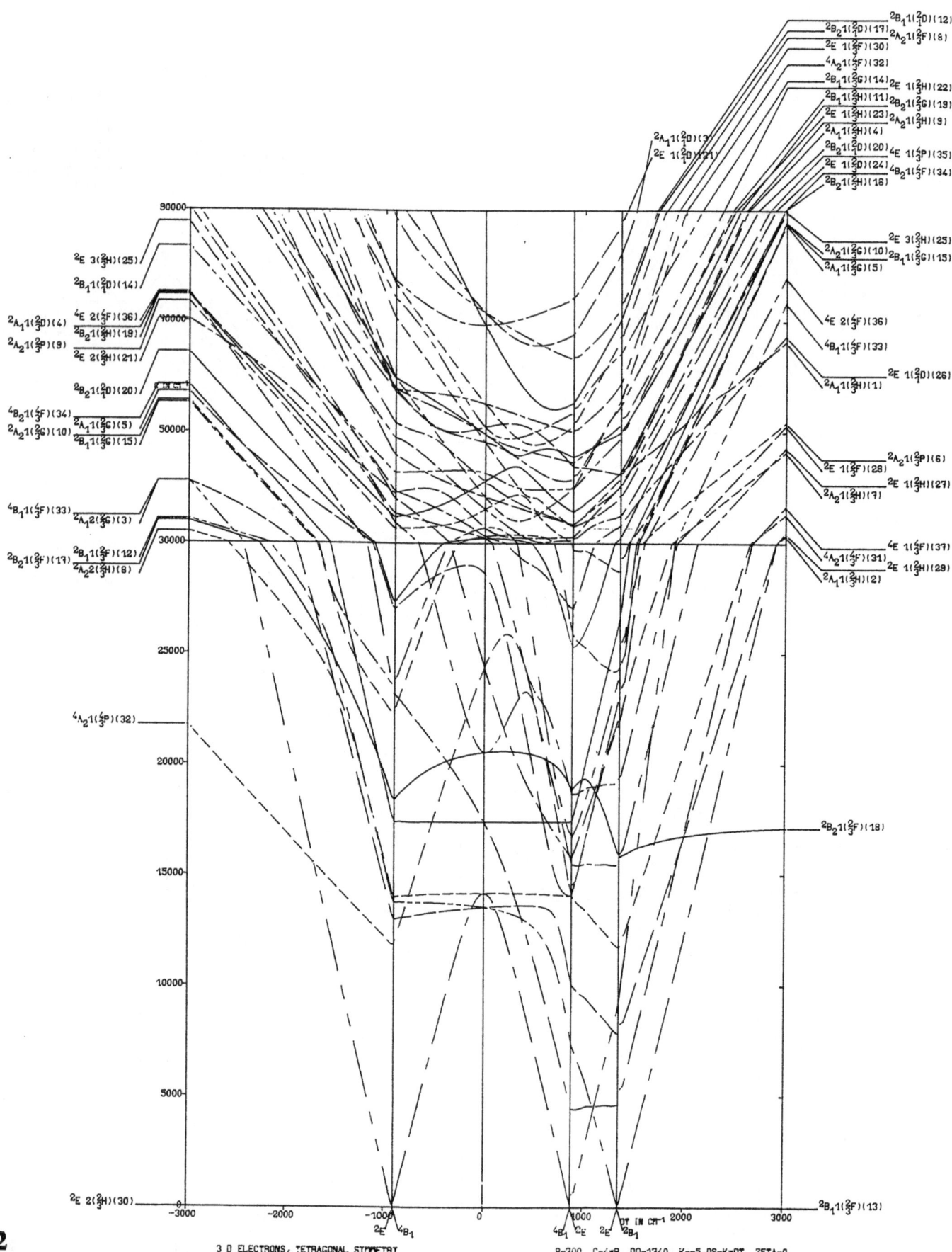

3 D ELECTRONS, TETRAGONAL SYMMETRY

B=700 C=4×B DQ=1740 K=-5 DS=K×DT ZETA=0

3 D ELECTRONS, TRIGONAL SYMMETRY

B=700   C=4*B   ZETA=0

ENERGY AS FUNCTION OF DQ
DT=0, 500, -500, 1000, -1000
K=1, -1, 5, -5

ENERGY AS FUNCTION OF DT
DQ=1230, 1740
K=1, -1, 5, -5

(33) ^4E $(E_E E_E(^3A_2)E_T)$
(32) ^4E $(E_T E_T(^3A_2)E_E)$
(31) ^4E $(A_1 E_T(^3E)E_E)$
(30) $^4A_2(E_E E_E(^3A_2)A_1)$
(29) $^4A_2(A_1 E_T(^3E)E_E)$
(28) $^4A_2(E_T E_T(^3A_2)A_1)$
(27) $^4A_1(A_1 E_T(^3E)E_E)$
(26) ^2E $(E_E E_E(^1A_1)E_E)$
(25) ^2E $(E_E E_E(^3A_2)E_T)$
(24) ^2E $(E_E E_E(^1E)E_T)$
(23) ^2E $(E_E E_E(^1A_1)E_T)$
(22) ^2E $(E_T E_T(^3A_2)E_E)$
(21) ^2E $(E_T E_T(^1E)E_E)$
(20) ^2E $(E_T E_T(^1A_1)E_E)$
(19) ^2E $(E_T E_T(^1A_1)E_T)$
(18) ^2E $(E_E E_E(^1E)A_1)$
(17) ^2E $(A_1 E_T(^3E)E_E)$
(16) ^2E $(A_1 E_T(^1E)E_E)$
(15) ^2E $(E_T E_T(^1E)A_1)$
(14) ^2E $(A_1 A_1(^1A_1)E_E)$
(13) ^2E $(A_1 A_1(^1A_1)E_T)$
(12) $^2A_2(E_E E_E(^1E)E_T)$
(11) $^2A_2(E_T E_T(^1E)E_E)$
(10) $^2A_2(E_E E_E(^3A_2)A_1)$
(9) $^2A_2(A_1 E_T(^3E)E_E)$
(8) $^2A_2(A_1 E_T(^1E)E_E)$
(7) $^2A_2(E_T E_T(^3A_2)A_1)$
(6) $^2A_1(E_E E_E(^1E)E_T)$
(5) $^2A_1(E_T E_T(^1E)E_E)$
(4) $^2A_1(E_E E_E(^1A_1)A_1)$
(3) $^2A_1(A_1 E_T(^3E)E_E)$
(2) $^2A_1(A_1 E_T(^1E)E_E)$
(1) $^2A_1(E_T E_T(^1A_1)A_1)$

143

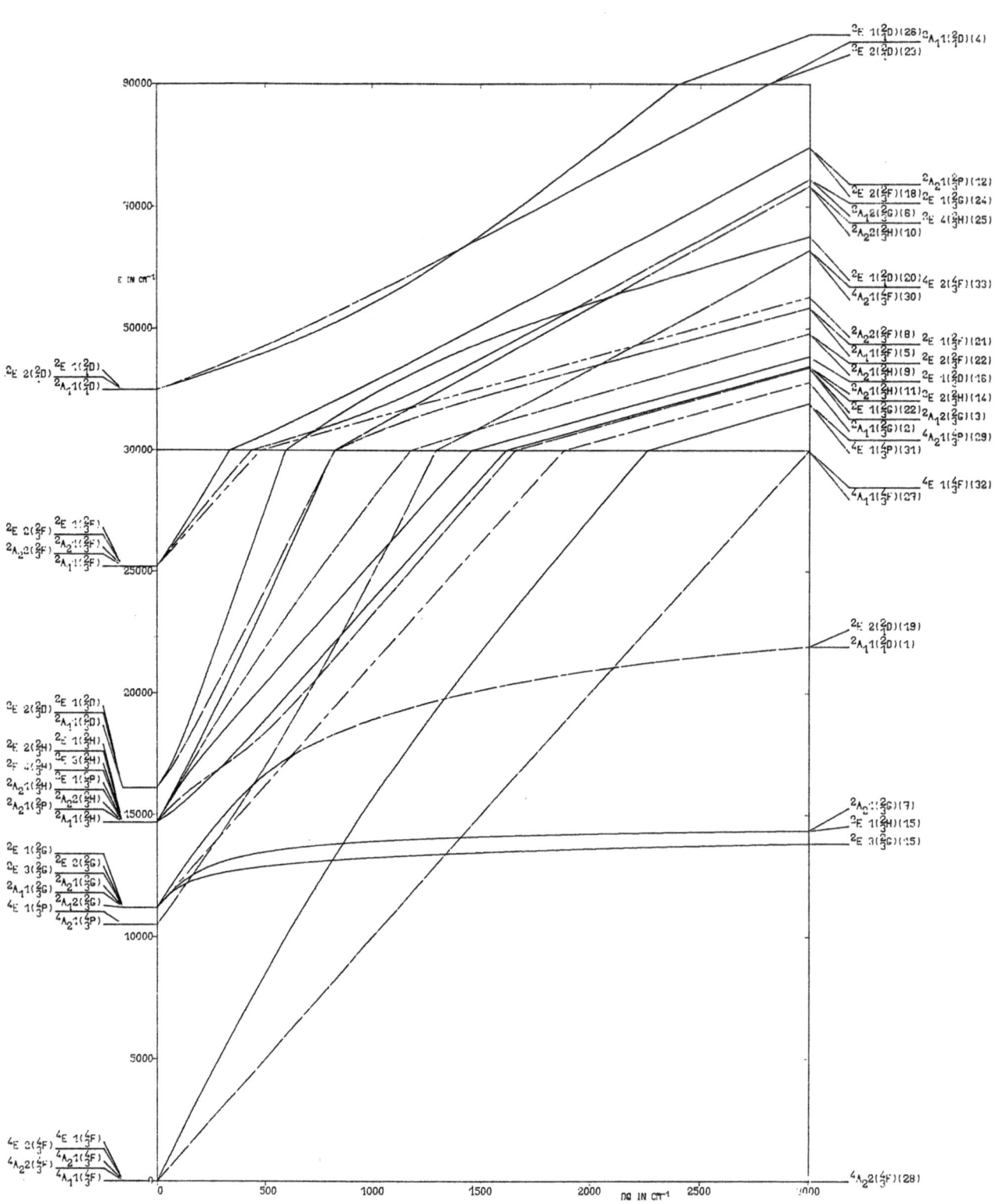

3 D ELECTRONS, TRIGONAL SYMMETRY   B=700 C=4.8 DT=0   K=0   ZETA=0

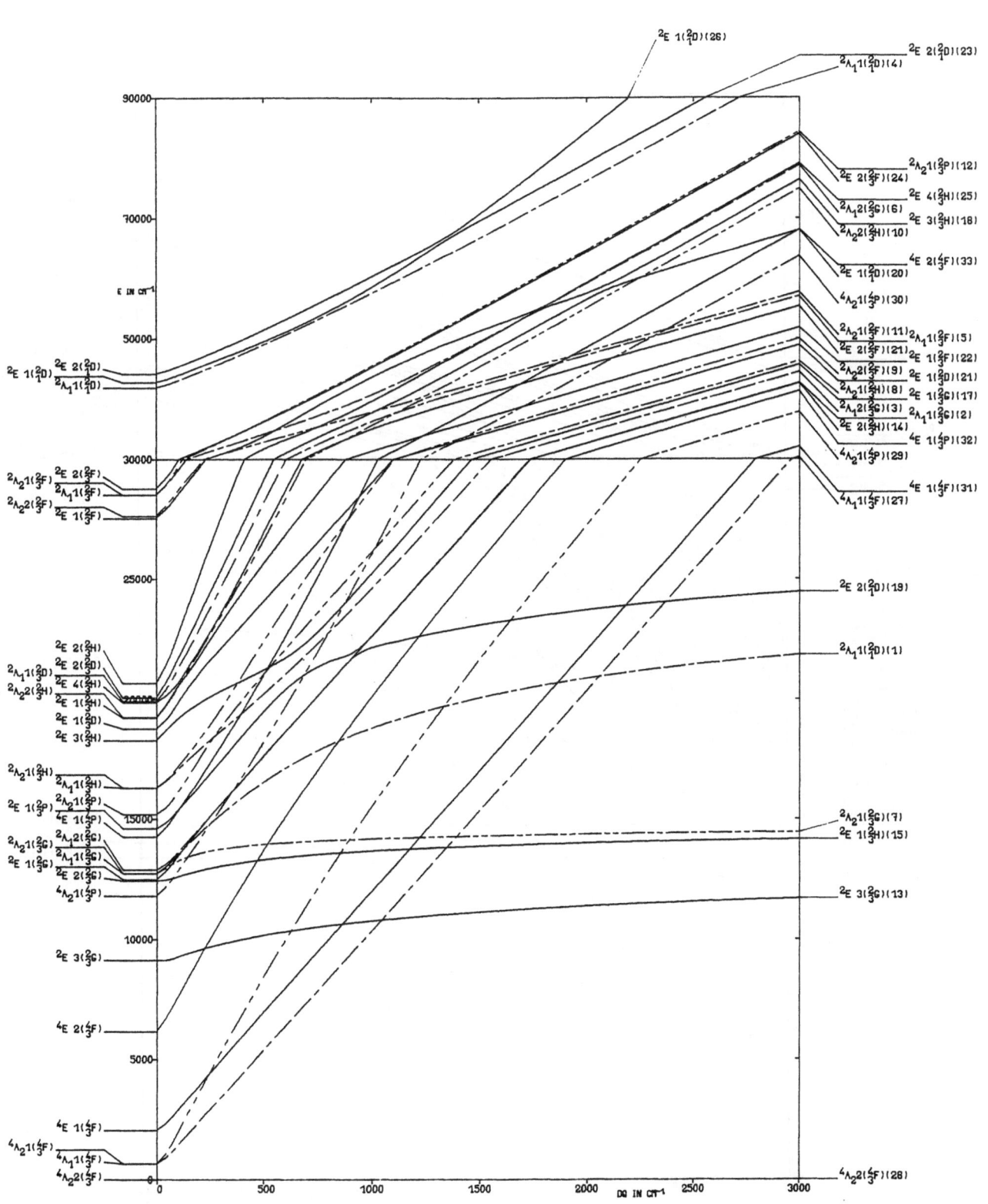

3 D ELECTRONS, TRIGONAL SYMMETRY   B=700 C=4×B DT=500 K=1 DS=K×DT ZETA=0

145

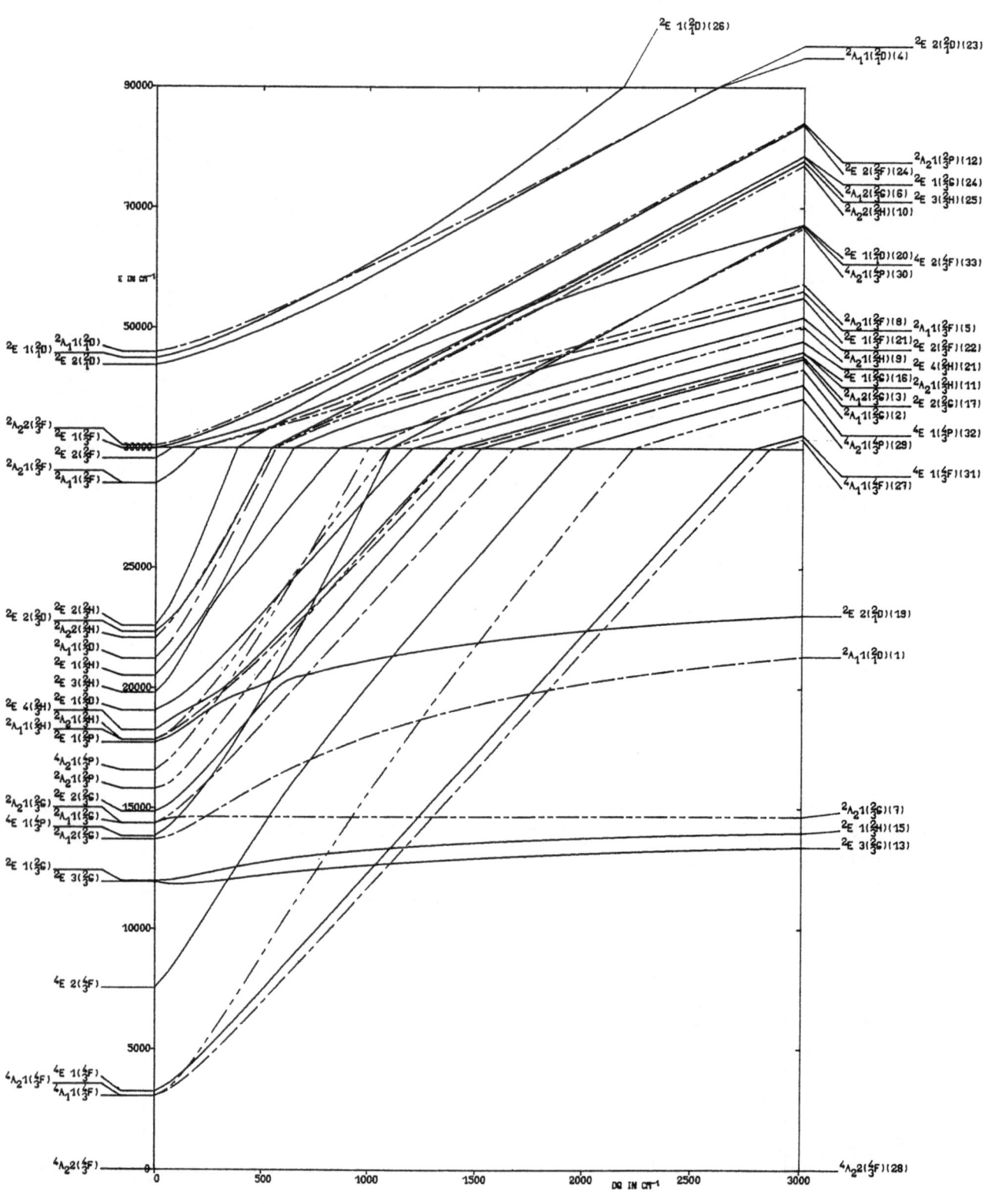

3 D ELECTRONS, TRIGONAL SYMMETRY

B=700 C=4*B DT=500 K=-1 DS=K*DT ZETA=0

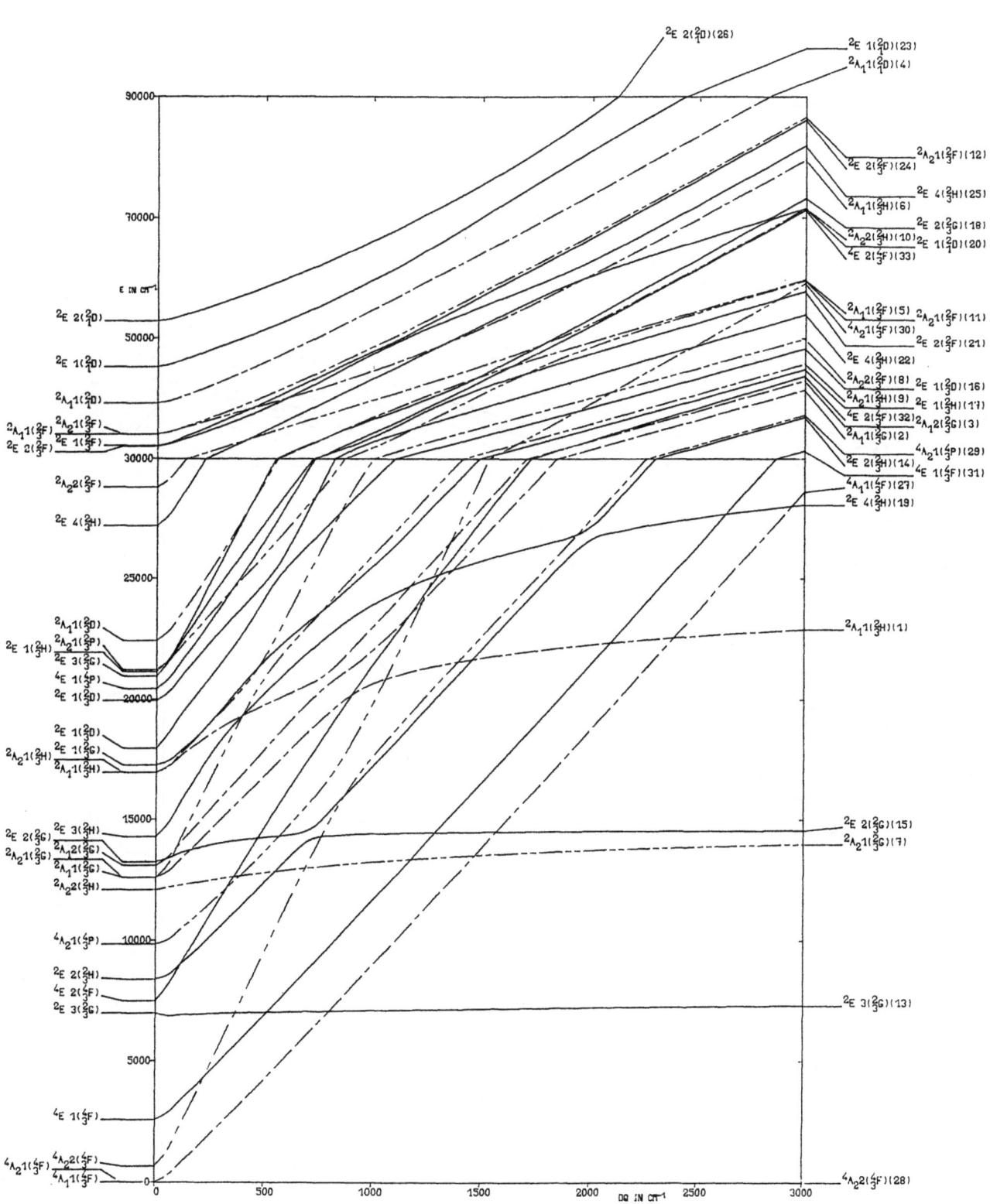

3 D ELECTRONS, TRIGONAL SYMMETRY   R=700 C=4xB DT=500 K=5 DS=KxDT ZETA=0

147

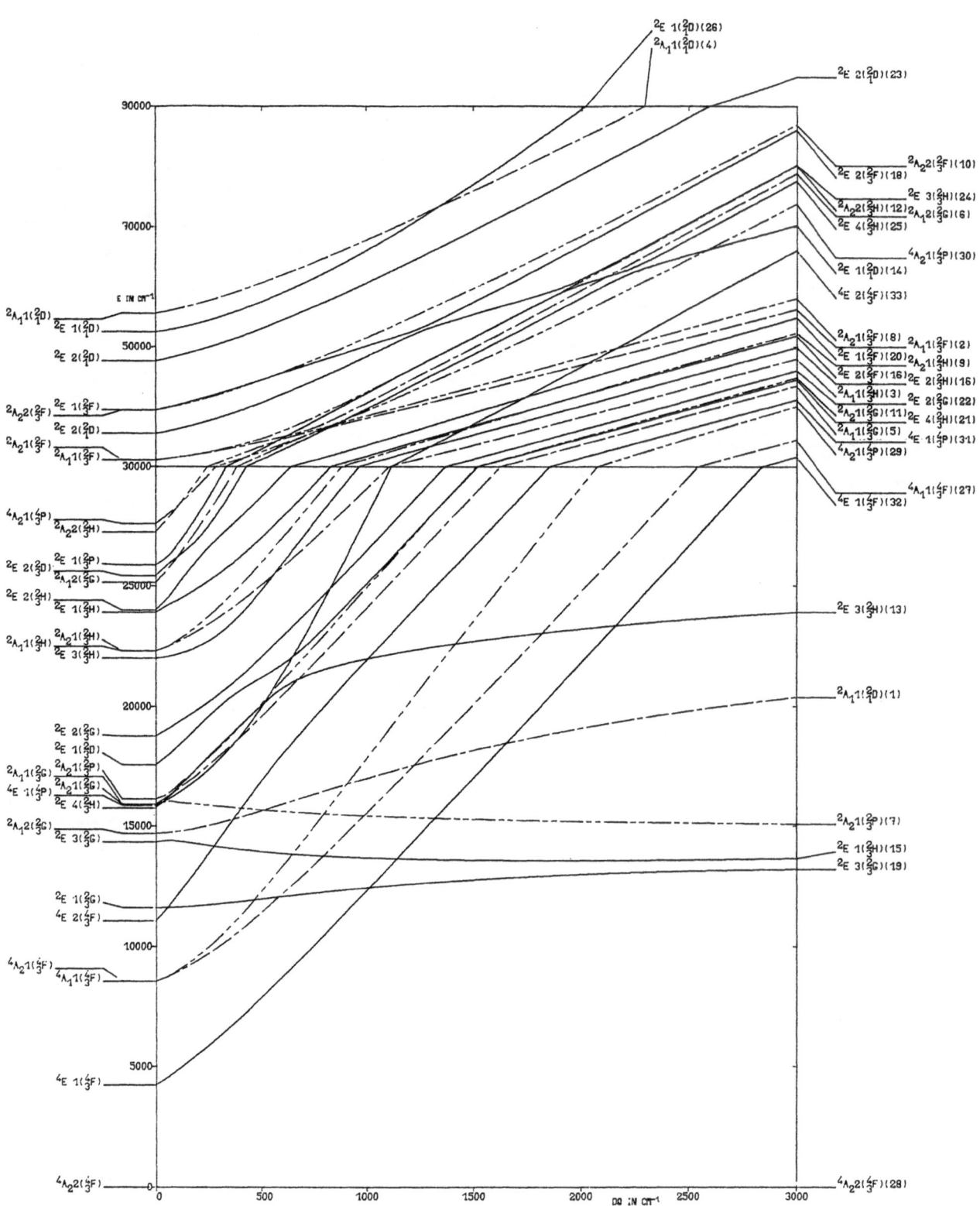

3 D ELECTRONS, TRIGONAL SYMMETRY   B=700 C=4×B DT=500 K=-5 DS=K×DT ZETA=0

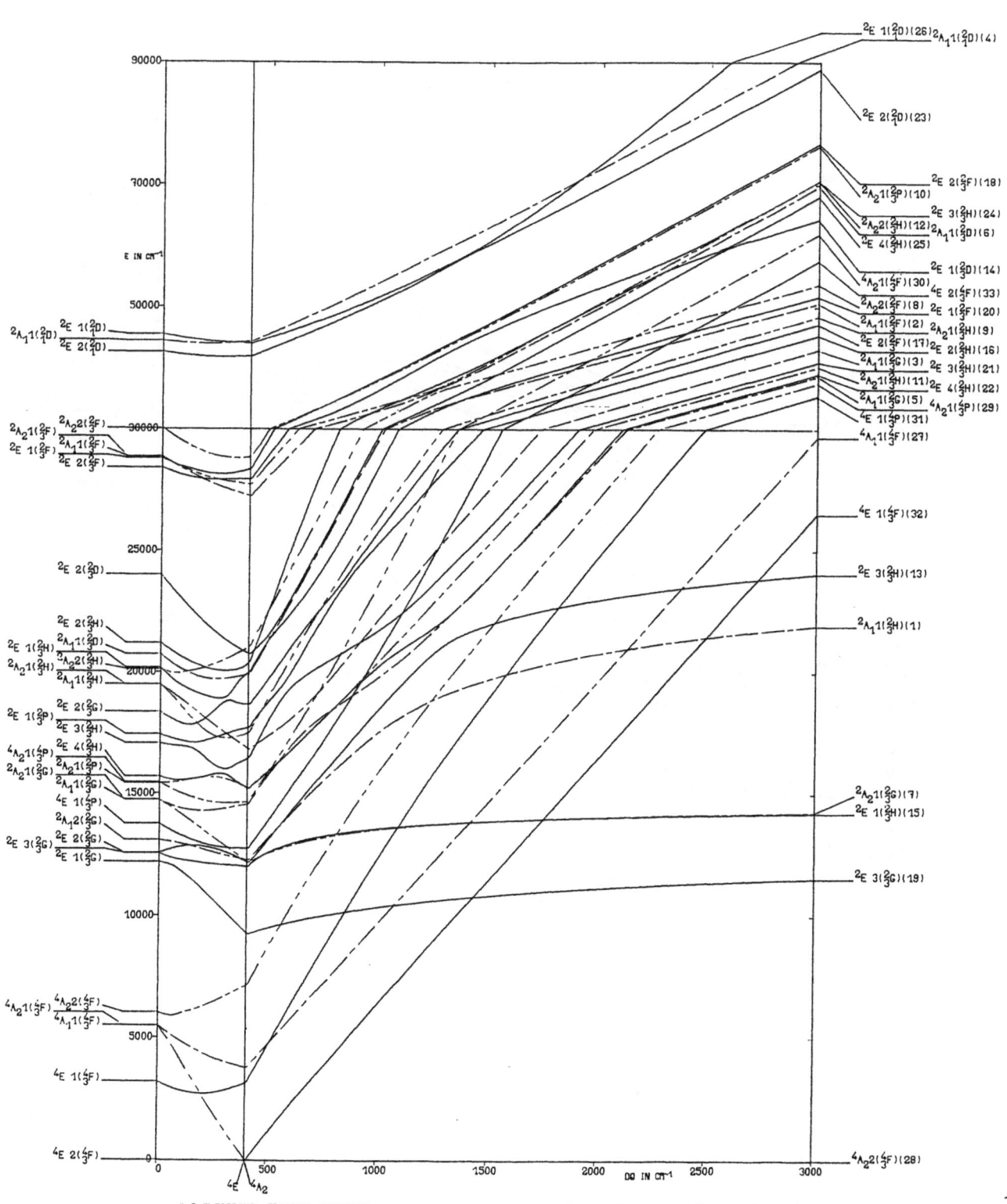

3 D ELECTRONS, TRIGONAL SYMMETRY

B=700 C=4×B DT=-500 K=1 DS=K×DT ZETA=0

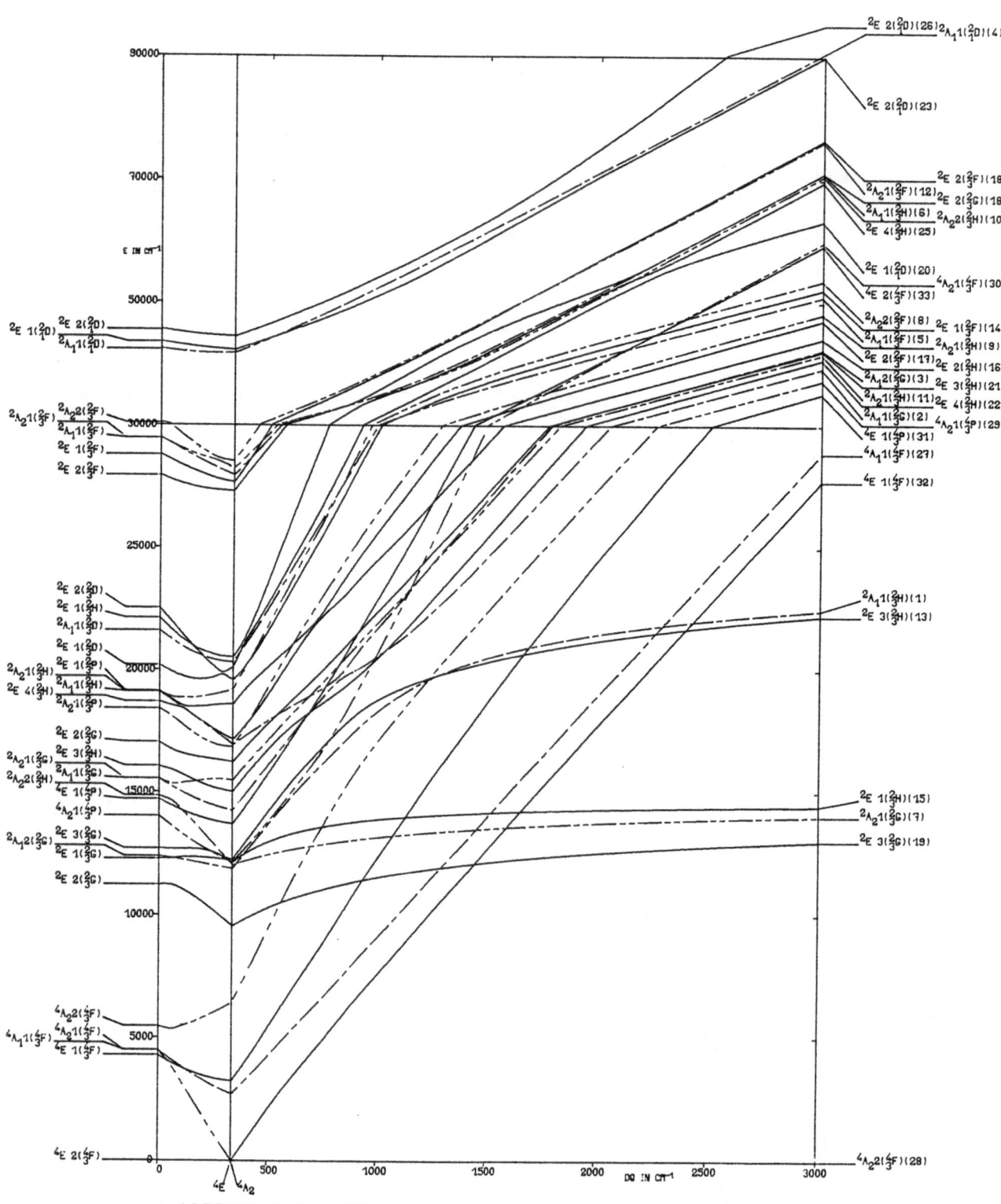

150

3 D ELECTRONS, TRIGONAL SYMMETRY

B=700 C=4×B DT=-500 K=-1 DS=K×DT ZETA=0

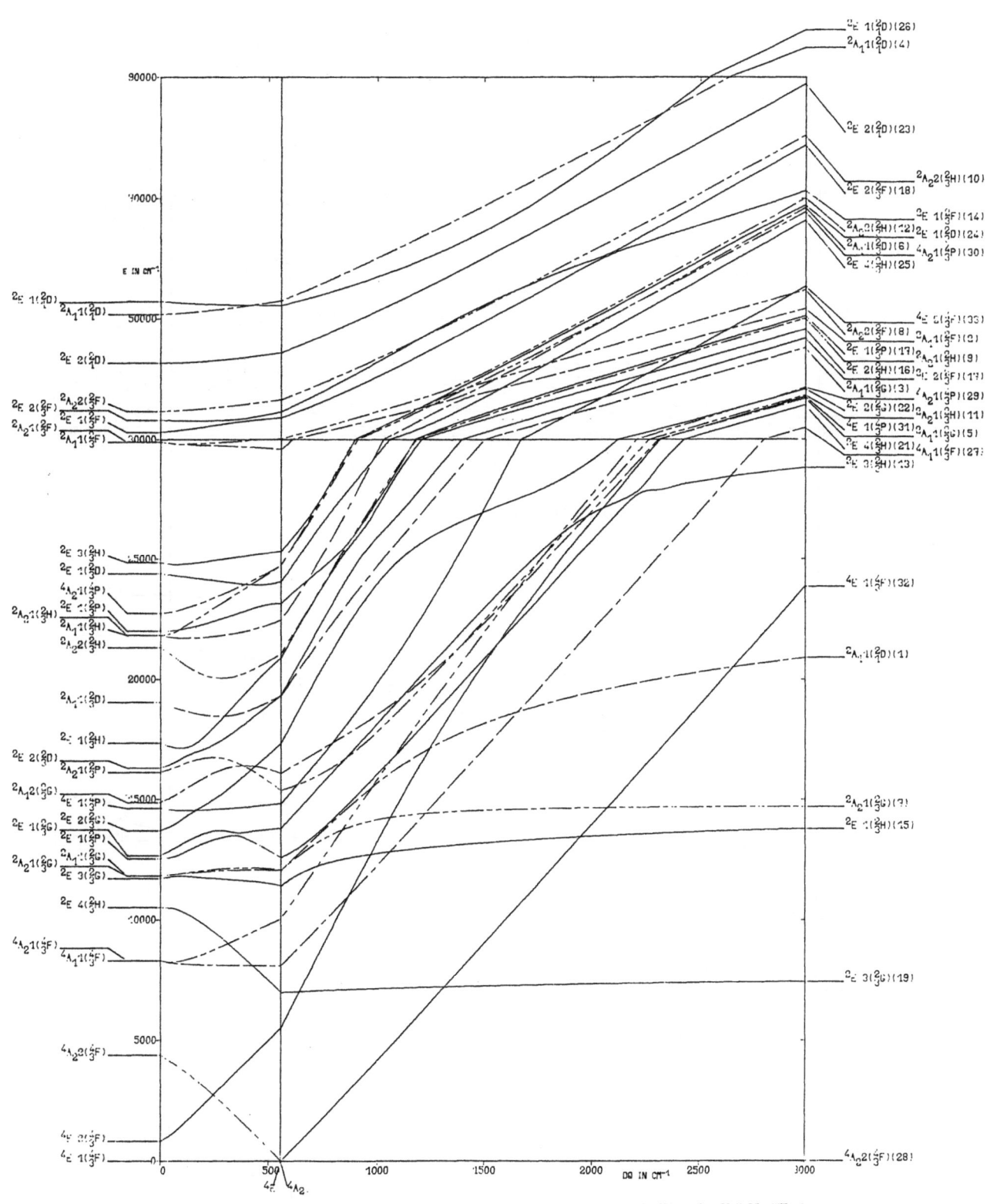

3 D ELECTRONS, TRIGONAL SYMMETRY   B=700 C=4*B DT=-500 K=5 DS=K*DT ZETA=0

151

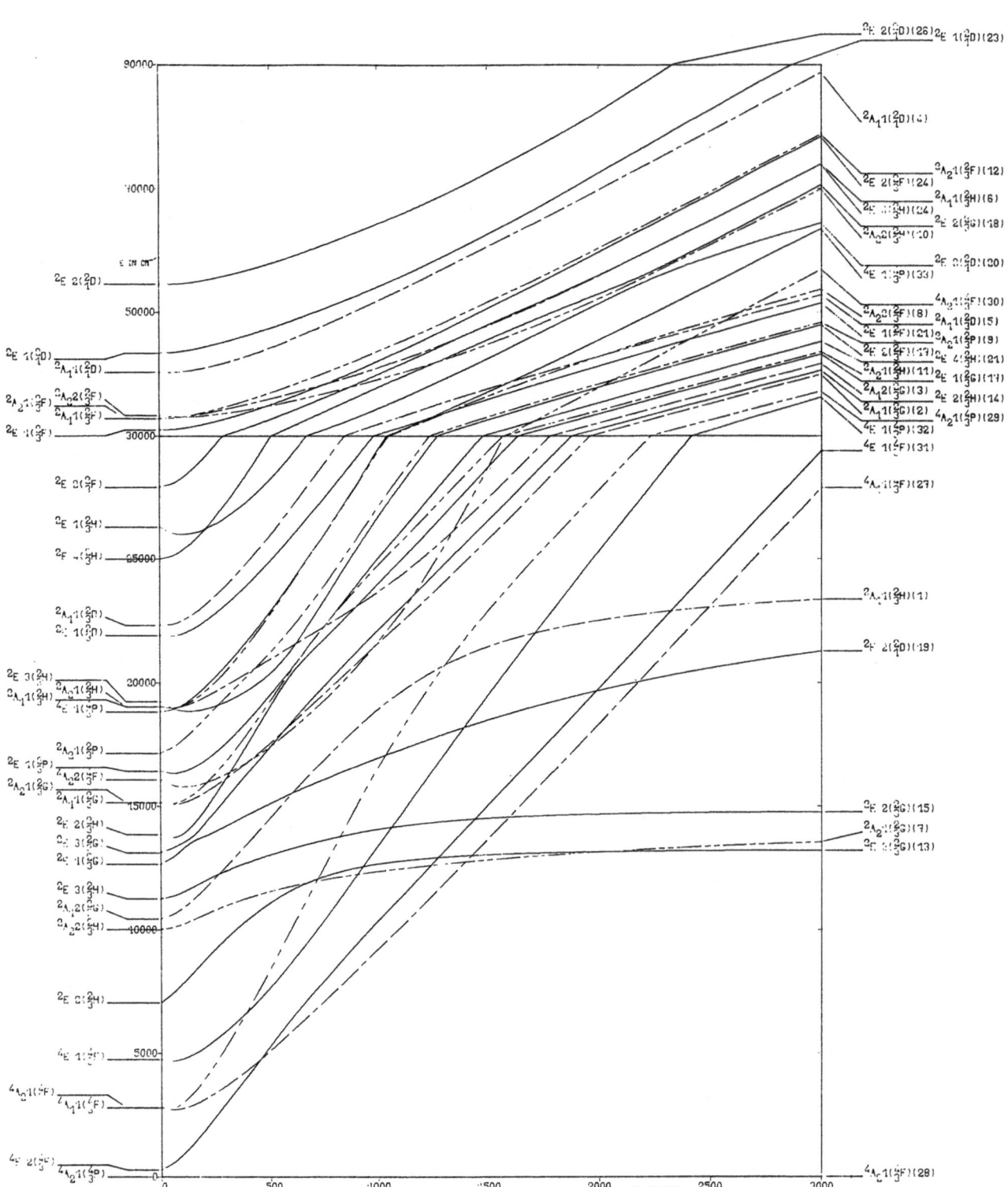

3 D ELECTRONS, TRIGONAL SYMMETRY   B=500 C=4*B DT=-500 K=-5 DS=K*DT ZETA=0

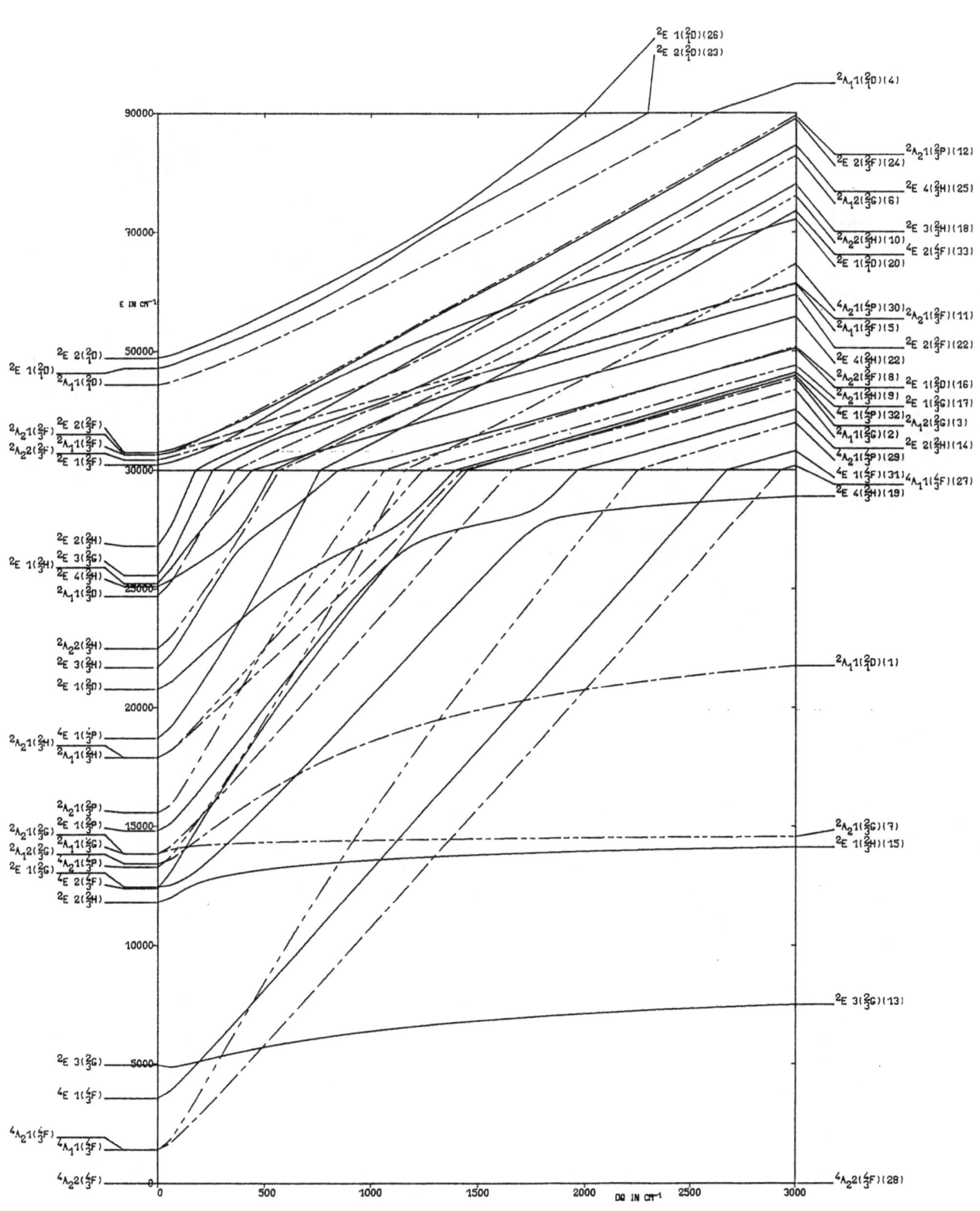

3 D ELECTRONS, TRIGONAL SYMMETRY   B=700 C=4×B DT=1000 K=1 DS=K×DT ZETA=0

153

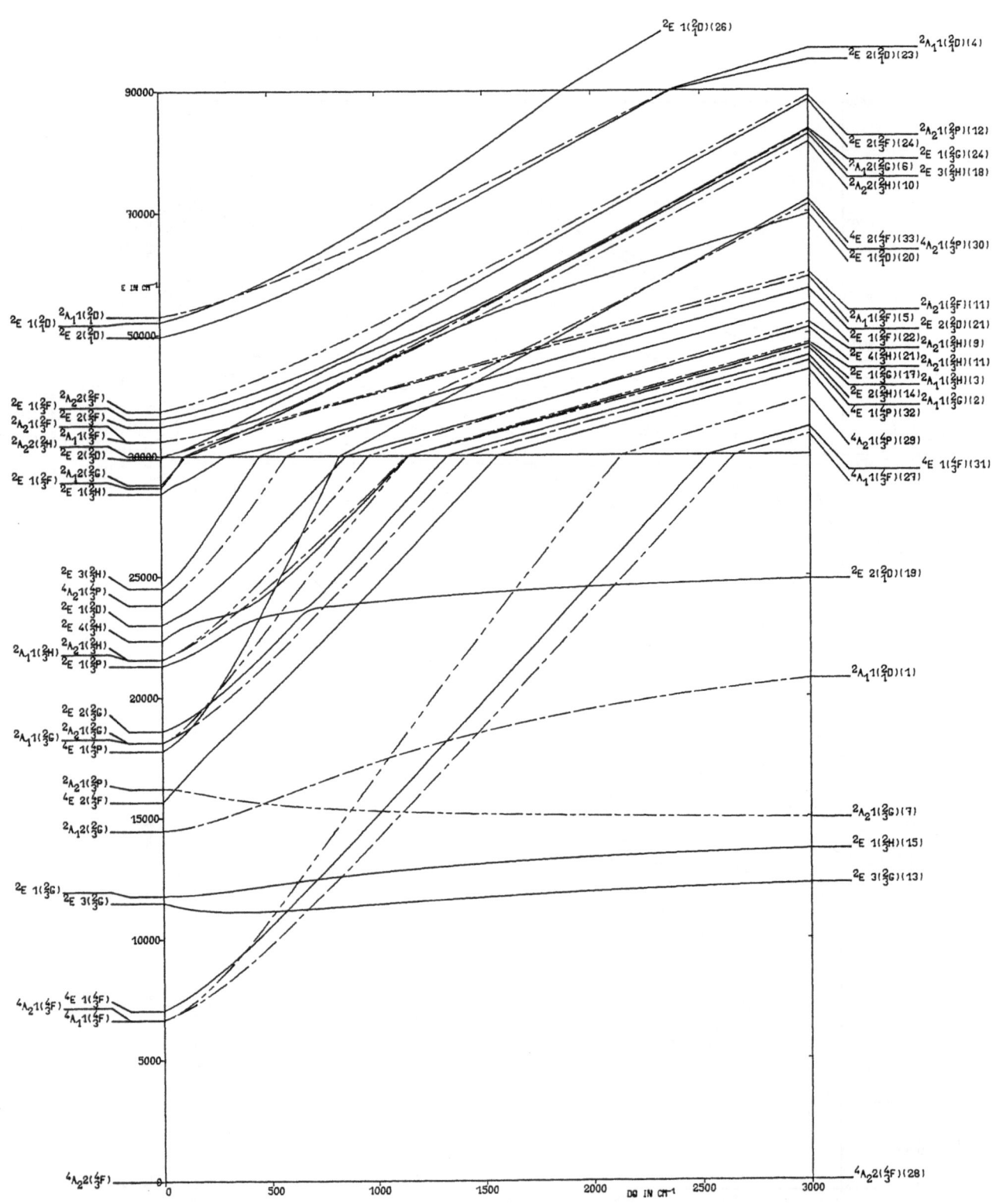

3 D ELECTRONS, TRIGONAL SYMMETRY

B=700 C=4*B DT=1000 K=-1 DS=K*DT ZETA=0

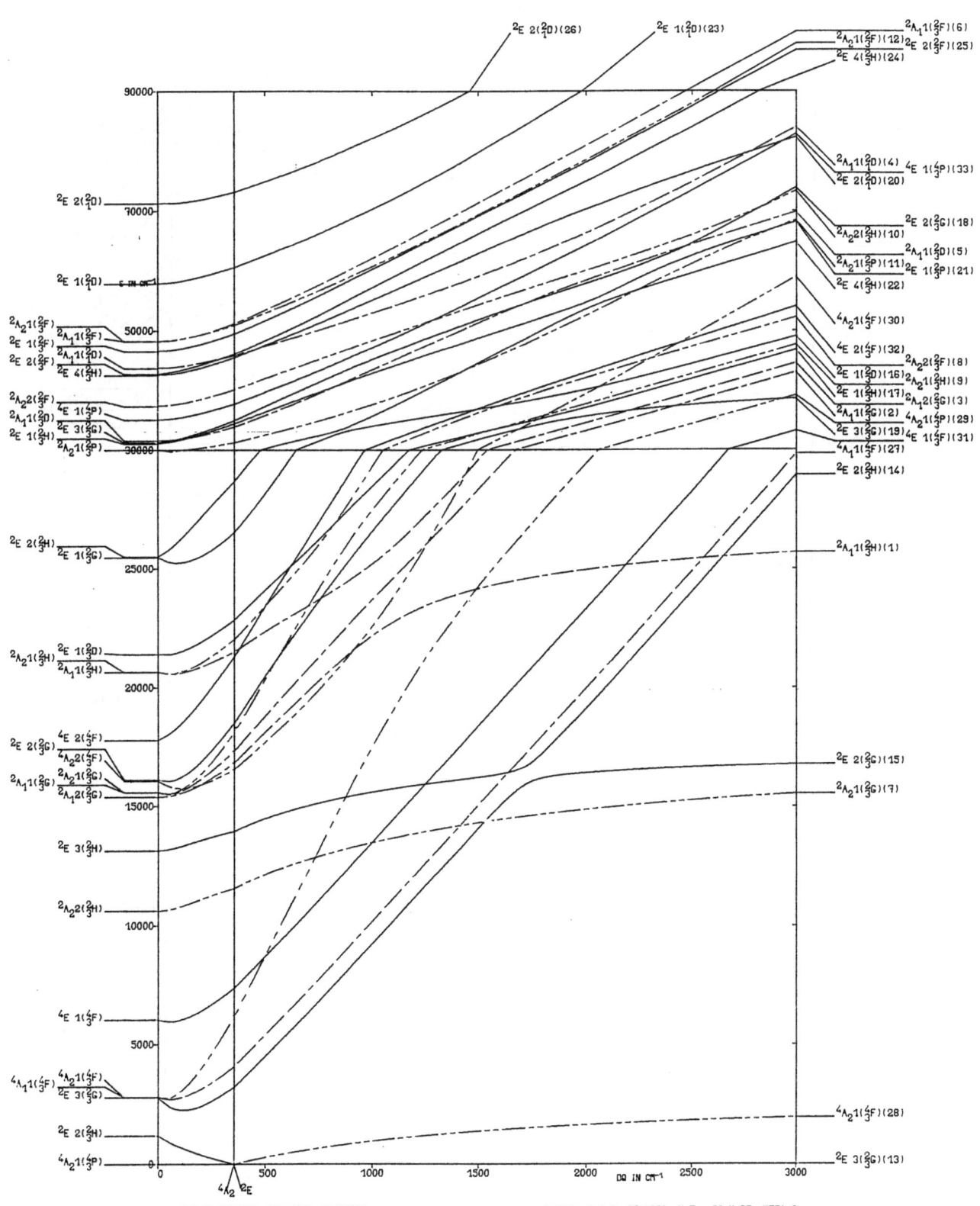

3 D ELECTRONS, TRIGONAL SYMMETRY   B=700   C=4×B   DT=1000   K=5   DS=K×DT   ZETA=0

155

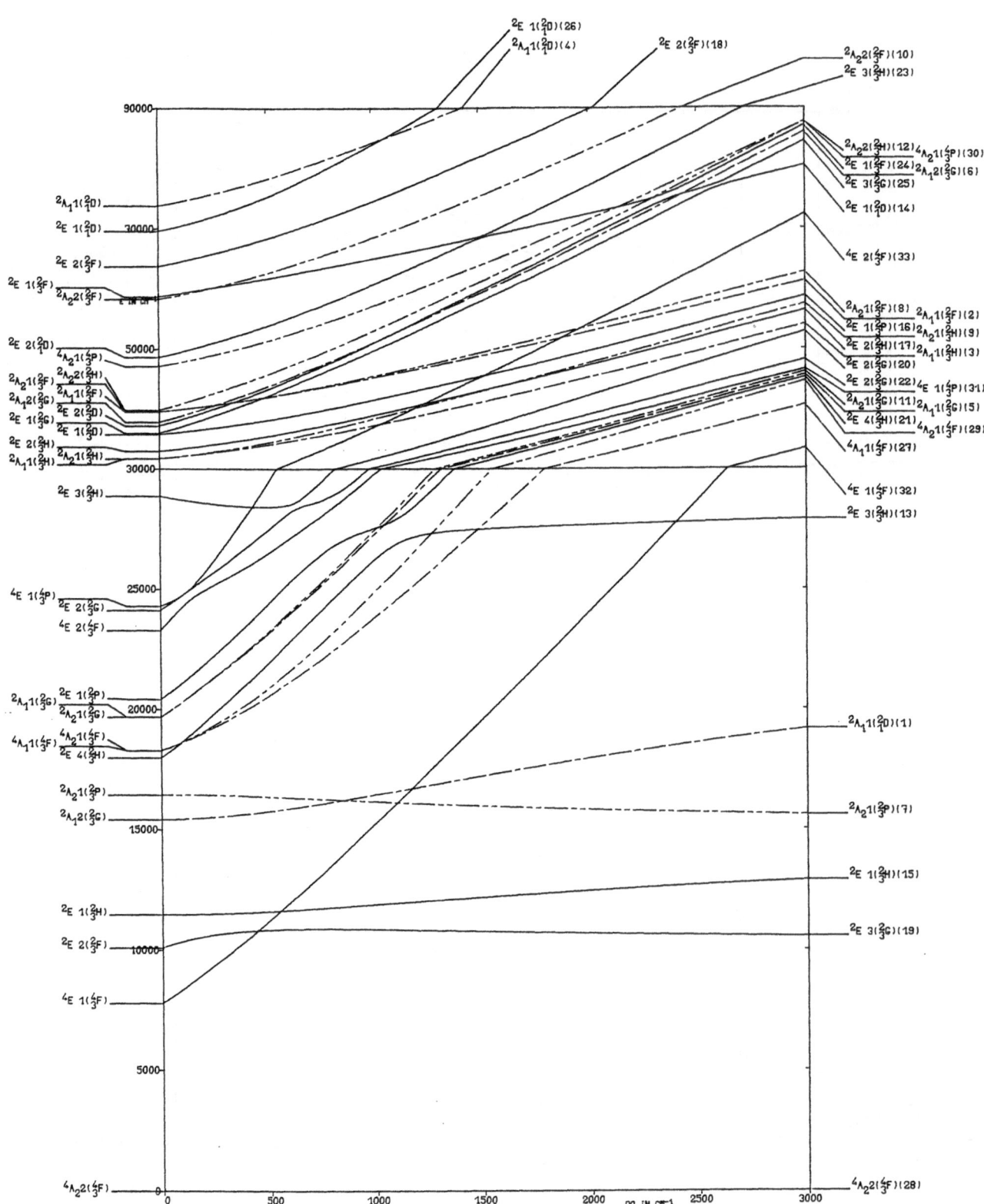

3 D ELECTRONS, TRIGONAL SYMMETRY

B=700   C=4*B   DT=1000   K=-5   DS=K*DT   ZETA=0

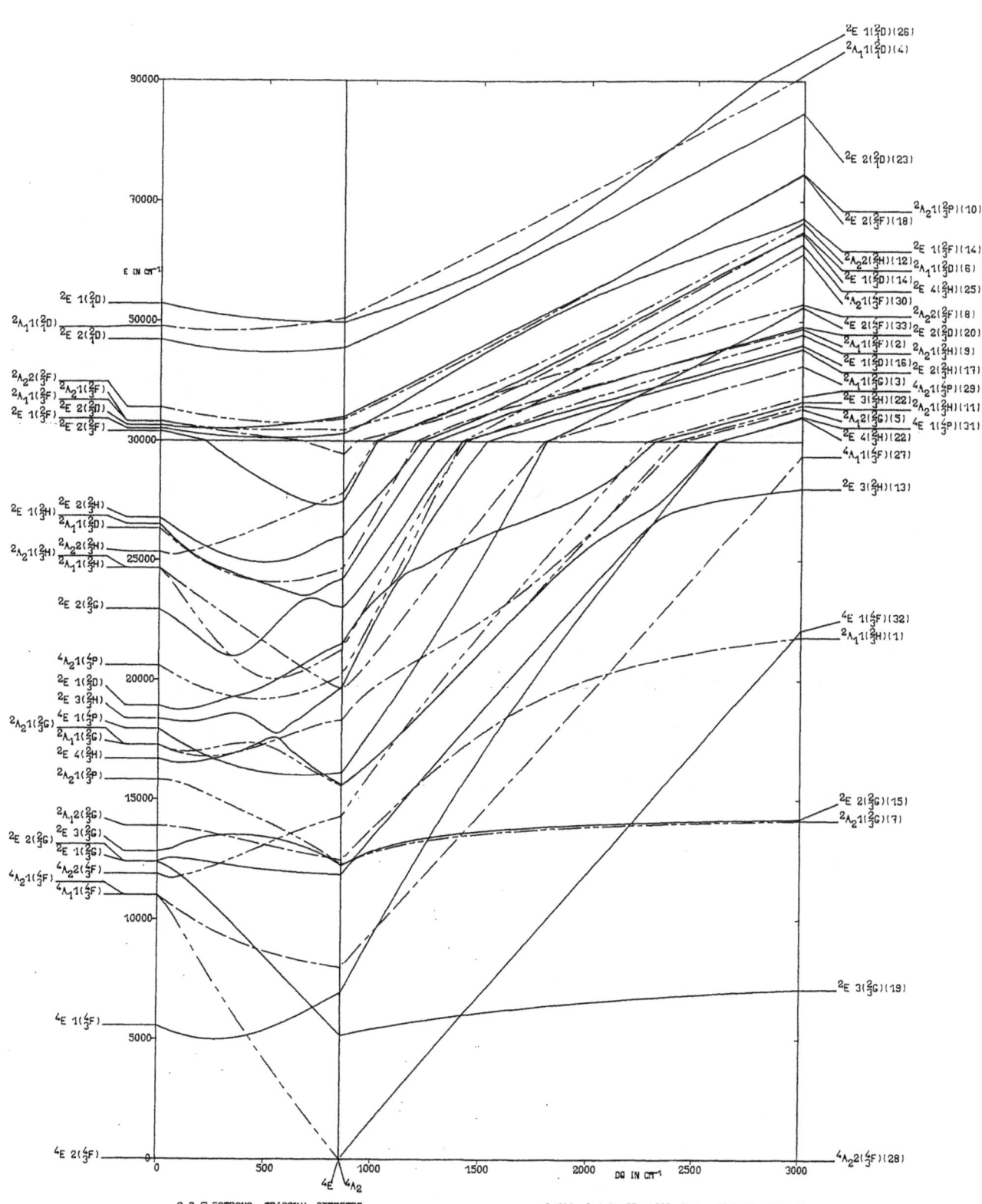

90000

70000

E IN CM⁻¹

$^2E\ 1(^2_1D)$
50000
$^2A_11(^2_1D)$
$^2E\ 2(^2_1D)$

$^2A_22(^2_3F)$
$^2A_11(^2_3D)\ ^2A_21(^2_3F)$
$^2E\ 1(^2_3F)\ ^2E\ 2(^2_3D)$
30000
$^2E\ 2(^2_3F)$

$^2E\ 1(^2_3H)\ ^2E\ 2(^2_3H)$
$^2A_11(^2_1D)$
$^2A_21(^2_3H)\ ^2A_22(^2_3H)$
25000
$^2A_11(^2_3H)$
$^2E\ 2(^2_3G)$

$^4A_21(^4_3P)$
$^2E\ 1(^2_1D)$
$^2E\ 3(^2_3H)$
$^2A_21(^2_3G)\ ^4E\ 1(^4_3P)$
$^2A_11(^2_3G)$
$^2E\ 4(^2_3H)$
$^2A_21(^2_3P)$
15000
$^2A_12(^2_3G)$
$^2E\ 2(^2_3G)\ ^2E\ 3(^2_3G)$
$^2E\ 1(^2_3G)$
$^4A_21(^4_3F)\ ^4A_22(^4_3F)$
$^4A_11(^4_3F)$
10000

$^4E\ 1(^4_3F)$
5000

$^4E\ 2(^4_3F)$
0

$^2E\ 1(^2_1D)(26)$
$^2A_11(^2_1D)(4)$

$^2E\ 2(^2_1D)(23)$

$^2A_21(^2_3P)(10)$
$^2E\ 2(^2_3F)(18)$

$^2E\ 1(^2_3F)(14)$
$^2A_22(^2_3H)(12)\ ^2A_11(^2_1D)(6)$
$^2E\ 1(^2_1D)(14)\ ^4E\ 4(^4_3H)(25)$
$^4A_21(^4_3F)(30)$
$^4E\ 2(^4_3F)(33)\ ^2E\ 2(^2_3D)(20)$
$^2A_11(^2_3F)(2)\ ^2A_21(^2_3H)(9)$
$^2E\ 1(^2_3D)(16)\ ^2E\ 2(^2_3H)(17)$
$^2A_11(^2_3G)(3)\ ^4A_21(^4P)(29)$
$^2E\ 3(^2_3H)(22)\ ^2A_21(^4P)(11)$
$^2A_12(^2_3G)(5)\ ^4E\ 1(^4P)(31)$
$^2E\ 4(^2_3H)(22)$
$^4A_11(^4_3F)(27)$

$^2E\ 3(^2_3H)(13)$

$^4E\ 1(^4_3F)(32)$
$^2A_11(^2_3H)(1)$

$^2E\ 2(^2_3G)(15)$
$^2A_12(^2_3G)(7)$

$^2E\ 3(^2_3G)(19)$

$^4A_22(^4_3F)(28)$

0   500   1000   1500   2000   2500   3000
DQ IN CM⁻¹

$^4E\ ^4A_2$

3 D ELECTRONS, TRIGONAL SYMMETRY

B=700   C=4×B   DT=-1000   K=1   DS=K×DT   ZETA=0

157

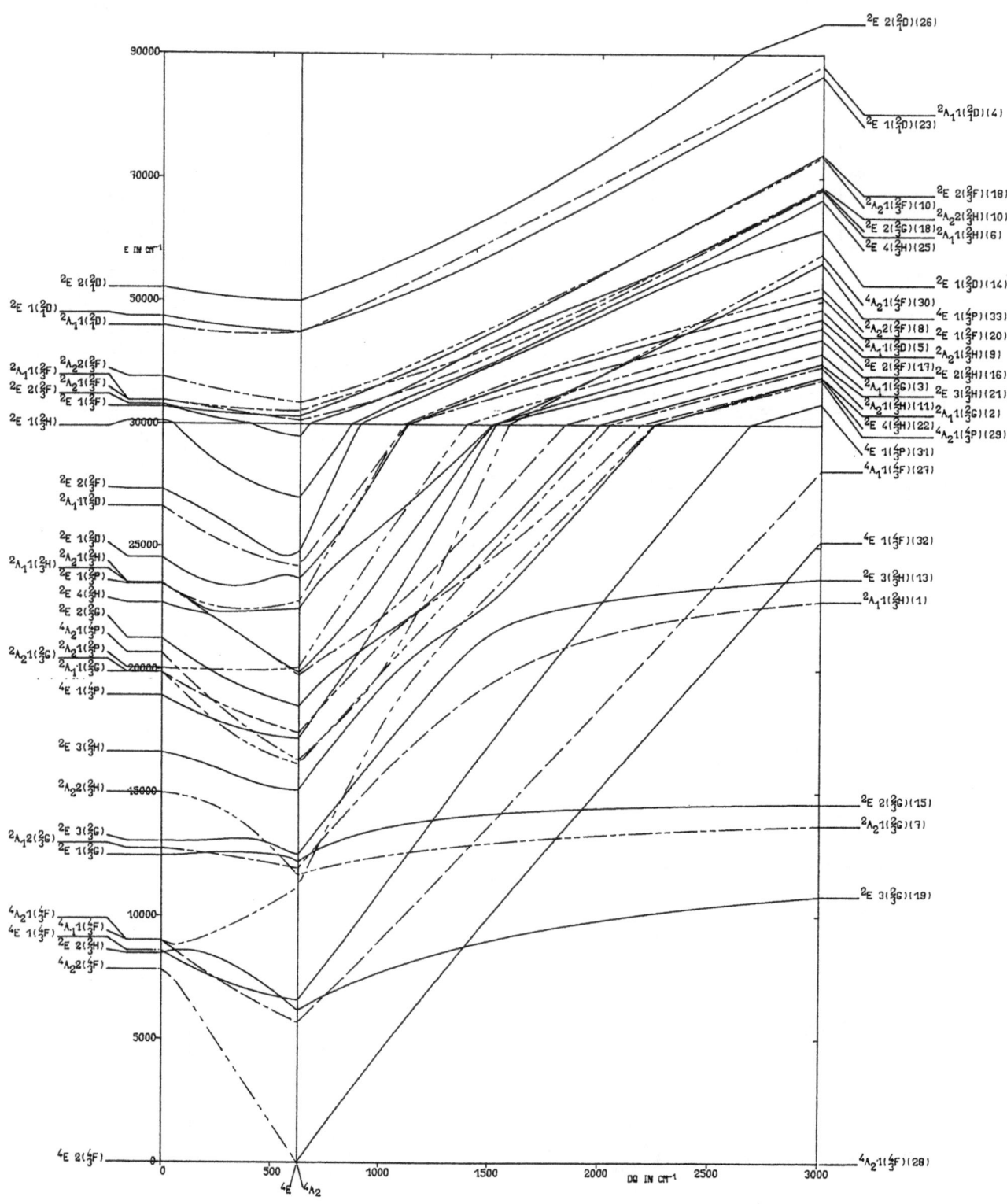

158

3 D ELECTRONS, TRIGONAL SYMMETRY

B=700 C=4×B DT=-1000 K=-1 DS=K×DT ZETA=0

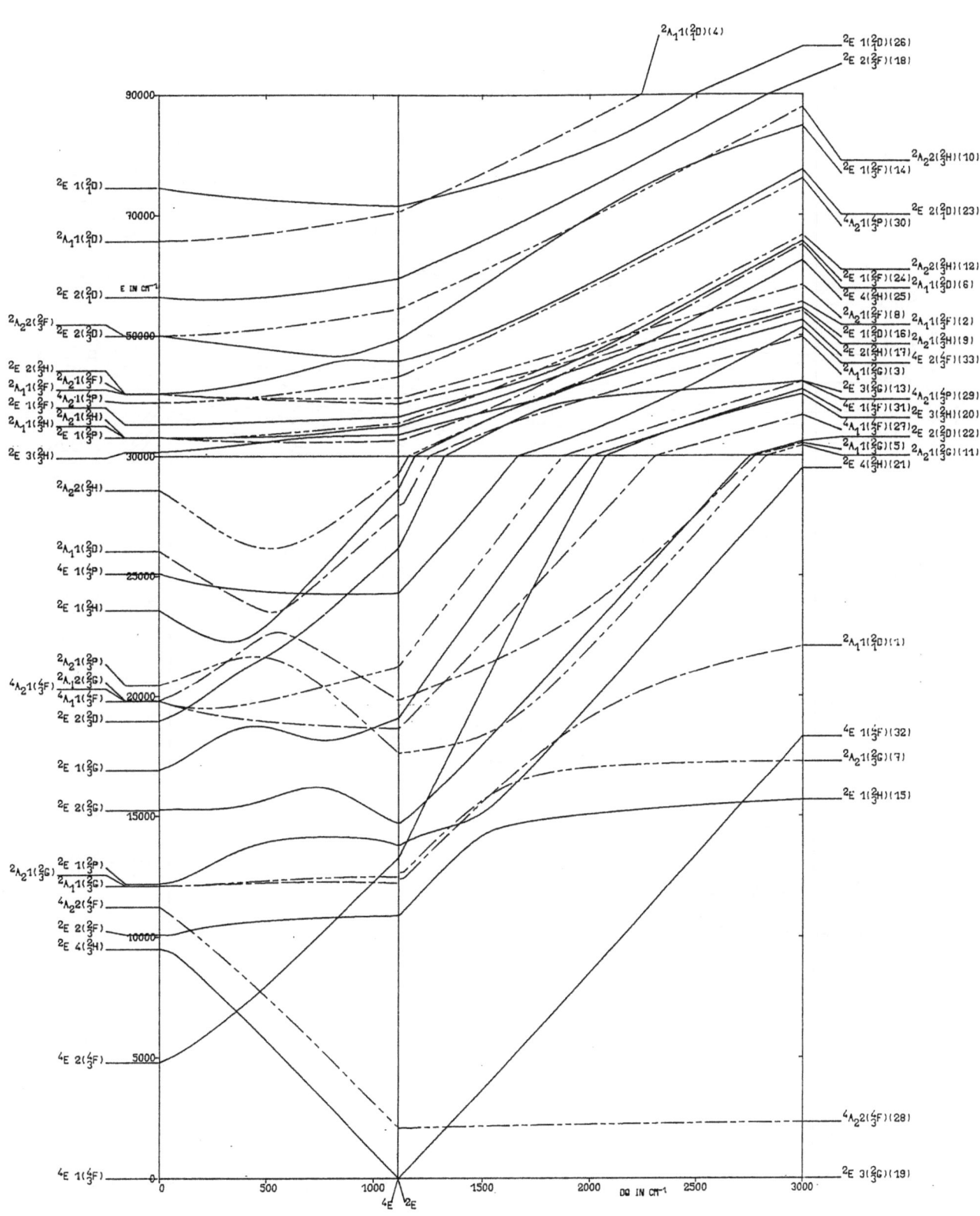

3 D ELECTRONS, TRIGONAL SYMMETRY   B=700 C=4×B DT=-1000 K=5   DS=K×DT ZETA=0

159

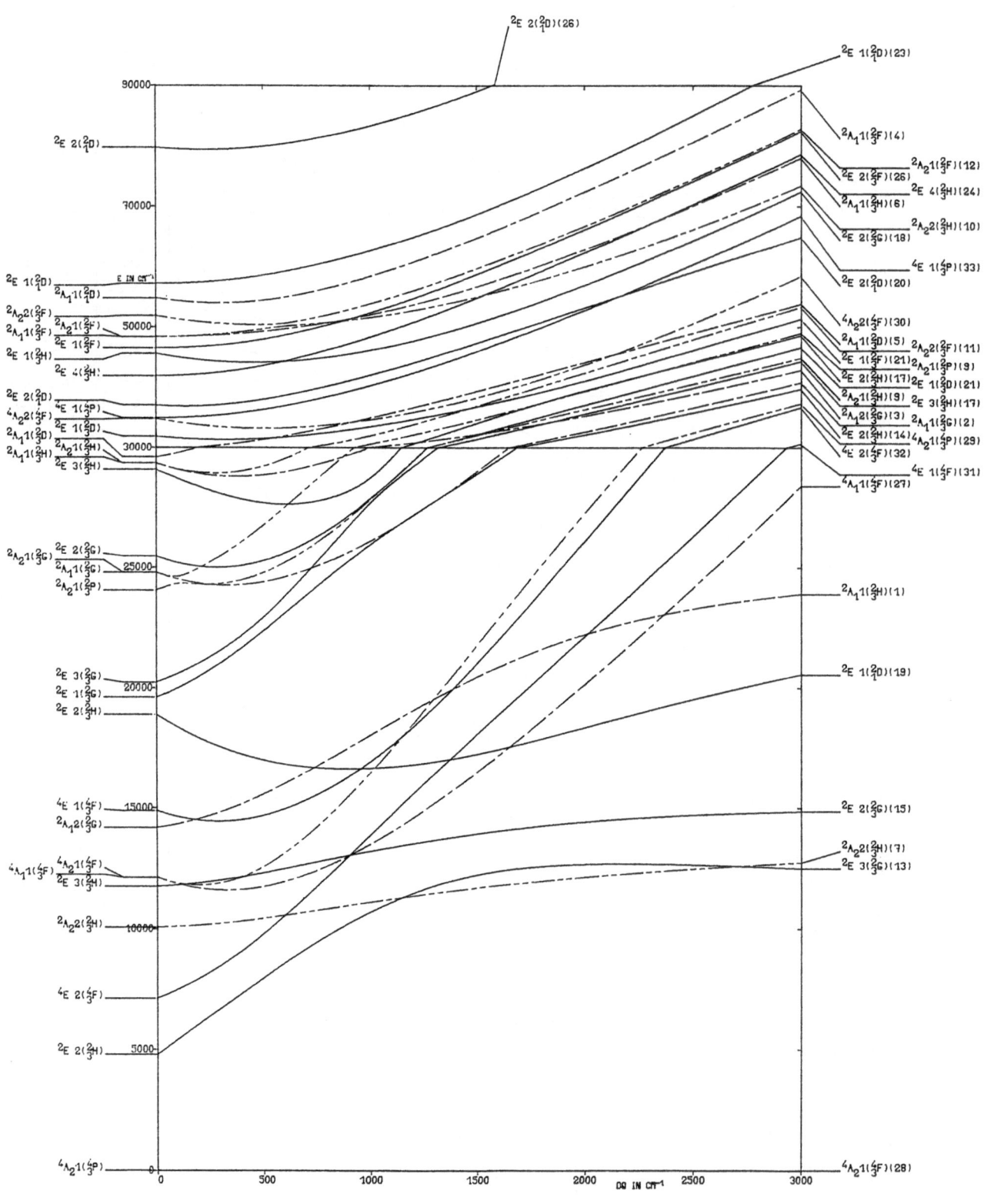

$^2E\ 2(^2_1D)(26)$

$^2E\ 1(^2_1D)(23)$

$^2E\ 2(^2_1D)$

$^2A_1 1(^2_3F)(4)$

$^2A_2 1(^2_3F)(12)$

$^2E\ 2(^2_3F)(26)$

$^2E\ 4(^2_3H)(24)$

$^2A_1 1(^2_3H)(6)$

$^2A_2 2(^2_3H)(10)$

$^2E\ 2(^2_3G)(18)$

$^4E\ 1(^4_3P)(33)$

$^2E\ 2(^2_1D)(20)$

$^2E\ 1(^2_1D)$
$^2A_1 1(^2_1D)$
E IN CM^{-1}

$^2A_2 2(^2_3F)$
$^2A_1 1(^2_3F)$
$^2E\ 1(^2_3H)$

$^2A_2 1(^2_3F)$
$^2E\ 1(^2_3F)$
$^2E\ 4(^2_3H)$

$^4A_2 2(^4_3F)(30)$
$^2A_1 1(^4_3F)(5)$
$^2A_2 2(^2_3F)(11)$
$^2E\ 1(^4_3F)(21)$
$^2A_2 1(^2_3F)(9)$
$^2E\ 2(^2_3H)(17)$
$^2E\ 1(^2_1D)(21)$
$^2A_2 1(^2_3H)(19)$
$^2E\ 3(^2_3H)(17)$
$^2A_1 2(^2_3G)(3)$
$^2A_1 1(^2_3G)(2)$
$^2E\ 2(^2_3H)(14)$
$^4A_2 1(^4_3P)(29)$
$^4E\ 2(^4_3F)(32)$

$^2E\ 2(^2_1D)$
$^4A_2 2(^4_3F)$
$^2A_1 1(^2_1D)$
$^2A_1 1(^2_3H)$

$^4E\ 1(^4_3P)$
$^2E\ 1(^2_1D)$
$^2A_2 1(^2_3H)$
$^2E\ 3(^2_3H)$

$^4E\ 1(^4_3F)(31)$
$^4A_1 1(^4_3F)(27)$

$^2A_2 1(^2_3G)$
$^2E\ 2(^2_3G)$
$^2A_1 1(^2_3G)$
$^2A_2 1(^2_3P)$

$^2A_1 1(^2_3H)(1)$

$^2E\ 3(^2_3G)$
$^2E\ 1(^2_3G)$
$^2E\ 2(^2_3H)$

$^2E\ 1(^2_1D)(19)$

$^4E\ 1(^4_3F)$
$^2A_1 2(^2_3G)$

$^2E\ 2(^2_3G)(15)$

$^4A_1 1(^4_3F)$
$^4A_2 1(^4_3F)$
$^2E\ 3(^2_3H)$

$^2A_2 2(^2_3H)(7)$
$^2E\ 3(^2_3G)(13)$

$^2A_2 2(^2_3H)$

$^4E\ 2(^4_3F)$

$^2E\ 2(^2_3H)$

$^4A_2 1(^4_3P)$

$^4A_2 1(^4_3F)(28)$

90000

70000

50000

30000

25000

20000

15000

10000

5000

0   500   1000   1500   2000   DQ IN CM^{-1} 2500   3000

160

3 D ELECTRONS, TRIGONAL SYMMETRY

B=700 C=4×B DT=-1000 K=-5 DS=K×DT ZETA=0

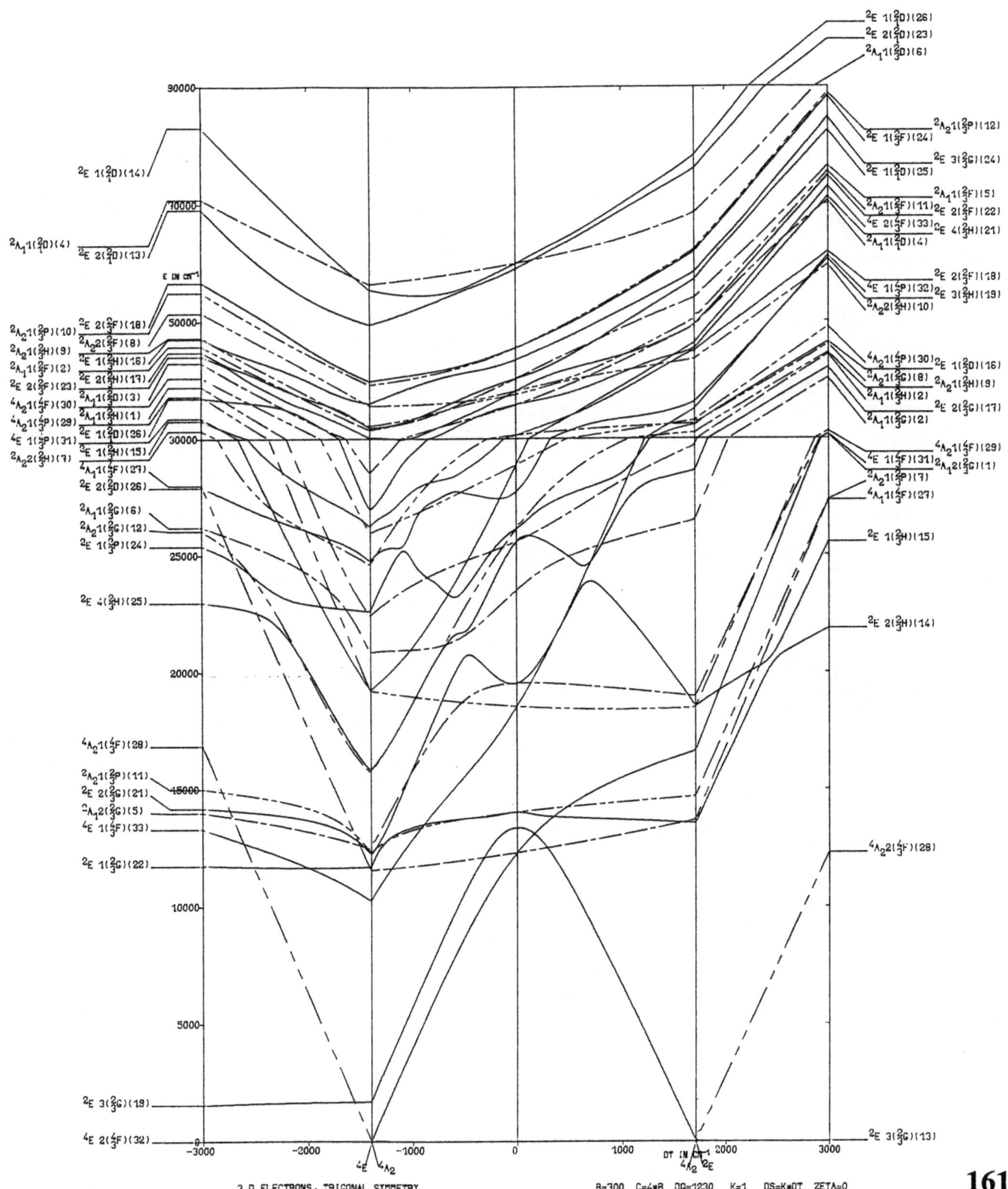

3 D ELECTRONS, TRIGONAL SYMMETRY   B=700   C=4×B   DQ=1230   K=1   DS=K×DT   ZETA=0

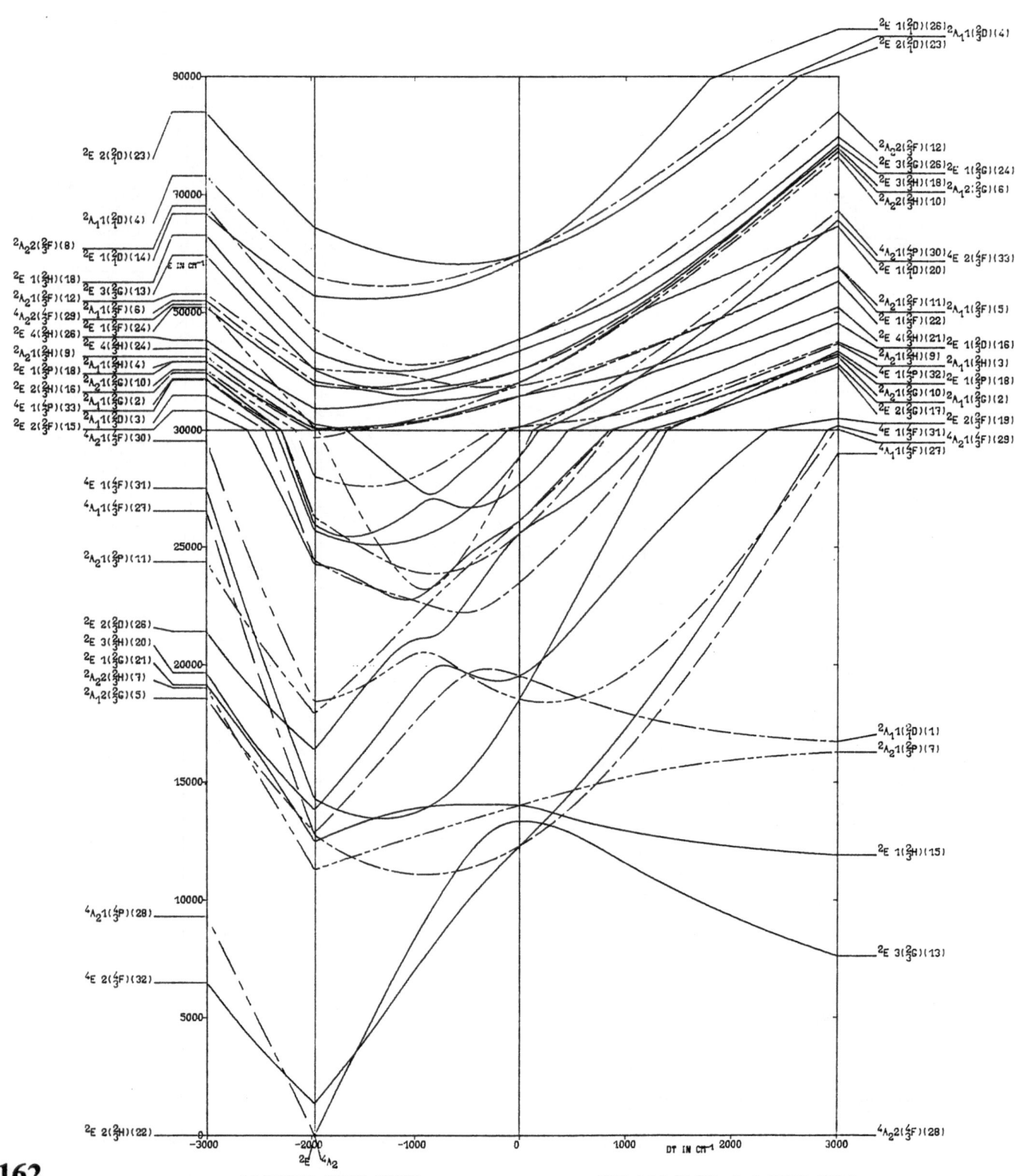

3 D ELECTRONS, TRIGONAL SYMMETRY

B=700 C=4×B DQ=1230 K=-1 DS=K×DT ZETA=0

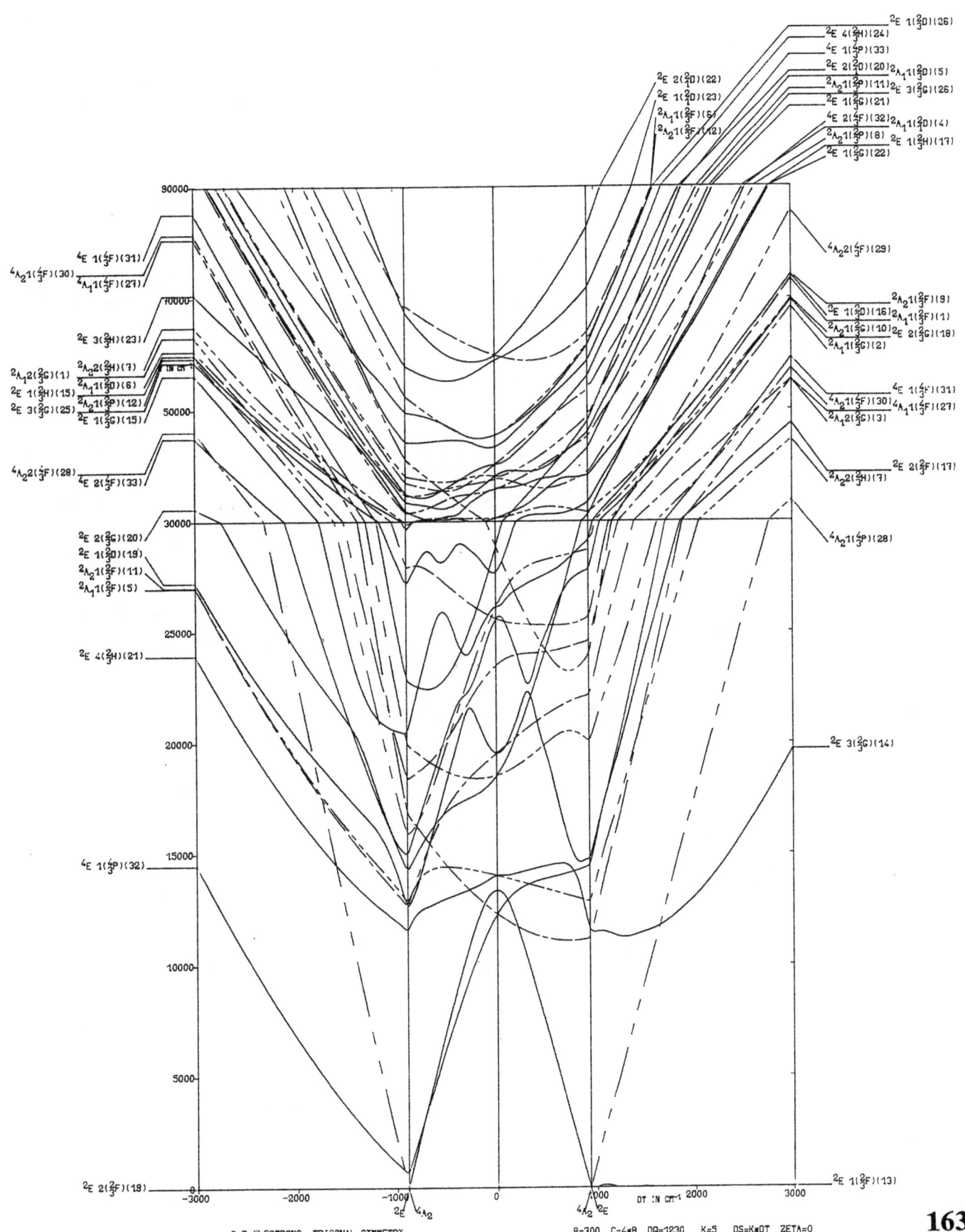

²E 1(²/₃D)(26)
²E 4(²/₃H)(24)
⁴E 1(⁴/₃P)(33)
²E 2(²/₃D)(20) 2_A₁1(²/₃D)(5)
²A₁1(²/₃F)(11) 2E 3(²/₃G)(26)
²E 1(²/₃G)(21)
⁴E 2(⁴/₃F)(32) 2_A₁1(²/₃D)(4)
²A₂1(⁴/₃P)(8) 2E 1(²/₃H)(17)
²E 1(²/₃G)(22)

²E 2(²/₃D)(22)
²E 1(²/₃D)(23)
²A₁1(²/₃F)(5)
²A₂1(²/₃F)(12)

⁴E 1(⁴/₃F)(31)
⁴A₂1(⁴/₃F)(30)
²A₁1(⁴/₃F)(27)

⁴A₂2(⁴/₃F)(29)

²A₁2(²/₃G)(1) 2_A₂2(²/₃H)(7)
²E 1(²/₃H)(15) 2_A₁1(²/₃D)(6)
²E 3(²/₃G)(25) 2_A₂1(⁴/₃P)(12)
              2E 1(²/₃G)(15)

²E 3(²/₃H)(23)

IN CM⁻¹

²A₁2(²/₃F)(9)
²E 1(²/₃D)(16) 2_A₁1(²/₃F)(1)
²A₁1(²/₃F)(10) 2E 2(²/₃G)(18)
²A₁1(²/₃G)(2)

⁴E 1(⁴/₃F)(31)
²A₁1(⁴/₃F)(30) 4_A₁1(⁴/₃F)(27)
²A₁2(⁴/₃F)(3)

⁴A₂2(⁴/₃F)(28) 4_E 2(⁴/₃F)(33)

²E 2(²/₃F)(17)

²A₂2(²/₃H)(7)

90000

²E 2(²/₃G)(20)
²E 1(²/₃D)(19)
²A₂1(²/₃F)(11)
²A₁1(²/₃F)(5)

30000

⁴A₂1(⁴/₃P)(28)

²E 4(²/₃H)(21)

25000

20000

²E 3(²/₃G)(14)

15000

⁴E 1(⁴/₃P)(32)

10000

5000

²E 2(²/₃F)(19)   0

        -3000   -2000   -1000   0   1000   DT IN CM⁻¹   2000   3000

²E   ⁴A₂   ⁴A₂ ²E

²E 1(²/₃F)(13)

3 D ELECTRONS, TRIGONAL SYMMETRY   B=700 C=4*B DQ=1230 K=5   DS=K*DT ZETA=0

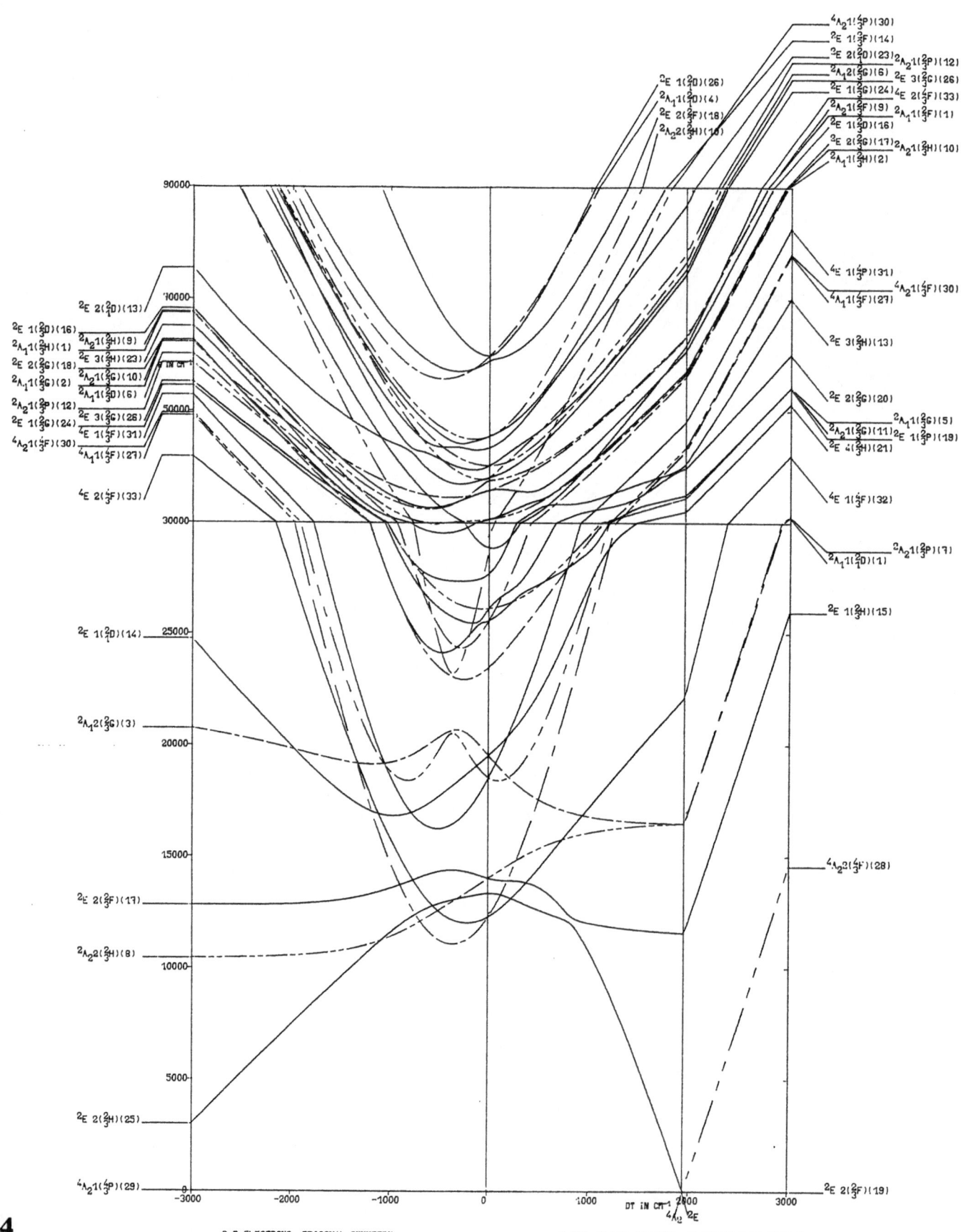

3 D ELECTRONS, TRIGONAL SYMMETRY

B=700 C=4∗B DQ=1230 K=-5 DS=K∗DT ZETA=0

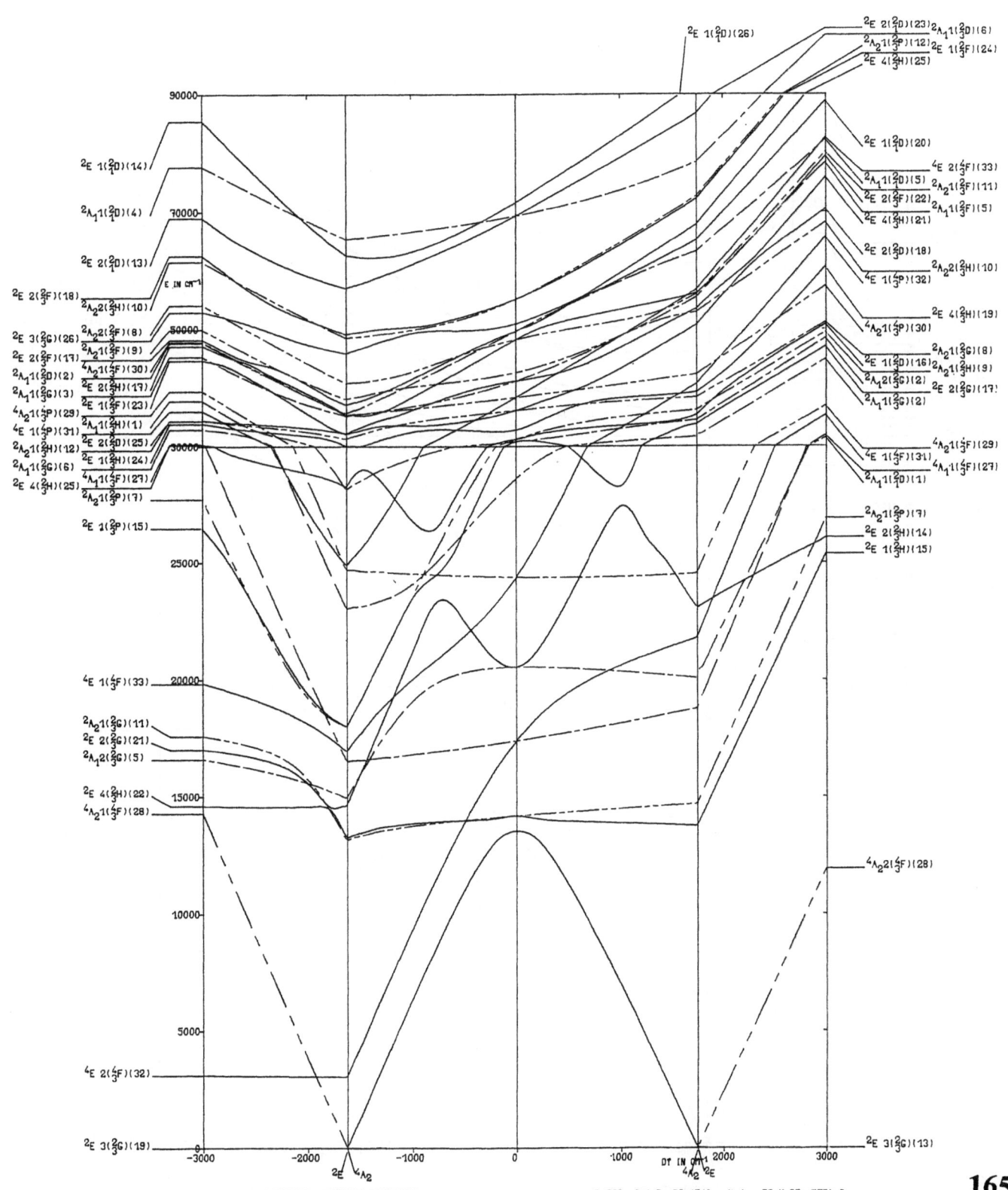

3 D ELECTRONS, TRIGONAL SYMMETRY

B=700   C=4×B   DQ=1740   K=1   DS=K×DT   ZETA=0

165

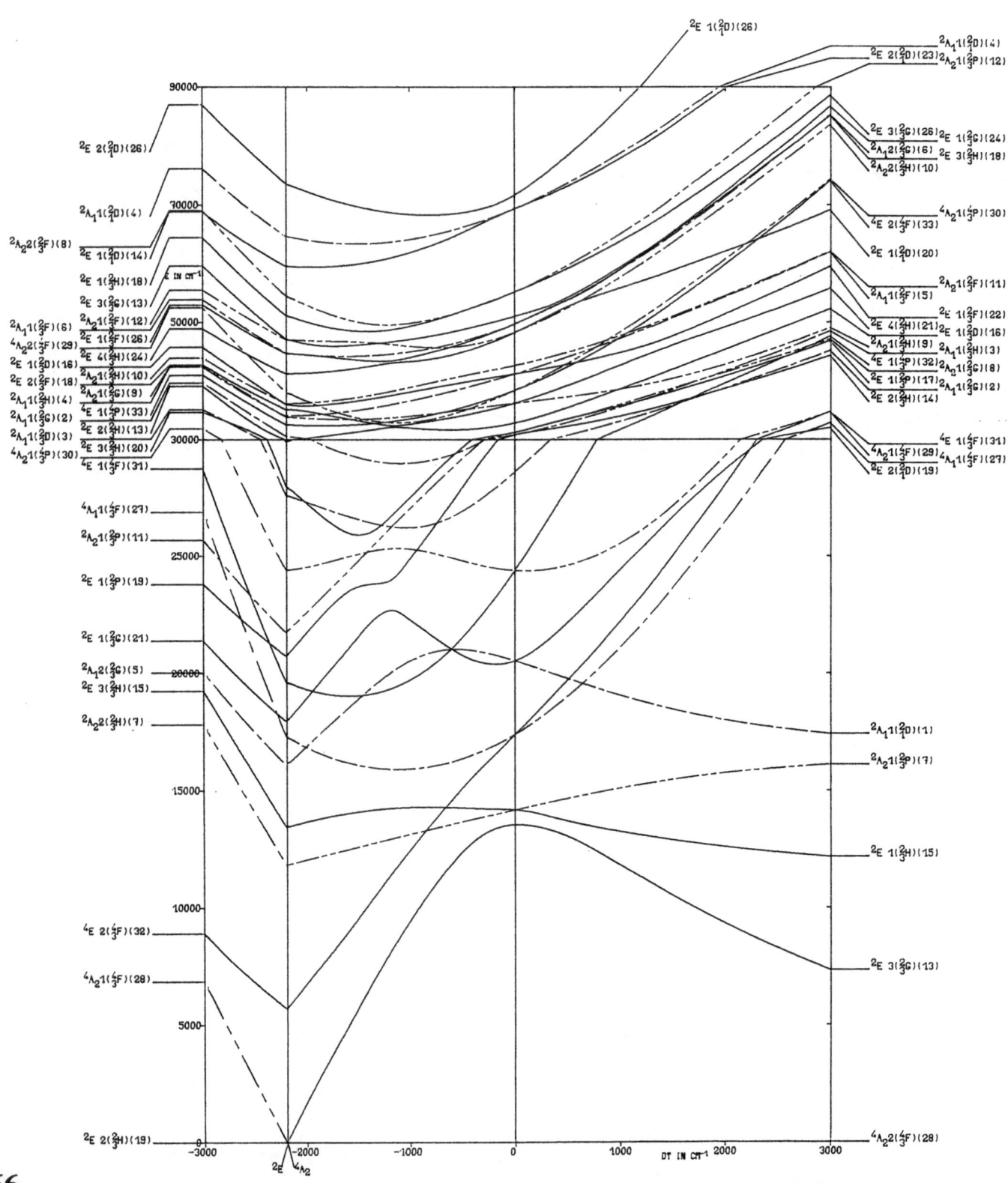

166

3 D ELECTRONS, TRIGONAL SYMMETRY

B=700 C=4×B DQ=1740 K=-1 DS=K×DT ZETA=0

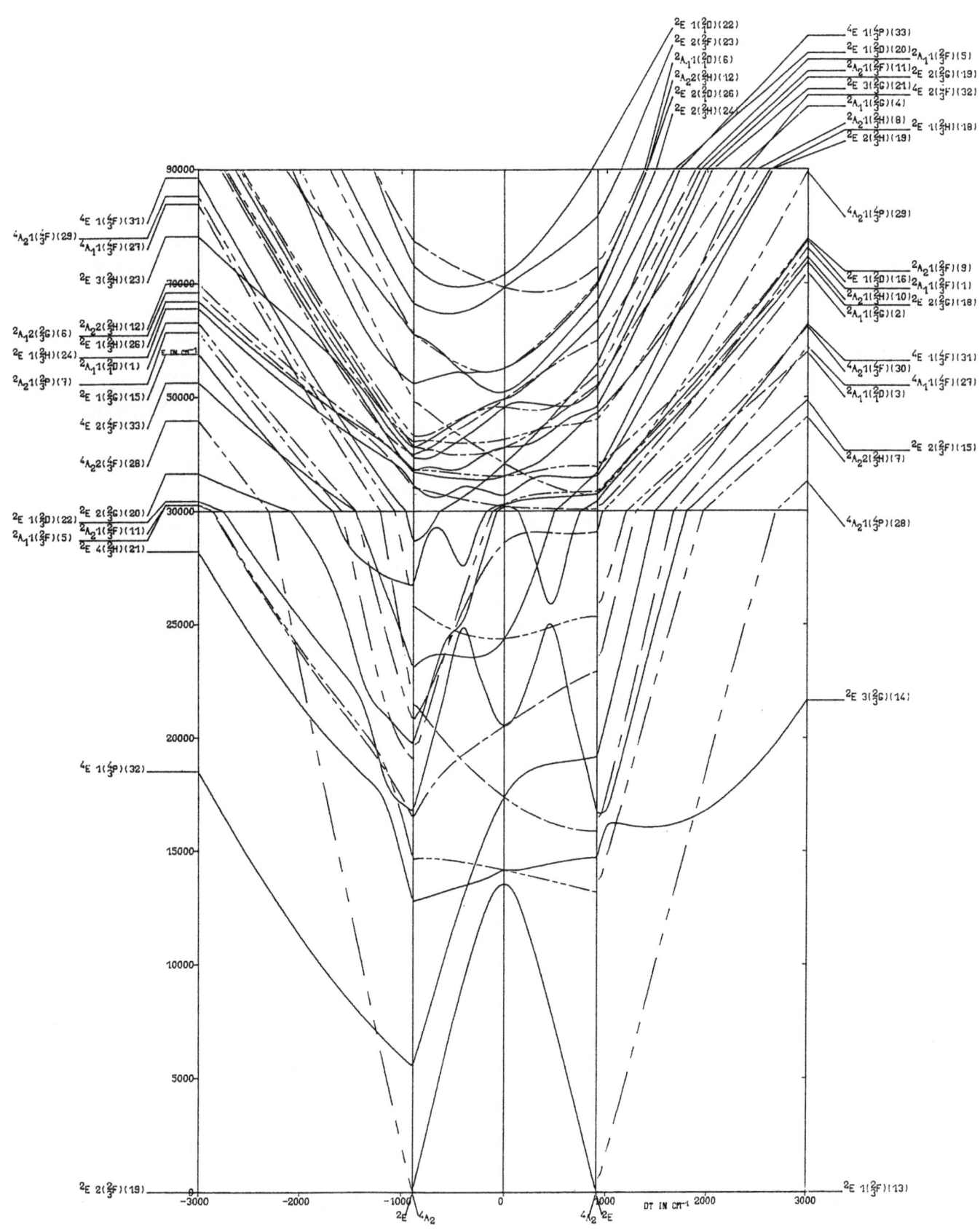

3 D ELECTRONS, TRIGONAL SYMMETRY   B=700 C=4×B DQ=1740 K=5 DS=K×DT ZETA=0

167

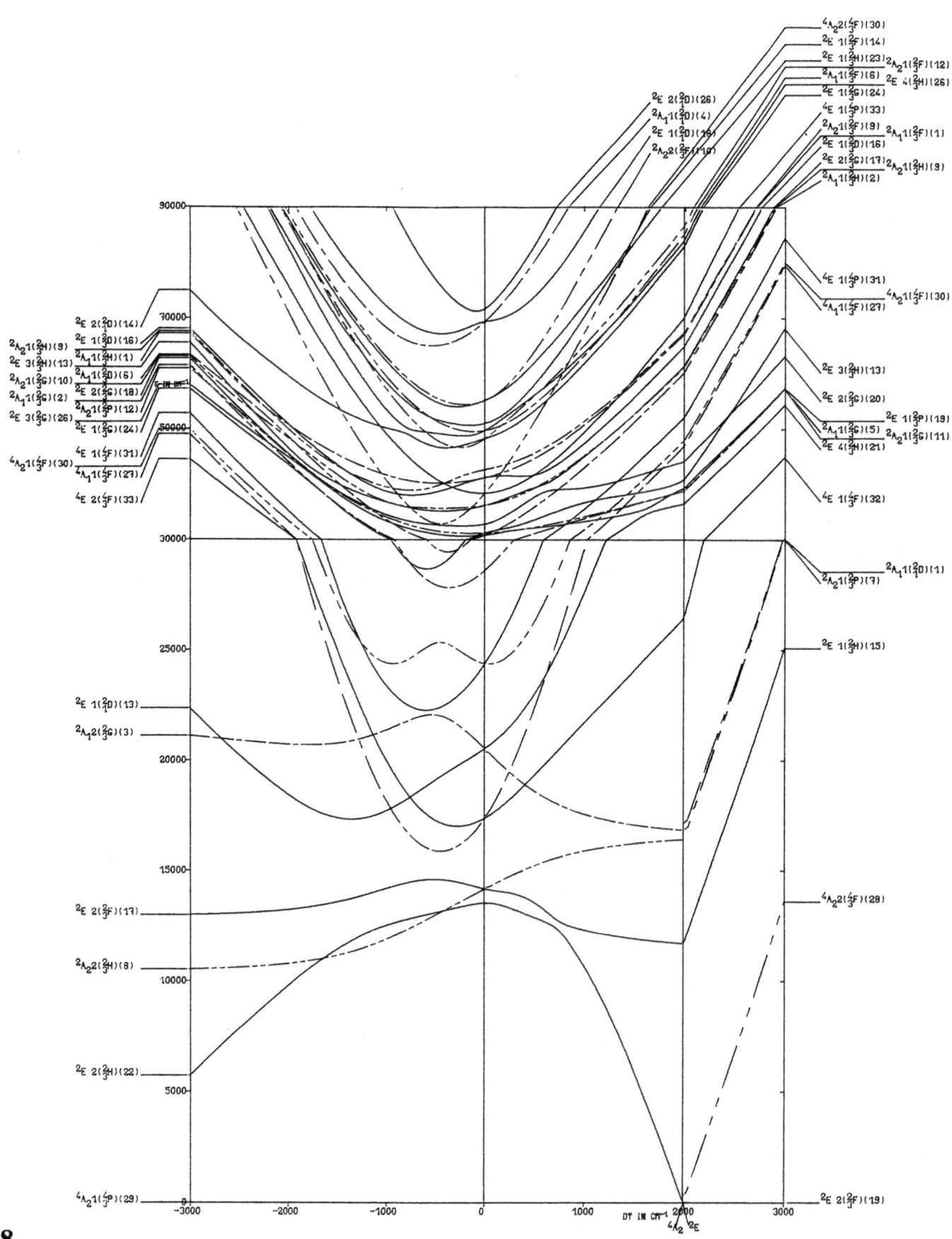

168

3 D ELECTRONS, TRIGONAL SYMMETRY

B=700   C=4×B   DQ=1740   K=-5   DS=K×DT   ZETA=0

3 D ELECTRONS, CYLINDRICAL SYMMETRY

B=700   C=4*B   ZETA=0

ENERGY AS FUNCTION OF DT
K= 1, -1, 5, -5

(29) $^2E_5(^1E_3(E_1E_2)E_2)$
(28) $^2E_4(^1E_2(E_1E_1)E_2)$
(27) $^2E_4(^1E_2(A_1E_2)E_2)$
(26) $^4E_3(^3E_1(A_1E_1)E_2)$
(25) $^2E_3(^1E_1(E_1E_2)E_2)$
(24) $^2E_3(^1E_2(A_1E_2)E_1)$
(23) $^2E_3(^1E_1(A_1E_1)E_2)$
(22) $^4E_2(^3A_2(E_1E_1)E_2)$
(21) $^2E_2(^1A_1(E_2E_2)E_2)$
(20) $^2E_2(^1E_1(E_1E_2)E_1)$
(19) $^2E_2(^1A_1(E_1E_1)E_2)$
(18) $^2E_2(^1E_1(A_1E_1)E_1)$
(17) $^2E_2(^1A_1(A_1A_1)E_2)$
(16) $^4E_1(^3E_1(E_1E_2)E_2)$
(15) $^4E_1(^3E_1(A_1E_1)E_2)$
(14) $^2E_1(^1E_3(E_1E_2)E_2)$
(13) $^2E_1(^1E_1(E_1E_2)E_2)$
(12) $^2E_1(^1A_1(E_1E_1)E_1)$
(11) $^2E_1(^1E_2(A_1E_2)E_1)$
(10) $^2E_1(^1E_1(A_1E_1)E_2)$
(9) $^2E_1(^1A_1(A_1A_1)E_1)$
(8) $^4A_2(^3E_2(A_1E_2)E_2)$
(7) $^4A_2(^3E_1(A_1E_1)E_1)$
(6) $^2A_2(^1E_2(E_1E_1)E_2)$
(5) $^2A_2(^1E_2(A_1E_2)E_2)$
(4) $^2A_2(^1E_1(A_1E_1)E_1)$
(3) $^2A_1(^1E_2(E_1E_1)E_2)$
(2) $^2A_1(^1E_2(A_1E_2)E_2)$
(1) $^2A_1(^1E_1(A_1E_1)E_1)$

169

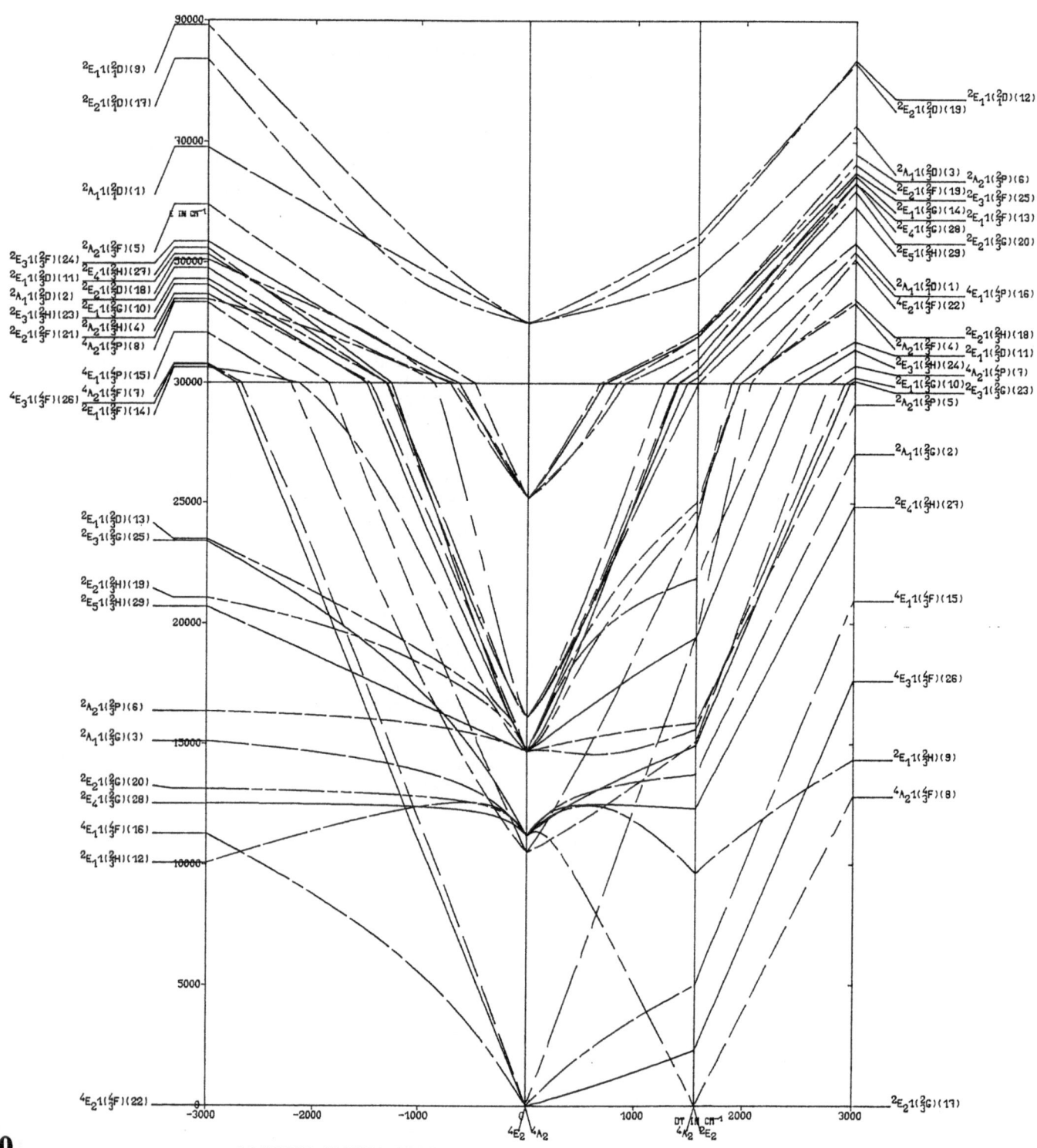

3 D ELECTRONS, CYLINDRICAL SYMMETRY   B=700   C=4×B   K=1   DS=K×DT   ZETA=0

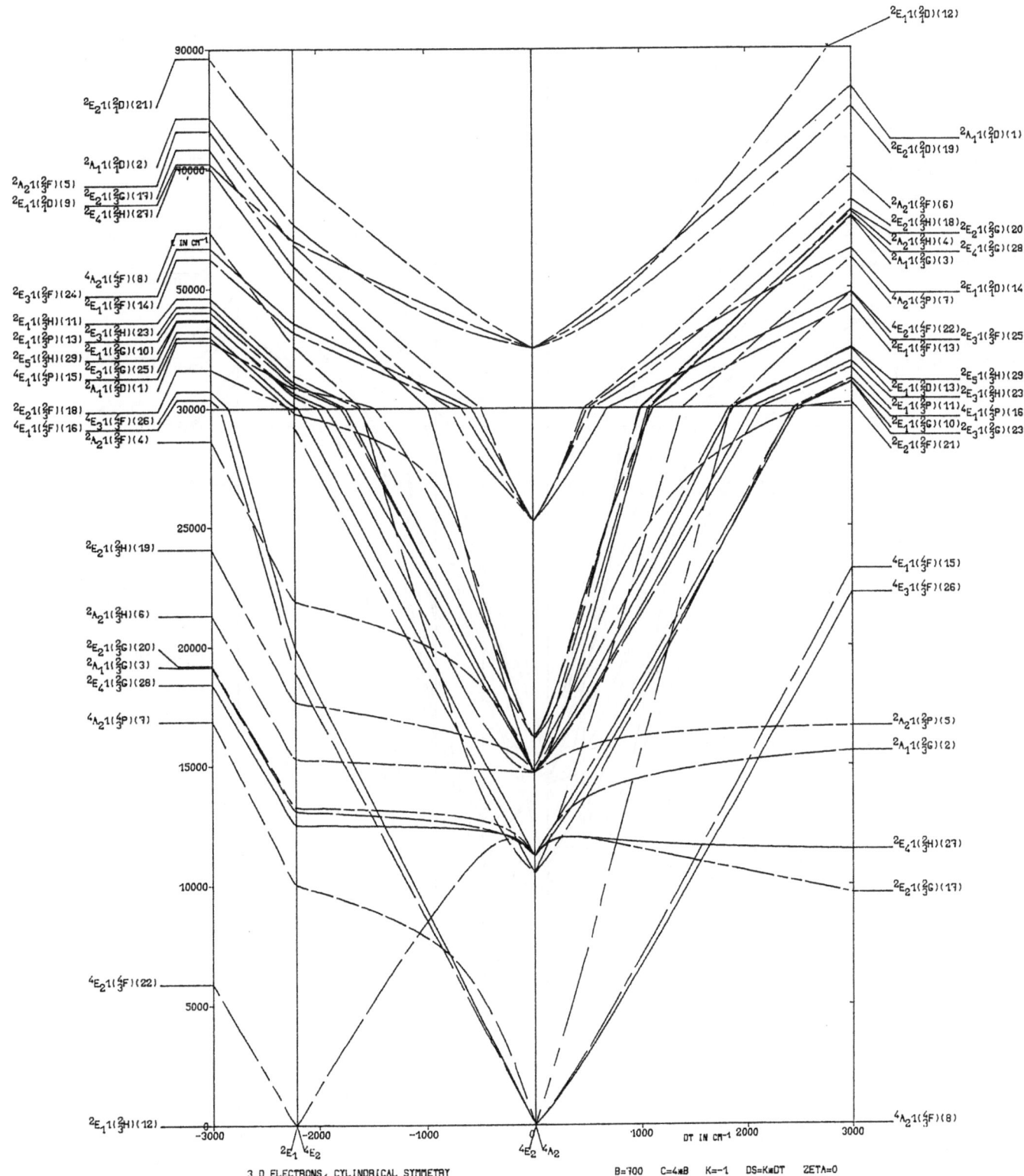

3 D ELECTRONS, CYLINDRICAL SYMMETRY   B=700   C=4×B   K=-1   DS=K×DT   2ETA=0

171

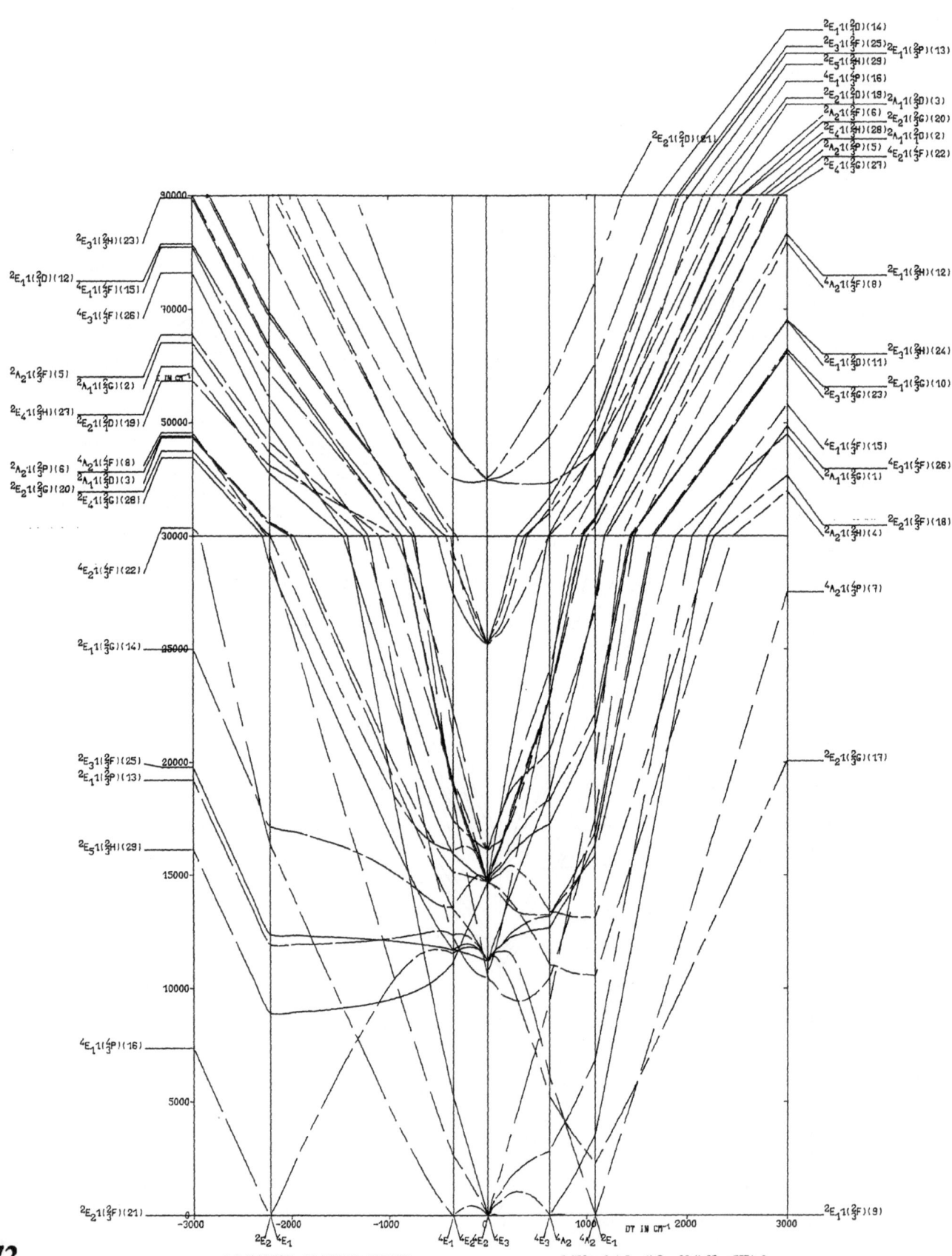

3 D ELECTRONS, CYLINDRICAL SYMMETRY   B=700   C=4×B   K=5   DS=K×DT   ZETA=0

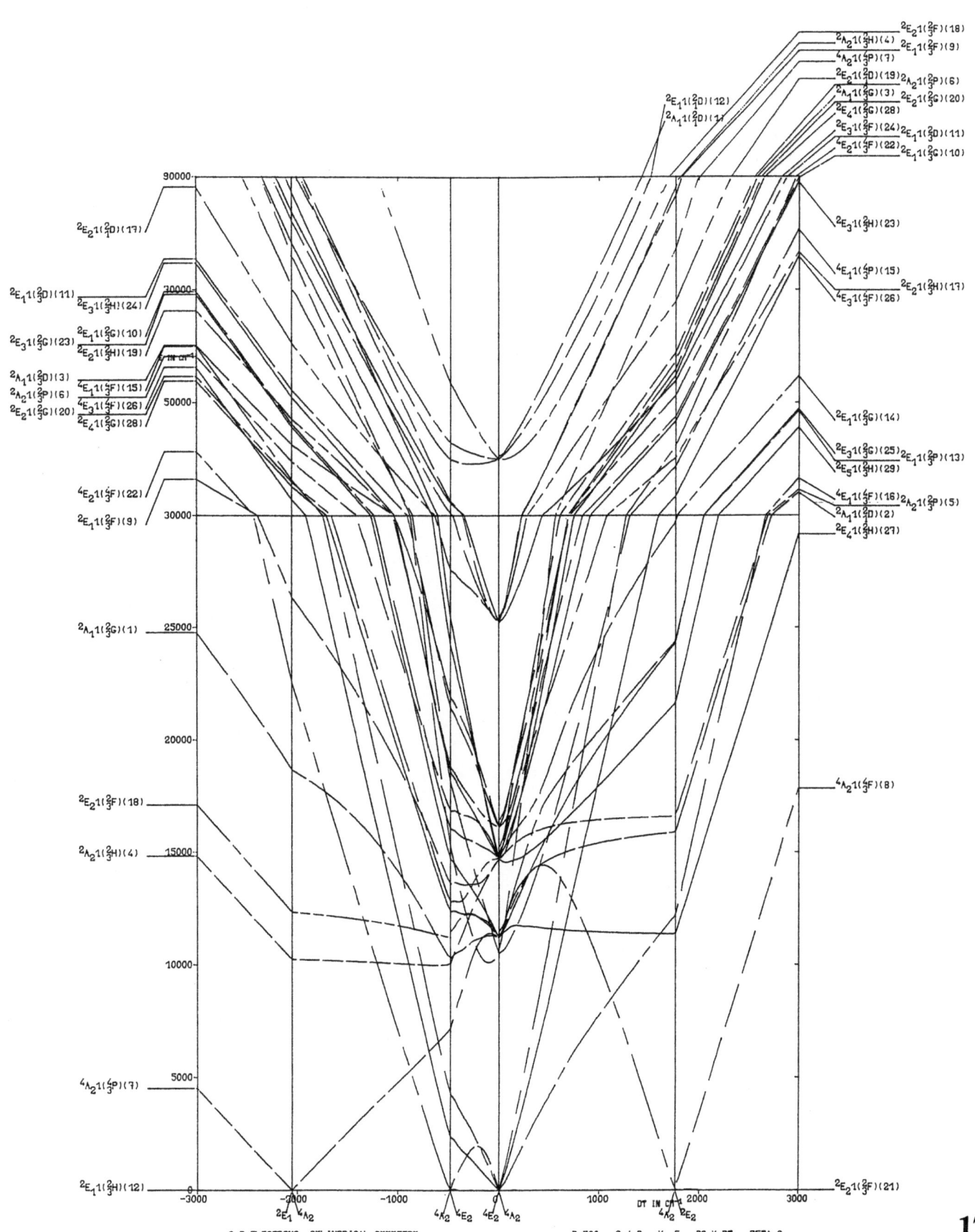

3 D ELECTRONS, CYLINDRICAL SYMMETRY   B=700   C=4mB   K=-5   DS=KmDT   ZETA=0

173

B=800   C=4×B   ZETA=0

ENERGY AS FUNCTION OF DQ
DT=0, 500, -500, 1000, -1000
K=1, -1, 5, -5

ENERGY AS FUNCTION OF DT
DQ=1390, 1930
K=1, -1, 5, -5

(76) $^5A_1(B_1B_2(^3A_2)E\ E\ (^3A_2))$
(75) $^5B_1(A_1B_2(^3B_2)E\ E\ (^3A_2))$
(74) $^5B_2(A_1B_1(^3B_1)E\ E\ (^3A_2))$
(73) $^5E\ (A_1B_1(^3B_1)B_2E\ (^3E\ ))$
(72) $^3E\ (B_2E\ (^1E\ )E\ E\ (^3A_2))$
(71) $^3E\ (B_1E\ (^1E\ )E\ E\ (^3A_2))$
(70) $^3E\ (B_1E\ (^3E\ )B_2B_2(^1A_1))$
(69) $^3E\ (B_1B_1(^1A_1)B_2E\ (^3E\ ))$
(68) $^3E\ (A_1E\ (^1E\ )E\ E\ (^3A_2))$
(67) $^3E\ (A_1E\ (^3E\ )B_2B_2(^1A_1))$
(66) $^3E\ (A_1B_1(^3B_1)B_2E\ (^3E\ ))$
(65) $^3E\ (A_1B_1(^3B_1)B_2E\ (^1E\ ))$
(64) $^3E\ (A_1B_1(^1B_1)B_2E\ (^3E\ ))$
(63) $^3E\ (A_1E\ (^3E\ )B_1B_1(^1A_1))$
(62) $^3E\ (A_1A_1(^1A_1)B_2E\ (^3E\ ))$
(61) $^3E\ (A_1A_1(^1A_1)B_1E\ (^3E\ ))$
(60) $^3B_2(B_1B_2(^3A_2)E\ E\ (^1B_1))$
(59) $^3B_2(A_1B_2(^3B_2)E\ E\ (^1A_1))$
(58) $^3B_2(A_1B_1(^3B_1)E\ E\ (^3A_2))$
(57) $^3B_2(A_1B_1(^1B_1)E\ E\ (^3A_2))$
(56) $^3B_2(A_1B_2(^3B_2)B_1B_1(^1A_1))$
(55) $^3B_1(B_1B_2(^3A_2)E\ E\ (^1B_2))$
(54) $^3B_1(A_1B_2(^3B_2)E\ E\ (^3A_2))$
(53) $^3B_1(A_1B_2(^1B_2)E\ E\ (^3A_2))$
(52) $^3B_1(A_1B_1(^3B_1)E\ E\ (^1A_1))$
(51) $^3B_1(A_1B_1(^3B_1)B_2B_2(^1A_1))$
(50) $^3A_2(B_2B_2(^1A_1)E\ E\ (^3A_2))$
(49) $^3A_2(B_1B_2(^3A_2)E\ E\ (^1A_1))$
(48) $^3A_2(B_1B_1(^1A_1)E\ E\ (^3A_2))$
(47) $^3A_2(A_1B_2(^3B_2)E\ E\ (^1B_1))$
(46) $^3A_2(A_1B_1(^3B_1)E\ E\ (^1B_2))$
(45) $^3A_2(A_1A_1(^1A_1)E\ E\ (^3A_2))$
(44) $^3A_2(A_1A_1(^1A_1)B_1B_2(^3A_2))$
(43) $^3A_1(B_1B_2(^3A_2)E\ E\ (^3A_2))$
(42) $^3A_1(B_1B_2(^1A_2)E\ E\ (^3A_2))$
(41) $^3A_1(A_1B_2(^3B_2)E\ E\ (^1B_2))$
(40) $^3A_1(A_1B_1(^3B_1)E\ E\ (^1B_1))$
(39) $^1E\ (B_2E\ (^1E\ )E\ E\ (^1A_1))$

(38) $^1E\ (B_1E\ (^1E\ )E\ E\ (^1A_1))$
(37) $^1E\ (B_1E\ (^1E\ )B_2B_2(^1A_1))$
(36) $^1E\ (B_1B_1(^1A_1)B_2E\ (^1E\ ))$
(35) $^1E\ (A_1E\ (^1E\ )E\ E\ (^1A_1))$
(34) $^1E\ (A_1E\ (^1E\ )B_2B_2(^1A_1))$
(33) $^1E\ (A_1B_1(^3B_1)B_2E\ (^3E\ ))$
(32) $^1E\ (A_1B_1(^1B_1)B_2E\ (^1E\ ))$
(31) $^1E\ (A_1E\ (^1E\ )B_1B_1(^1A_1))$
(30) $^1E\ (A_1A_1(^1A_1)B_2E\ (^1E\ ))$
(29) $^1E\ (A_1A_1(^1A_1)B_1E\ (^1E\ ))$
(28) $^1B_2(B_2B_2(^1A_1)E\ E\ (^1B_2))$
(27) $^1B_2(B_1B_2(^1A_2)E\ E\ (^1B_1))$
(26) $^1B_2(B_1B_1(^1A_1)E\ E\ (^1B_2))$
(25) $^1B_2(A_1B_2(^1B_2)B_2B_2(^1A_1))$
(24) $^1B_2(A_1B_2(^3B_1)E\ E\ (^3A_2))$
(23) $^1B_2(A_1B_2(^1B_2)B_1B_1(^1A_1))$
(22) $^1B_2(A_1A_1(^1A_1)E\ E\ (^1B_2))$
(21) $^1B_1(B_2B_2(^1A_1)E\ E\ (^1B_1))$
(20) $^1B_1(B_1B_2(^1A_2)E\ E\ (^1B_2))$
(19) $^1B_1(B_1B_1(^1A_1)E\ E\ (^1B_1))$
(18) $^1B_1(A_1B_2(^3B_2)E\ E\ (^3A_2))$
(17) $^1B_1(A_1B_1(^1B_1)E\ E\ (^1A_1))$
(16) $^1B_1(A_1B_1(^1B_1)B_2B_2(^1A_1))$
(15) $^1B_1(A_1A_1(^1A_1)E\ E\ (^1B_1))$
(14) $^1A_2(B_1B_2(^1A_2)E\ E\ (^1A_1))$
(13) $^1A_2(A_1B_2(^1B_2)E\ E\ (^1B_1))$
(12) $^1A_2(A_1B_1(^1B_1)E\ E\ (^1B_2))$
(11) $^1A_2(A_1A_1(^1A_1)B_1B_2(^1A_2))$
(10) $^1A_1(E\ E\ (^1A_1)E\ E\ (^1A_1))$
(9) $^1A_1(B_2B_2(^1A_1)E\ E\ (^1A_1))$
(8) $^1A_1(B_1B_2(^3A_2)E\ E\ (^3A_2))$
(7) $^1A_1(B_1B_1(^1A_1)E\ E\ (^1A_1))$
(6) $^1A_1(B_1B_1(^1A_1)B_2B_2(^1A_1))$
(5) $^1A_1(A_1B_2(^1B_2)E\ E\ (^1B_2))$
(4) $^1A_1(A_1B_1(^1B_1)E\ E\ (^1B_1))$
(3) $^1A_1(A_1A_1(^1A_1)E\ E\ (^1A_1))$
(2) $^1A_1(A_1A_1(^1A_1)B_2B_2(^1A_1))$
(1) $^1A_1(A_1A_1(^1A_1)B_1B_1(^1A_1))$

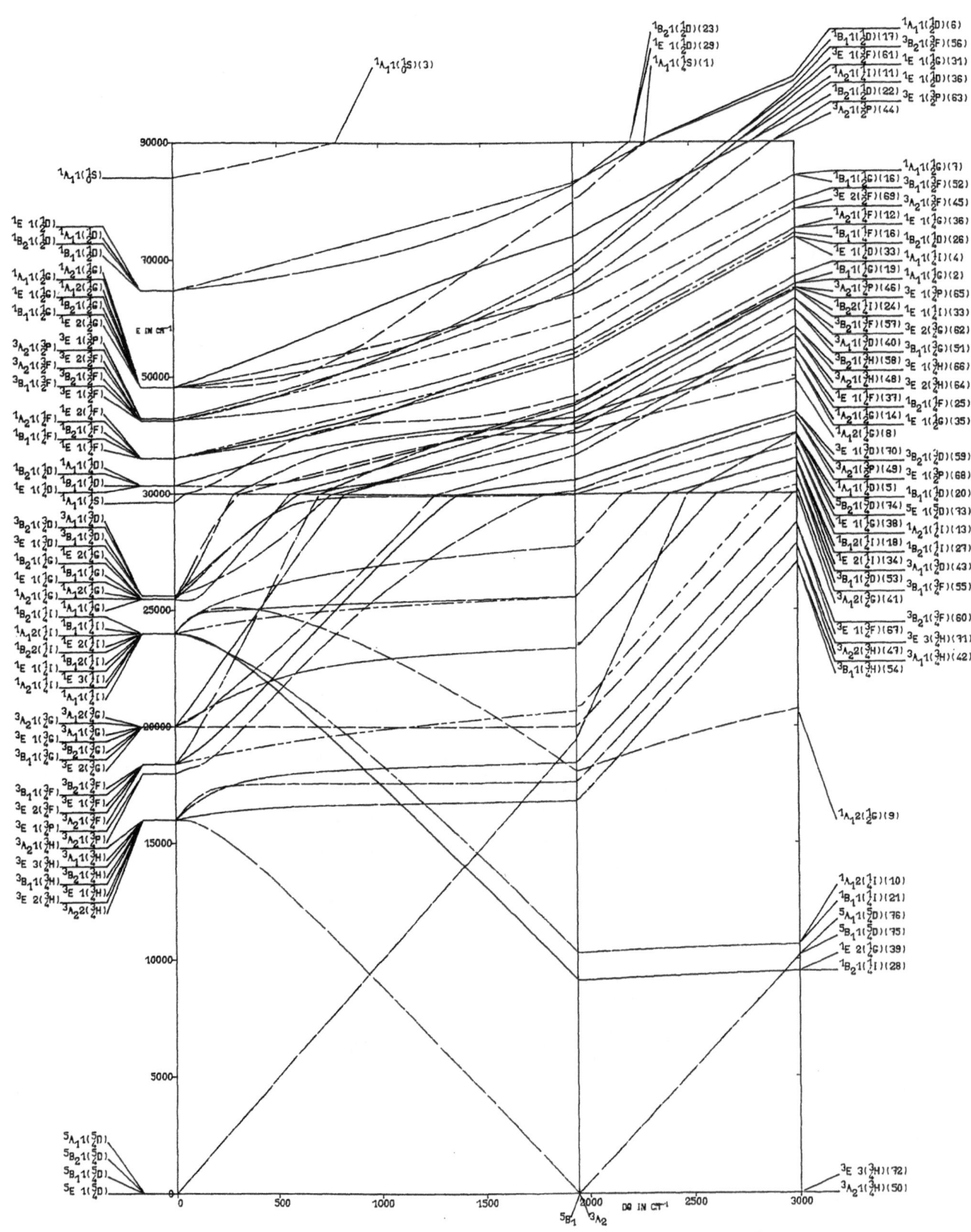

4 D ELECTRONS, TETRAGONAL SYMMETRY

B=800 C=4×B DT=0 K=0 DS=K×DT ZETA=0

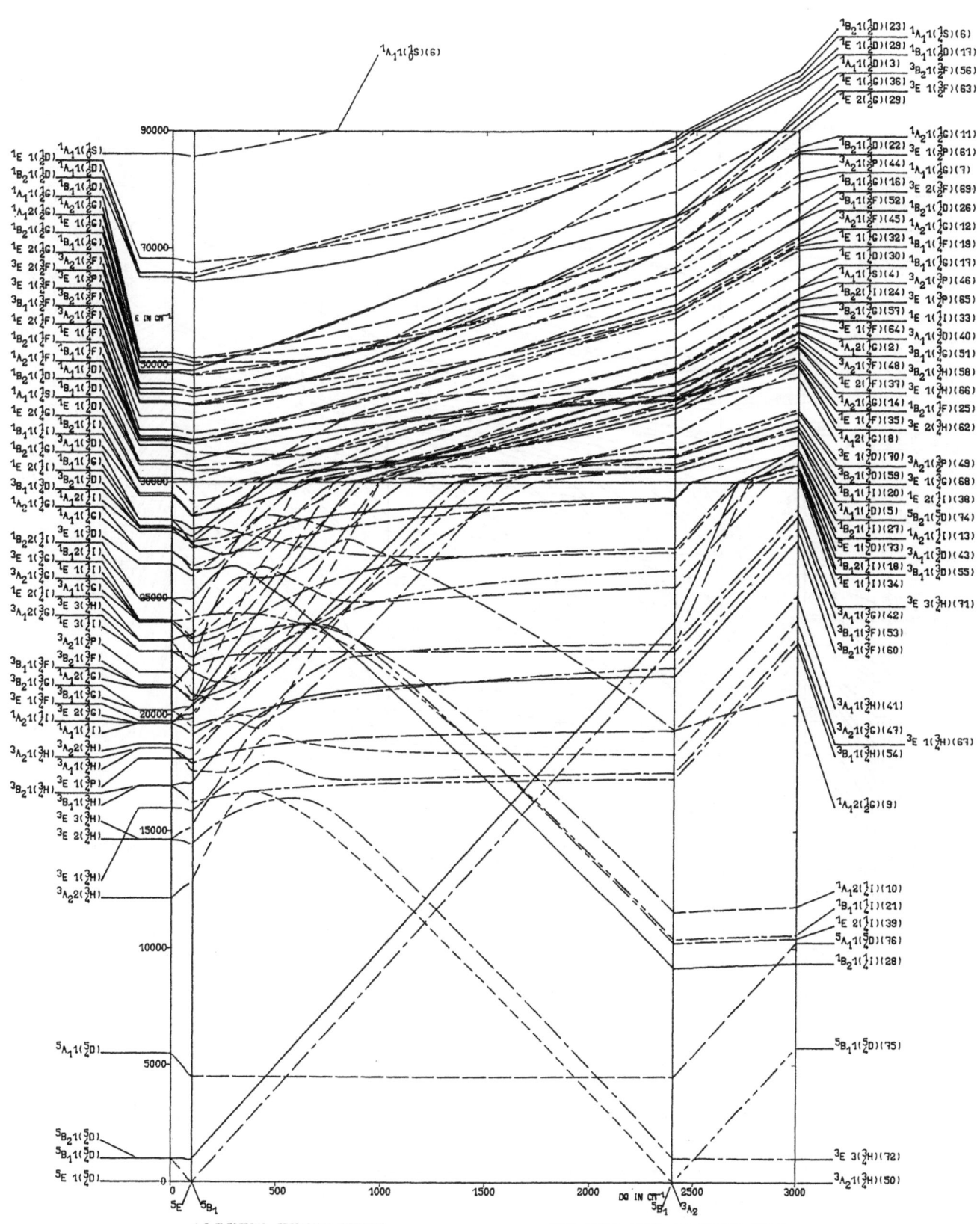

4 D ELECTRONS, TETRAGONAL SYMMETRY

B=800   C=4×B   DT=500   K=1   DS=K×DT   ZETA=0

177

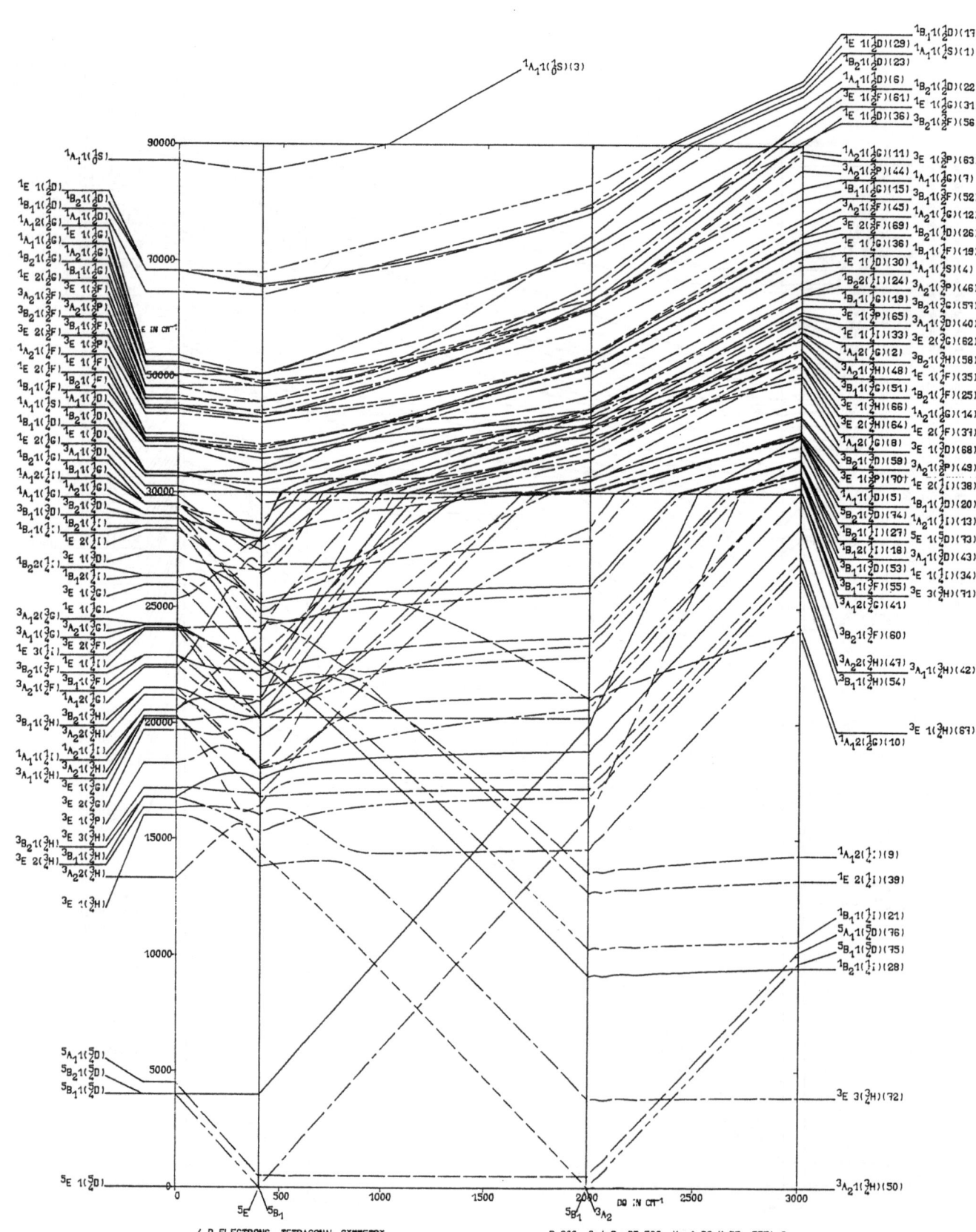

178

4 D ELECTRONS, TETRAGONAL SYMMETRY

B=800 C=4×B DT=500 K=−1 DS=K×DT ZETA=0

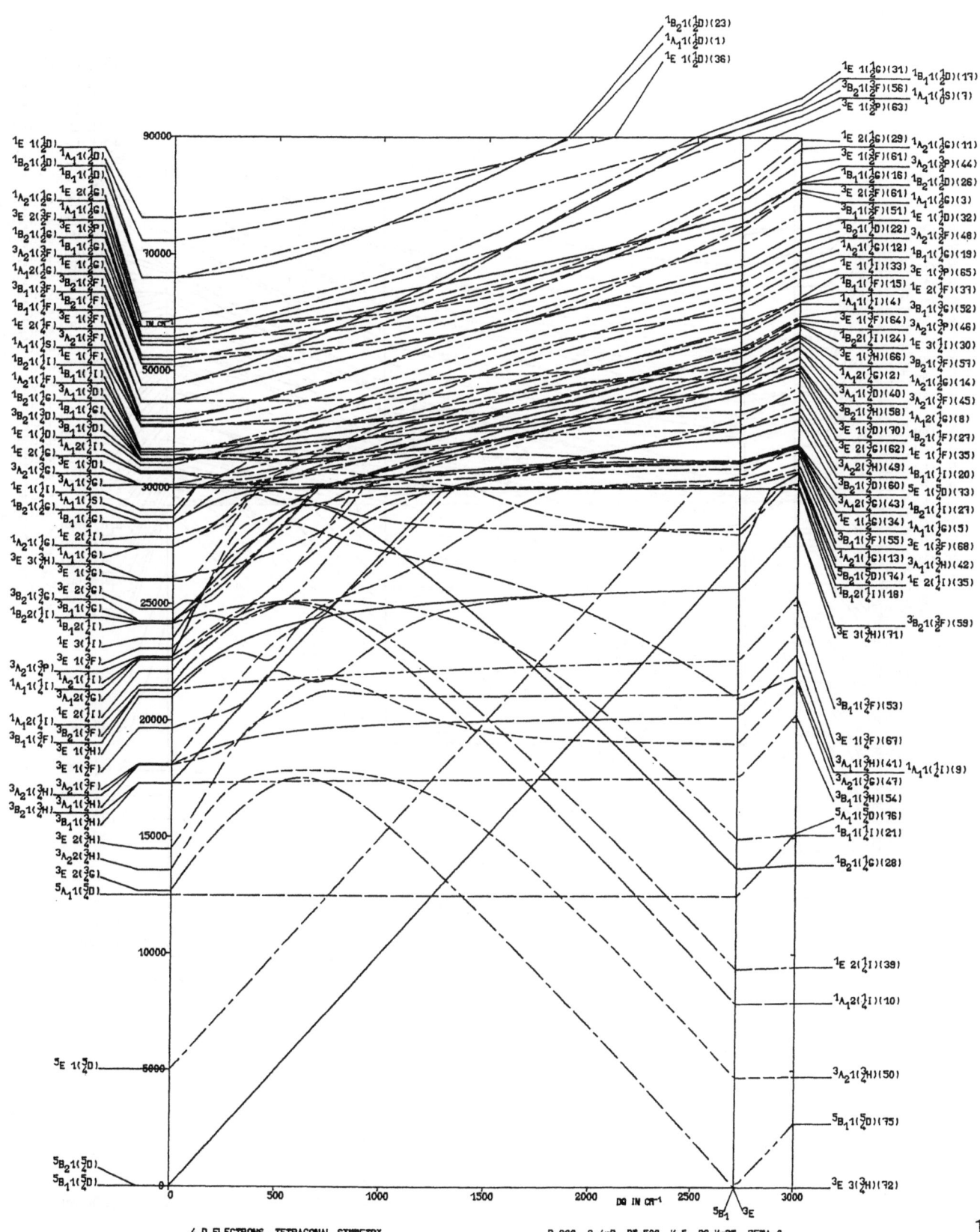

4 D ELECTRONS, TETRAGONAL SYMMETRY   B=800 C=4×B DT=500 K=5 DS=K×DT ZETA=0

179

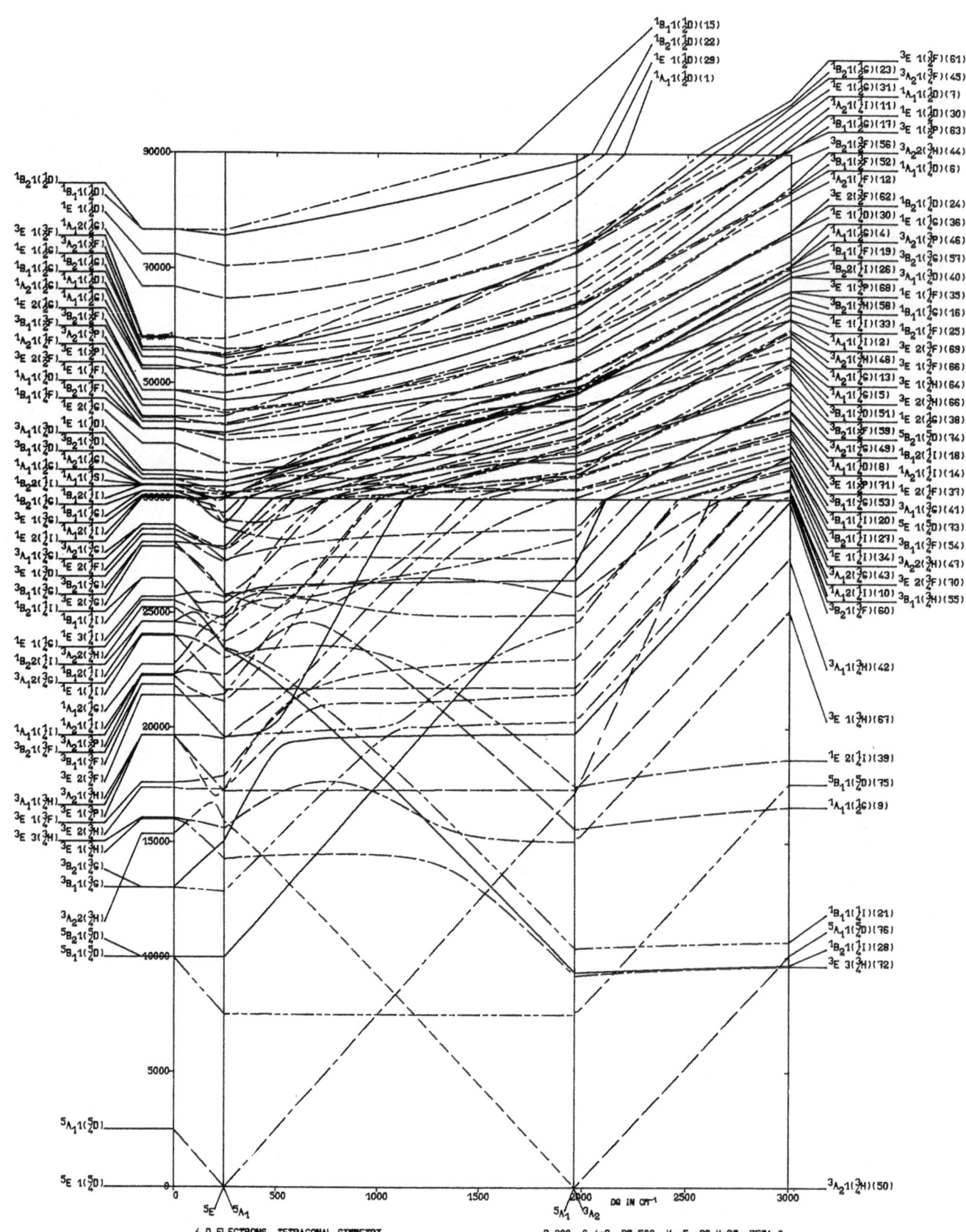

180

4 D ELECTRONS, TETRAGONAL SYMMETRY

B=800   C=4×B   DT=500   K=-5   DS=K×DT   ZETA=0

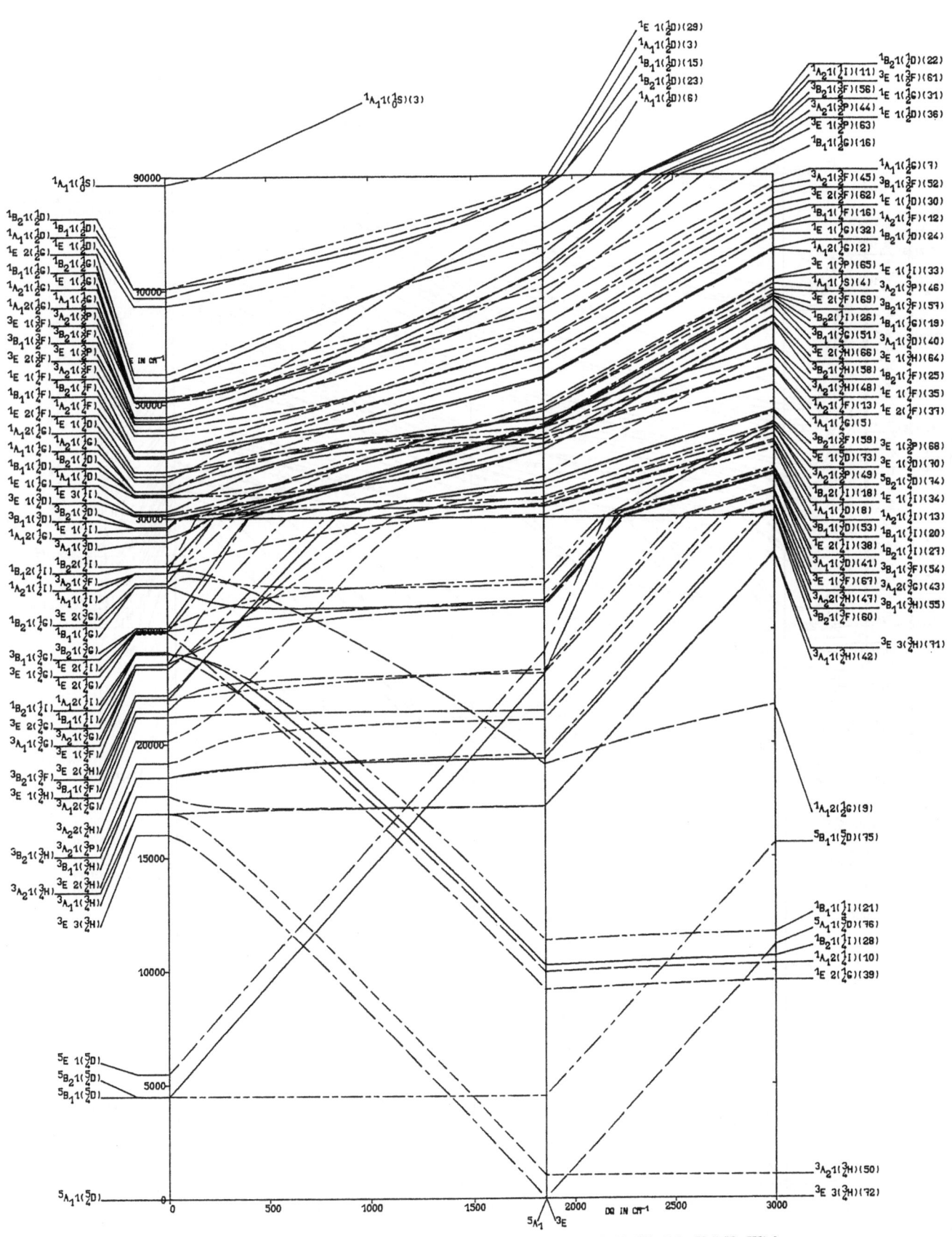

4 D ELECTRONS, TETRAGONAL SYMMETRY

B=800 C=4×B DT=-500 K=1 DS=K×DT ZETA=0

181

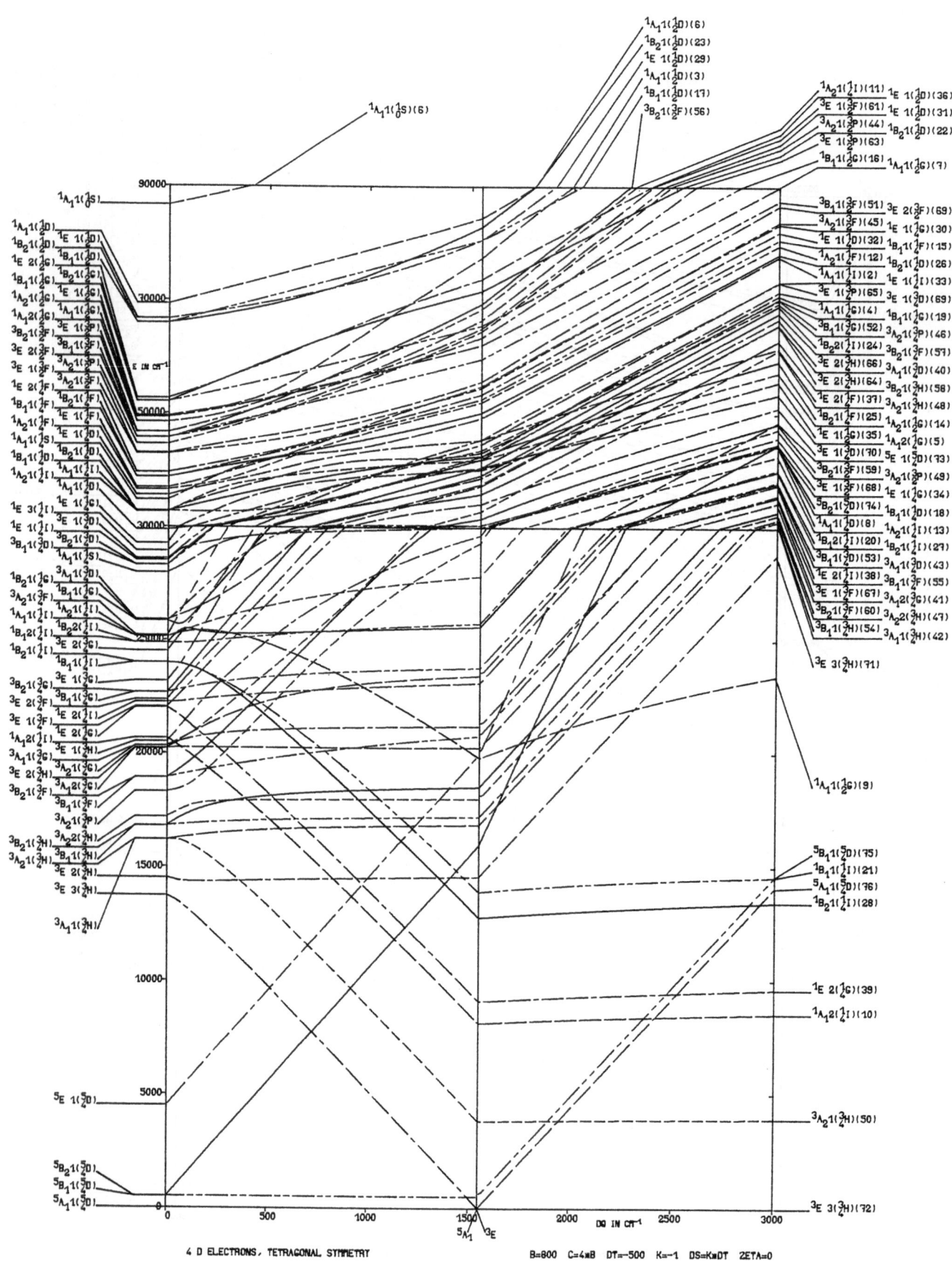

182

4 D ELECTRONS, TETRAGONAL SYMMETRY

B=800 C=4×B DT=-500 K=-1 DS=K×DT ZETA=0

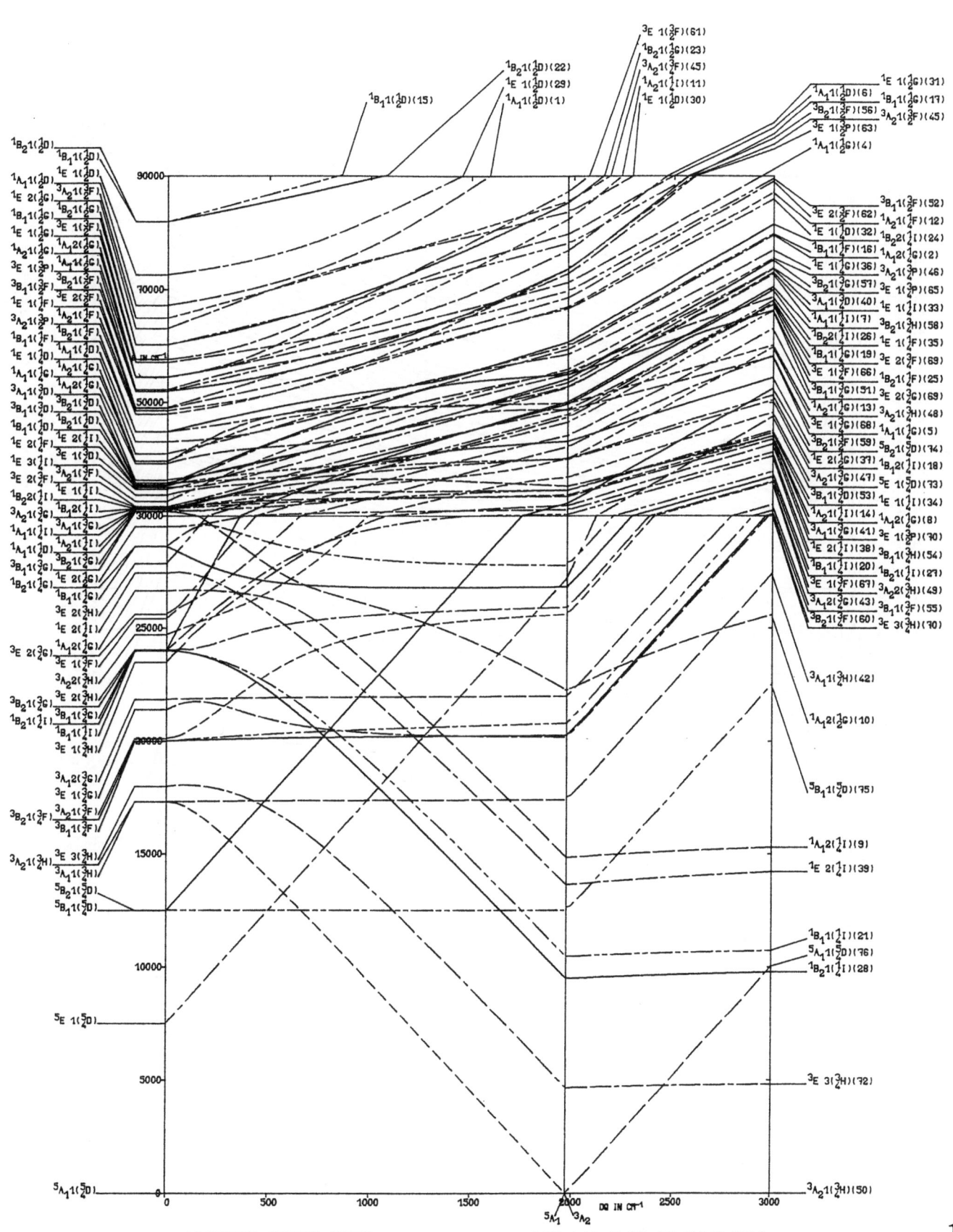

4 D ELECTRONS, TETRAGONAL SYMMETRY

B=800 C=4×B DT=-500 K=5 DS=K×DT ZETA=0

183

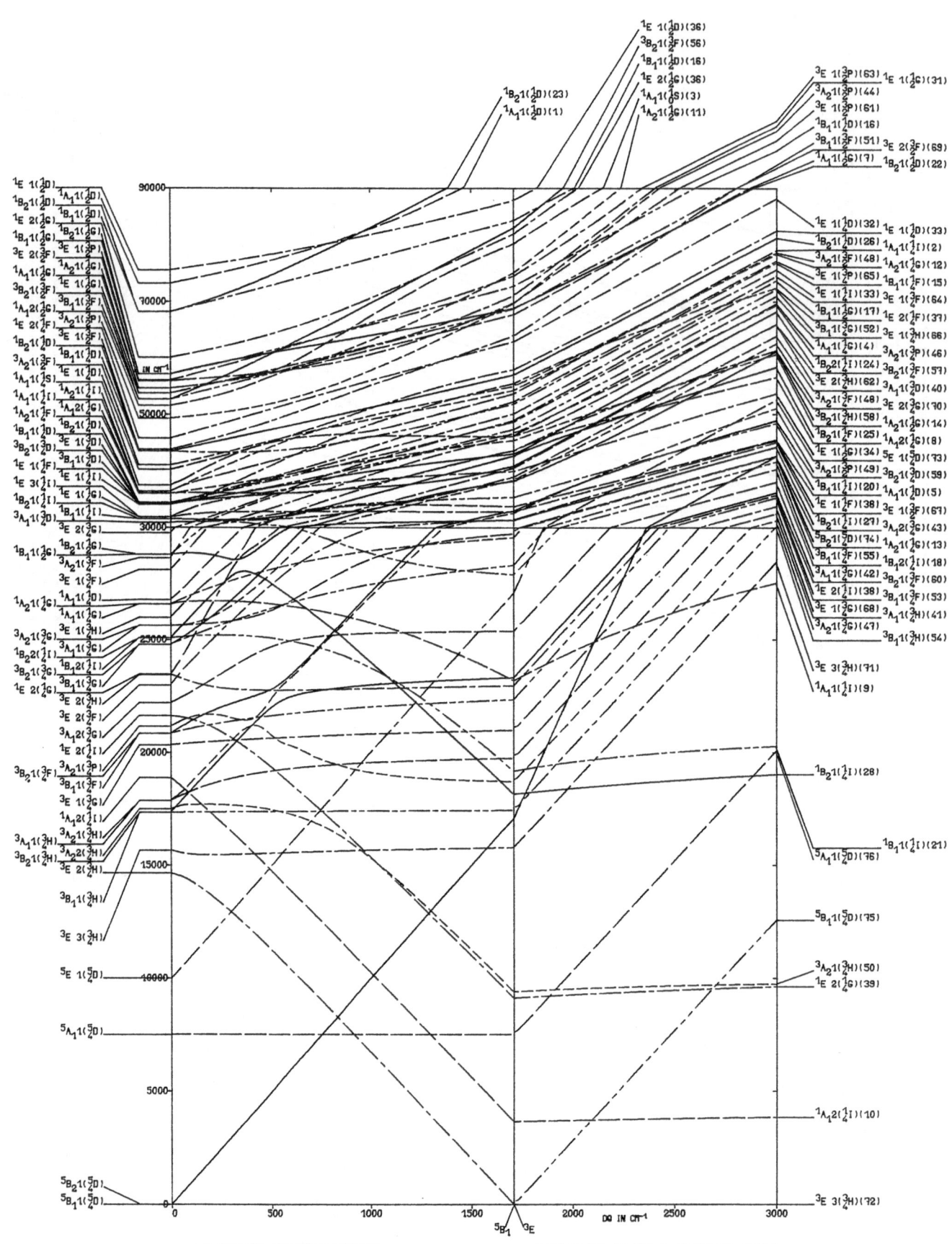

4 D ELECTRONS, TETRAGONAL SYMMETRY

B=800   C=4×B   DT=-500   K=-5   DS=K×DT   ZETA=0

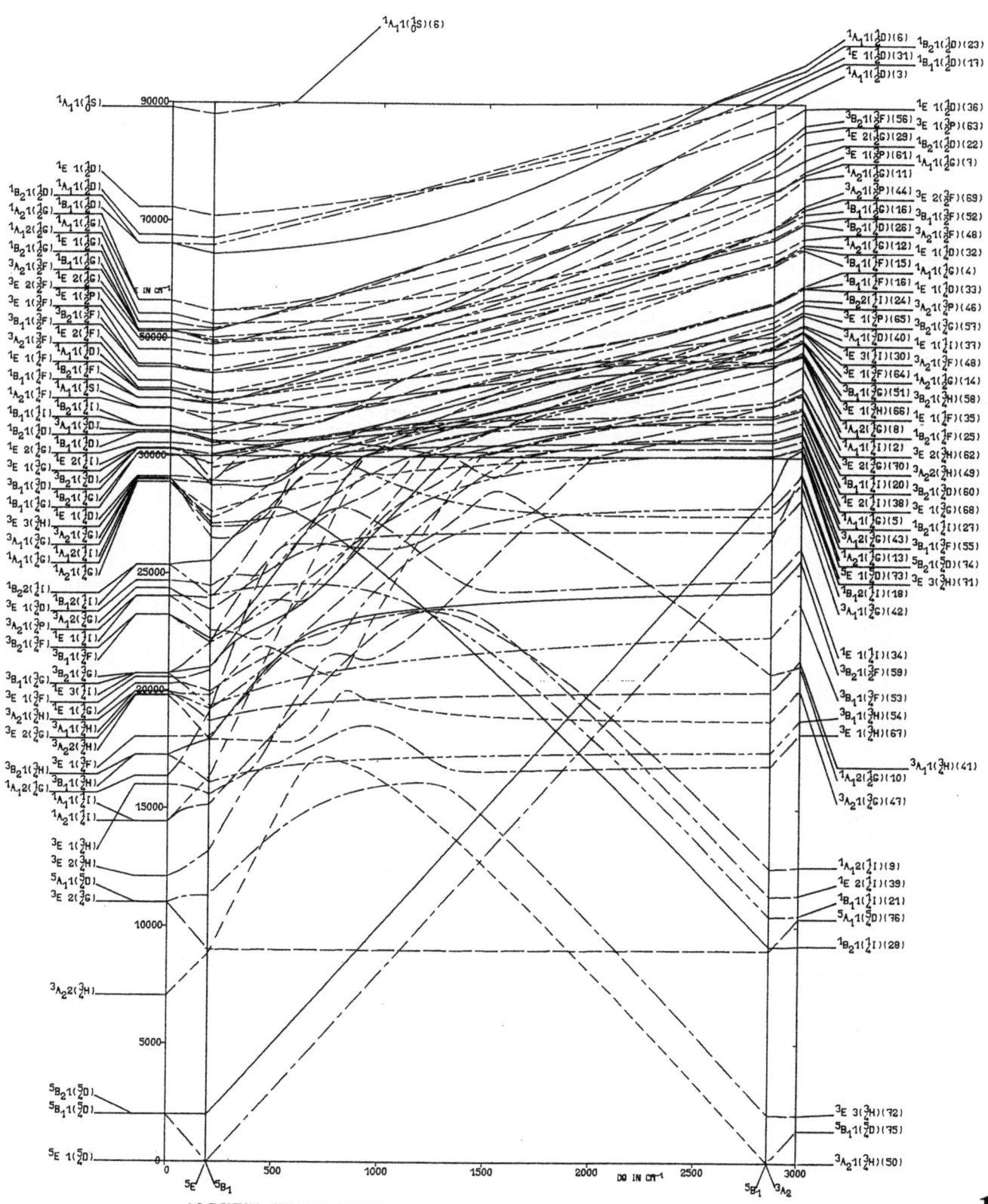

4 D ELECTRONS, TETRAGONAL SYMMETRY

B=800 C=4×B DT=1000 K=1 DS=K×DT ZETA=0

185

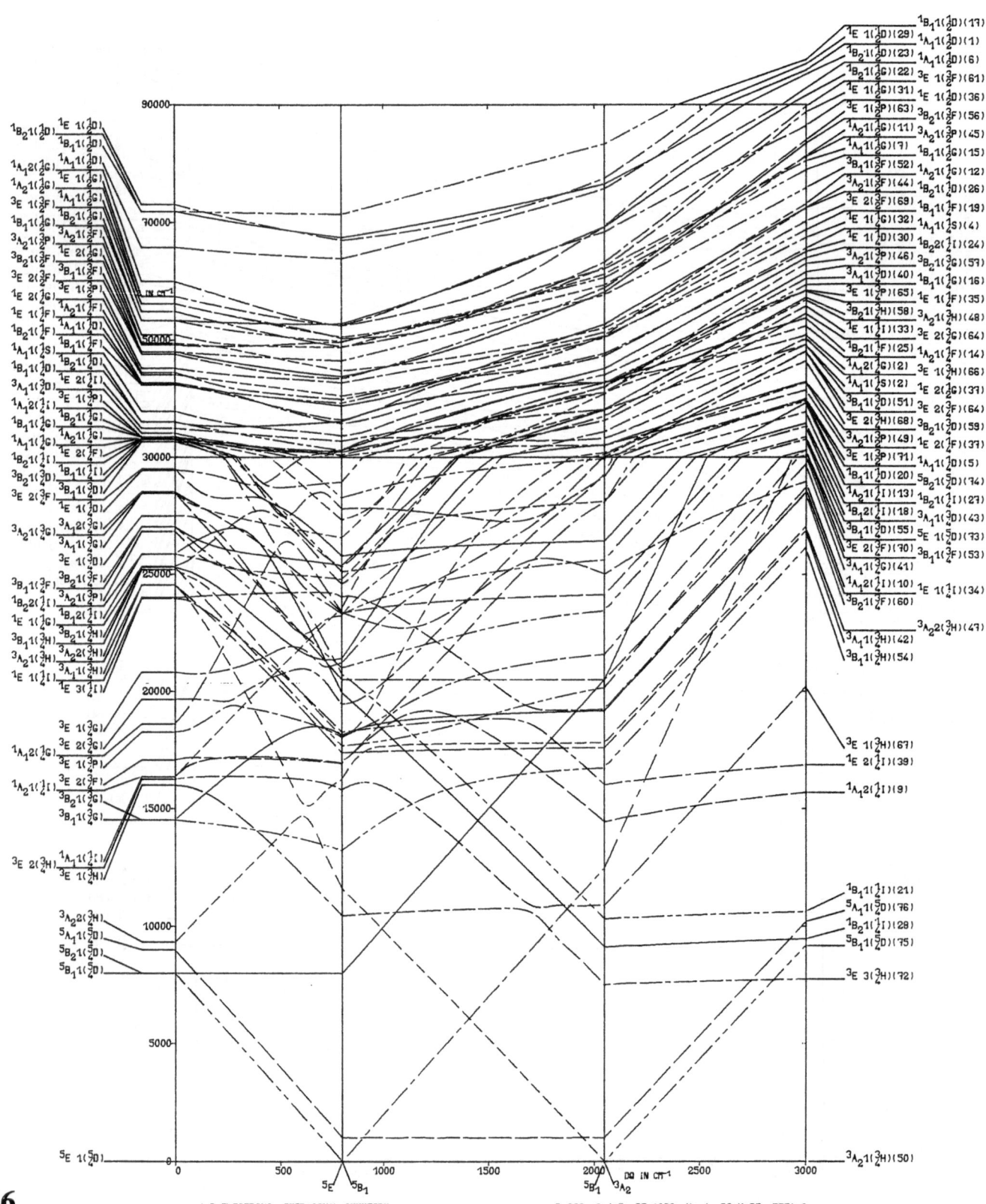

4 D ELECTRONS, TETRAGONAL SYMMETRY

B=800 C=4×B DT=1000 K=-1 DS=K×DT ZETA=0

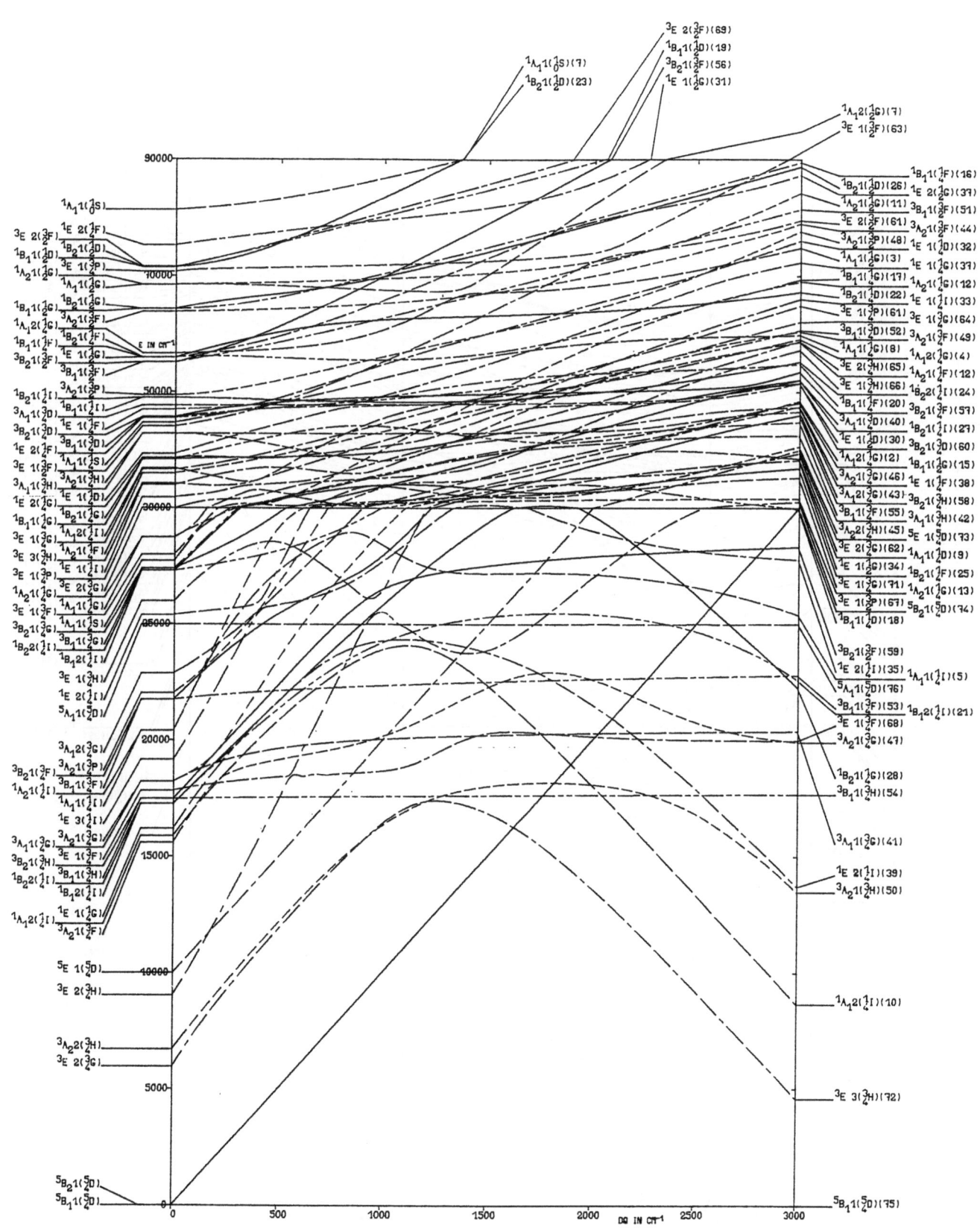

4 D ELECTRONS, TETRAGONAL SYMMETRY   B=800 C=4×B DT=1000 K=5 DS=K×DT ZETA=0

187

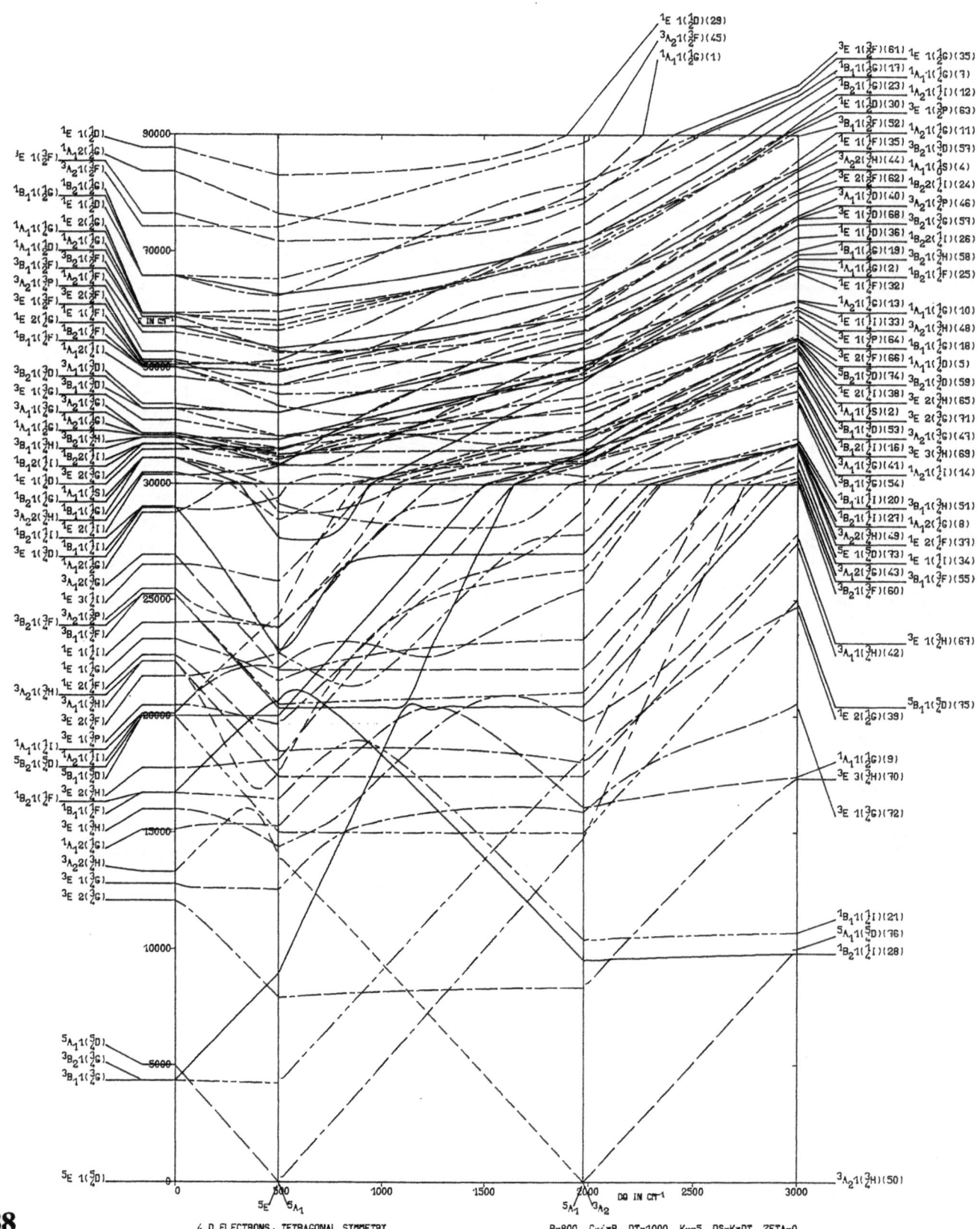

188

4 D ELECTRONS, TETRAGONAL SYMMETRY

B=800 C=4×B DT=1000 K=-5 DS=K×DT ZETA=0

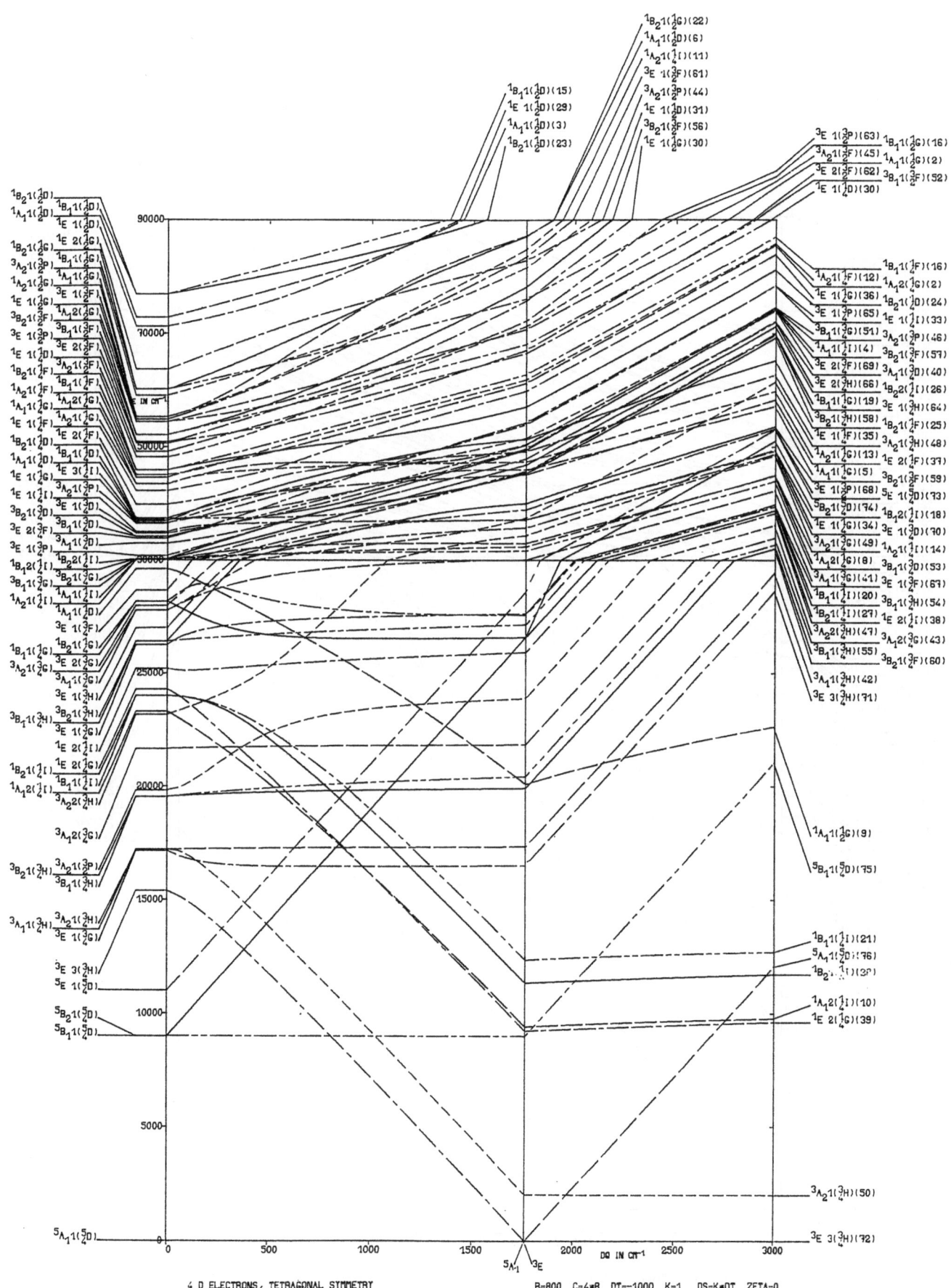

4 D ELECTRONS, TETRAGONAL SYMMETRY   B=800 C=4×B DT=-1000 K=1   DS=K×DT ZETA=0

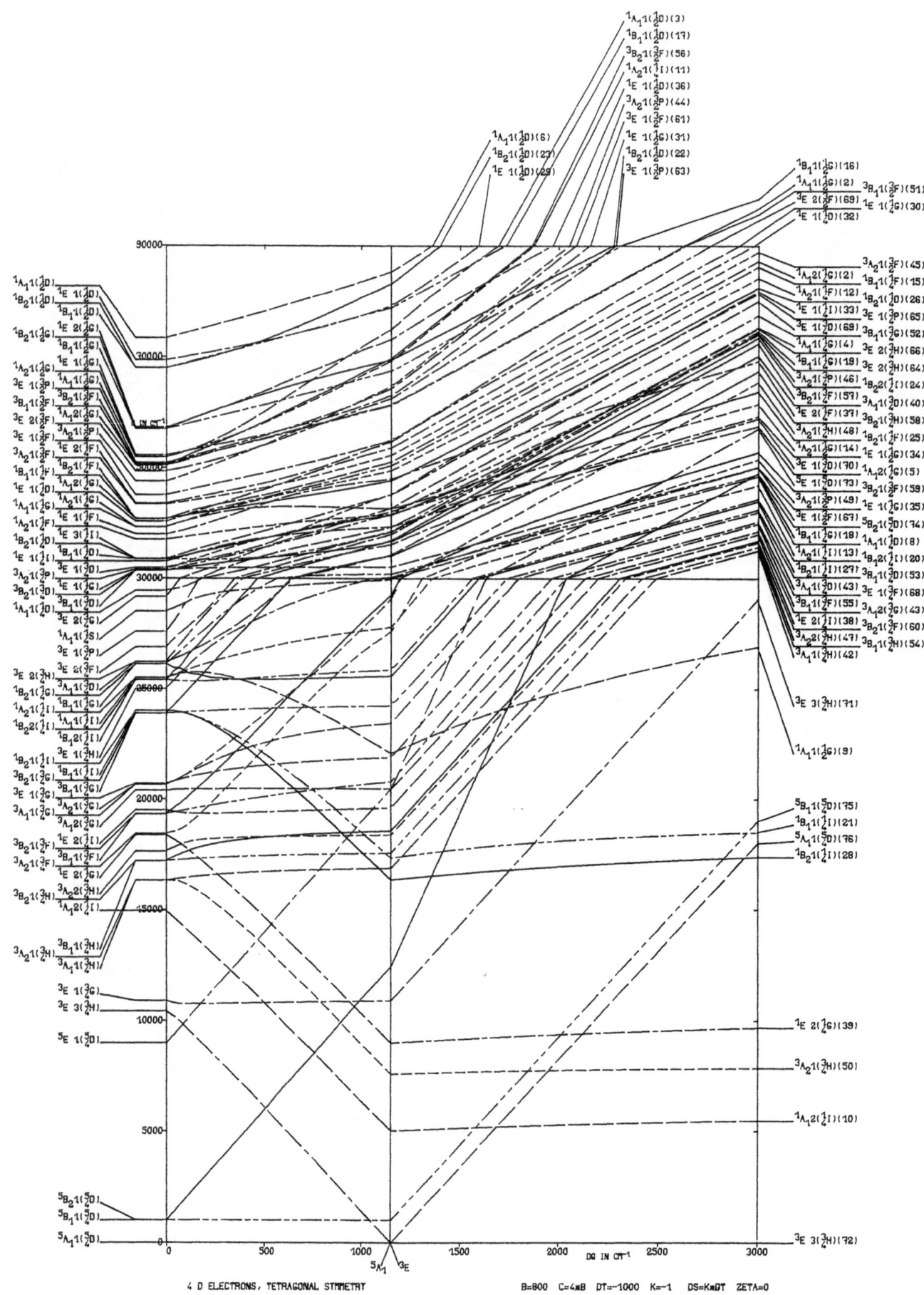

4 D ELECTRONS, TETRAGONAL SYMMETRY   B=800 C=4×B DT=-1000 K=-1   DS=K×DT ZETA=0

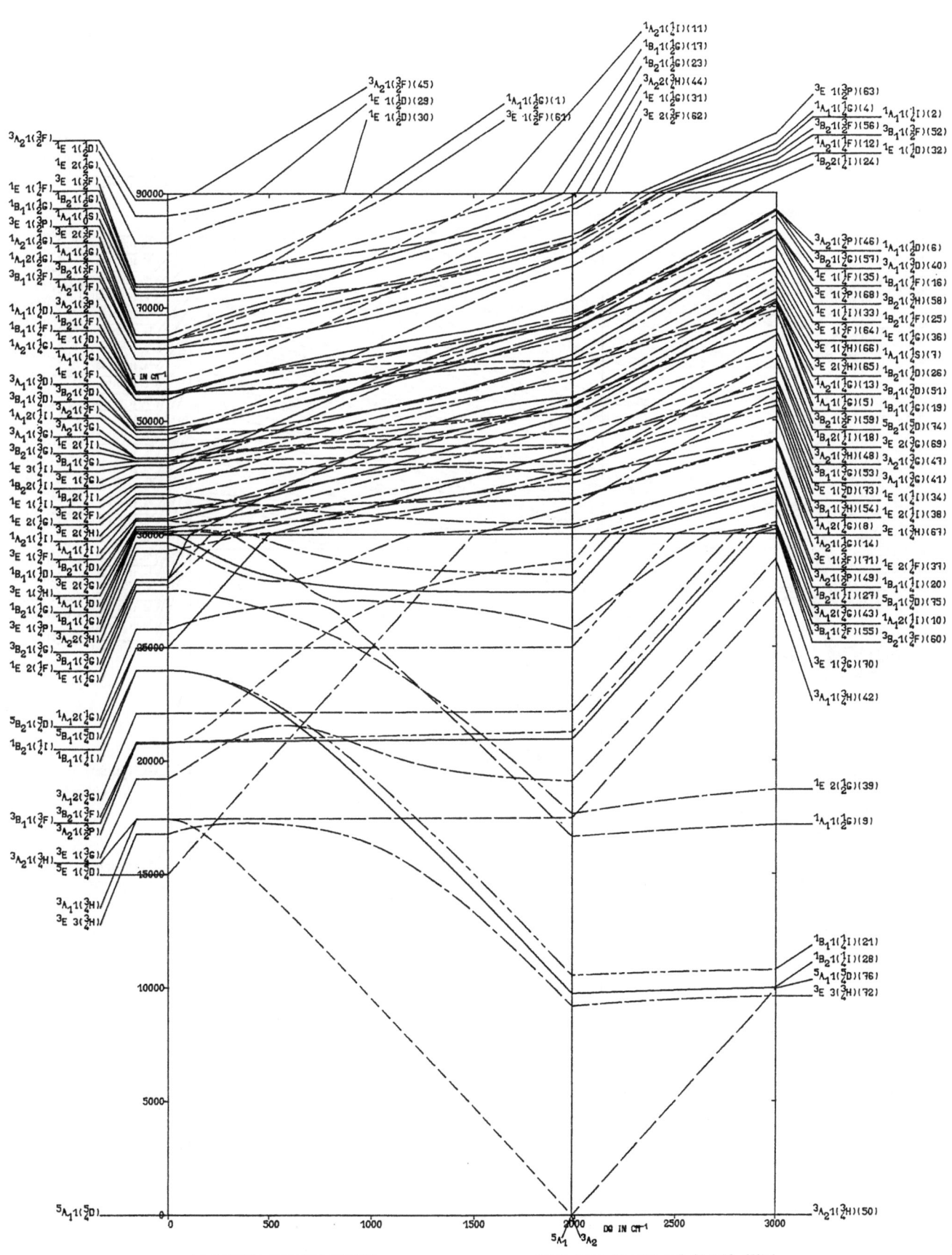

4 D ELECTRONS, TETRAGONAL SYMMETRY   B=800 C=4×B DT=-1000 K=5 DS=K×DT ZETA=0

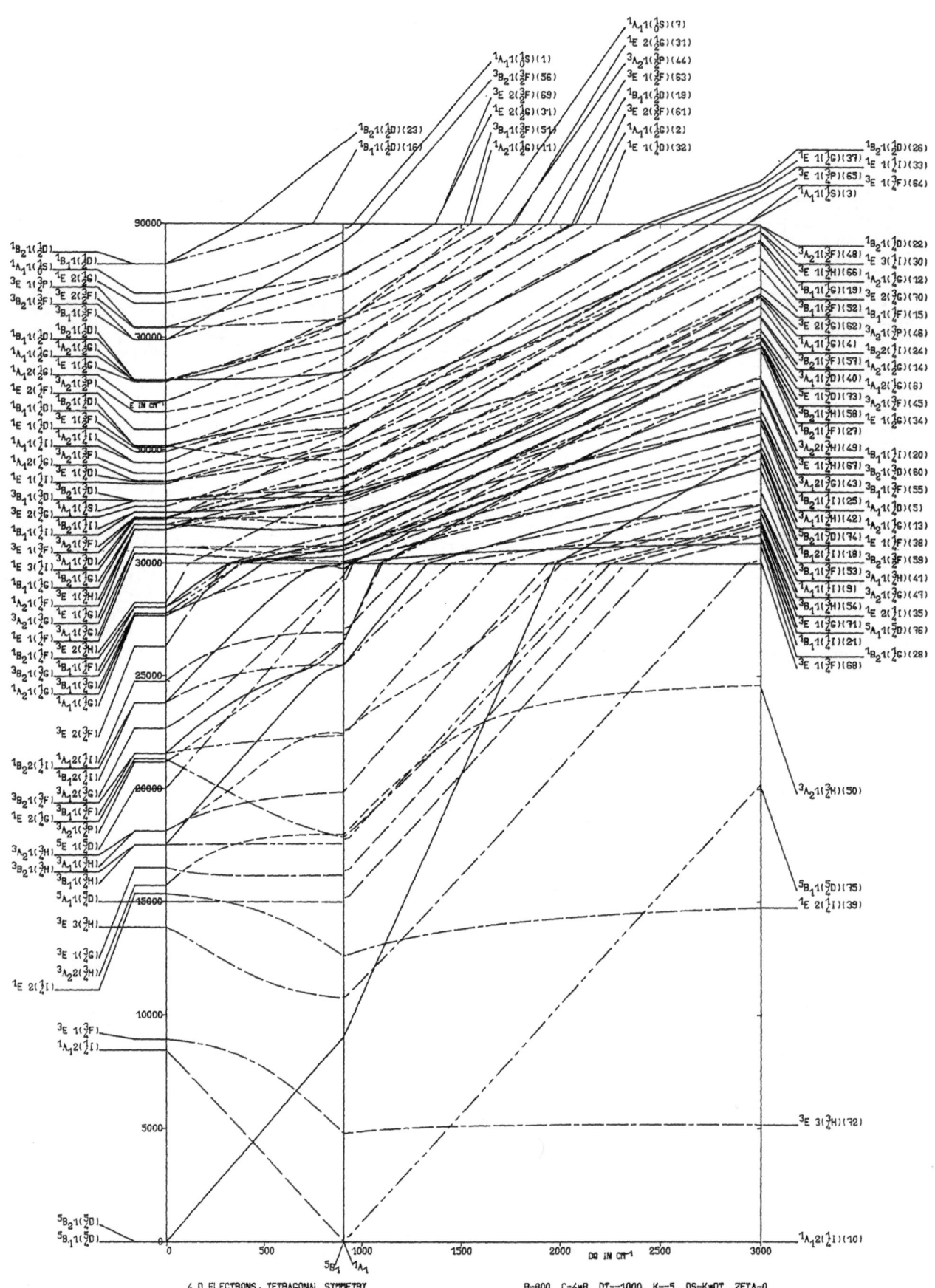

4 D ELECTRONS, TETRAGONAL SYMMETRY

B=800 C=4×B DT=-1000 K=-5 DS=K×DT ZETA=0

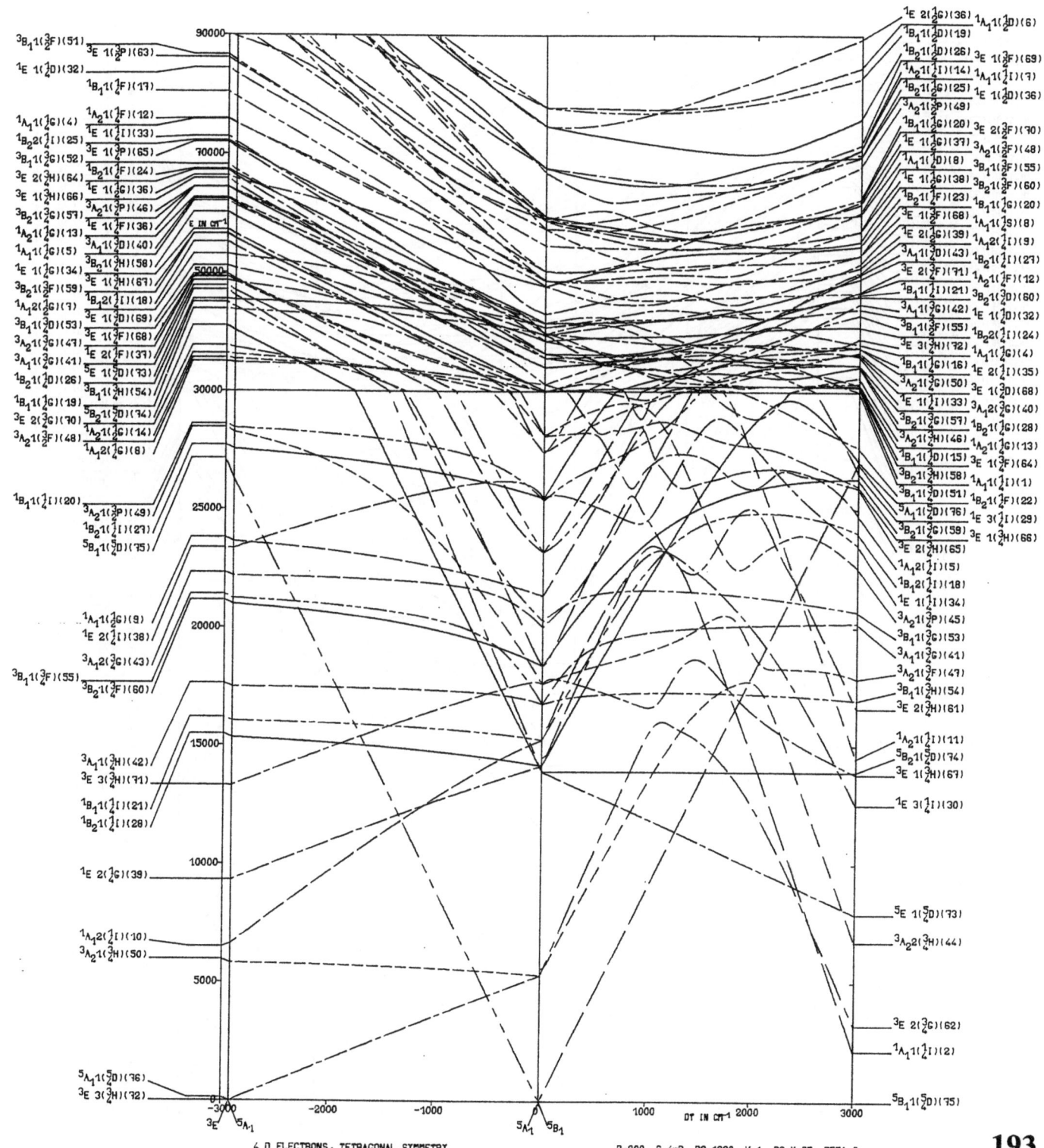

4 D ELECTRONS, TETRAGONAL SYMMETRY

B=800 C=4×B DQ=1390 K=1 DS=K×DT ZETA=0

193

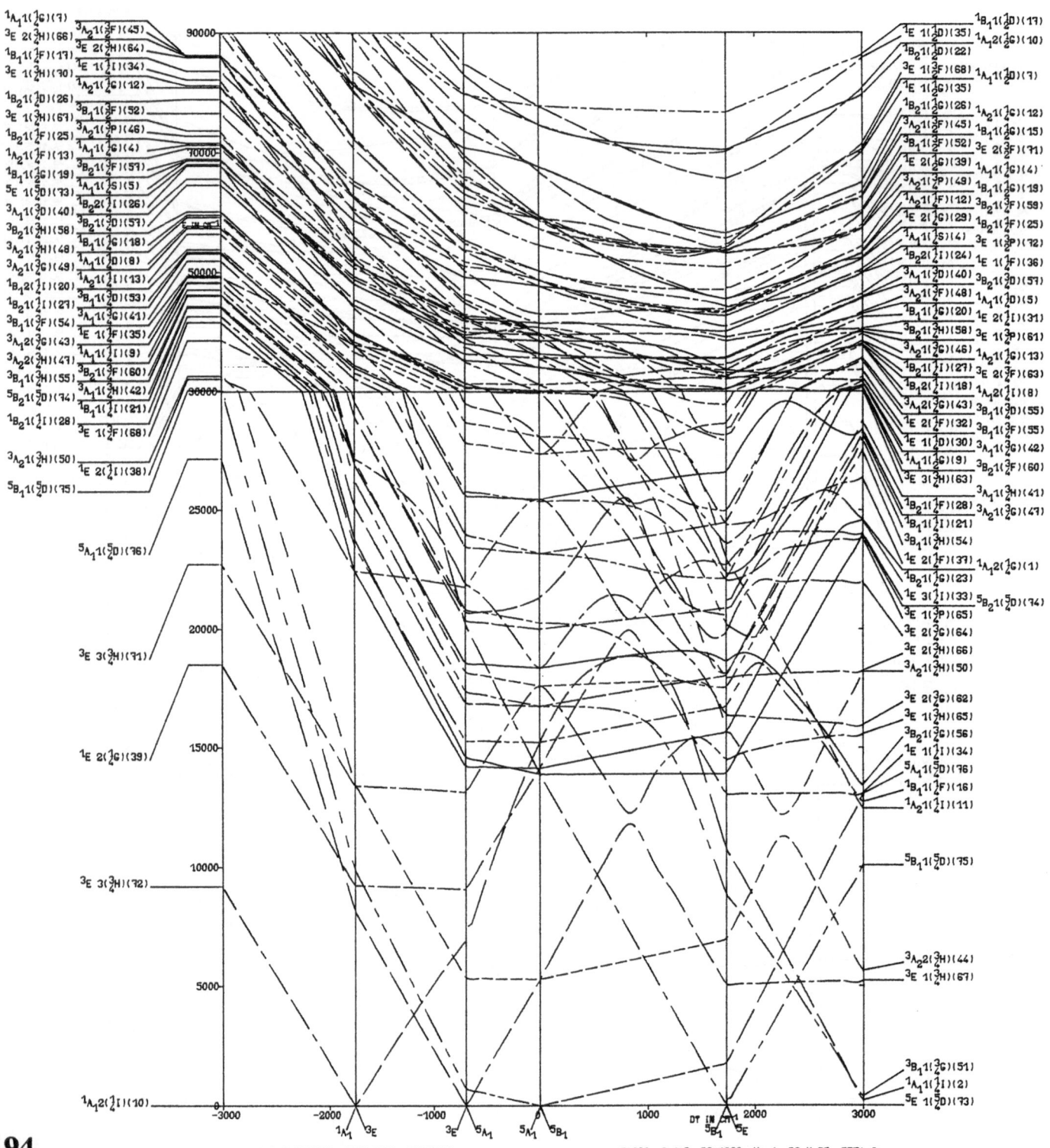

4 D ELECTRONS, TETRAGONAL SYMMETRY

B=800 C=4×B DQ=1390 K=-1 DS=K×DT ZETA=0

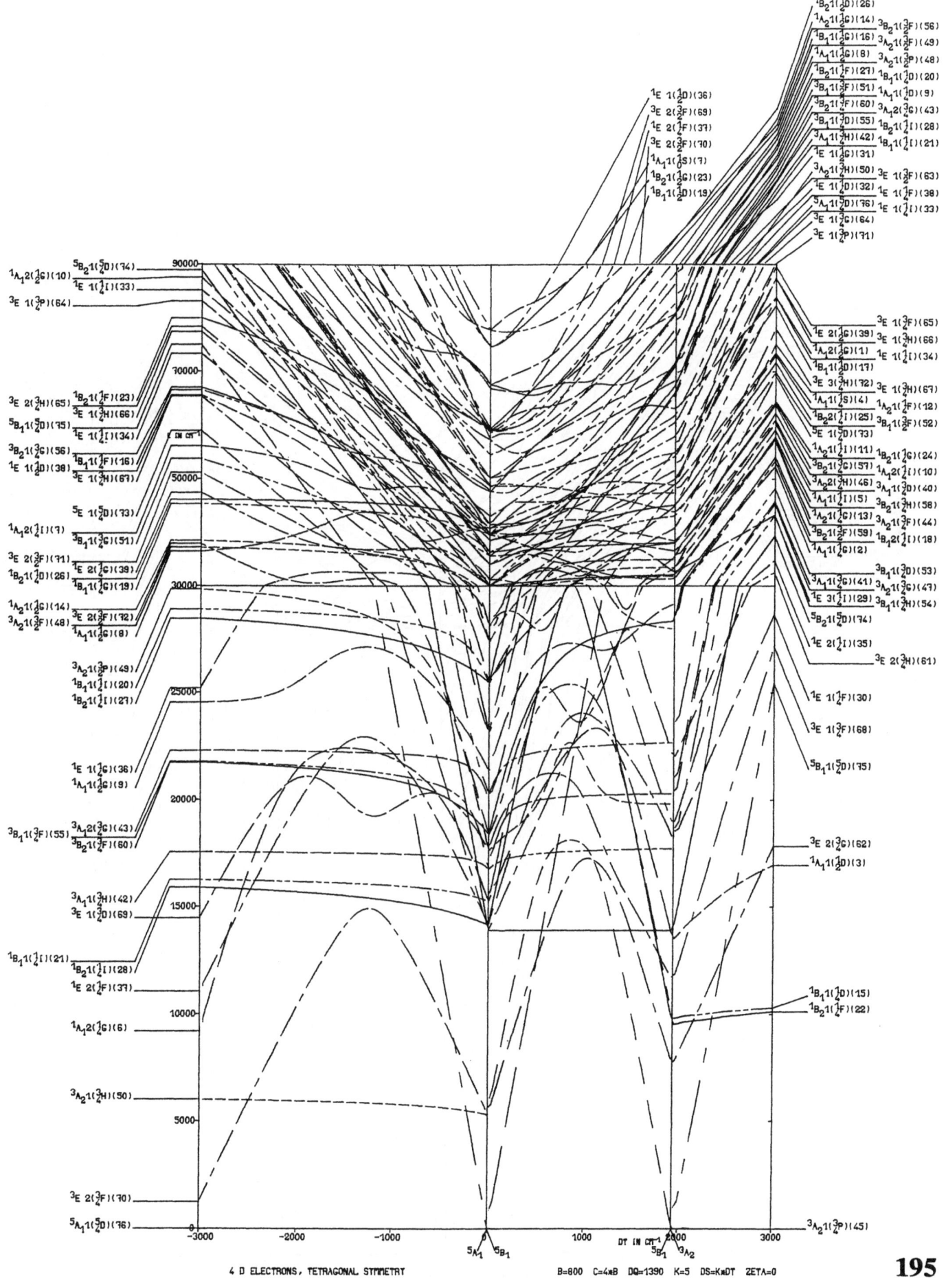

4 D ELECTRONS, TETRAGONAL SYMMETRY   B=800 C=4×B DQ=1390 K=5 DS=K×DT ZETA=0

195

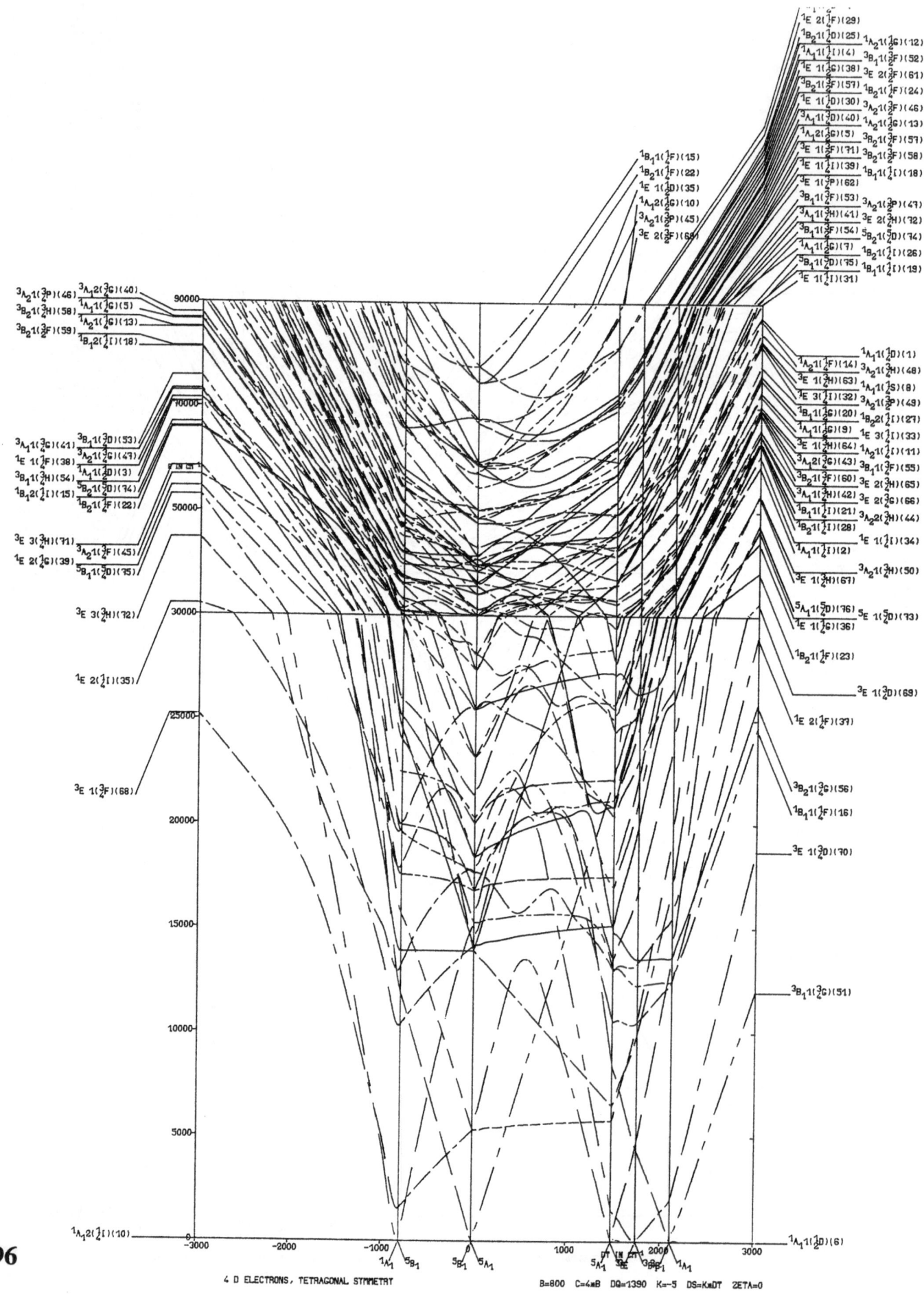

196

4 D ELECTRONS, TETRAGONAL SYMMETRY

B=800 C=4×B DQ=1390 K=-5 DS=K×DT ZETA=0

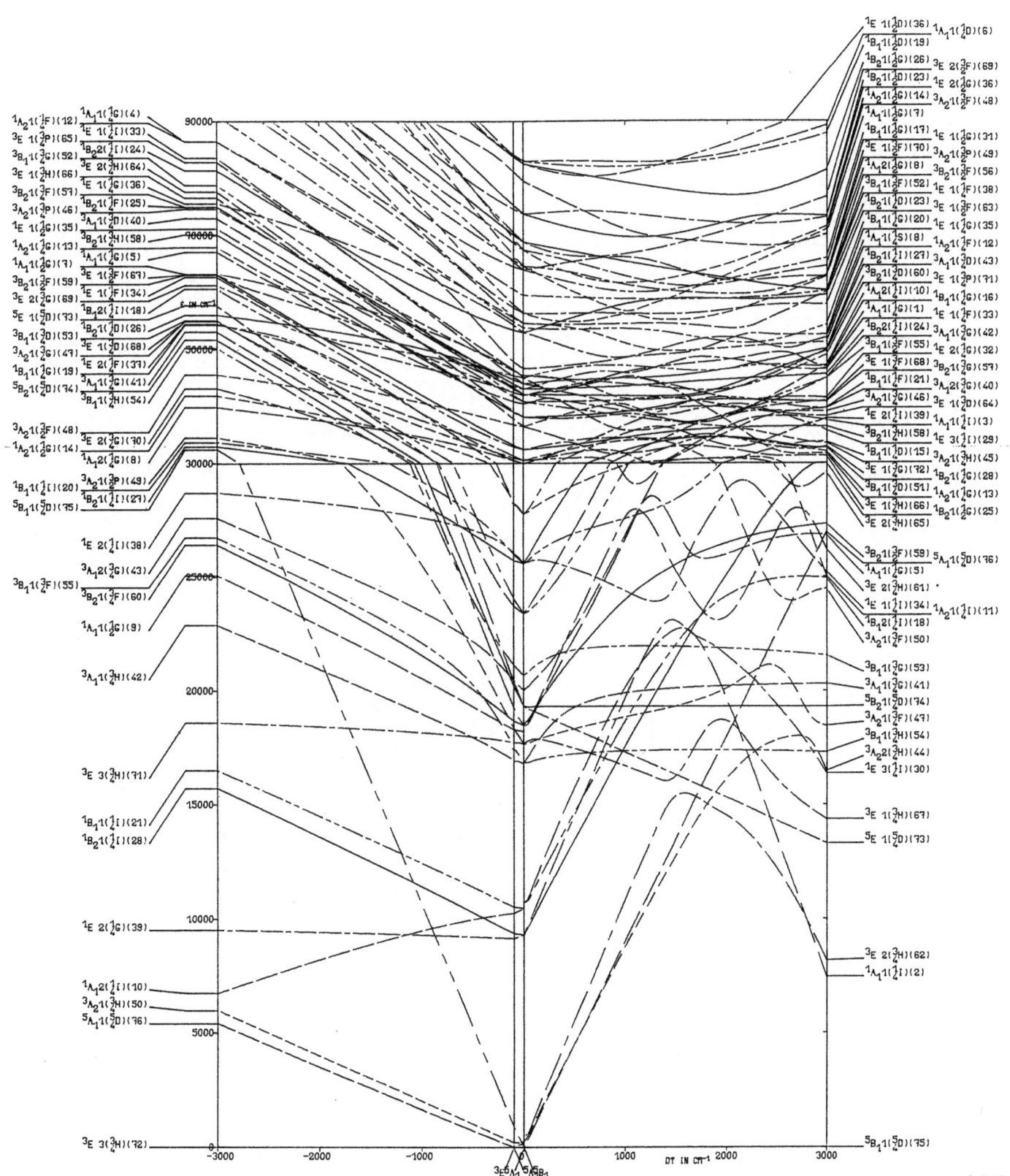

4 D ELECTRONS, TETRAGONAL SYMMETRY   B=800 C=4×B DQ=1930 K=1   DS=K×DT ZETA=0

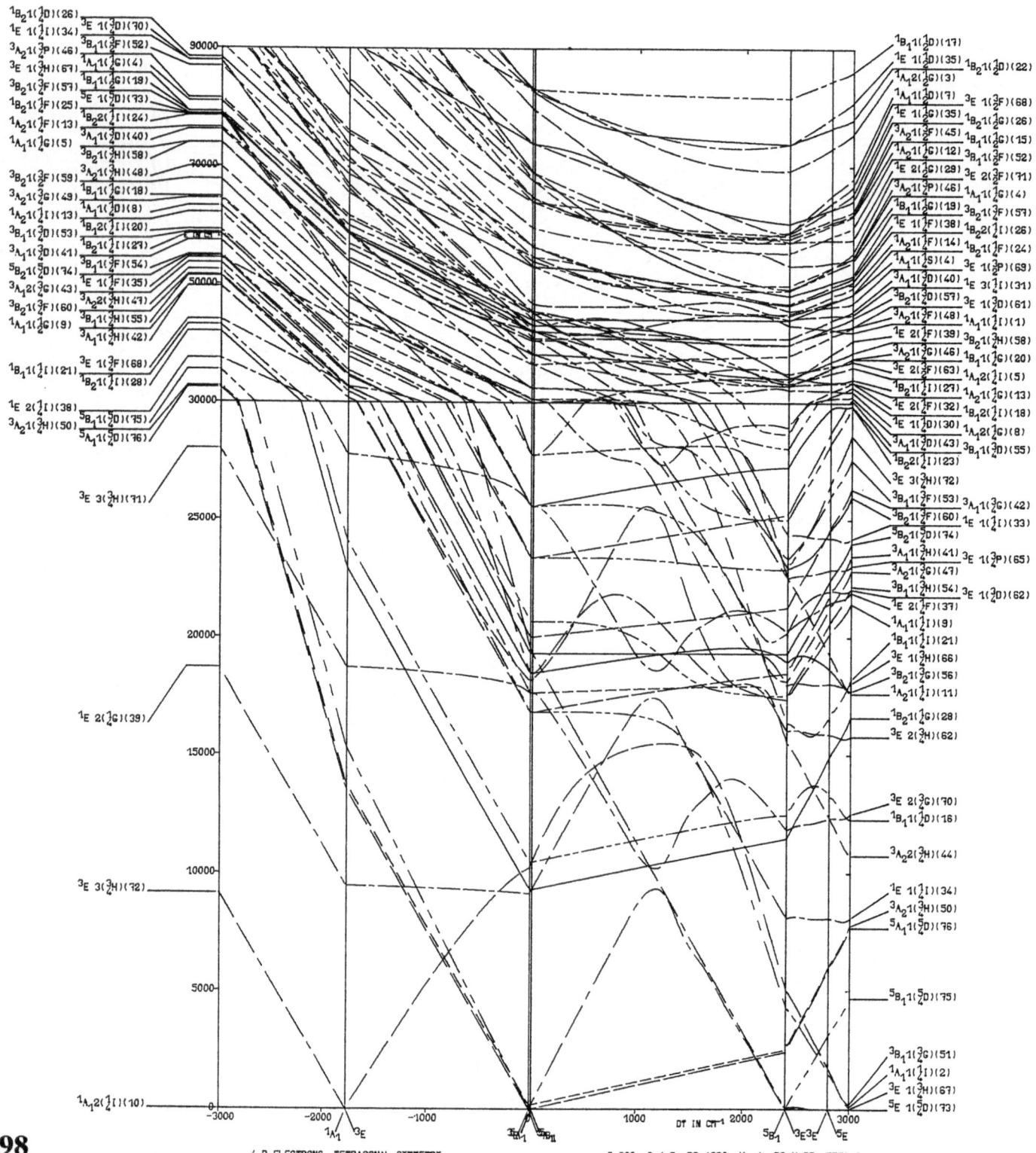

4 D ELECTRONS, TETRAGONAL SYMMETRY

B=800 C=4×B DQ=1930 K=−¼ DS=K×DT ZETA=0

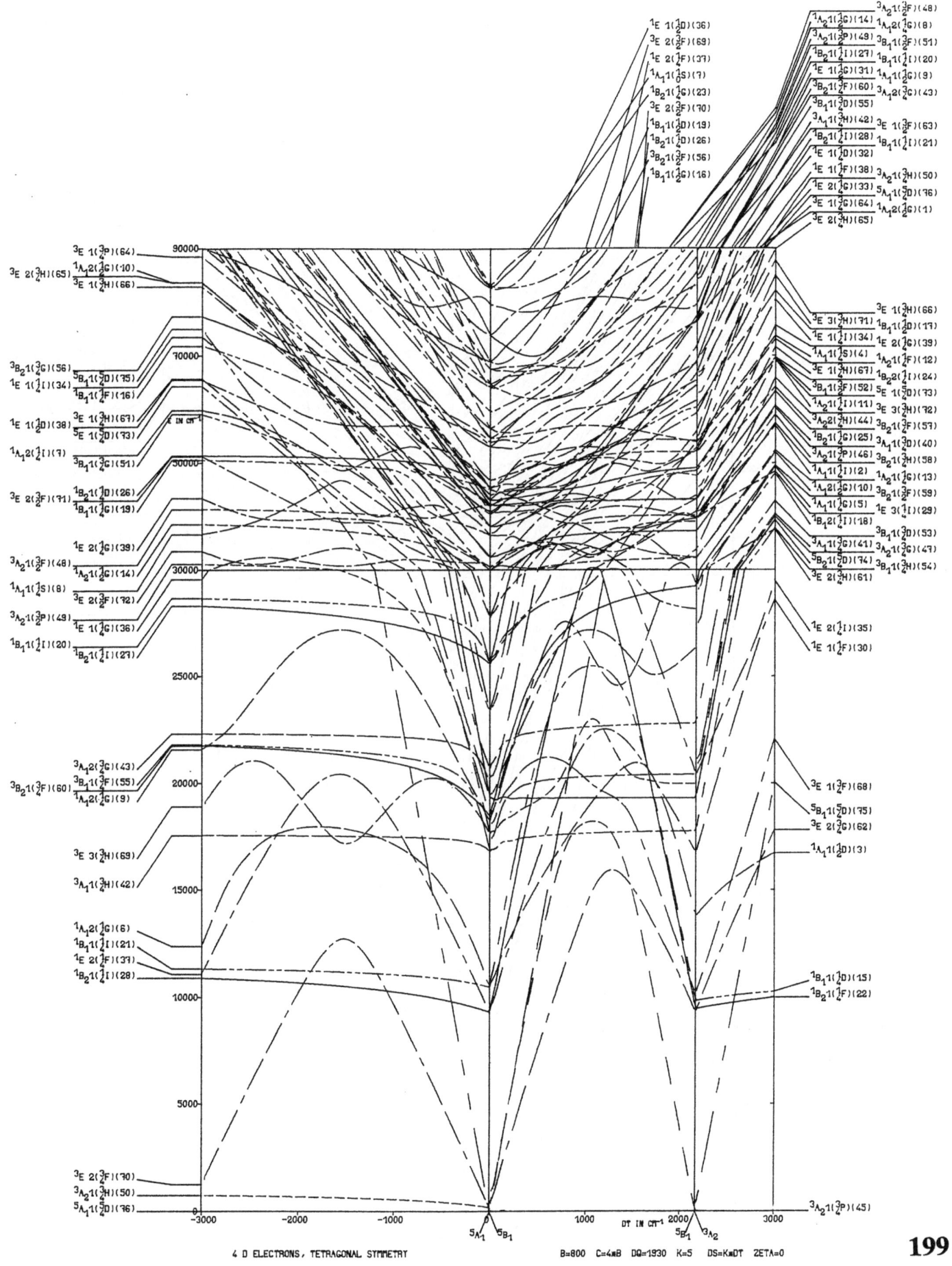

4 D ELECTRONS, TETRAGONAL SYMMETRY   B=800 C=4×B DQ=1930 K=5 DS=K×DT ZETA=0

199

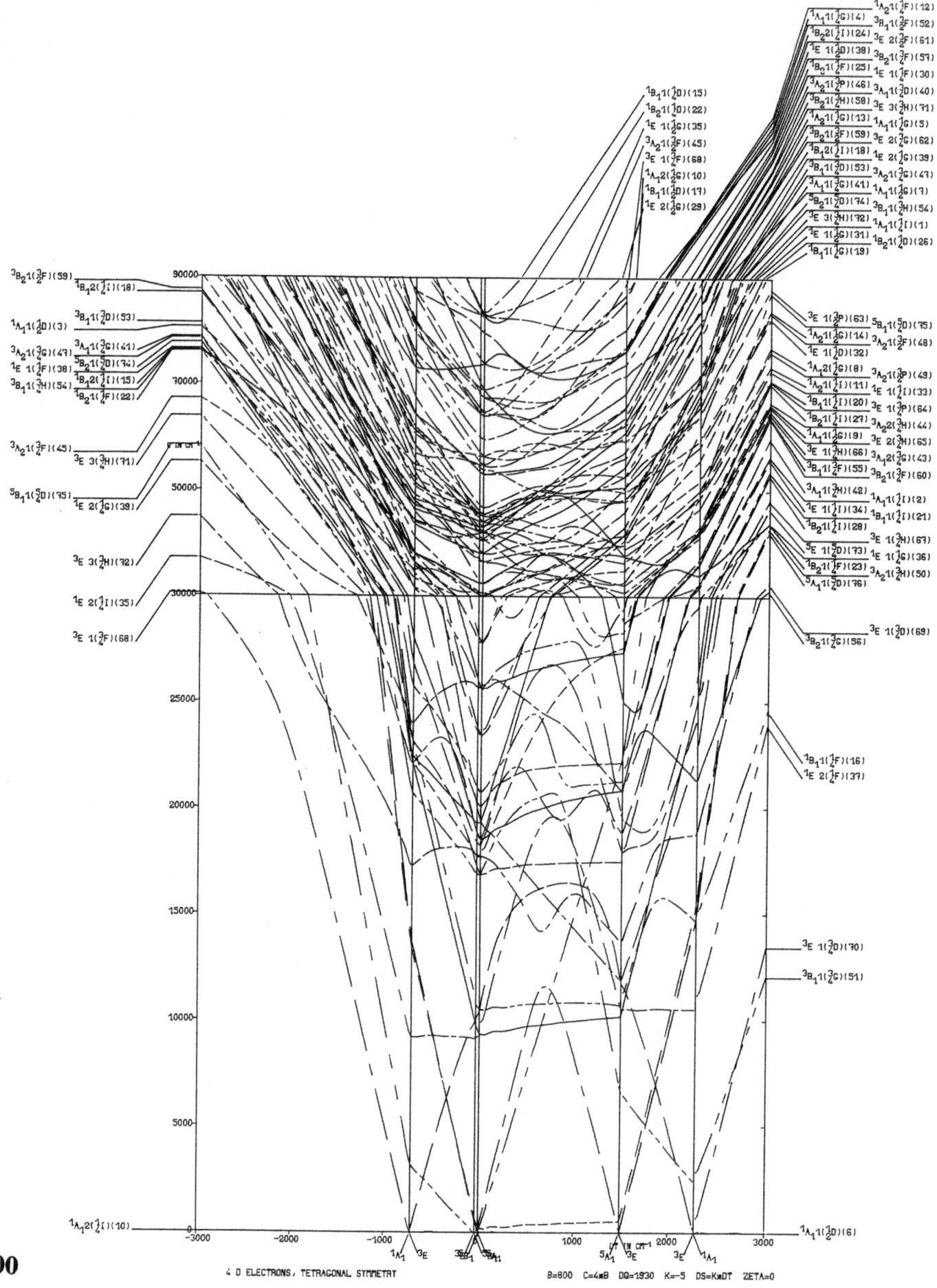

4 D ELECTRONS, TETRAGONAL SYMMETRY

B=800 C=4×B DQ=1930 K=-5 DS=K×DT ZETA=0

4 D ELECTRONS, TRIGONAL SYMMETRY

B=800   C=4×B   ZETA=0

ENERGY AS FUNCTION OF DQ
DT=0, 500, -500, 1000, -1000
K=1, -1, 5, -5

ENERGY AS FUNCTION OF DT
DQ=1390, 1930
K=1, -1, 5, -5

(67) $^5A_1(E_TE_T(^3A_2)E_EE_E(^3A_2))$
(66) $^5E (A_1E_T(^3E)E_EE_E(^3A_2))$
(65) $^5E (A_1E_E(^3E)E_TE_T(^3A_2))$
(64) $^3E (E_TE_E(^1E)E_EE_E(^3A_2))$
(63) $^3E (E_TE_T(^3A_2)E_EE_E(^1E))$
(62) $^3E (E_TE_E(^1E)E_EE_E(^3A_2))$
(61) $^3E (E_TE_E(^1E)E_TE_T(^3A_2))$
(60) $^3E (A_1E_E(^1E)E_EE_E(^3A_2))$
(59) $^3E (A_1E_T(^3E)E_EE_E(^3A_2))$
(58) $^3E (A_1E_T(^3E)E_EE_E(^1E))$
(57) $^3E (A_1E_T(^3E)E_EE_E(^1A_1))$
(56) $^3E (A_1E_T(^1E)E_EE_E(^3A_2))$
(55) $^3E (A_1E_E(^3E)E_TE_T(^3A_2))$
(54) $^3E (A_1E_E(^3E)E_TE_T(^1E))$
(53) $^3E (A_1E_E(^3E)E_TE_T(^1A_1))$
(52) $^3E (A_1E_E(^1E)E_TE_T(^3A_2))$
(51) $^3E (A_1E_T(^1E)E_TE_T(^3A_2))$
(50) $^3E (A_1A_1(^1A_1)E_TE_E(^3E))$
(49) $^3A_2(E_TE_E(^1A_1)E_EE_E(^3A_2))$
(48) $^3A_2(E_TE_T(^3A_2)E_EE_E(^1A_1))$
(47) $^3A_2(E_TE_T(^1A_1)E_EE_E(^3A_2))$
(46) $^3A_2(E_TE_E(^1A_1)E_TE_T(^3A_2))$
(45) $^3A_2(A_1E_T(^3E)E_EE_E(^1E))$
(44) $^3A_2(A_1E_E(^3E)E_TE_T(^1E))$
(43) $^3A_2(A_1A_1(^1A_1)E_EE_E(^3A_2))$
(42) $^3A_2(A_1A_1(^1A_1)E_TE_E(^3A_2))$
(41) $^3A_2(A_1A_1(^1A_1)E_TE_T(^3A_2))$
(40) $^3A_1(E_TE_E(^1A_2)E_EE_E(^3A_2))$
(39) $^3A_1(E_TE_T(^3A_2)E_EE_E(^3A_2))$
(38) $^3A_1(E_TE_E(^1A_2)E_TE_T(^3A_2))$
(37) $^3A_1(A_1E_T(^3E)E_EE_E(^1E))$
(36) $^3A_1(A_1E_E(^3E)E_TE_T(^1E))$
(35) $^3A_1(A_1A_1(^1A_1)E_TE_E(^3A_1))$
(34) $^1E (E_TE_E(^1A_1)E_EE_E(^1E))$

(33) $^1E (E_TE_T(^1E)E_EE_E(^1E))$
(32) $^1E (E_TE_T(^1E)E_EE_E(^1A_1))$
(31) $^1E (E_TE_T(^1A_1)E_EE_E(^1E))$
(30) $^1E (E_TE_E(^1A_1)E_TE_T(^1E))$
(29) $^1E (A_1E_E(^1E)E_EE_E(^1A_1))$
(28) $^1E (A_1E_T(^3E)E_EE_E(^3A_2))$
(27) $^1E (A_1E_T(^1E)E_EE_E(^1E))$
(26) $^1E (A_1E_T(^1E)E_EE_E(^1A_1))$
(25) $^1E (A_1E_E(^3E)E_TE_T(^3A_2))$
(24) $^1E (A_1E_E(^1E)E_TE_T(^1E))$
(23) $^1E (A_1E_E(^1E)E_TE_T(^1A_1))$
(22) $^1E (A_1E_T(^1E)E_TE_T(^1A_1))$
(21) $^1E (A_1A_1(^1A_1)E_EE_E(^1E))$
(20) $^1E (A_1A_1(^1A_1)E_TE_E(^1E))$
(19) $^1E (A_1A_1(^1A_1)E_TE_T(^1E))$
(18) $^1A_2(E_TE_T(^1A_2)E_EE_E(^1A_1))$
(17) $^1A_2(E_TE_T(^1E)E_EE_E(^1E))$
(16) $^1A_2(E_TE_E(^1A_2)E_TE_T(^1A_1))$
(15) $^1A_2(A_1E_T(^1E)E_EE_E(^1E))$
(14) $^1A_2(A_1E_E(^1E)E_TE_T(^1E))$
(13) $^1A_2(A_1A_1(^1A_1)E_TE_E(^1A_2))$
(12) $^1A_1(E_EE_E(^1A_1)E_EE_E(^1A_1))$
(11) $^1A_1(E_TE_E(^1A_1)E_EE_E(^1A_1))$
(10) $^1A_1(E_TE_T(^3A_2)E_EE_E(^3A_2))$
(9) $^1A_1(E_TE_T(^1E)E_EE_E(^1E))$
(8) $^1A_1(E_TE_T(^1A_1)E_EE_E(^1A_1))$
(7) $^1A_1(E_TE_T(^1A_1)E_TE_T(^1A_1))$
(6) $^1A_1(E_TE_T(^1A_1)E_TE_T(^1A_1))$
(5) $^1A_1(A_1E_T(^1E)E_EE_E(^1E))$
(4) $^1A_1(A_1E_E(^1E)E_TE_T(^1E))$
(3) $^1A_1(A_1A_1(^1A_1)E_EE_E(^1A_1))$
(2) $^1A_1(A_1A_1(^1A_1)E_TE_E(^1A_1))$
(1) $^1A_1(A_1A_1(^1A_1)E_TE_T(^1A_1))$

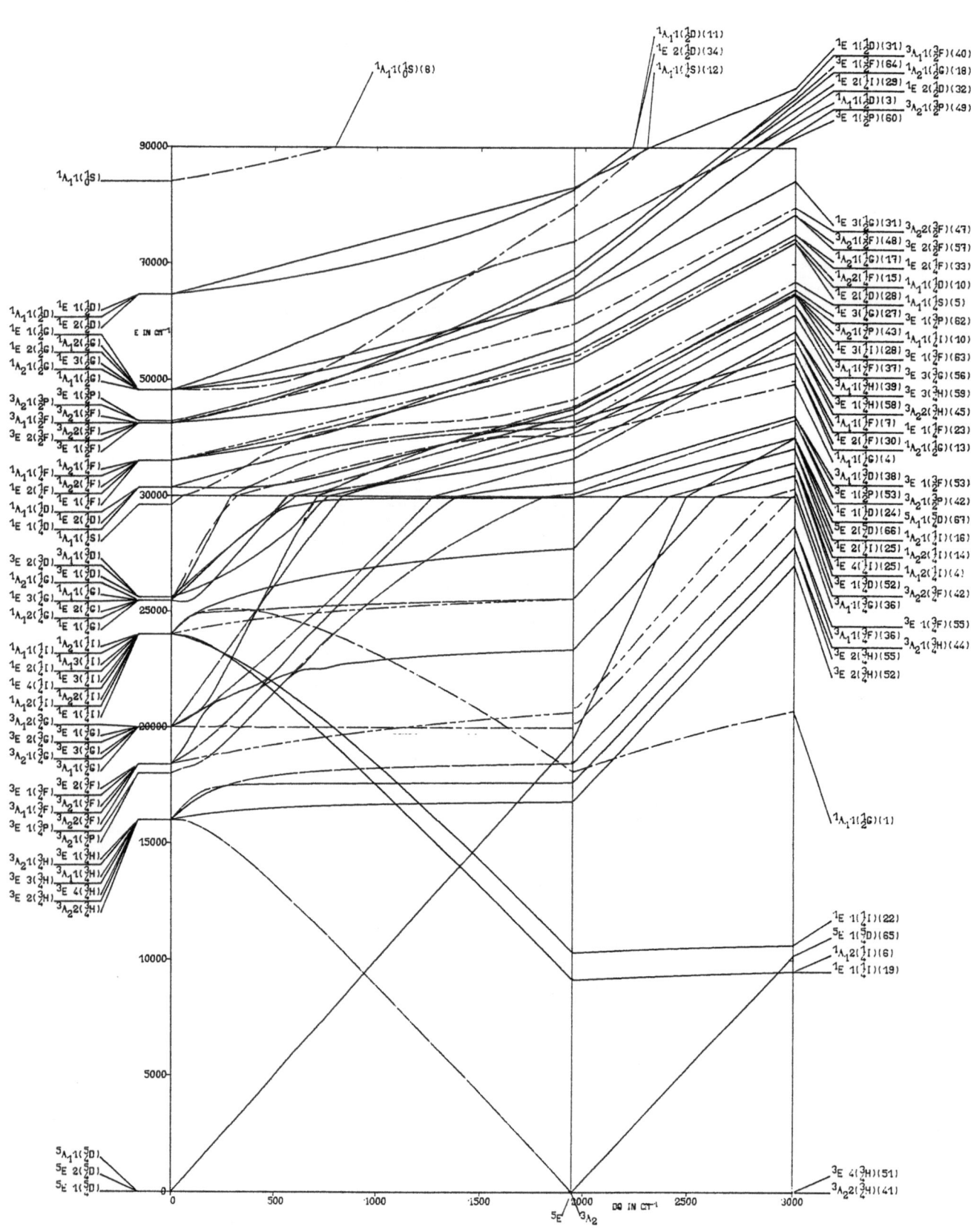

4 D ELECTRONS, TRIGONAL SYMMETRY

B=600 C=4×B DT=0 K=0 DS=K×DT ZETA=0

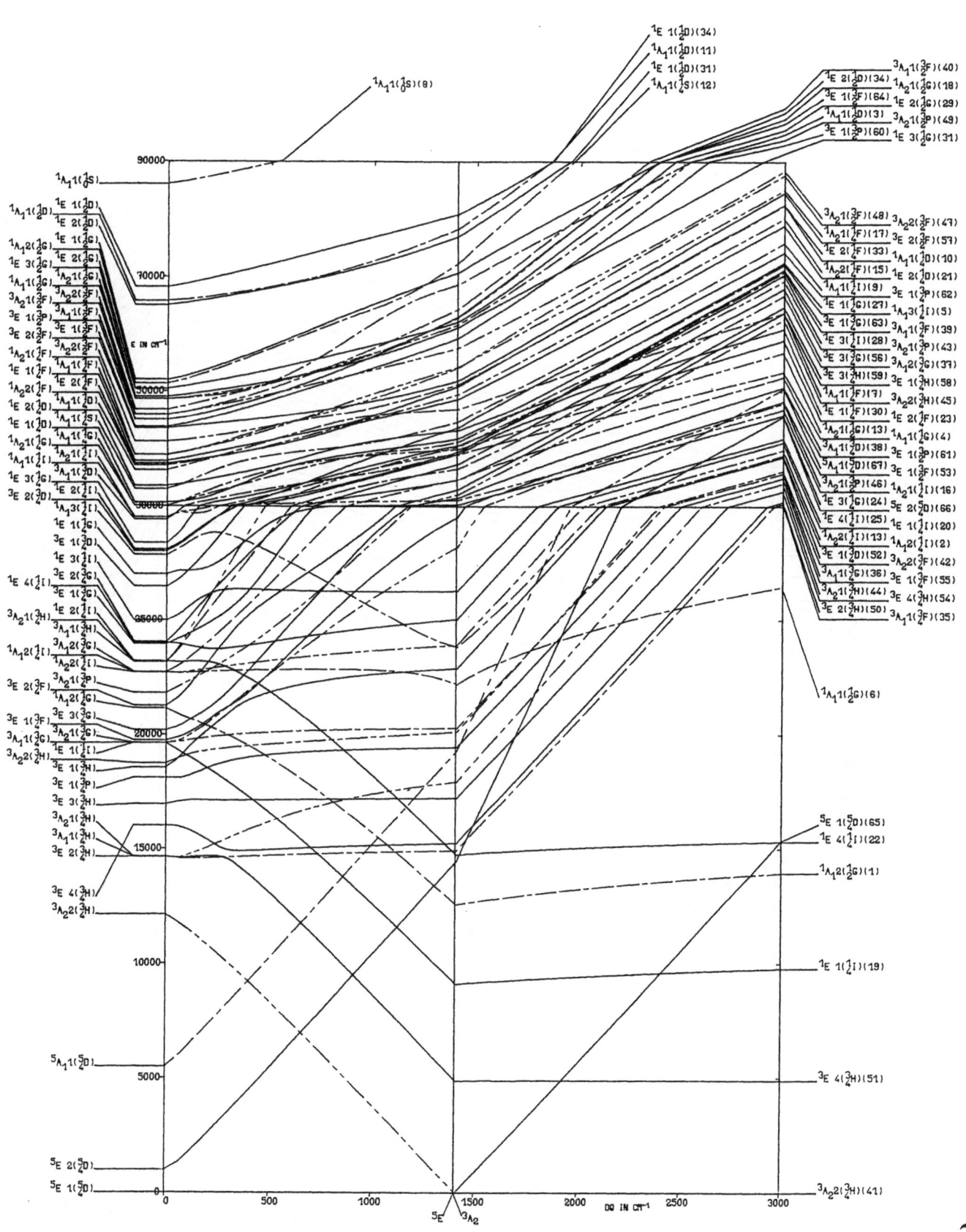

4 D ELECTRONS, TRIGONAL SYMMETRY

B=800 C=4×B DT=500 K=1 DS=K×DT ZETA=0

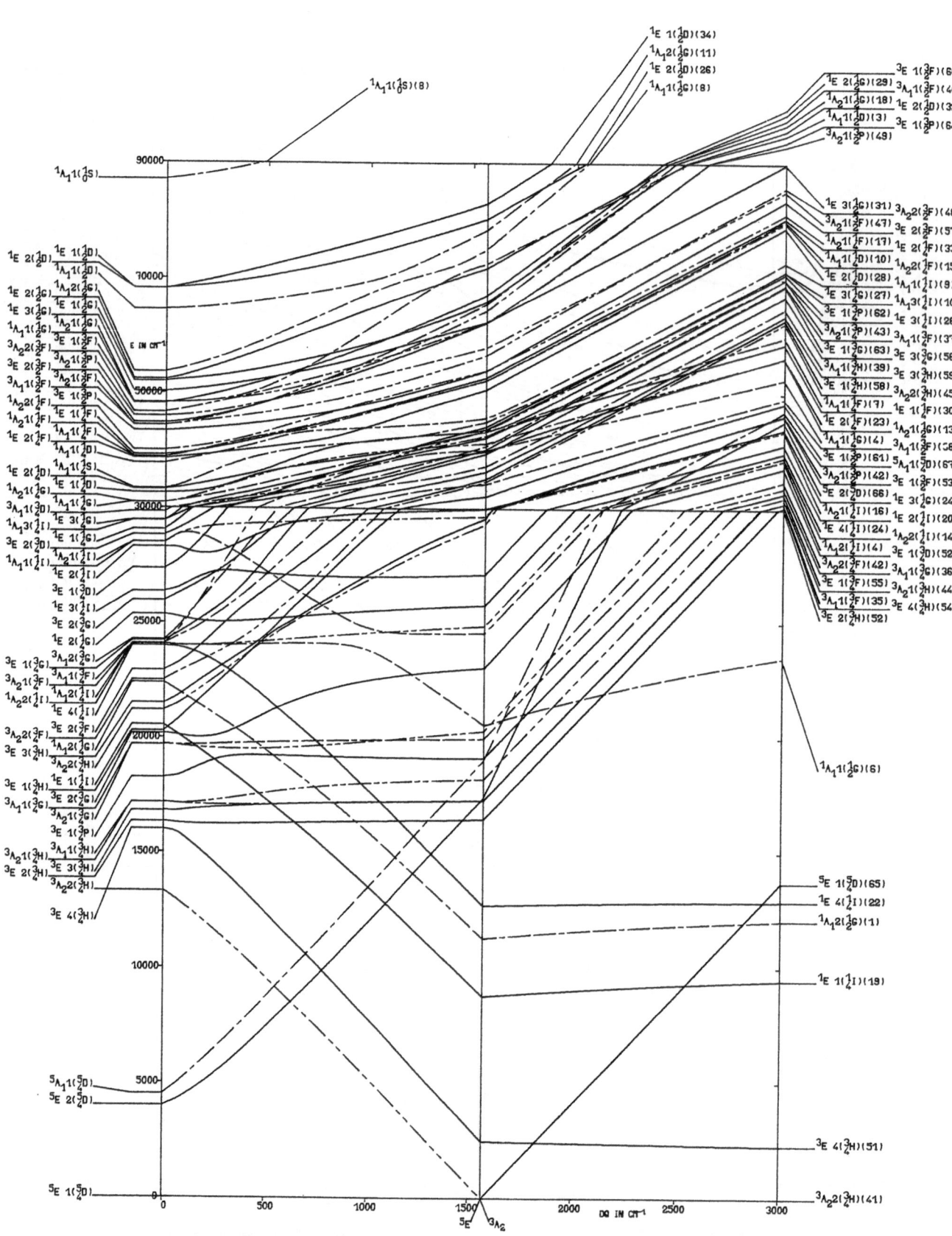

4 D ELECTRONS, TRIGONAL SYMMETRY

B=800 C=4×B DT=500 K=−1 DS=K×DT ZETA=0

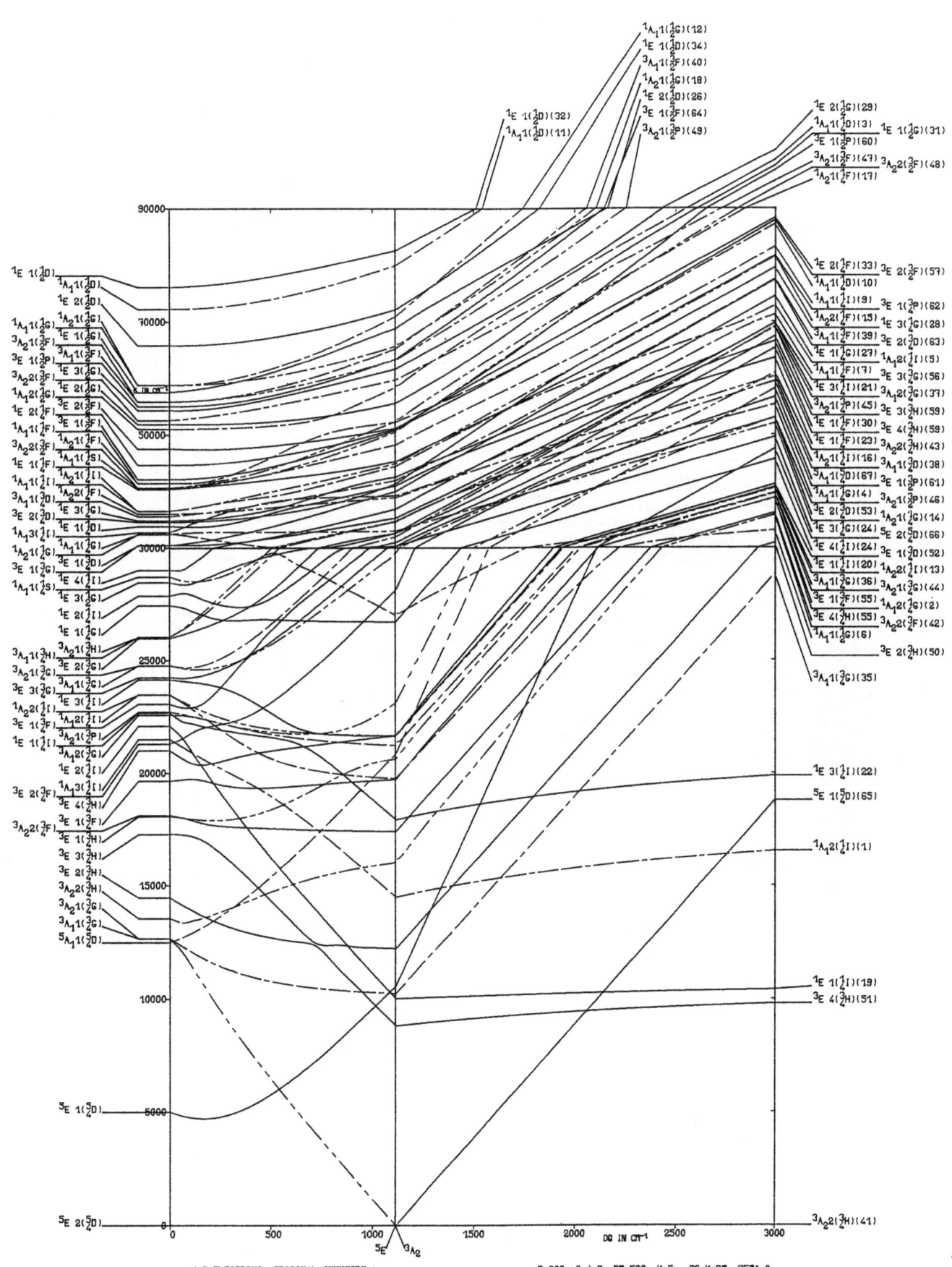

4 D ELECTRONS, TRIGONAL SYMMETRY   B=800   C=4×B   DT=500   K=5   DS=K×DT   ZETA=0

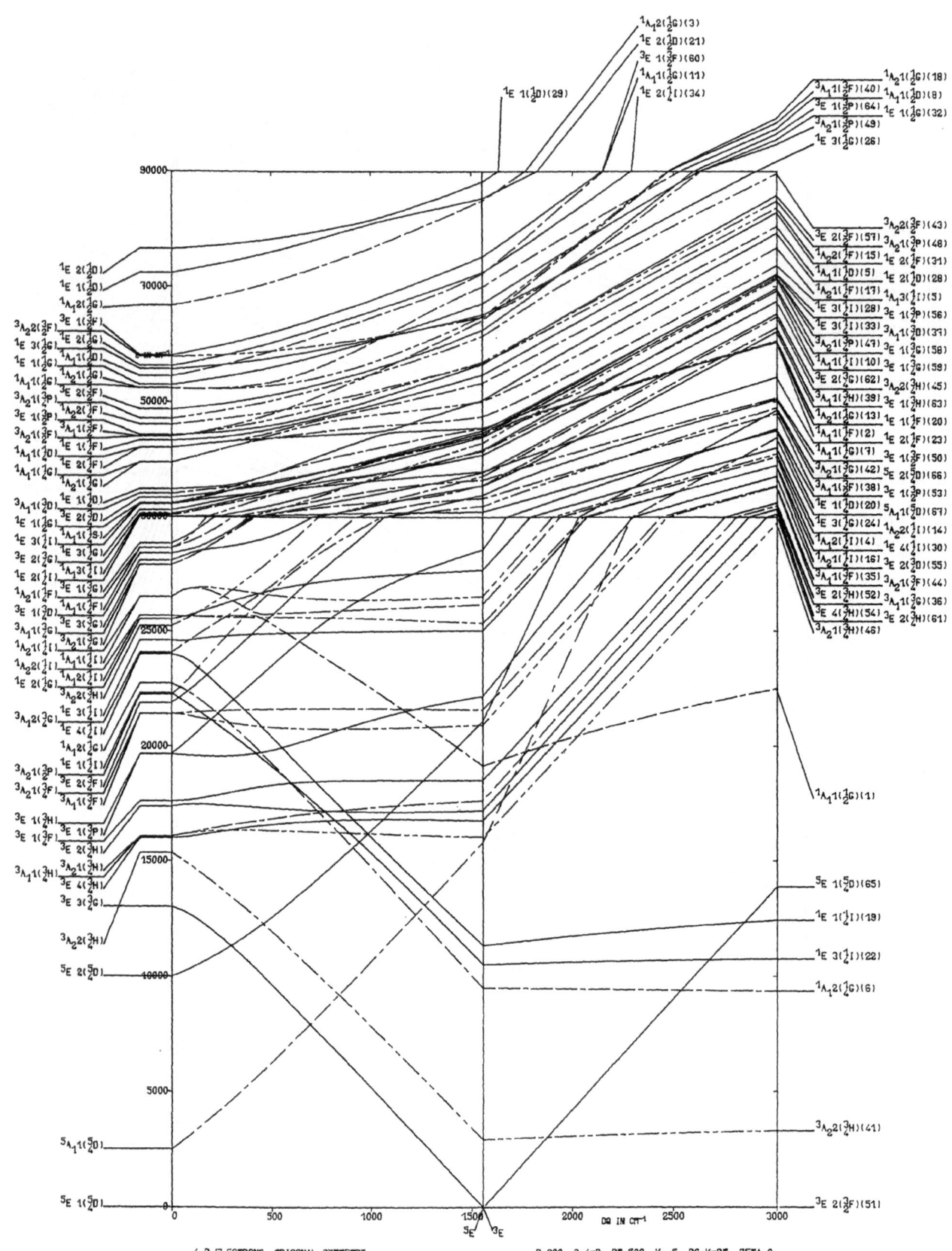

4 D ELECTRONS, TRIGONAL SYMMETRY

B=800 C=4×B DT=500 K=-5 DS=K×DT ZETA=0

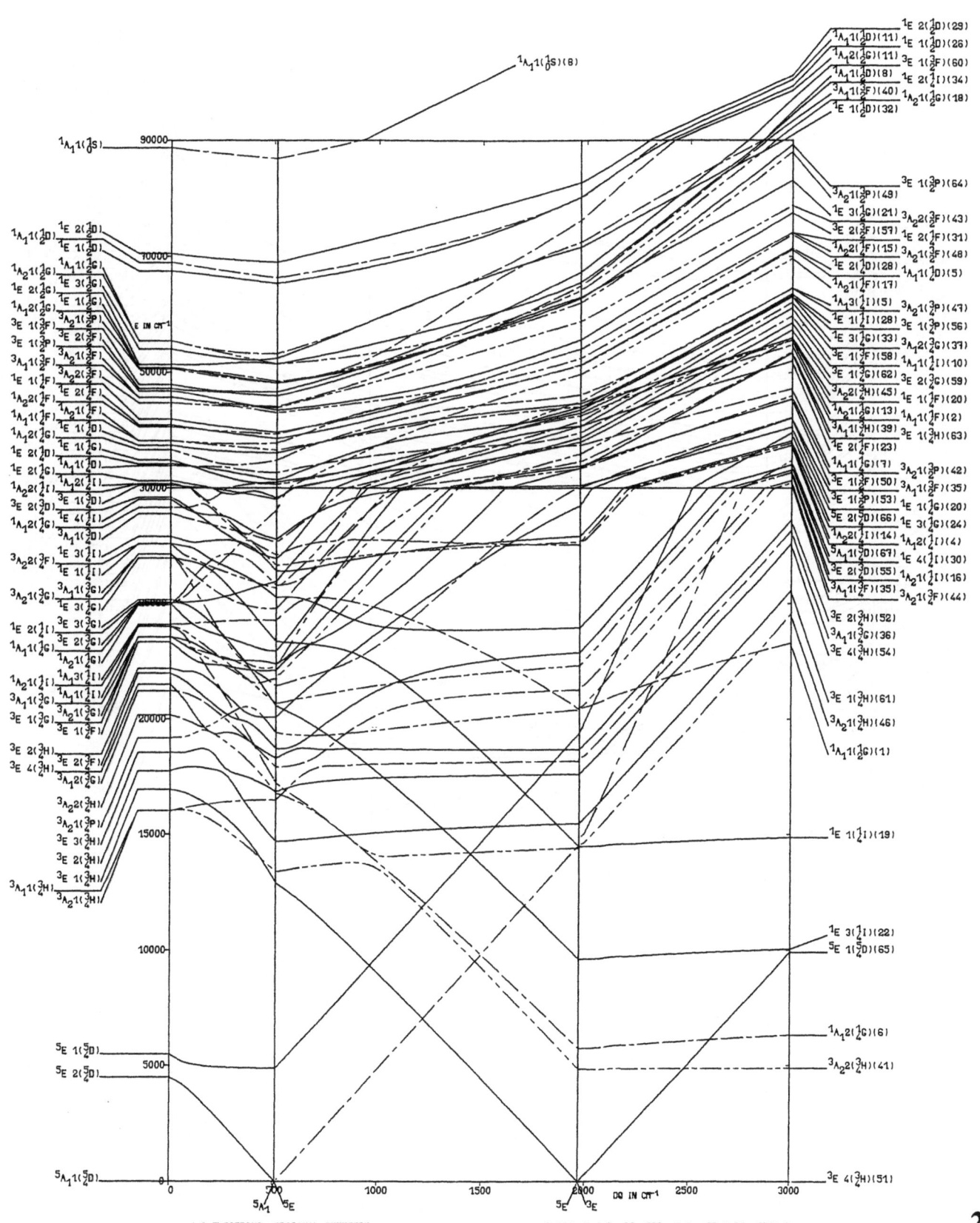

4 D ELECTRONS, TRIGONAL SYMMETRY   B=800   C=4×B   DT=-500   K=1   DS=K×DT   ZETA=0

207

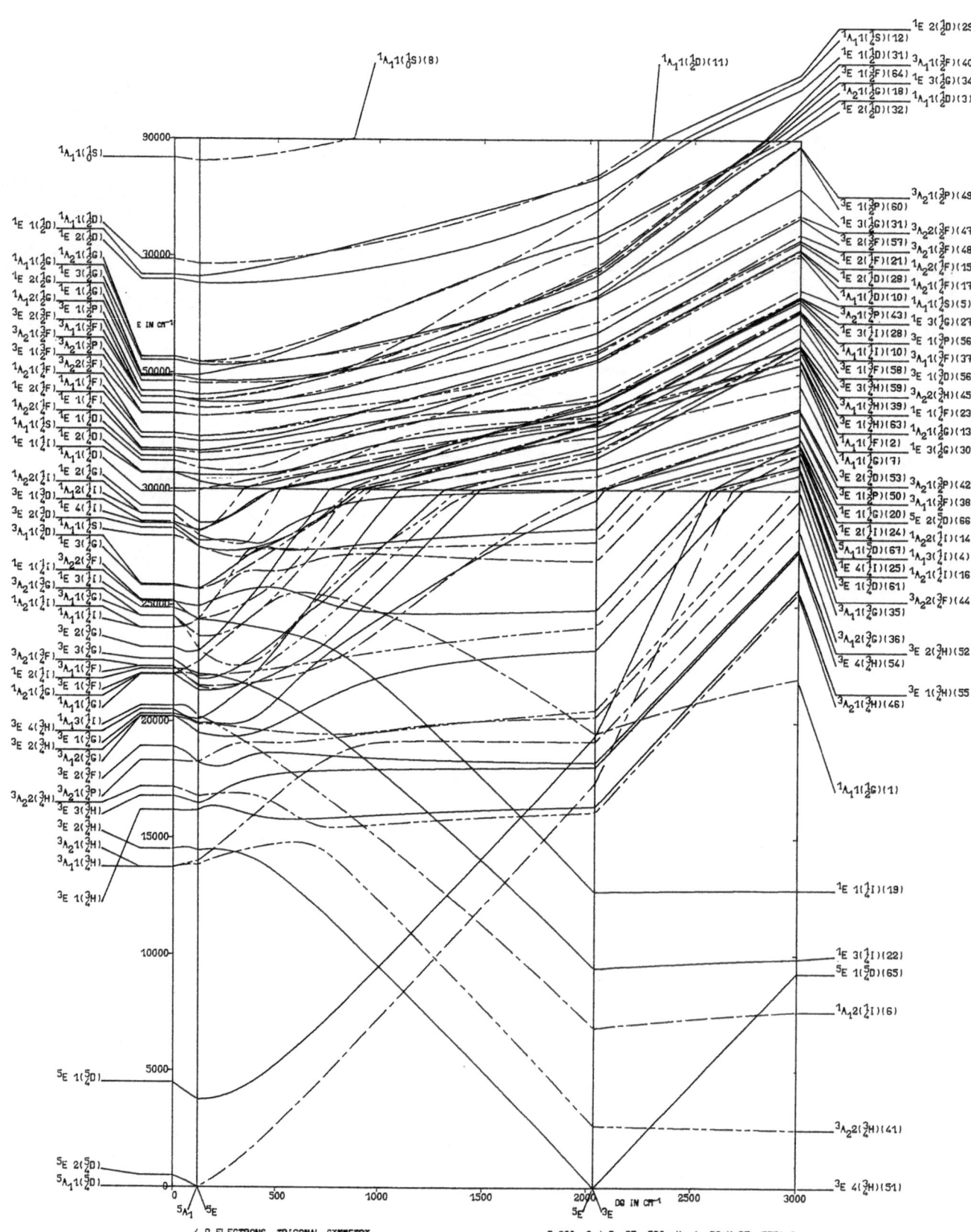

4 D ELECTRONS, TRIGONAL SYMMETRY

B=800 C=4×B DT=-500 K=-1 DS=K×DT ZETA=0

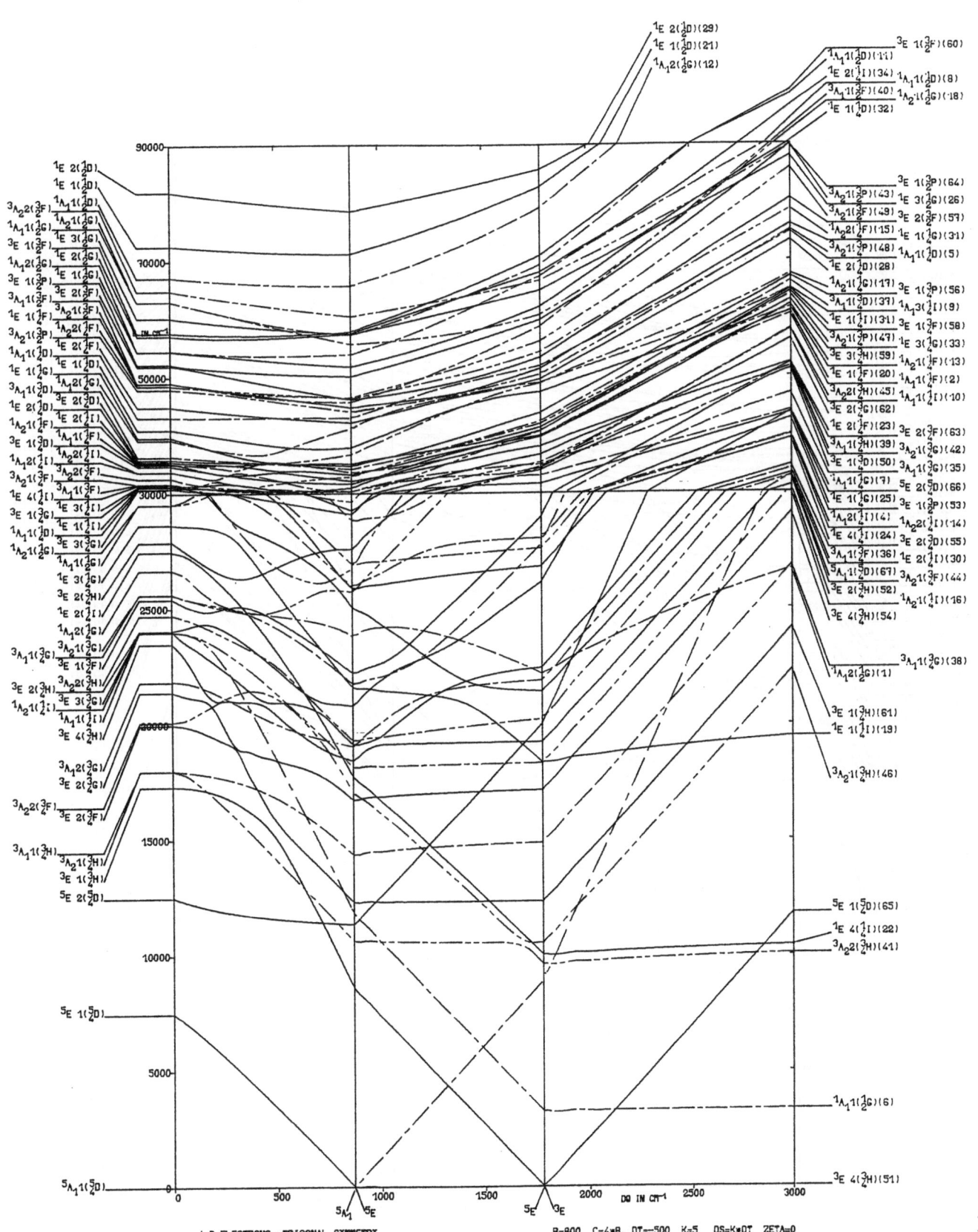

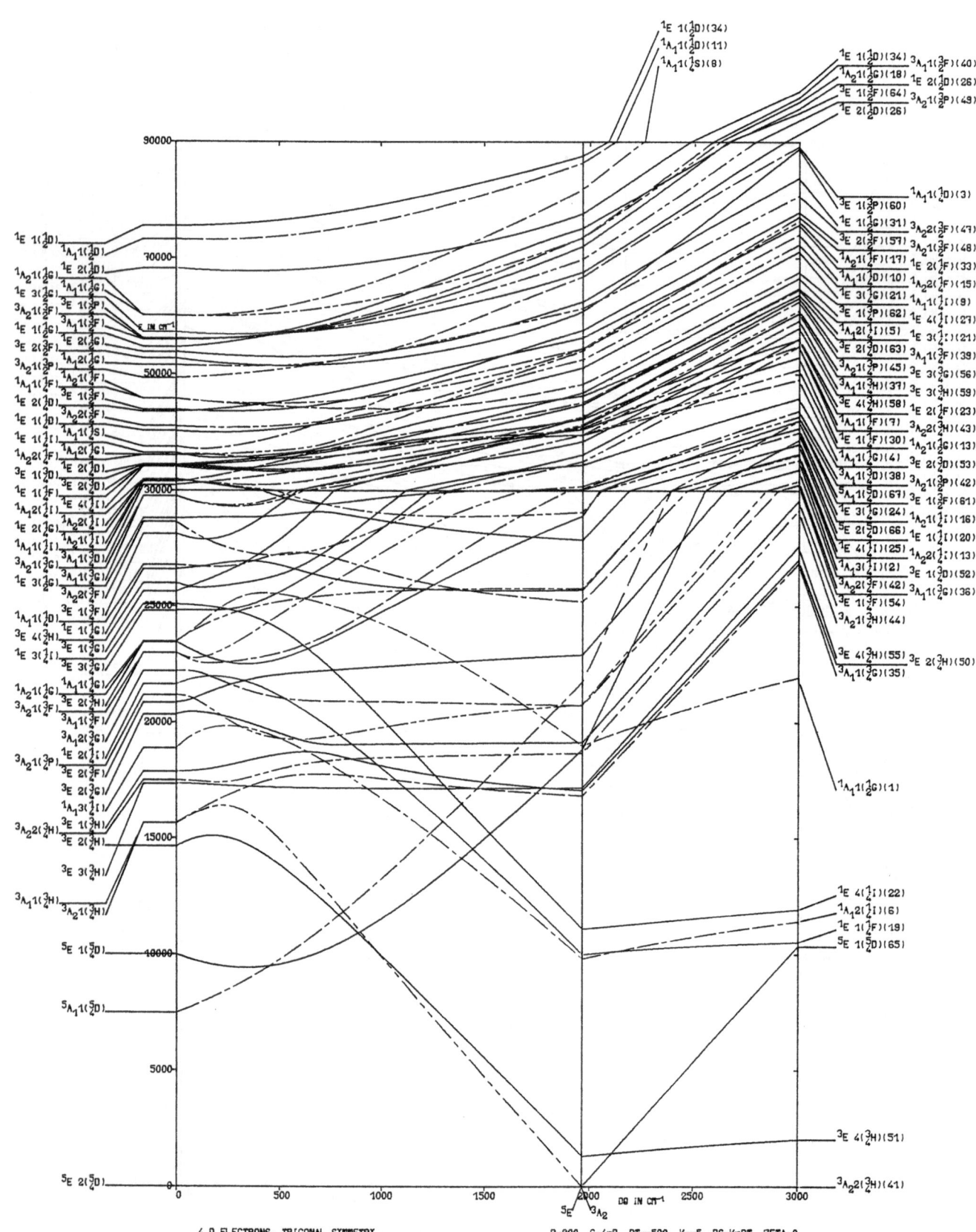

4 D ELECTRONS, TRIGONAL SYMMETRY

B=800 C=4×B DT=-500 K=-5 DS=K×DT ZETA=0

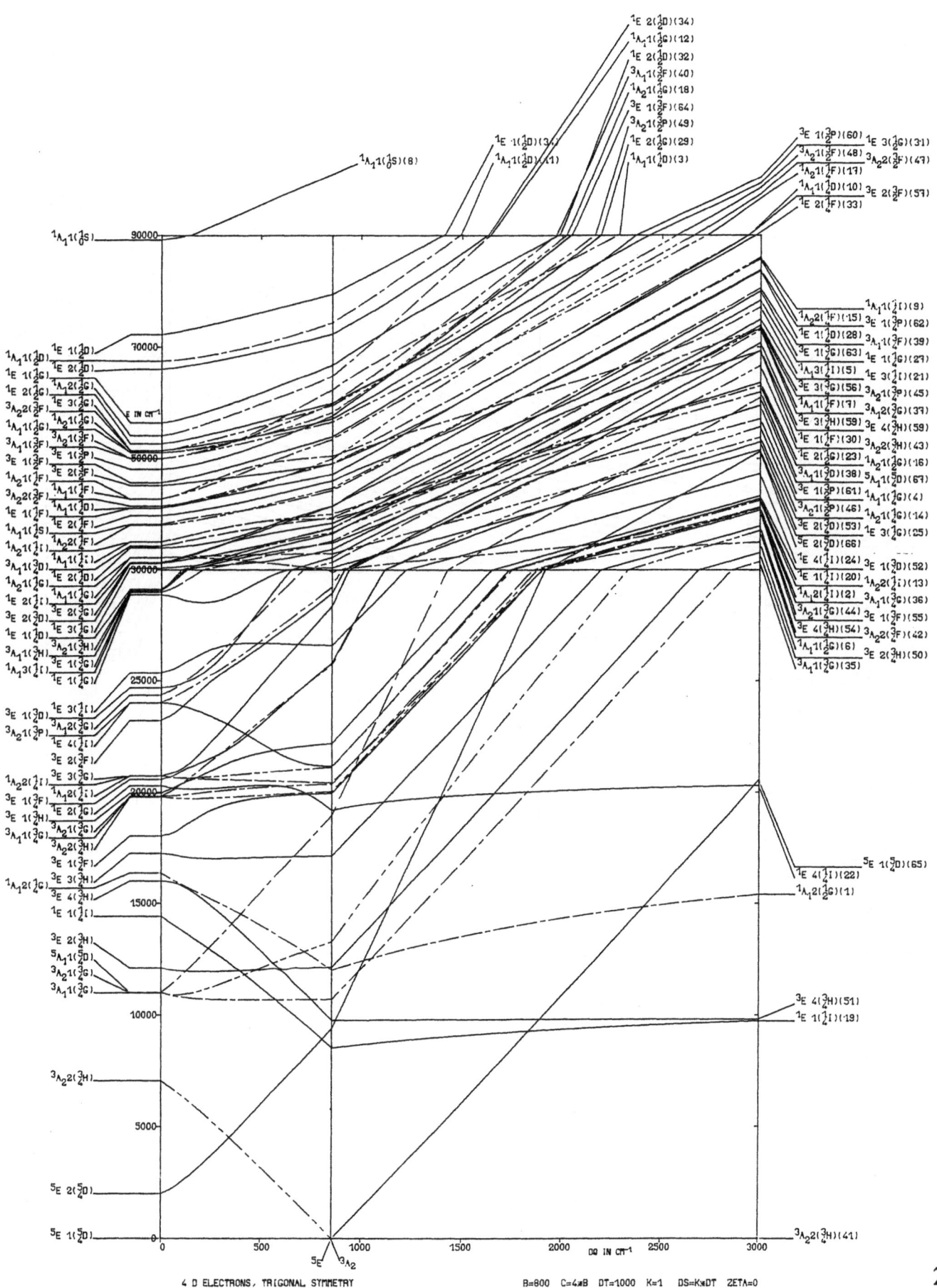

4 D ELECTRONS, TRIGONAL SYMMETRY   B=800   C=4×B   DT=1000   K=1   DS=K×DT   ZETA=0

211

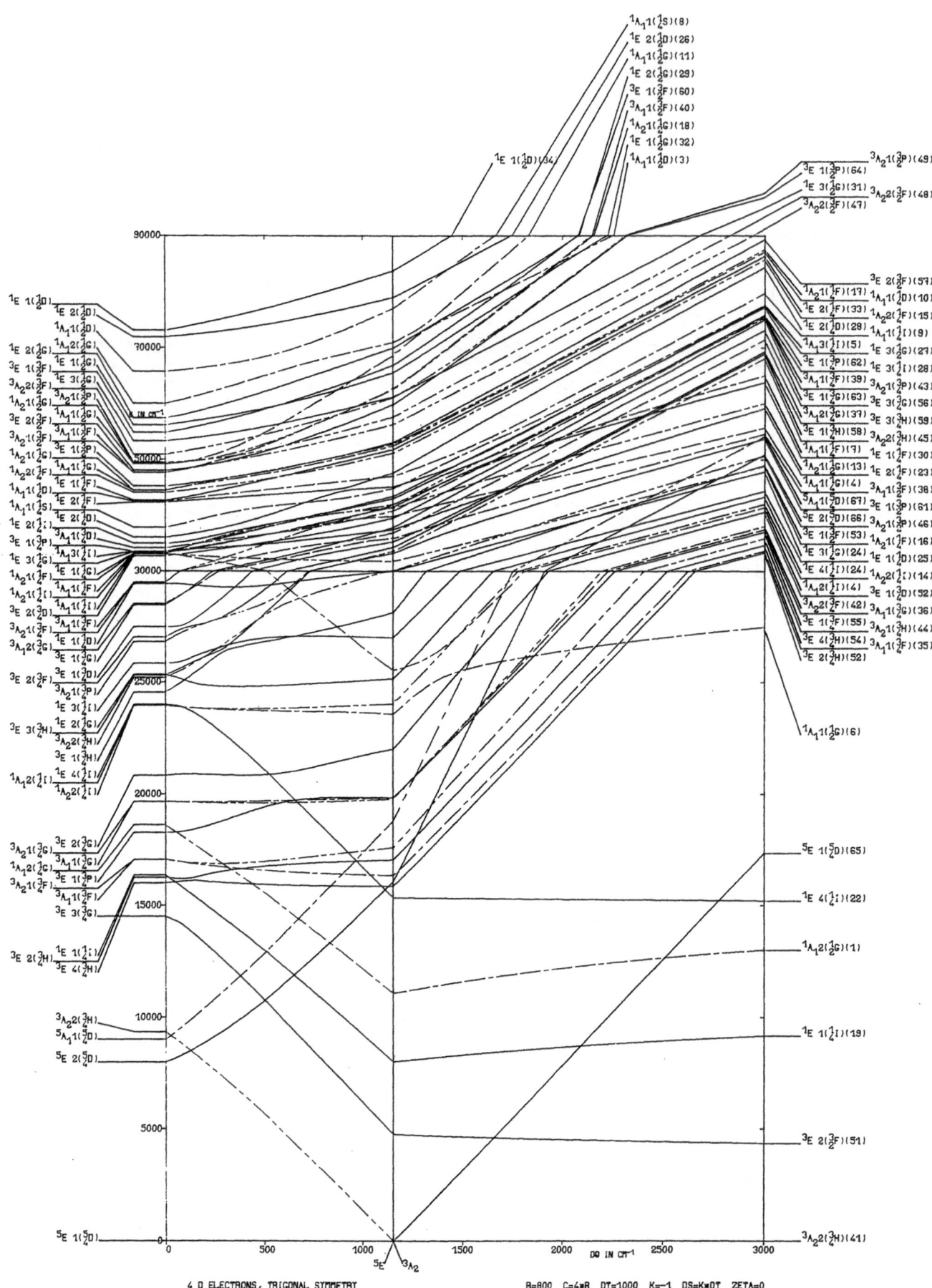

4 D ELECTRONS, TRIGONAL SYMMETRY

B=800 C=4×B DT=1000 K=−1 DS=K×DT ZETA=0

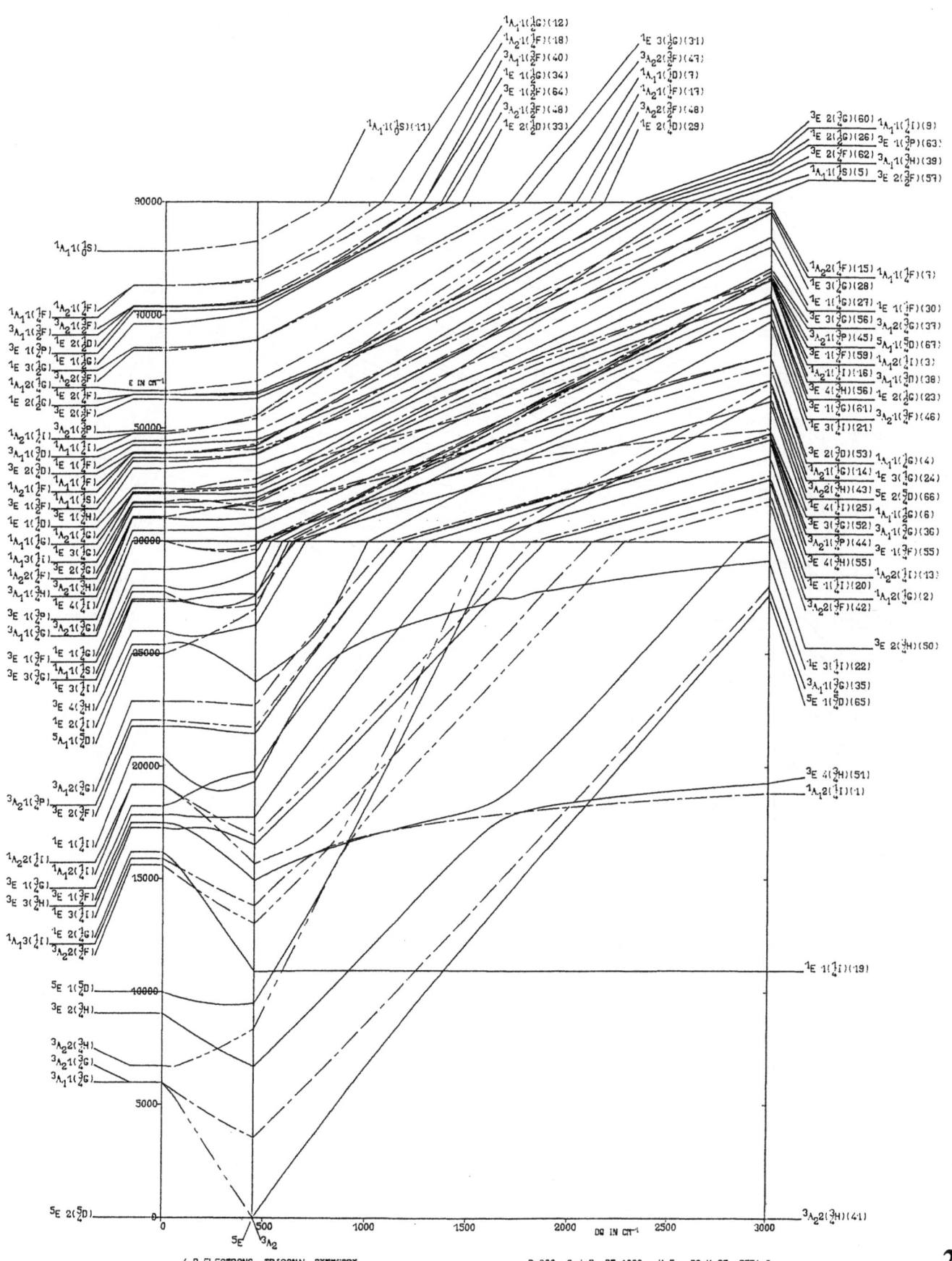

4 D ELECTRONS, TRIGONAL SYMMETRY

B=800 C=4×B DT=1000 K=5 DS=K×DT ZETA=0

213

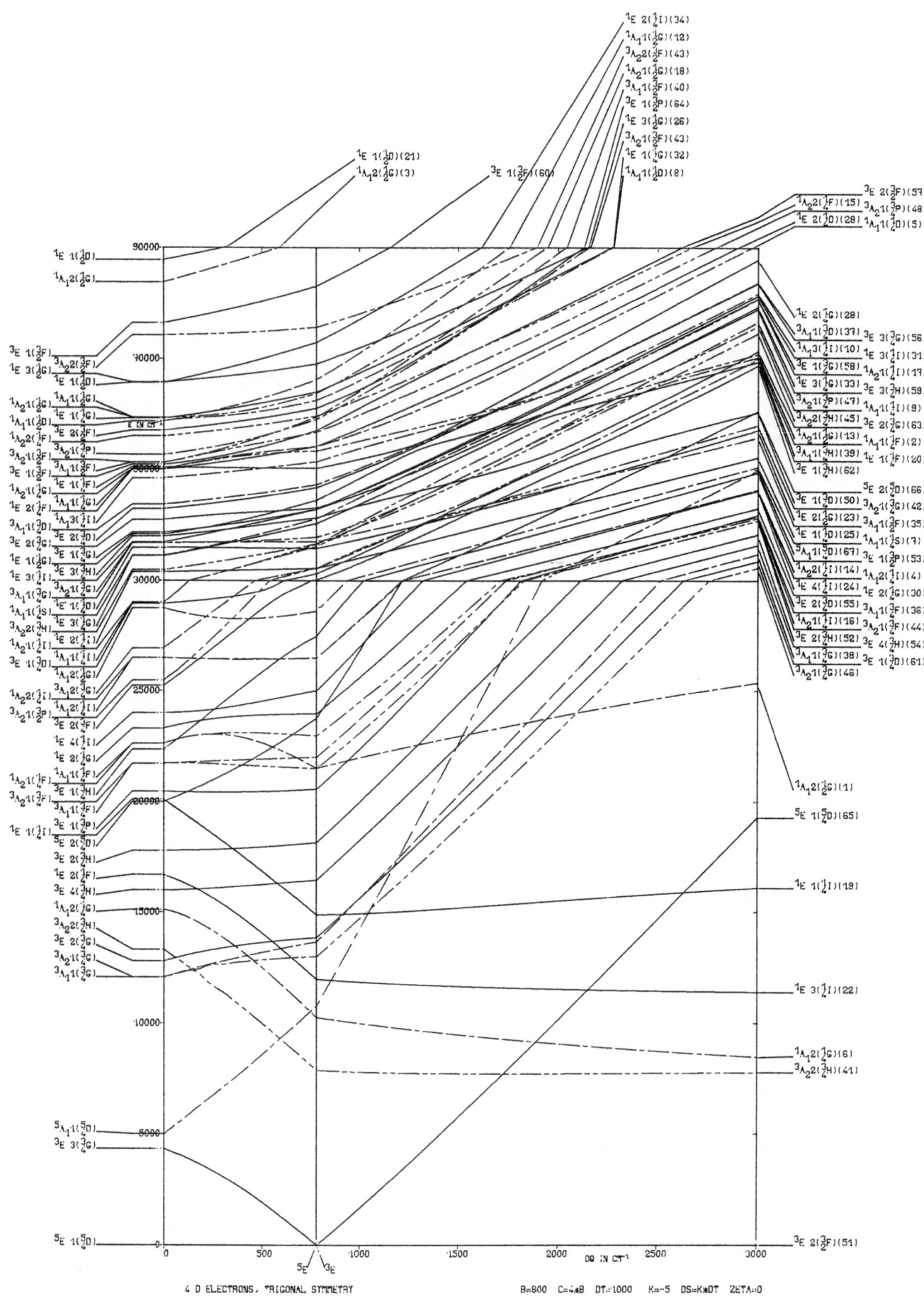

214

4 D ELECTRONS, TRIGONAL SYMMETRY   B=800   C=4.8B   DT.=1000   K=-5   DS=K×DT   ZETA=0

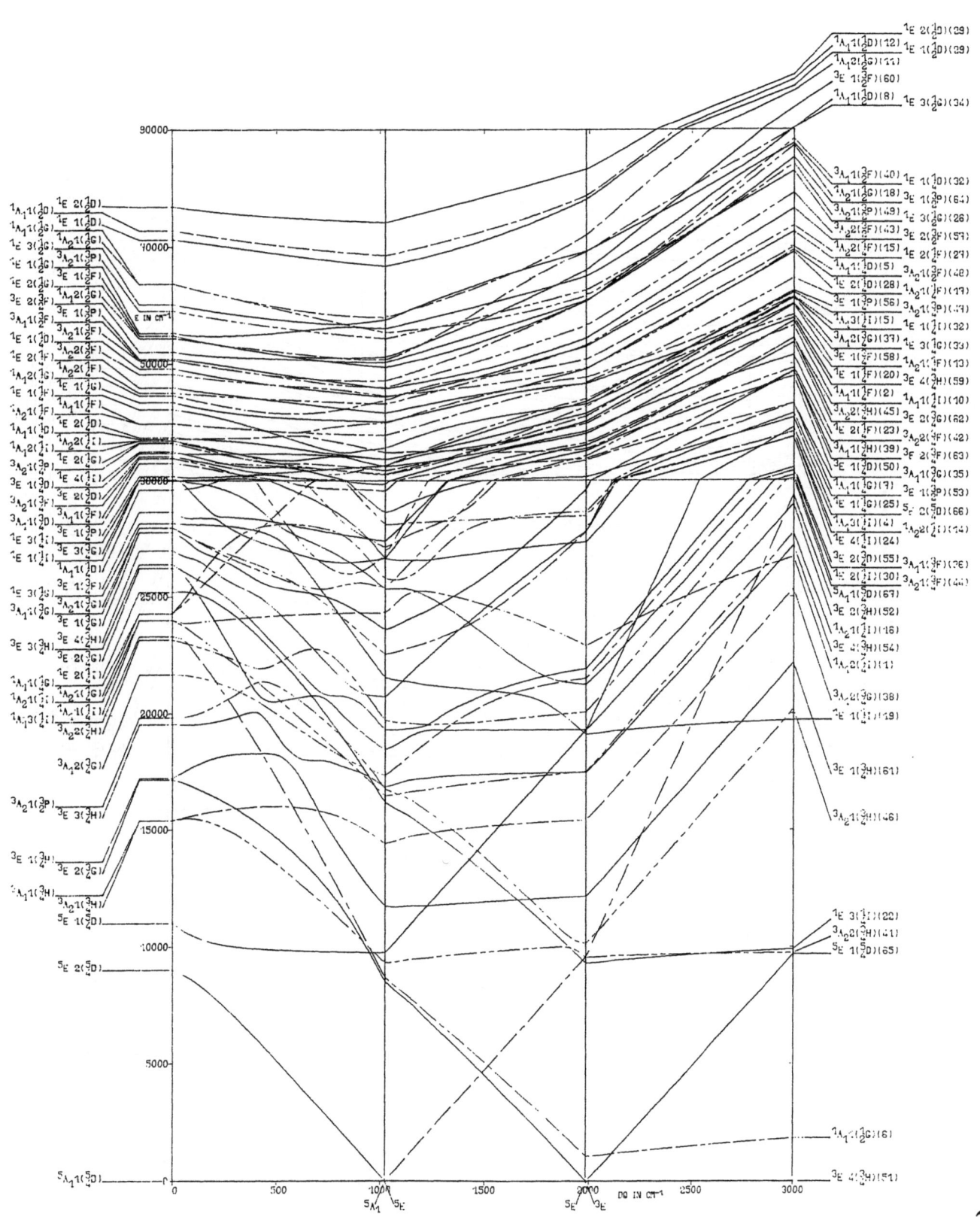

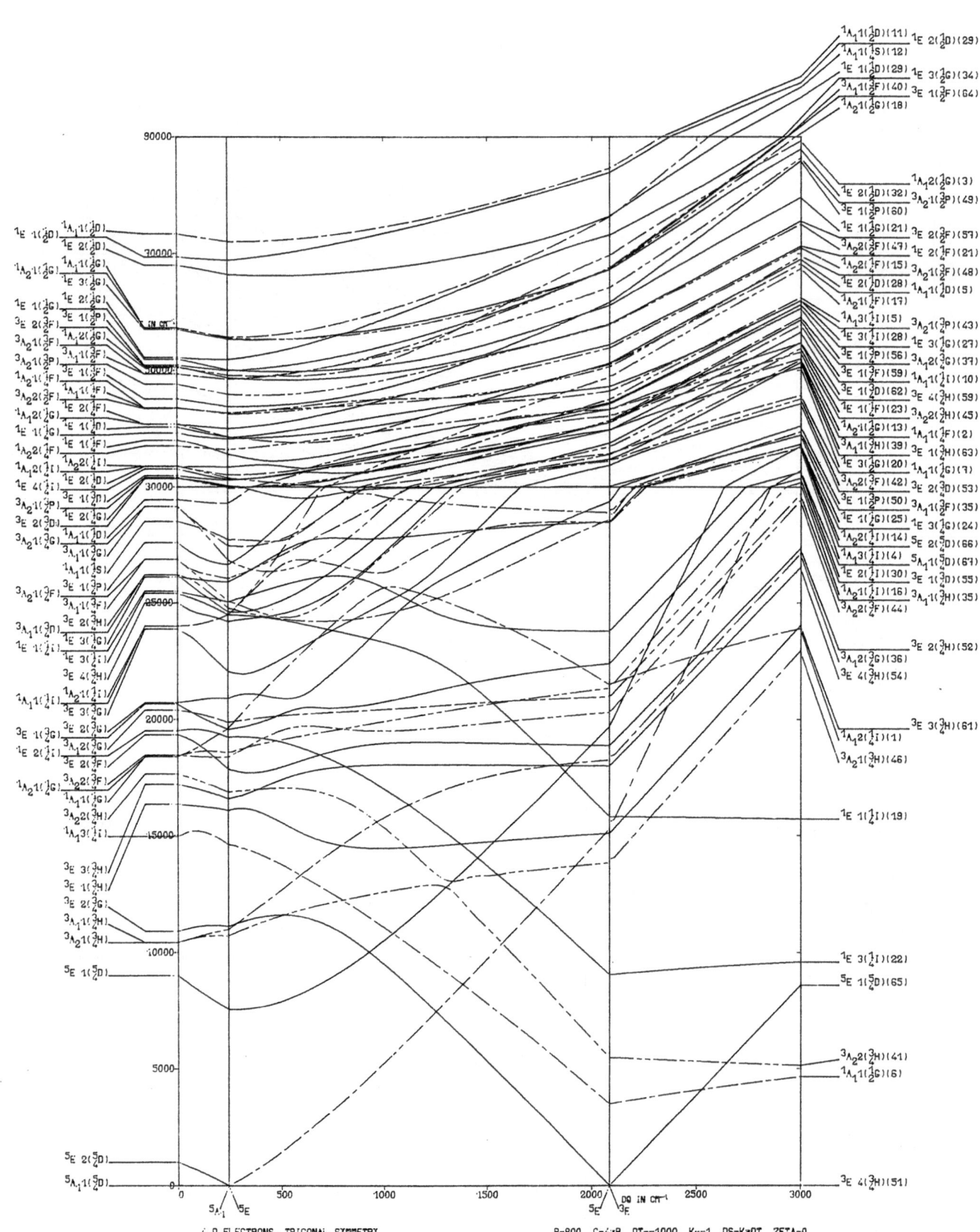

4 D ELECTRONS, TRIGONAL SYMMETRY   B=800   C=4*B   DT=-1000   K=-1   DS=K*DT   ZETA=0

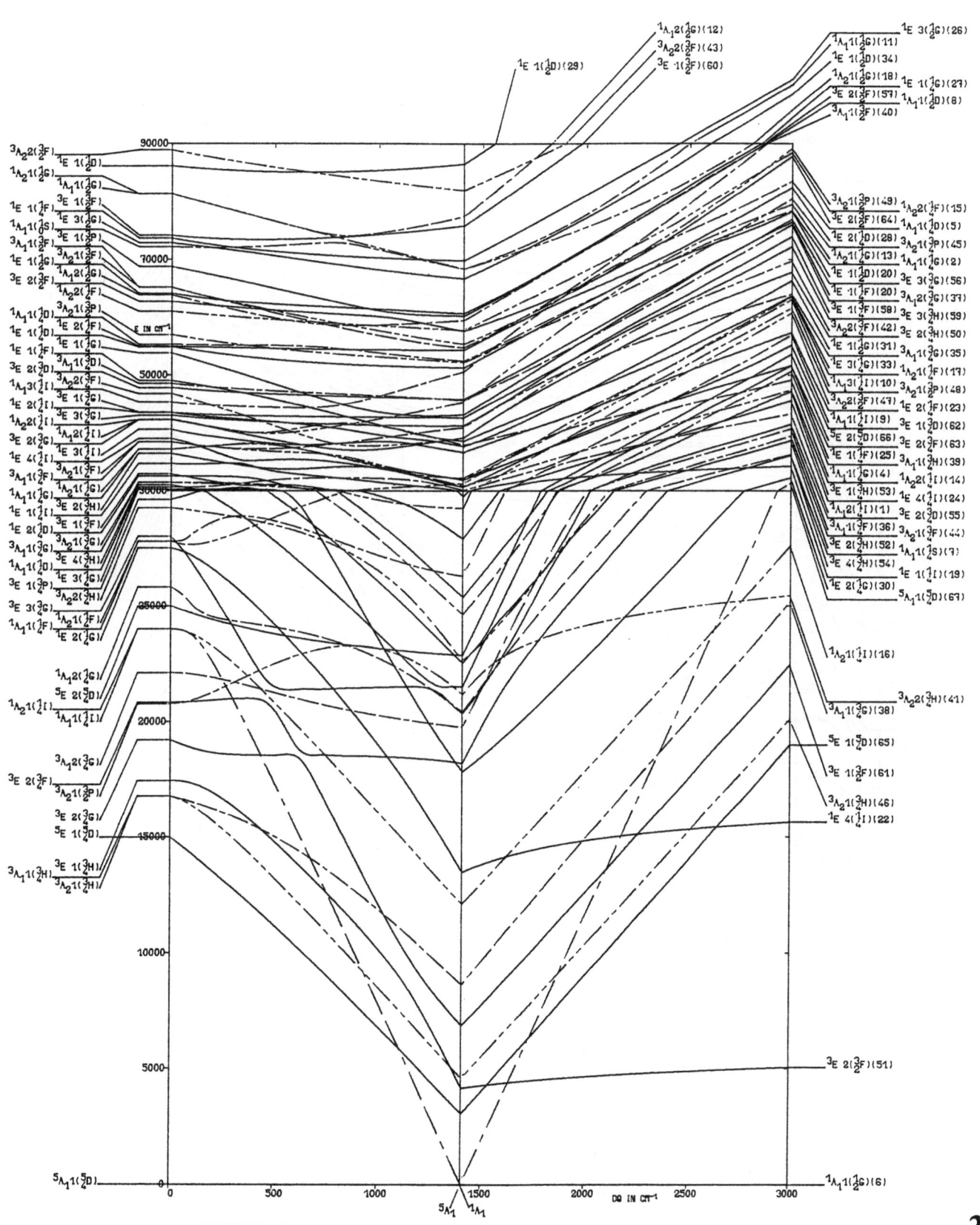

4 D ELECTRONS, TRIGONAL SYMMETRY   B=800 C=4×B DT=−1000 K=5 DS=K×DT ZETA=0

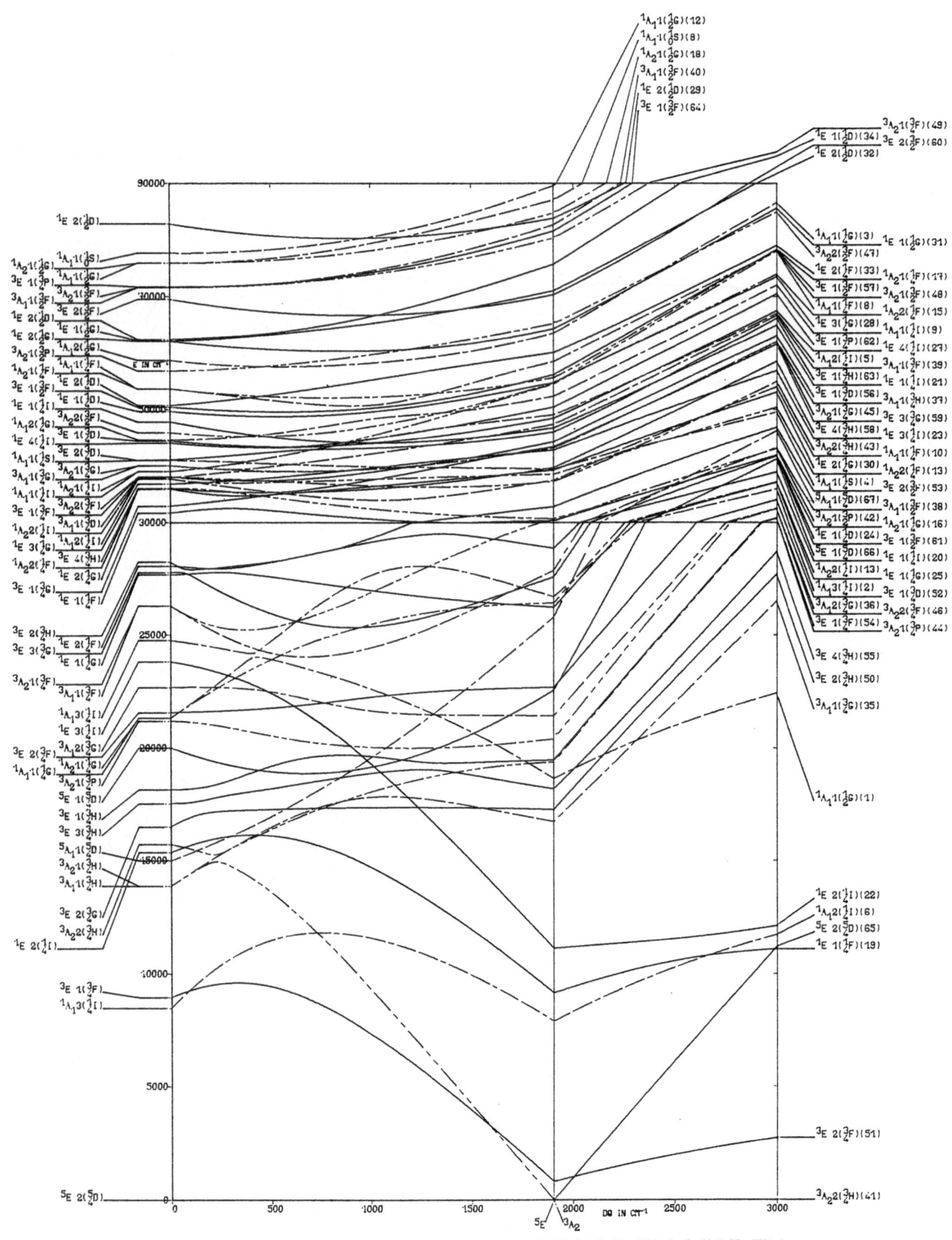

4 D ELECTRONS, TRIGONAL SYMMETRY

B=800 C=4×B DT=-1000 K=-5 DS=K×DT ZETA=0

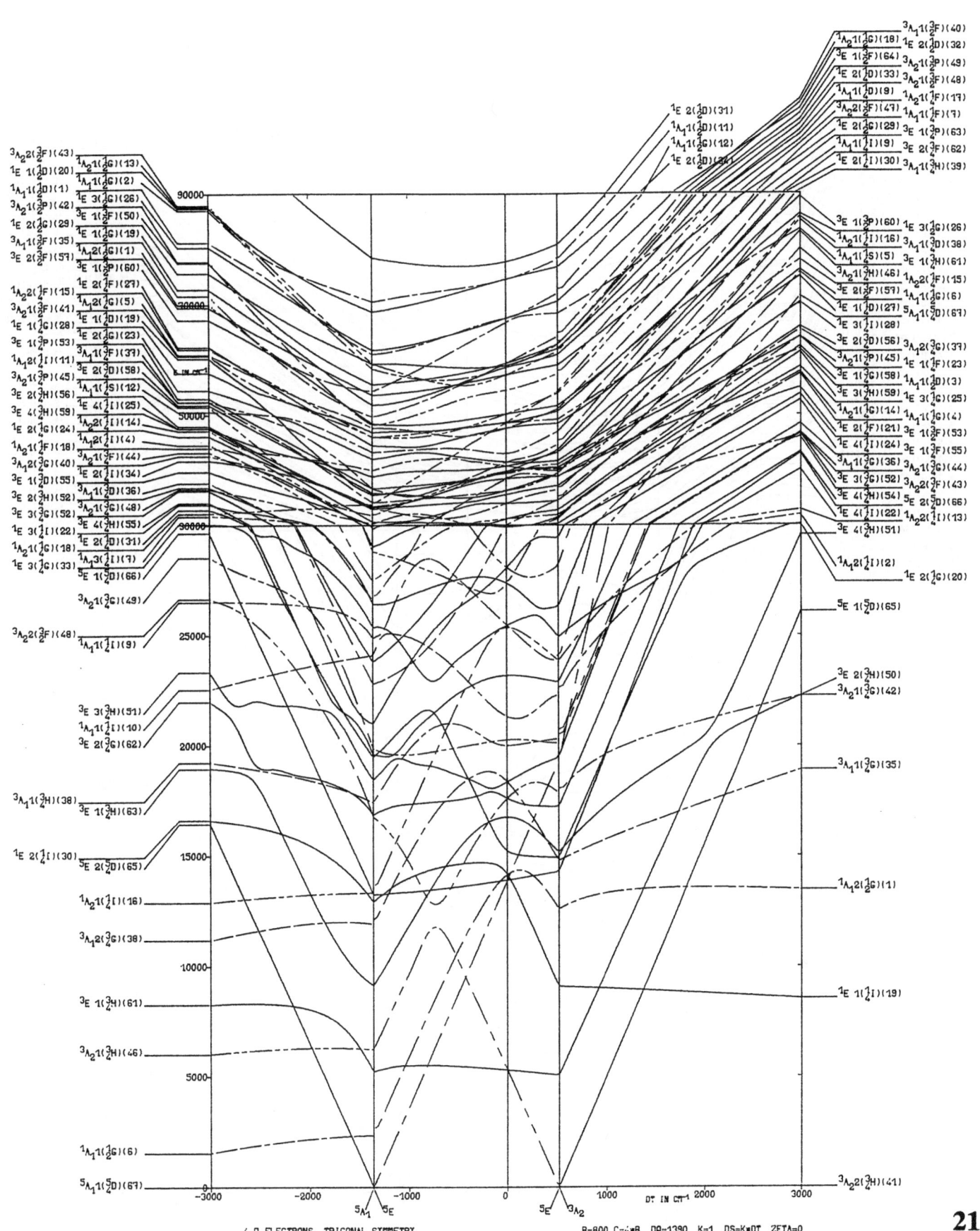

4 D ELECTRONS, TRIGONAL SYMMETRY   B=800 C=4×B   DQ=1390   K=1   DS=K×DT   ZETA=0

219

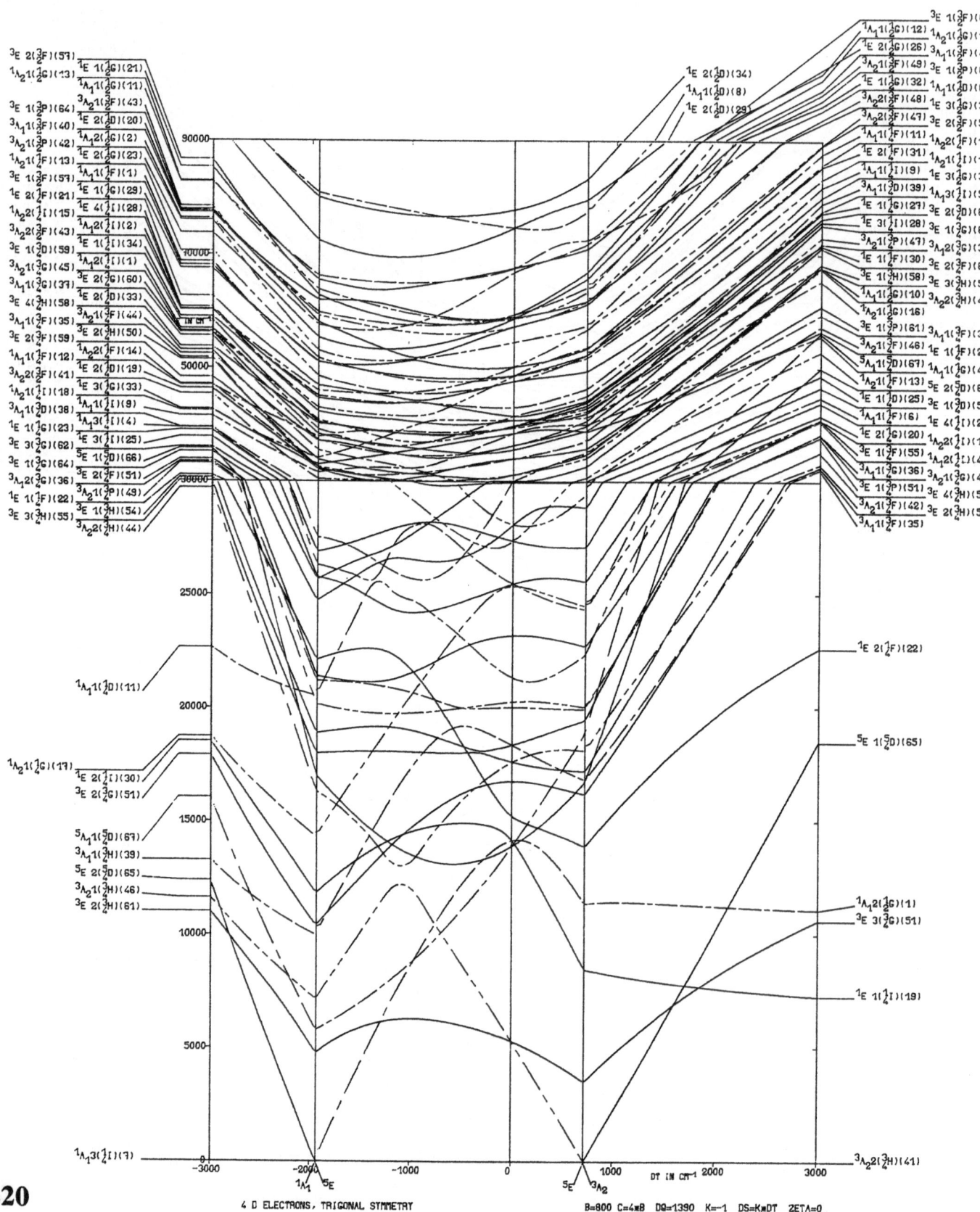

4 D ELECTRONS, TRIGONAL SYMMETRY

B=800 C=4*B DQ=1390 K=-1 DS=K*DT ZETA=0

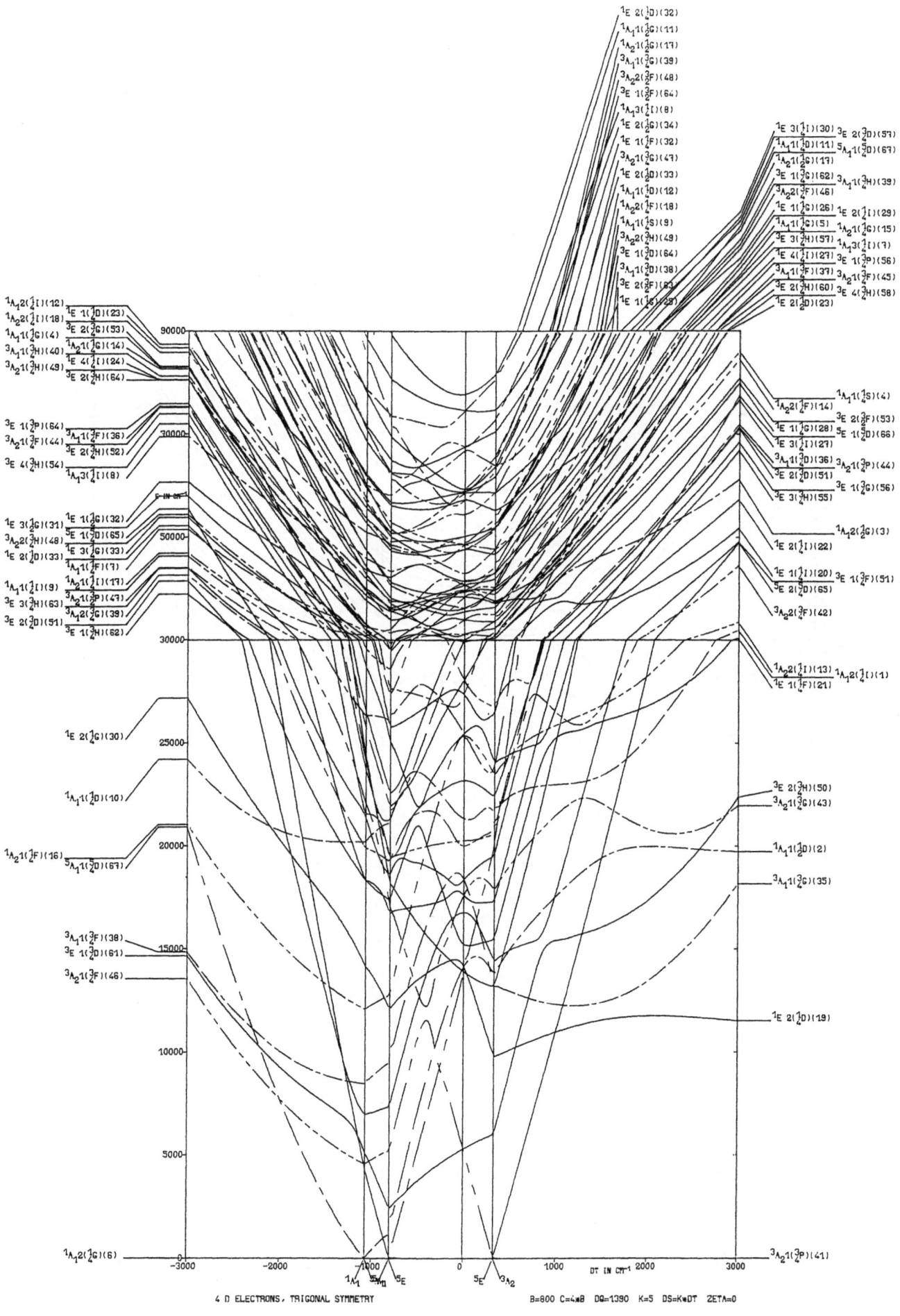

$^1A_2 2(\frac{1}{4}I)(12)$
$^1A_2 2(\frac{1}{4}I)(18)$ $^1E\ 1(\frac{1}{4}D)(23)$
$^1A_1 1(\frac{1}{4}G)(4)$ $^3E\ 2(\frac{2}{4}G)(53)$
$^3A_1 1(\frac{3}{4}H)(40)$ $^1A_2 1(\frac{1}{4}G)(14)$
$^3A_2 1(\frac{3}{4}H)(49)$ $^1E\ 4(\frac{1}{4}I)(24)$
$^3E\ 2(\frac{2}{4}H)(64)$

$^3E\ 1(\frac{2}{4}P)(64)$ $^1A_1 1(\frac{1}{4}F)(36)$
$^3A_2 1(\frac{2}{4}F)(44)$ $^3E\ 2(\frac{2}{4}F)(52)$
$^3E\ 4(\frac{3}{4}H)(54)$ $^1A_1 3(\frac{1}{4}I)(8)$

$^1E\ 3(\frac{1}{4}G)(31)$ $^1E\ 1(\frac{1}{4}G)(32)$
$^3A_2 2(\frac{3}{4}H)(48)$ $^1E\ 1(\frac{1}{4}D)(65)$
$^1E\ 2(\frac{1}{4}D)(33)$
$^1A_1 1(\frac{1}{4}F)(7)$
$^1A_1 1(\frac{1}{4}I)(9)$ $^1A_2 1(\frac{1}{4}I)(17)$
$^3E\ 3(\frac{3}{4}H)(63)$ $^3A_2 1(\frac{2}{4}P)(47)$
$^3E\ 2(\frac{2}{4}D)(51)$ $^1A_2 1(\frac{1}{4}G)(39)$
$^3E\ 1(\frac{1}{4}H)(62)$

$^1E\ 2(\frac{1}{4}G)(30)$

$^1A_1 1(\frac{1}{4}D)(10)$

$^1A_2 1(\frac{1}{4}F)(16)$
$^5A_1 1(\frac{5}{4}D)(67)$

$^3A_1 1(\frac{3}{4}F)(38)$
$^3E\ 1(\frac{3}{4}D)(61)$
$^3A_2 1(\frac{3}{4}F)(46)$

$^1A_1 2(\frac{1}{4}G)(6)$

$^1E\ 2(\frac{1}{4}D)(32)$
$^1A_1 1(\frac{1}{4}G)(11)$
$^1A_2 1(\frac{1}{4}G)(17)$
$^3A_1 1(\frac{2}{4}G)(39)$
$^3A_2 2(\frac{3}{4}F)(48)$
$^3E\ 1(\frac{2}{4}F)(64)$
$^1A_1 3(\frac{1}{4}I)(8)$
$^1E\ 2(\frac{1}{4}G)(34)$
$^1E\ 1(\frac{1}{4}F)(32)$
$^3A_2 1(\frac{2}{4}G)(47)$
$^1E\ 2(\frac{1}{4}D)(33)$
$^1A_1 1(\frac{1}{4}D)(12)$
$^1A_2 2(\frac{1}{4}F)(18)$
$^1A_1 1(\frac{1}{4}S)(9)$
$^3A_2 2(\frac{3}{4}H)(49)$
$^3E\ 1(\frac{2}{4}D)(64)$
$^3A_1 1(\frac{3}{4}D)(38)$
$^3E\ 2(\frac{3}{4}F)(83)$
$^1E\ 1(\frac{1}{4}G)(23)$

$^1E\ 3(\frac{1}{4}I)(30)$ $^3E\ 2(\frac{3}{4}D)(57)$
$^1A_1 1(\frac{1}{4}D)(11)$ $^5A_1 1(\frac{5}{4}D)(67)$
$^1A_2 1(\frac{1}{4}G)(17)$
$^3E\ 1(\frac{2}{4}G)(62)$ $^3A_1 1(\frac{3}{4}H)(39)$
$^3A_2 2(\frac{3}{4}F)(46)$
$^1E\ 1(\frac{1}{4}G)(26)$ $^1E\ 2(\frac{1}{4}I)(29)$
$^1A_1 1(\frac{1}{4}G)(5)$ $^1A_2 1(\frac{1}{4}G)(15)$
$^3E\ 3(\frac{2}{4}H)(57)$ $^1A_1 3(\frac{1}{4}I)(7)$
$^1E\ 4(\frac{1}{4}I)(27)$ $^3E\ 1(\frac{3}{4}P)(56)$
$^3A_1 1(\frac{3}{4}F)(37)$ $^3A_2 1(\frac{2}{4}F)(45)$
$^3E\ 2(\frac{3}{4}H)(60)$ $^3E\ 4(\frac{3}{4}H)(58)$
$^1E\ 2(\frac{1}{4}D)(23)$

$^1A_1 1(\frac{1}{4}S)(4)$
$^1A_2 2(\frac{1}{4}F)(14)$
$^1E\ 2(\frac{3}{4}F)(53)$
$^1E\ 1(\frac{1}{4}G)(28)$ $^5E\ 1(\frac{5}{4}D)(66)$
$^1E\ 3(\frac{1}{4}I)(27)$
$^3A_1 1(\frac{3}{4}D)(36)$ $^3A_2 1(\frac{3}{4}P)(44)$
$^3E\ 2(\frac{3}{4}D)(51)$
$^3E\ 1(\frac{3}{4}G)(56)$
$^3E\ 3(\frac{3}{4}H)(55)$

$^1A_1 2(\frac{1}{4}G)(3)$

$^1E\ 2(\frac{1}{4}I)(22)$

$^1E\ 1(\frac{1}{4}I)(20)$ $^3E\ 1(\frac{3}{4}F)(51)$
$^5E\ 2(\frac{5}{4}D)(65)$

$^3A_2 2(\frac{3}{4}F)(42)$

$^1A_2 2(\frac{1}{4}I)(13)$ $^1A_1 2(\frac{1}{4}I)(1)$
$^1E\ 1(\frac{1}{4}F)(21)$

$^3E\ 2(\frac{3}{4}H)(50)$
$^3A_2 1(\frac{3}{4}G)(43)$

$^1A_1 1(\frac{1}{4}D)(2)$

$^3A_1 1(\frac{3}{4}G)(35)$

$^1E\ 2(\frac{1}{4}D)(19)$

$^3A_2 1(\frac{3}{4}P)(41)$

90000

30000

F IN cm⁻¹

50000

30000

25000

20000

15000

10000

5000

0

-3000   -2000   -1000   0   1000   DT IN CM⁻¹ 2000   3000

$^1A_1$ $^5_A\Pi$ ^5E ^5E $^3A_2$

4 D ELECTRONS, TRIGONAL SYMMETRY   B=800 C=4×B DQ=1390 K=5 DS=K×DT ZETA=0

221

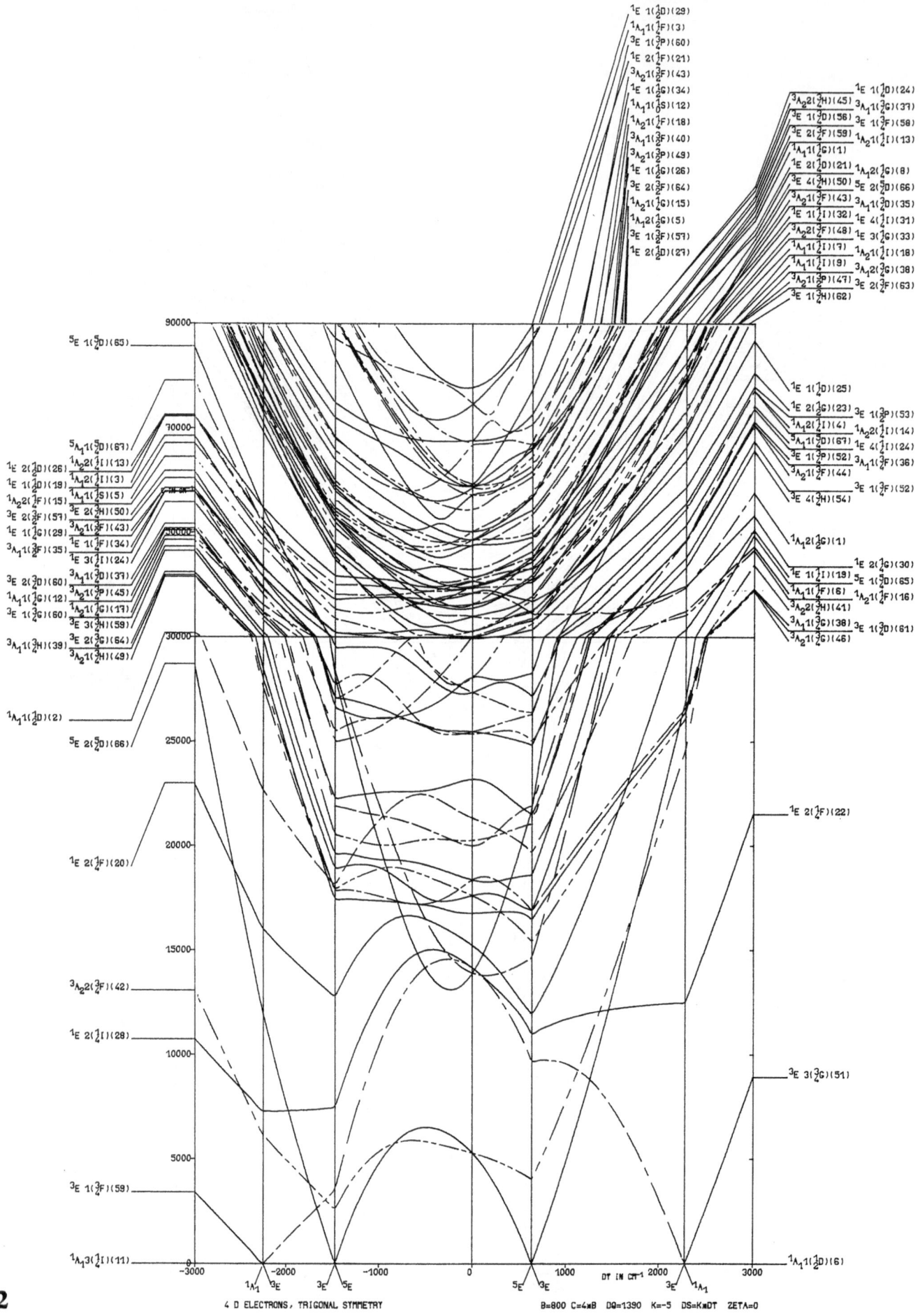

4 D ELECTRONS, TRIGONAL SYMMETRY

B=800 C=4×B DQ=1390 K=-5 DS=K×DT ZETA=0

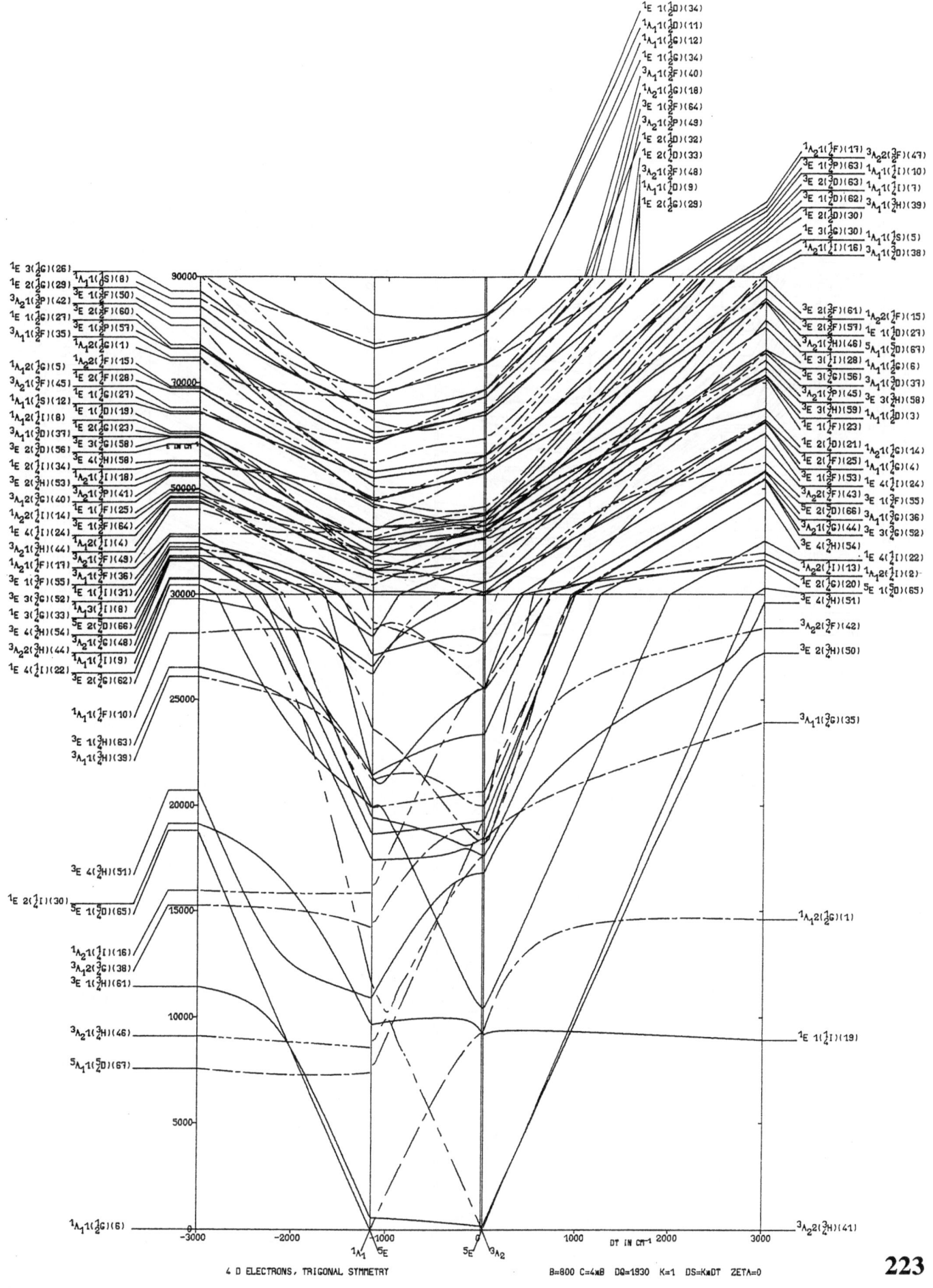

4 D ELECTRONS, TRIGONAL SYMMETRY   B=800 C=4×B DQ=1930 K=1 DS=K×DT ZETA=0

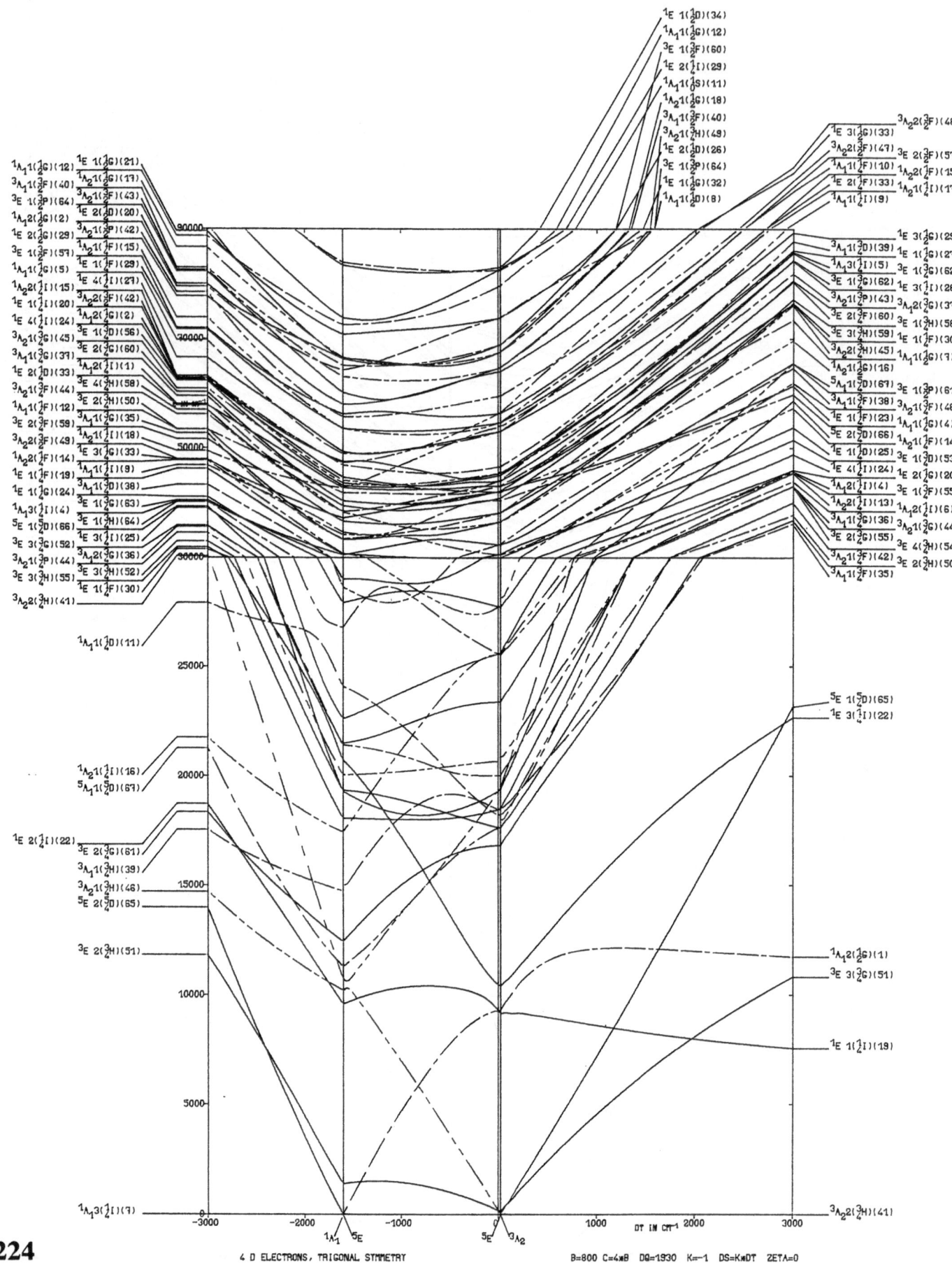

4 D ELECTRONS, TRIGONAL SYMMETRY

B=800 C=4×B DQ=1930 K=-1 DS=K×DT ZETA=0

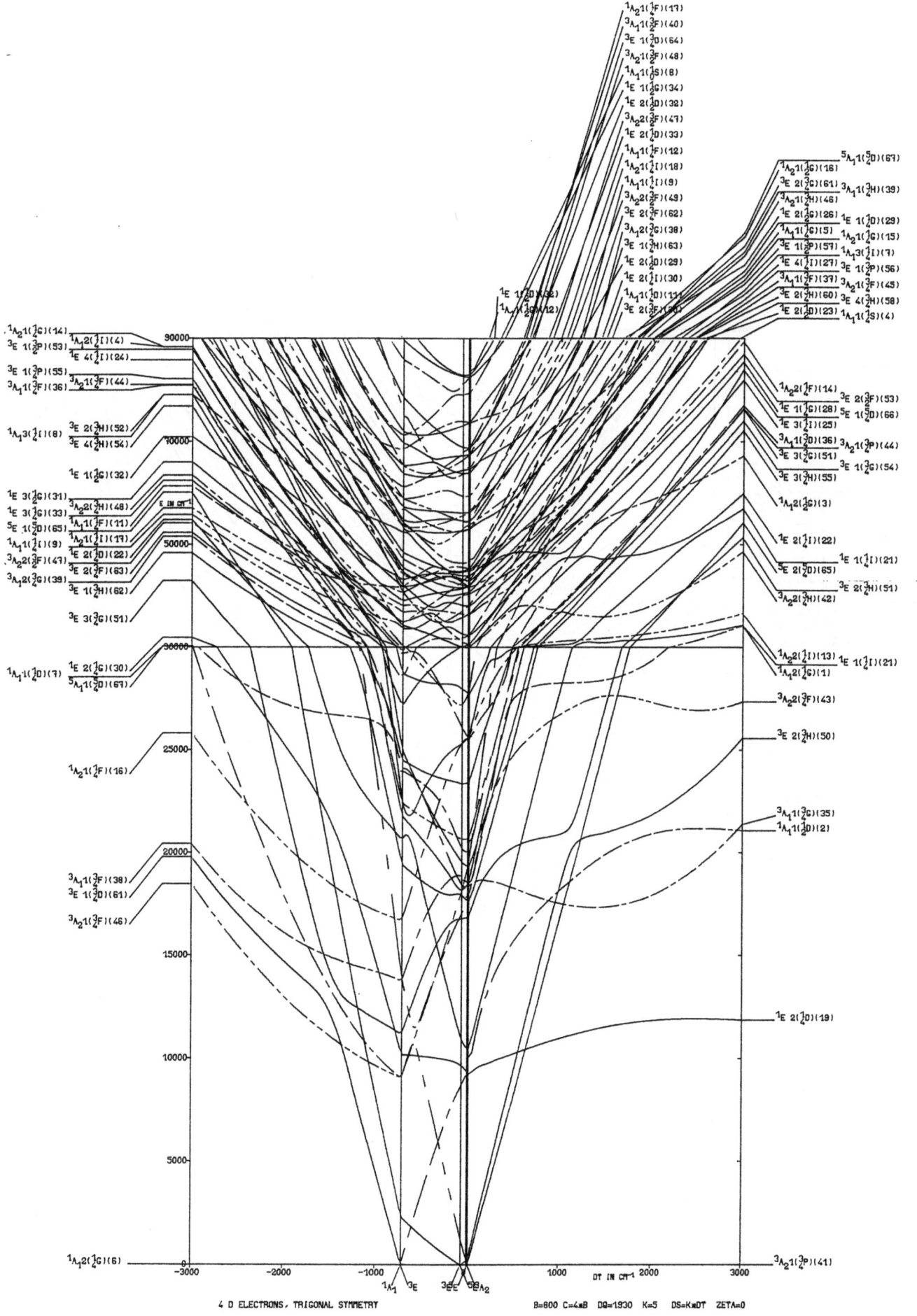

4 D ELECTRONS, TRIGONAL SYMMETRY

B=800 C=4xB DQ=1930 K=5 DS=KxDT ZETA=0

225

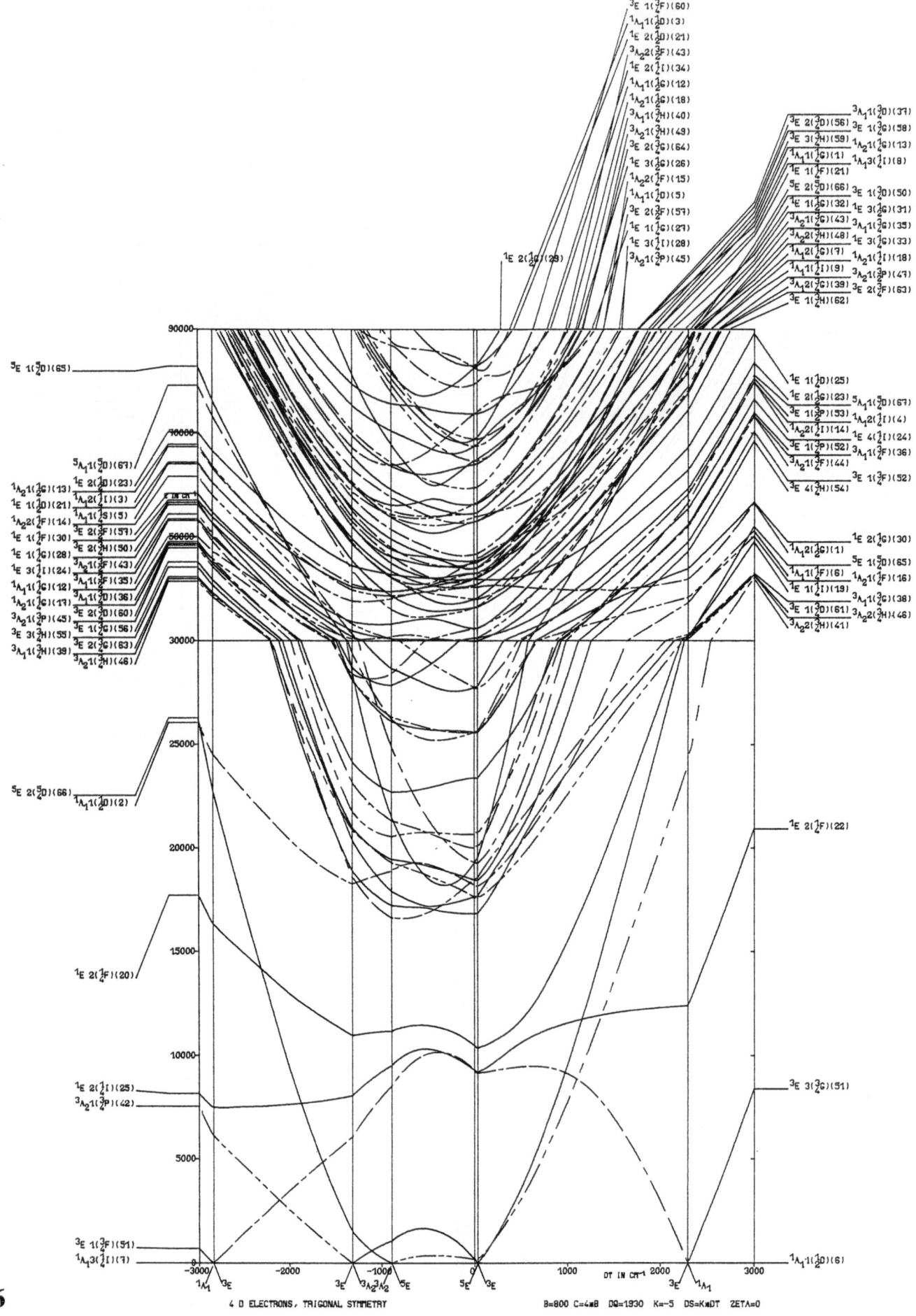

4 D ELECTRONS, TRIGONAL SYMMETRY

B=800 C=4×B DQ=1930 K=-5 DS=K×DT ZETA=0

B=800   C=4×B   ZETA=0

ENERGY AS FUNCTION OF DT
K=1, -1, 5, -5

(58) $^1E_6(^2E_4(^1E_2(E_1E_1)E_2)E_2)$

(57) $^3E_5(^2E_3(^1E_2(A_1E_2)E_1)E_2)$

(56) $^1E_5(^2E_3(^1E_1(A_1E_1)E_2)E_2)$

(55) $^3E_4(^2E_2(^1E_1(E_1E_2)E_1)E_2)$

(54) $^3E_4(^2E_2(^1E_1(A_1E_1)E_1)E_2)$

(53) $^1E_4(^2E_2(^1A_1(E_1E_1)E_2)E_2)$

(52) $^1E_4(^2E_2(^1E_1(A_1E_1)E_1)E_2)$

(51) $^1E_4(^2E_2(^1A_1(A_1A_1)E_2)E_2)$

(50) $^3E_3(^2E_1(^1E_3(E_1E_2)E_2)E_2)$

(49) $^3E_3(^2E_1(^1A_1(E_1E_1)E_1)E_2)$

(48) $^3E_3(^2E_1(^1E_2(A_1E_2)E_1)E_1)$'

(47) $^3E_3(^2E_1(^1A_1(A_1A_1)E_1)E_2)$

(46) $^1E_3(^2E_1(^1E_1(E_1E_2)E_2)E_2)$

(45) $^1E_3(^2E_1(^1A_1(E_1E_1)E_1)E_2)$

(44) $^1E_3(^2E_1(^1E_1(A_1E_1)E_2)E_2)$

(43) $^1E_3(^2E_1(^1A_1(A_1A_1)E_1)E_2)$

(42) $^5E_2(^4A_2(^3E_1(A_1E_1)E_1)E_2)$

(41) $^3E_2(^2A_1(^1E_2(E_1E_1)E_2)E_2)$

(40) $^3E_2(^2A_1(^1E_2(A_1E_2)E_2)E_2)$

(39) $^3E_2(^2E_1(^1E_2(A_1E_2)E_1)E_1)$

(38) $^3E_2(^2A_2(^1E_1(A_1E_1)E_1)E_2)$

(37) $^3E_2(^2A_1(^1E_1(A_1E_1)E_1)E_2)$

(36) $^1E_2(^2A_2(^1E_2(E_1E_1)E_2)E_2)$

(35) $^1E_2(^2A_1(^1E_2(E_1E_1)E_2)E_2)$

(34) $^1E_2(^2A_1(^1E_2(A_1E_2)E_2)E_2)$

(33) $^1E_2(^2A_2(^1E_1(A_1E_1)E_1)E_2)$

(32) $^1E_2(^2A_1(^1E_1(A_1E_1)E_1)E_2)$

(31) $^1E_2(^2E_1(^1A_1(A_1A_1)E_1)E_1)$

(30) $^5E_1(^4E_1(^3E_1(A_1E_1)E_2)E_2)$

(29) $^3E_1(^2E_1(^1E_1(E_1E_2)E_2)E_2)$

(28) $^3E_1(^2E_1(^1A_1(E_1E_1)E_1)E_2)$

(27) $^3E_1(^2E_3(^1E_2(A_1E_2)E_1)E_2)$

(26) $^3E_1(^2E_1(^1E_2(A_1E_2)E_1)E_2)$

(25) $^3E_1(^2E_1(^1E_1(A_1E_1)E_2)E_2)$

(24) $^3E_1(^2A_1(^1E_1(A_1E_1)E_1)E_1)$

(23) $^3E_1(^2E_1(^1A_1(A_1A_1)E_1)E_2)$

(22) $^1E_1(^2E_1(^1E_1(E_1E_2)E_2)E_2)$

(21) $^1E_1(^2E_1(^1A_1(E_1E_1)E_1)E_2)$

(20) $^1E_1(^2E_1(^1E_2(A_1E_2)E_1)E_2)$

(19) $^1E_1(^2E_1(^1E_1(A_1E_1)E_2)E_2)$

(18) $^1E_1(^2A_1(^1E_1(A_1E_1)E_1)E_1)$

(17) $^1E_1(^2E_1(^1A_1(A_1A_1)E_1)E_2)$

(16) $^3A_2(^2E_2(^1E_1(E_1E_2)E_1)E_2)$

(15) $^3A_2(^2E_2(^1A_1(E_1E_1)E_2)E_2)$

(14) $^3A_2(^2E_2(^1E_1(A_1E_1)E_1)E_2)$

(13) $^3A_2(^2E_2(^1A_1(A_1A_1)E_2)E_2)$

(12) $^3A_2(^2E_1(^1A_1(A_1A_1)E_1)E_1)$

(11) $^1A_2(^2E_2(^1E_1(A_1E_1)E_1)E_2)$

(10) $^5A_1(^4E_2(^3A_2(E_1E_1)E_2)E_2)$

(9) $^3A_1(^2E_2(^1E_1(E_1E_2)E_1)E_2)$

(8) $^3A_1(^2E_2(^1E_1(A_1E_1)E_1)E_2)$

(7) $^1A_1(^2E_2(^1A_1(E_2E_2)E_2)E_2)$

(6) $^1A_1(^2E_2(^1E_1(E_1E_2)E_1)E_2)$

(5) $^1A_1(^2E_2(^1A_1(E_1E_1)E_2)E_2)$

(4) $^1A_1(^2E_1(^1A_1(E_1E_1)E_1)E_1)$

(3) $^1A_1(^2E_2(^1E_1(A_1E_1)E_1)E_2)$

(2) $^1A_1(^2E_2(^1A_1(A_1A_1)E_2)E_2)$

(1) $^1A_1(^2E_1(^1A_1(A_1A_1)E_1)E_1)$

227

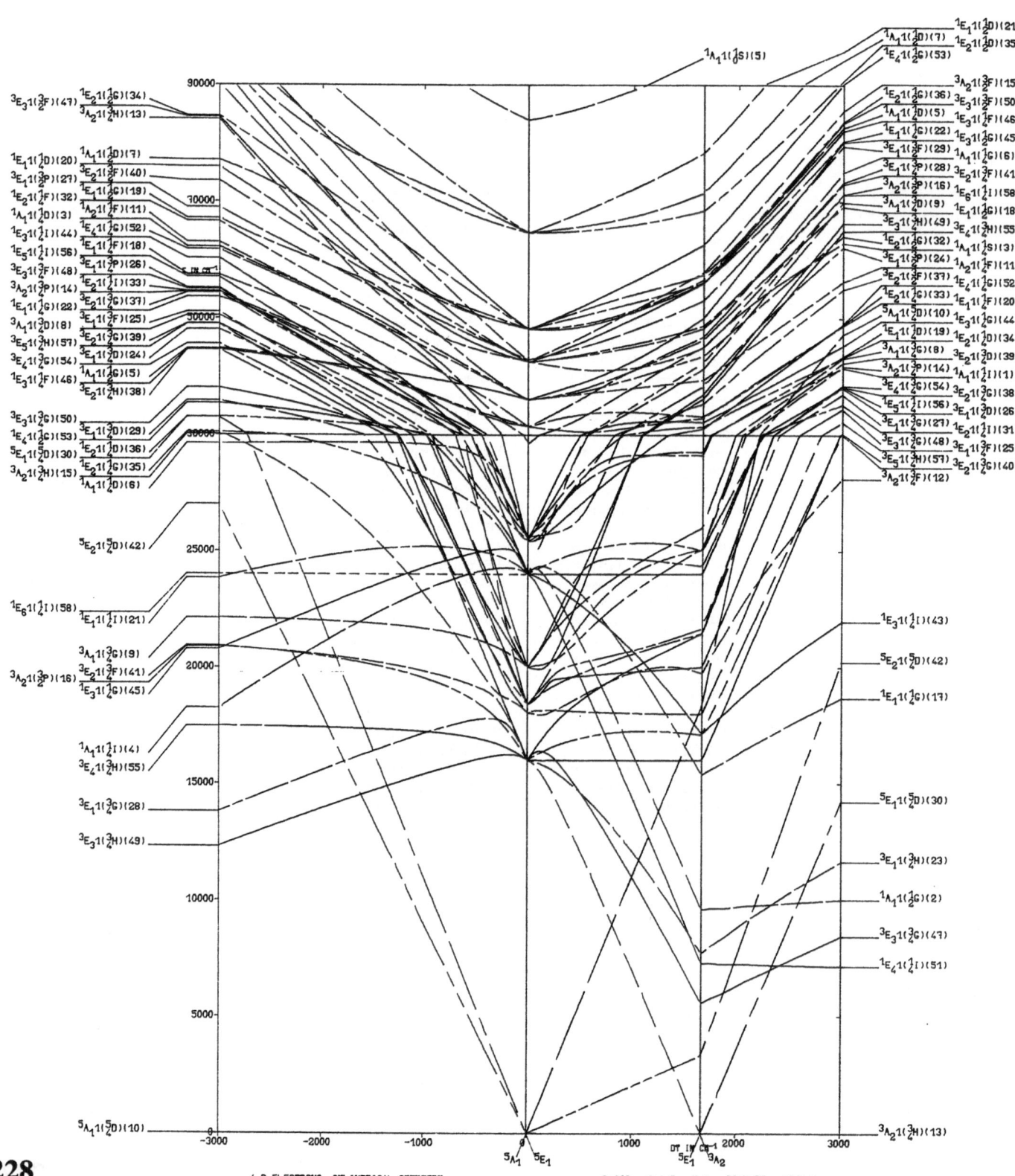

4 D ELECTRONS, CYLINDRICAL SYMMETRY

B=800   C=4×B   K=1   DS=K×DT   ZETA=0

228

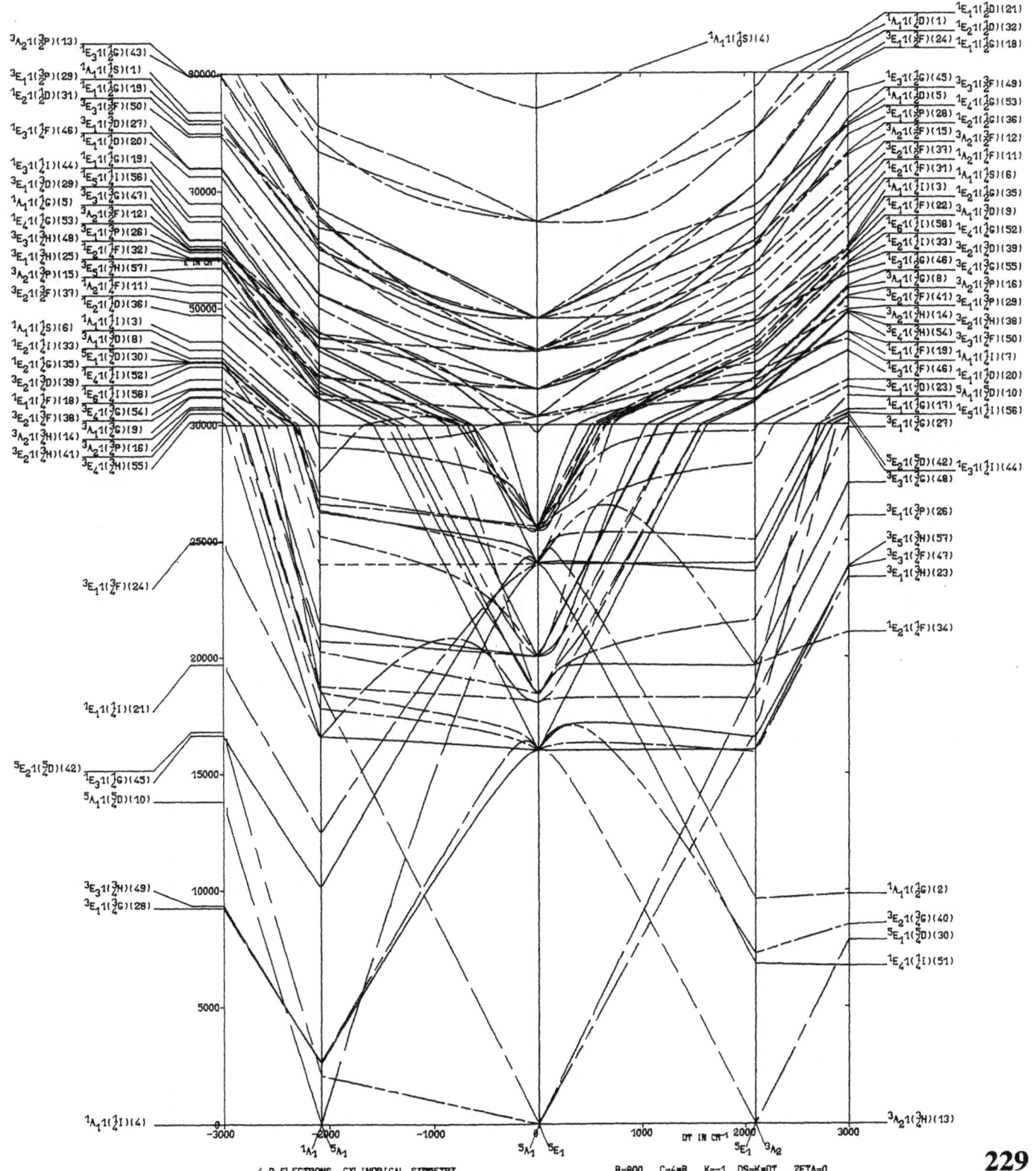

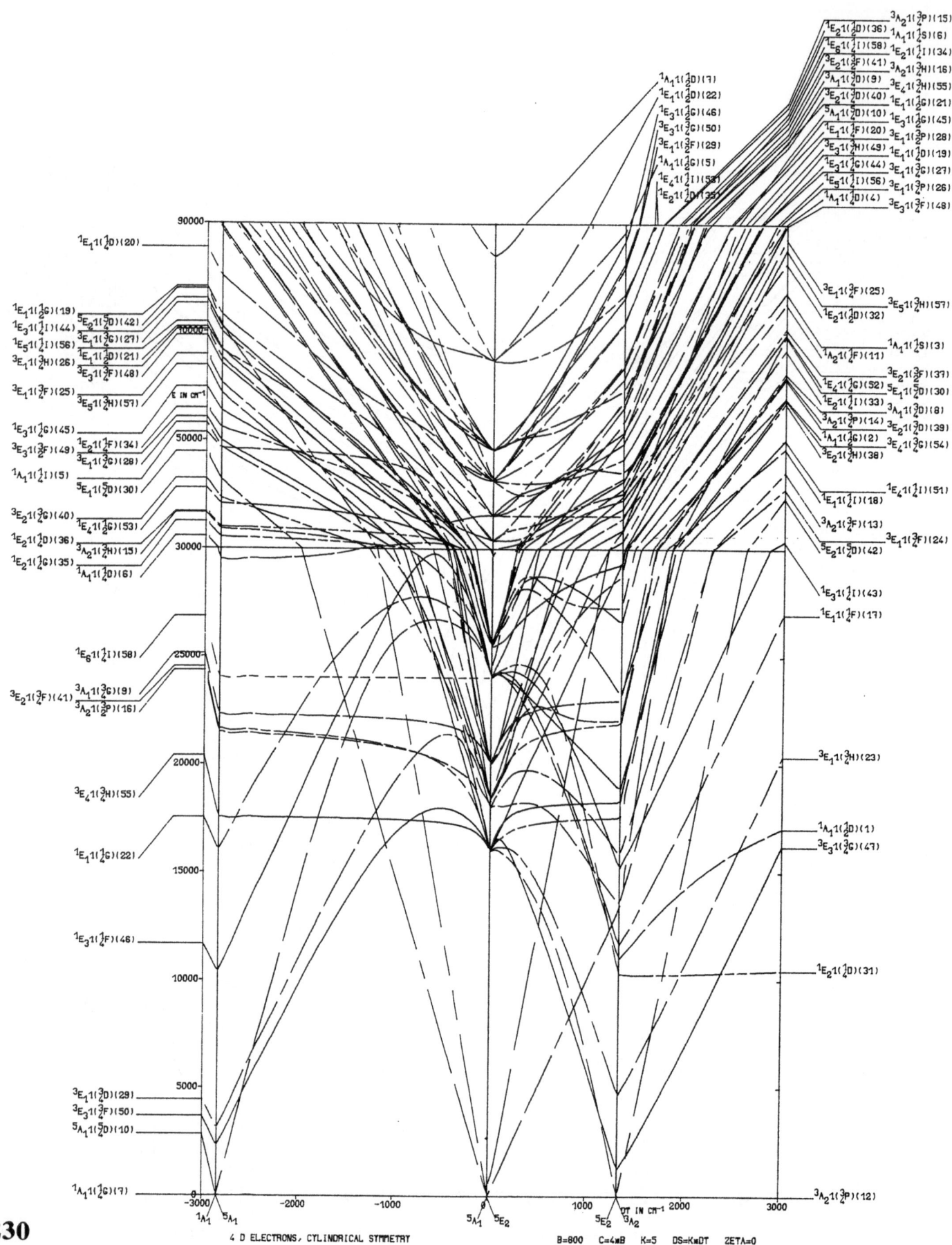

4 D ELECTRONS, CYLINDRICAL SYMMETRY   B=800   C=4×B   K=5   DS=K×DT   ZETA=0

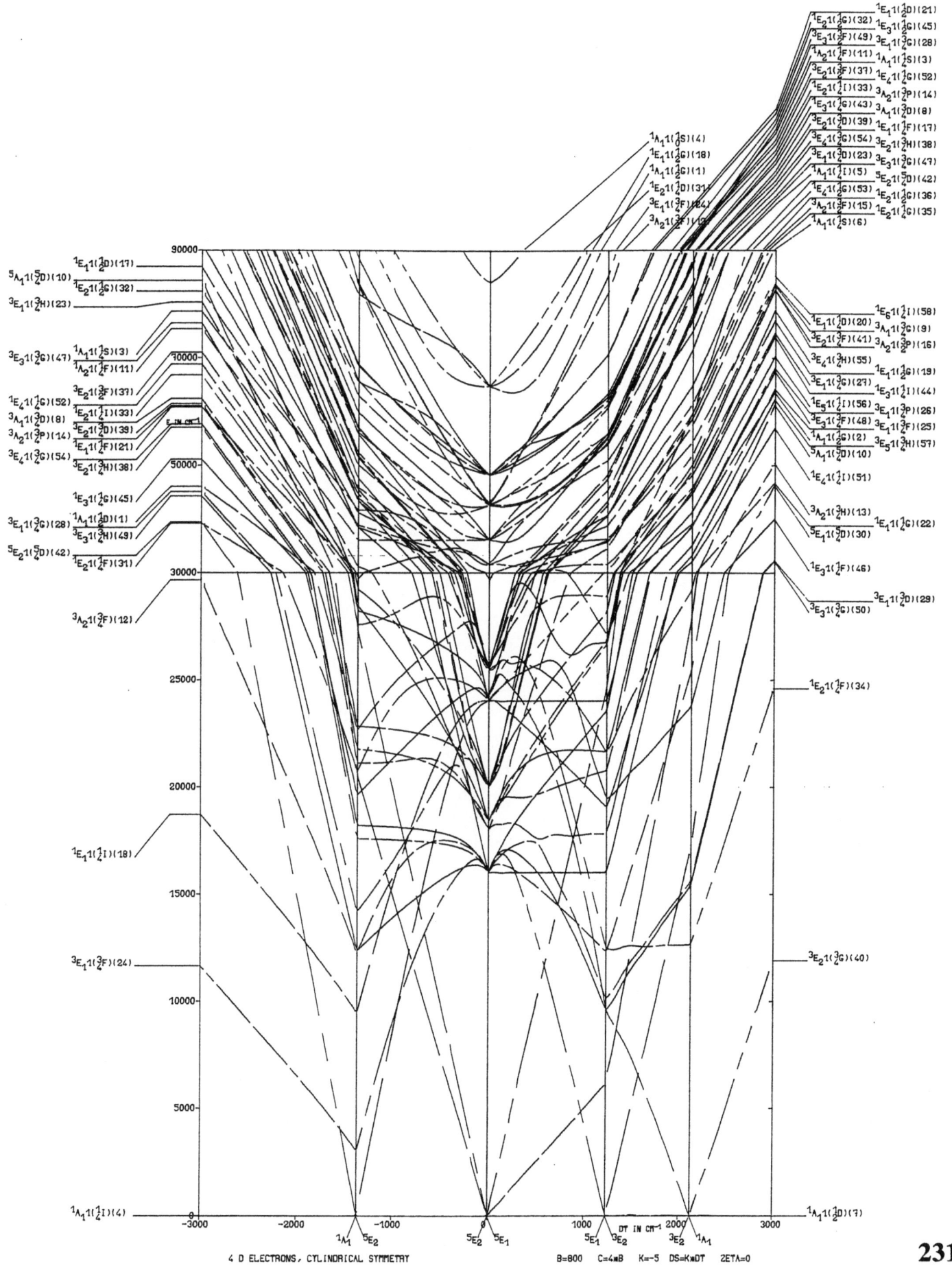

4 D ELECTRONS, CYLINDRICAL SYMMETRY   B=800   C=4×B   K=-5   DS=K×DT   2ETA=0

231

5 D ELECTRONS, TETRAGONAL SYMMETRY

B=825   C=4×B   ZETA=0

ENERGY AS FUNCTION OF DQ
DT=0, 500, -500, 1000, -1000
K=1, -1, 5, -5

ENERGY AS FUNCTION OF DT
DQ=850, 1430
K=1, -1, 5, -5

(76) $^6A_1 1(A_1B_1B_2(^4A_2)E\ E\ (^3A_2))$

(75) $^4E\ 1(B_1B_2E\ (^4E\ )E\ E\ (^1A_1))$

(74) $^4E\ 1(A_1B_2E\ (^4E\ )E\ E\ (^1A_1))$

(73) $^4E\ 1(A_1B_1E\ (^4E\ )E\ E\ (^1A_1))$

(72) $^4E\ 1(A_1B_1B_2(^4A_2)B_2E\ (^1E\ ))$

(71) $^4E\ 1(A_1B_2B_1(^2A_1)B_1E\ (^3E\ ))$

(70) $^4E\ 1(A_1A_1B_1(^2B_1)B_2E\ (^3E\ ))$

(69) $^4B_21(B_2E\ B_2(^2B_1)B_2E\ (^3A_2))$

(68) $^4B_21(A_1B_1B_2(^4A_2)E\ E\ (^1B_1))$

(67) $^4B_21(A_1A_1B_1(^2B_1)E\ E\ (^3A_2))$

(66) $^4B_11(B_1B_1B_2(^2B_2)E\ E\ (^3A_2))$

(65) $^4B_11(A_1B_1B_2(^4A_2)E\ E\ (^1B_2))$

(64) $^4B_11(A_1A_1B_2(^2B_2)E\ E\ (^3A_2))$

(63) $^4A_21(A_1E\ B_2(^2A_1)B_2E\ (^3A_2))$

(62) $^4A_21(A_1B_1B_2(^4A_2)E\ E\ (^1A_1))$

(61) $^4A_21(A_1E\ B_1(^2A_1)B_1E\ (^3A_2))$

(60) $^4A_12(A_1B_1B_2(^2A_2)E\ E\ (^3A_2))$

(59) $^4A_11(A_1B_1B_2(^4A_2)E\ E\ (^3A_2))$

(58) $^4A_11(A_1B_1B_2(^2A_2)E\ E\ (^3A_2))$

(57) $^2E\ 1(B_2B_2E\ (^2E\ )E\ E\ (^1A_1))$

(56) $^2E\ 1(B_1B_2E\ (^4E\ )E\ E\ (^3A_2))$

(55) $^2E\ 1(B_1B_2E\ (^2E\ )E\ E\ (^1A_1))$

(54) $^2E\ 1(B_1B_1E\ (^2E\ )E\ E\ (^1A_1))$

(53) $^2E\ 1(B_1B_1B_2(^2B_2)B_2E\ (^1E\ ))$

(52) $^2E\ 1(A_1B_2E\ (^4E\ )E\ E\ (^3A_2))$

(51) $^2E\ 1(A_1B_2E\ (^2E\ )E\ E\ (^1A_1))$

(50) $^2E\ 1(A_1B_1E\ (^4E\ )E\ E\ (^3A_2))$

(49) $^2E\ 1(A_1B_1E\ (^2E\ )E\ E\ (^1A_1))$

(48) $^2E\ 1(A_1B_1B_2(^4A_2)B_2E\ (^3E\ ))$

(47) $^2E\ 1(A_1B_1B_2(^2A_2)B_2E\ (^1E\ ))$

(46) $^2E\ 1(A_1B_2B_1(^2A_1)B_1E\ (^3E\ ))$

(45) $^2E\ 1(A_1B_2B_1(^2A_1)B_1E\ (^1E\ ))$

(44) $^2E\ 1(A_1A_1E\ (^2E\ )E\ E\ (^1A_1))$

(43) $^2E\ 1(A_1A_1B_2(^2B_2)B_2E\ (^1E\ ))$

(42) $^2E\ 1(A_1A_1B_1(^2B_1)B_2E\ (^3E\ ))$

(41) $^2E\ 1(A_1A_1B_1(^2B_1)B_2E\ (^1E\ ))$

(40) $^2E\ 1(A_1A_1B_1(^2B_1)B_1E\ (^1E\ ))$

(39) $^2B_21(B_2E\ E\ (^2A_1)E\ E\ (^1B_2))$

(38) $^2B_21(B_1E\ B_2(^2B_1)B_2E\ (^3A_2))$

(37) $^2B_21(B_1B_1B_2(^2B_2)E\ E\ (^1A_1))$

(36) $^2B_21(A_1E\ B_2(^2A_1)B_2E\ (^1B_2))$

(35) $^2B_22(A_1B_1B_2(^2A_2)E\ E\ (^1B_1))$

(34) $^2B_21(A_1B_1B_2(^2A_2)E\ E\ (^1B_1))$

(33) $^2B_21(A_1E\ B_1(^2A_1)B_1E\ (^1B_2))$

(32) $^2B_21(A_1A_1B_2(^2B_2)E\ E\ (^1A_1))$

(31) $^2B_21(A_1A_1B_1(^2B_1)E\ E\ (^3A_2))$

(30) $^2B_21(A_1A_1B_1(^2B_1)B_1B_2(^1A_2))$

(29) $^2B_11(B_1E\ E\ (^2A_1)E\ E\ (^1B_1))$

(28) $^2B_11(B_1E\ B_2(^2B_1)B_2E\ (^1A_1))$

(27) $^2B_11(B_1B_1B_2(^2B_2)E\ E\ (^3A_2))$

(26) $^2B_11(A_1E\ B_2(^2A_1)B_2E\ (^1B_1))$

(25) $^2B_12(A_1B_1B_2(^2A_2)E\ E\ (^1B_2))$

(24) $^2B_11(A_1B_1B_2(^2A_2)E\ E\ (^1B_2))$

(23) $^2B_11(A_1E\ B_1(^2A_1)B_1E\ (^1B_1))$

(22) $^2B_11(A_1A_1B_2(^2B_2)E\ E\ (^3A_2))$

(21) $^2B_11(A_1A_1B_1(^2B_1)E\ E\ (^1A_1))$

(20) $^2B_11(A_1A_1B_1(^2B_1)B_2B_2(^1A_1))$

(19) $^2A_21(B_1E\ B_2(^2B_1)B_2E\ (^1B_2))$

(18) $^2A_21(B_1B_1B_2(^2B_2)E\ E\ (^1B_1))$

(17) $^2A_21(A_1E\ B_2(^2A_1)B_2E\ (^3A_2))$

(16) $^2A_22(A_1B_1B_2(^2A_2)E\ E\ (^1A_1))$

(15) $^2A_21(A_1B_1B_2(^2A_2)E\ E\ (^1A_1))$

(14) $^2A_21(A_1E\ B_1(^2A_1)B_1E\ (^3A_2))$

(13) $^2A_21(A_1A_1B_2(^2B_2)E\ E\ (^1B_1))$

(12) $^2A_21(A_1A_1B_1(^2B_1)E\ E\ (^1B_2))$

(11) $^2A_11(B_1E\ B_2(^2B_1)B_2E\ (^1B_1))$

(10) $^2A_11(B_1B_1B_2(^2B_2)E\ E\ (^1B_2))$

(9) $^2A_11(A_1E\ E\ (^2A_1)E\ E\ (^1A_1))$

(8) $^2A_11(A_1E\ B_2(^2A_1)B_2E\ (^1A_1))$

(7) $^2A_12(A_1B_1B_2(^2A_2)E\ E\ (^3A_2))$

(6) $^2A_11(A_1B_1B_2(^4A_2)E\ E\ (^3A_2))$

(5) $^2A_11(A_1B_1B_2(^2A_2)E\ E\ (^3A_2))$

(4) $^2A_11(A_1E\ B_1(^2A_1)B_1E\ (^1A_1))$

(3) $^2A_11(A_1B_2B_1(^2A_1)B_1B_2(^1A_1))$

(2) $^2A_11(A_1A_1B_2(^2B_2)E\ E\ (^1B_2))$

(1) $^2A_11(A_1A_1B_1(^2B_1)E\ E\ (^1B_1))$

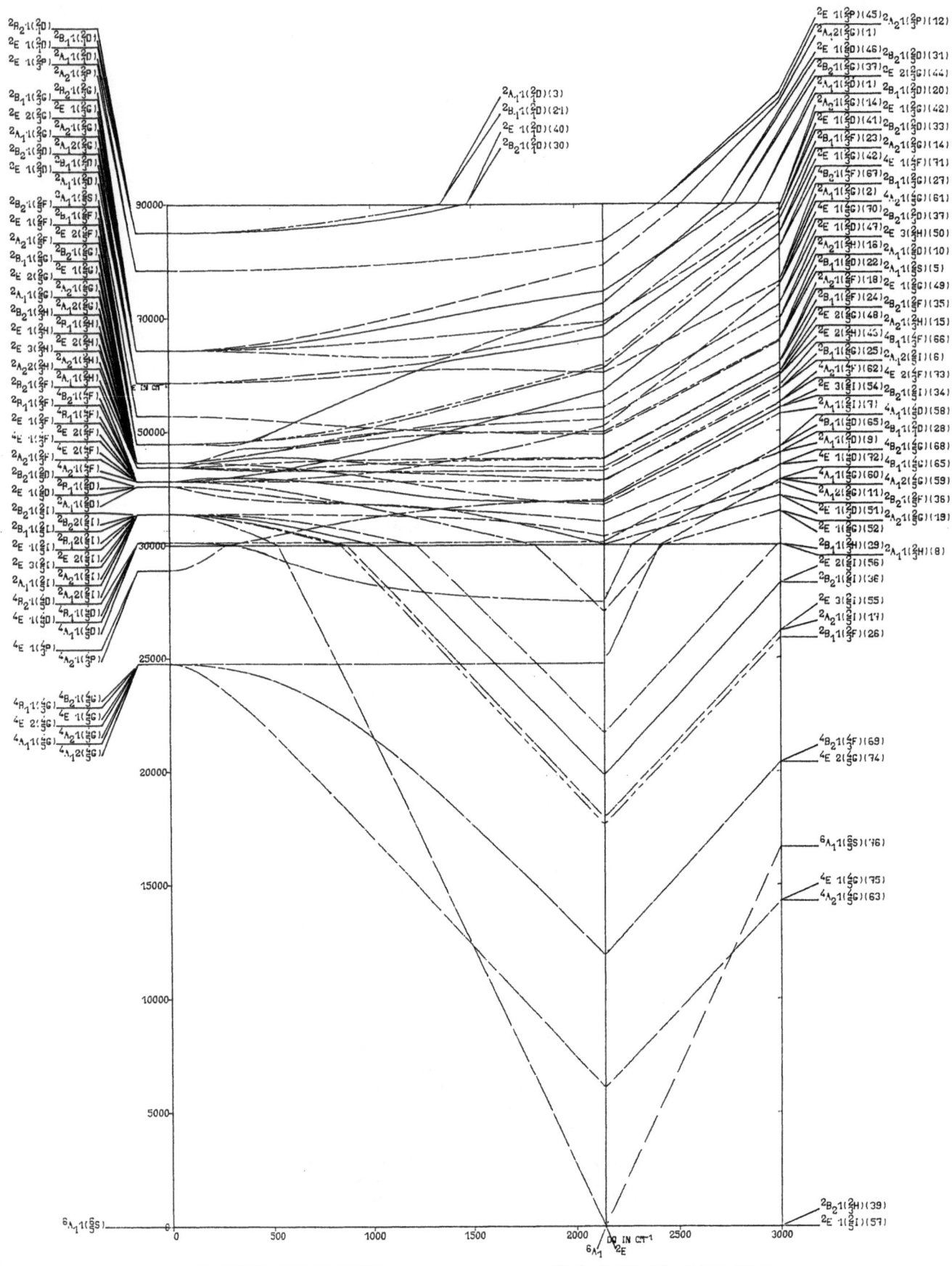

234

5 D ELECTRONS, TETRAGONAL SYMMETRY

G=825 C=4xB DT=0 K=0 DS=KxDT ZETA=0

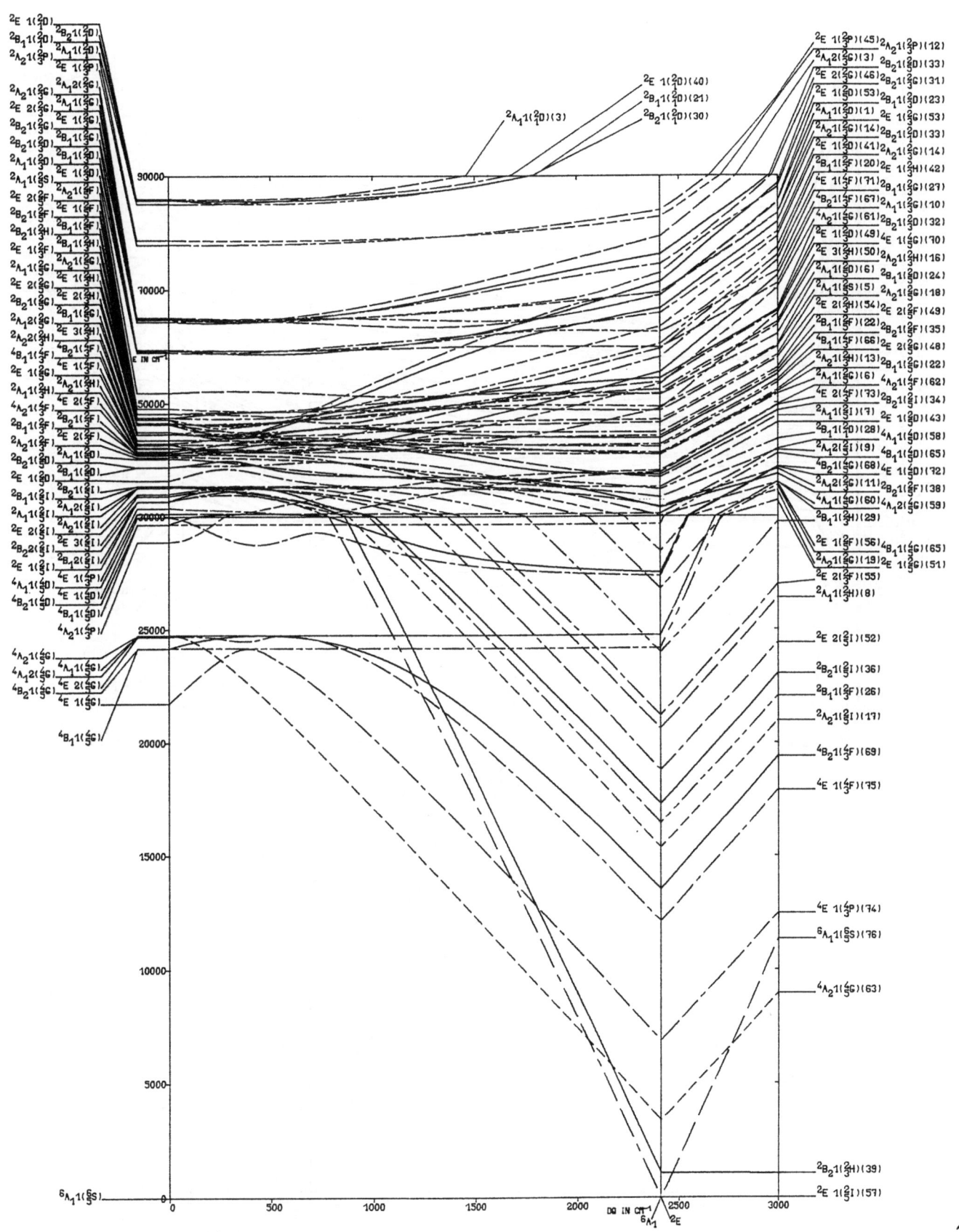

5 D ELECTRONS, TETRAGONAL SYMMETRY

B=825 C=4×B DT=500 K=1 DS=K×DT ZETA=0

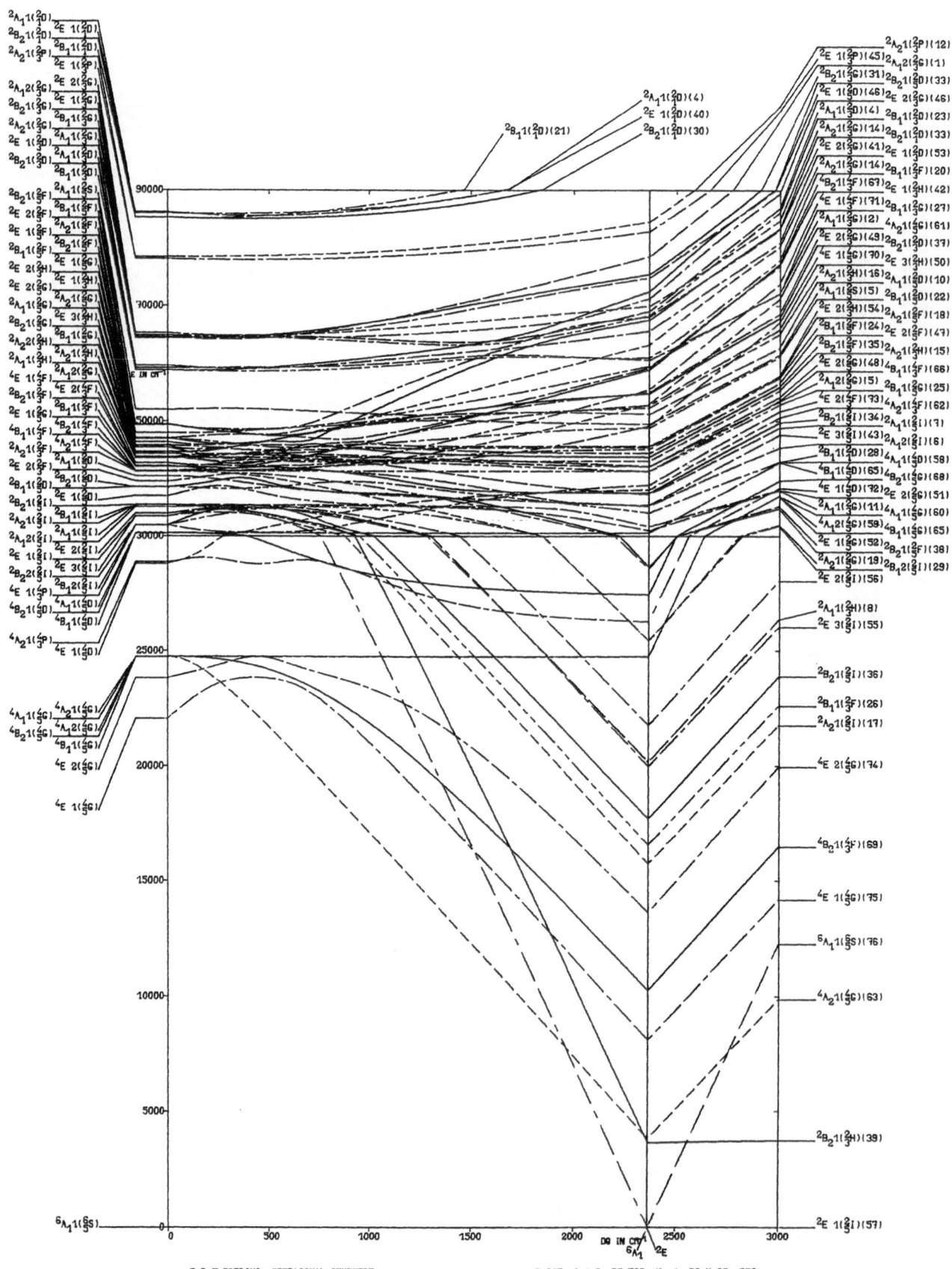

5 D ELECTRONS, TETRAGONAL SYMMETRY   B=825 C=4×B DT=500 K=-1 DS=K×DT ZETA=0

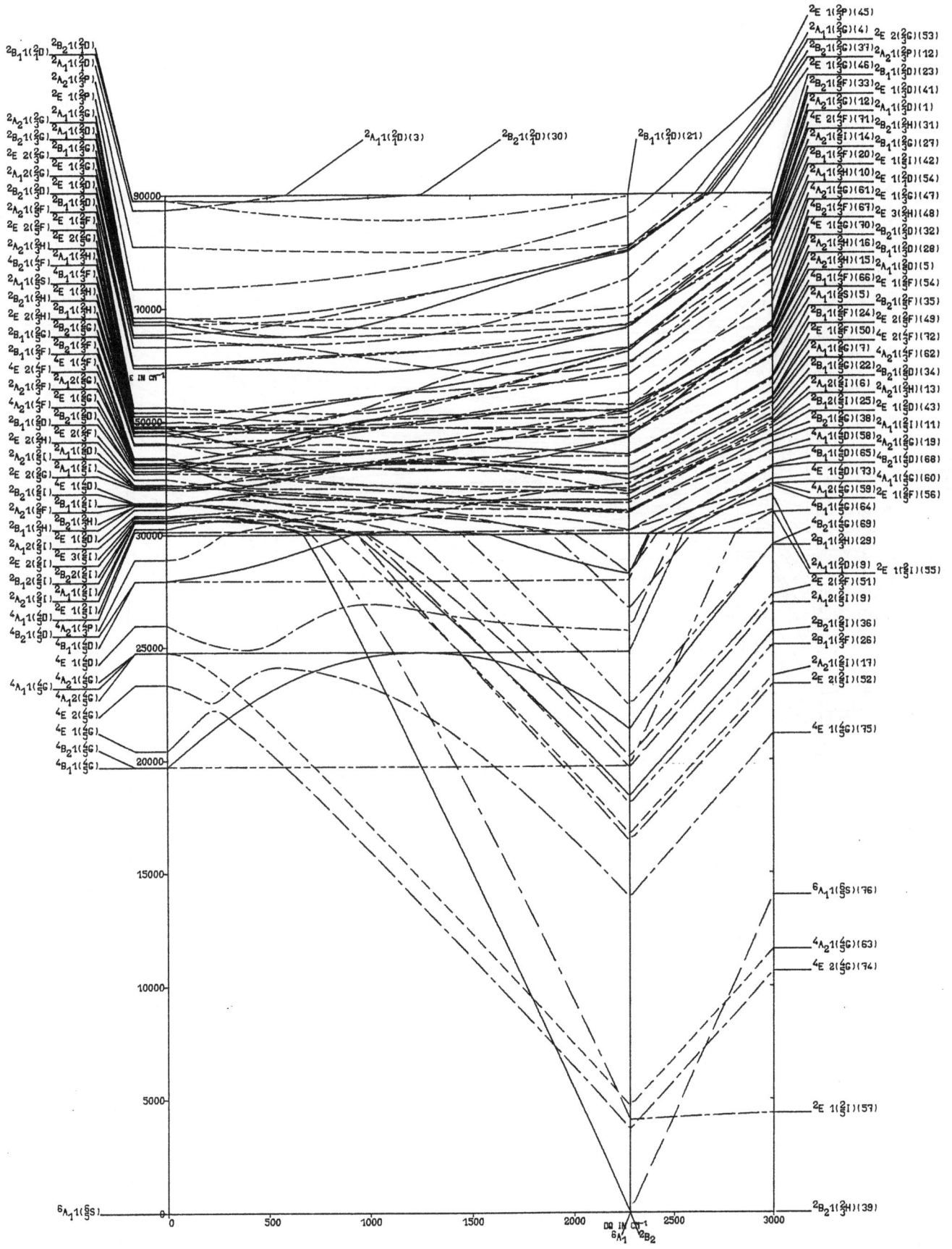

5 D ELECTRONS, TETRAGONAL SYMMETRY   G=825 C=4×B DT=500 K=5   DS=K×DT ZETA=0

237

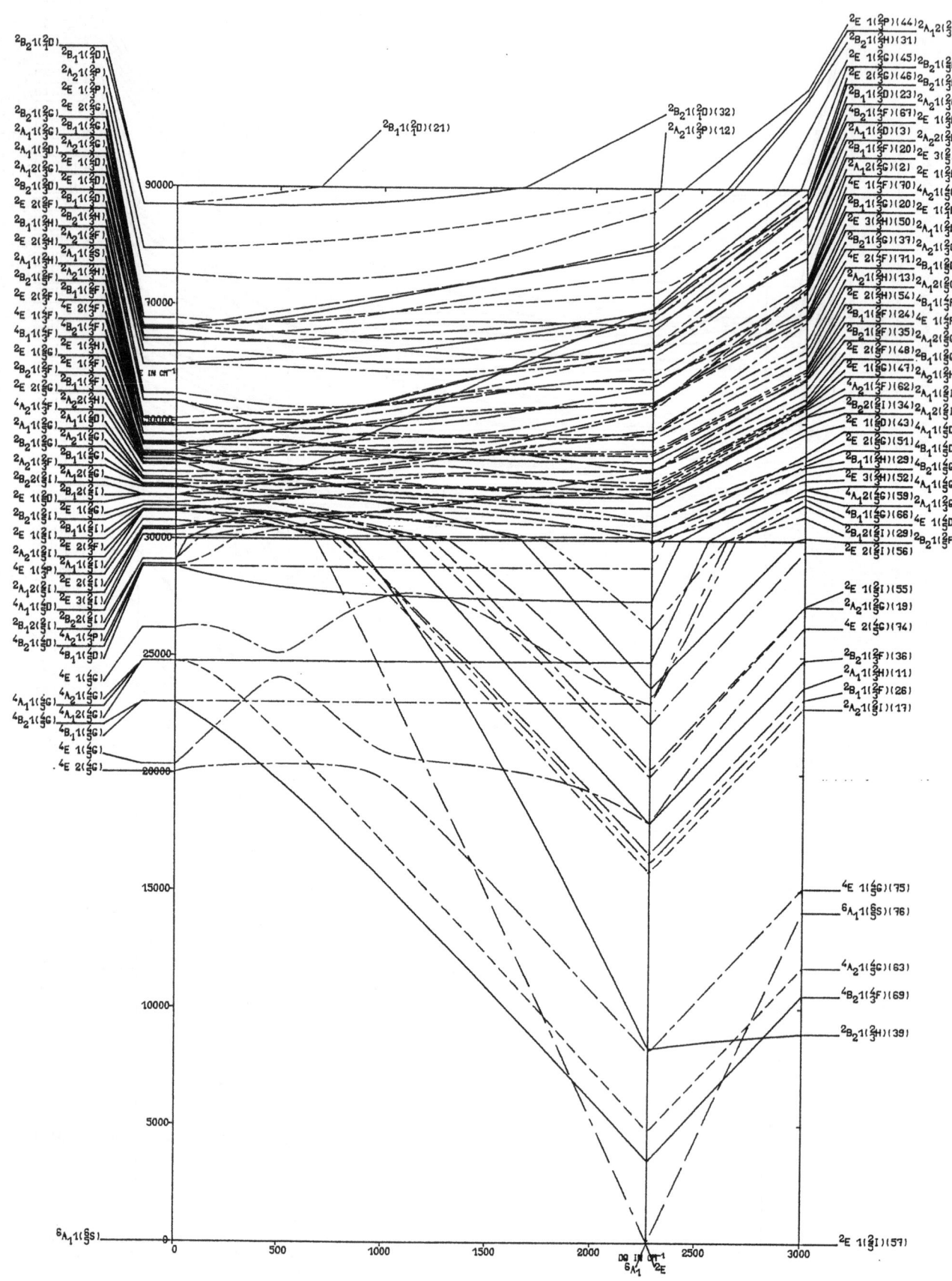

5 D ELECTRONS, TETRAGONAL SYMMETRY

B=825 C=4*B DT=500 K=-5 DS=K*DT ZETA=0

238

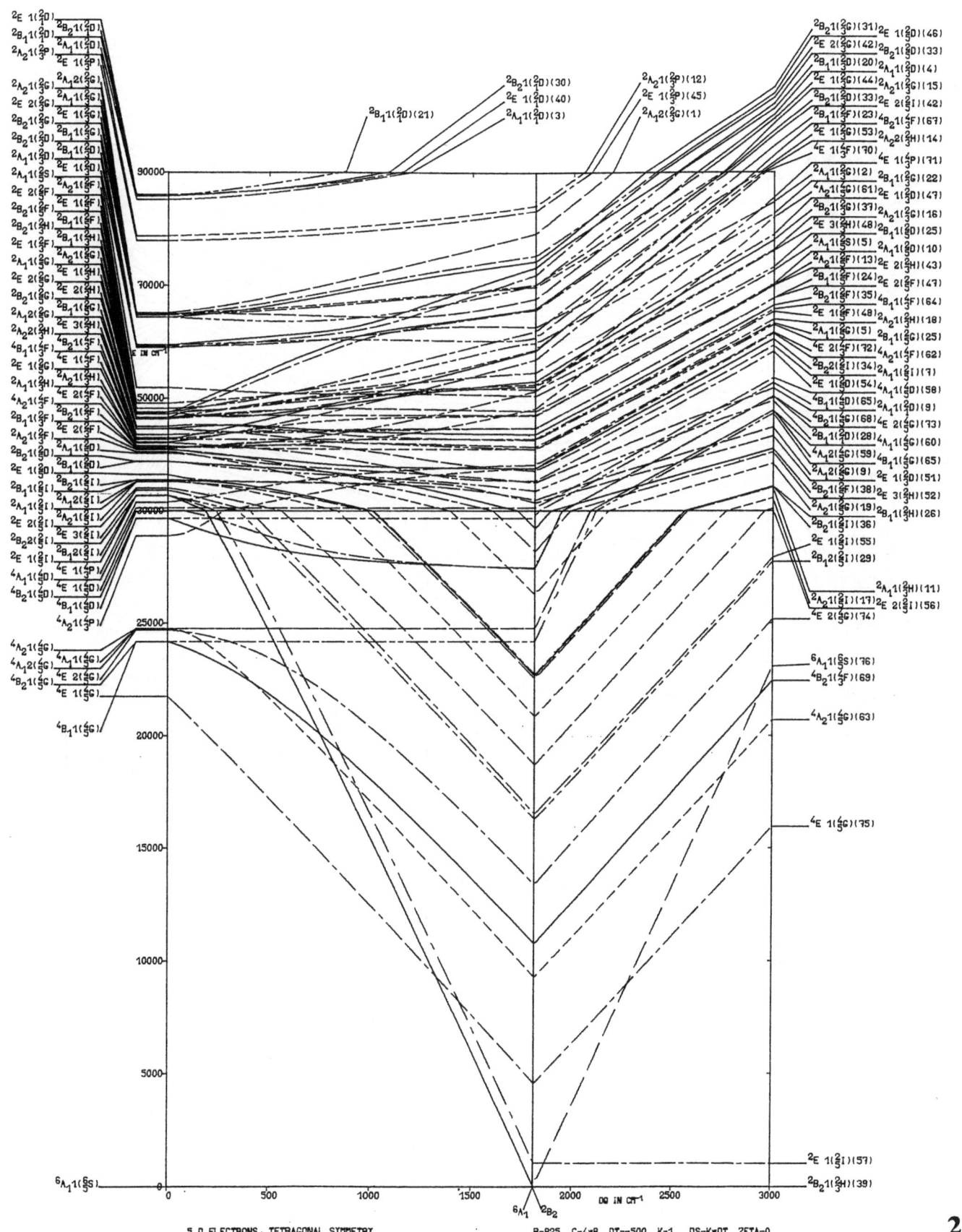

5 D ELECTRONS, TETRAGONAL SYMMETRY   B=825 C=4×B DT=-500 K=1   DS=K×DT ZETA=0

239

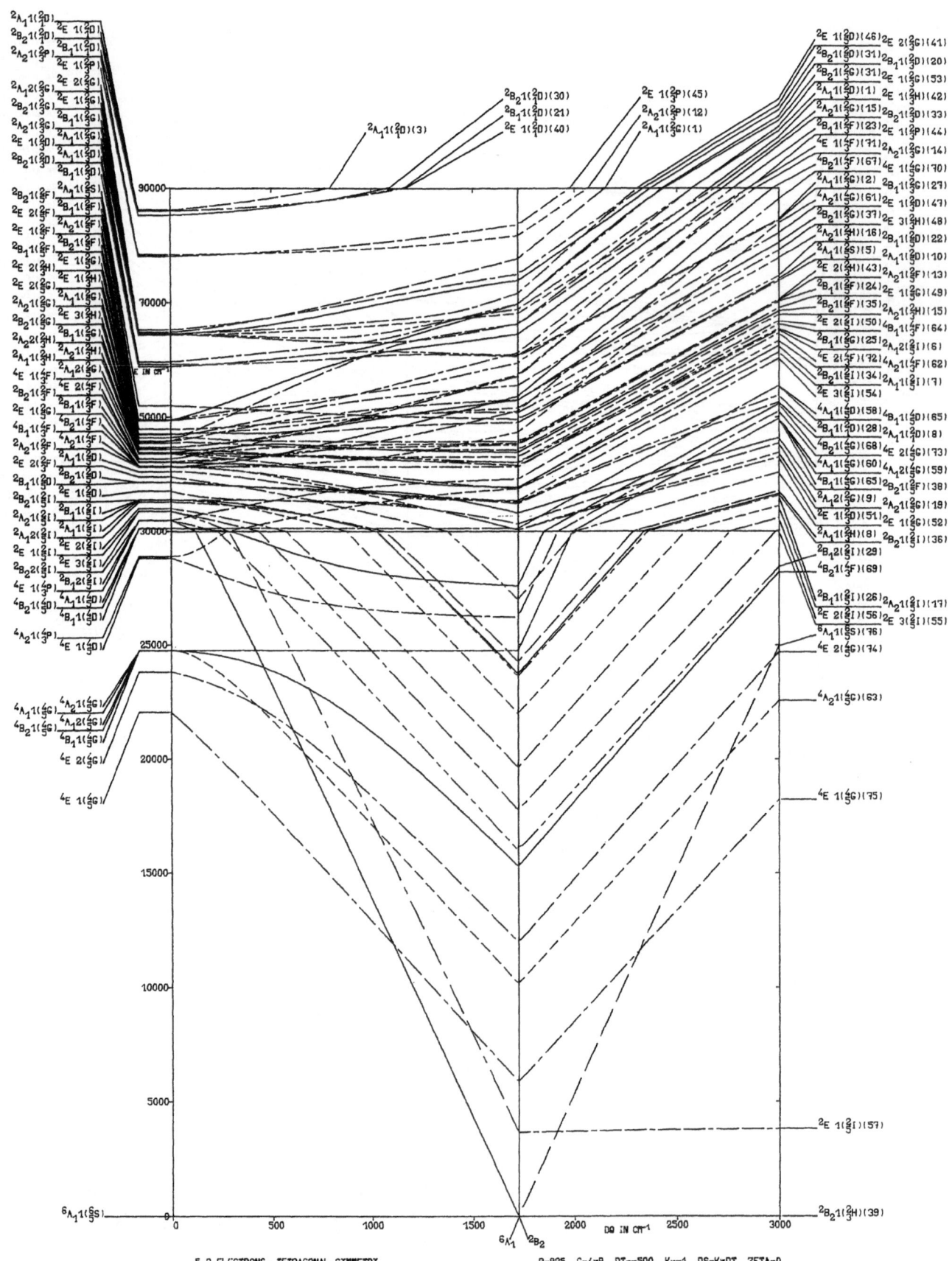

5 D ELECTRONS, TETRAGONAL SYMMETRY

B=825 C=4×B DT=-500 K=-1 DS=K×DT ZETA=0

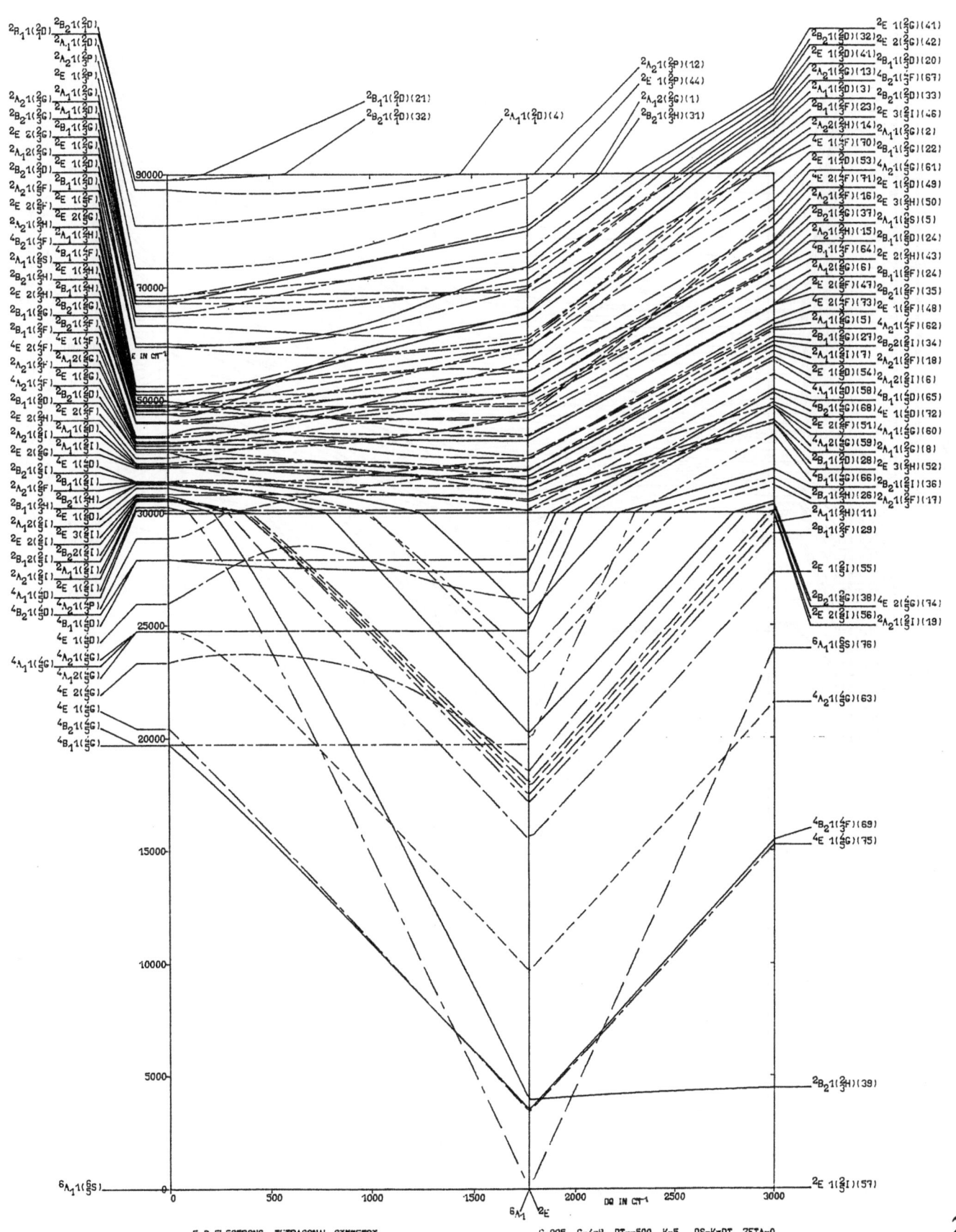

5 D ELECTRONS, TETRAGONAL SYMMETRY

G=825   C=4×B   DT=-500   K=5   DS=K×DT   ZETA=0

241

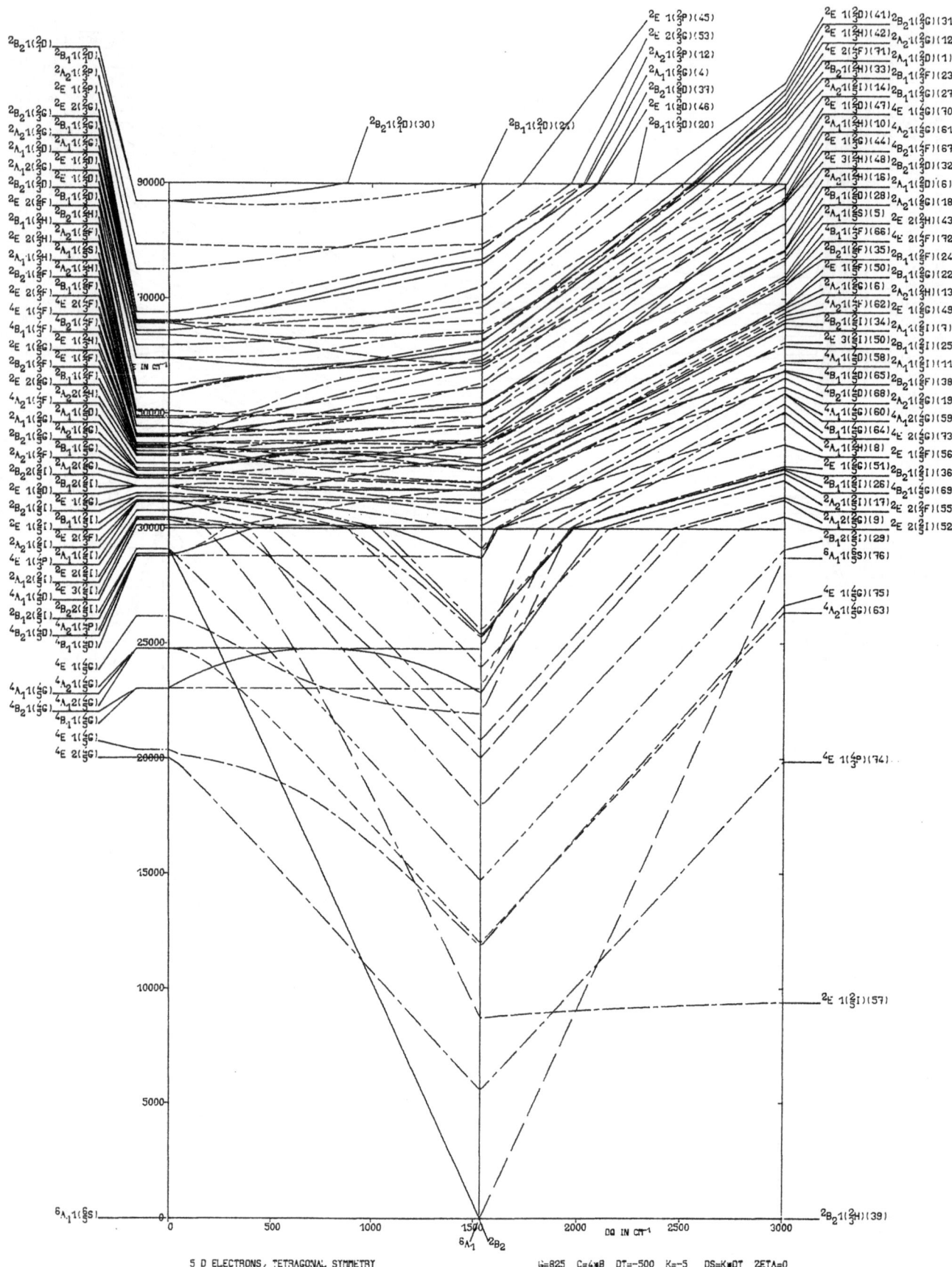

242

5 D ELECTRONS, TETRAGONAL SYMMETRY

G=825 C=4×B DT=-500 K=-5 DS=K×DT ZETA=0

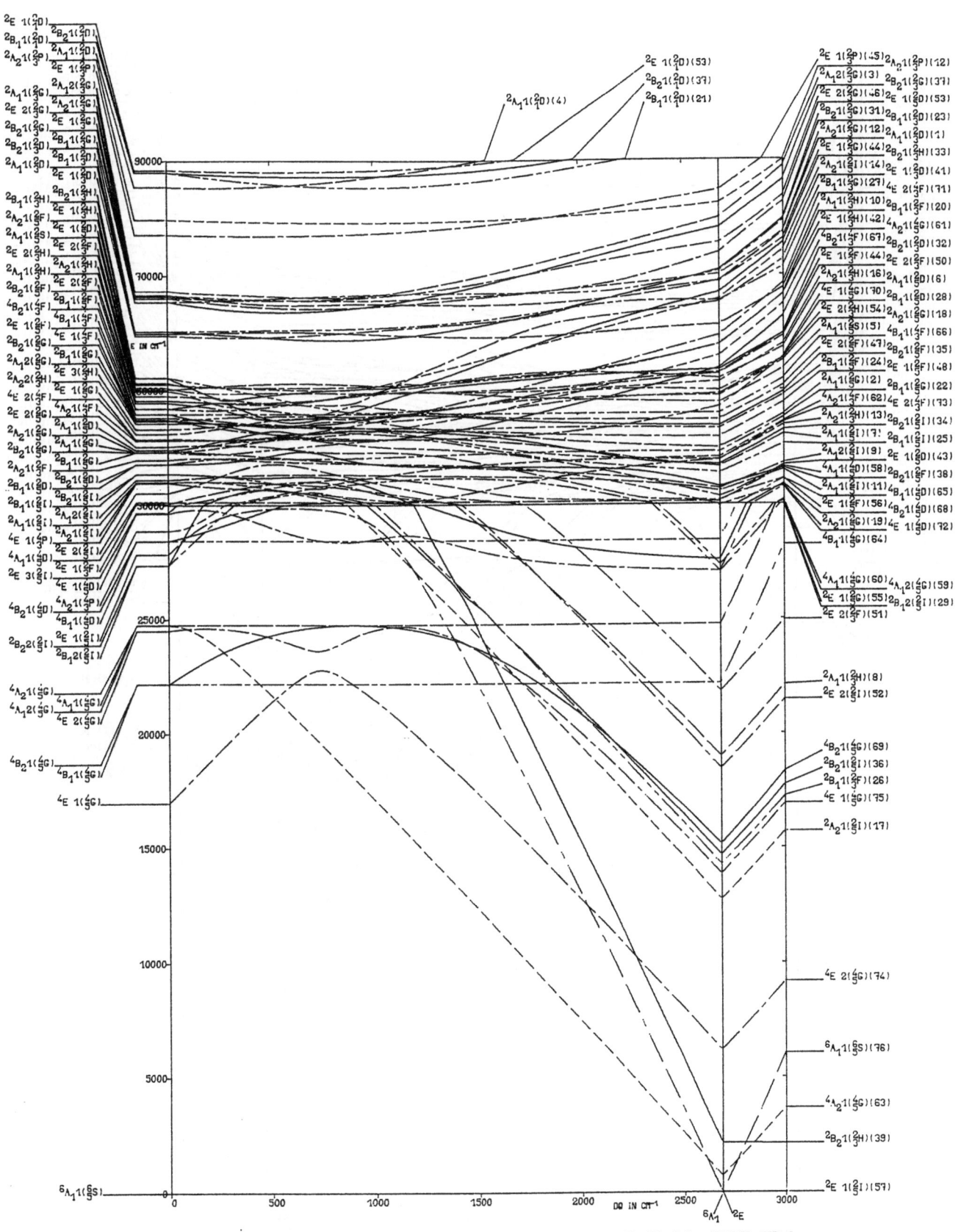

5 D ELECTRONS, TETRAGONAL SYMMETRY

G=625 C=4*B DT=1000 K=1 DS=K*DT ZETA=0

243

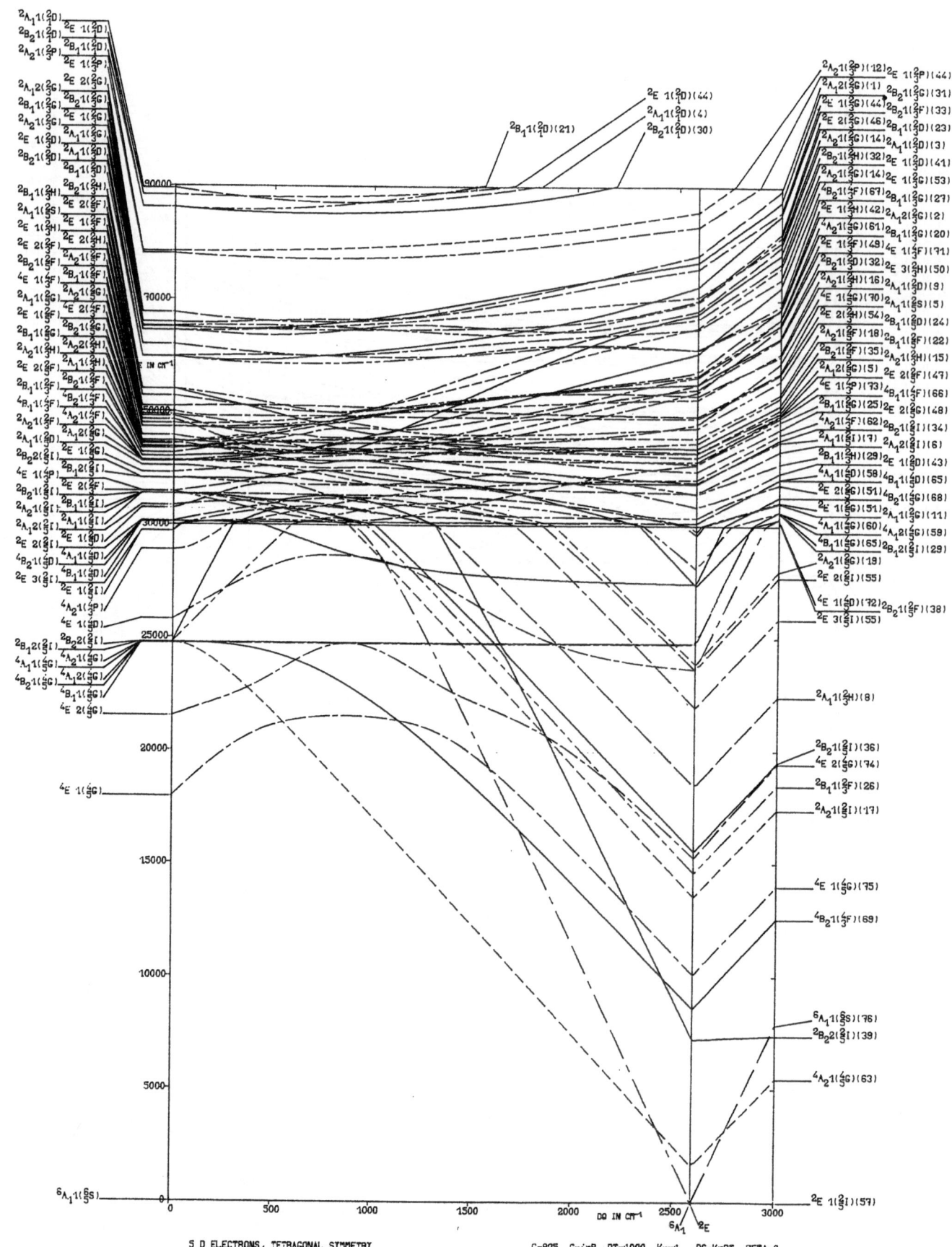

5 D ELECTRONS, TETRAGONAL SYMMETRY

DQ IN CM⁻¹

G=825 C=4×B DT=1000 K=-1 DS=K×DT ZETA=0

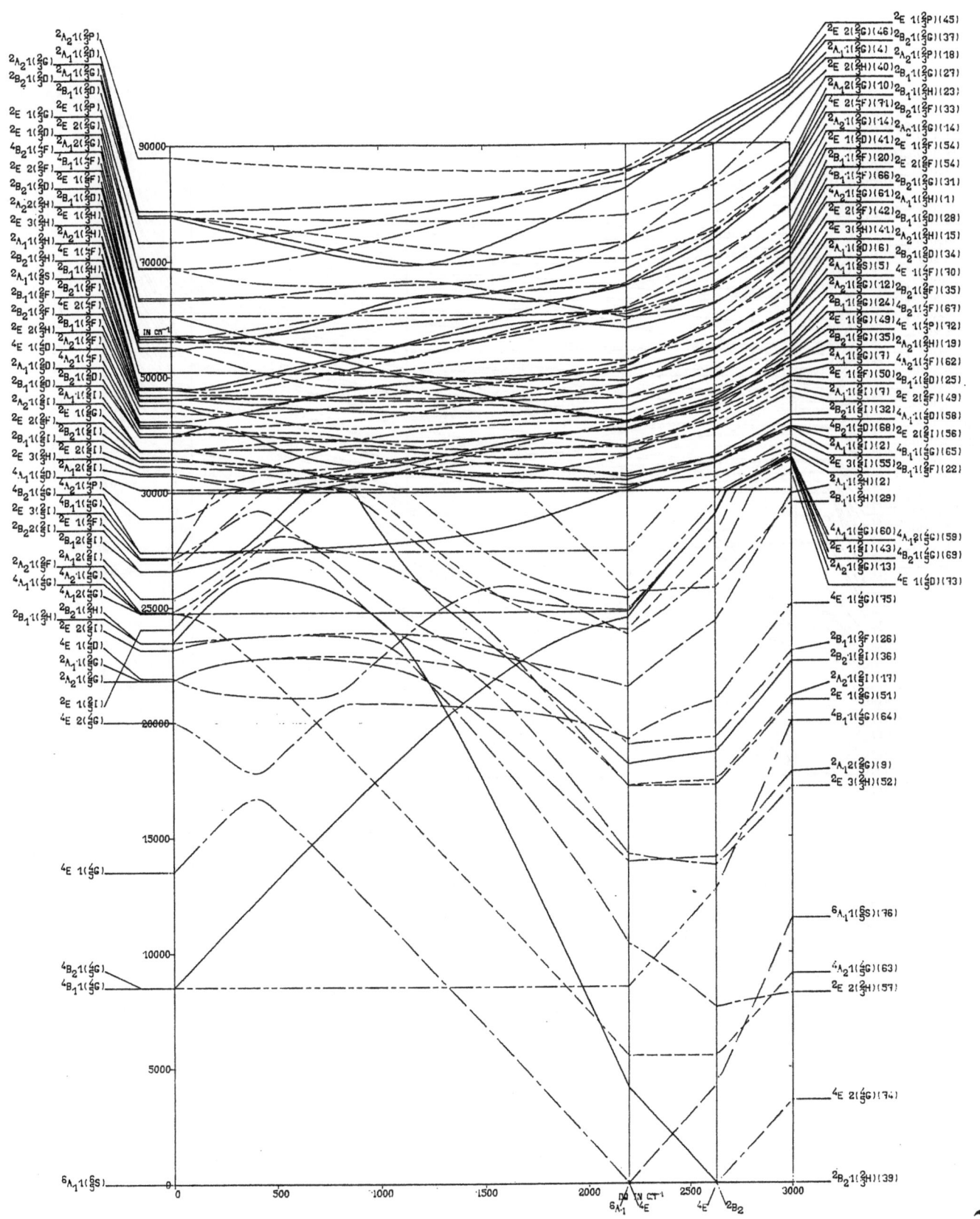

5 D ELECTRONS, TETRAGONAL SYMMETRY   G=825 C=4*B DT=1000 K=5   DS=K*DT ZETA=0

245

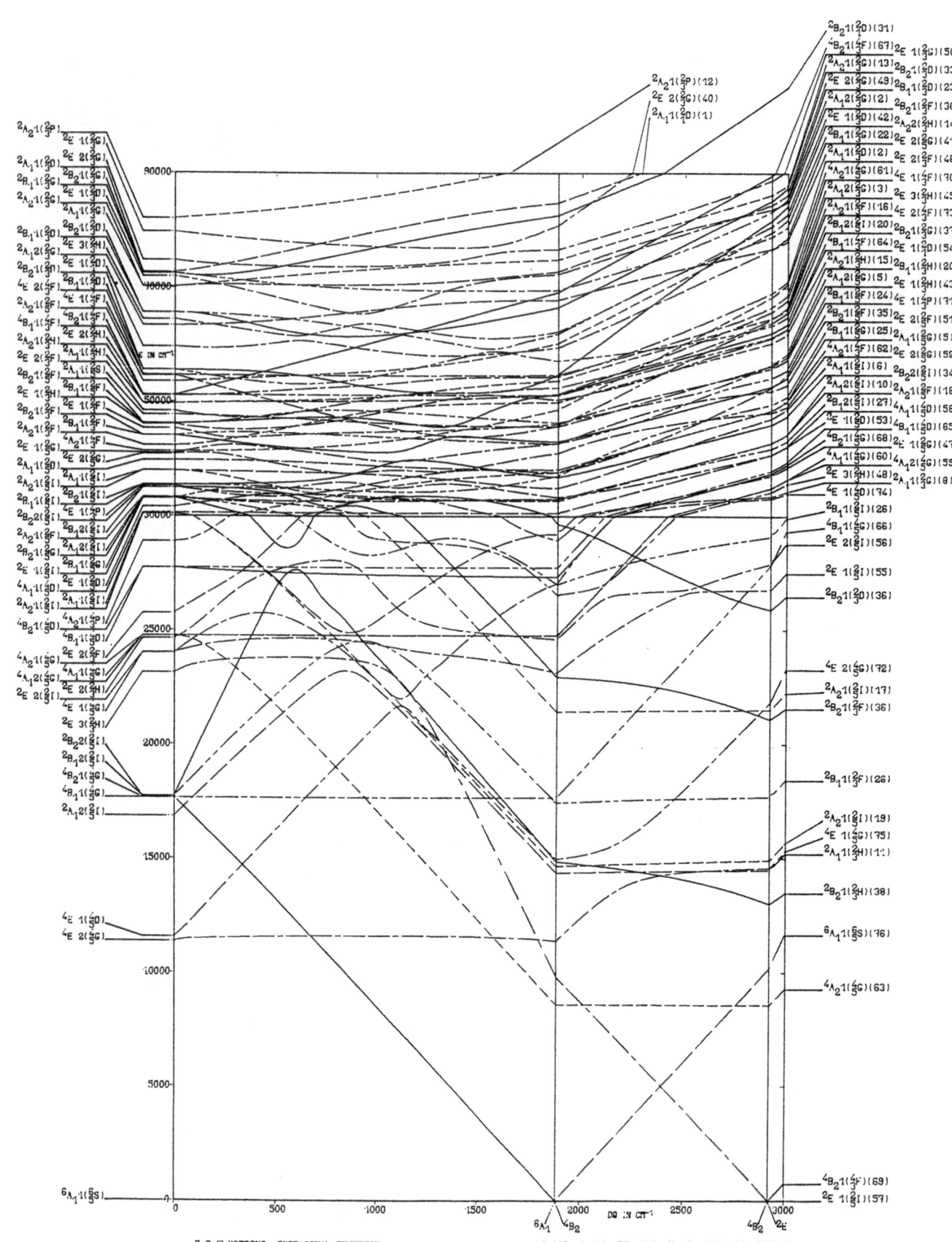

5 D ELECTRONS, TETRAGONAL SYMMETRY

G=825   C=4×B   DT=1000   K=-5   DS=K×DT   ZETA=0

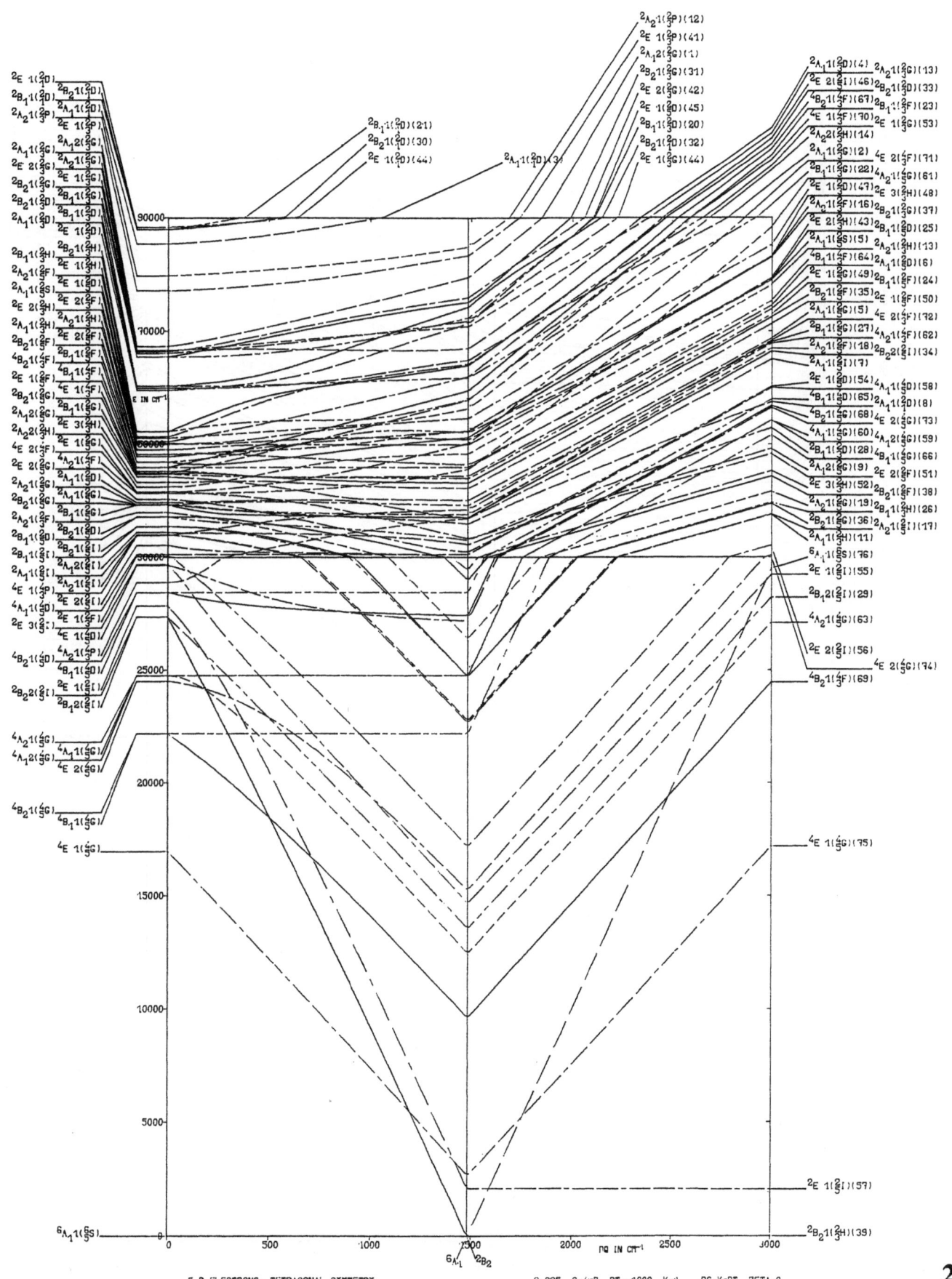

5 D ELECTRONS, TETRAGONAL SYMMETRY   G=825   C=4×B   DT=-1000   K=1   DS=K×DT   ZETA=0

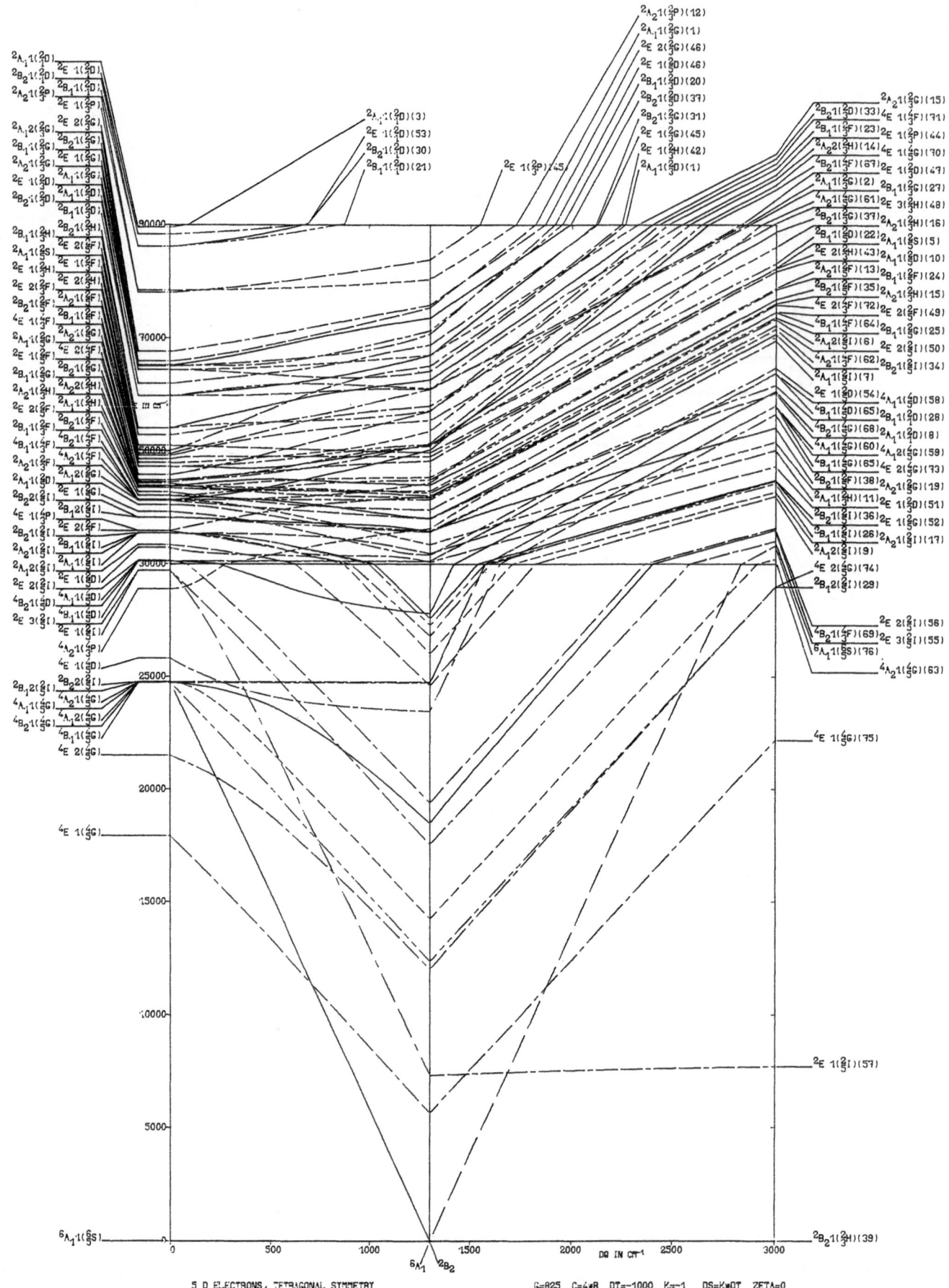

5 D ELECTRONS, TETRAGONAL SYMMETRY

G=825 C=4×B DT=-1000 K=-1 DS=K×DT ZETA=0

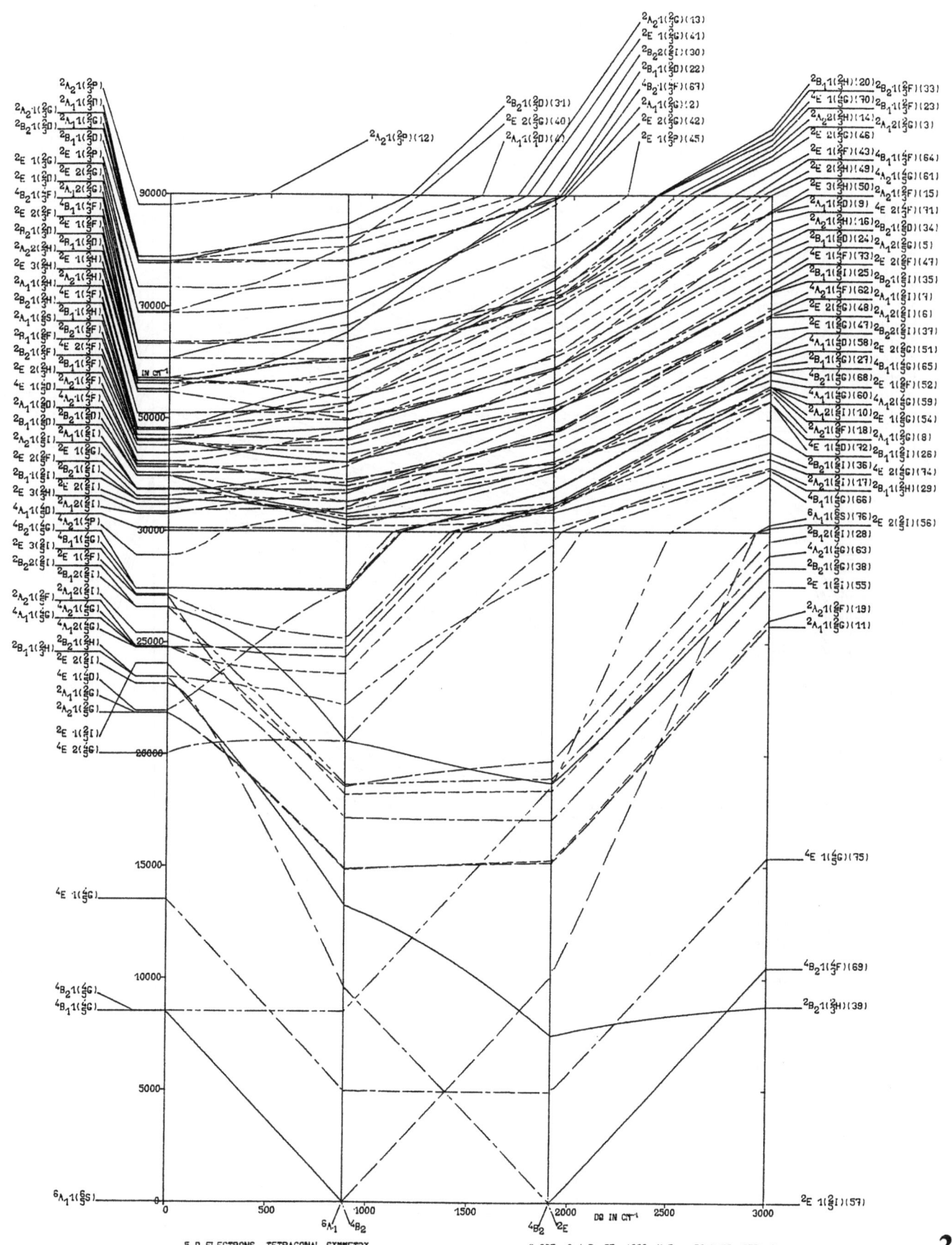

5 D ELECTRONS, TETRAGONAL SYMMETRY

G=825   C=4.8B   DT=-1000   K=5   DS=K×DT   ZETA=0

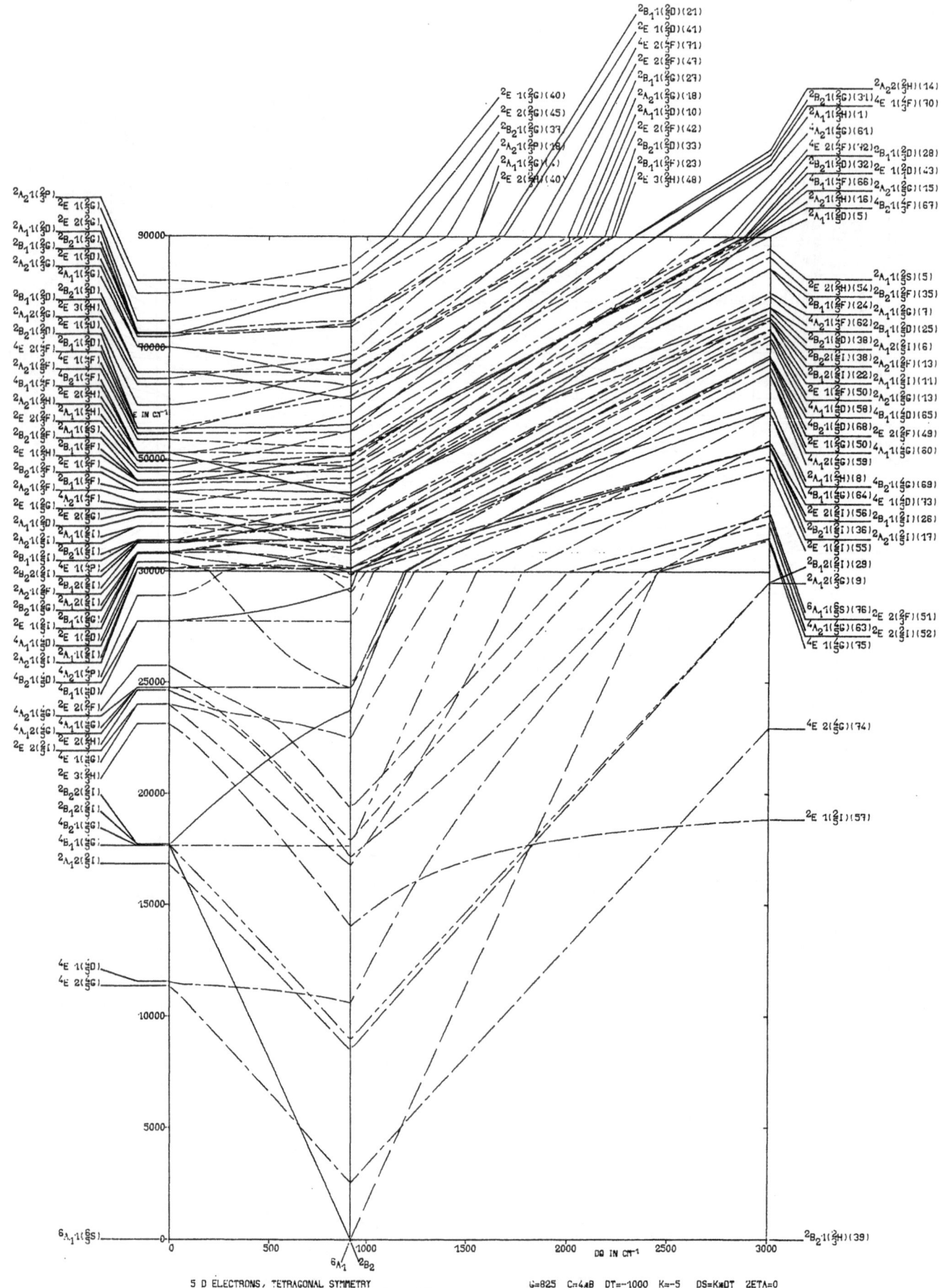

250

5 D ELECTRONS, TETRAGONAL SYMMETRY   G=825   C=4.B   DT=-1000   K=-5   DS=K×DT   ZETA=0

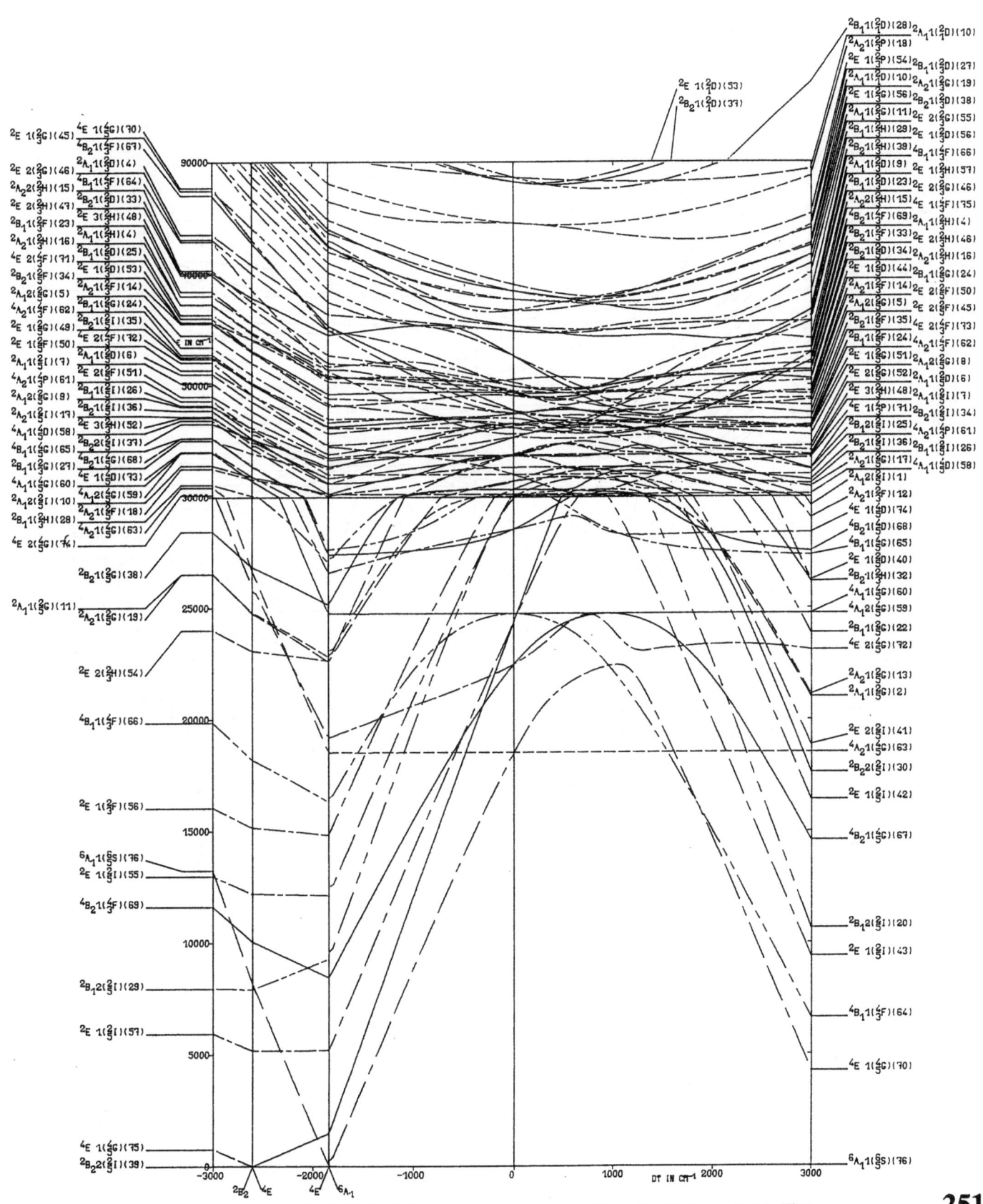

5 D ELECTRONS, TETRAGONAL SYMMETRY

B=825 C=4*B DQ=850 K=1 DS=K*DT ZETA=0

251

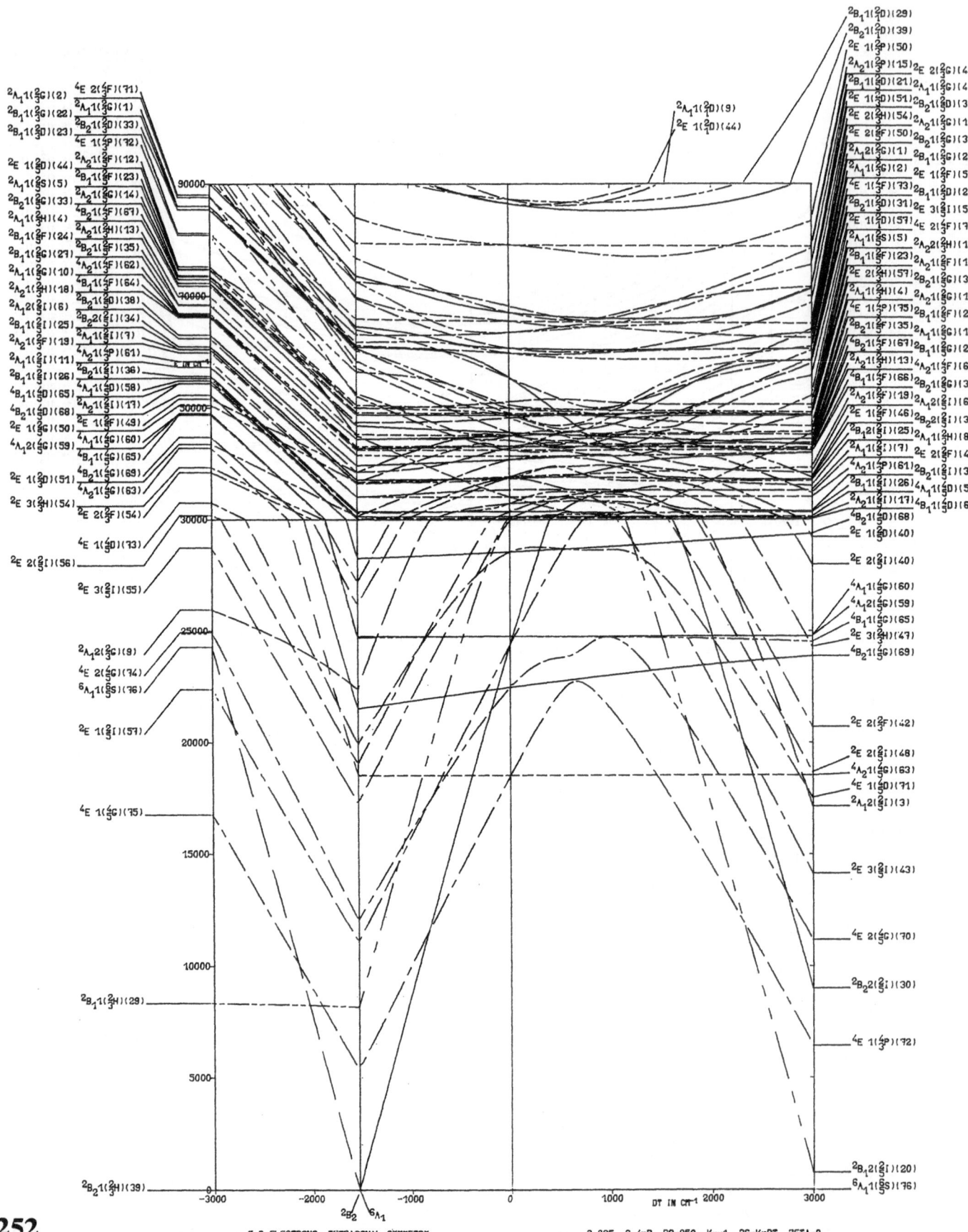

5 D ELECTRONS, TETRAGONAL SYMMETRY

B=825   C=4*B   DQ=850   K=-1   DS=K*DT   2ETA=0

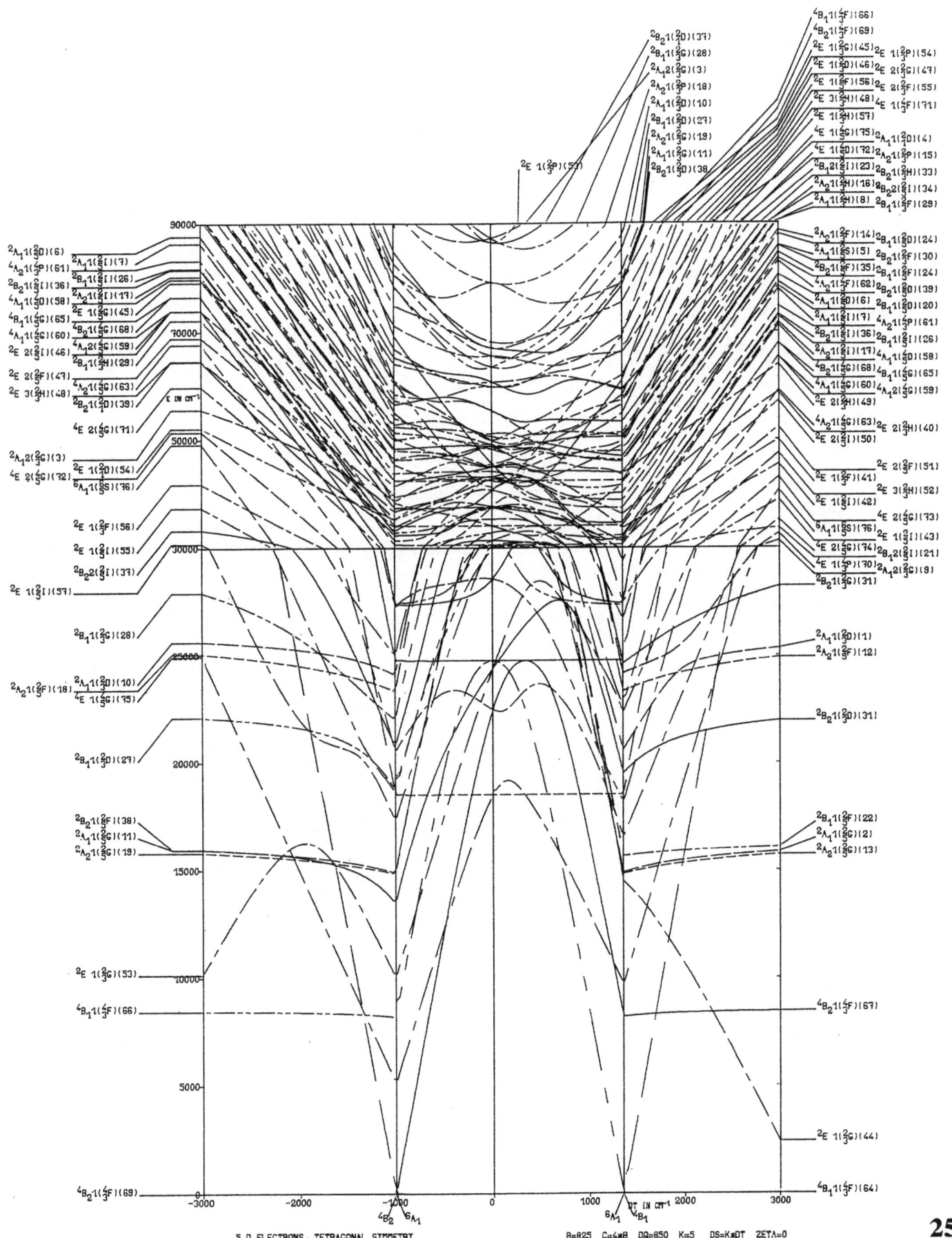

5 D ELECTRONS, TETRAGONAL SYMMETRY

B=825   C=4×B   DQ=850   K=5   DS=K×DT   ZETA=0

253

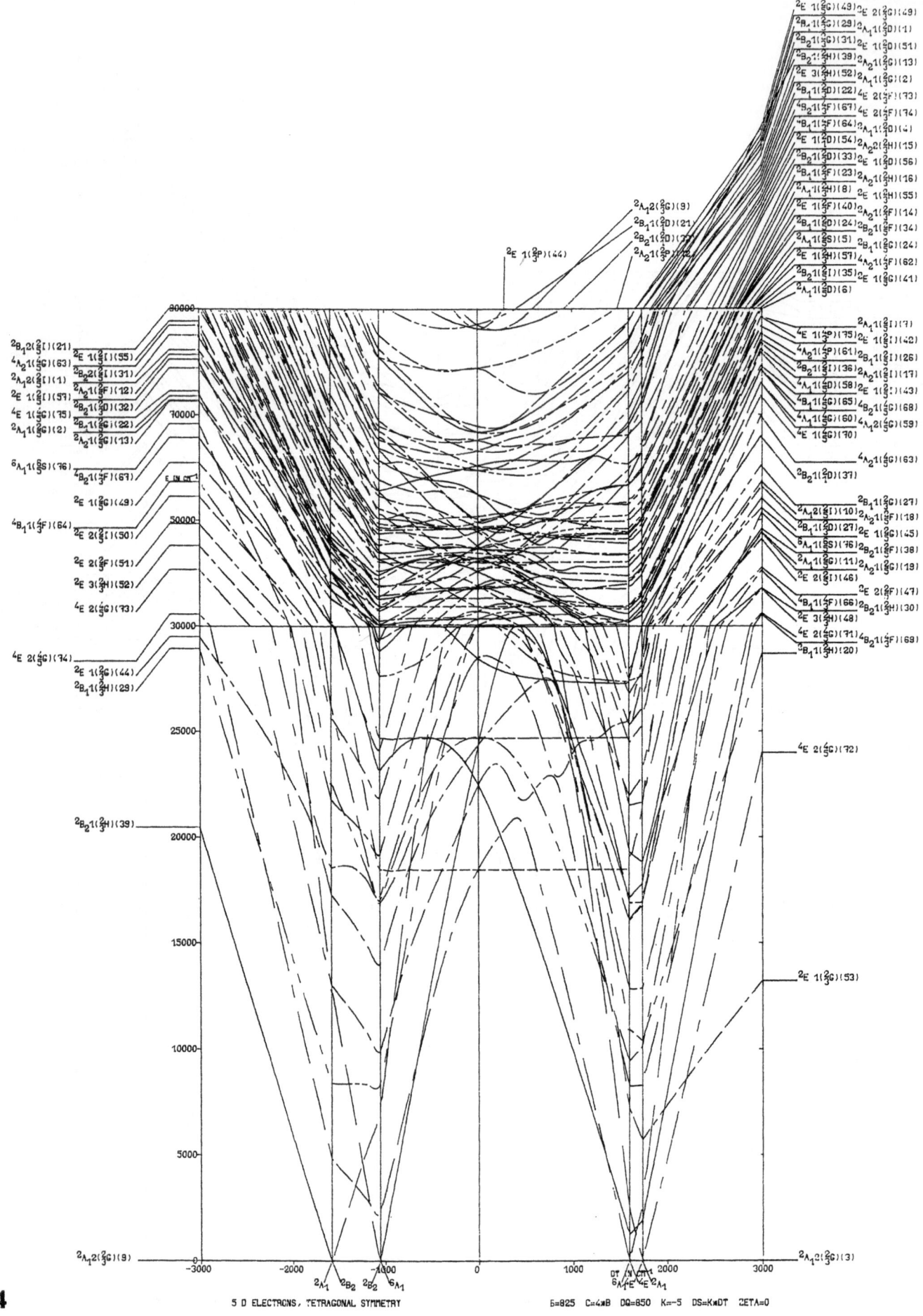

5 D ELECTRONS, TETRAGONAL SYMMETRY

B=825 C=4×B DQ=850 K=-5 DS=K×DT ZETA=0

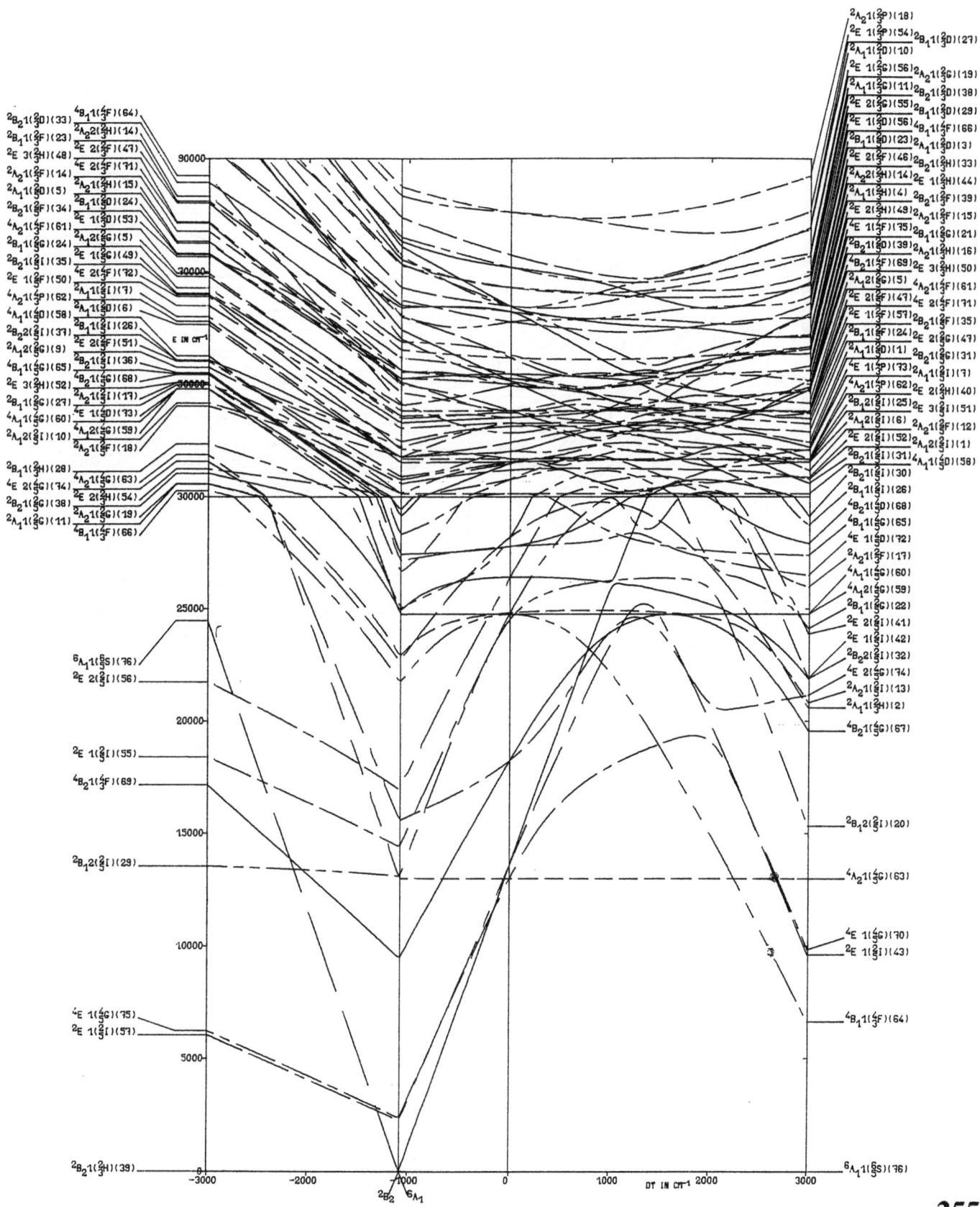

5 D ELECTRONS, TETRAGONAL SYMMETRY   B=825 C=4×B DQ=1430 K=1 DS=K×DT ZETA=0

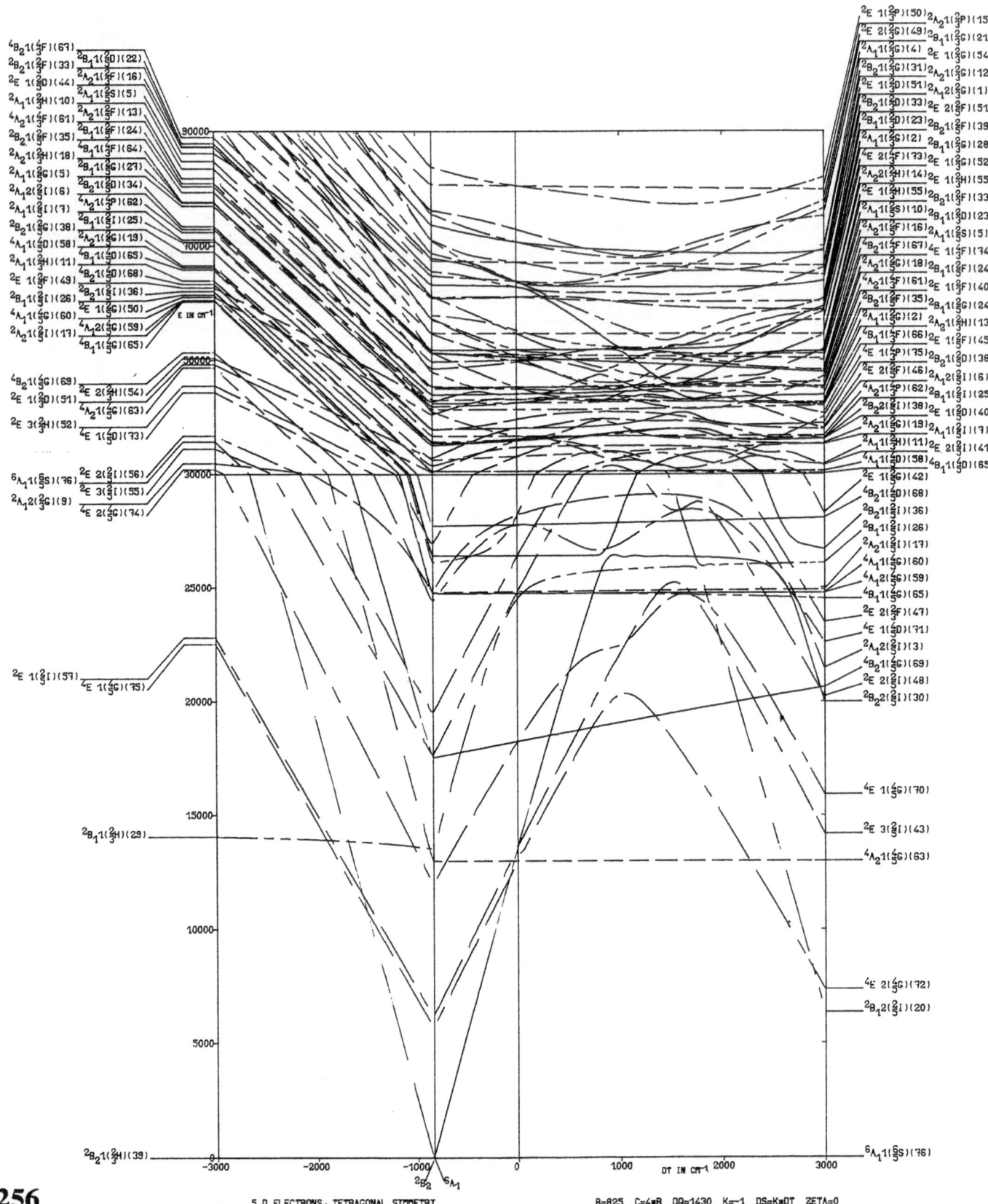

5 D ELECTRONS, TETRAGONAL SYMMETRY

B=825 C=4×B DQ=1430 K=-1 DS=K×DT ZETA=0

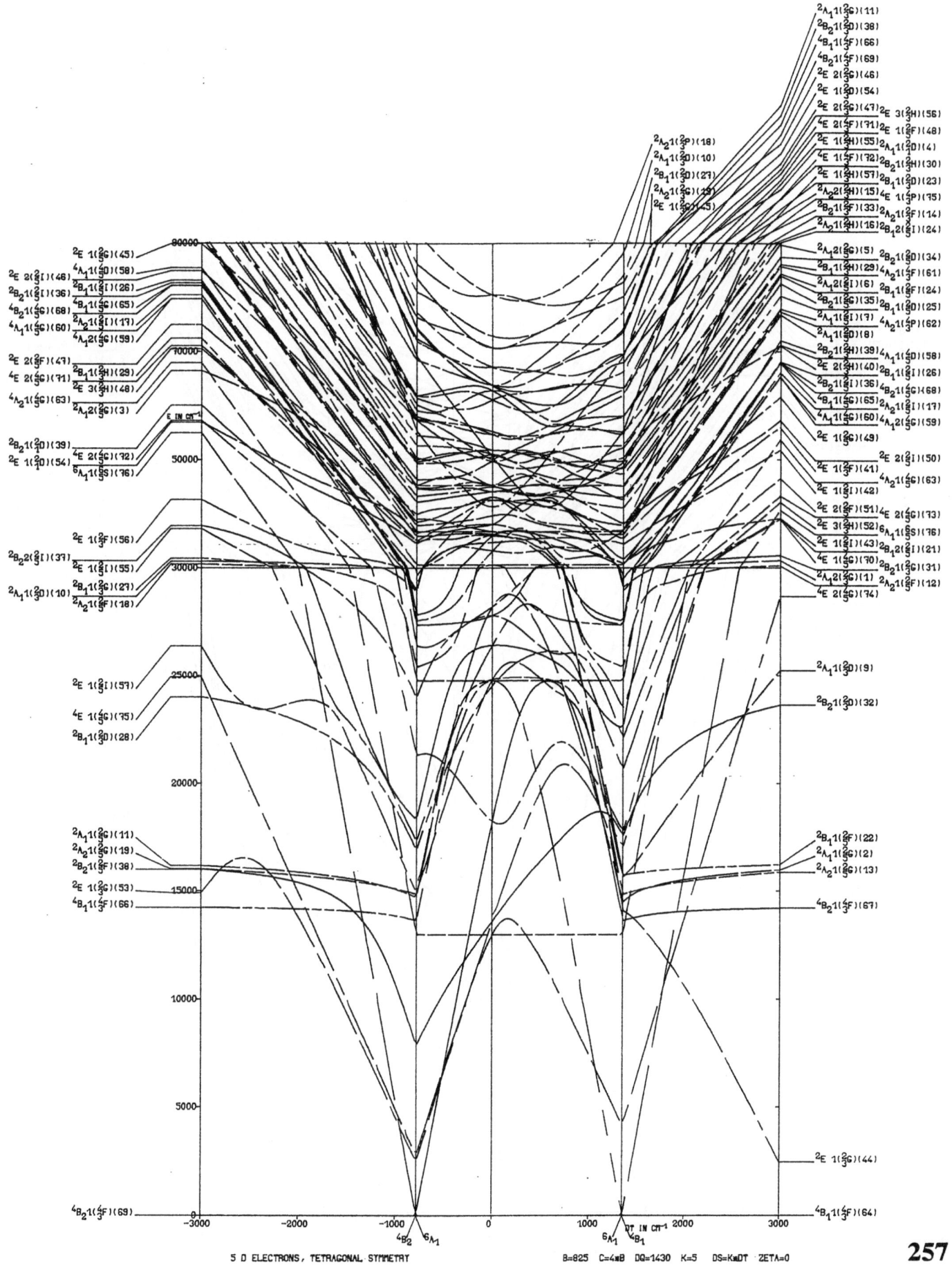

5 D ELECTRONS, TETRAGONAL SYMMETRY   B=825   C=4∗B   DQ=1430   K=5   DS=K∗DT   ZETA=0

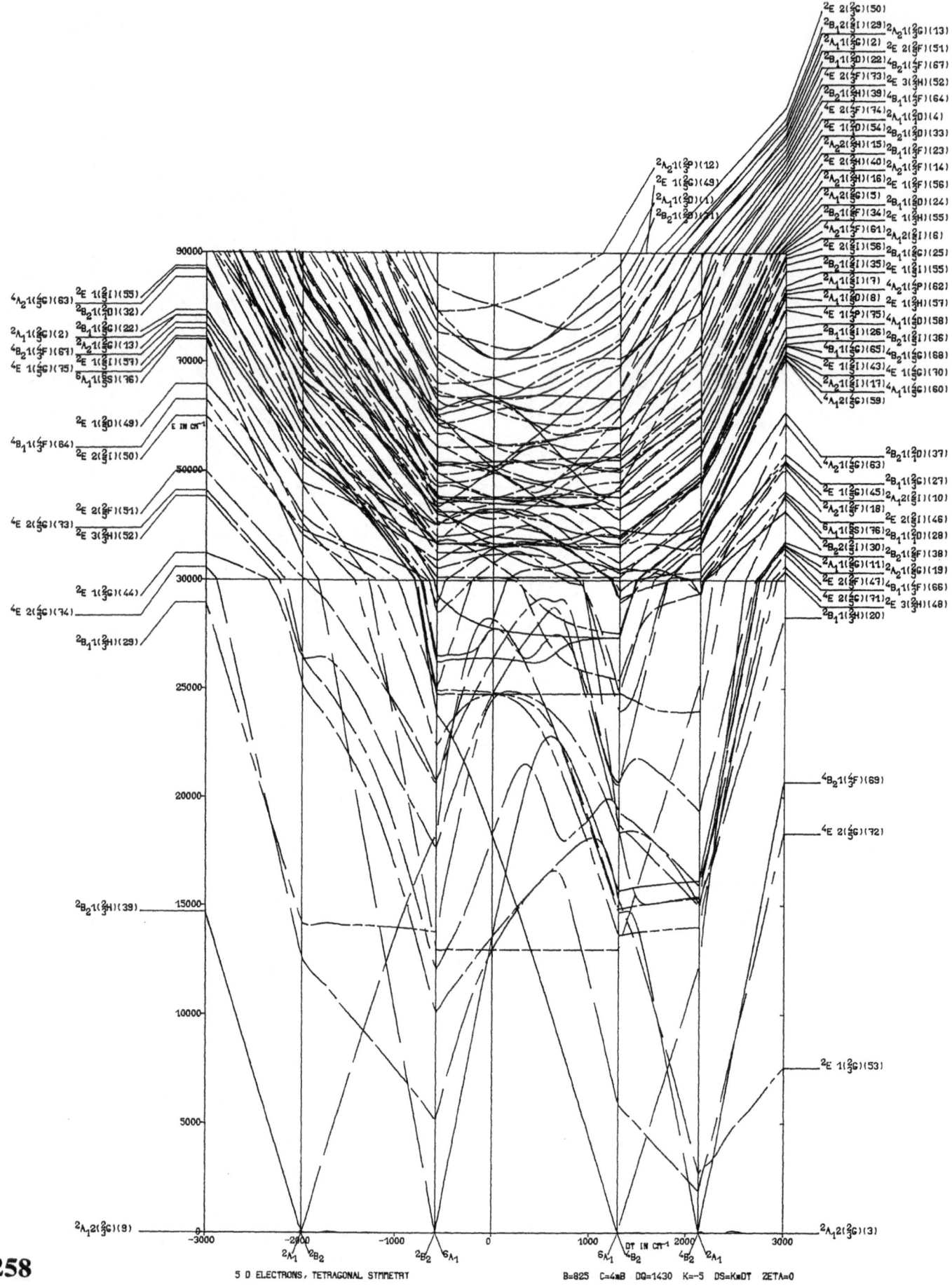

5 D ELECTRONS, TETRAGONAL SYMMETRY   B=825 C=4xB DQ=1430 K=-5 DS=KxDT ZETA=0

B=825 C=4×B ZETA=0

ENERGY AS FUNCTION OF DQ
DT=0, 500, -500, 1000, -1000
K=1, -1, 5, -5

ENERGY AS FUNCTION OF DT
DQ=850, 1430
K=1, -1, 5, -5

(67) $^6A_11(E_EE_EA_1(^4A_2)E_TE_T(^3A_2))$
(66) $^4E\ 1(E_EE_EE_E(^2E\ )E_TE_T(^3A_2))$
(65) $^4E\ 1(E_EE_EE_T(^4E\ )E_TE_T(^1A_1))$
(64) $^4E\ 1(E_EE_EE_E(^2E\ )A_1E_T(^3E\ ))$
(63) $^4E\ 1(E_EE_EA_1(^4A_2)E_TE_T(^1E\ ))$
(62) $^4E\ 1(E_EE_EA_1(^2E\ )E_TE_T(^3A_2))$
(61) $^4E\ 1(A_1E_TE_E(^2E\ )E_TE_T(^3A_2))$
(60) $^4E\ 1(E_EE_EE_T(^4E\ )A_1A_1(^1A_1))$
(59) $^4E\ 1(E_TE_TE_E(^4E\ )A_1A_1(^1A_1))$
(58) $^4A_21(E_EE_EE_E(^2E\ )A_1E_T(^3E\ ))$
(57) $^4A_21(E_EE_EA_1(^4A_2)E_TE_T(^1A_1))$
(56) $^4A_21(E_EE_EA_1(^2A_1)E_TE_T(^3A_2))$
(55) $^4A_21(A_1E_TE_E(^2A_1)E_TE_T(^3A_2))$
(54) $^4A_11(E_EE_EE_E(^2E\ )A_1E_T(^3E\ ))$
(53) $^4A_11(E_EE_EA_1(^4A_2)E_TE_T(^3A_2))$
(52) $^4A_11(E_EE_EA_1(^2A_2)E_TE_T(^3A_2))$
(51) $^4A_11(A_1E_TE_E(^2A_2)E_TE_T(^3A_2))$
(50) $^2E\ 1(E_EE_EE_T(^2A_1)E_EE_E(^1E\ ))$
(49) $^2E\ 1(E_EE_EE_E(^2E\ )E_TE_T(^3A_2))$
(48) $^2E\ 1(E_EE_EE_E(^2E\ )E_TE_T(^1E\ ))$
(47) $^2E\ 1(E_EE_EE_E(^2E\ )E_TE_T(^1A_1))$
(46) $^2E\ 1(E_EE_EE_T(^4E\ )E_TE_T(^3A_2))$
(45) $^2E\ 1(E_EE_EE_T(^2E\ )E_TE_T(^1A_1))$
(44) $^2E\ 1(E_EE_EE_T(^2A_1)E_TE_T(^1E\ ))$
(43) $^2E\ 1(E_TE_TE_E(^2A_1)E_TE_T(^1E\ ))$
(42) $^2E\ 1(E_EE_EE_E(^2E\ )A_1E_T(^3E\ ))$
(41) $^2E\ 1(E_EE_EE_E(^2E\ )A_1E_T(^1E\ ))$
(40) $^2E\ 1(E_EE_EA_1(^2E\ )E_TE_T(^3A_2))$
(39) $^2E\ 1(E_EE_EA_1(^2E\ )E_TE_T(^1E\ ))$
(38) $^2E\ 1(E_EE_EA_1(^2E\ )E_TE_T(^1A_1))$
(37) $^2E\ 1(E_EE_EA_1(^2A_2)E_TE_T(^1E\ ))$
(36) $^2E\ 1(E_EE_EA_1(^2A_1)E_TE_T(^1E\ ))$
(35) $^2E\ 1(A_1E_TE_E(^2E\ )E_TE_T(^3A_2))$

(34) $^2E\ 1(A_1E_TE_E(^2A_1)E_TE_T(^1E\ ))$
(33) $^2E\ 1(E_EE_EE_E(^2E\ )A_1A_1(^1A_1))$
(32) $^2E\ 3(E_EE_EE_T(^2E\ )A_1A_1(^1A_1))$
(31) $^2E\ 2(E_EE_EE_T(^2E\ )A_1A_1(^1A_1))$
(30) $^2E\ 1(E_EE_EE_T(^2E\ )A_1A_1(^1A_1))$
(29) $^2E\ 3(E_TE_TE_E(^2E\ )A_1A_1(^1A_1))$
(28) $^2E\ 2(E_TE_TE_E(^2E\ )A_1A_1(^1A_1))$
(27) $^2E\ 1(E_TE_TE_E(^2E\ )A_1A_1(^1A_1))$
(26) $^2E\ 1(E_TE_TE_T(^2E\ )A_1A_1(^1A_1))$
(25) $^2A_21(E_EE_EE_E(^2E\ )E_TE_T(^1E\ ))$
(24) $^2A_21(E_EE_EE_T(^2A_2)E_TE_T(^1A_1))$
(23) $^2A_22(E_EE_EE_E(^2E\ )A_1E_T(^3E\ ))$
(22) $^2A_22(E_EE_EE_E(^2E\ )A_1E_T(^1E\ ))$
(21) $^2A_21(E_EE_EA_1(^2A_1)E_TE_T(^3A_2))$
(20) $^2A_21(E_EE_EA_1(^2E\ )E_TE_T(^1E\ ))$
(19) $^2A_21(E_EE_EA_1(^2A_2)E_TE_T(^1A_1))$
(18) $^2A_21(A_1E_TE_E(^2A_1)E_TE_T(^3A_2))$
(17) $^2A_21(A_1E_TE_E(^2A_2)E_TE_T(^1A_1))$
(16) $^2A_21(E_EE_EE_T(^2A_2)A_1A_1(^1A_1))$
(15) $^2A_21(E_TE_TE_E(^2A_2)A_1A_1(^1A_1))$
(14) $^2A_11(E_EE_EE_E(^2E\ )E_TE_T(^1E\ ))$
(13) $^2A_11(E_EE_EE_T(^2A_1)E_TE_T(^1A_1))$
(12) $^2A_11(E_EE_EA_1(^2A_1)E_EE_E(^1A_1))$
(11) $^2A_11(E_EE_EE_E(^2E\ )A_1E_T(^3E\ ))$
(10) $^2A_11(E_EE_EE_E(^2E\ )A_1E_T(^1E\ ))$
(9) $^2A_11(E_EE_EA_1(^4A_2)E_TE_T(^3A_2))$
(8) $^2A_11(E_EE_EA_1(^2A_2)E_TE_T(^3A_2))$
(7) $^2A_11(E_EE_EA_1(^2E\ )E_TE_T(^1E\ ))$
(6) $^2A_11(E_EE_EA_1(^2A_1)E_TE_T(^1A_1))$
(5) $^2A_11(A_1E_TE_E(^2A_2)E_TE_T(^3A_2))$
(4) $^2A_11(A_1E_TE_E(^2A_1)E_TE_T(^1A_1))$
(3) $^2A_11(E_TE_TA_1(^2A_1)E_TE_T(^1A_1))$
(2) $^2A_11(E_EE_EE_T(^2A_1)A_1A_1(^1A_1))$
(1) $^2A_11(E_TE_TE_E(^2A_1)A_1A_1(^1A_1))$

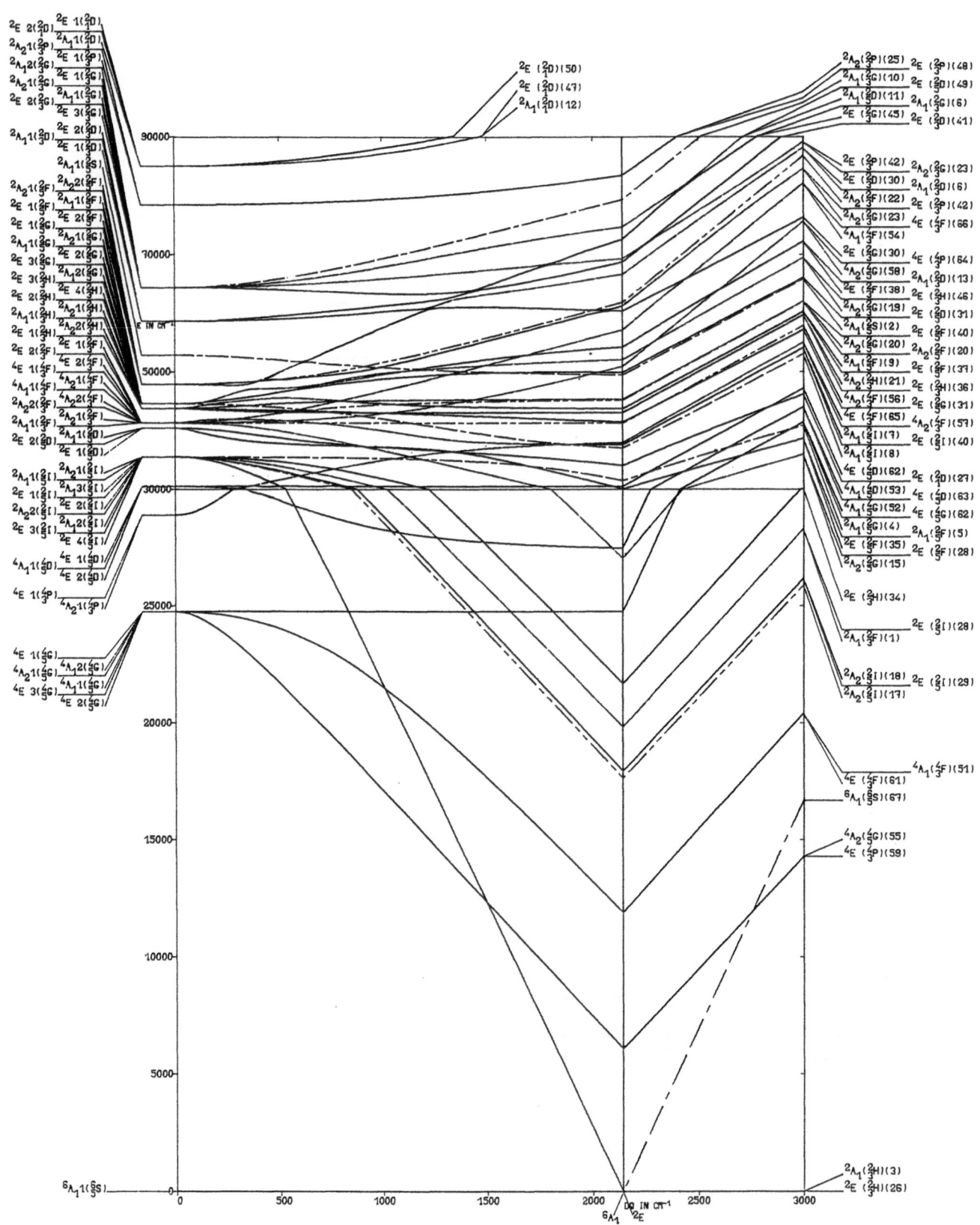

$^2E\ 2(^2D)$ $^2E\ 1(^2D)$
$^2A_21(^2P)$ $^2A_11(^2D)$
$^2A_12(^2G)$ $^2E\ 1(^2P)$
$^2A_21(^2G)$ $^2E\ 1(^2G)$
$^2E\ 2(^2G)$ $^2A_11(^2G)$
   $^2E\ 3(^2G)$
$^2A_11(^2D)$ $^2E\ 1(^2D)$

   $^2A_11(^2S)$

$^2A_21(^2F)$ $^2A_22(^2F)$
$^2E\ 1(^2F)$ $^2A_11(^2F)$
$^2E\ 1(^2G)$ $^2E\ 2(^2G)$
$^2A_11(^2G)$ $^2E\ 2(^2G)$
$^2E\ 3(^2H)$ $^2A_12(^2G)$
$^2E\ 2(^2H)$ $^2E\ 4(^2H)$
$^2A_11(^2H)$ $^2A_21(^2H)$
$^2E\ 1(^2H)$ $^2A_22(^2H)$
$^2E\ 2(^2F)$ $^2E\ 1(^2F)$
$^4E\ 1(^4F)$ $^2E\ 2(^2F)$
$^4A_11(^4F)$ $^2A_21(^2F)$
$^2A_22(^2F)$ $^2A_22(^2F)$
$^2A_11(^2F)$ $^2A_21(^2F)$
$^2E\ 2(^2D)$ $^2A_11(^2D)$
   $^2E\ 1(^2D)$

$^2A_11(^2I)$ $^2A_22(^2I)$
$^2E\ 1(^2I)$ $^2A_23(^2I)$
$^2A_22(^2I)$ $^2E\ 2(^2I)$
$^2E\ 3(^2I)$ $^2A_12(^2I)$
   $^2E\ 4(^2I)$

$^4E\ 1(^4D)$ $^4E\ 2(^4D)$
$^4E\ 1(^4P)$ $^4A_21(^4P)$

$^4E\ 1(^4G)$
$^4A_21(^4G)$ $^4A_12(^4G)$
$^4E\ 3(^4G)$ $^4A_11(^4G)$
   $^4E\ 2(^4G)$

$^6A_11(^6S)$

$^2A_2(^2D)(50)$
$^2E\ (^2D)(47)$
$^2A_1(^2D)(12)$

$^2A_2(^2P)(25)$ $^2E\ (^2P)(48)$
$^2A_1(^2G)(10)$ $^2E\ (^2D)(49)$
$^2A_1(^2D)(11)$ $^2A_1(^2G)(6)$
$^2E\ (^2G)(45)$ $^2E\ (^2D)(41)$

$^2E\ (^2P)(42)$ $^2A_2(^2G)(23)$
$^2E\ (^2D)(30)$ $^2A_1(^2D)(6)$
$^2A_2(^2F)(22)$ $^2E\ (^2F)(42)$
$^2A_2(^2G)(23)$ $^4A_1(^4F)(66)$
$^4A_1(^4F)(54)$
$^2E\ (^2F)(30)$ $^4E\ (^4P)(64)$
$^4E\ (^4G)(58)$ $^2A_1(^2D)(13)$
$^2E\ (^2F)(38)$ $^2E\ (^2H)(46)$
$^2A_2(^2G)(19)$ $^2E\ (^2D)(31)$
$^2A_1(^2S)(2)$ $^2E\ (^2F)(40)$
$^2A_2(^2G)(20)$ $^2A_2(^2F)(20)$
$^2A_1(^2F)(9)$ $^2E\ (^2F)(37)$
$^2A_1(^2F)(9)$ $^2E\ (^2H)(36)$
$^4A_2(^4F)(56)$ $^2E\ (^2G)(31)$
$^4E\ (^4F)(65)$ $^4A_2(^4F)(57)$
$^2A_1(^2I)(7)$ $^2E\ (^2I)(40)$
$^2A_1(^2I)(8)$
$^4E\ (^4D)(62)$ $^2E\ (^2D)(27)$
$^4A_1(^4D)(53)$ $^4E\ (^4D)(63)$
$^4A_1(^4G)(52)$ $^4E\ (^4G)(62)$
$^2A_2(^2G)(4)$ $^2A_1(^2F)(5)$
$^2E\ (^2F)(35)$ $^2E\ (^2F)(28)$
$^2A_2(^2G)(15)$

$^2E\ (^2H)(34)$

$^2E\ (^2I)(28)$
$^2A_1(^2F)(1)$

$^2A_2(^2I)(18)$ $^2E\ (^2I)(29)$
$^2A_2(^2I)(17)$

$^4E\ (^4F)(61)$ $^4A_1(^4F)(51)$
$^6A_1(^6S)(67)$
$^4A_2(^4G)(55)$
$^4E\ (^4P)(59)$

$^2A_1(^2H)(3)$
$^2E\ (^2H)(26)$

$E\ IN\ CM^{-1}$

90000

70000

50000

30000

25000

20000

15000

10000

5000

0

Dq IN CM^{-1}

$^6A_11$
^2E

0   500   1000   1500   2000   2500   3000

5 D ELECTRONS, TRIGONAL SYMMETRY

B=825   C=4×B   DT=0   K=0   DS=K×DT   ZETA=0

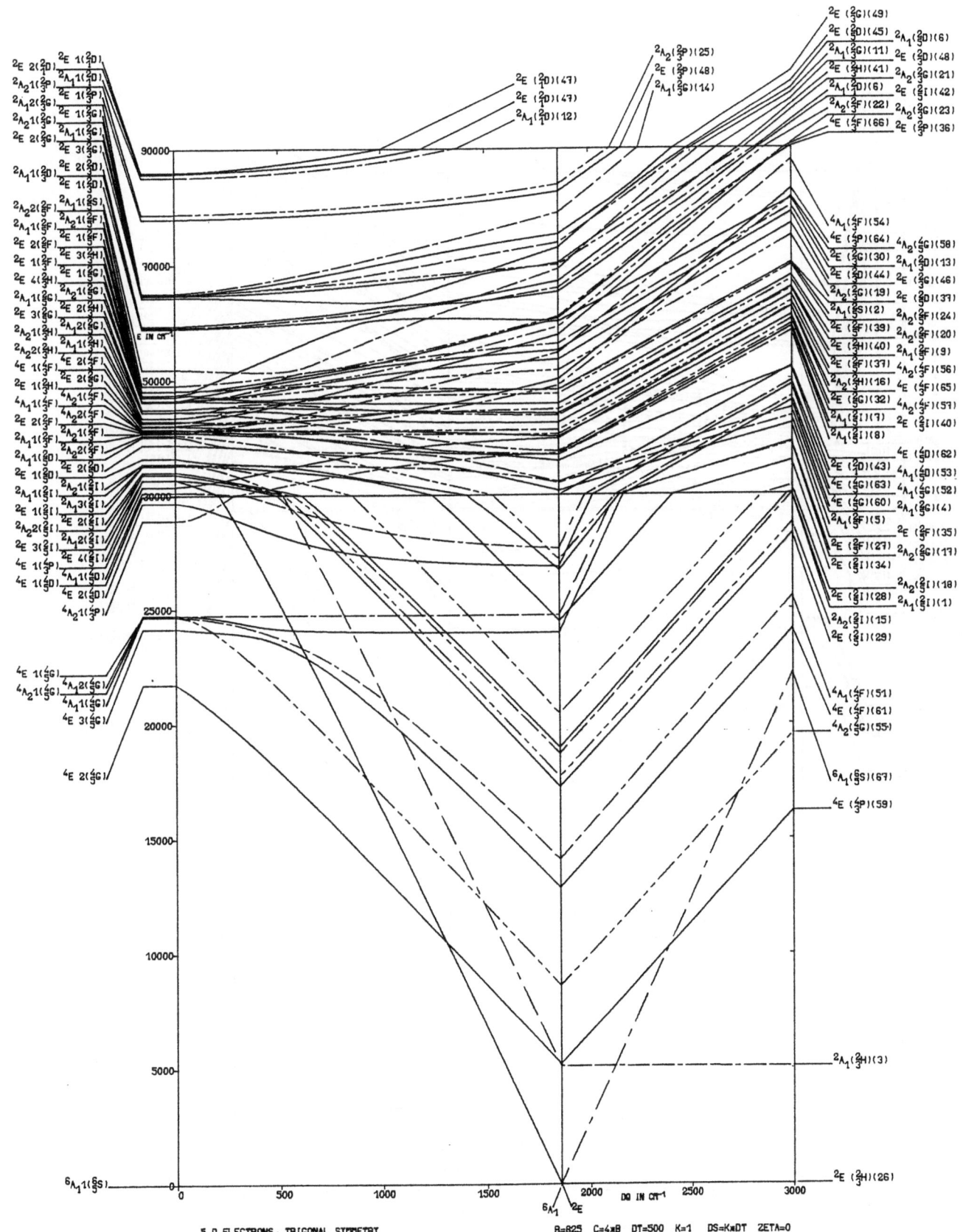

5 D ELECTRONS, TRIGONAL SYMMETRY

B=825 C=4×B DT=500 K=1 DS=K×DT ZETA=0

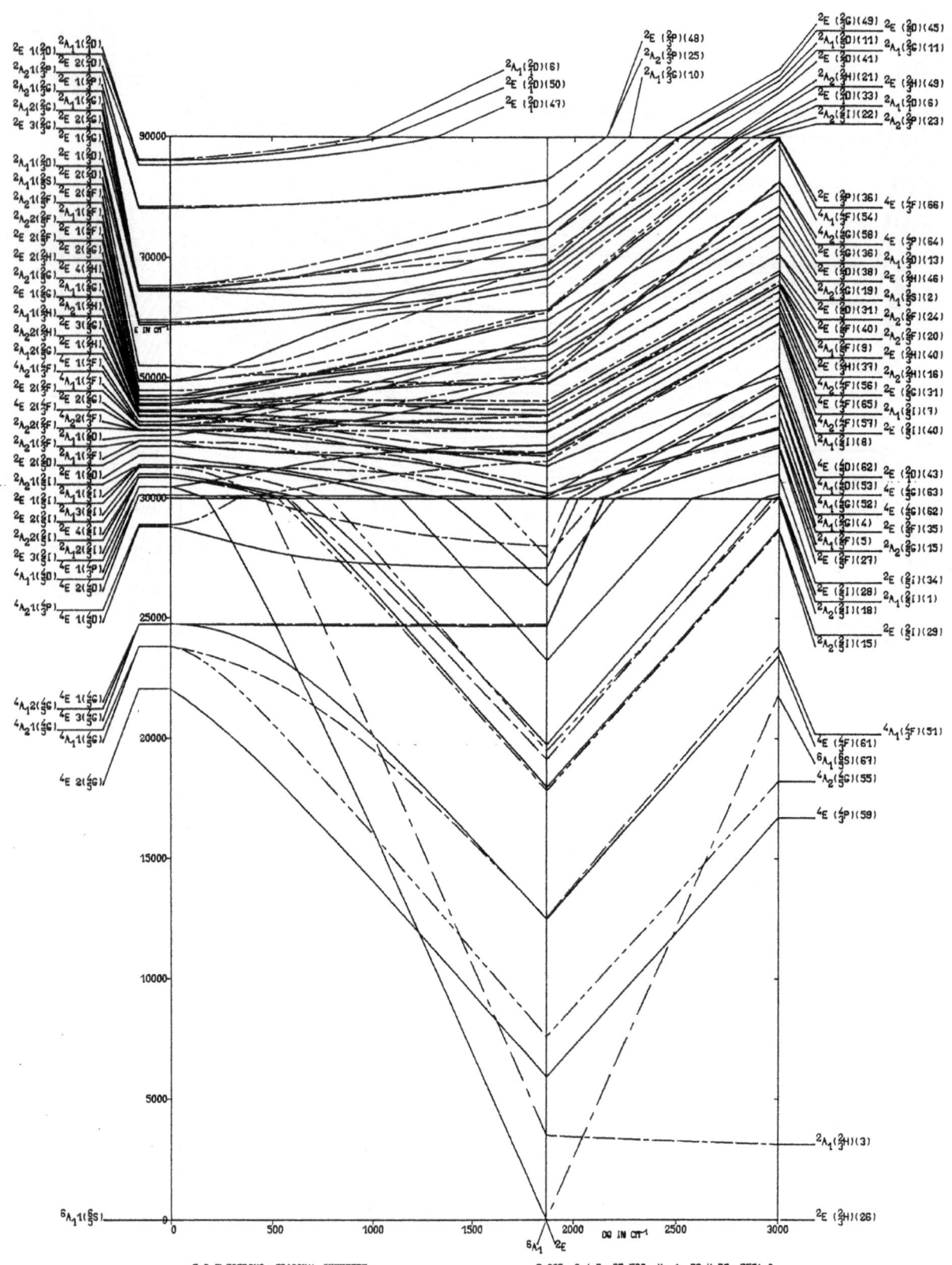

262

5 D ELECTRONS, TRIGONAL SYMMETRY

B=825 C=4×B DT=500 K=-1 DS=K×DT ZETA=0

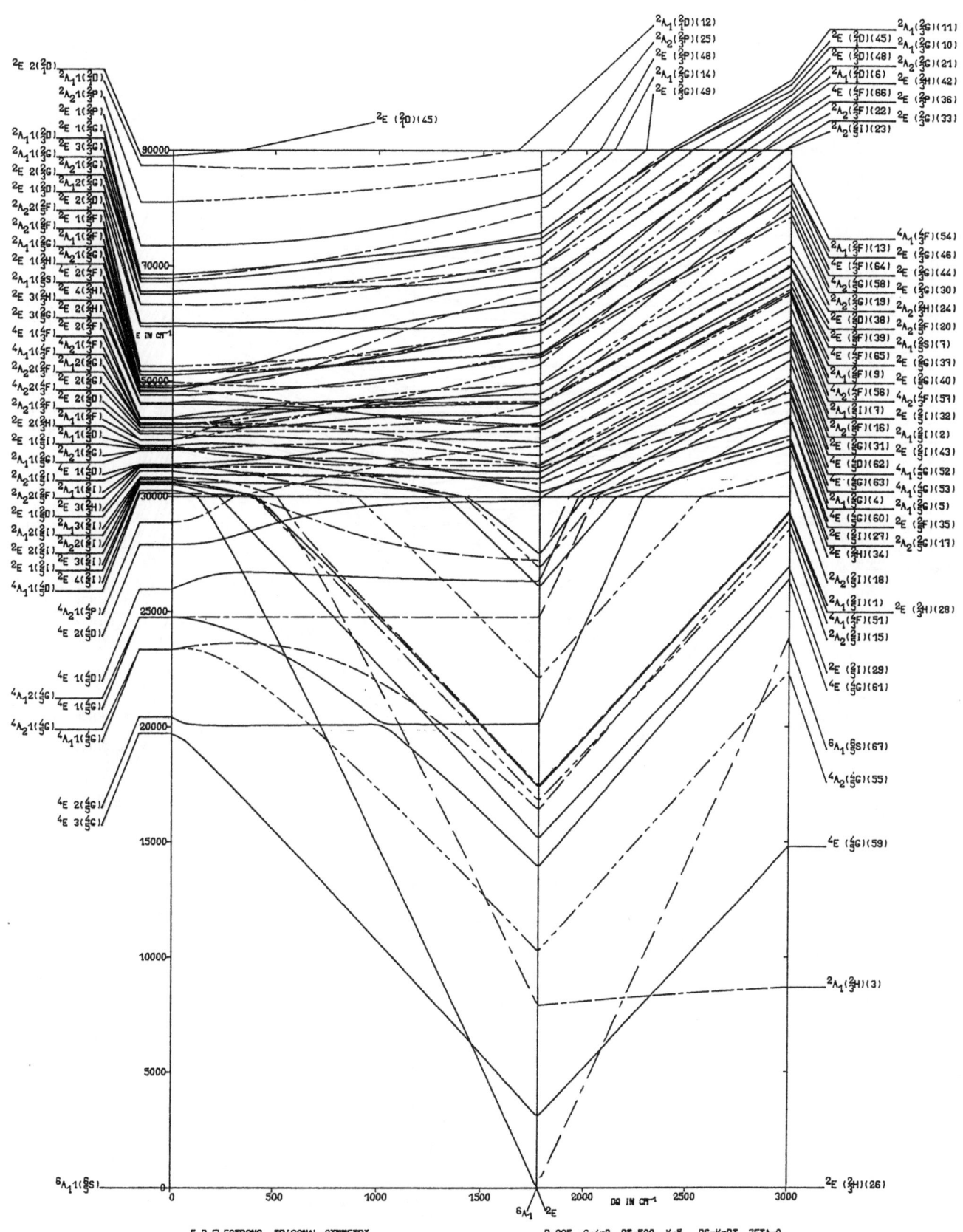

5 D ELECTRONS, TRIGONAL SYMMETRY   B=825 C=4×B DT=500 K=5   DS=K×DT 2ETA=0

263

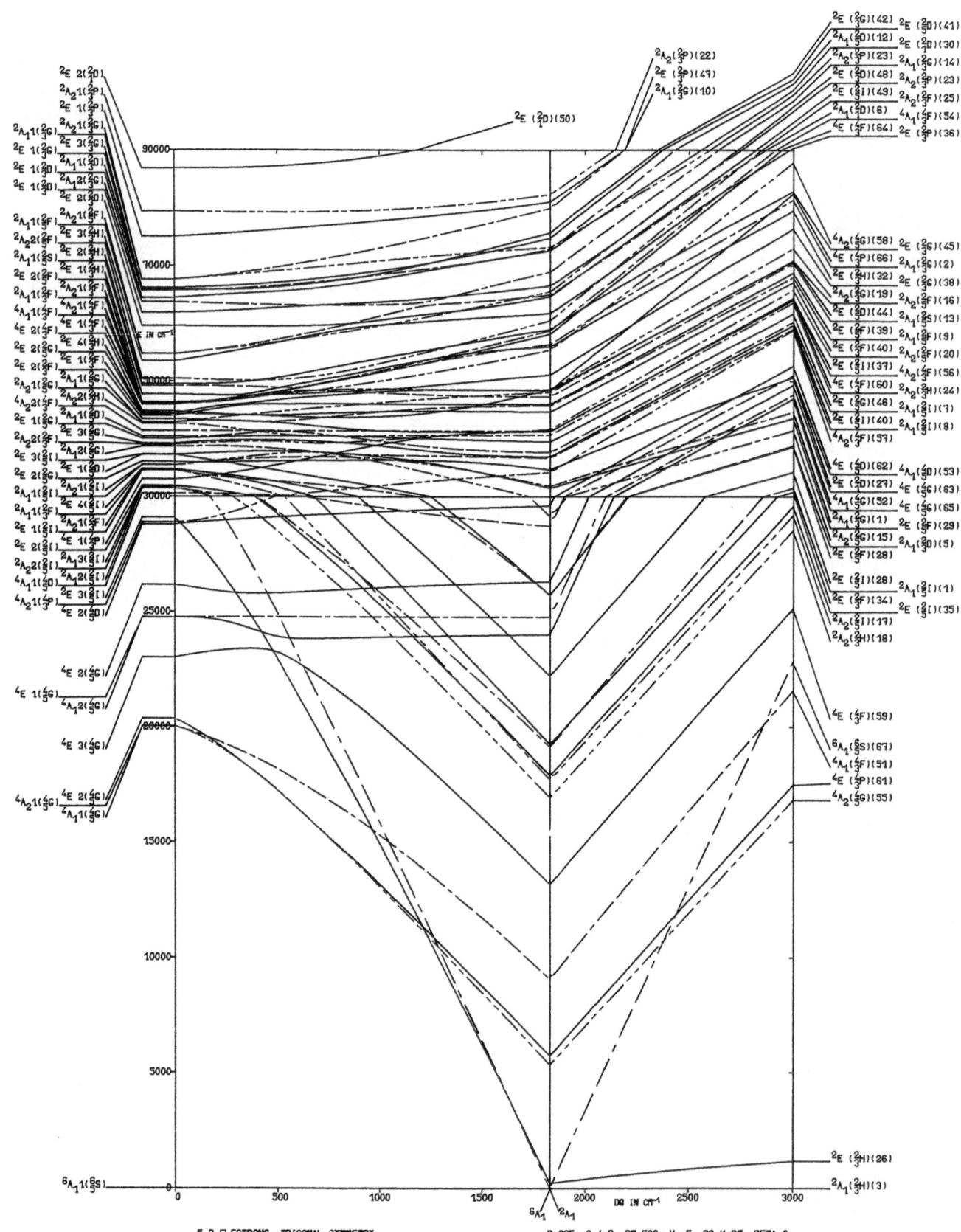

5 D ELECTRONS, TRIGONAL SYMMETRY

B=825 C=4×B DT=500 K=-5 DS=K×DT ZETA=0

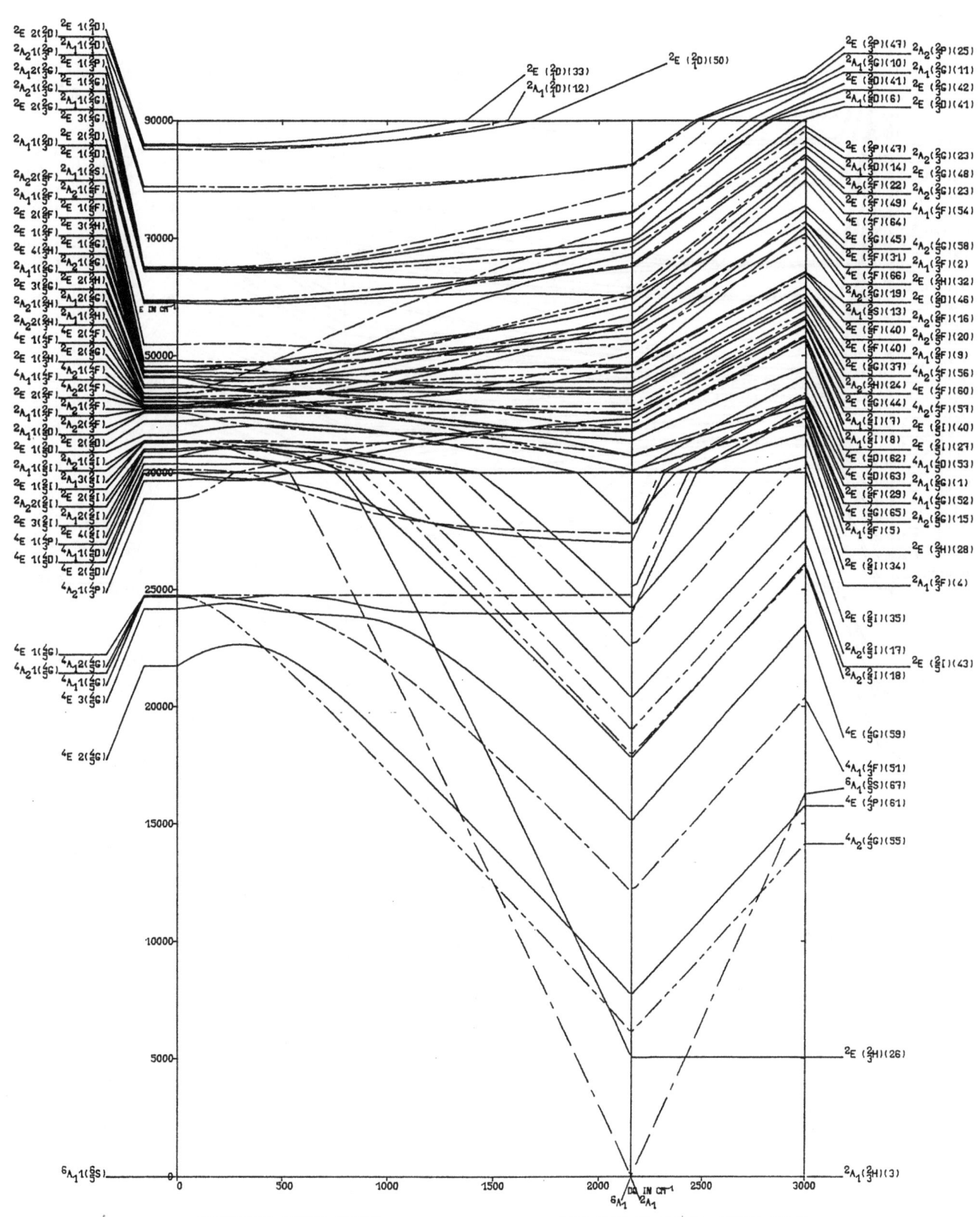

5 D ELECTRONS, TRIGONAL SYMMETRY   B=825   C=4*B   DT=-500   K=1   DS=K*DT   ZETA=0

265

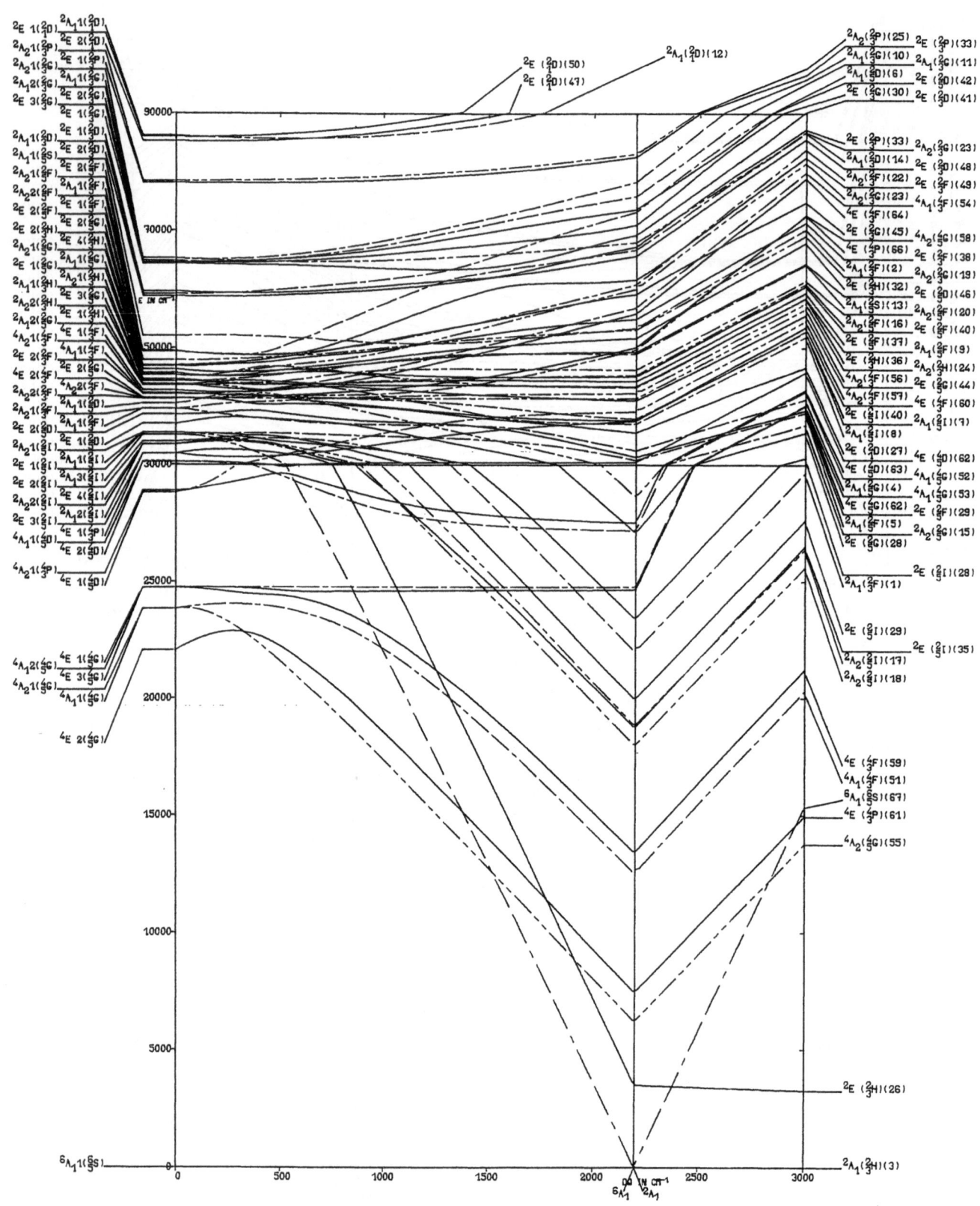

5 D ELECTRONS, TRIGONAL SYMMETRY

B=825   C=4×B   DT=-500   K=-1   DS=K×DT   ZETA=0

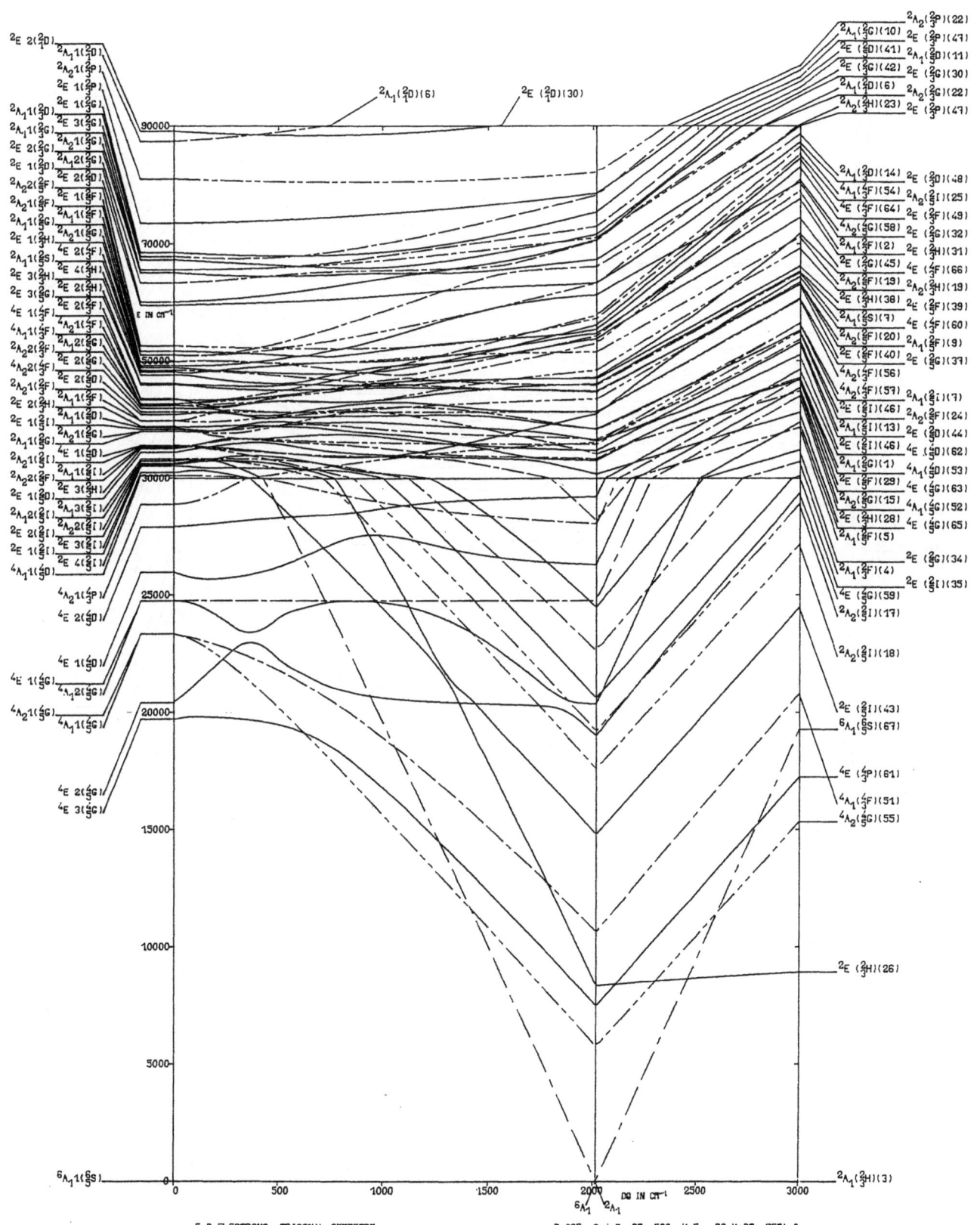

5 D ELECTRONS, TRIGONAL SYMMETRY   B=825 C=4×B DT=-500 K=5 DS=K×DT ZETA=0

267

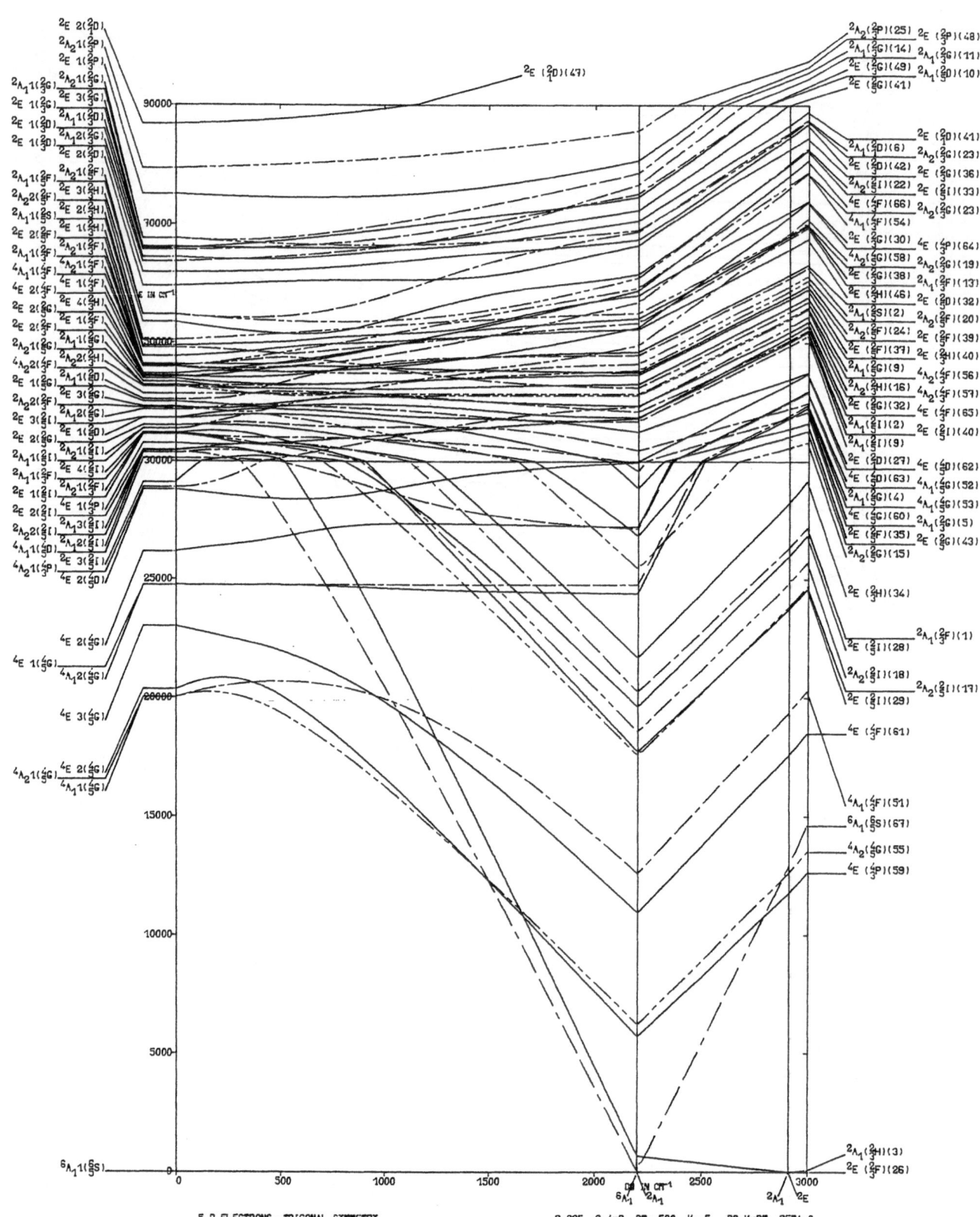

5 D ELECTRONS, TRIGONAL SYMMETRY

B=825   C=4×B   DT=-500   K=-5   DS=K×DT   ZETA=0

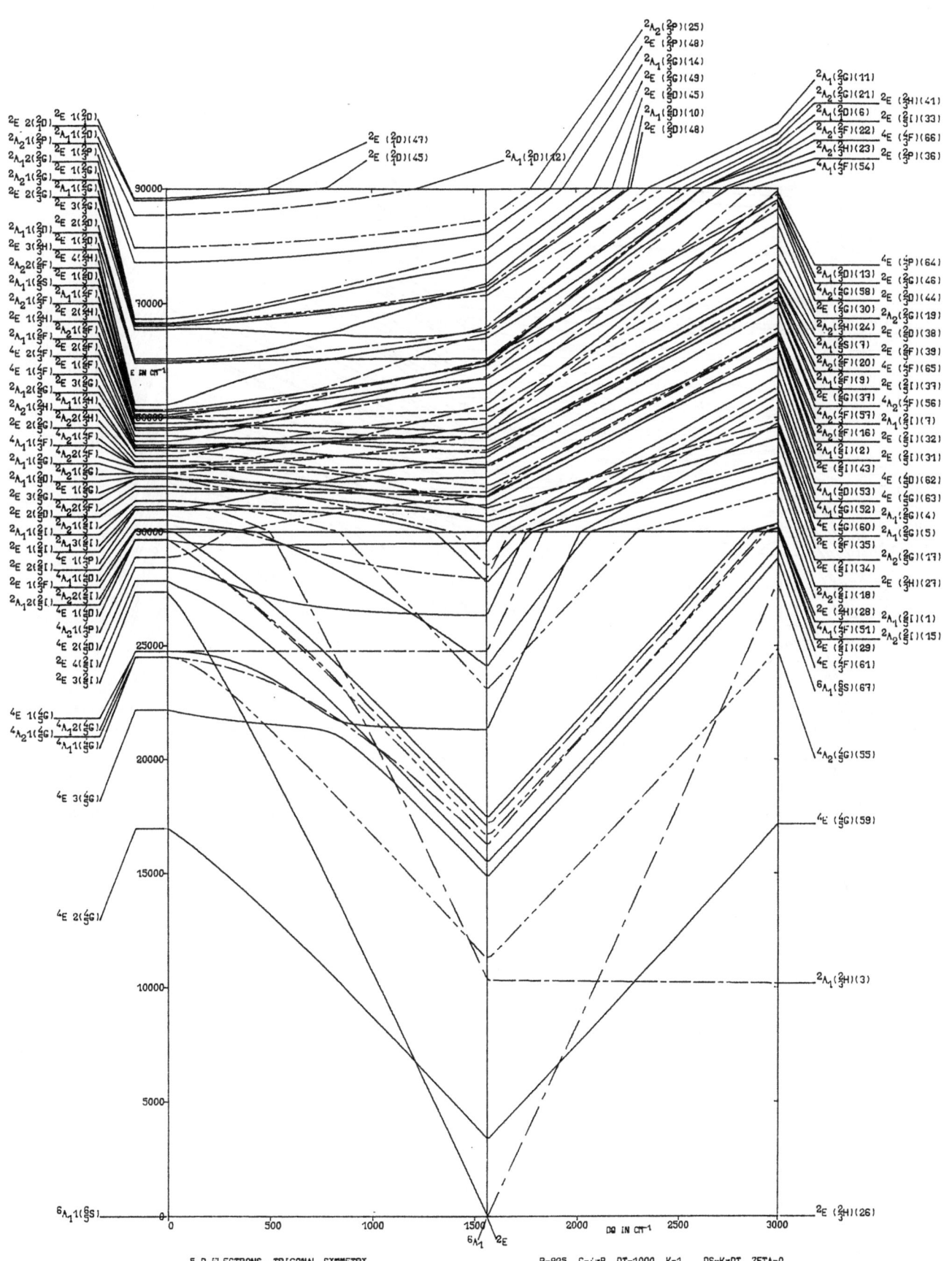

5 D ELECTRONS, TRIGONAL SYMMETRY   B=825 C=4×B DT=1000 K=1   DS=K×DT ZETA=0

269

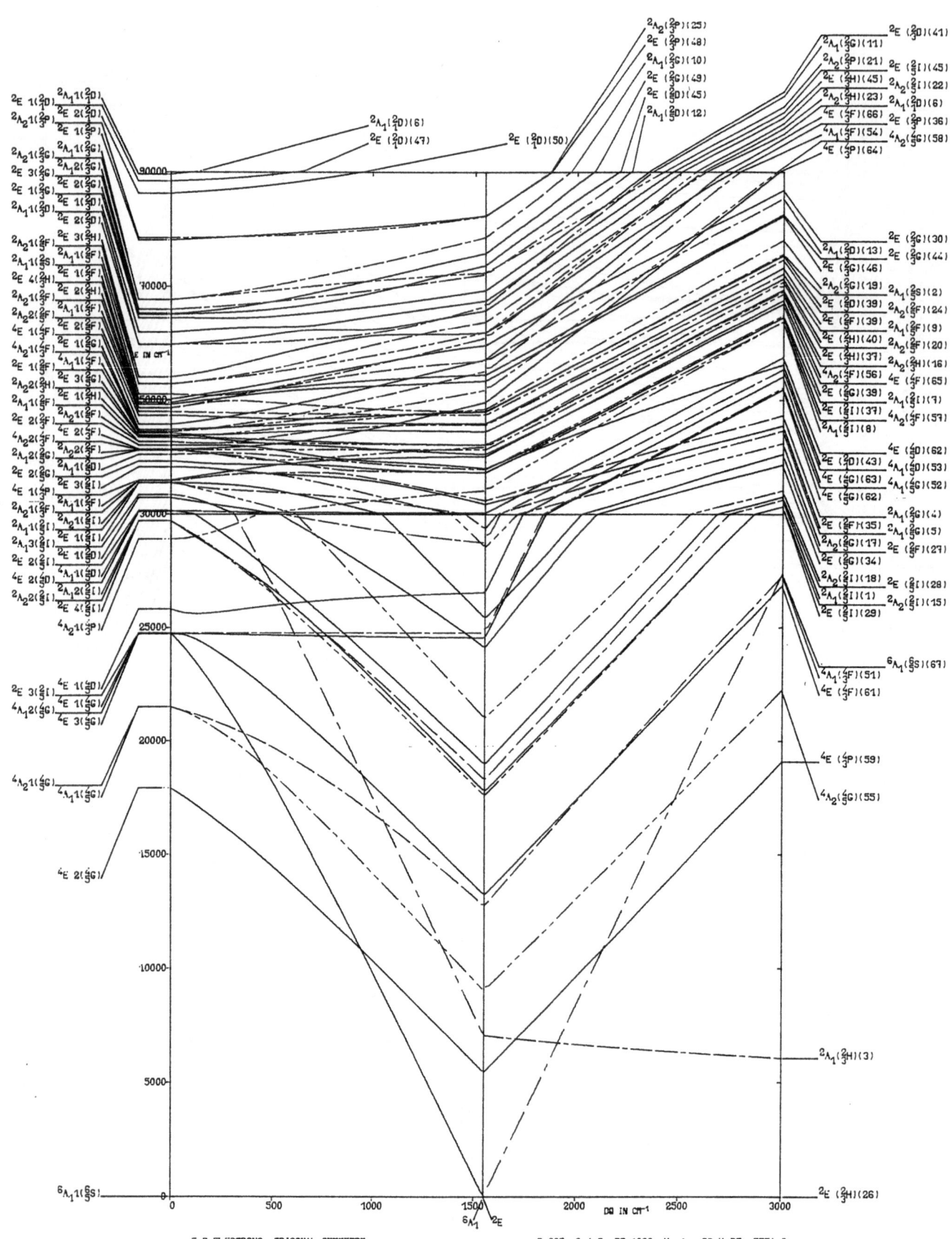

270

5 D ELECTRONS, TRIGONAL SYMMETRY

B=825 C=4×B DT=1000 K=-1 DS=K×DT ZETA=0

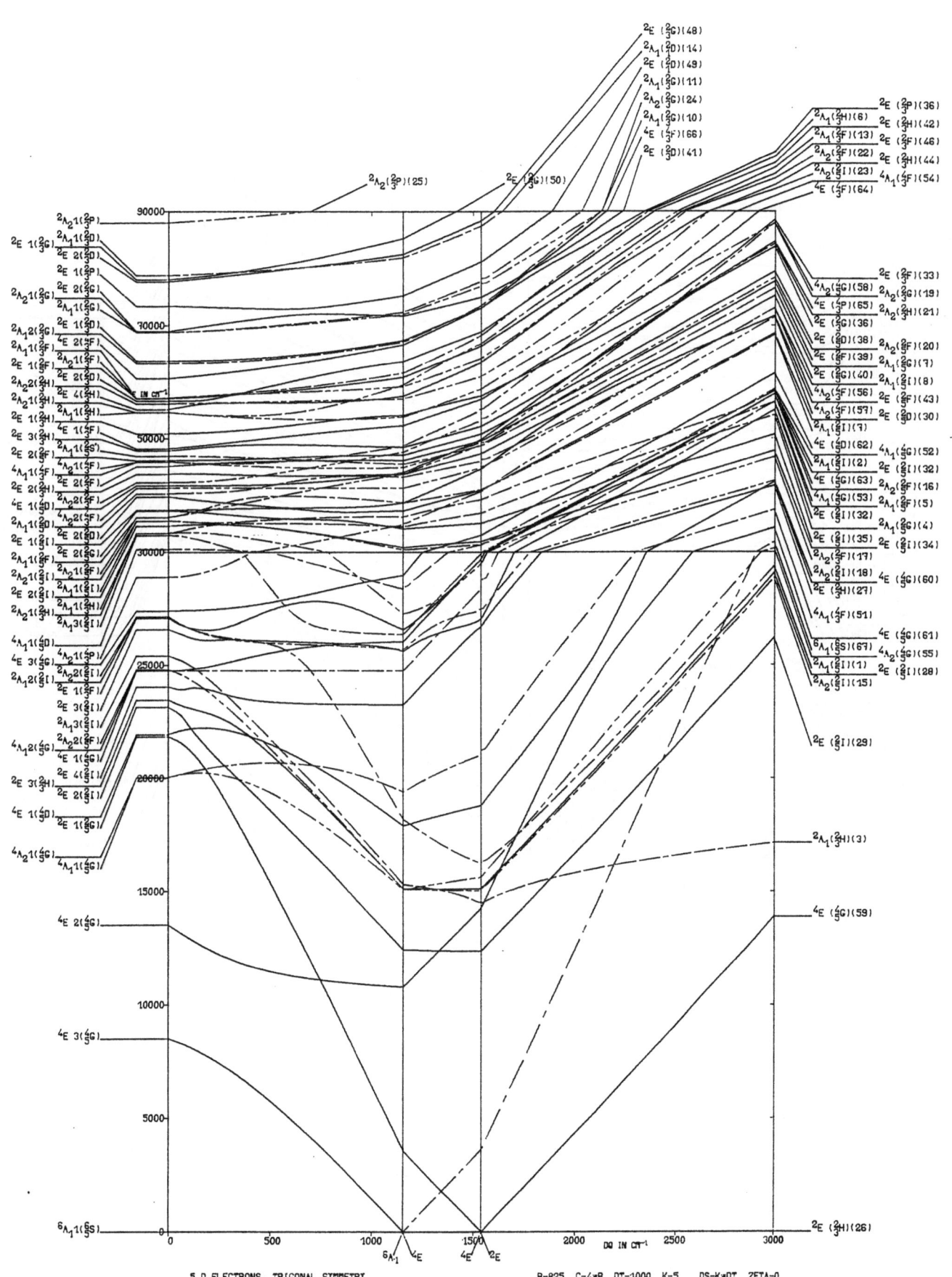

5 D ELECTRONS, TRIGONAL SYMMETRY   B=825 C=4×B DT=1000 K=5   DS=K×DT ZETA=0

271

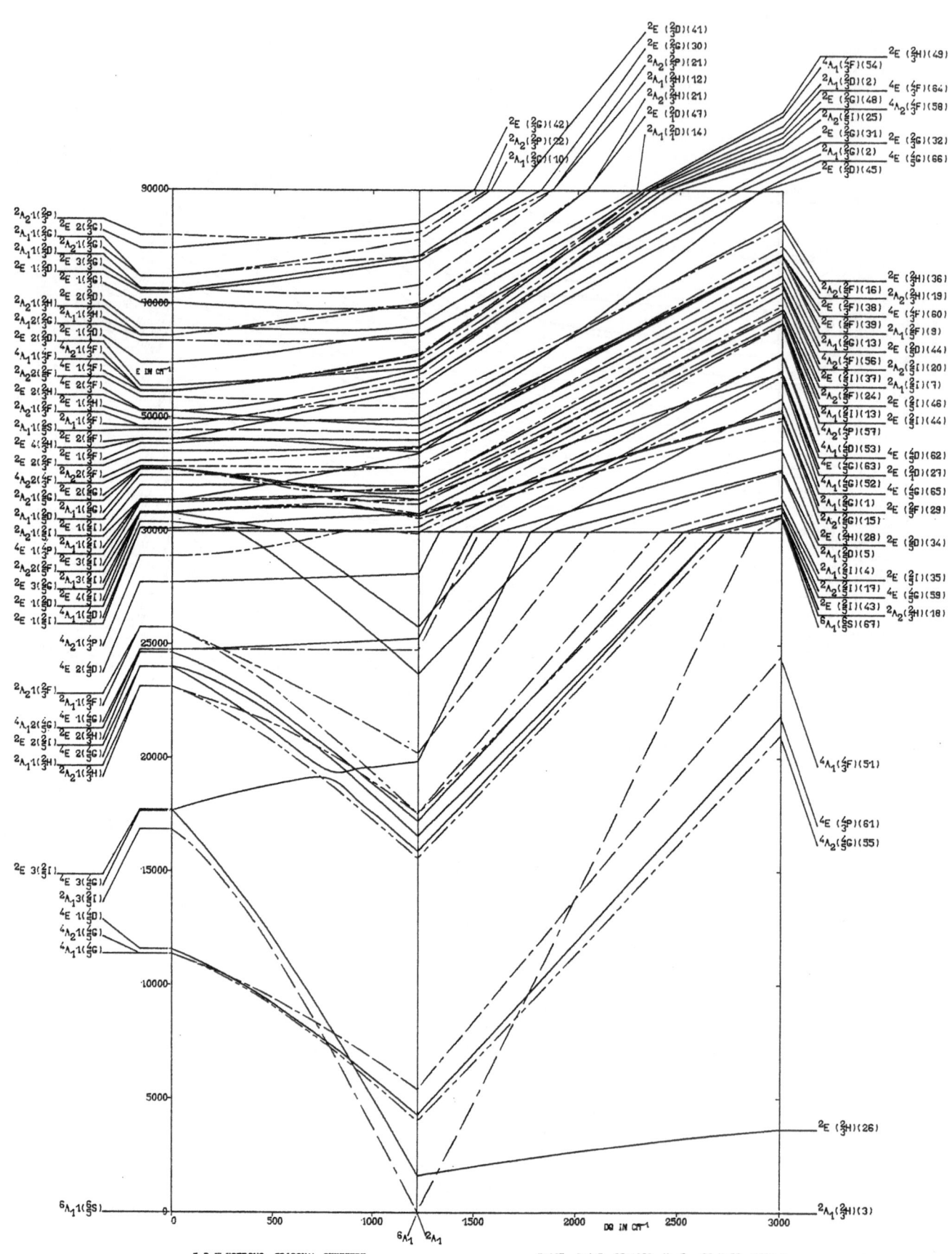

5 D ELECTRONS, TRIGONAL SYMMETRY

B=825   C=4×B   DT=1000   K=-5   DS=K×DT   ZETA=0

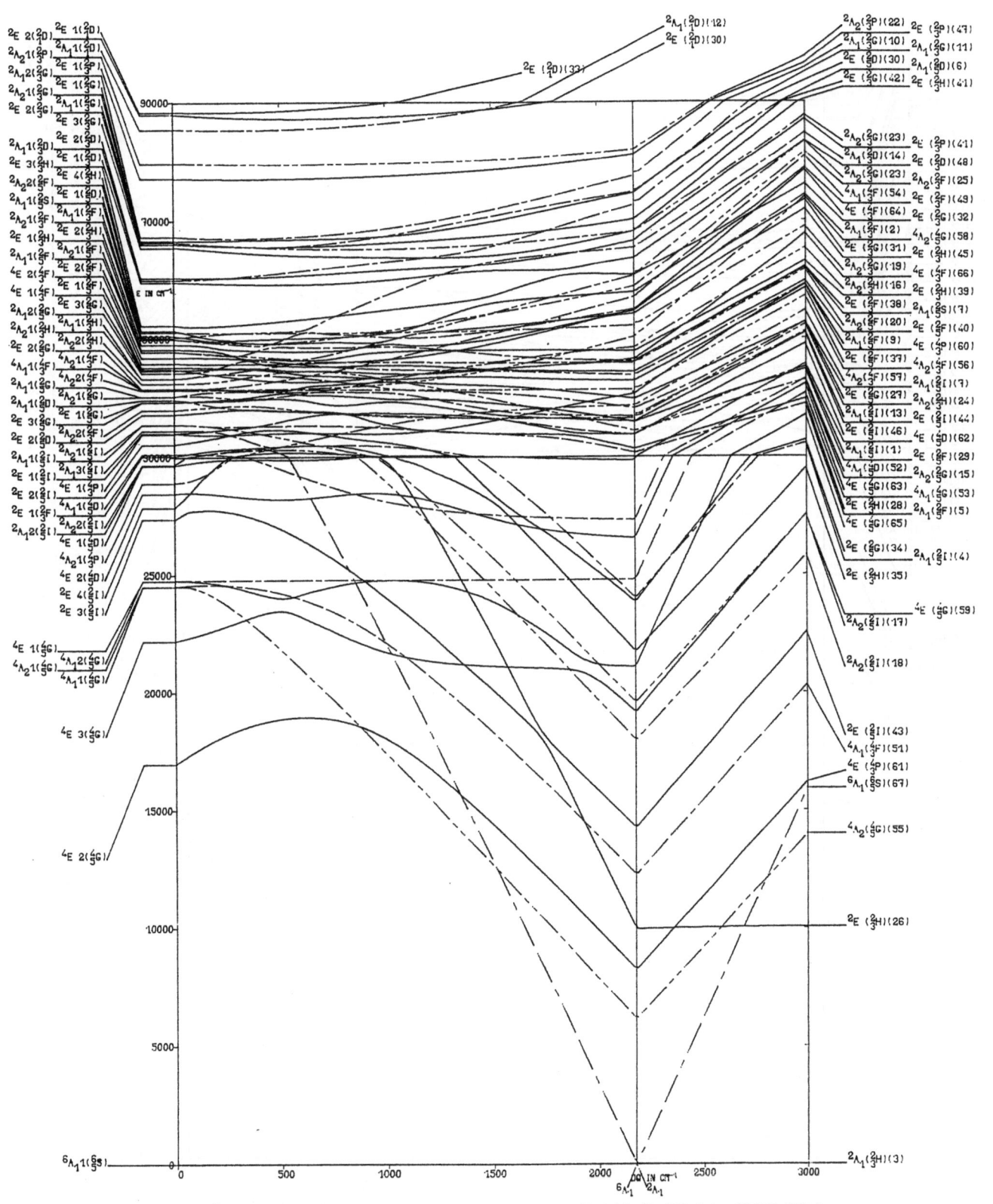

5 Ô ELECTRONS, TRIGONAL SYMMETRY   B=825 C=4×B DT=-1000 K=1   DS=K×DT ZETA=0

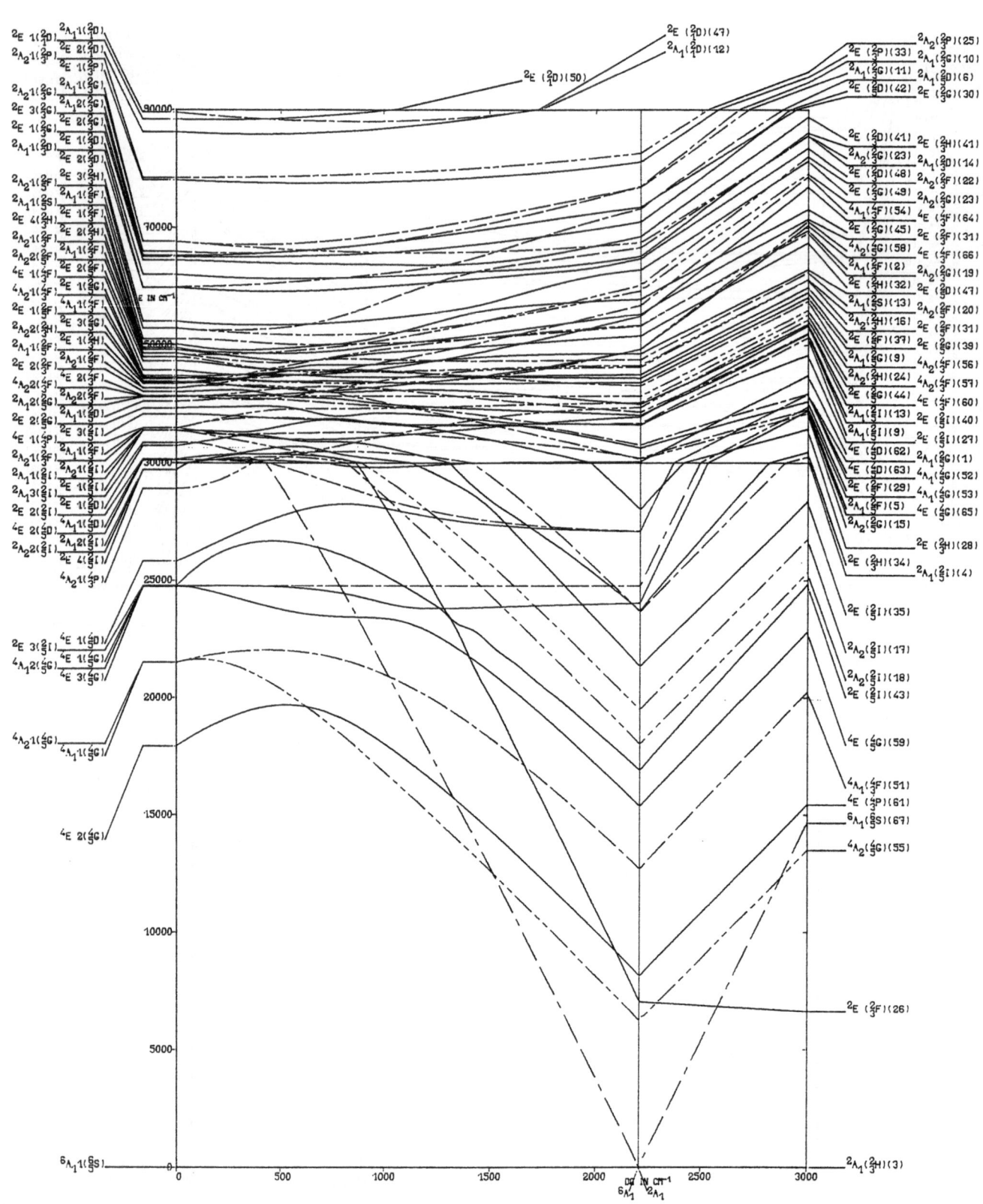

274

5 D ELECTRONS, TRIGONAL SYMMETRY

B=825 C=4*B DT=-1000 K=-1 DS=K*DT ZETA=0

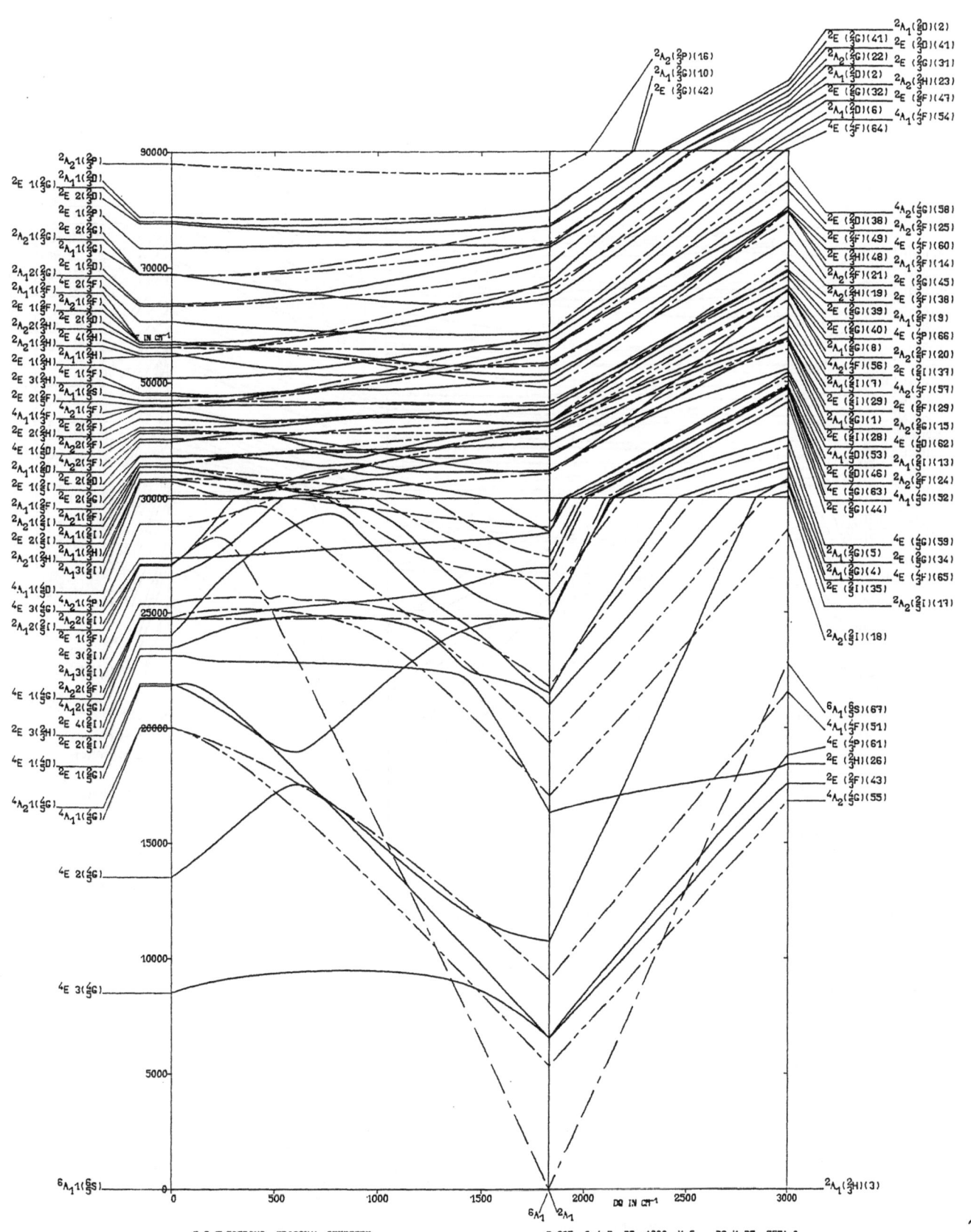

5 D ELECTRONS, TRIGONAL SYMMETRY   B=825 C=4×B DT=-1000 K=5   DS=K×DT ZETA=0

275

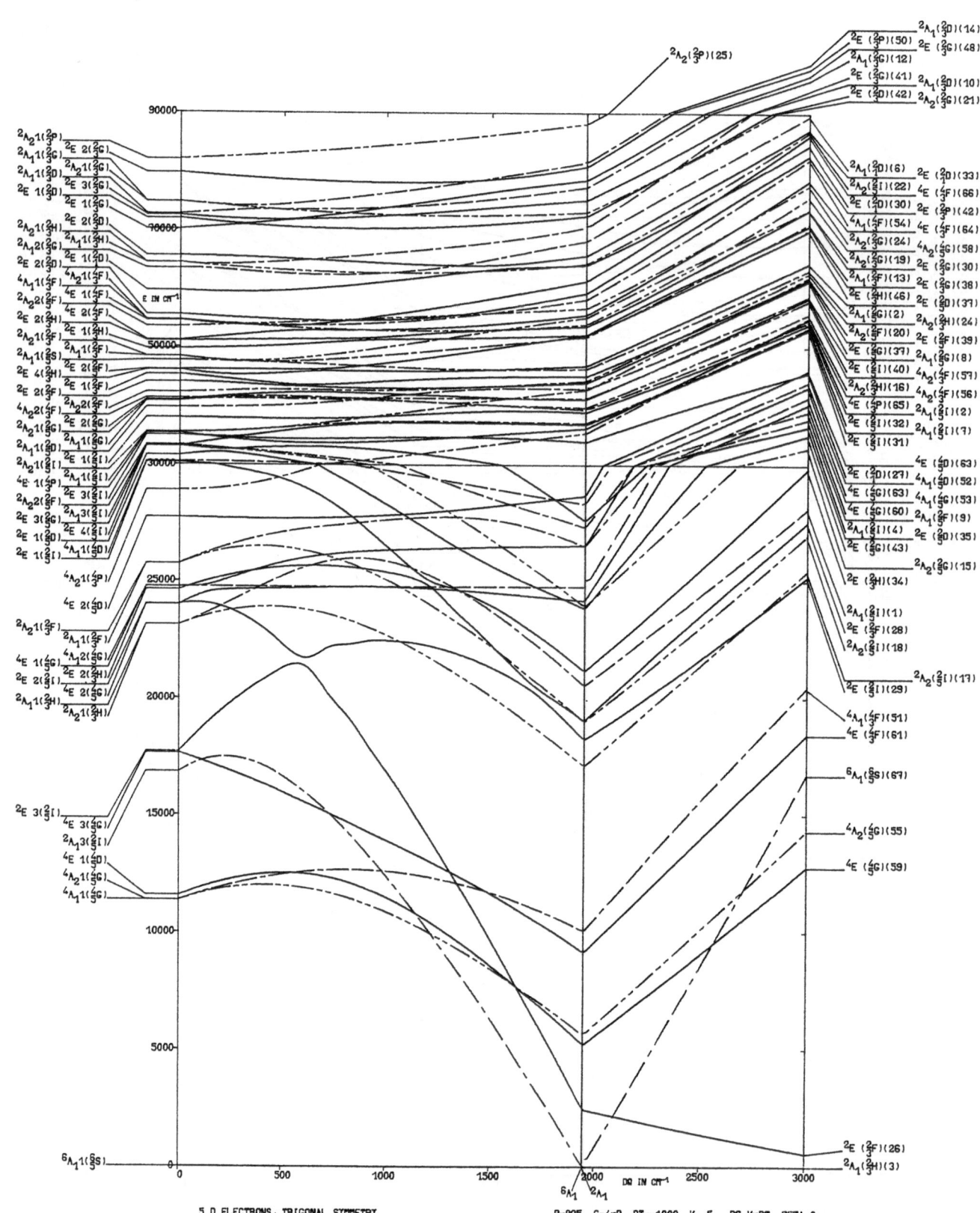

5 D ELECTRONS, TRIGONAL SYMMETRY

B=825   C=4×B   DT=-1000   K=-5   DS=K×DT   ZETA=0

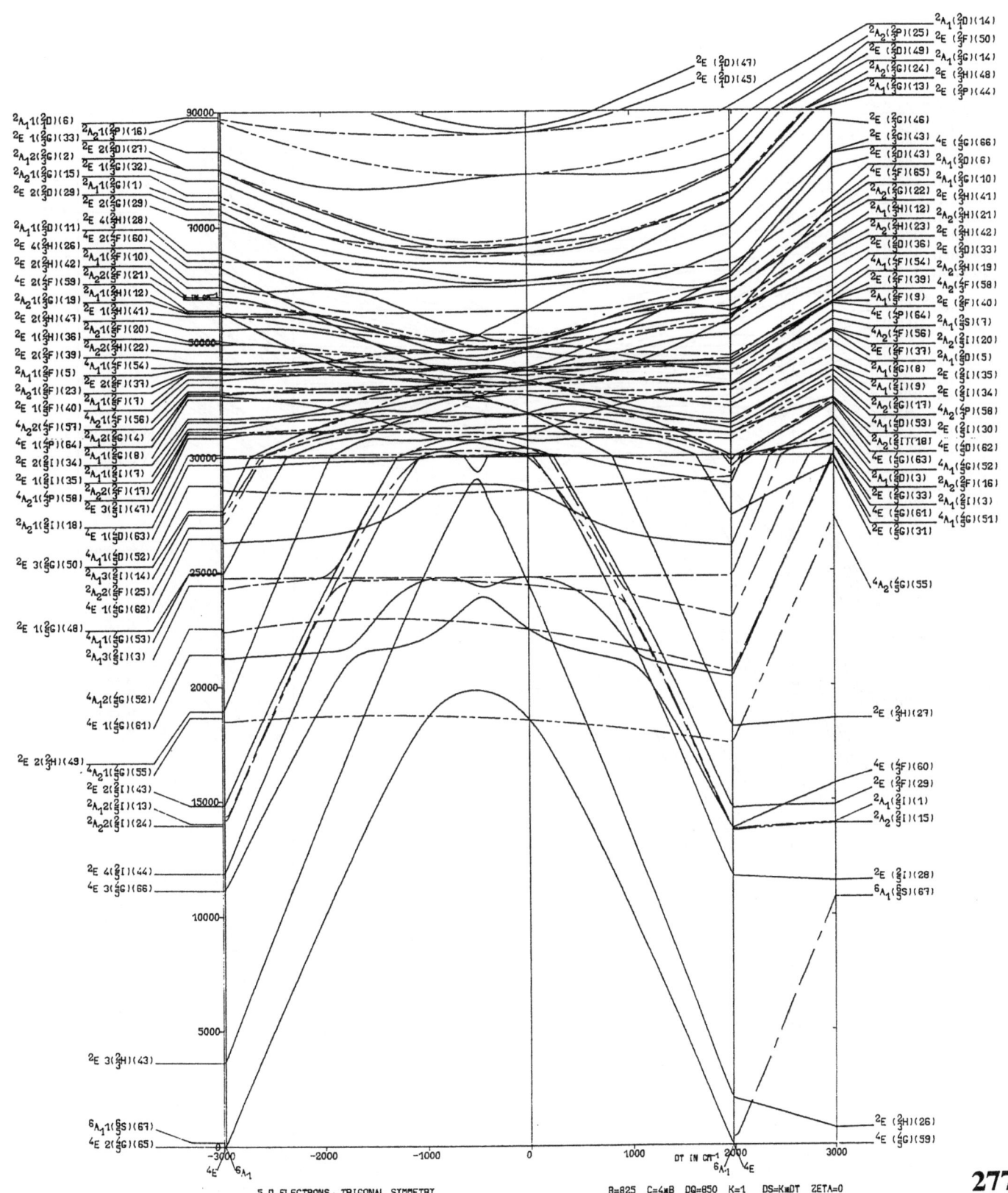

5 D ELECTRONS, TRIGONAL SYMMETRY

B=825 C=4×B DQ=850 K=1 DS=K×DT ZETA=0

277

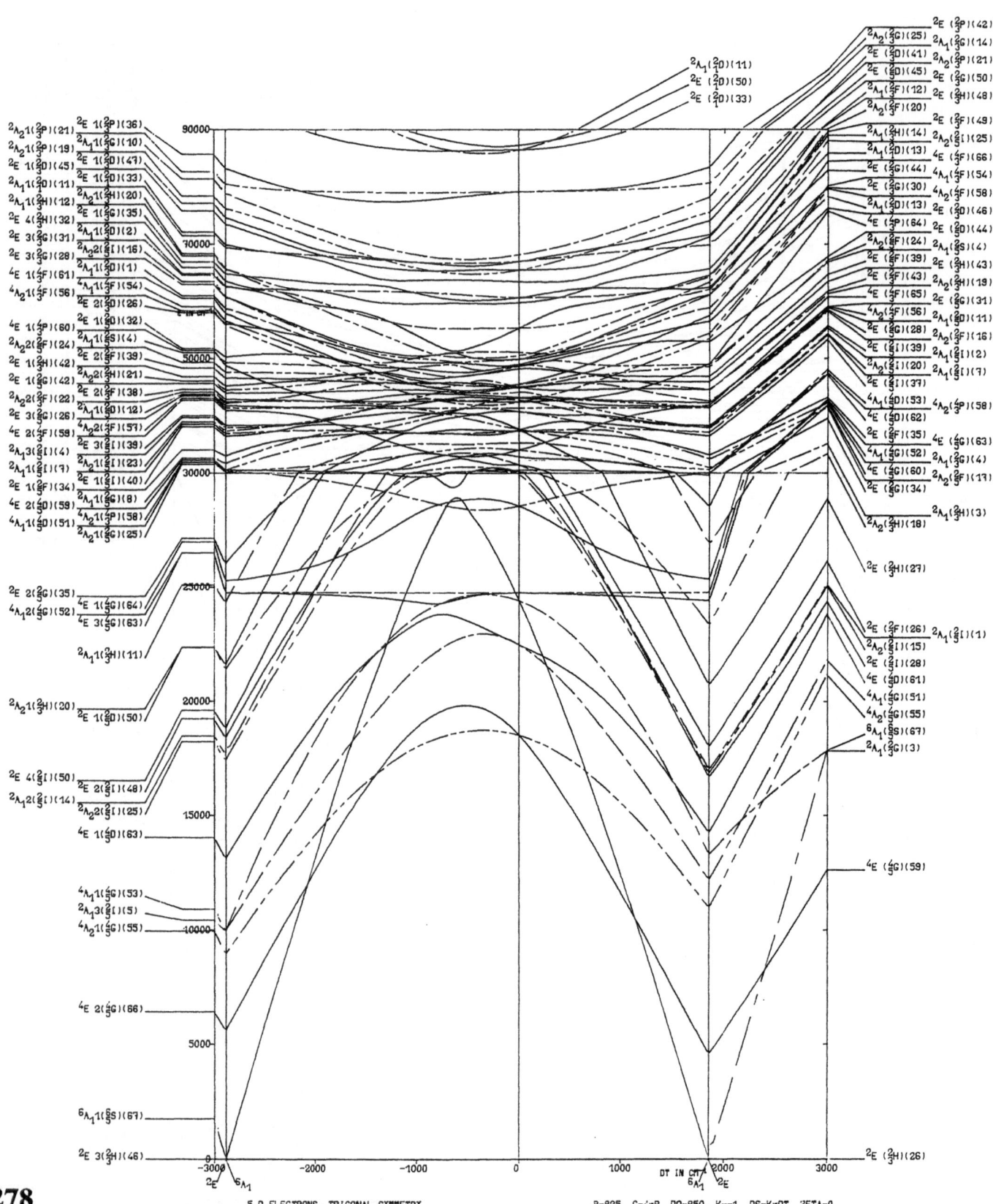

$^{2}A_{2}1(^{2}_{1}P)(21)$ $^{2}E\ 1(^{2}_{1}P)(36)$
$^{2}A_{2}1(^{2}_{1}P)(19)$ $^{2}A_{1}1(^{2}_{2}G)(10)$
$^{2}E\ 1(^{2}_{2}D)(45)$ $^{2}E\ 1(^{2}_{1}D)(47)$
$^{2}A_{1}1(^{2}_{2}D)(11)$ $^{2}E\ 1(^{2}_{4}D)(33)$
$^{2}A_{1}1(^{2}_{4}H)(12)$ $^{2}A_{2}1(^{2}_{4}H)(20)$
$^{2}E\ 4(^{2}_{3}H)(32)$ $^{2}E\ 1(^{2}_{5}H)(35)$
$^{2}E\ 3(^{2}_{5}G)(31)$ $^{2}A_{1}1(^{2}_{3}D)(2)$
$^{2}E\ 3(^{2}_{3}G)(28)$ $^{2}A_{2}2(^{2}_{2}I)(16)$
$^{4}E\ 1(^{2}_{3}F)(61)$ $^{2}A_{1}1(^{2}_{2}D)(1)$
$^{4}A_{2}1(^{2}_{4}F)(56)$ $^{2}A_{1}1(^{2}_{5}F)(54)$
$^{2}E\ 2(^{2}_{2}D)(26)$
$^{4}E\ 1(^{4}_{1}P)(60)$ $^{2}E\ 1(^{2}_{6}D)(32)$
$^{2}A_{2}2(^{2}_{2}F)(24)$ $^{2}A_{1}1(^{2}_{2}S)(4)$
$^{2}E\ 1(^{2}_{2}H)(42)$ $^{2}A_{2}2(^{2}_{4}F)(39)$
$^{2}E\ 1(^{2}_{6}G)(42)$ $^{2}A_{2}2(^{2}_{3}H)(38)$
$^{2}A_{2}2(^{2}_{5}F)(22)$ $^{2}A_{1}1(^{2}_{6}D)(12)$
$^{2}E\ 3(^{2}_{6}G)(26)$ $^{2}E\ 3(^{2}_{3}I)(39)$
$^{4}E\ 2(^{2}_{2}F)(59)$ $^{2}A_{2}2(^{2}_{3}F)(57)$
$^{2}A_{1}3(^{2}_{2}I)(14)$ $^{2}A_{2}1(^{2}_{2}I)(23)$
$^{2}A_{1}1(^{2}_{1}I)(7)$ $^{2}E\ 1(^{2}_{1}I)(40)$
$^{2}E\ 1(^{2}_{6}F)(34)$ $^{2}E\ 2(^{2}_{6}D)(8)$
$^{2}E\ 2(^{2}_{5}D)(59)$ $^{2}A_{1}1(^{4}_{1}P)(58)$
$^{4}A_{1}1(^{2}_{5}D)(51)$ $^{2}A_{2}1(^{2}_{1}G)(25)$

$^{2}E\ 2(^{2}_{3}G)(35)$
$^{4}A_{1}2(^{4}_{1}G)(52)$ $^{4}E\ 1(^{4}_{1}G)(64)$
$^{4}E\ 3(^{4}_{5}G)(63)$

$^{2}A_{1}1(^{2}_{3}H)(11)$

$^{2}A_{2}1(^{2}_{3}H)(20)$ $^{2}E\ 1(^{2}_{5}D)(50)$

$^{2}E\ 4(^{2}_{2}I)(50)$ $^{2}E\ 2(^{2}_{4}I)(48)$
$^{2}A_{1}2(^{2}_{2}I)(14)$ $^{2}A_{2}2(^{2}_{6}I)(25)$

$^{4}E\ 1(^{4}_{1}D)(63)$

$^{4}A_{1}1(^{4}_{4}G)(53)$
$^{2}A_{1}3(^{2}_{2}I)(5)$
$^{4}A_{2}1(^{4}_{5}G)(55)$

$^{4}E\ 2(^{4}_{5}G)(66)$

$^{6}A_{1}1(^{6}_{6}S)(67)$

$^{2}E\ 3(^{2}_{3}H)(46)$

$^{2}A_{1}(^{2}_{1}D)(11)$
$^{2}E\ (^{2}_{1}D)(50)$
$^{2}E\ (^{2}_{1}D)(33)$

$^{2}E\ (^{2}_{1}P)(42)$
$^{2}A_{2}(^{2}_{1}G)(25)$ $^{2}A_{1}(^{2}_{2}G)(14)$
$^{2}E\ (^{2}_{2}D)(41)$ $^{2}A_{2}(^{2}_{2}P)(21)$
$^{2}E\ (^{2}_{1}D)(45)$ $^{2}E\ (^{2}_{1}G)(50)$
$^{2}A_{1}(^{2}_{2}F)(12)$ $^{2}E\ (^{2}_{4}H)(48)$
$^{2}A_{1}(^{2}_{1}F)(20)$ $^{2}E\ (^{2}_{2}F)(49)$
$^{2}A_{1}(^{2}_{4}H)(14)$ $^{2}A_{2}(^{2}_{2}I)(25)$
$^{2}A_{1}(^{2}_{5}D)(13)$ $^{4}E\ (^{2}_{3}F)(66)$
$^{2}E\ (^{2}_{6}D)(44)$ $^{4}A_{1}(^{2}_{3}F)(54)$
$^{2}E\ (^{2}_{2}G)(30)$ $^{2}E\ (^{2}_{5}H)(58)$
$^{2}A_{1}(^{2}_{4}D)(13)$ $^{2}E\ (^{2}_{6}G)(46)$
$^{4}E\ (^{4}_{1}P)(64)$ $^{2}E\ (^{2}_{3}D)(44)$
$^{2}A_{2}(^{2}_{5}F)(24)$ $^{2}A_{1}(^{2}_{2}S)(4)$
$^{2}E\ (^{2}_{2}F)(39)$ $^{2}E\ (^{2}_{4}H)(43)$
$^{4}E\ (^{4}_{1}F)(43)$ $^{4}A_{2}(^{2}_{2}H)(19)$
$^{4}E\ (^{4}_{5}F)(65)$ $^{2}E\ (^{2}_{6}G)(31)$
$^{4}A_{2}(^{4}_{4}F)(56)$ $^{2}A_{1}(^{2}_{6}H)(11)$
$^{2}E\ (^{2}_{6}G)(28)$ $^{2}A_{1}(^{2}_{2}F)(16)$
$^{2}E\ (^{2}_{3}I)(39)$ $^{2}A_{1}(^{2}_{3}I)(2)$
$^{2}A_{2}(^{2}_{1}I)(20)$ $^{2}A_{1}(^{2}_{1}I)(7)$
$^{2}E\ (^{2}_{2}I)(37)$
$^{4}A_{1}(^{2}_{5}D)(53)$ $^{4}A_{2}(^{4}_{1}P)(58)$
$^{2}E\ (^{2}_{6}D)(62)$
$^{2}E\ (^{2}_{5}D)(35)$ $^{4}E\ (^{4}_{4}G)(63)$
$^{4}A_{1}(^{4}_{5}F)(52)$ $^{2}A_{1}(^{2}_{2}G)(4)$
$^{4}E\ (^{4}_{5}F)(60)$ $^{2}A_{2}(^{2}_{5}F)(13)$
$^{2}E\ (^{2}_{6}G)(34)$

$^{2}A_{2}(^{2}_{3}H)(18)$ $^{2}A_{1}(^{2}_{3}H)(3)$

$^{2}E\ (^{2}_{3}H)(27)$

$^{2}E\ (^{2}_{2}F)(26)$ $^{2}A_{1}(^{2}_{1}I)(1)$
$^{2}A_{2}(^{2}_{2}I)(15)$
$^{2}E\ (^{2}_{1}I)(28)$
$^{4}E\ (^{4}_{2}D)(61)$
$^{4}A_{1}(^{4}_{5}S)(51)$
$^{4}A_{2}(^{4}_{1}G)(55)$
$^{6}A_{1}(^{6}_{6}S)(67)$
$^{2}A_{1}(^{2}_{3}G)(3)$

$^{4}E\ (^{4}_{5}G)(59)$

$^{2}E\ (^{2}_{3}H)(26)$

$^{-3000}$ $^{-2000}$ $^{-1000}$ 0 1000 DT IN CM 2000 3000

^{2}E $^{6}A_{1}$ $^{6}A_{1}$ ^{2}E

278

5 D ELECTRONS, TRIGONAL SYMMETRY   B=825 C=4*B DQ=850 K=-1 DS=K*DT ZETA=0

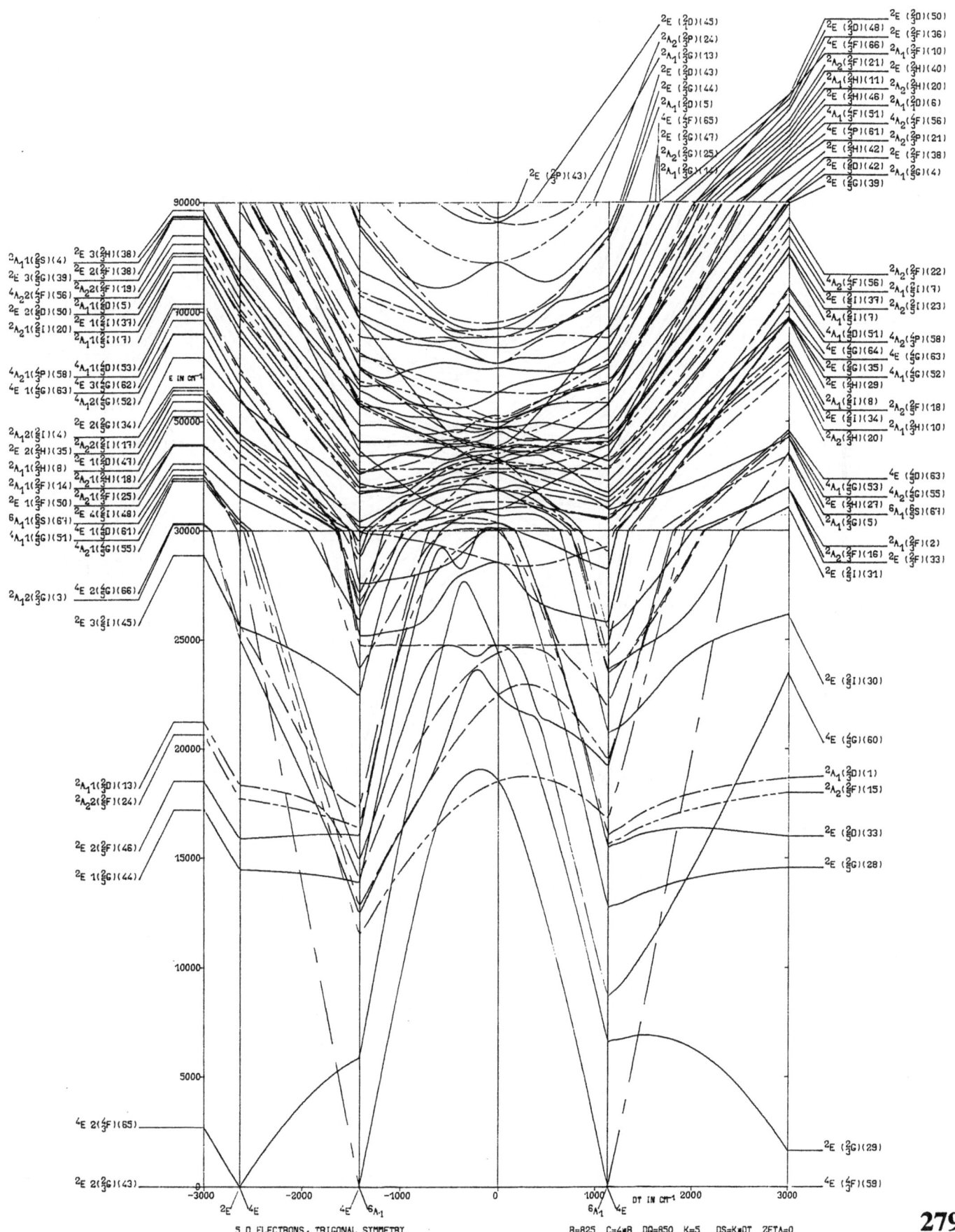

5 D ELECTRONS, TRIGONAL SYMMETRY

B=825 C=4×B DQ=850 K=5   DS=K×DT ZETA=0

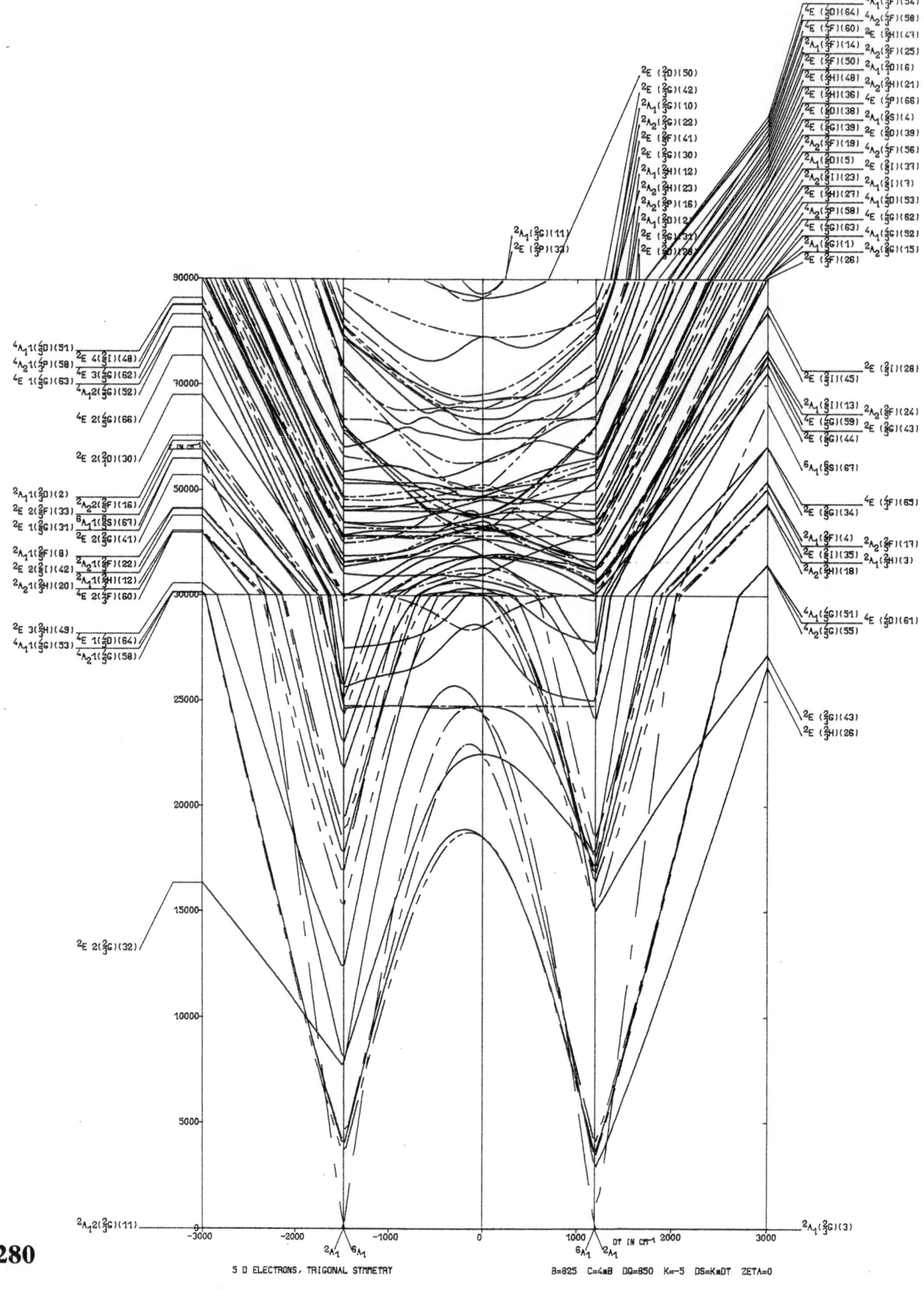

5 D ELECTRONS, TRIGONAL SYMMETRY

B=825 C=4×B DQ=850 K=-5 DS=K×DT ZETA=0

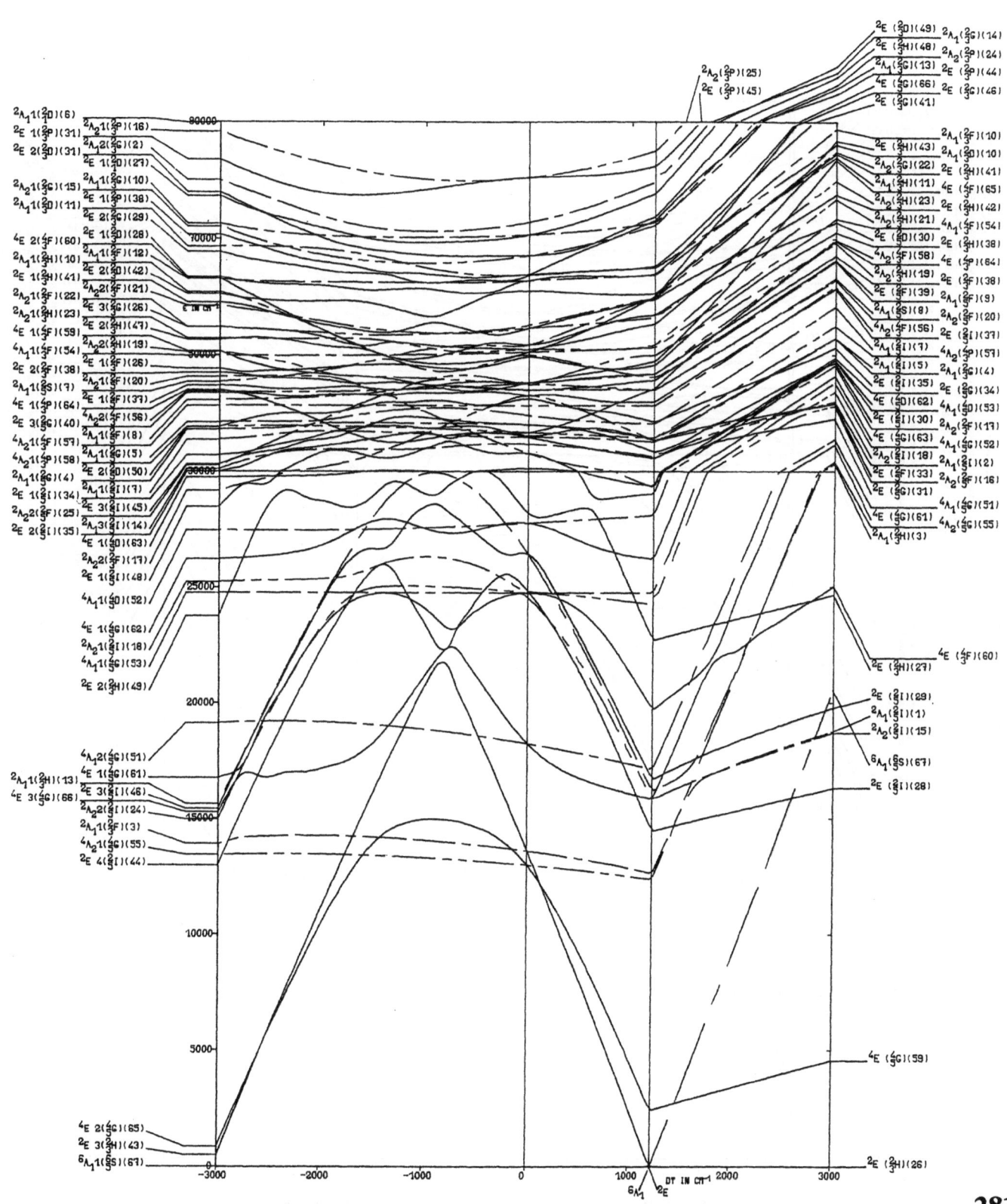

5 D ELECTRONS, TRIGONAL SYMMETRY

B=825 C=4×B DQ=1430 K=1 DS=K×DT ZETA=0

281

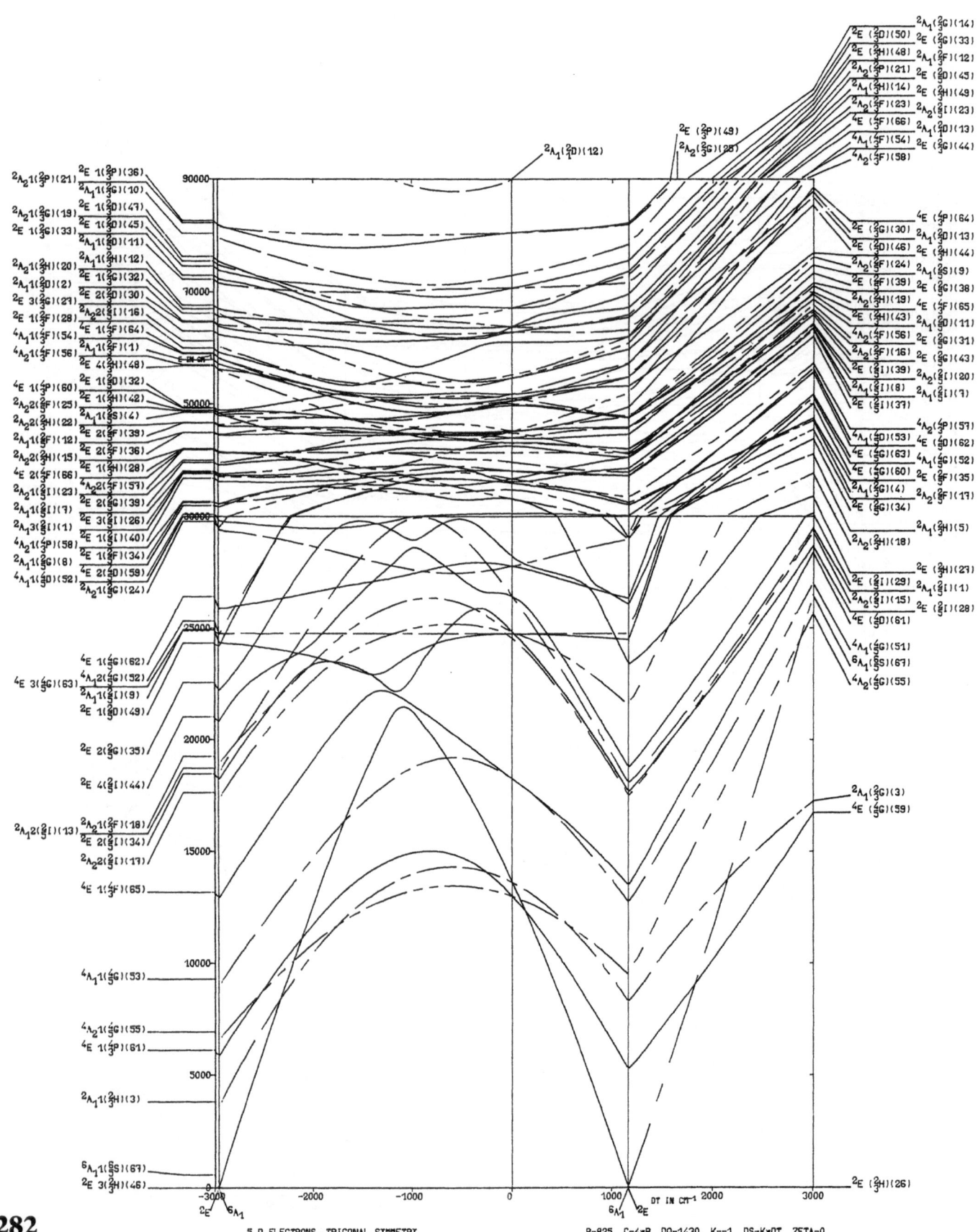

282

5 D ELECTRONS, TRIGONAL SYMMETRY

B=825 C=4xB DQ=1430 K=-1 DS=KxDT ZETA=0

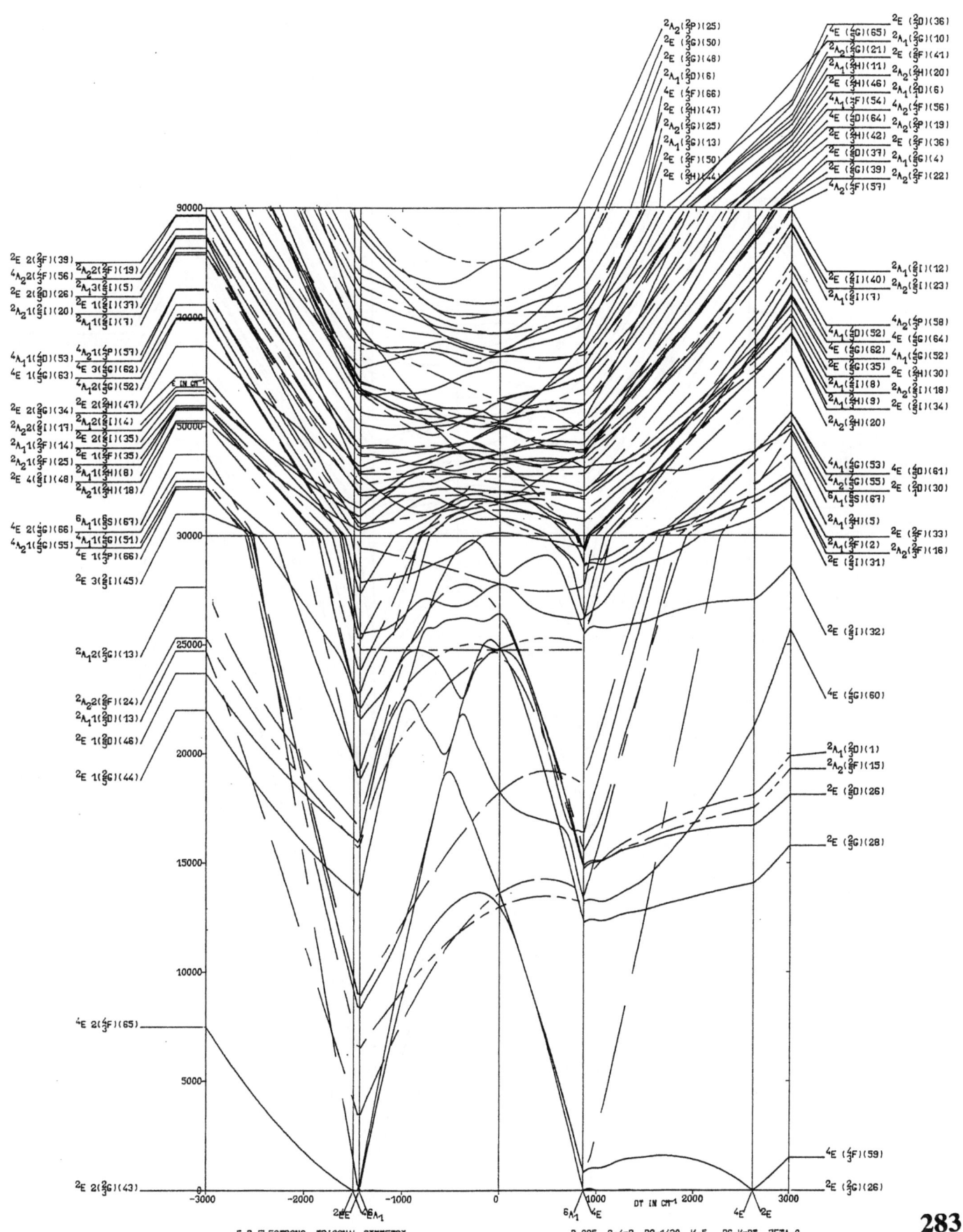

5 D ELECTRONS, TRIGONAL SYMMETRY

B=825 C=4*B DQ=1430 K=5 DS=K*DT ZETA=0

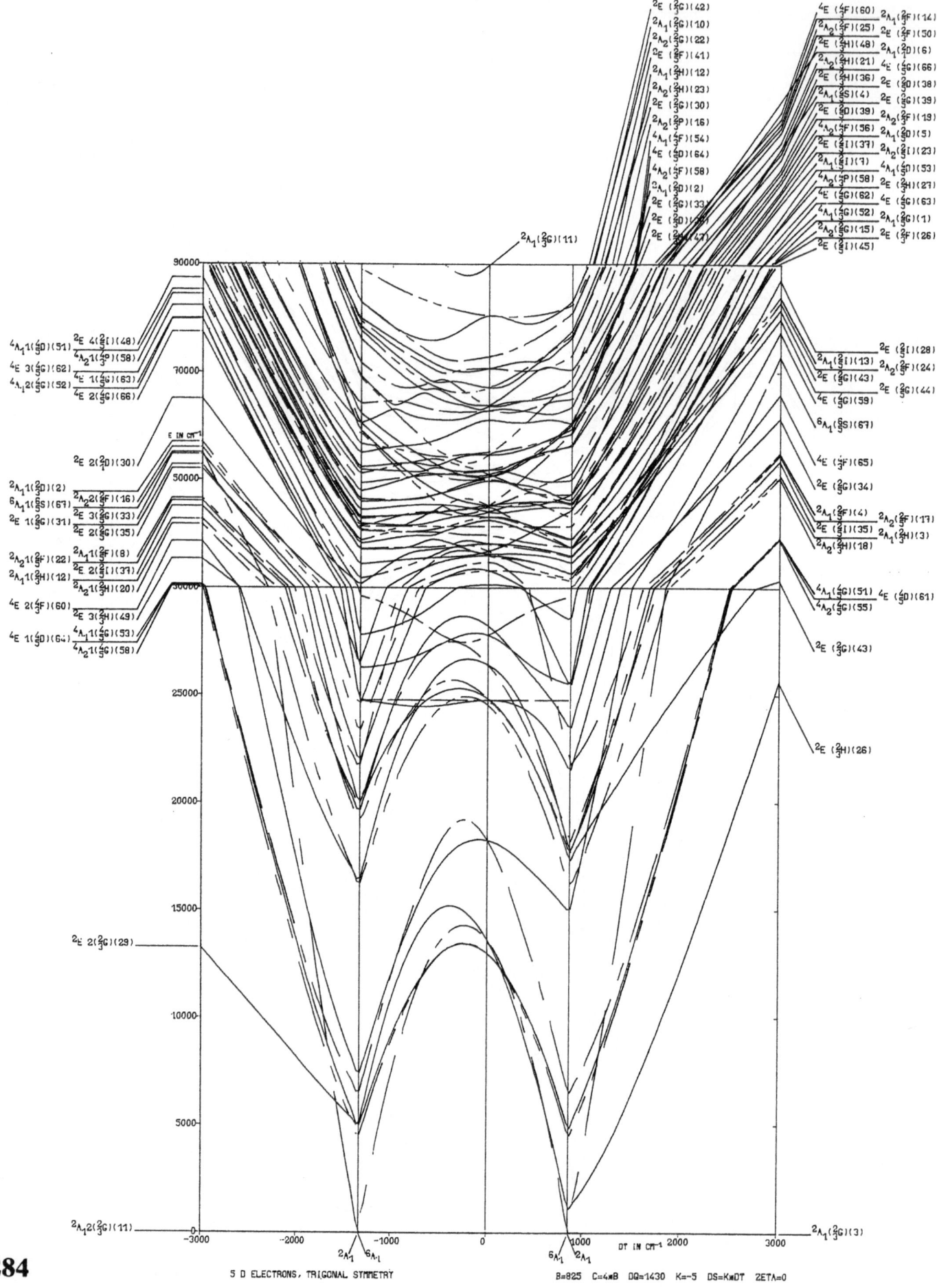

5 D ELECTRONS, TRIGONAL SYMMETRY

B=825 C=4*B DQ=1430 K=-5 DS=K*DT ZETA=0

B=825   C=4∗B   ZETA=0

ENERGY AS FUNCTION OF DT
K=1, -1, 5, -5

(58) $^2E_6(^1E_4(^2E_2(^1E_1(A_1E_1)E_1)E_2)E_2)$

(57) $^2E_5(^1E_3(^2E_1(^1A_1(E_1E_1)E_1)E_2)E_2)$

(56) $^2E_5(^1E_3(^2E_1(^1A_1(A_1A_1)E_1)E_2)E_2)$

(55) $^4E_4(^3E_2(^2E_1(^1E_2(A_1E_2)E_1)E_1)E_1)$

(54) $^2E_4(^1E_2(^2A_1(^1E_2(E_1E_1)E_2)E_2)E_2)$

(53) $^2E_4(^1E_2(^2A_2(^1E_1(A_1E_1)E_1)E_2)E_2)$

(52) $^2E_4(^1E_2(^2A_1(^1E_1(A_1E_1)E_1)E_2)E_2)$

(51) $^2E_4(^1E_2(^2E_1(^1A_1(A_1A_1)E_1)E_1)E_2)$

(50) $^4E_3(^3E_1(^2E_3(^1E_2(A_1E_2)E_1)E_2)E_2)$

(49) $^4E_3(^3E_1(^2A_1(^1E_1(A_1E_1)E_1)E_1)E_2)$

(48) $^2E_3(^1E_1(^2E_1(^1A_1(E_1E_1)E_1)E_2)E_2)$

(47) $^2E_3(^1E_1(^2E_1(^1E_2(A_1E_2)E_2)E_2)E_2)$

(46) $^2E_3(^1E_1(^2E_1(^1E_1(A_1E_1)E_2)E_2)E_2)$

(45) $^2E_3(^1E_2(^2A_1(^1E_1(A_1E_1)E_1)E_2)E_1)$

(44) $^2E_3(^1E_1(^2A_1(^1E_1(A_1E_1)E_1)E_1)E_2)$

(43) $^2E_3(^1E_2(^2E_1(^1A_1(A_1A_1)E_1)E_2)E_2)$

(42) $^4E_2(^3A_1(^2E_2(^1E_1(E_1E_2)E_1)E_2)E_2)$

(41) $^4E_2(^3A_1(^2E_2(^1E_1(A_1E_1)E_1)E_2)E_2)$

(40) $^4E_2(^3A_2(^2E_1(^1A_1(A_1A_1)E_1)E_1)E_2)$

(39) $^2E_2(^1A_1(^2E_2(^1E_1(E_1E_2)E_1)E_2)E_2)$

(38) $^2E_2(^1A_1(^2E_2(^1A_1(E_1E_1)E_2)E_2)E_2)$

(37) $^2E_2(^1A_1(^2E_2(^1A_1(E_1E_1)E_1)E_2)E_2)$

(36) $^2E_2(^1E_4(^2E_2(^1E_1(A_1E_1)E_1)E_2)E_2)$

(35) $^2E_2(^1A_2(^2E_2(^1E_1(A_1E_1)E_1)E_2)E_2)$

(34) $^2E_2(^1A_1(^2E_2(^1E_1(A_1E_1)E_1)E_2)E_2)$

(33) $^2E_2(^1A_1(^2E_2(^1A_1(A_1A_1)E_2)E_2)E_2)$

(32) $^2E_2(^1E_1(^2E_1(^1A_1(A_1A_1)E_1)E_2)E_1)$

(31) $^2E_2(^1A_1(^2E_1(^1A_1(A_1A_1)E_1)E_1)E_2)$

(30) $^4E_1(^3E_1(^2E_1(^1A_1(E_1E_1)E_1)E_2)E_2)$

(29) $^4E_1(^3E_1(^2E_1(^1E_2(A_1E_2)E_1)E_2)E_2)$

(28) $^4E_1(^3E_1(^2A_1(^1E_1(A_1E_1)E_1)E_1)E_2)$

(27) $^4E_1(^3E_1(^2E_1(^1A_1(A_1A_1)E_1)E_2)E_1)$

(26) $^2E_1(^1E_1(^2E_1(^1E_1(E_1E_2)E_2)E_2)E_2)$

(25) $^2E_1(^1E_3(^2E_1(^1A_1(E_1E_1)E_1)E_2)E_2)$

(24) $^2E_1(^1E_1(^2E_1(^1A_1(E_1E_1)E_1)E_2)E_2)$

(23) $^2E_1(^1E_1(^2E_1(^1E_2(A_1E_2)E_1)E_2)E_2)$

(22) $^2E_1(^1E_1(^2E_1(^1E_1(A_1E_1)E_2)E_2)E_2)$

(21) $^2E_1(^1E_2(^2A_1(^1E_1(A_1E_1)E_1)E_2)E_1)$

(20) $^2E_1(^1E_1(^2A_1(^1E_1(A_1E_1)E_1)E_2)E_2)$

(19) $^2E_1(^1E_3(^2E_1(^1A_1(A_1A_1)E_1)E_2)E_2)$

(18) $^2E_1(^1E_1(^2E_1(^1A_1(A_1A_1)E_1)E_2)E_2)$

(17) $^2E_1(^1A_1(^2E_1(^1A_1(A_1A_1)E_1)E_1)E_1)$

(16) $^4A_2(^3E_2(^2E_1(^1E_2(A_1E_2)E_1)E_1)E_2)$

(15) $^4A_2(^3E_2(^2A_1(^1E_1(A_1E_1)E_1)E_2)E_2)$

(14) $^2A_2(^1E_2(^2A_1(^1E_2(E_1E_1)E_2)E_2)E_2)$

(13) $^2A_2(^1E_2(^2A_2(^1E_1(A_1E_1)E_1)E_2)E_2)$

(12) $^2A_2(^1E_2(^2A_1(^1E_1(A_1E_1)E_1)E_2)E_2)$

(11) $^2A_2(^1E_2(^2E_1(^1A_1(A_1A_1)E_1)E_1)E_2)$

(10) $^6A_1(^5E_2(^4A_2(^3E_1(A_1E_1)E_1)E_2)E_2)$

(9) $^4A_1(^3E_2(^2E_1(^1E_2(A_1E_2)E_1)E_1)E_2)$

(8) $^4A_1(^3E_2(^2A_2(^1E_1(A_1E_1)E_1)E_2)E_2)$

(7) $^2A_1(^1E_2(^2A_1(^1E_2(E_1E_1)E_2)E_2)E_2)$

(6) $^2A_1(^1E_2(^2A_1(^1E_2(A_1E_2)E_2)E_2)E_2)$

(5) $^2A_1(^1E_1(^2E_1(^1E_2(A_1E_2)E_2)E_2)E_1)$

(4) $^2A_1(^1E_2(^2A_2(^1E_1(A_1E_1)E_1)E_2)E_2)$

(3) $^2A_1(^1E_2(^2A_1(^1E_1(A_1E_1)E_1)E_2)E_2)$

(2) $^2A_1(^1E_1(^2A_1(^1E_1(A_1E_1)E_1)E_1)E_1)$

(1) $^2A_1(^1E_2(^2E_1(^1A_1(A_1A_1)E_1)E_1)E_2)$

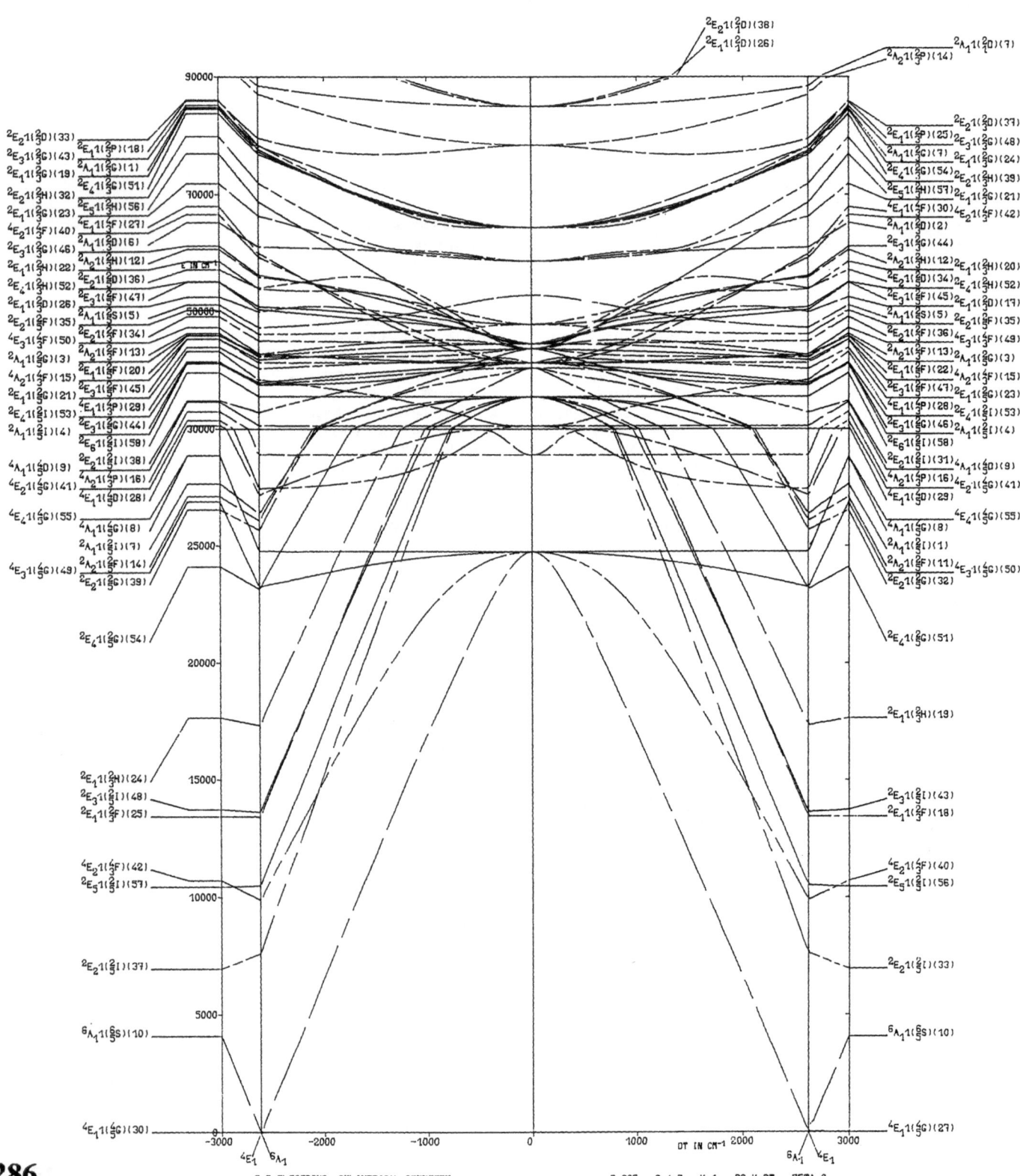

5 D ELECTRONS, CYLINDRICAL SYMMETRY

B=825   C=4*B   K=1   DS=K*DT   ZETA=0

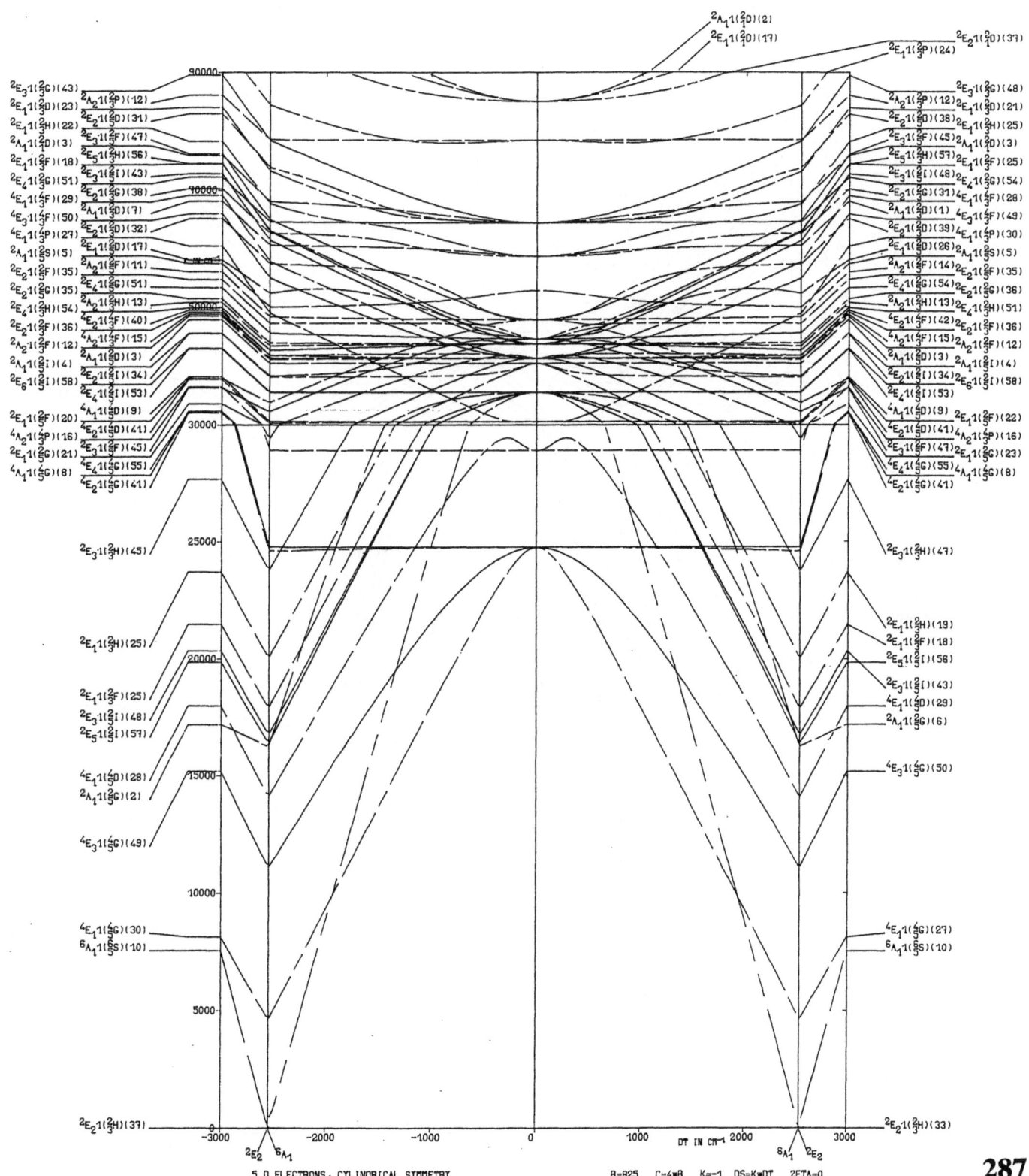

5 D ELECTRONS, CYLINDRICAL SYMMETRY   B=825   C=4×B   K=-1   DS=K×DT   ZETA=0

287

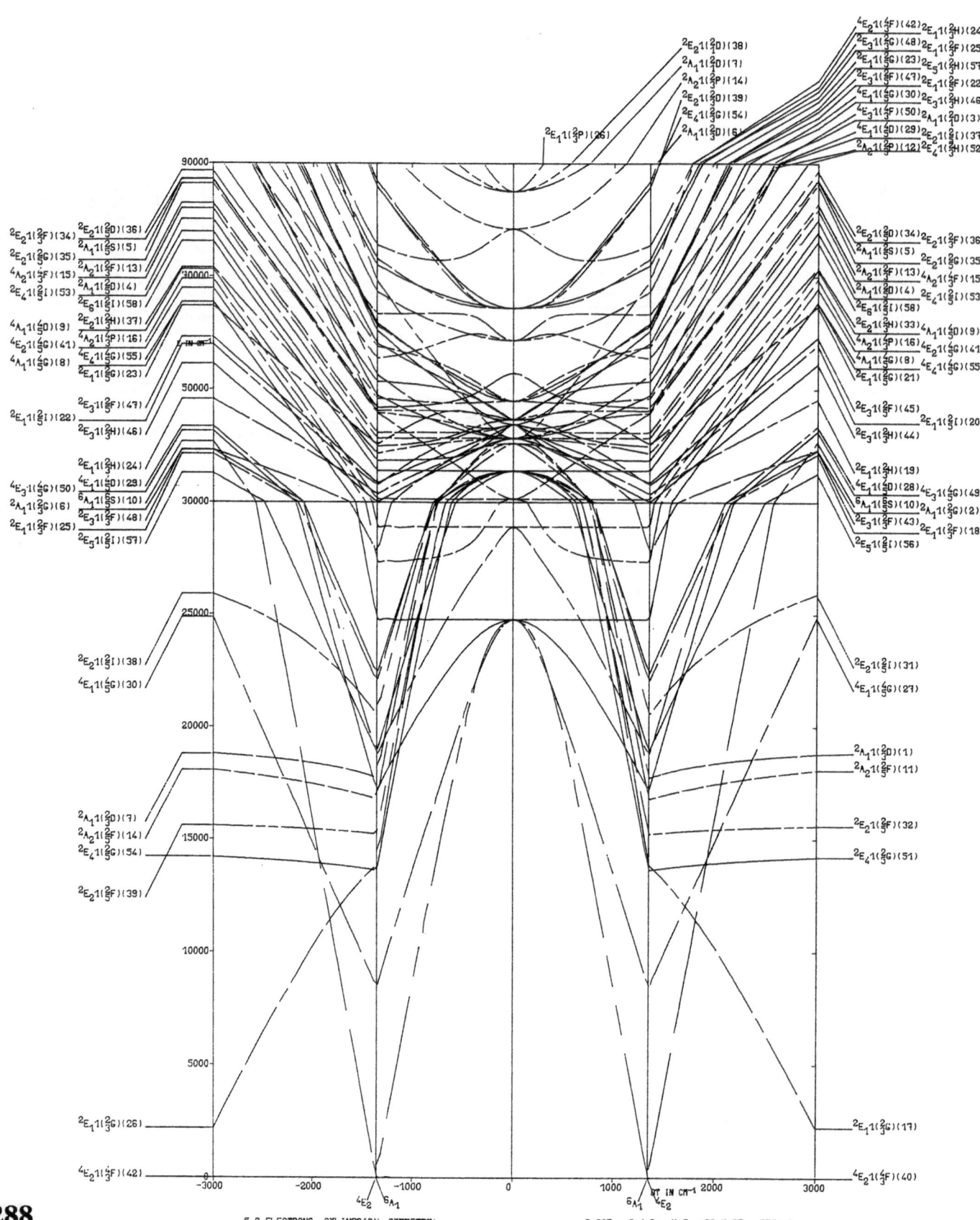

5 D ELECTRONS, CYLINDRICAL SYMMETRY

B=825   C=4×B   K=5   DS=K×DT   ZETA=0

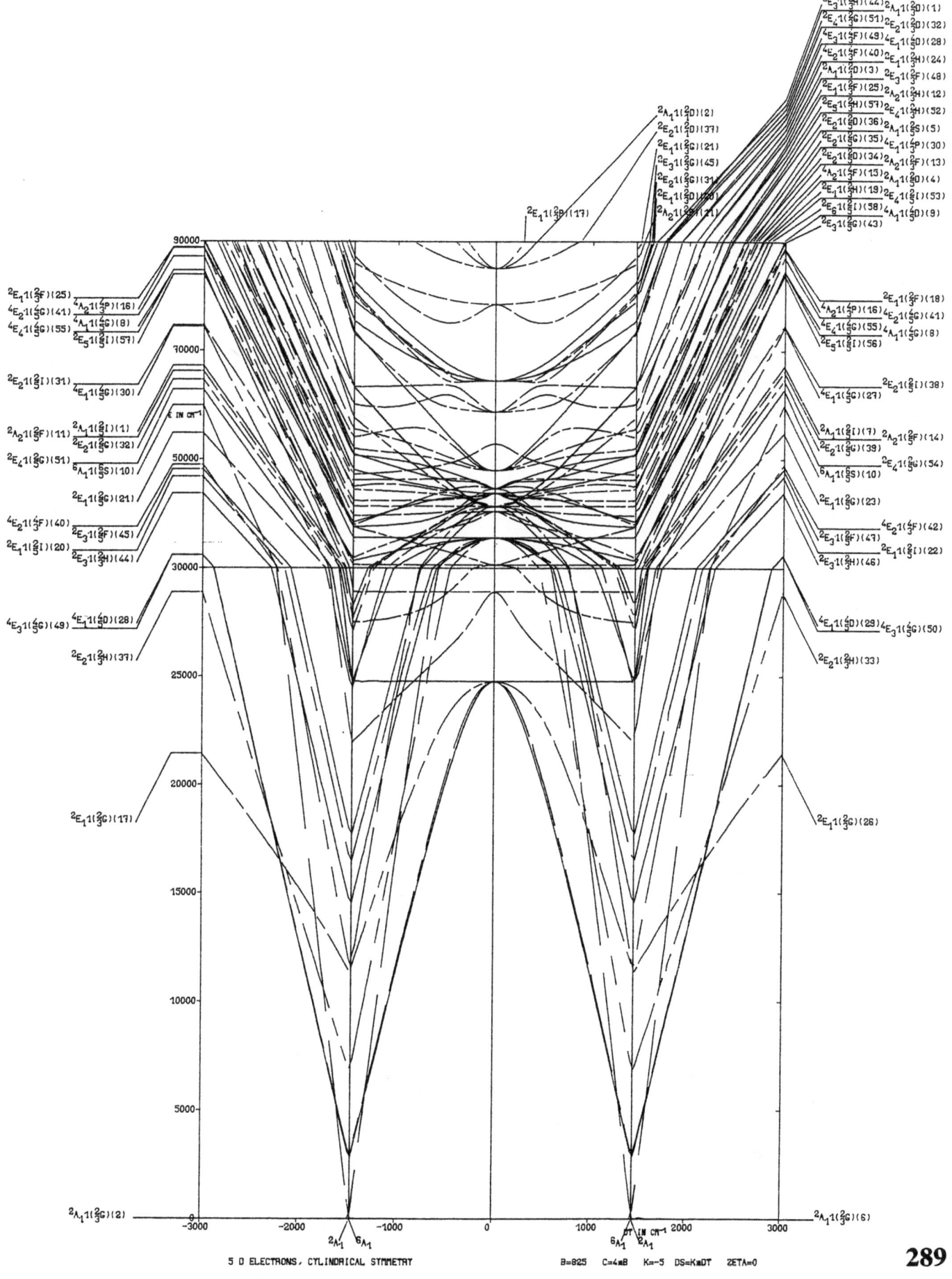

$^{2}E_{1}1(^{2}_{2}G)(17)$

$^{2}A_{1}1(^{2}_{3}G)(2)$

5 D ELECTRONS, CYLINDRICAL SYMMETRY

B=825   C=4·B   K=-5   DS=K·DT   ZETA=0

289

B=806   C=4×B   ZETA=0

ENERGY AS FUNCTION OF DQ
DT=0, 500, -500, 1000, -1000
K=1, -1, 5, -5

ENERGY AS FUNCTION OF DT
B=806   DQ=940 / B=670   DQ=1820
K=1, -1, 5, -5

(76) $^5A_1(B_1B_2(^3A_2)E\ E\ (^3A_2))$
(75) $^5B_1(A_1B_2(^3B_2)E\ E\ (^3A_2))$
(74) $^5B_2(A_1B_1(^3B_1)E\ E\ (^3A_2))$
(73) $^5E\ (A_1B_1(^3B_1)B_2E\ (^3E\ ))$
(72) $^3E\ (B_2E\ (^1E\ )E\ E\ (^3A_2))$
(71) $^3E\ (B_1E\ (^1E\ )E\ E\ (^3A_2))$
(70) $^3E\ (B_1E\ (^3E\ )B_2B_2(^1A_1))$
(69) $^3E\ (B_1B_1(^1A_1)B_2E\ (^3E\ ))$
(68) $^3E\ (A_1E\ (^1E\ )E\ E\ (^3A_2))$
(67) $^3E\ (A_1E\ (^3E\ )B_2B_2(^1A_1))$
(66) $^3E\ (A_1B_1(^3B_1)B_2E\ (^3E\ ))$
(65) $^3E\ (A_1B_1(^3B_1)B_2E\ (^1E\ ))$
(64) $^3E\ (A_1B_1(^1B_1)B_2E\ (^3E\ ))$
(63) $^3E\ (A_1E\ (^3E\ )B_1B_1(^1A_1))$
(62) $^3E\ (A_1A_1(^1A_1)B_2E\ (^3E\ ))$
(61) $^3E\ (A_1A_1(^1A_1)B_1E\ (^3E\ ))$
(60) $^3B_2(B_1B_2(^3A_2)E\ E\ (^1B_1))$
(59) $^3B_2(A_1B_2(^3B_2)E\ E\ (^1A_1))$
(58) $^3B_2(A_1B_1(^3B_1)E\ E\ (^3A_2))$
(57) $^3B_2(A_1B_1(^1B_1)E\ E\ (^3A_2))$
(56) $^3B_2(A_1B_2(^3B_2)B_1B_1(^1A_1))$
(55) $^3B_1(B_1B_2(^3A_2)E\ E\ (^1B_2))$
(54) $^3B_1(A_1B_2(^3B_2)E\ E\ (^3A_2))$
(53) $^3B_1(A_1B_2(^1B_2)E\ E\ (^3A_2))$
(52) $^3B_1(A_1B_1(^3B_1)E\ E\ (^1A_1))$
(51) $^3B_1(A_1B_1(^3B_1)B_2B_2(^1A_1))$
(50) $^3A_2(B_2B_2(^1A_1)E\ E\ (^3A_2))$
(49) $^3A_2(B_1B_2(^3A_2)E\ E\ (^1A_1))$
(48) $^3A_2(B_1B_1(^1A_1)E\ E\ (^3A_2))$
(47) $^3A_2(A_1B_2(^3B_2)E\ E\ (^1B_1))$
(46) $^3A_2(A_1B_1(^3B_1)E\ E\ (^1B_2))$
(45) $^3A_2(A_1A_1(^1A_1)E\ E\ (^3A_2))$
(44) $^3A_2(A_1A_1(^1A_1)B_1B_2(^3A_2))$
(43) $^3A_1(B_1B_2(^3A_2)E\ E\ (^3A_2))$
(42) $^3A_1(B_1B_2(^1A_2)E\ E\ (^3A_2))$
(41) $^3A_1(A_1B_2(^3B_2)E\ E\ (^1B_2))$
(40) $^3A_1(A_1B_1(^3B_1)E\ E\ (^1B_1))$
(39) $^1E\ (B_2E\ (^1E\ )E\ E\ (^1A_1))$

(38) $^1E\ (B_1E\ (^1E\ )E\ E\ (^1A_1))$
(37) $^1E\ (B_1E\ (^1E\ )B_2B_2(^1A_1))$
(36) $^1E\ (B_1B_1(^1A_1)B_2E\ (^1E\ ))$
(35) $^1E\ (A_1E\ (^1E\ )E\ E\ (^1A_1))$
(34) $^1E\ (A_1E\ (^1E\ )B_2B_2(^1A_1))$
(33) $^1E\ (A_1B_1(^3B_1)B_2E\ (^3E\ ))$
(32) $^1E\ (A_1B_1(^1B_1)B_2E\ (^1E\ ))$
(31) $^1E\ (A_1E\ (^1E\ )B_1B_1(^1A_1))$
(30) $^1E\ (A_1A_1(^1A_1)B_2E\ (^1E\ ))$
(29) $^1E\ (A_1A_1(^1A_1)B_1E\ (^1E\ ))$
(28) $^1B_2(B_2B_2(^1A_1)E\ E\ (^1B_2))$
(27) $^1B_2(B_2B_2(^1A_2)E\ E\ (^1B_1))$
(26) $^1B_2(B_1B_1(^1A_1)E\ E\ (^1B_2))$
(25) $^1B_2(A_1B_2(^1B_2)E\ E\ (^1A_1))$
(24) $^1B_2(A_1B_1(^3B_1)E\ E\ (^3A_2))$
(23) $^1B_2(A_1B_2(^1B_2)B_1B_1(^1A_1))$
(22) $^1B_2(A_1A_1(^1A_1)E\ E\ (^1B_2))$
(21) $^1B_1(B_2B_2(^1A_1)E\ E\ (^1B_1))$
(20) $^1B_1(B_1B_2(^1A_2)E\ E\ (^1B_2))$
(19) $^1B_1(B_1B_1(^1A_1)E\ E\ (^1B_1))$
(18) $^1B_1(A_1B_2(^3B_2)E\ E\ (^3A_2))$
(17) $^1B_1(A_1B_1(^1B_1)E\ E\ (^1A_1))$
(16) $^1B_1(A_1B_1(^1B_1)B_2B_2(^1A_1))$
(15) $^1B_1(A_1A_1(^1A_1)E\ E\ (^1B_1))$
(14) $^1A_2(B_1B_2(^1A_2)E\ E\ (^1A_1))$
(13) $^1A_2(A_1B_2(^1B_2)E\ E\ (^1B_1))$
(12) $^1A_2(A_1B_1(^1B_1)E\ E\ (^1B_2))$
(11) $^1A_2(A_1A_1(^1A_1)B_1B_2(^1A_2))$
(10) $^1A_1(E\ E\ (^1A_1)E\ E\ (^1A_1))$
(9) $^1A_1(B_2B_2(^1A_1)E\ E\ (^1A_1))$
(8) $^1A_1(B_1B_2(^3A_2)E\ E\ (^3A_2))$
(7) $^1A_1(B_1B_1(^1A_1)E\ E\ (^1A_1))$
(6) $^1A_1(B_1B_1(^1A_1)B_2B_2(^1A_1))$
(5) $^1A_1(A_1B_2(^1B_2)E\ E\ (^1B_2))$
(4) $^1A_1(A_1B_1(^1B_1)E\ E\ (^1B_1))$
(3) $^1A_1(A_1A_1(^1A_1)E\ E\ (^1A_1))$
(2) $^1A_1(A_1A_1(^1A_1)B_2B_2(^1A_1))$
(1) $^1A_1(A_1A_1(^1A_1)B_1B_1(^1A_1))$

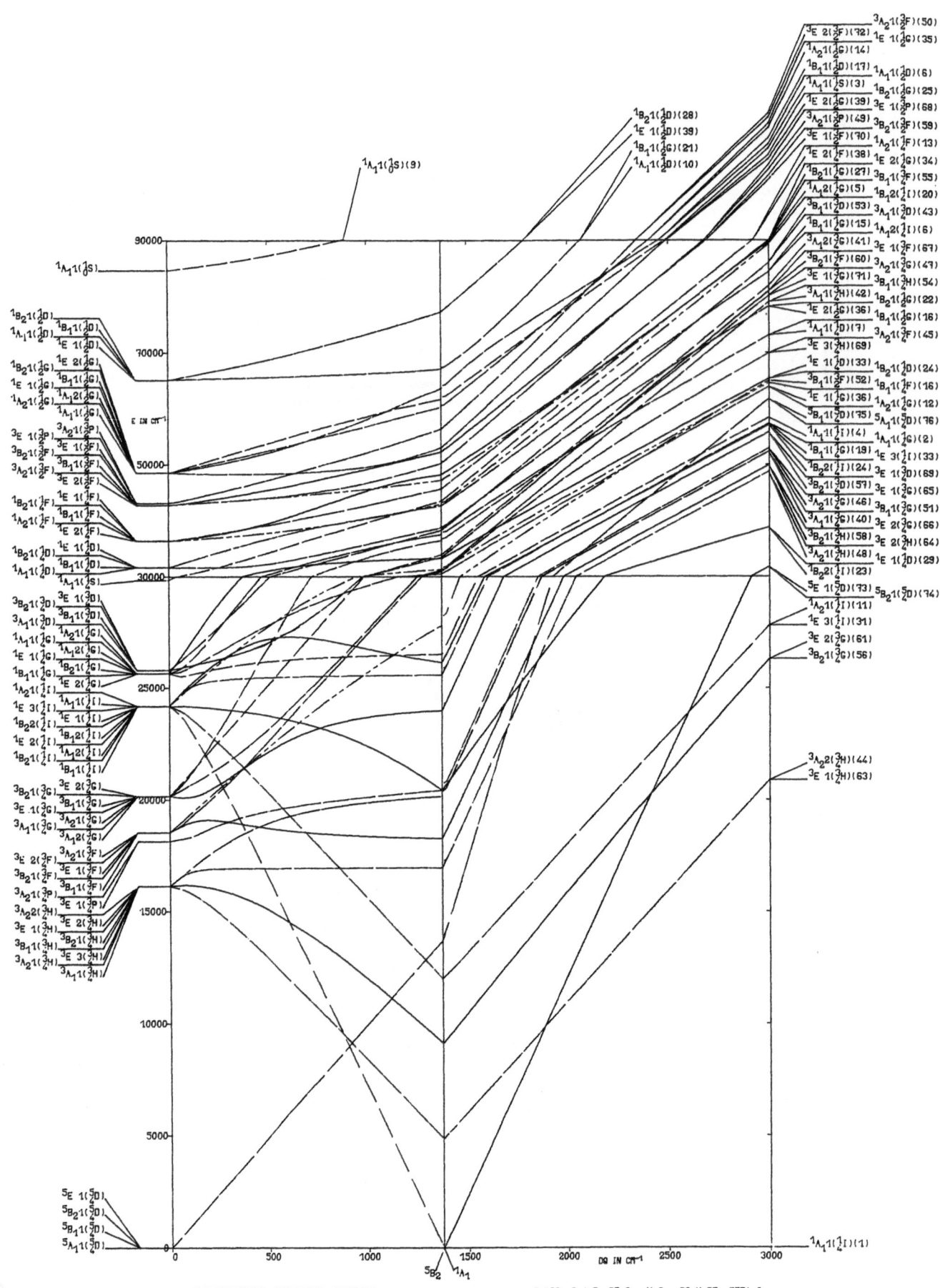

6 D ELECTRONS, TETRAGONAL SYMMETRY

B=806 C=4×B DT=0 K=0 DS=K×DT ZETA=0

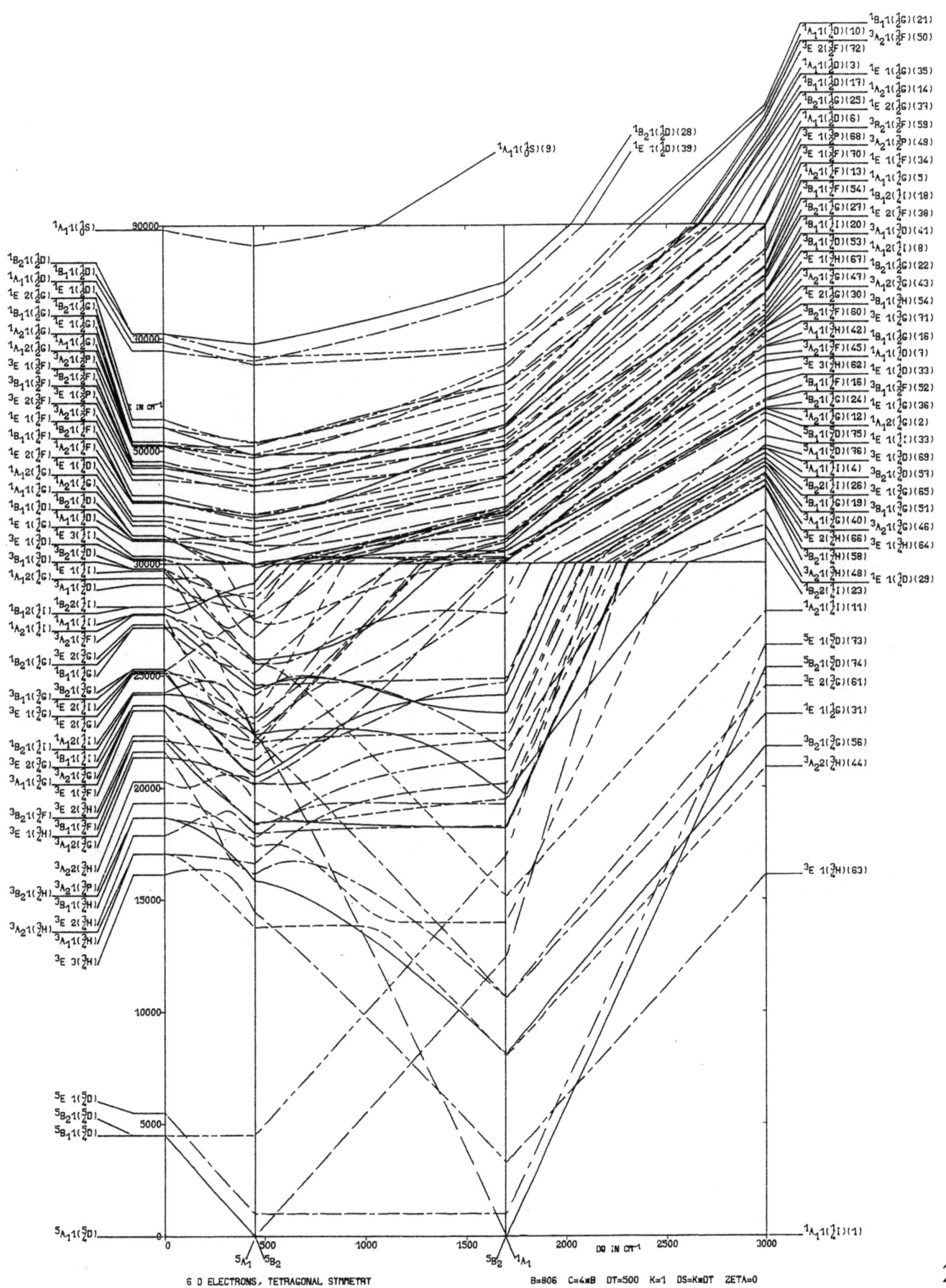

6 D ELECTRONS, TETRAGONAL SYMMETRY   B=806 C=4×B DT=500 K=1 DS=K×DT ZETA=0

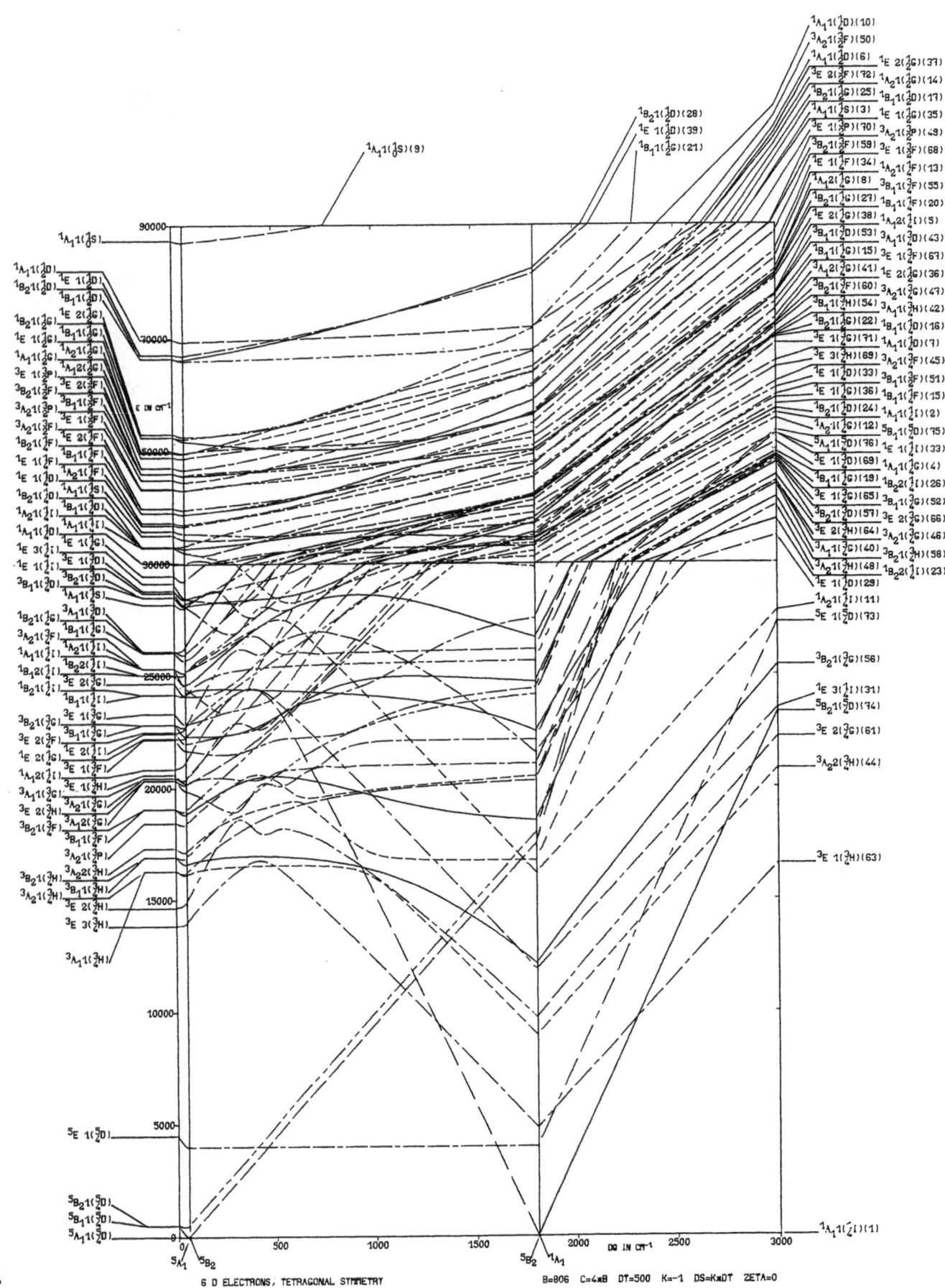

6 D ELECTRONS, TETRAGONAL SYMMETRY

B=806 C=4×B DT=500 K=-1 DS=K×DT ZETA=0

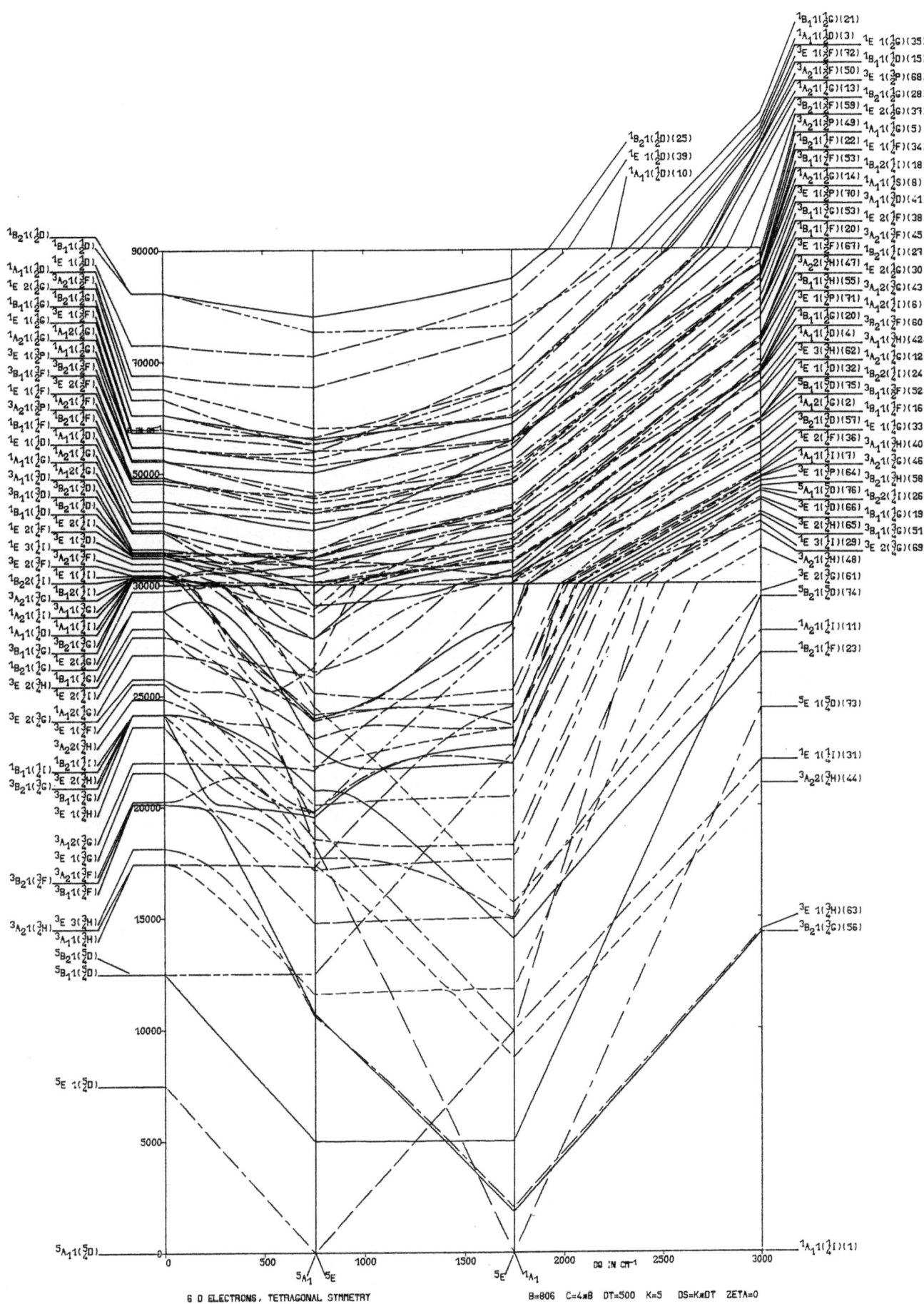

6 D ELECTRONS, TETRAGONAL SYMMETRY   B=806 C=4×B DT=500 K=5 DS=K×DT ZETA=0

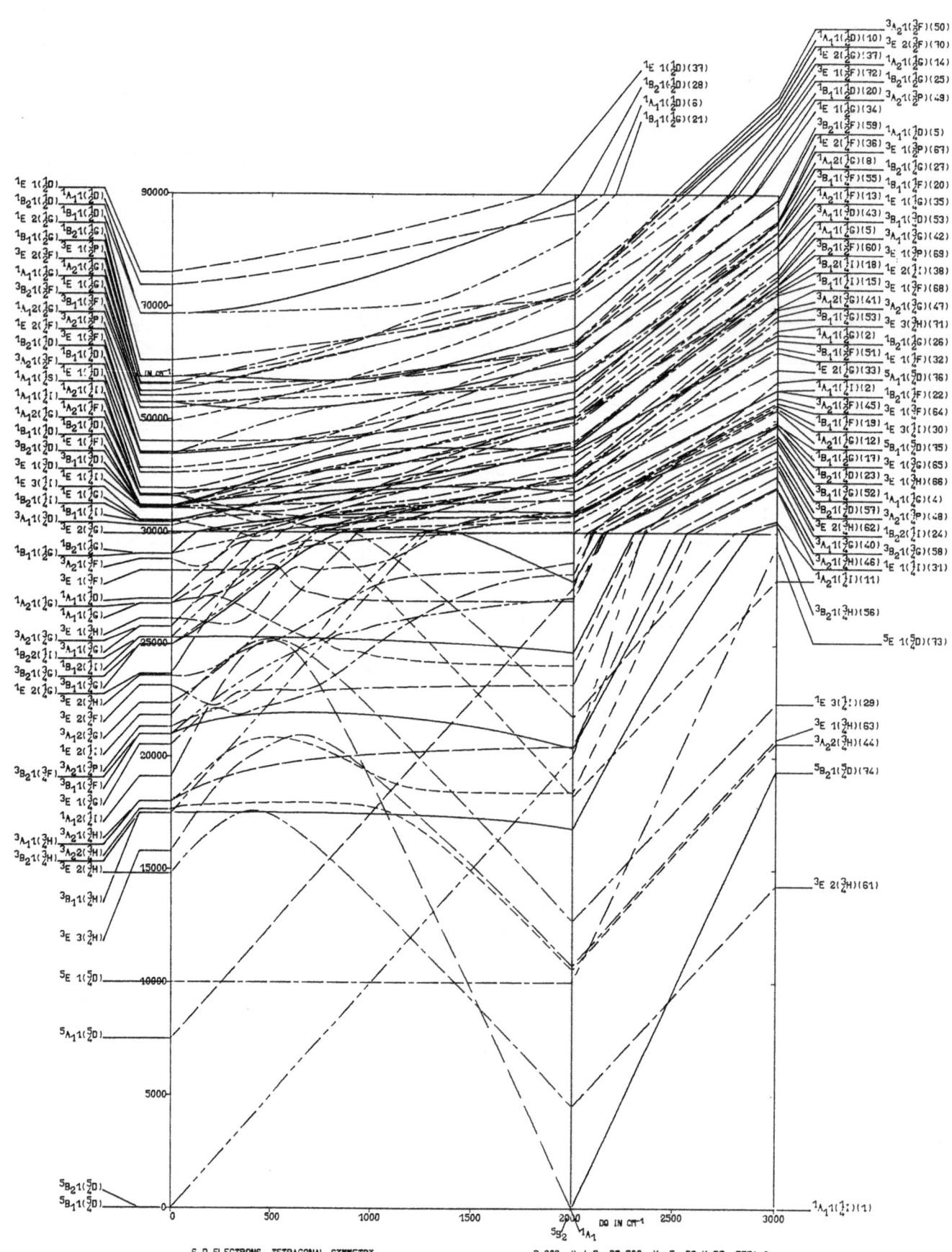

6 D ELECTRONS, TETRAGONAL SYMMETRY

B=806 C=4*B DT=500 K=-5 DS=K*DT ZETA=0

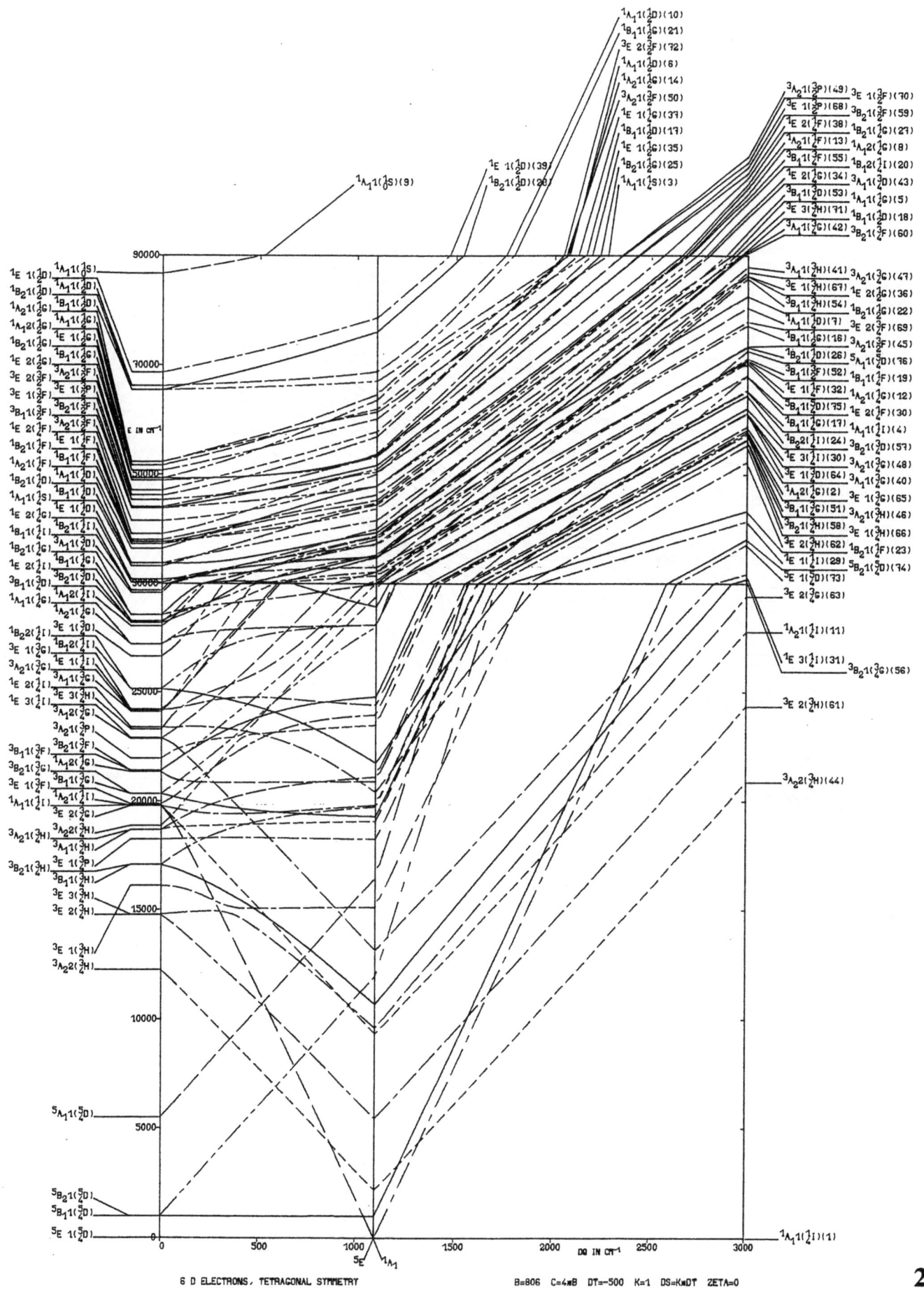

6 D ELECTRONS, TETRAGONAL SYMMETRY

B=806 C=4×B DT=-500 K=1 DS=K×DT ZETA=0

297

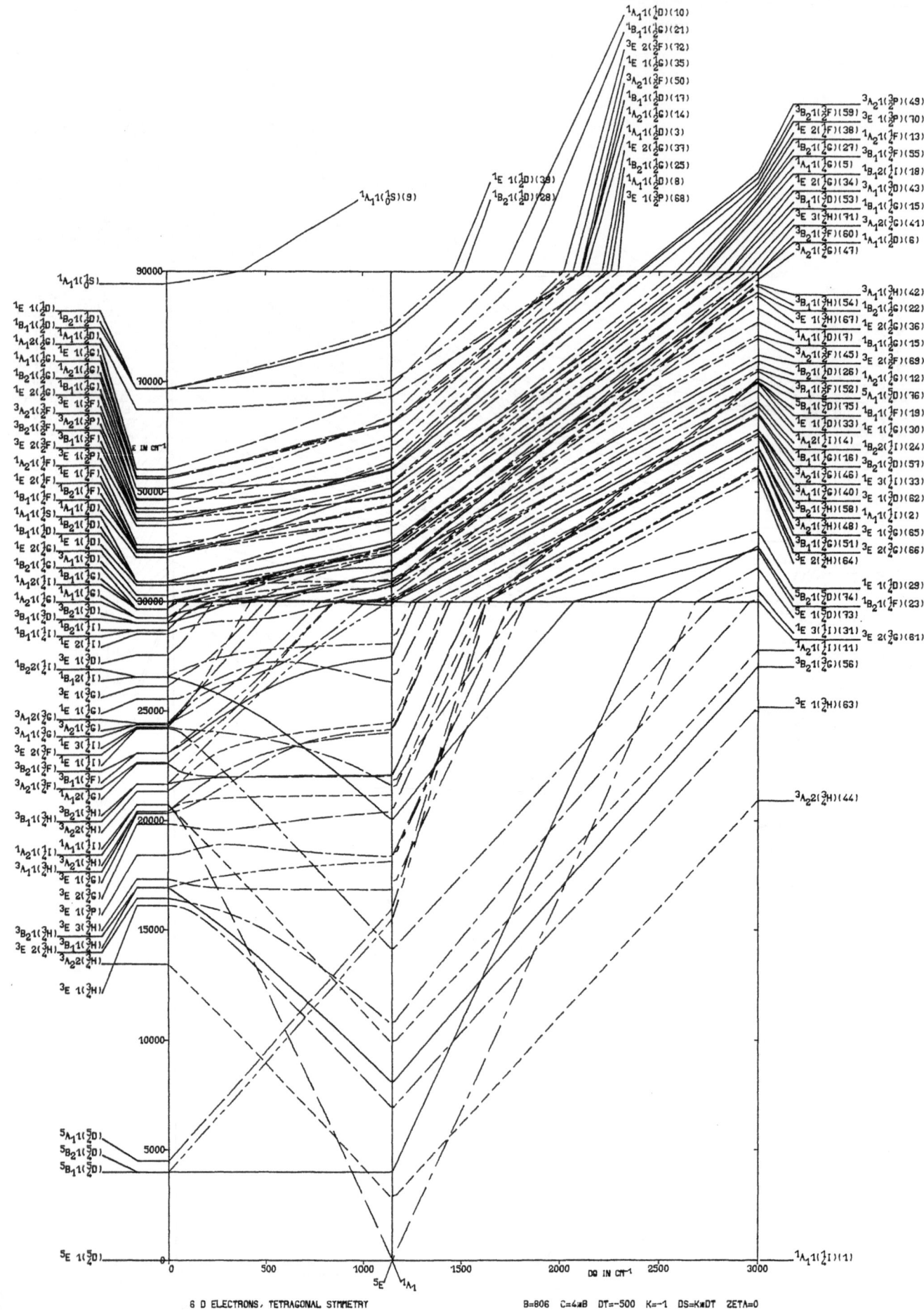

6 D ELECTRONS, TETRAGONAL SYMMETRY   B=806 C=4×B DT=-500 K=-1 DS=K×DT ZETA=0

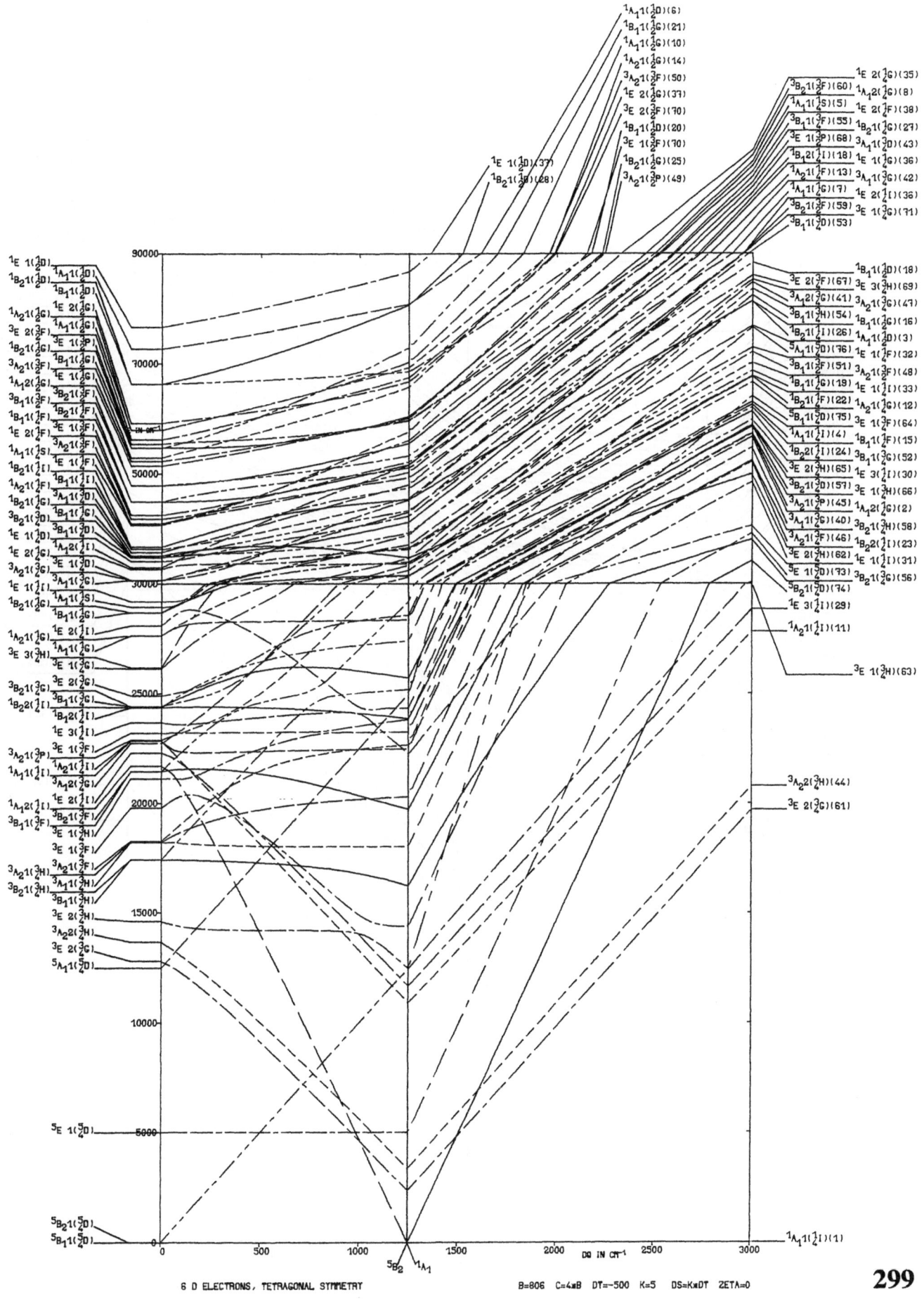

6 D ELECTRONS, TETRAGONAL SYMMETRY   B=806 C=4xB DT=-500 K=5 DS=KxDT ZETA=0

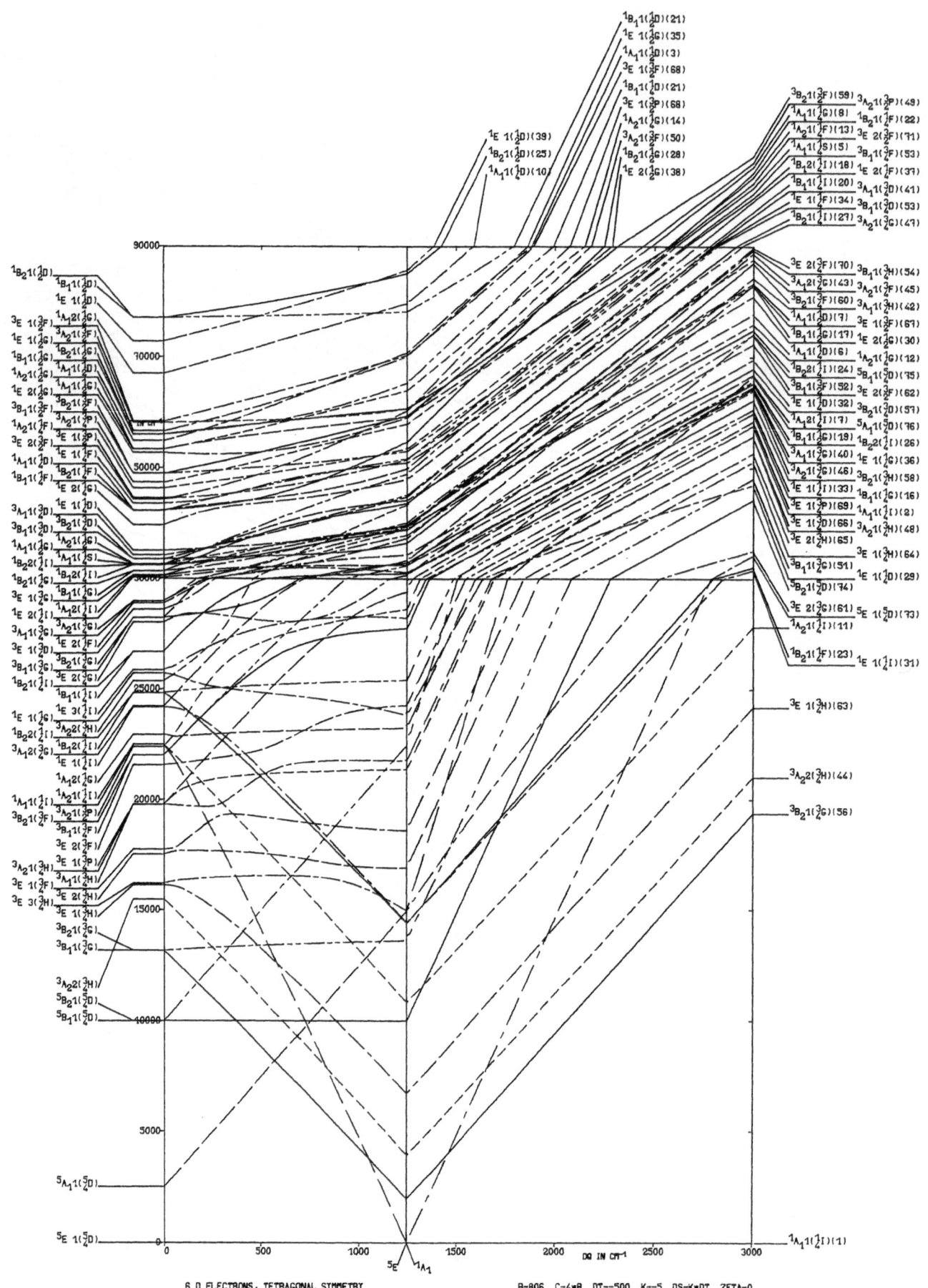

300

6 D ELECTRONS, TETRAGONAL SYMMETRY   B=806 C=4×B DT=-500 K=-5 DS=K×DT ZETA=0

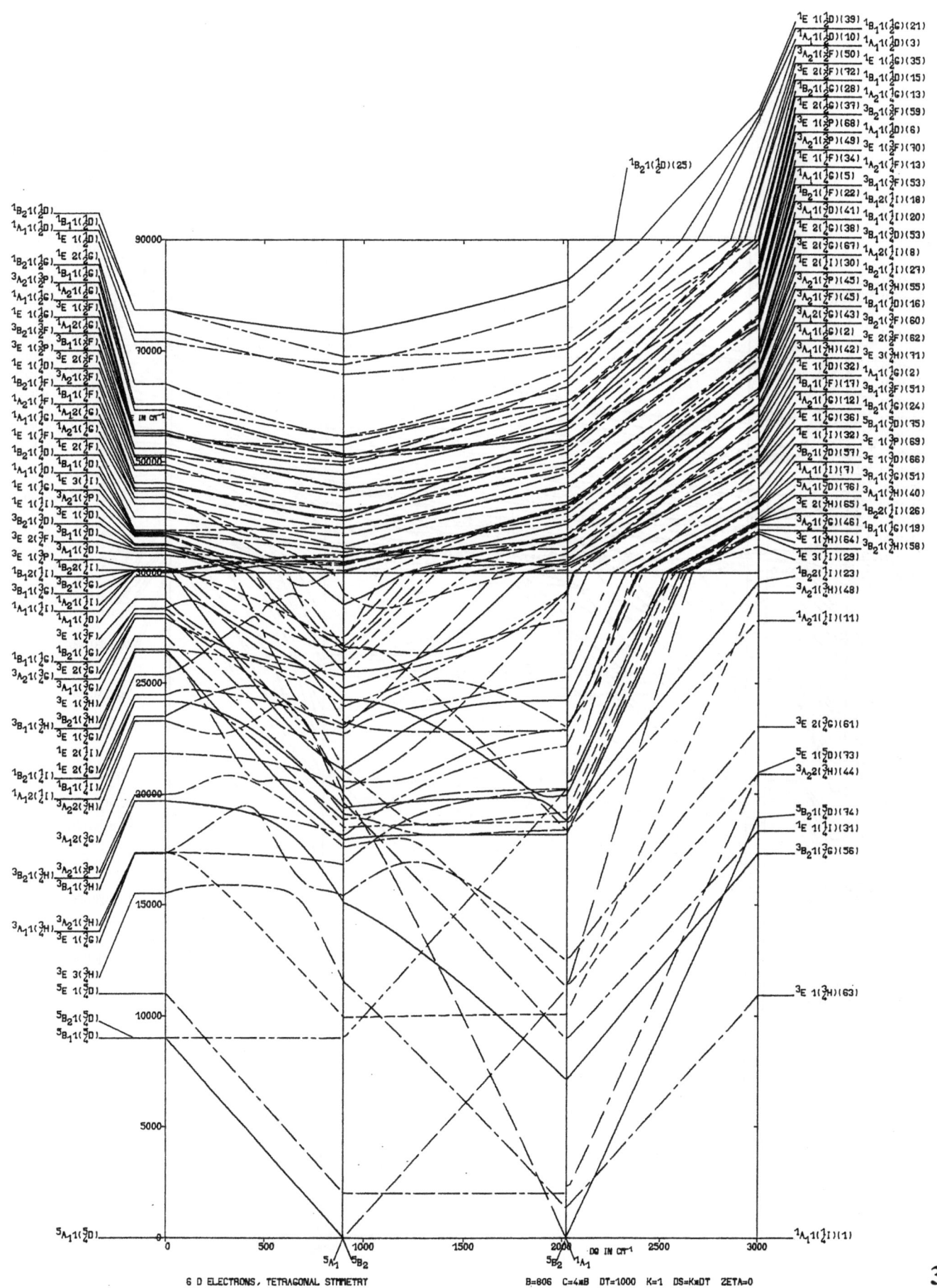

6 D ELECTRONS, TETRAGONAL SYMMETRY   B=806   C=4×B   DT=1000   K=1   DS=K×DT   ZETA=0

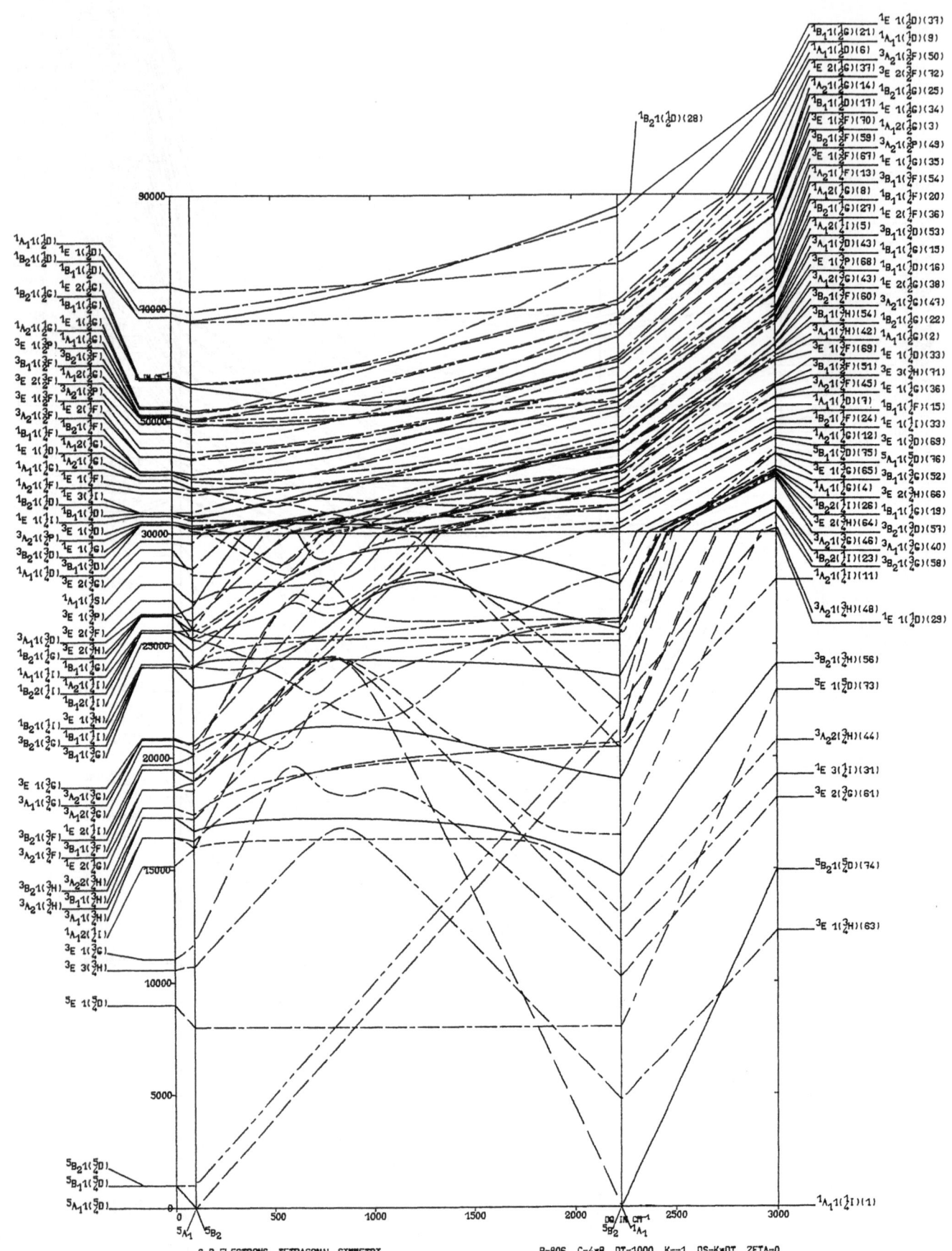

6 D ELECTRONS, TETRAGONAL SYMMETRY

B=806 C=4×B DT=1000 K=-1 DS=K×DT ZETA=0

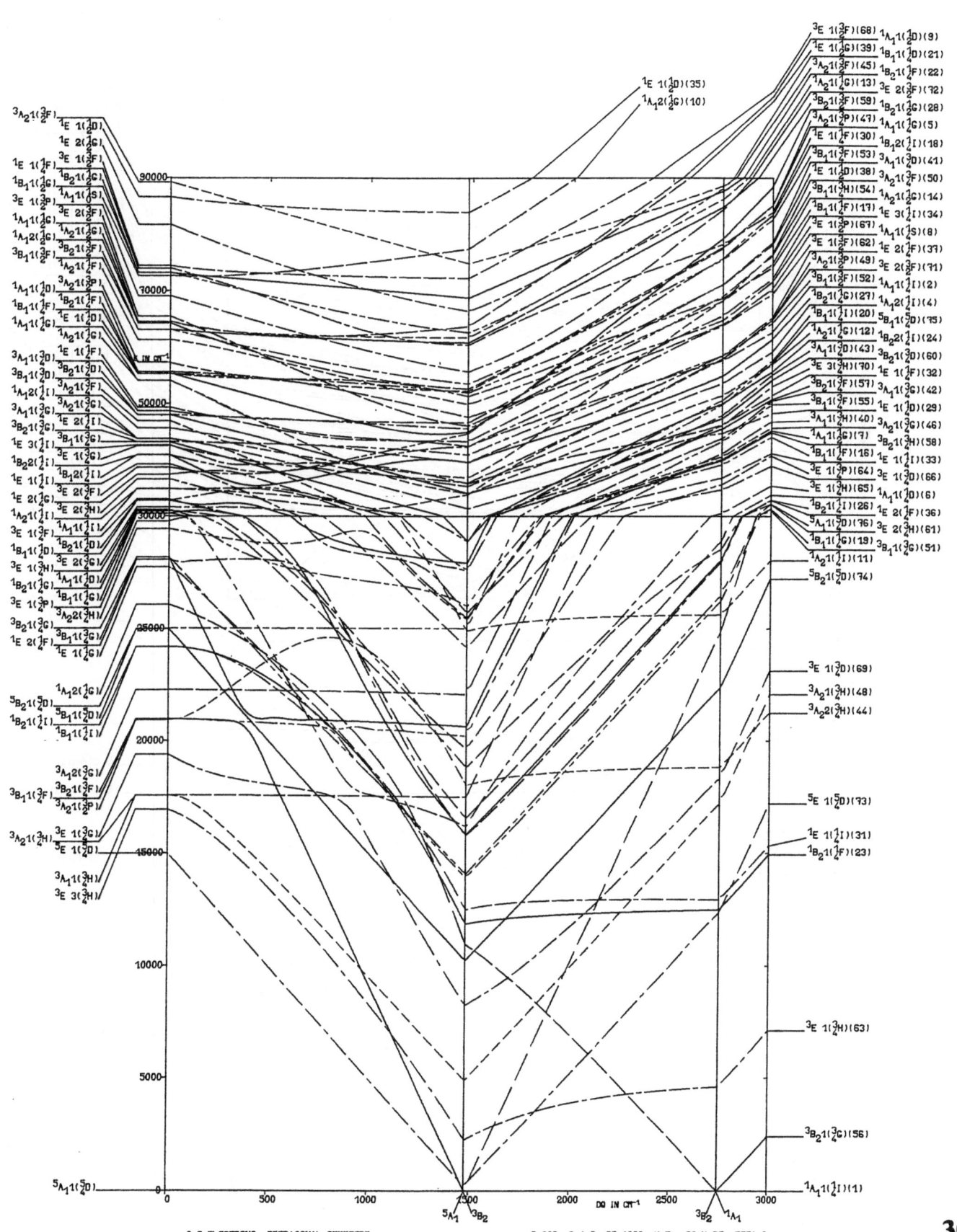

6 D ELECTRONS, TETRAGONAL SYMMETRY

B=806 C=4×B DT=1000 K=5 DS=K×DT ZETA=0

303

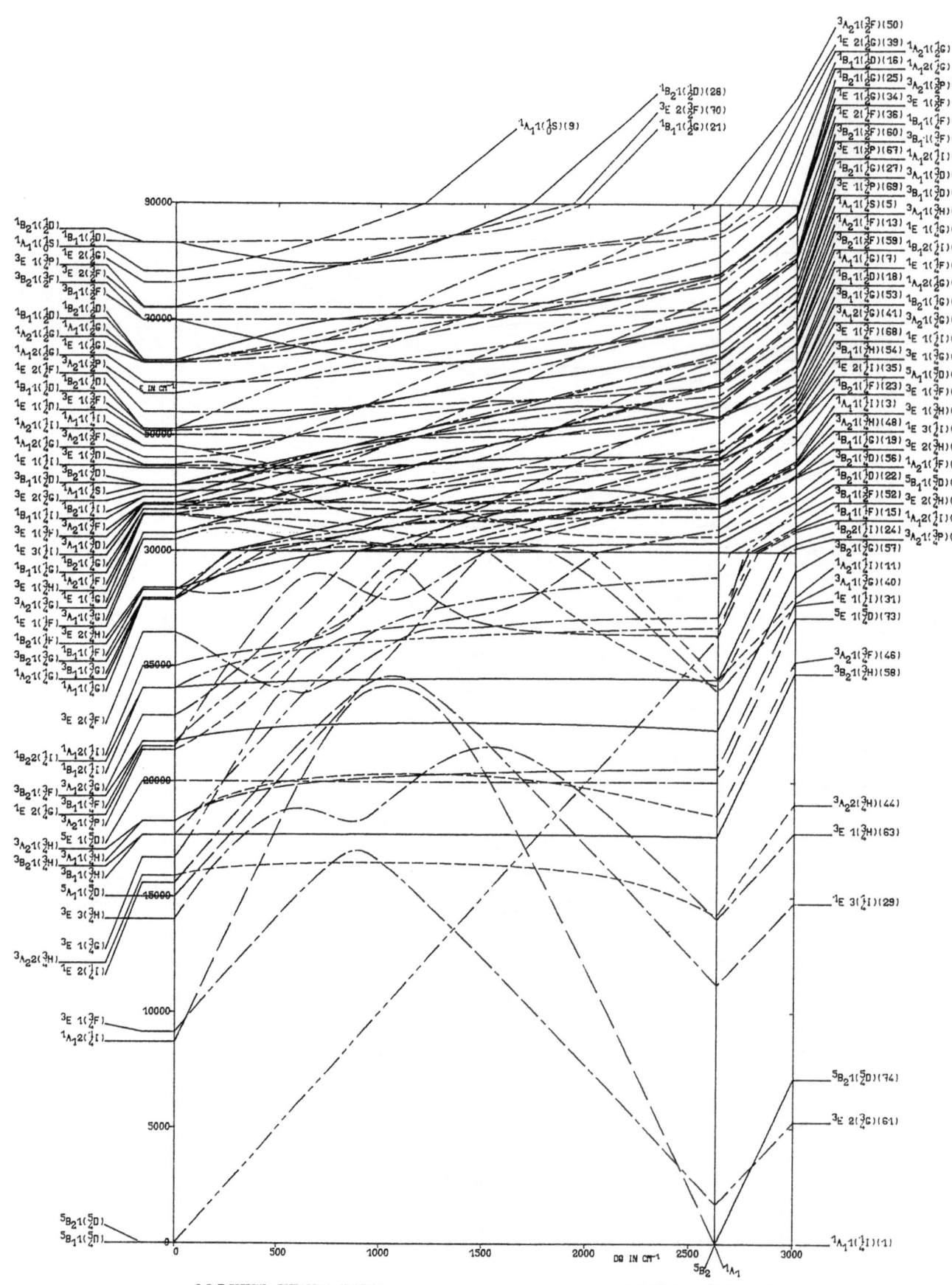

G D ELECTRONS, TETRAGONAL SYMMETRY

B=806 C=4×B DT=1000 K=-5 DS=K×DT ZETA=0

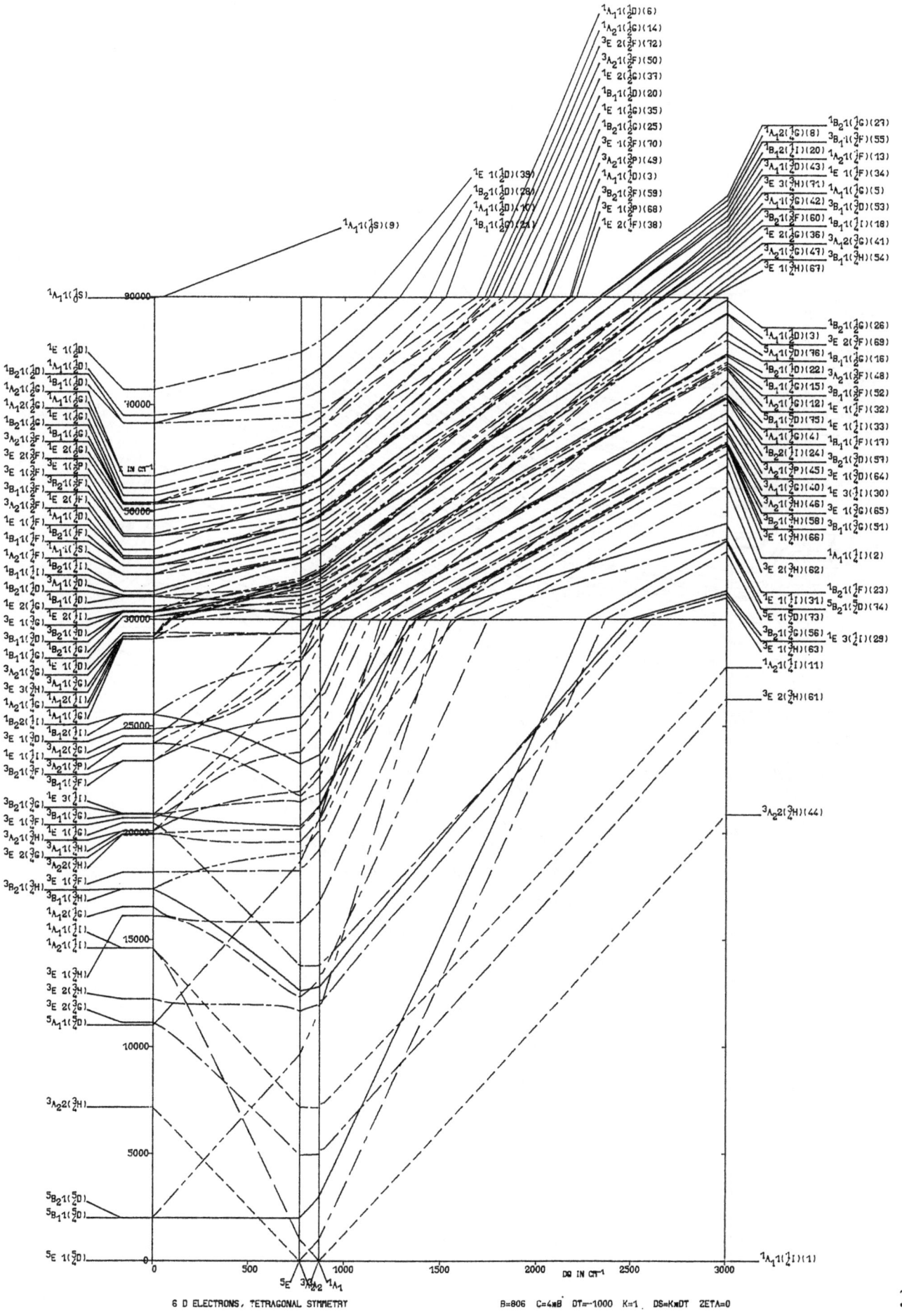

6 D ELECTRONS, TETRAGONAL SYMMETRY   B=806 C=4×B DT=-1000 K=1 DS=K×DT ZETA=0

305

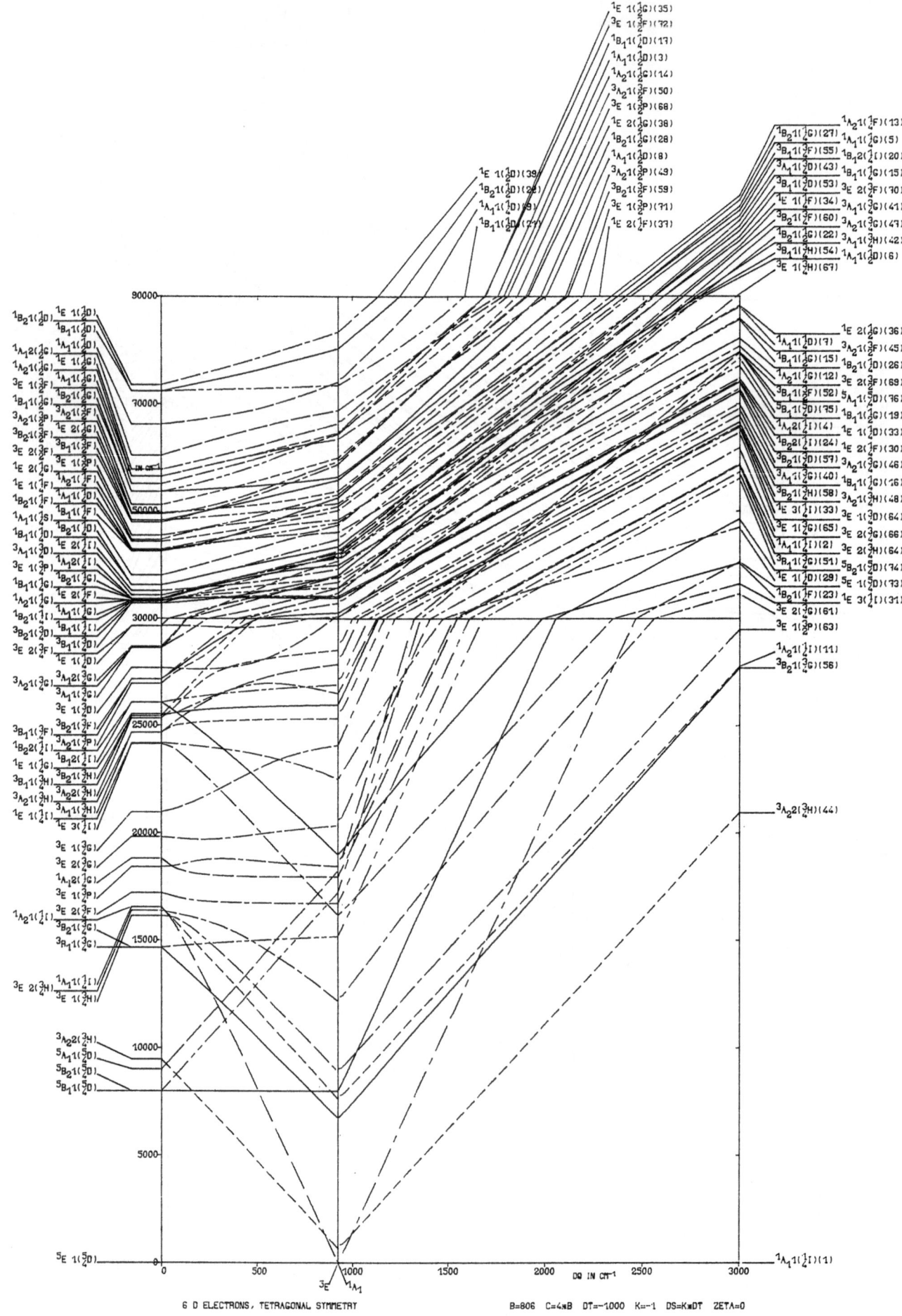

6 D ELECTRONS, TETRAGONAL SYMMETRY

B=806 C=4×B DT=-1000 K=-1 DS=K×DT ZETA=0

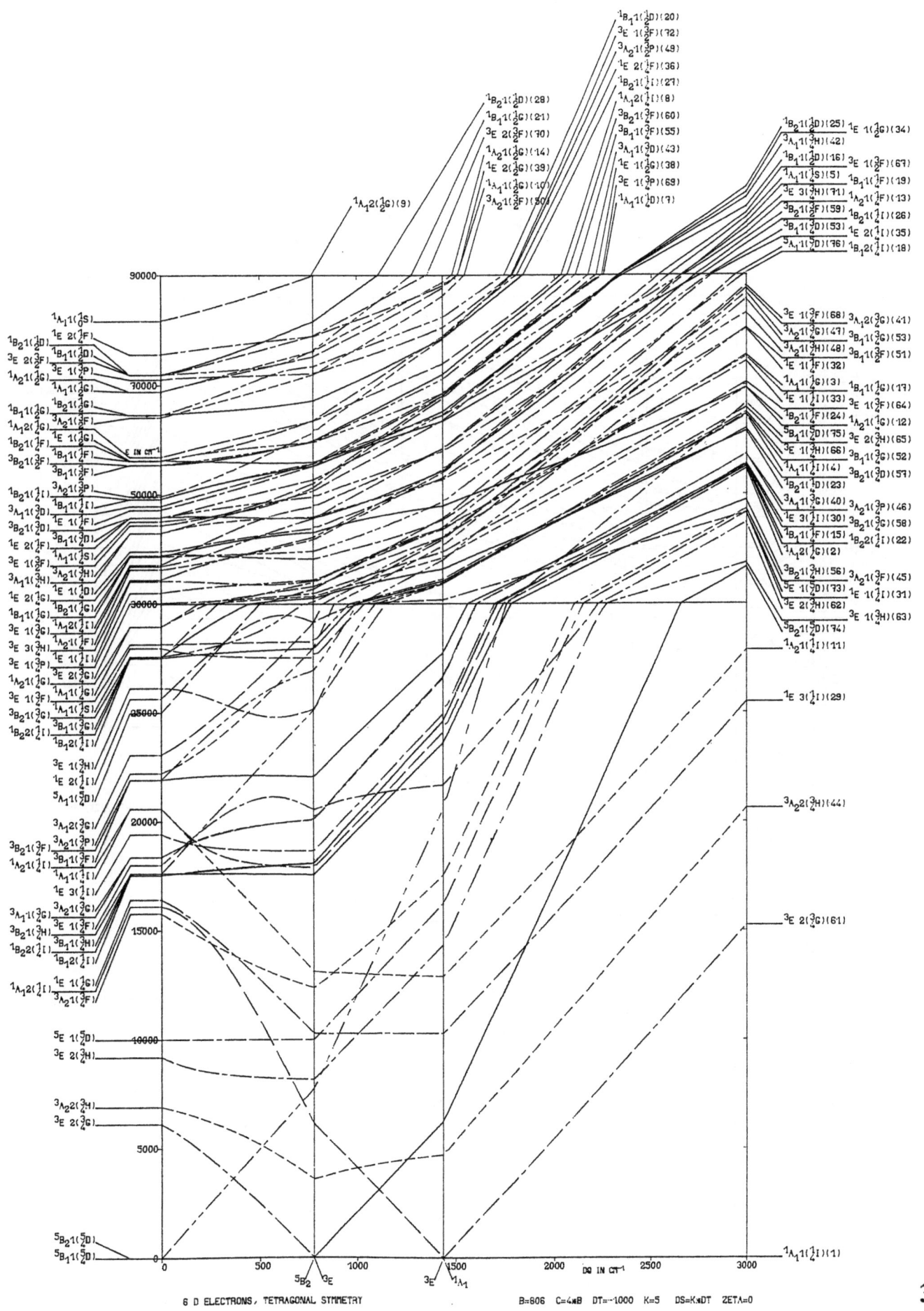

6 D ELECTRONS, TETRAGONAL SYMMETRY   B=806 C=4xB DT=-1000 K=5 DS=KxDT ZETA=0

307

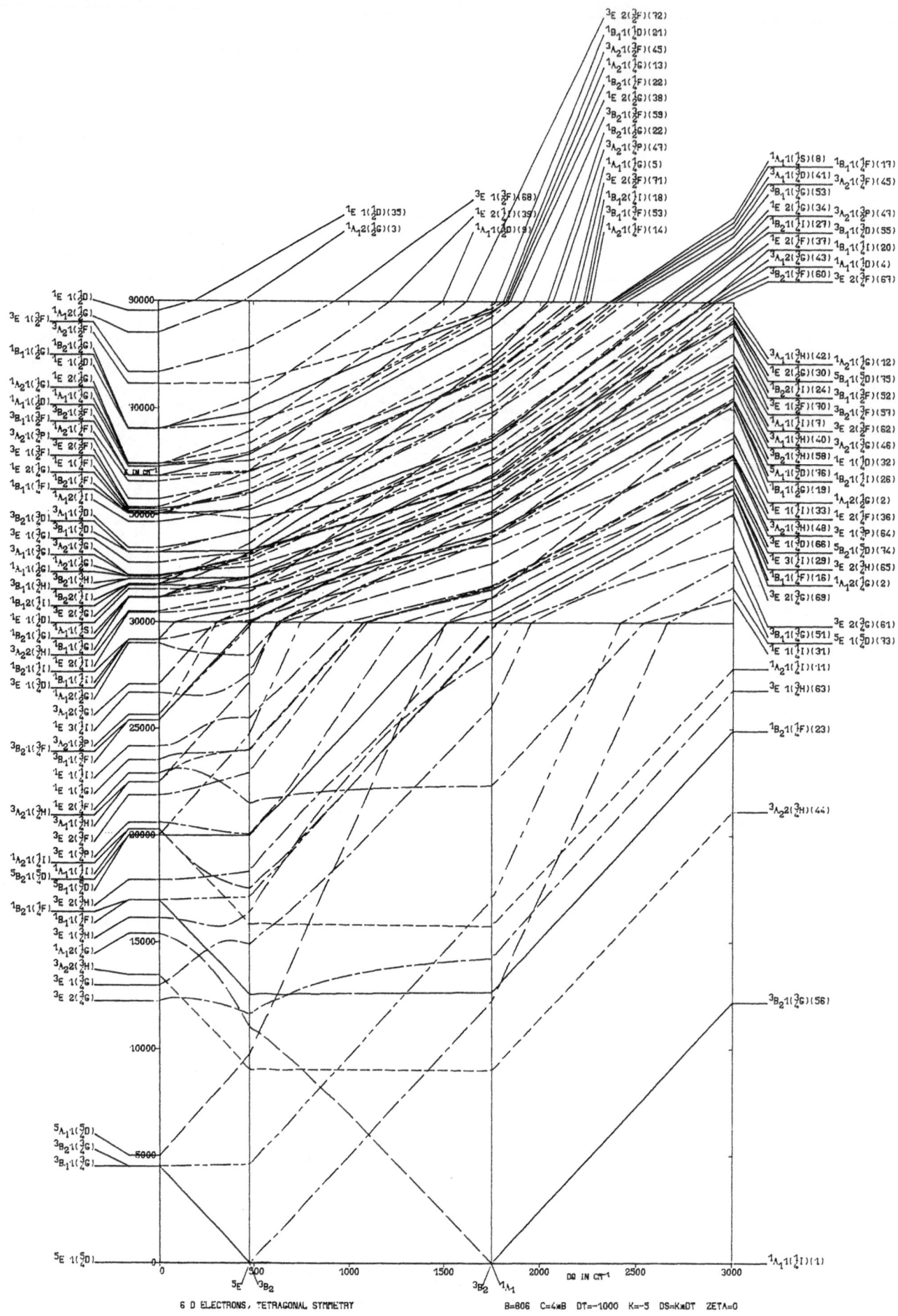

6 D ELECTRONS, TETRAGONAL SYMMETRY

B=806   C=4×B   DT=-1000   K=-5   DS=K×DT   ZETA=0

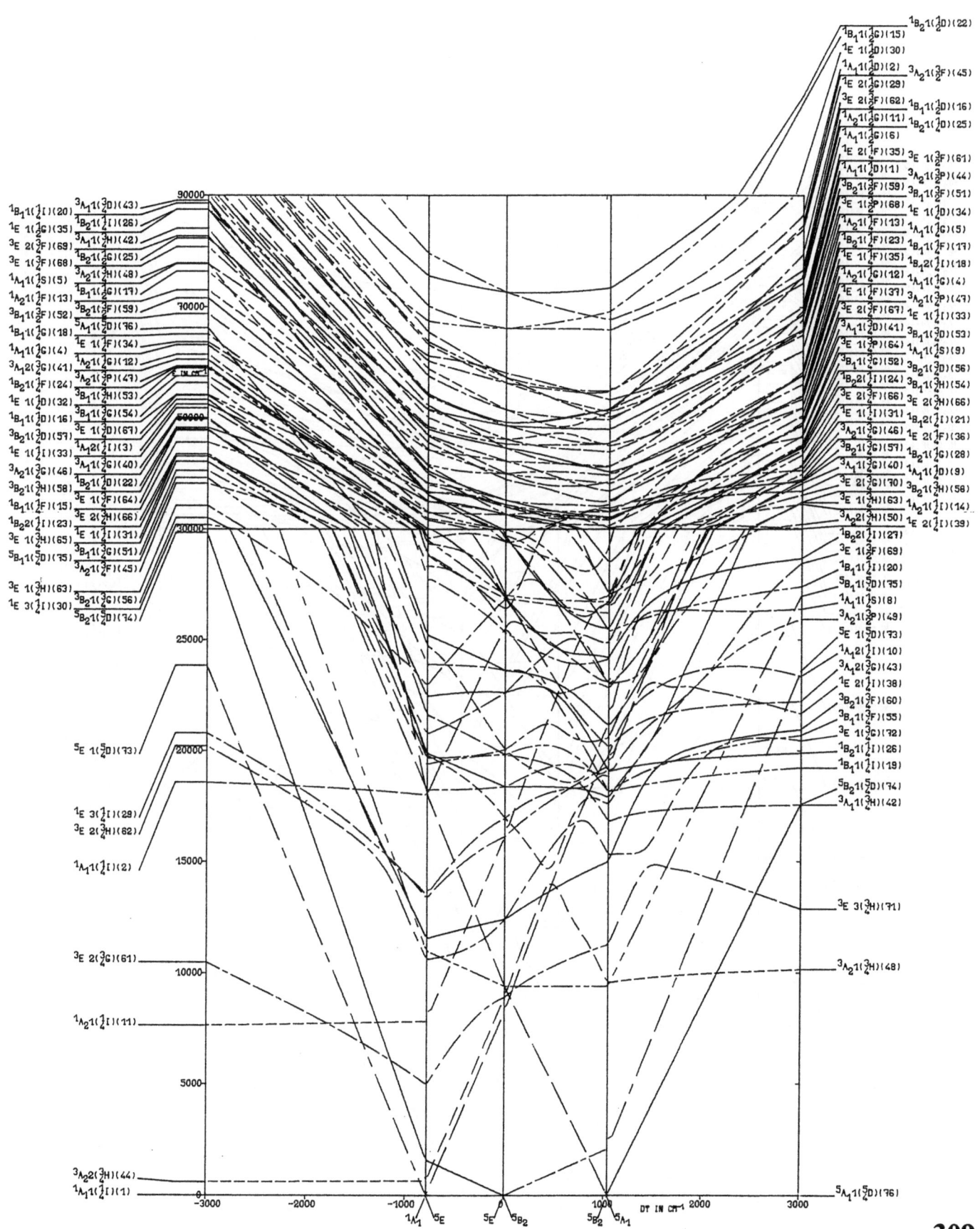

6 D ELECTRONS, TETRAGONAL SYMMETRY   B=806 C=4×B DQ=940 K=1 DS=K×DT ZETA=0

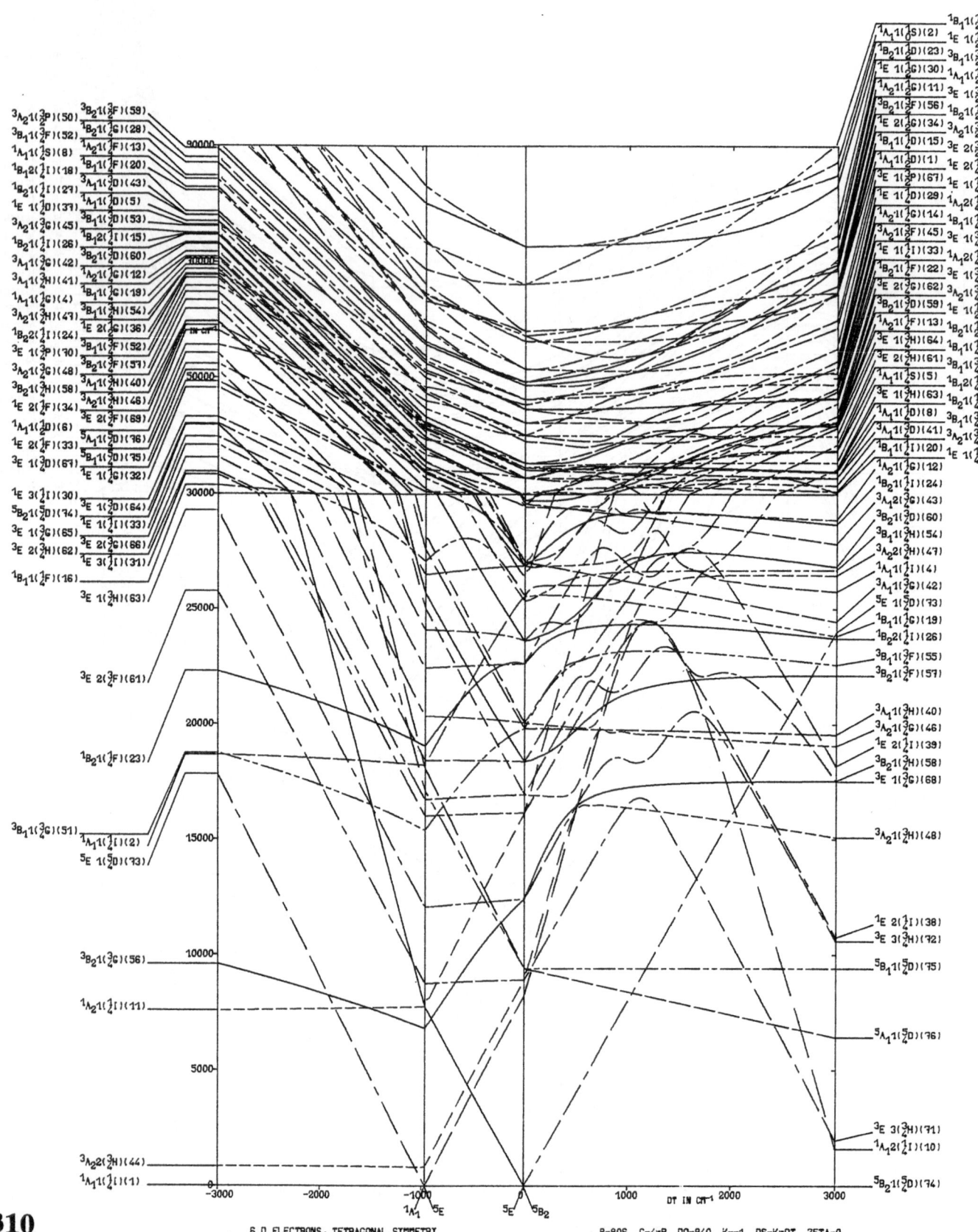

6 D ELECTRONS, TETRAGONAL SYMMETRY

B=806 C=4×B DQ=940 K=-1 DS=K×DT ZETA=0

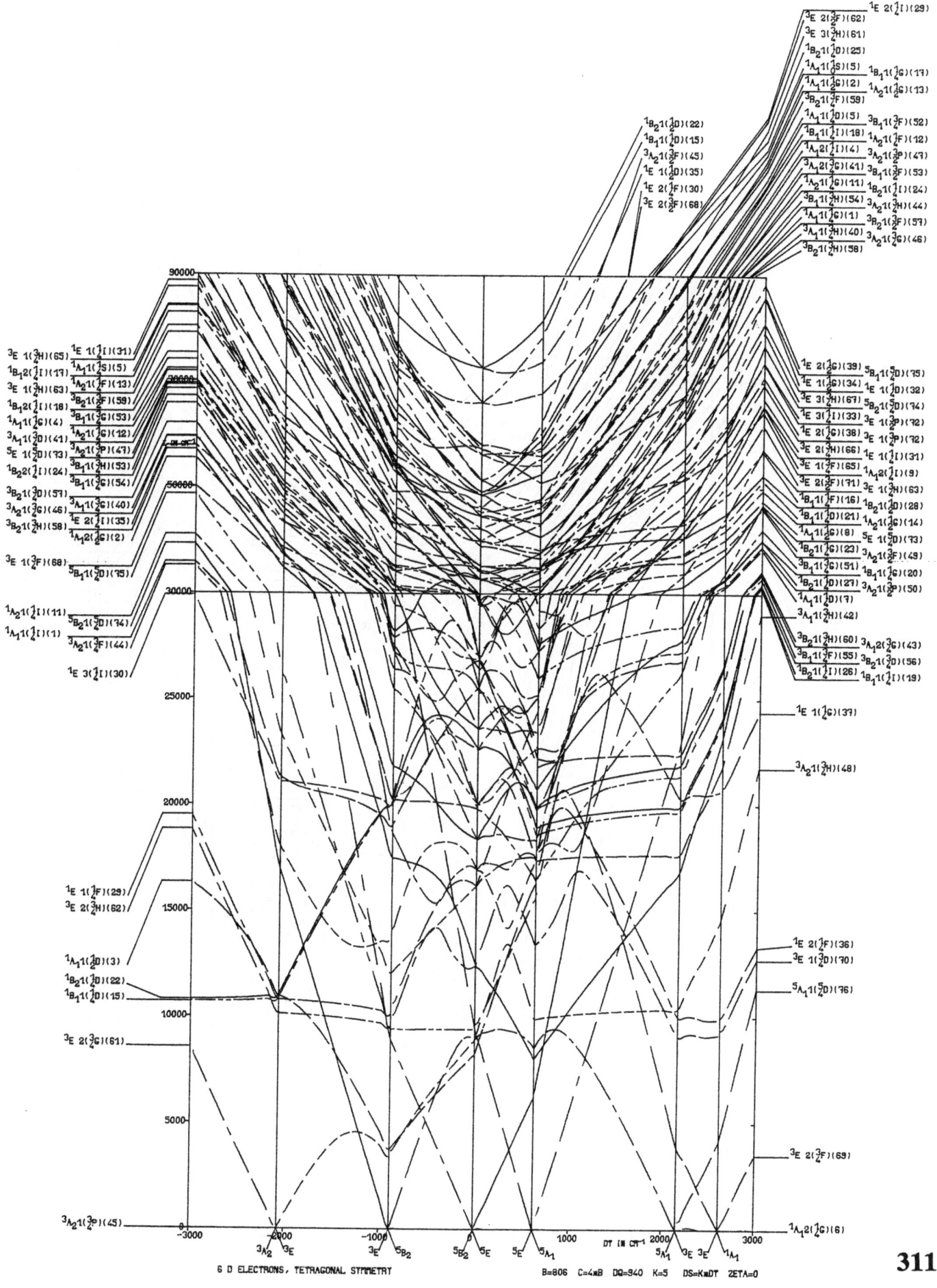

6 D ELECTRONS, TETRAGONAL SYMMETRY   B=806 C=4mB DQ=940 K=5   DS=KmDT ZETA=0

311

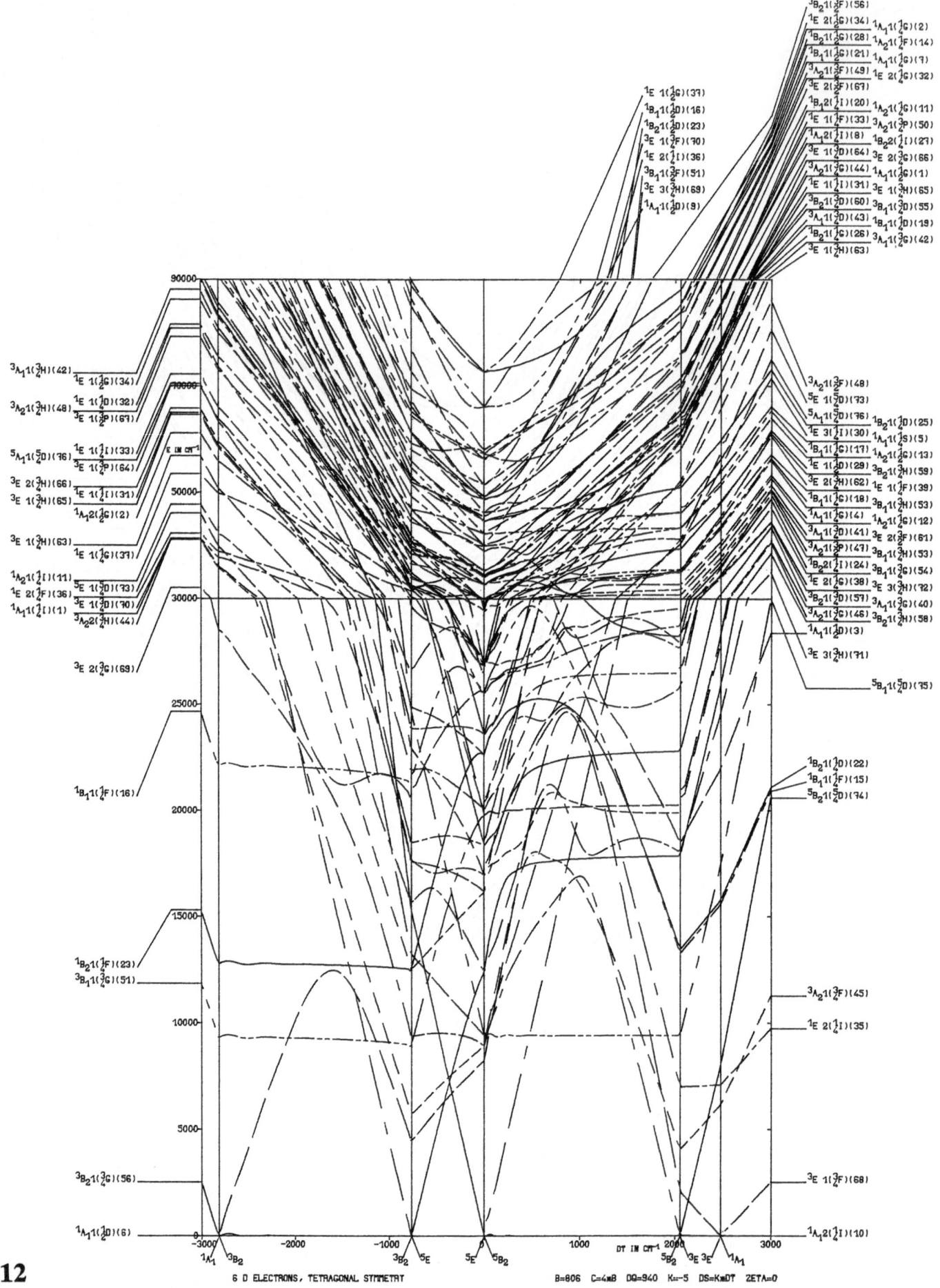

312

6 D ELECTRONS, TETRAGONAL SYMMETRY   B=806 C=4×B DQ=940 K=-5 DS=K×DT ZETA=0

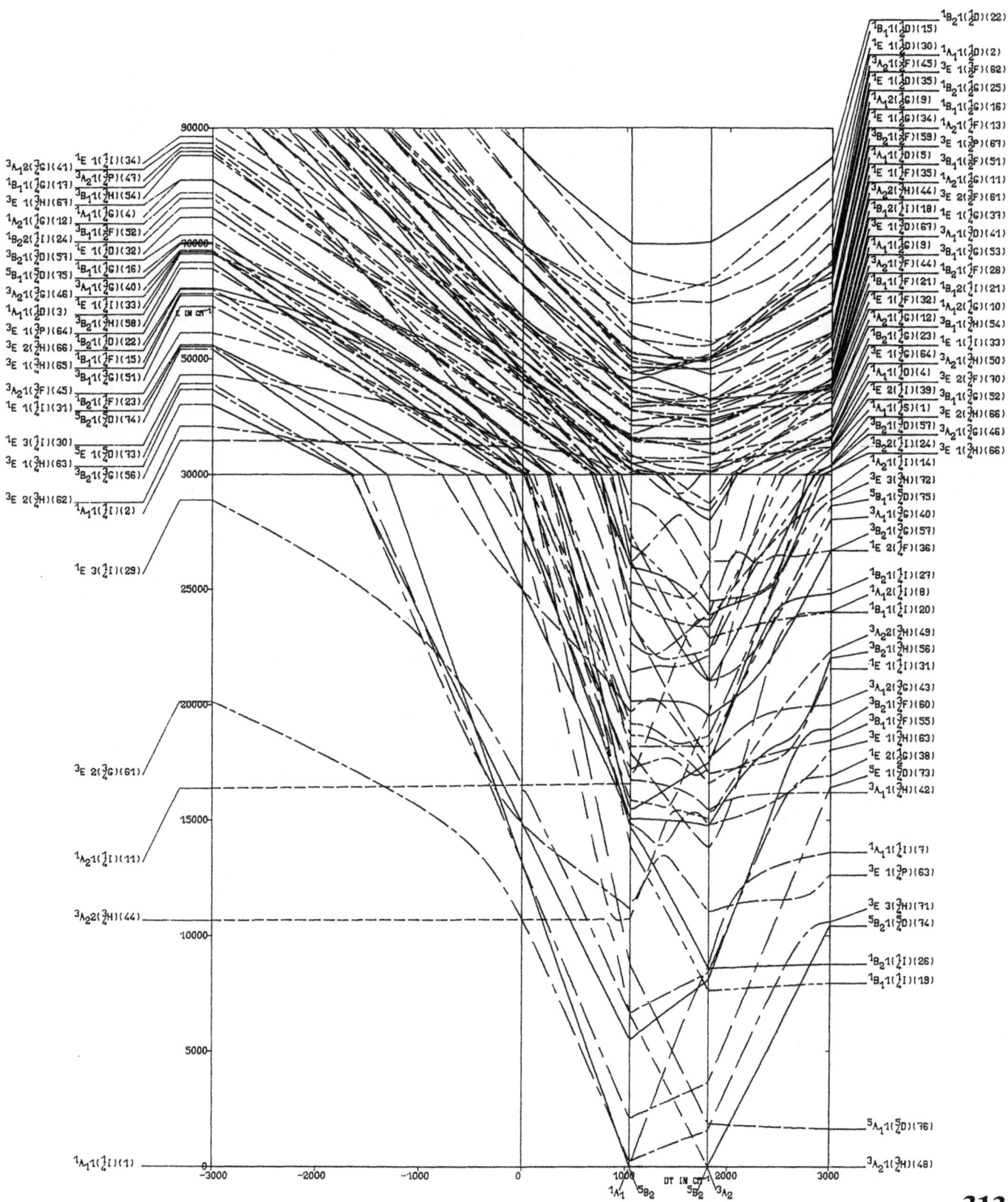

6 D ELECTRONS, TETRAGONAL SYMMETRY

B=670 C=4×B DQ=1820 K=1 DS=K×DT ZETA=0

313

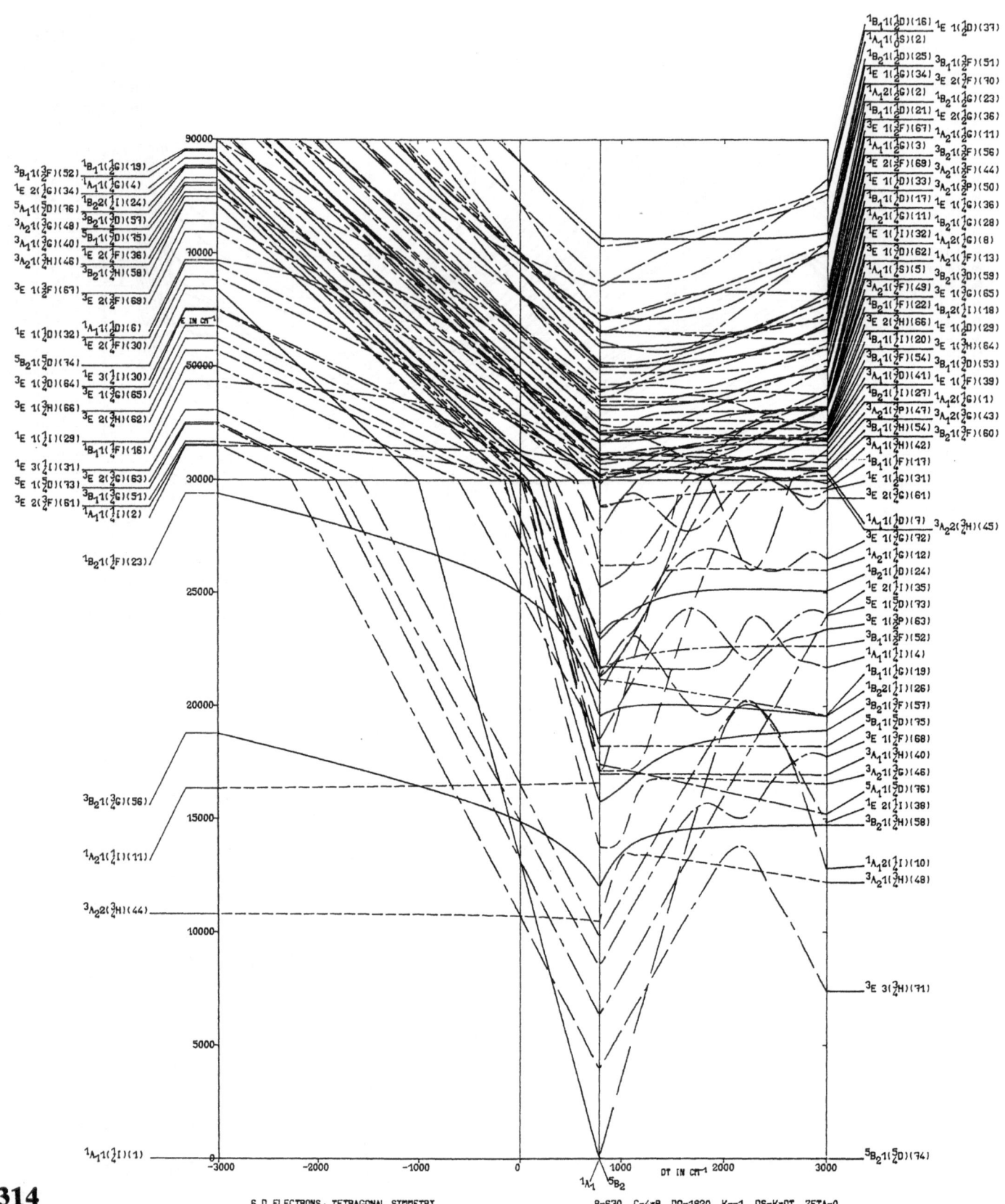

6 D ELECTRONS, TETRAGONAL SYMMETRY

B=670   C=4×B   DQ=1820   K=-1   DS=K×DT   ZETA=0

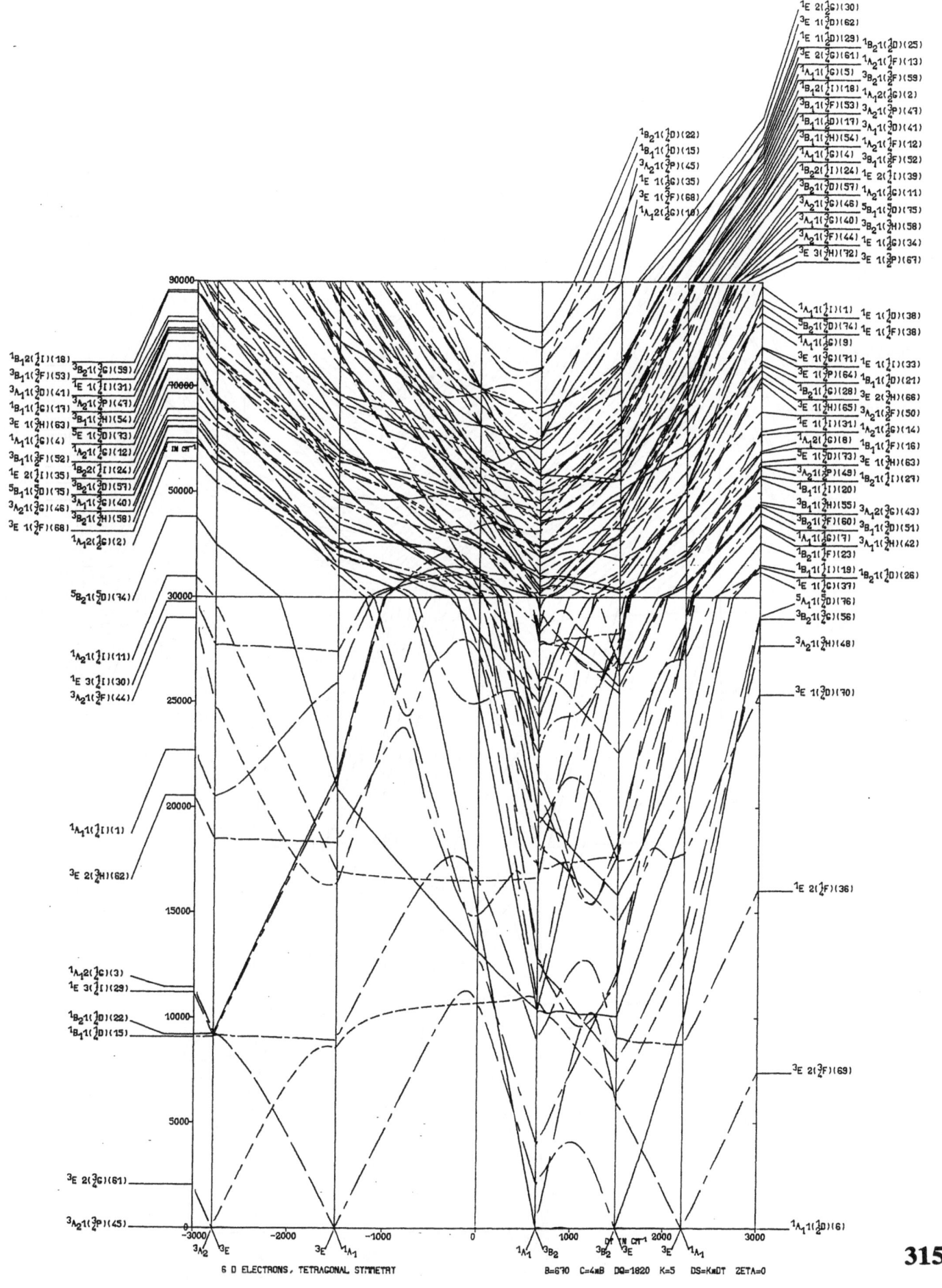

6 D ELECTRONS, TETRAGONAL SYMMETRY   B=670   C=4×B   DQ=1820   K=5   DS=K×DT   ZETA=0

315

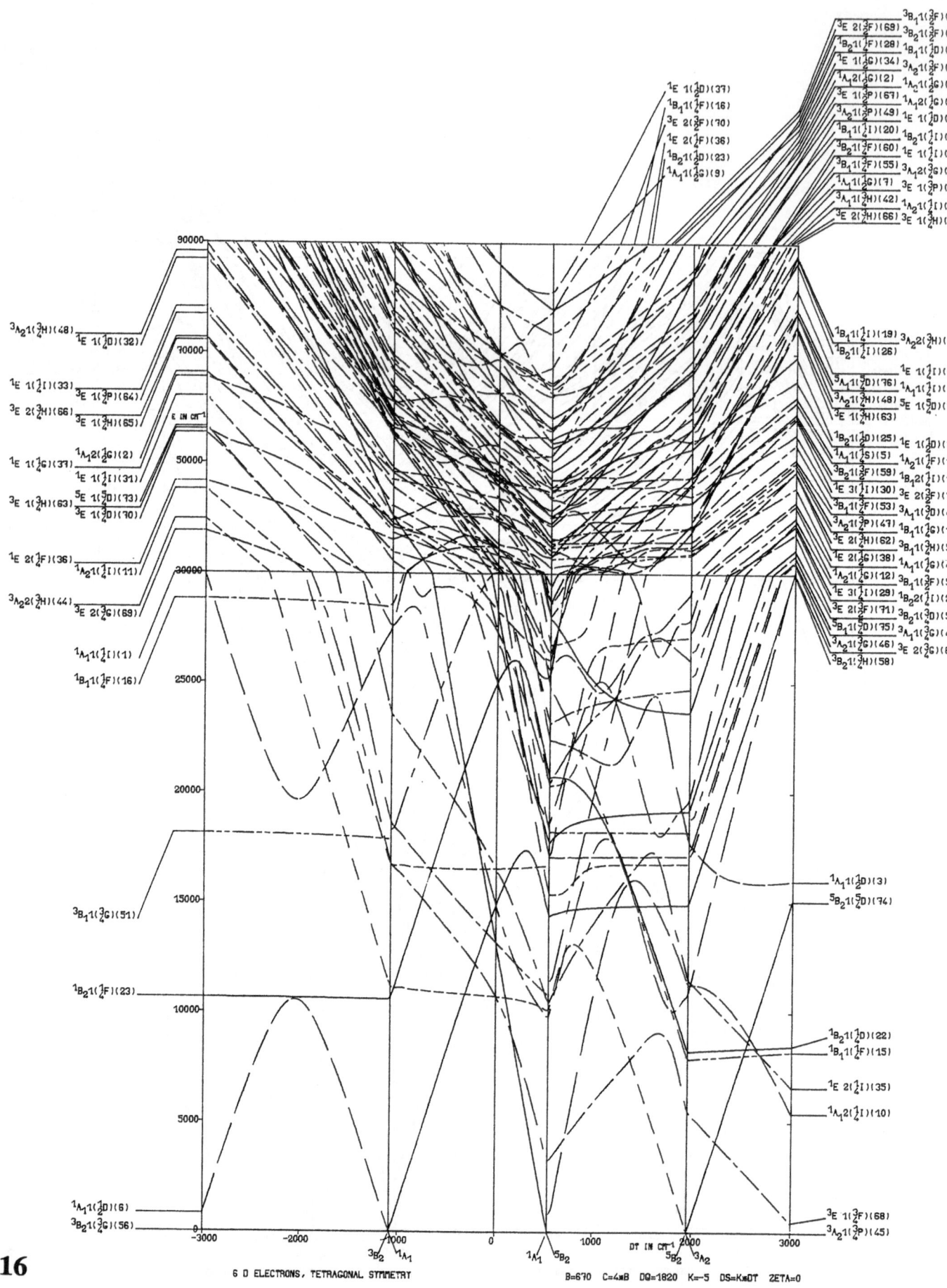

316

6 D ELECTRONS, TETRAGONAL SYMMETRY   B=670 C=4×B DQ=1820 K=-5 DS=K×DT ZETA=0

6 D ELECTRONS, TRIGONAL SYMMETRY

$B=806$ $C=4 \times B$ $ZETA=0$

ENERGY AS FUNCTION OF DQ
$DT=0, 500, -500, 1000, -1000$
$K=1, -1, 5, -5$

ENERGY AS FUNCTION OF DT
$B=806$ $DQ=940$ / $B=670$ $DQ=1820$
$K=1, -1, 5, -5$

(67) $^5A_1(E_TE_T(^3A_2)E_EE_E(^3A_2))$
(66) $^5E (A_1E_T(^3E)E_EE_E(^3A_2))$
(65) $^5E (A_1E_E(^3E)E_TE_T(^3A_2))$
(64) $^3E (E_TE_E(^1E)E_EE_E(^3A_2))$
(63) $^3E (E_TE_T(^3A_2)E_EE_E(^1E))$
(62) $^3E (E_TE_T(^1E)E_EE_E(^3A_2))$
(61) $^3E (E_TE_E(^1E)E_TE_T(^3A_2))$
(60) $^3E (A_1E_E(^1E)E_EE_E(^3A_2))$
(59) $^3E (A_1E_T(^3E)E_EE_E(^3A_2))$
(58) $^3E (A_1E_T(^3E)E_EE_E(^1E))$
(57) $^3E (A_1E_T(^3E)E_EE_E(^1A_1))$
(56) $^3E (A_1E_T(^1E)E_EE_E(^3A_2))$
(55) $^3E (A_1E_E(^3E)E_TE_T(^3A_2))$
(54) $^3E (A_1E_E(^3E)E_TE_T(^1E))$
(53) $^3E (A_1E_E(^3E)E_TE_T(^1A_1))$
(52) $^3E (A_1E_E(^1E)E_TE_T(^3A_2))$
(51) $^3E (A_1E_T(^1E)E_TE_T(^3A_2))$
(50) $^3E (A_1A_1(^1A_1)E_TE_T(^3E))$
(49) $^3A_2(E_TE_E(^1A_1)E_EE_E(^3A_2))$
(48) $^3A_2(E_TE_T(^3A_2)E_EE_E(^1A_1))$
(47) $^3A_2(E_TE_T(^1A_1)E_EE_E(^3A_2))$
(46) $^3A_2(E_TE_E(^1A_1)E_TE_T(^3A_2))$
(45) $^3A_2(A_1E_T(^3E)E_EE_E(^1E))$
(44) $^3A_2(A_1E_E(^3E)E_TE_T(^1E))$
(43) $^3A_2(A_1A_1(^1A_1)E_EE_E(^3A_2))$
(42) $^3A_2(A_1A_1(^1A_1)E_TE_E(^3A_2))$
(41) $^3A_2(A_1A_1(^1A_1)E_TE_T(^3A_2))$
(40) $^3A_1(E_TE_E(^1A_2)E_EE_E(^3A_2))$
(39) $^3A_1(E_TE_T(^3A_2)E_EE_E(^3A_2))$
(38) $^3A_1(E_TE_E(^1A_2)E_TE_T(^3A_2))$
(37) $^3A_1(A_1E_T(^3E)E_EE_E(^1E))$
(36) $^3A_1(A_1E_E(^3E)E_TE_T(^1E))$
(35) $^3A_1(A_1A_1(^1A_1)E_TE_E(^3A_1))$
(34) $^1E (E_TE_E(^1A_1)E_EE_E(^1E))$

(33) $^1E (E_TE_T(^1E)E_EE_E(^1E))$
(32) $^1E (E_TE_T(^1E)E_EE_E(^1A_1))$
(31) $^1E (E_TE_T(^1A_1)E_EE_E(^1E))$
(30) $^1E (E_TE_E(^1A_1)E_TE_T(^1E))$
(29) $^1E (A_1E_E(^1E)E_EE_E(^1A_1))$
(28) $^1E (A_1E_T(^3E)E_EE_E(^3A_2))$
(27) $^1E (A_1E_T(^1E)E_EE_E(^1E))$
(26) $^1E (A_1E_T(^1E)E_EE_E(^1A_1))$
(25) $^1E (A_1E_E(^3E)E_TE_T(^3A_2))$
(24) $^1E (A_1E_E(^1E)E_TE_T(^1E))$
(23) $^1E (A_1E_E(^1E)E_TE_T(^1A_1))$
(22) $^1E (A_1E_T(^1E)E_TE_T(^1A_1))$
(21) $^1E (A_1A_1(^1A_1)E_EE_E(^1E))$
(20) $^1E (A_1A_1(^1A_1)E_TE_T(^1E))$
(19) $^1E (A_1A_1(^1A_1)E_TE_T(^1E))$
(18) $^1A_2(E_TE_T(^1A_2)E_EE_E(^1A_1))$
(17) $^1A_2(E_TE_T(^1E)E_EE_E(^1E))$
(16) $^1A_2(E_TE_T(^1A_2)E_TE_T(^1A_1))$
(15) $^1A_2(A_1E_T(^1E)E_EE_E(^1E))$
(14) $^1A_2(A_1E_E(^1E)E_TE_T(^1E))$
(13) $^1A_2(A_1A_1(^1A_1)E_TE_E(^1A_2))$
(12) $^1A_1(E_EE_E(^1A_1)E_EE_E(^1A_1))$
(11) $^1A_1(E_TE_E(^1A_1)E_EE_E(^1A_1))$
(10) $^1A_1(E_TE_T(^3A_2)E_EE_E(^3A_2))$
(9) $^1A_1(E_TE_T(^1E)E_EE_E(^1E))$
(8) $^1A_1(E_TE_T(^1A_1)E_EE_E(^1A_1))$
(7) $^1A_1(E_TE_E(^1A_1)E_TE_T(^1A_1))$
(6) $^1A_1(E_TE_T(^1A_1)E_TE_T(^1A_1))$
(5) $^1A_1(A_1E_T(^1E)E_EE_E(^1E))$
(4) $^1A_1(A_1E_E(^1E)E_TE_T(^1E))$
(3) $^1A_1(A_1A_1(^1A_1)E_EE_E(^1A_1))$
(2) $^1A_1(A_1A_1(^1A_1)E_TE_E(^1A_1))$
(1) $^1A_1(A_1A_1(^1A_1)E_TE_T(^1A_1))$

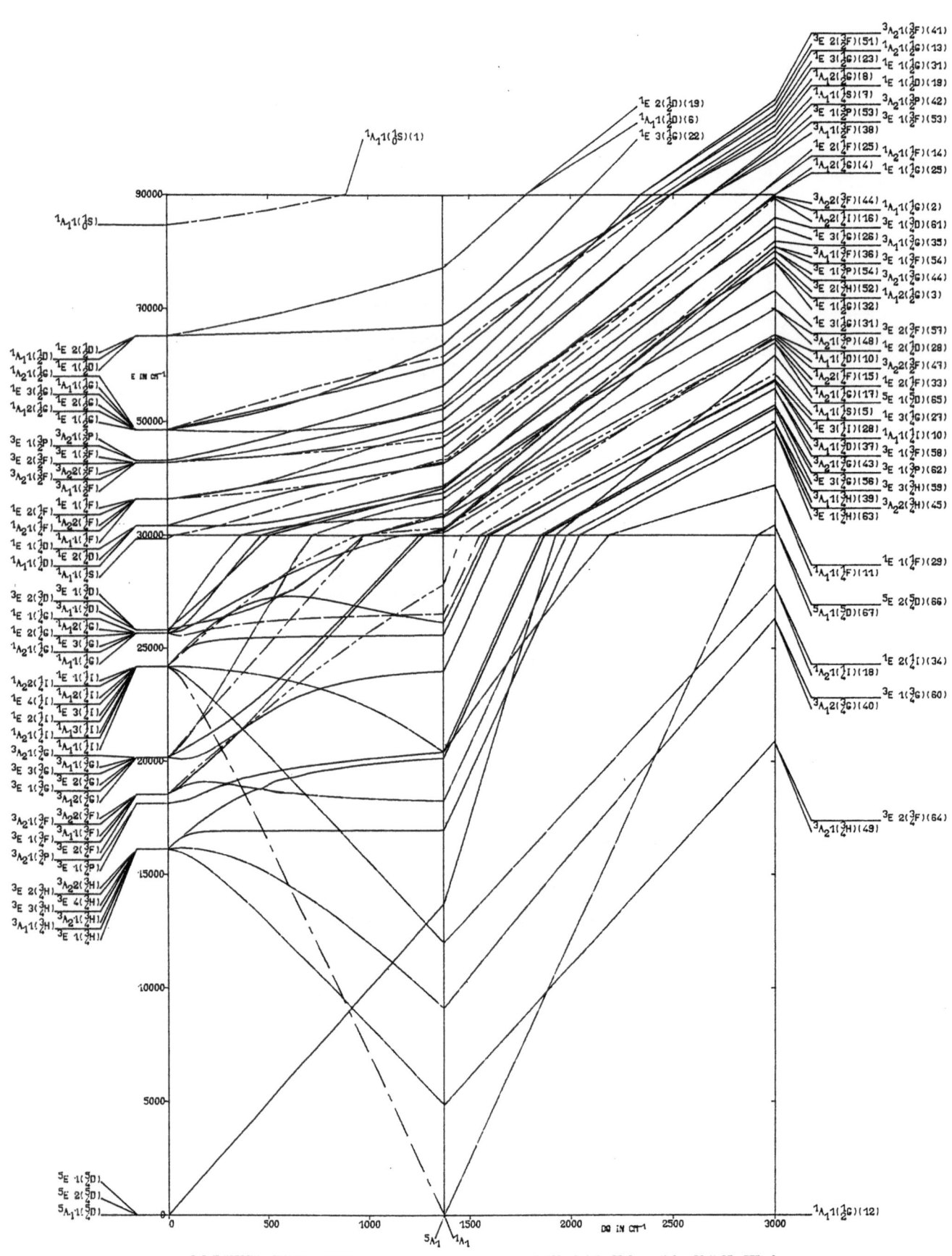

6 D ELECTRONS, TRIGONAL SYMMETRY   B=806 C=4*B DT=0   K=0   DS=K*DT ZETA=0

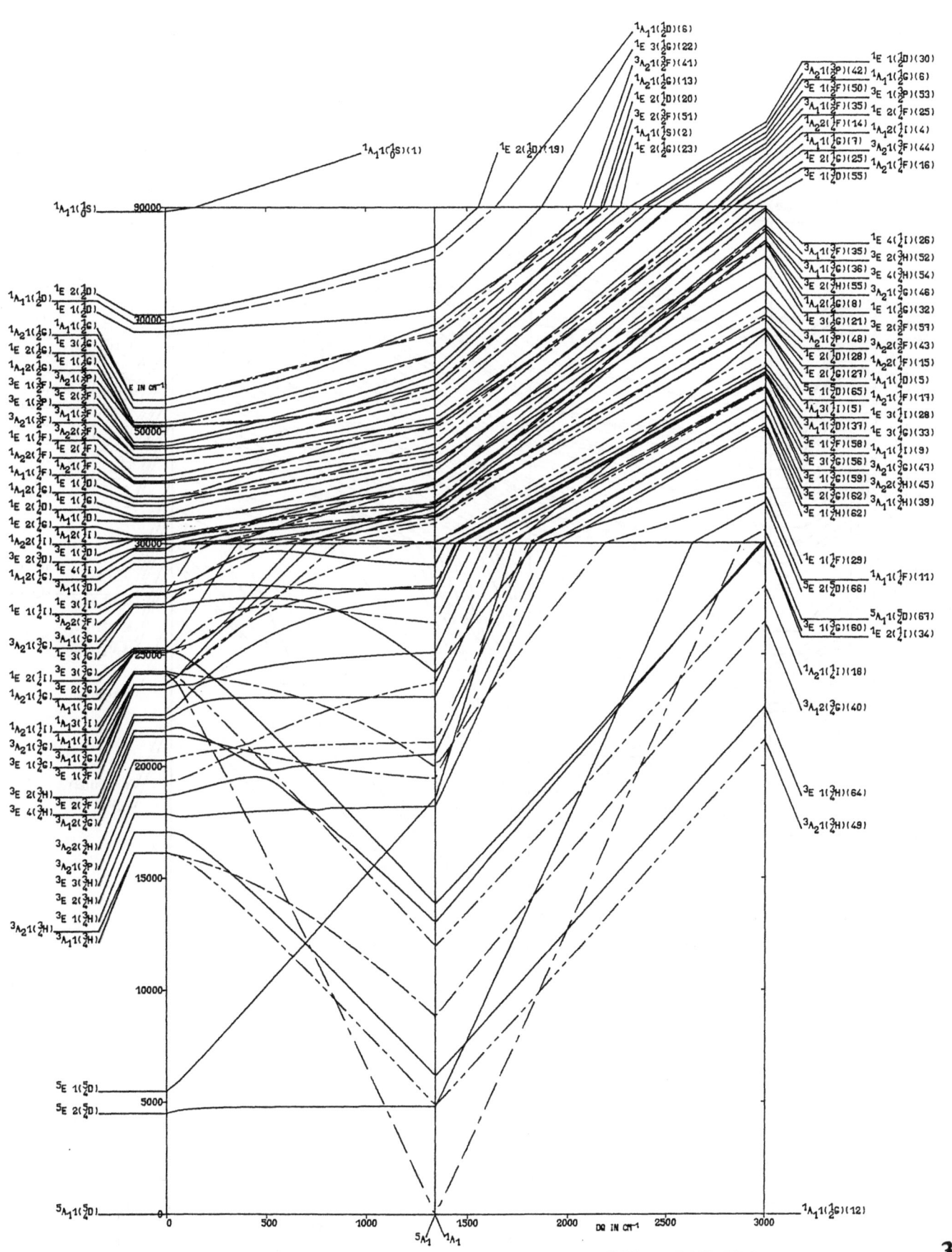

6 D ELECTRONS, TRIGONAL SYMMETRY   B=806   C=4×B   DT=500   K=1   DS=K×DT   ZETA=0

319

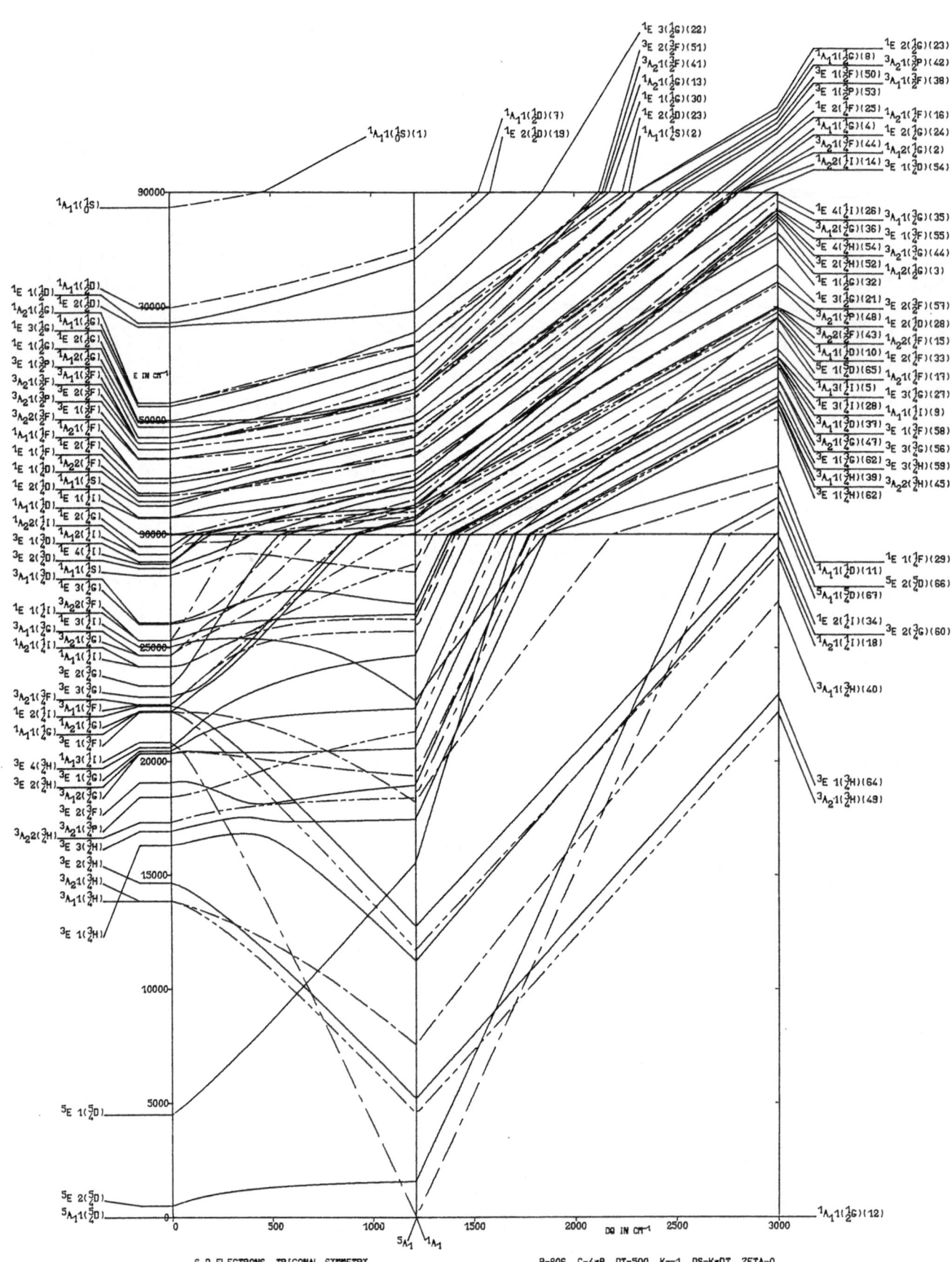

6 D ELECTRONS, TRIGONAL SYMMETRY   B=806 C=4×B DT=500 K=-1 DS=K×DT ZETA=0

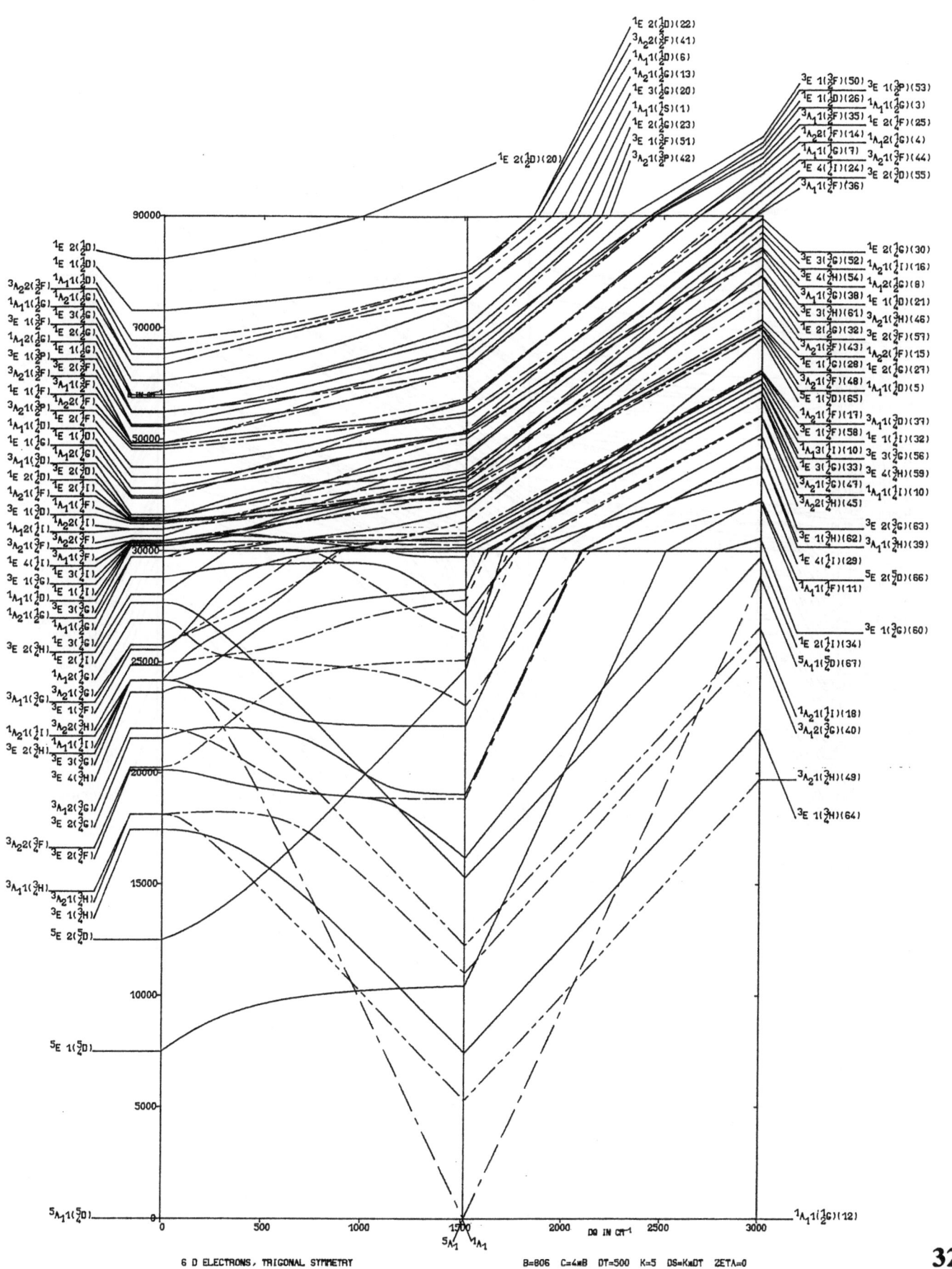

6 D ELECTRONS, TRIGONAL SYMMETRY   B=806 C=4*B DT=500 K=5 DS=K*DT ZETA=0

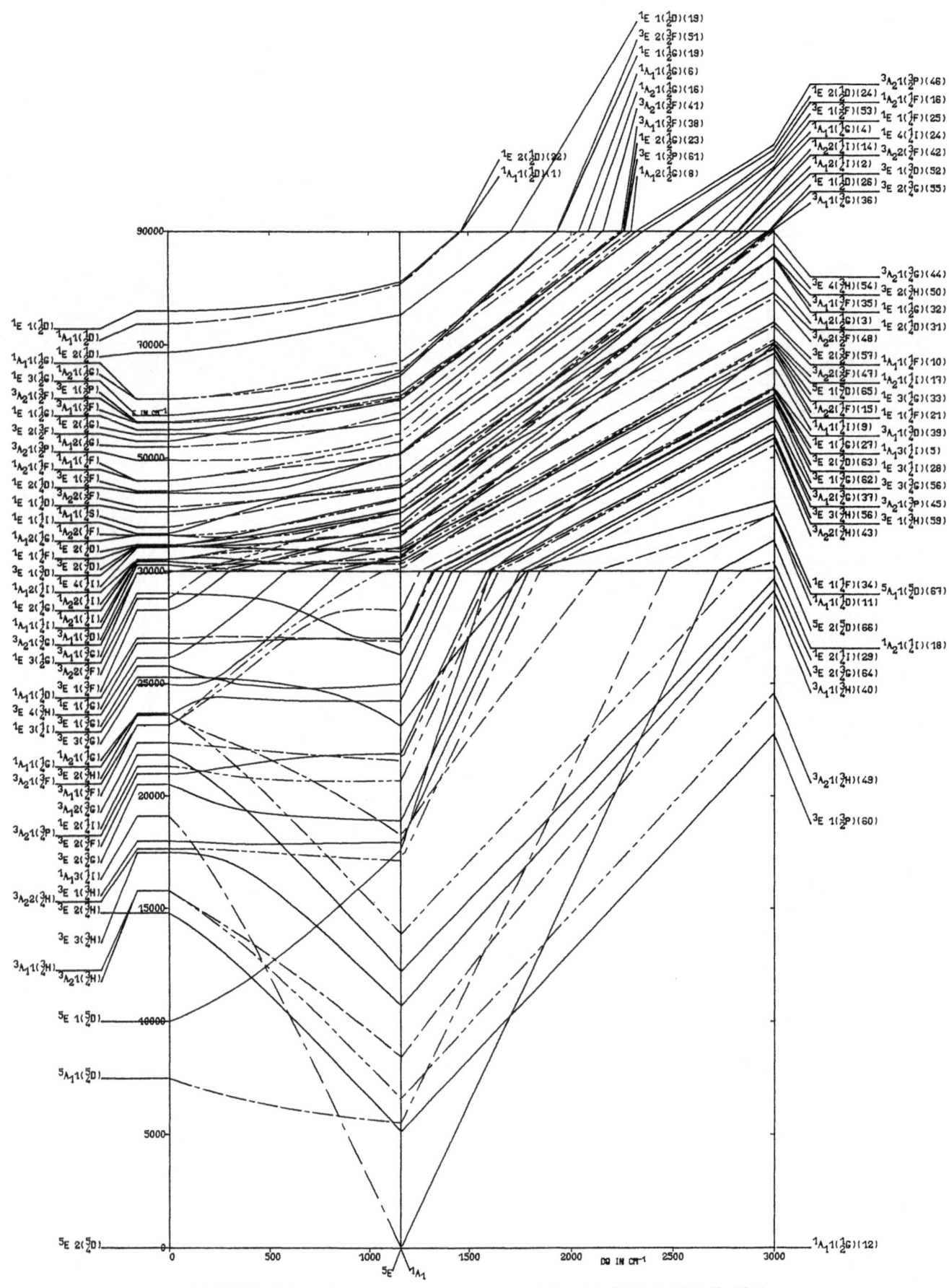

6 D ELECTRONS, TRIGONAL SYMMETRY

B=806 C=4*B DT=500 K=-5 DS=K*DT 2ETA=0

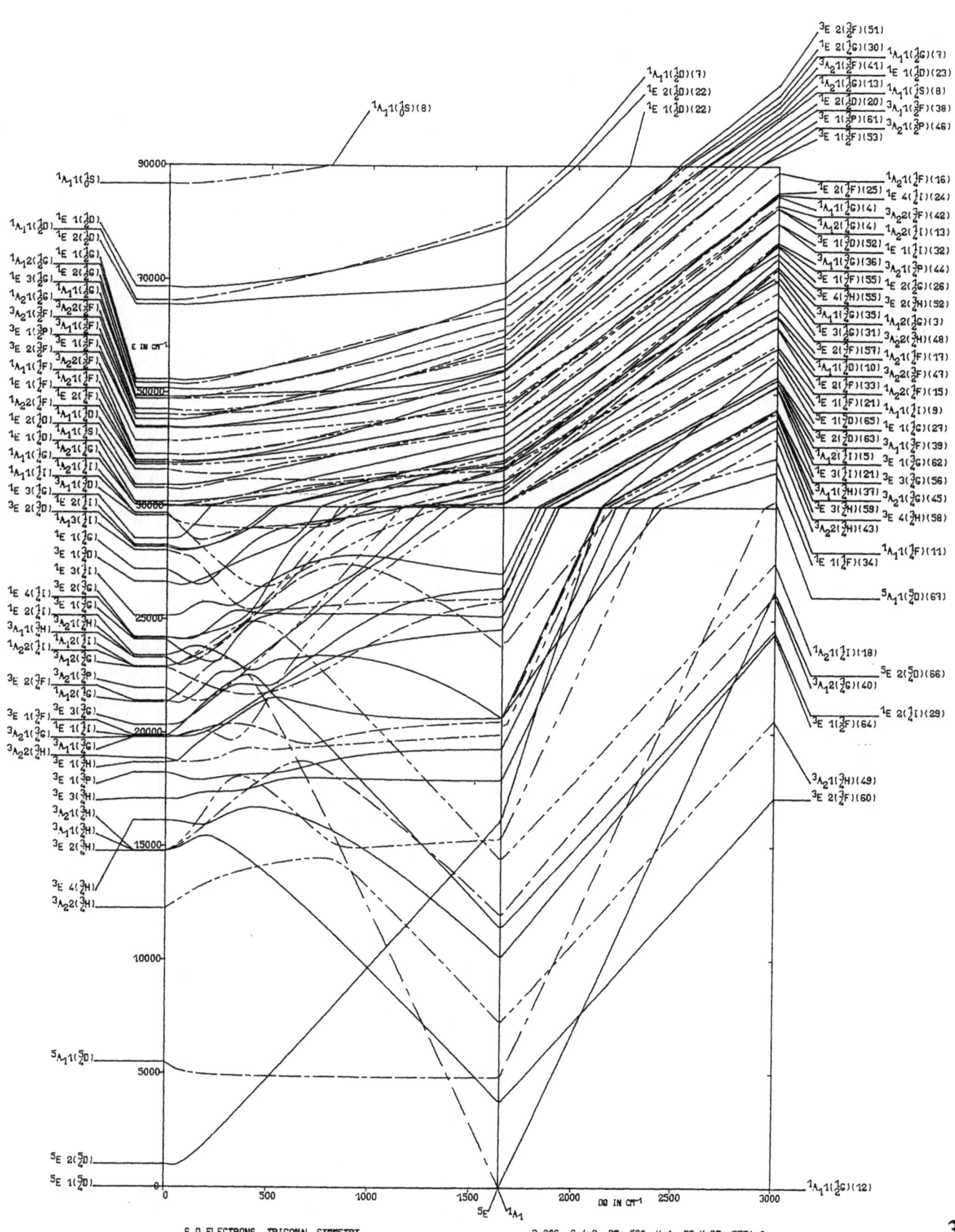

6 D ELECTRONS, TRIGONAL SYMMETRY   B=806 C=4×B DT=-500 K=1 DS=K×DT ZETA=0

323

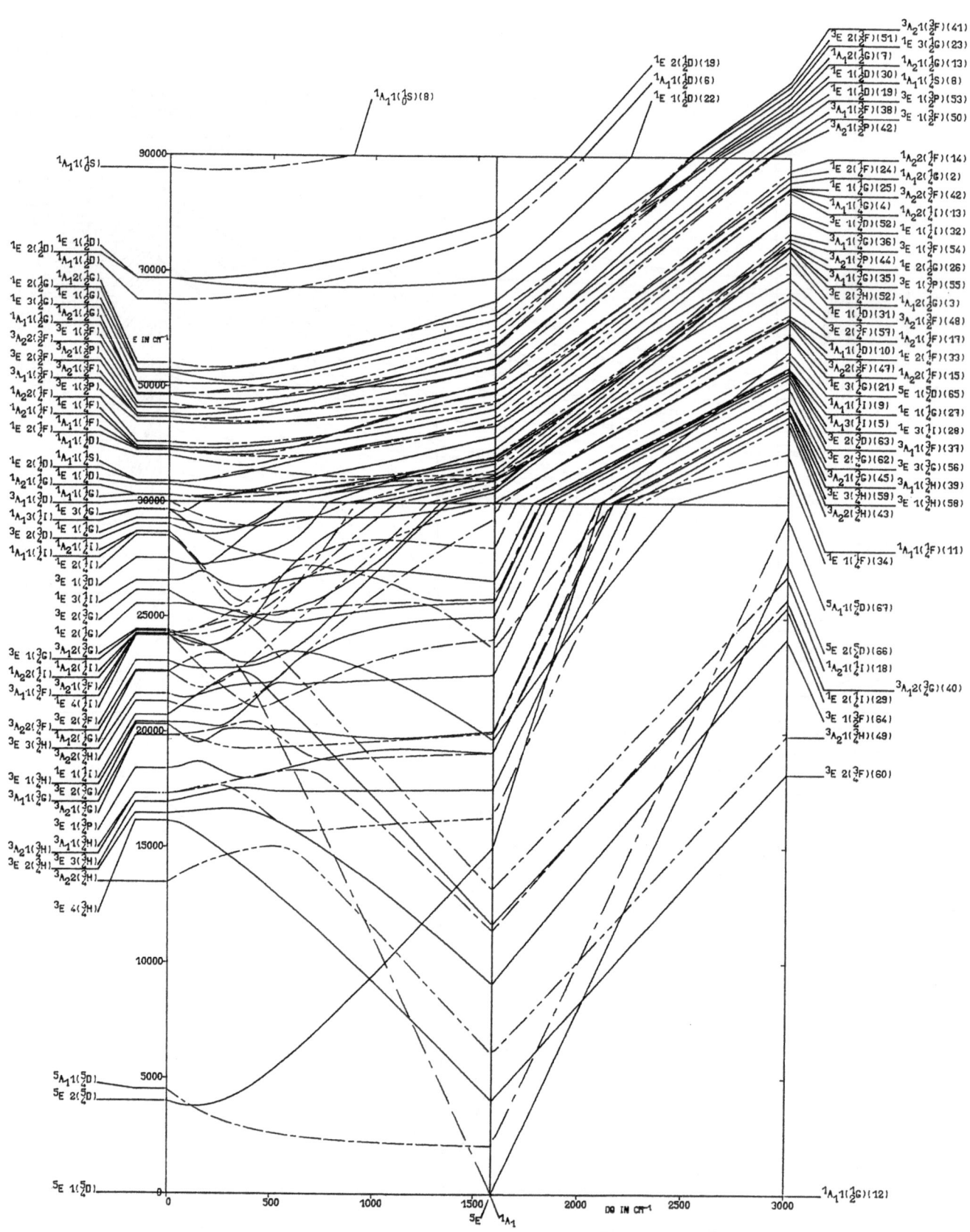

6 D ELECTRONS, TRIGONAL SYMMETRY

B=806   C=4×B   DT=-500   K=-1   DS=K×DT   ZETA=0

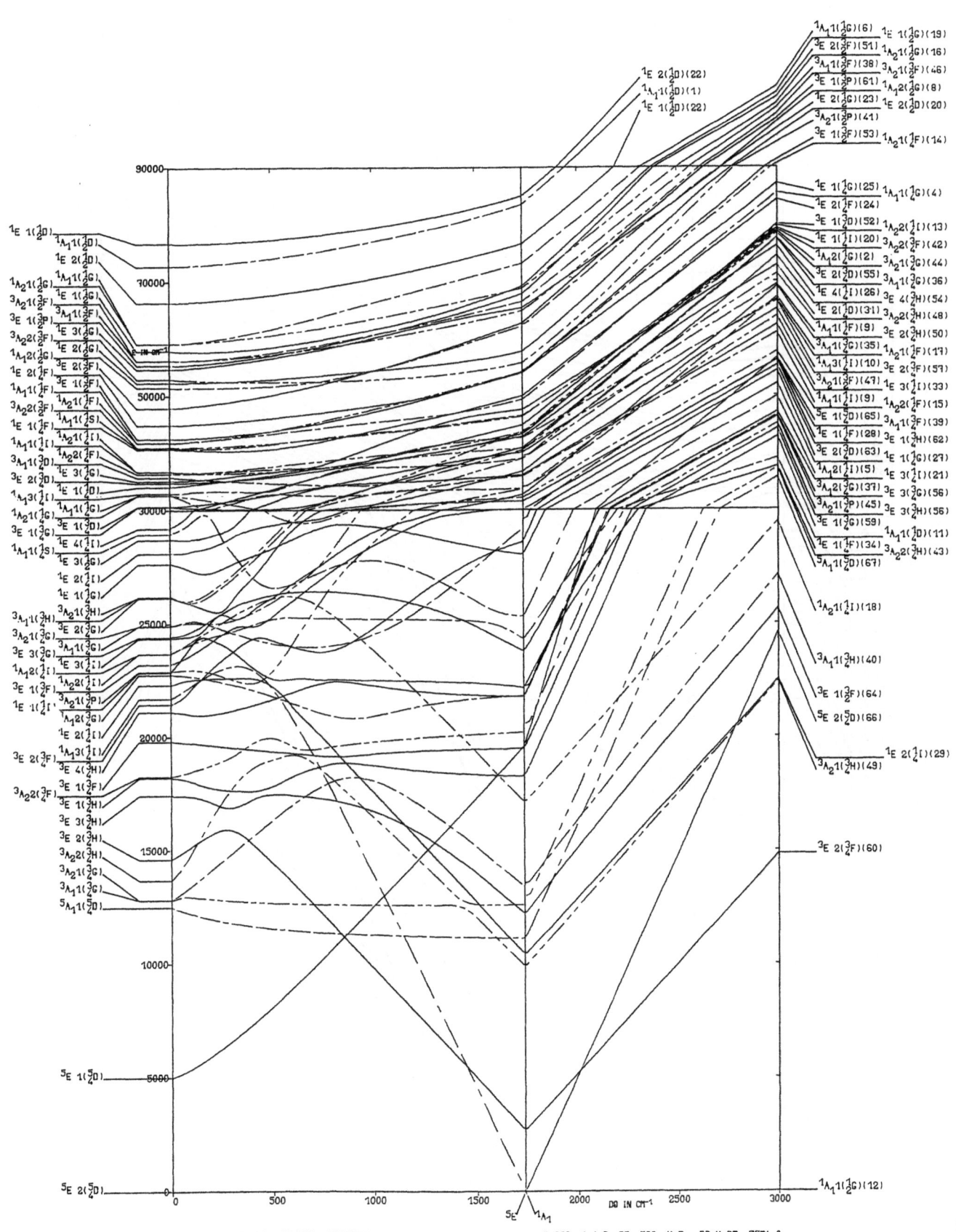

6 D ELECTRONS, TRIGONAL SYMMETRY   B=806 C=4*B DT=-500 K=5   DS=K*DT ZETA=0

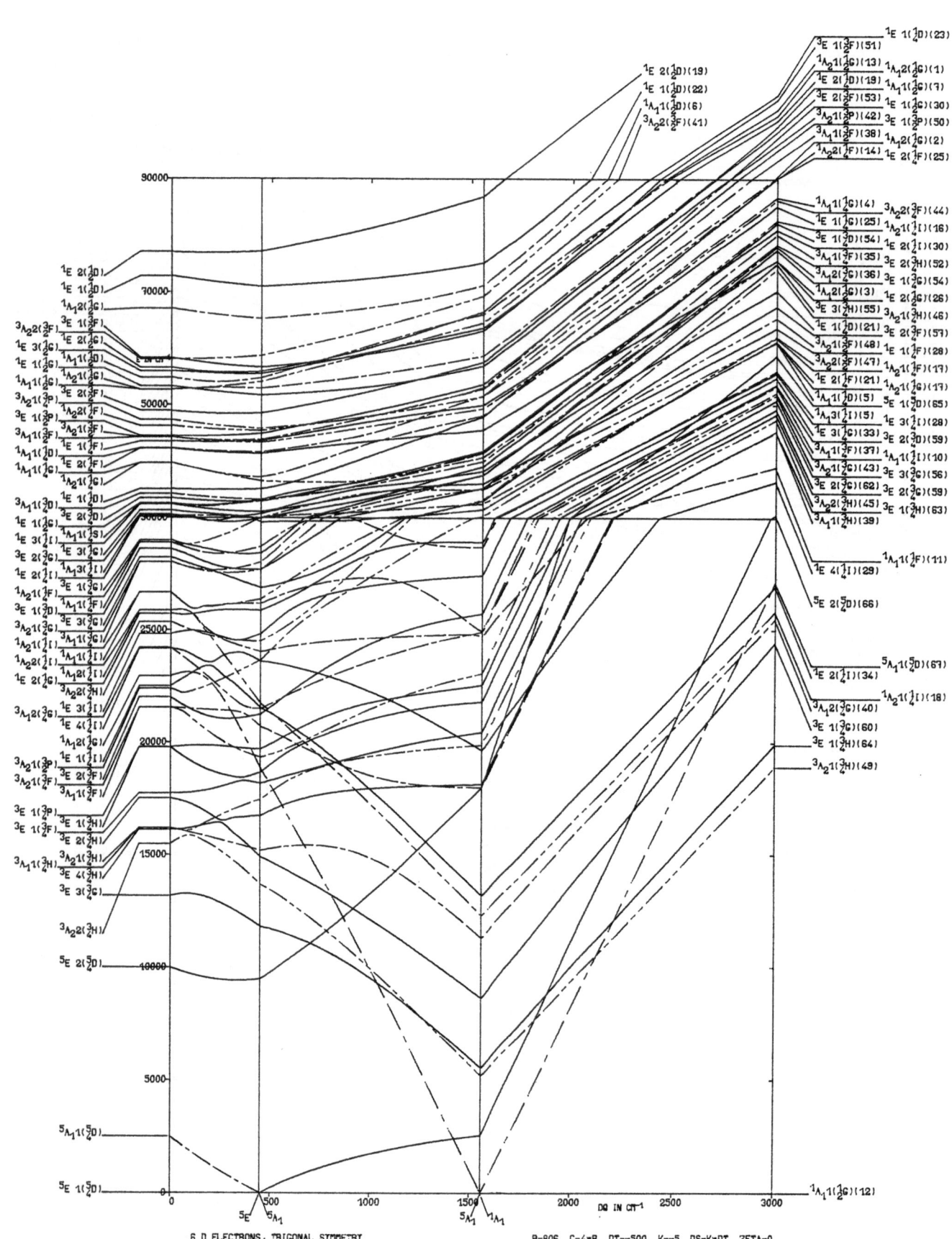

6 D ELECTRONS, TRIGONAL SYMMETRY

B=806 C=4×B DT=-500 K=-5 DS=K×DT ZETA=0

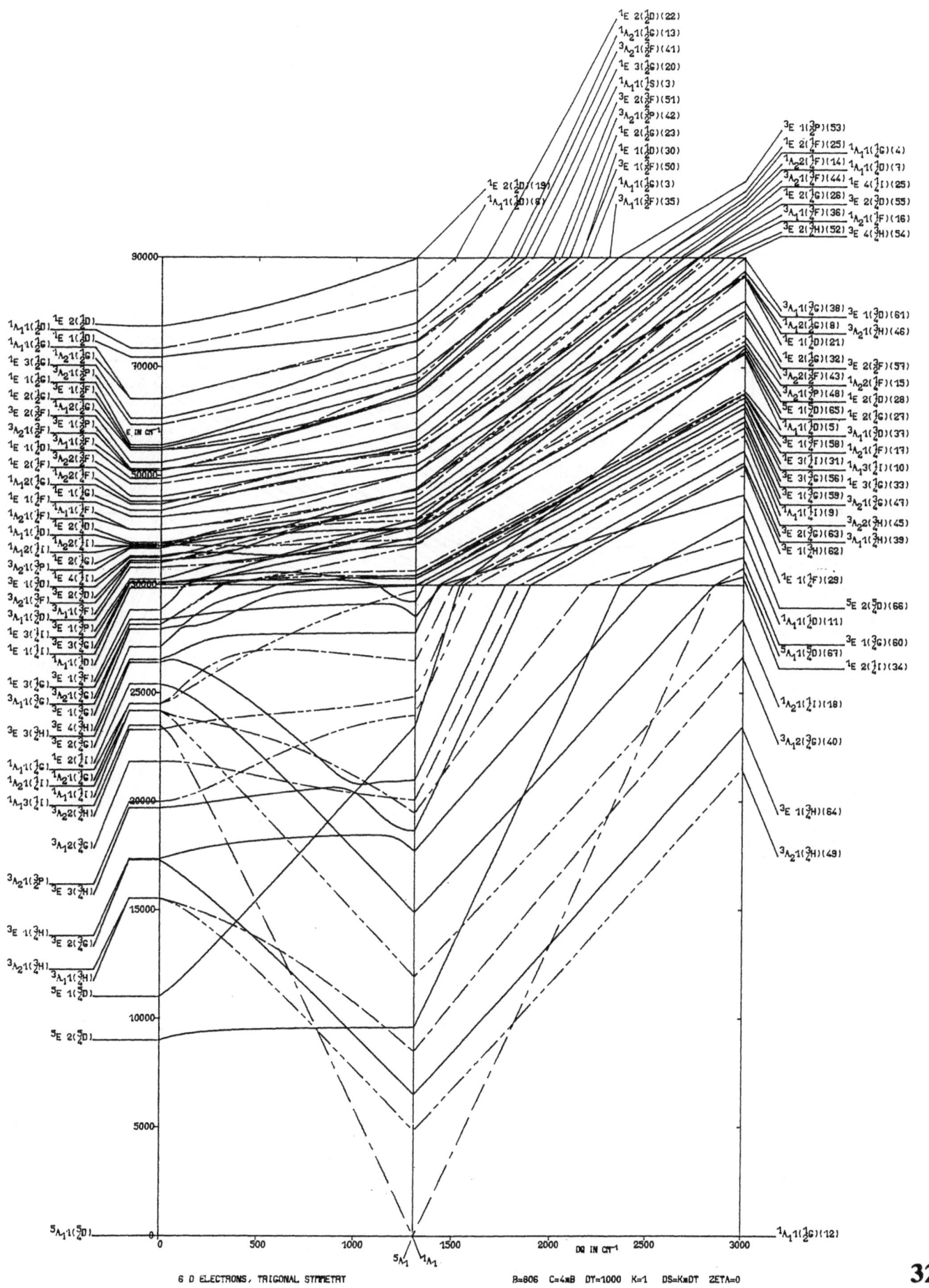

6 D ELECTRONS, TRIGONAL SYMMETRY   B=806 C=4×B DT=1000 K=1   DS=K×DT ZETA=0

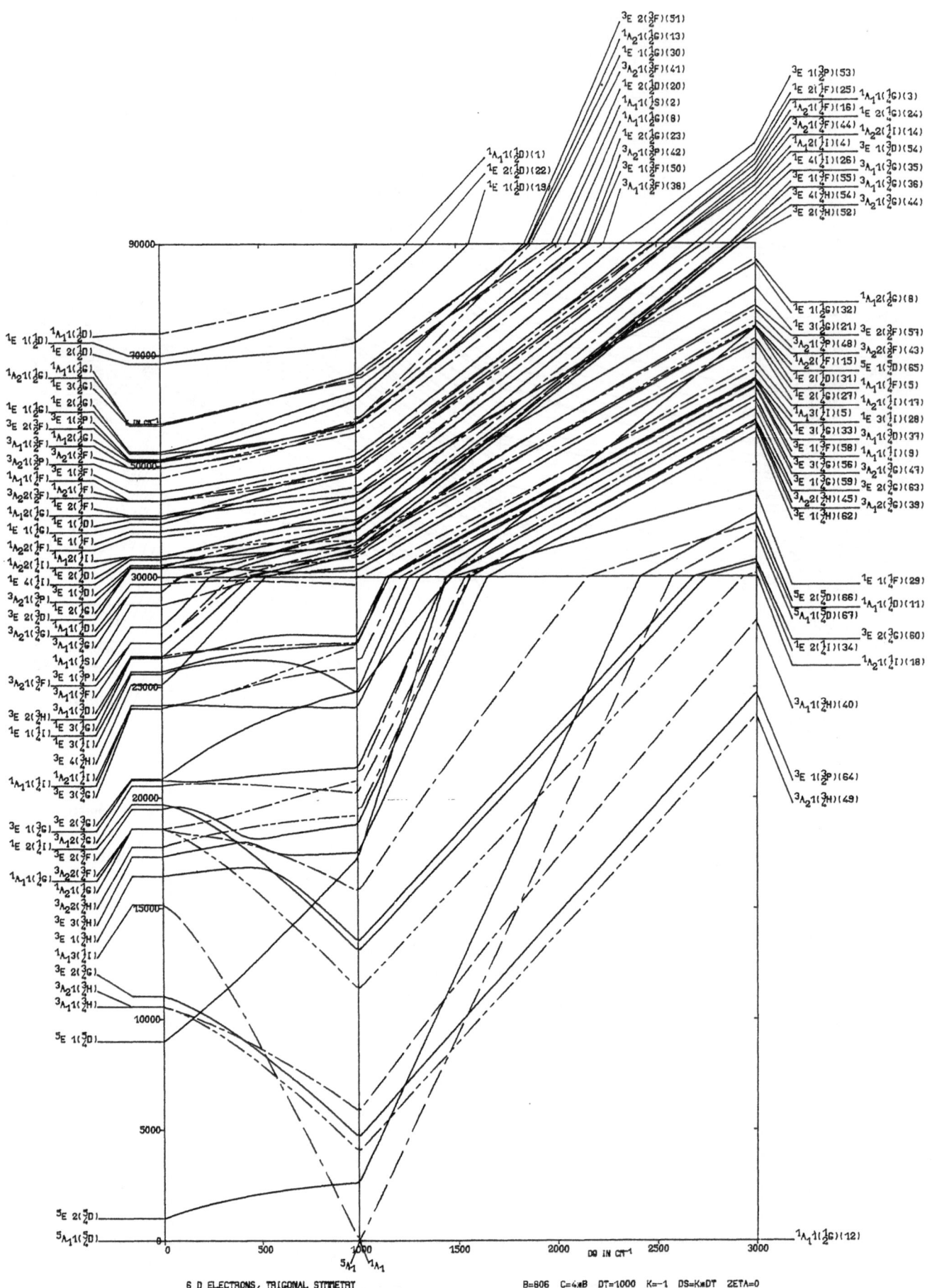

6 D ELECTRONS, TRIGONAL SYMMETRY

B=806 C=4.xB DT=1000 K=-1 DS=KxDT ZETA=0

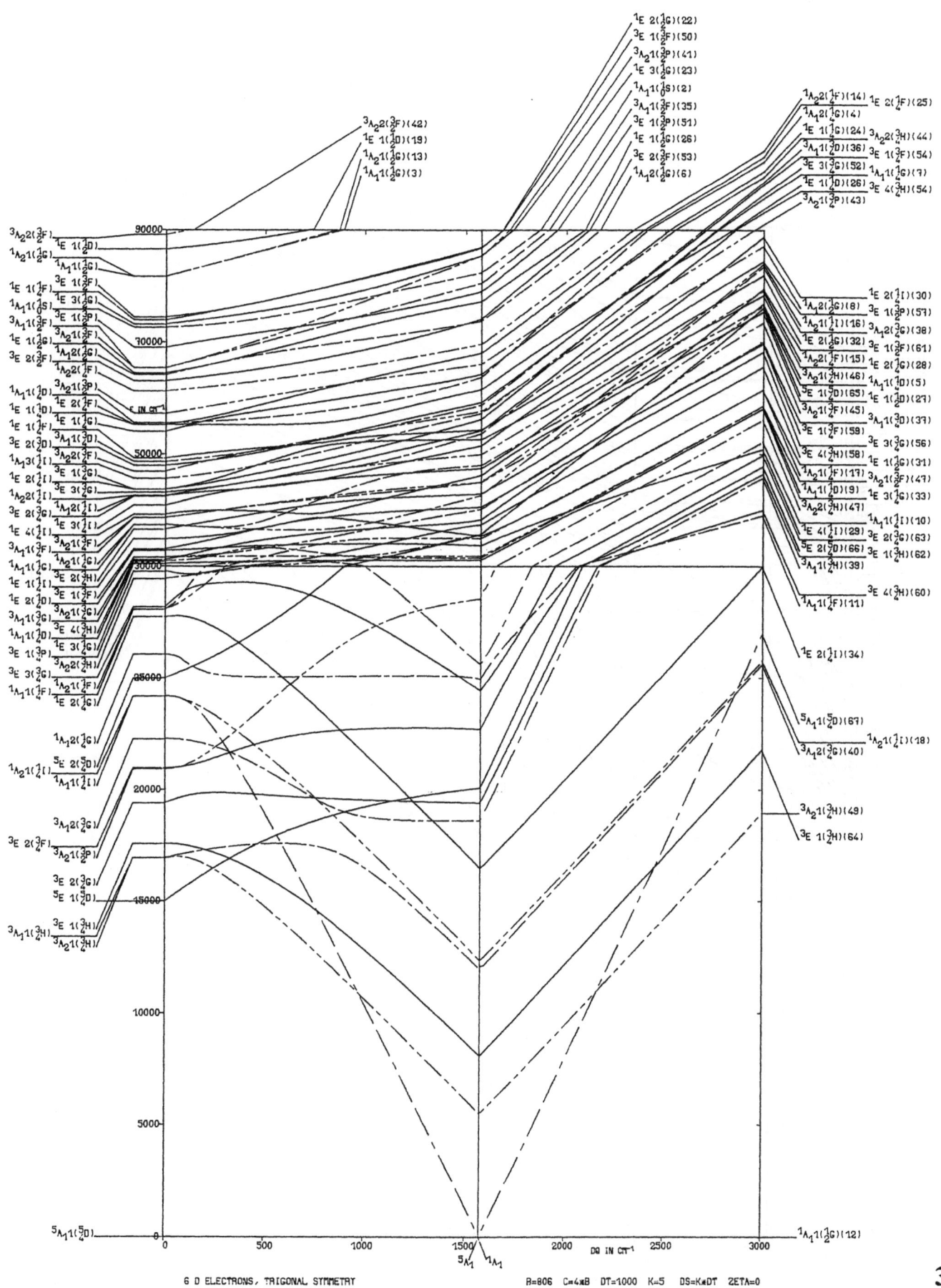

6 D ELECTRONS, TRIGONAL SYMMETRY

B=806 C=4*B DT=1000 K=5 DS=K*DT ZETA=0

DQ IN CM⁻¹

329

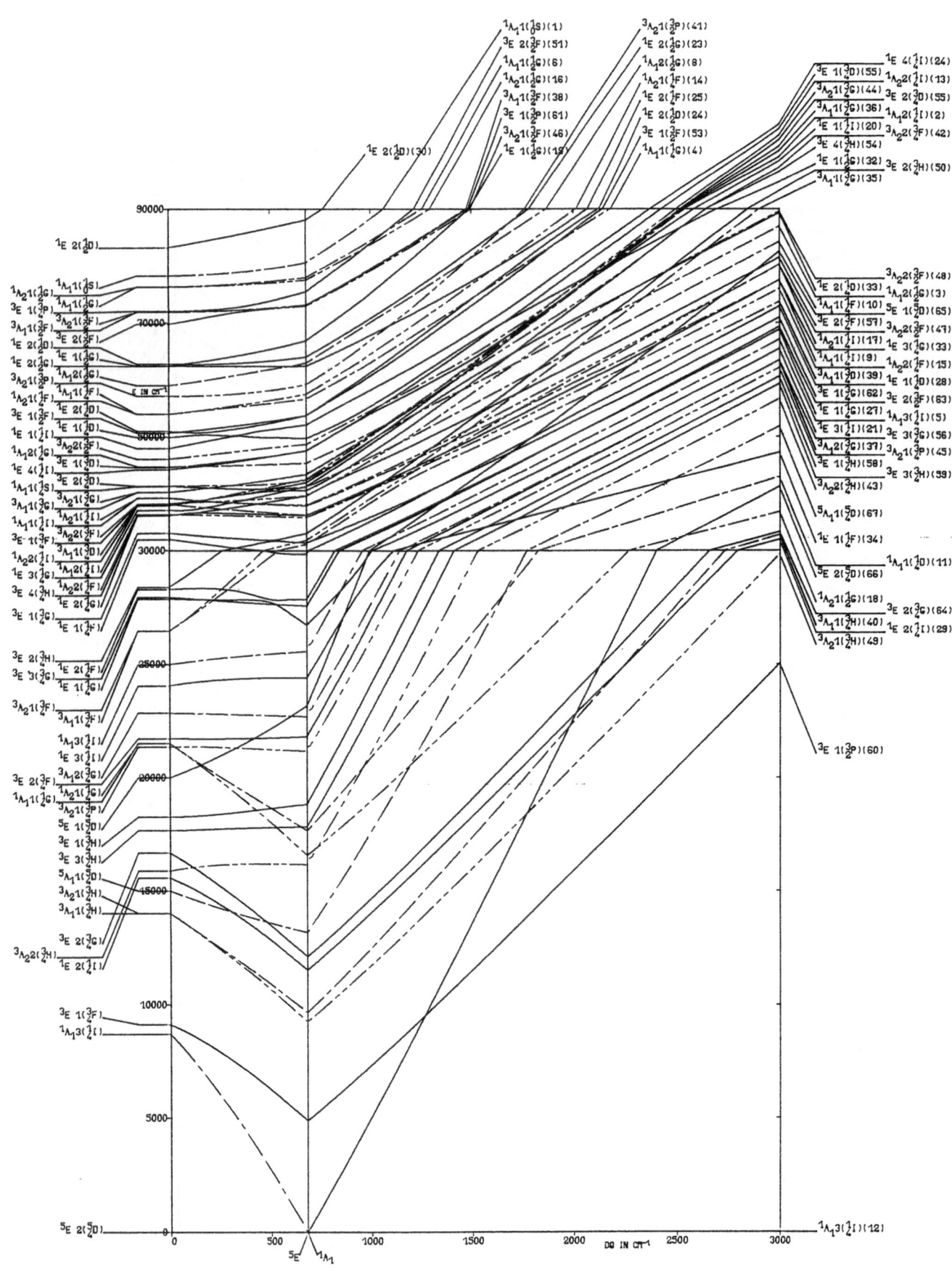

6 D ELECTRONS, TRIGONAL SYMMETRY

B=806   C=4*B   DT=1000   K=-5   DS=K*DT   ZETA=0

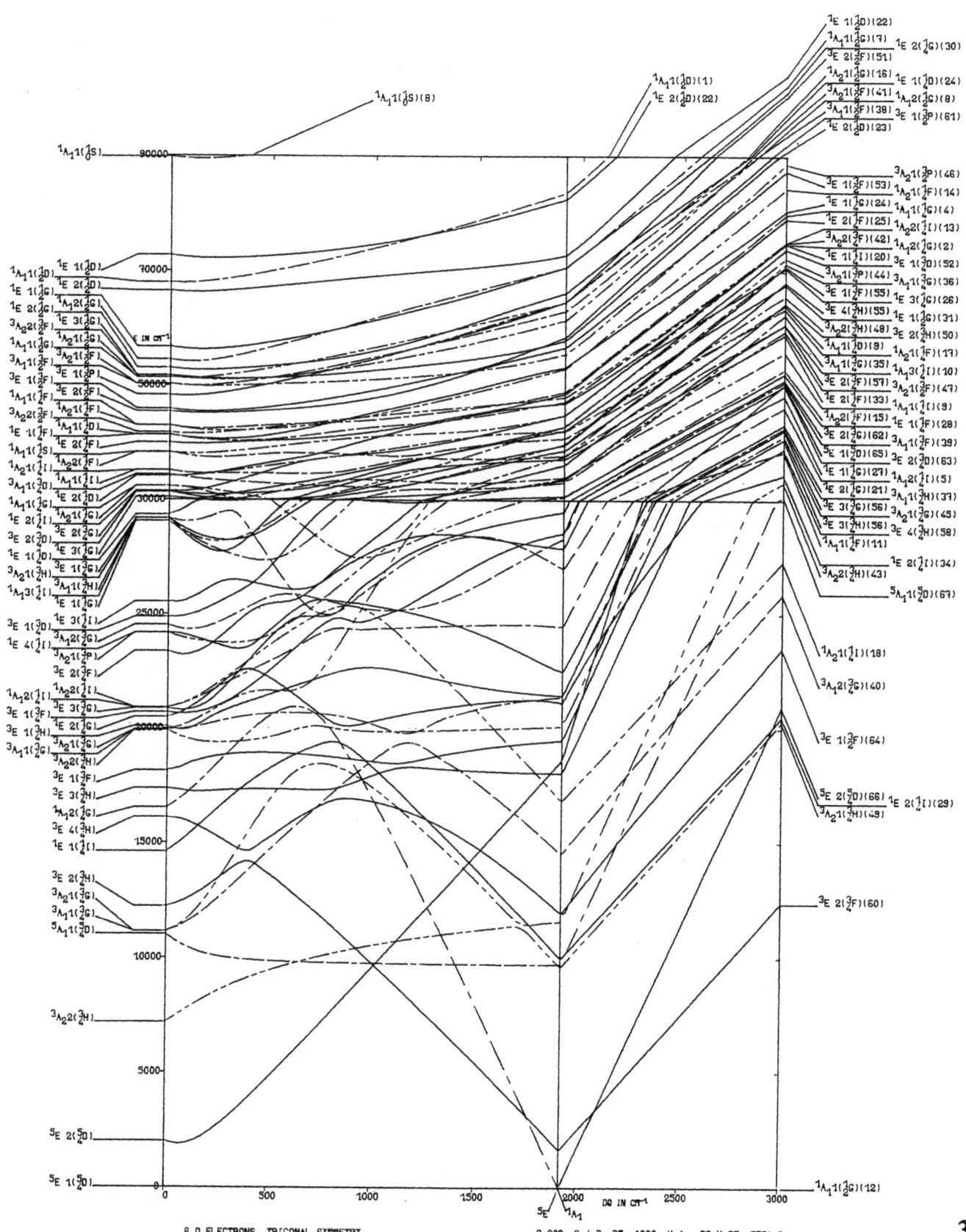

6 D ELECTRONS, TRIGONAL SYMMETRY

B=806 C=4×B DT=-1000 K=1 DS=K×DT ZETA=0

331

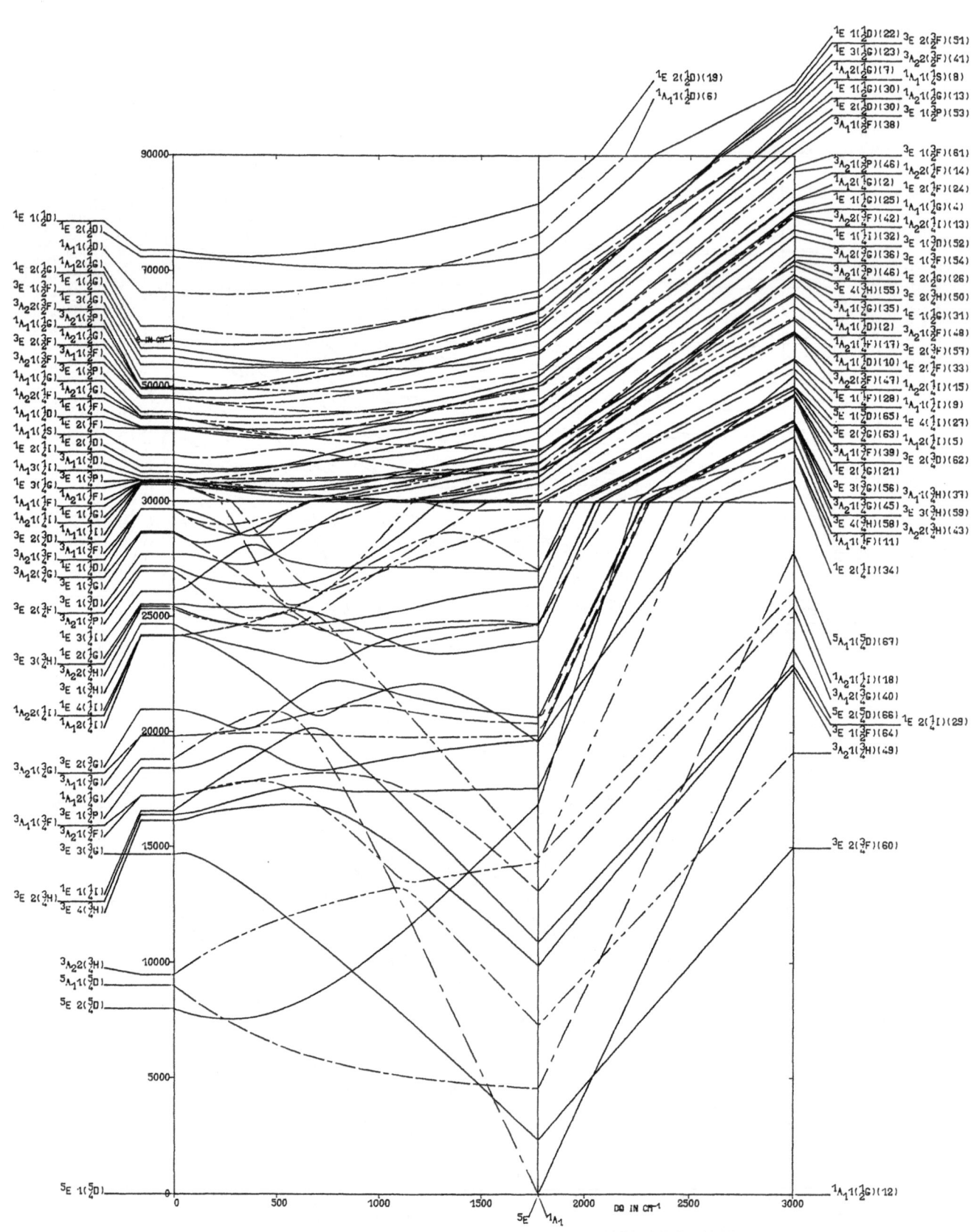

6 D ELECTRONS, TRIGONAL SYMMETRY

B=806 C=4*B DT=-1000 K=-1 DS=K*DT ZETA=0

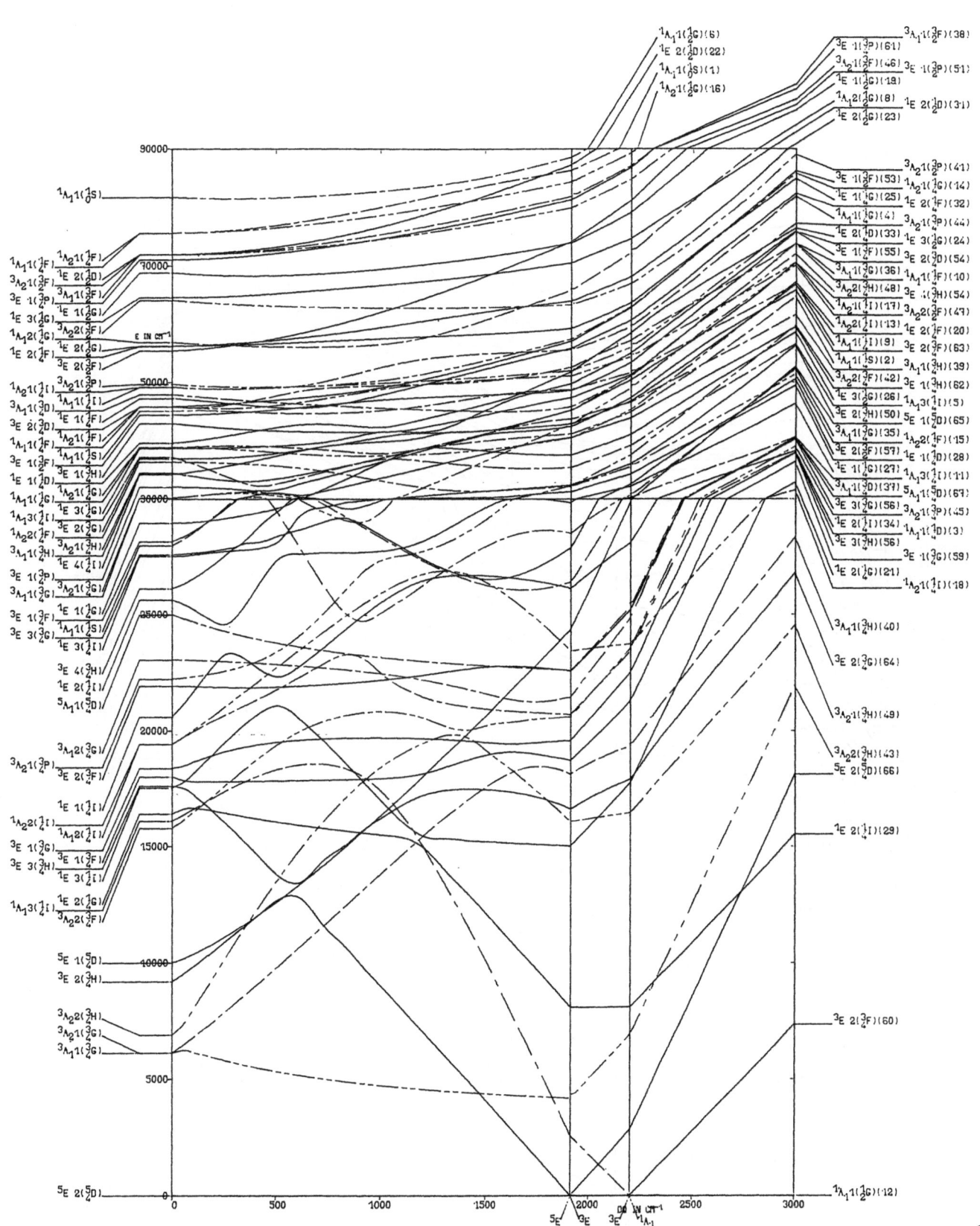

6 D ELECTRONS, TRIGONAL SYMMETRY
B=806 C=4×B DT=-1000 K=5 DS=K×DT ZETA=0

333

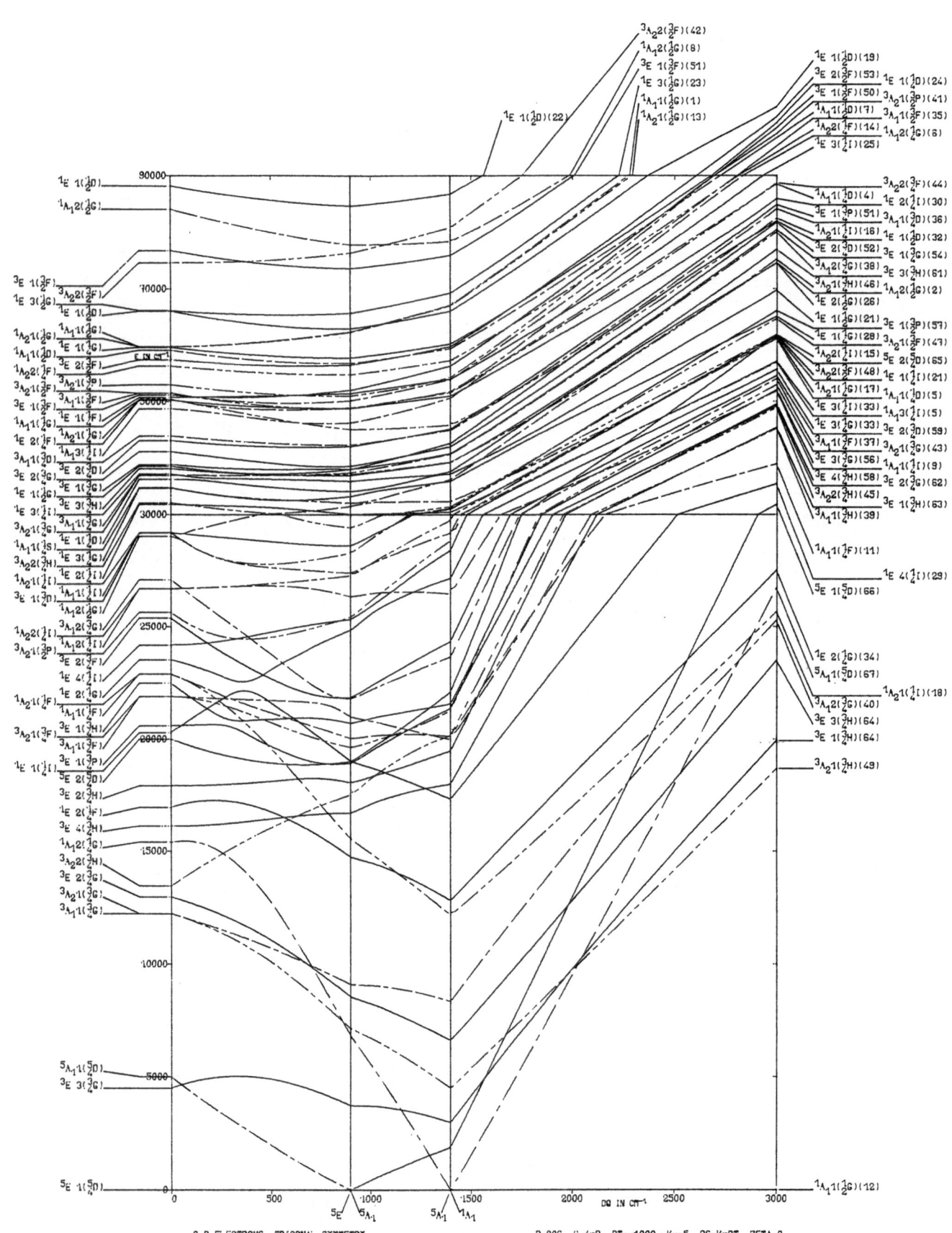

6 D ELECTRONS, TRIGONAL SYMMETRY   B=806   C=4×B   DT=-1000   K=-5   DS=K×DT   ZETA=0

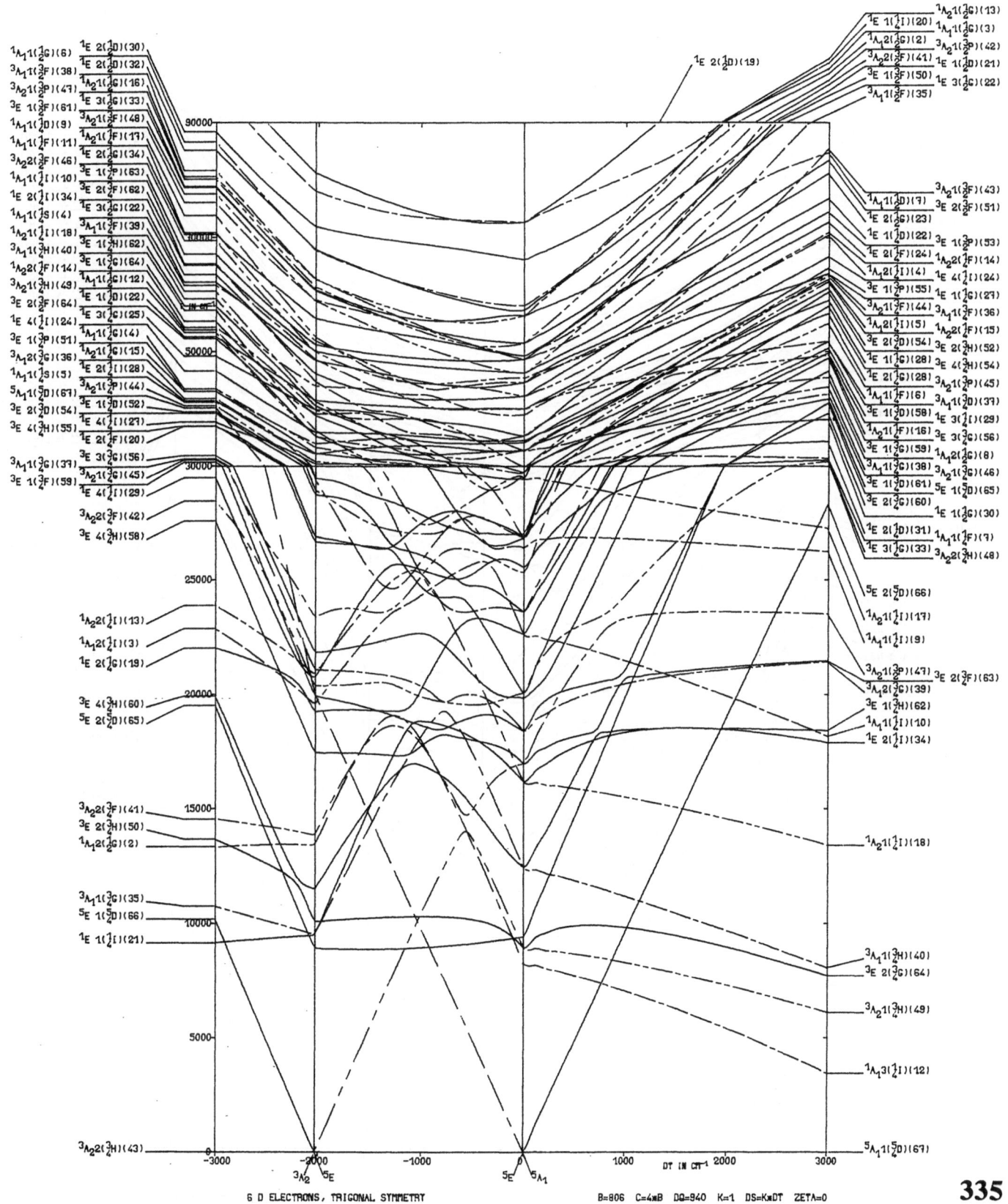

6 D ELECTRONS, TRIGONAL SYMMETRY   B=806 C=4⋆B DQ=940 K=1 DS=K⋆DT ZETA=0

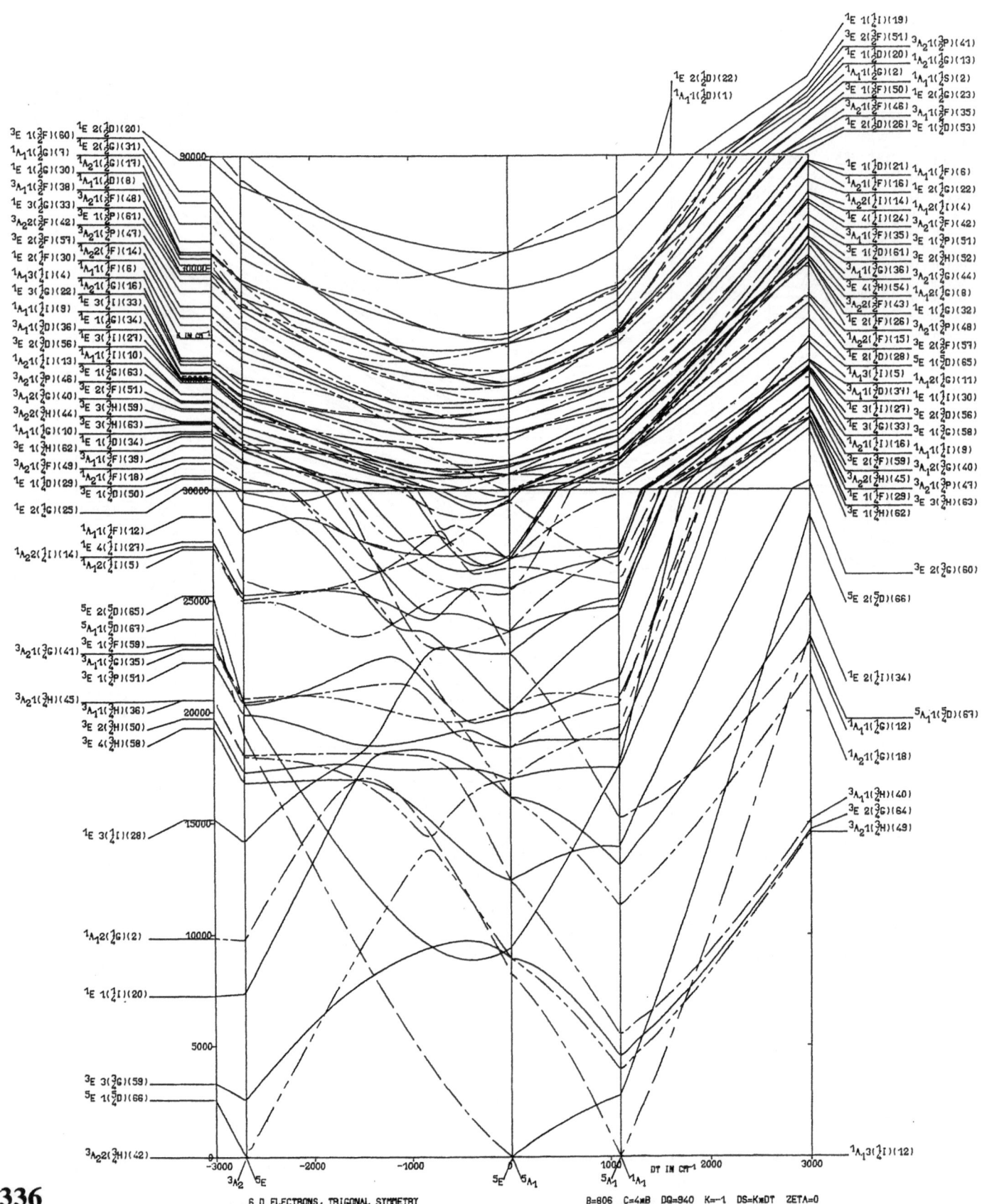

6 D ELECTRONS, TRIGONAL SYMMETRY

B=806   C=4*B   DQ=940   K=-1   DS=K*DT   ZETA=0

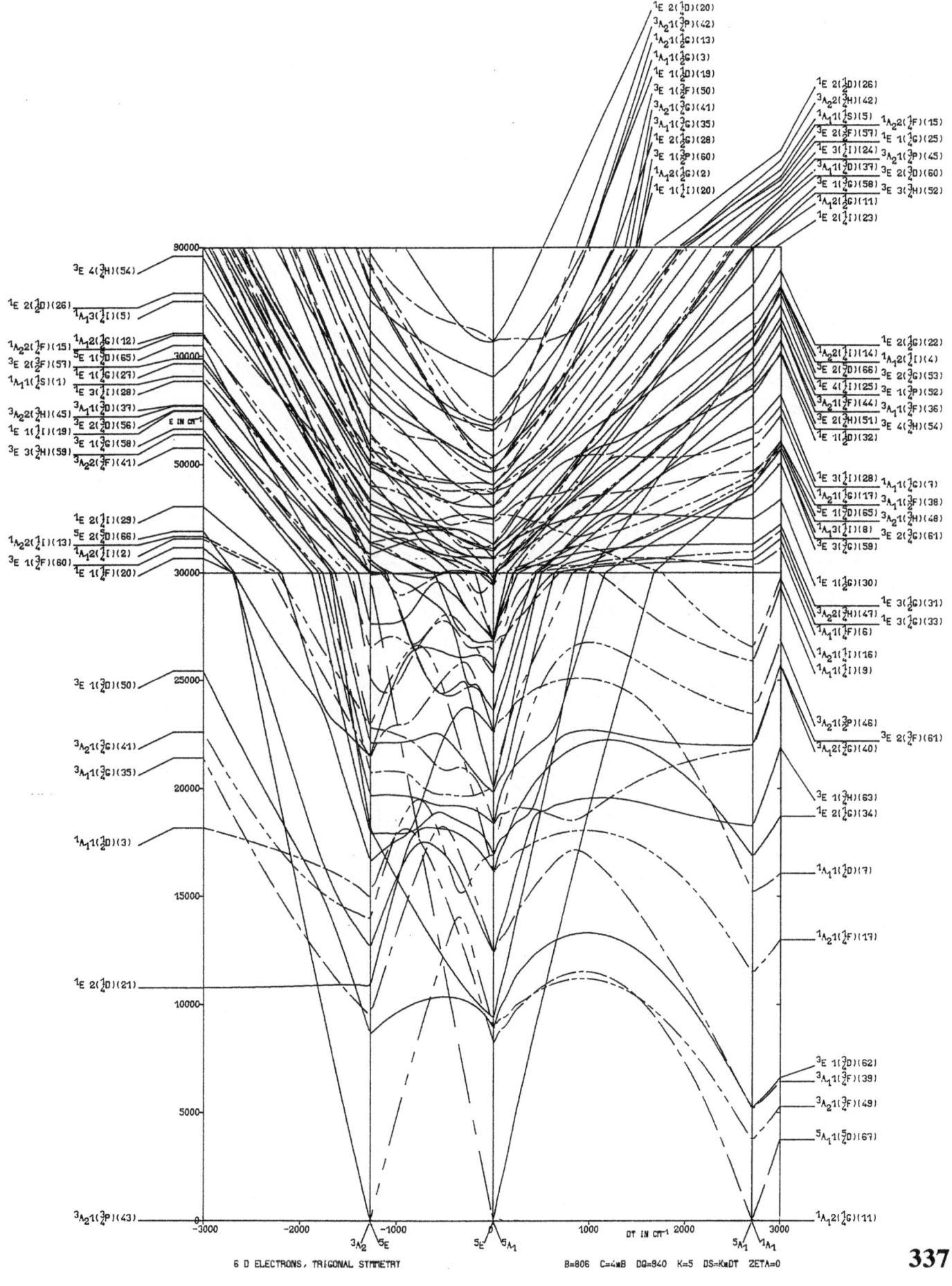

6 D ELECTRONS, TRIGONAL SYMMETRY   B=806 C=4×B DQ=940 K=5 DS=K×DT ZETA=0

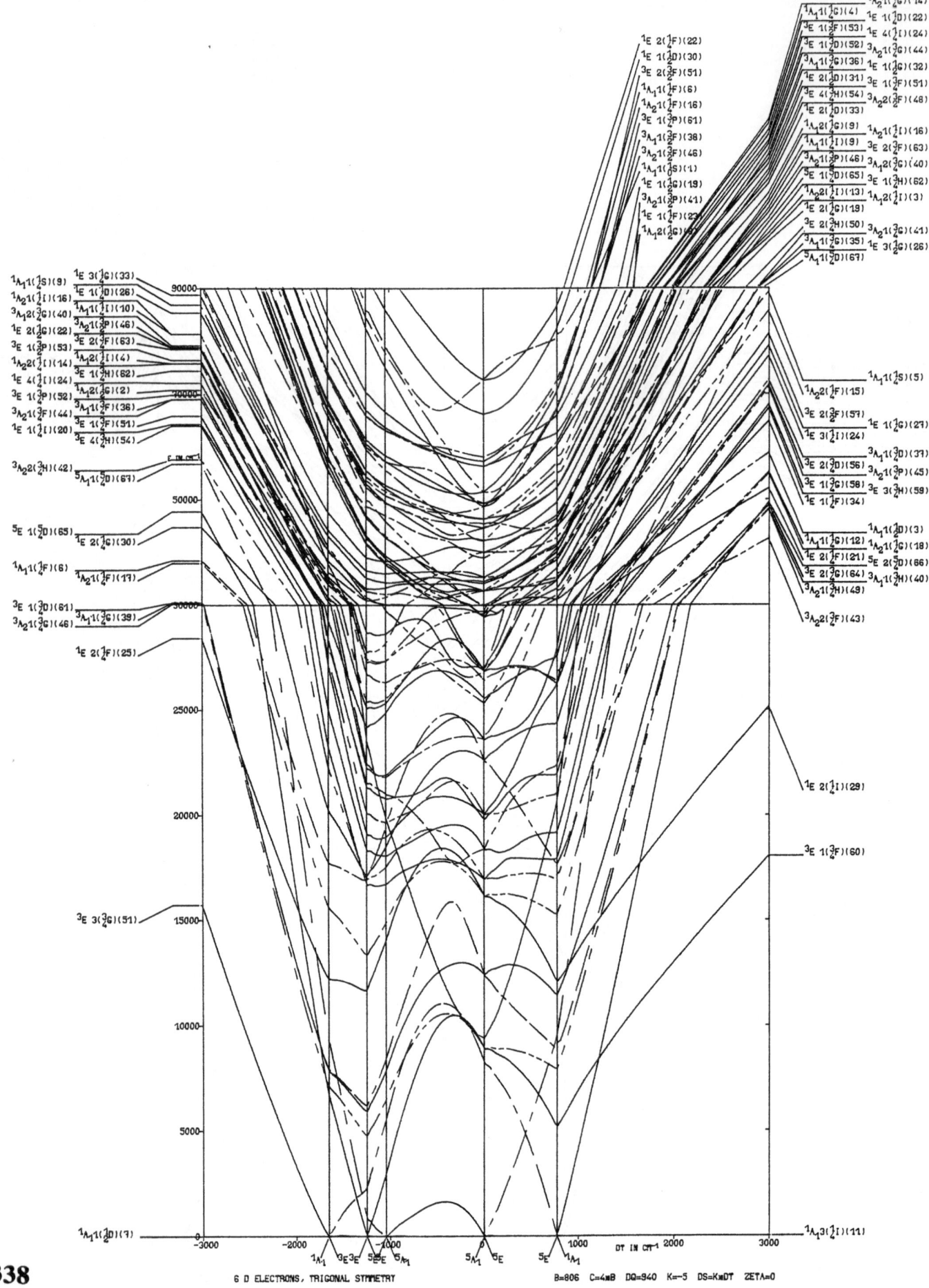

6 D ELECTRONS, TRIGONAL SYMMETRY

B=806 C=4×B DQ=940 K=-5 DS=K×DT ZETA=0

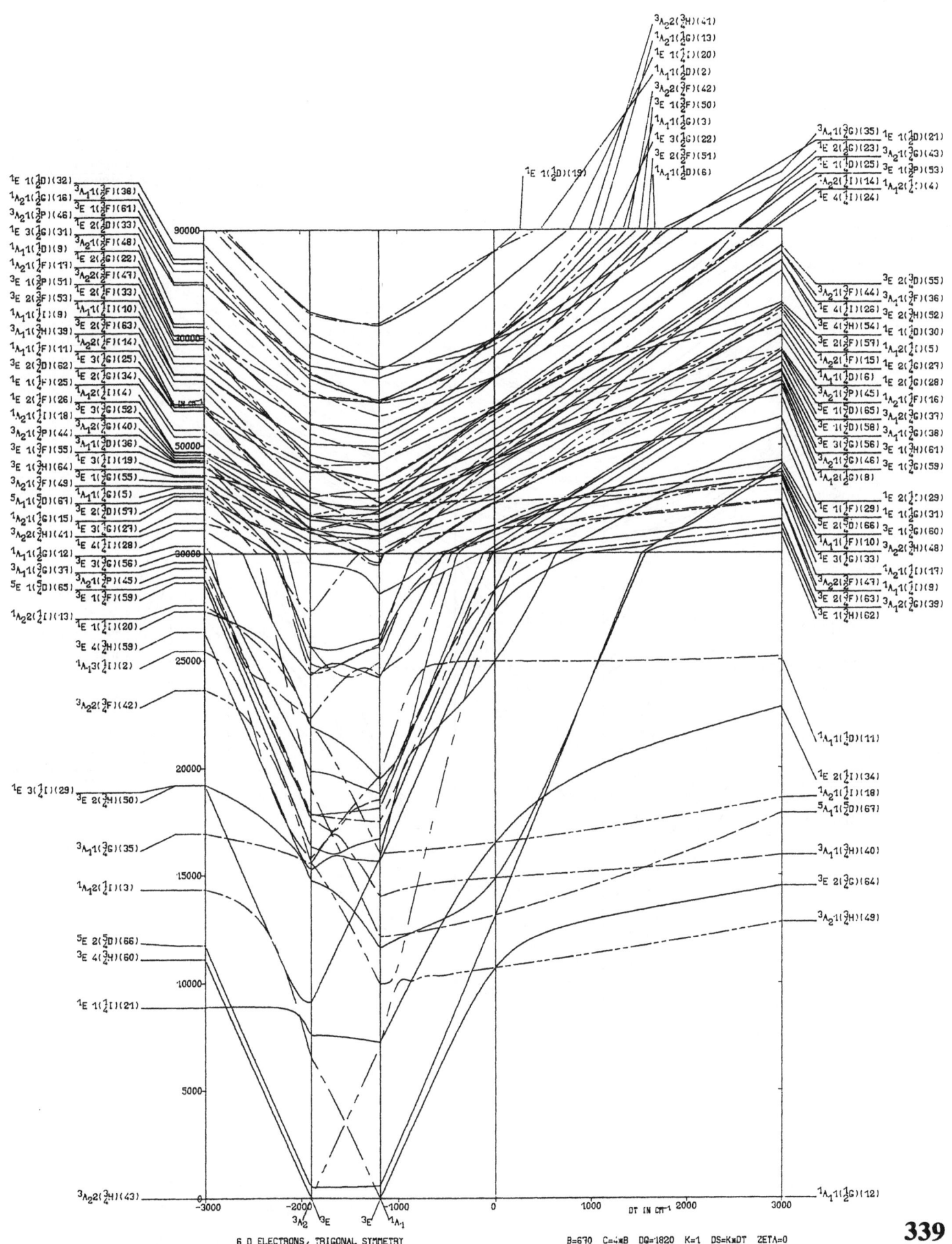

6 D ELECTRONS, TRIGONAL SYMMETRY   B=670 C=4×B DQ=1820 K=1 DS=K×DT ZETA=0

339

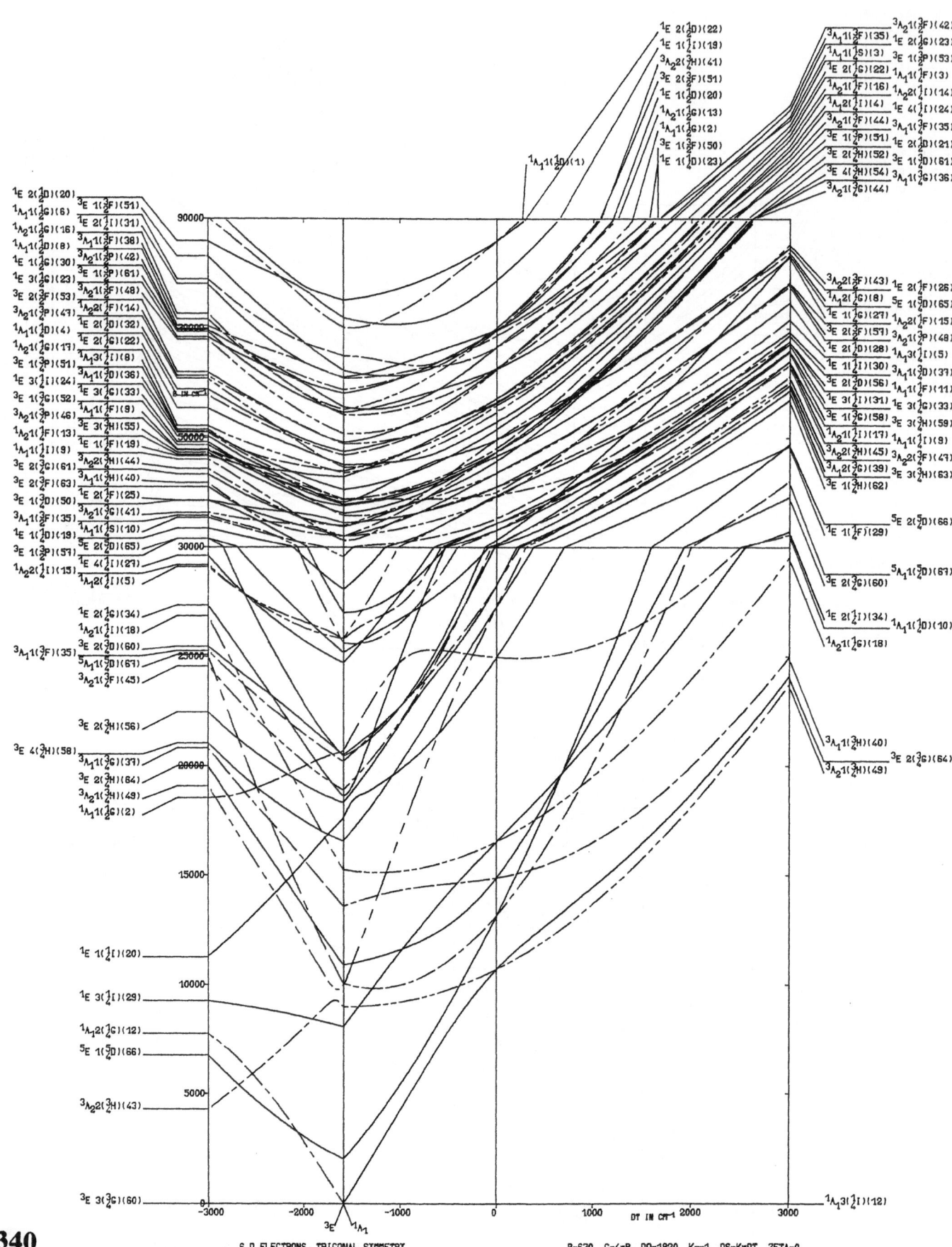

1E 2(½D)(20)   3E 1(³/₂F)(51)   1E 2(½D)(22)
1A₁1(³/₂G)(6)   1E 2(½I)(31)   1E 1(½I)(19)   3A₁1(³/₂F)(35)   3A₂1(³/₂F)(42)
1A₂1(³/₂G)(16)   3A₁1(³/₂F)(38)   3A₂2(³/₂H)(41)   1A₁1(³/₂G)(3)   1E 2(³/₂G)(23)
1A₁1(³/₂G)(8)   3A₂1(³/₂P)(42)   3E 2(³/₂F)(51)   1A₁1(½I)(3)   3E 1(½P)(53)
1E 1(³/₂G)(30)   3E 1(³/₂P)(61)   1E 1(½F)(16)   1A₂1(½F)(16)   1A₁1(½F)(3)
1E 3(³/₂G)(23)   3A₂1(³/₂F)(48)   3E 1(³/₂D)(20)   1A₁2(½I)(4)   1A₂2(½I)(14)
3E 2(³/₂F)(53)   1A₂2(³/₂F)(14)   1A₂1(³/₂G)(13)   3A₂1(³/₂F)(44)   1E 4(½I)(24)
3A₂1(³/₂P)(47)   1E 2(½D)(32)   1A₁1(³/₂G)(2)   3E 1(½P)(51)   3A₁1(³/₂F)(35)
1A₁1(½D)(4)   1E 2(³/₂G)(22)   3E 1(³/₂F)(50)   3E 2(³/₂H)(52)   1E 2(½D)(21)
1A₂1(³/₂D)(17)   1A₁3(½I)(8)   1E 1(½D)(23)   3E 4(³/₂H)(54)   3A₁1(³/₂G)(36)
3E 1(³/₂P)(51)   1A₁1(½D)(36)   3A₂1(½G)(44)
1E 3(½I)(24)   3E 3(³/₂G)(33)
3E 1(³/₂G)(52)   1A₁1(½F)(9)
3A₂1(³/₂P)(46)   3E 3(³/₂H)(55)   3A₂2(³/₂F)(43)   1E 2(½F)(26)
1A₂1(½F)(13)   1E 1(½F)(19)   1A₁2(½G)(18)   5E 1(½D)(65)
1A₁1(½I)(9)   3A₂2(³/₂H)(44)   1E 1(³/₂G)(27)   1A₂2(½F)(15)
3E 2(³/₂G)(61)   3A₁1(³/₂H)(40)   3E 2(³/₂F)(57)   3A₂1(½P)(48)
3E 2(³/₂F)(63)   1E 2(½F)(25)   1E 2(½D)(28)   1A₁3(½I)(5)
3E 1(½D)(50)   3A₂1(³/₂G)(41)   1E 1(½I)(30)   3A₁1(½D)(37)
3A₁1(½D)(35)   1A₁1(½S)(10)   3E 2(½D)(56)   1A₁1(½F)(11)
1E 1(½D)(19)   1E 2(½D)(65)   3E 1(½I)(31)   1E 3(³/₂G)(33)
3E 1(³/₂P)(57)   1E 4(½I)(27)   3E 1(³/₂G)(58)   3E 3(³/₂H)(59)
1A₂2(½I)(15)   1A₁2(½I)(5)   1A₂1(½I)(17)   1A₁1(½I)(9)
                                                                                                                                           3A₂2(½F)(45)   3A₂2(½F)(47)
1E 2(³/₂G)(34)   3A₂1(½G)(39)   3E 3(³/₂H)(63)
1A₂1(½I)(18)   5E 1(½H)(62)
3E 2(½D)(60)
5A₁1(½D)(67)   1E 1(½F)(29)   5E 2(½D)(66)
3A₂1(³/₂F)(45)

                                                                                                                                           5A₁1(½D)(67)

3E 2(³/₂H)(56)   3E 2(³/₂G)(60)

3E 4(³/₂H)(58)   3A₁1(³/₂G)(37)   1E 2(½I)(34)   1A₁1(½D)(10)
3E 2(³/₂H)(64)   1A₂1(½G)(18)
3A₂1(³/₂H)(49)
1A₁1(½G)(2)

                                                                                                                                           3A₁1(³/₂H)(40)
1E 1(½I)(20)   3A₂1(½H)(49)   3E 2(³/₂G)(64)

1E 3(½I)(29)

1A₁2(½G)(12)
5E 1(³/₂D)(66)

3A₂2(³/₂H)(43)

3E 3(³/₂G)(60)   1A₁3(½I)(12)

        -3000   -2000   -1000   0   1000   DT IN CM-1 2000   3000
                              3E   1A₁

340

6 D ELECTRONS, TRIGONAL SYMMETRY   B=670 C=4×B DQ=1820 K=-1 DS=K×DT ZETA=0

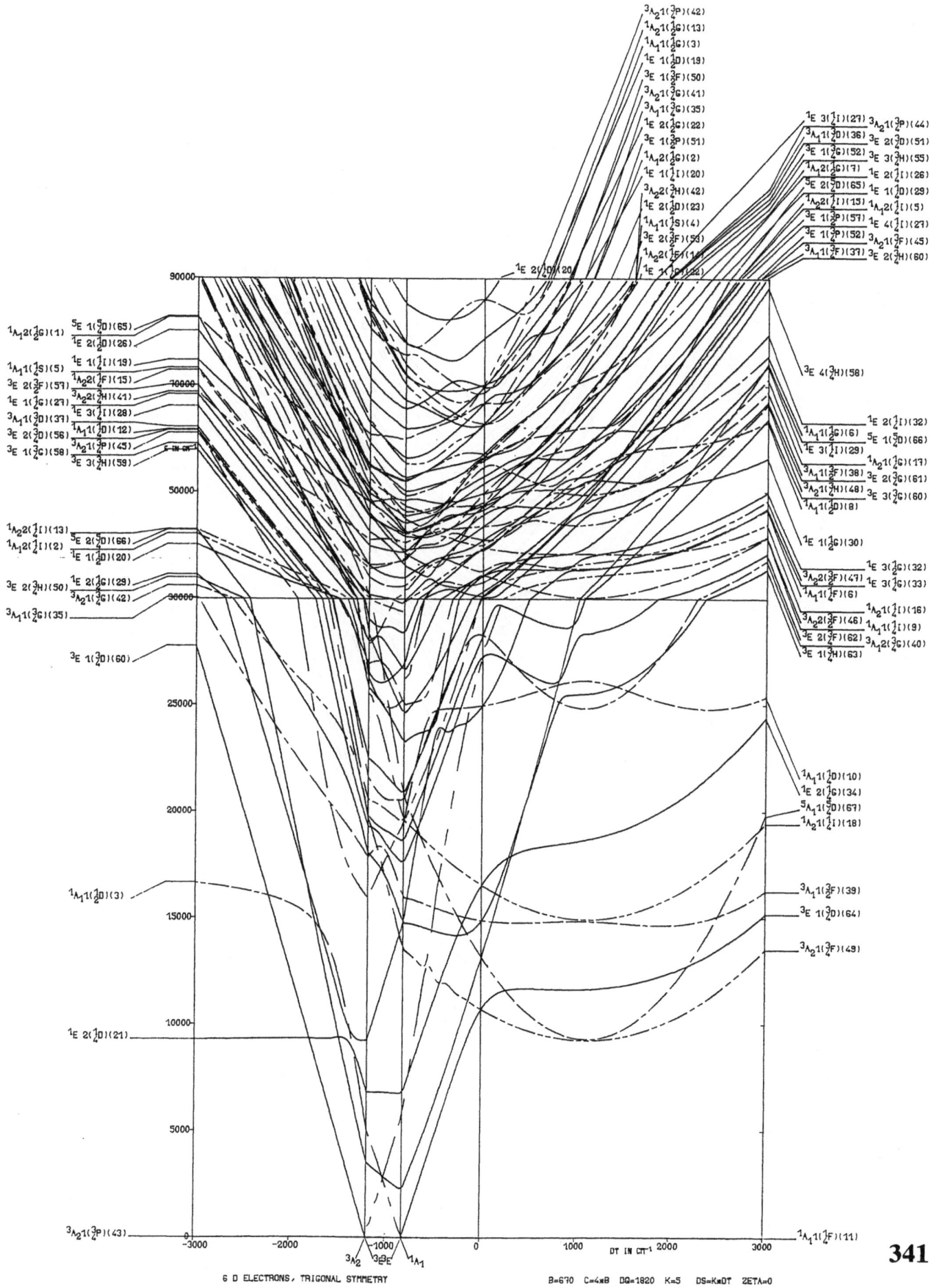

6 D ELECTRONS, TRIGONAL SYMMETRY   B=670 C=4×B DQ=1820 K=5   DS=K×DT ZETA=0

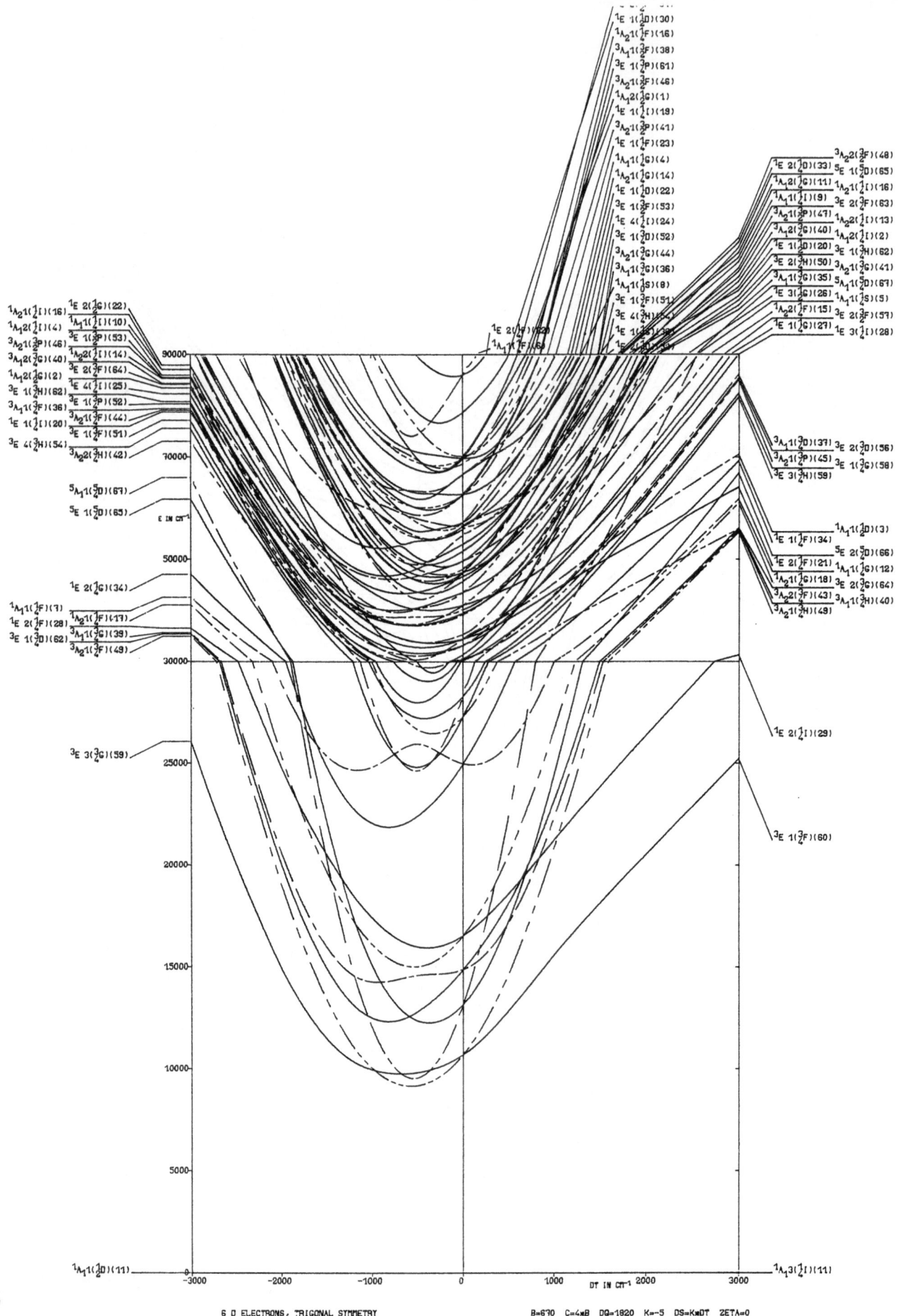

6 D ELECTRONS, TRIGONAL SYMMETRY   B=670   C=4×B   DQ=1820   K=-5   DS=K×DT   ZETA=0

B=806, 670   C=4×B   2ETA=0

ENERGY AS FUNCTION OF DT
K=1, -1, 5, -5

(58) $^1E_6(^2E_4(^1E_2(E_1E_1)E_2)E_2)$

(57) $^3E_5(^2E_3(^1E_2(A_1E_2)E_1)E_2)$

(56) $^1E_5(^2E_3(^1E_1(A_1E_1)E_2)E_2)$

(55) $^3E_4(^2E_2(^1E_1(E_1E_2)E_1)E_2)$

(54) $^3E_4(^2E_2(^1E_1(A_1E_1)E_1)E_2)$

(53) $^1E_4(^2E_2(^1A_1(E_1E_1)E_2)E_2)$

(52) $^1E_4(^2E_2(^1E_1(A_1E_1)E_1)E_2)$

(51) $^1E_4(^2E_2(^1A_1(A_1A_1)E_2)E_2)$

(50) $^3E_3(^2E_1(^1E_3(E_1E_2)E_2)E_2)$

(49) $^3E_3(^2E_1(^1A_1(E_1E_1)E_1)E_2)$

(48) $^3E_3(^2E_1(^1E_2(A_1E_2)E_1)E_2)$

(47) $^3E_3(^2E_1(^1A_1(A_1A_1)E_1)E_2)$

(46) $^1E_3(^2E_1(^1E_1(E_1E_2)E_2)E_2)$

(45) $^1E_3(^2E_1(^1A_1(E_1E_1)E_1)E_2)$

(44) $^1E_3(^2E_1(^1E_1(A_1E_1)E_2)E_2)$

(43) $^1E_3(^2E_1(^1A_1(A_1A_1)E_1)E_2)$

(42) $^5E_2(^4A_2(^3E_1(A_1E_1)E_1)E_2)$

(41) $^3E_2(^2A_1(^1E_2(E_1E_1)E_2)E_2)$

(40) $^3E_2(^2A_1(^1E_2(A_1E_2)E_2)E_2)$

(39) $^3E_2(^2E_1(^1E_2(A_1E_2)E_1)E_1)$

(38) $^3E_2(^2A_2(^1E_1(A_1E_1)E_1)E_2)$

(37) $^3E_2(^2A_1(^1E_1(A_1E_1)E_1)E_2)$

(36) $^1E_2(^2A_2(^1E_2(E_1E_1)E_2)E_2)$

(35) $^1E_2(^2A_1(^1E_2(E_1E_1)E_2)E_2)$

(34) $^1E_2(^2A_1(^1E_2(A_1E_2)E_2)E_2)$

(33) $^1E_2(^2A_2(^1E_1(A_1E_1)E_1)E_2)$

(32) $^1E_2(^2A_1(^1E_1(A_1E_1)E_1)E_2)$

(31) $^1E_2(^2E_1(^1A_1(A_1A_1)E_1)E_1)$

(30) $^5E_1(^4E_1(^3E_1(A_1E_1)E_2)E_2)$

(29) $^3E_1(^2E_1(^1E_1(E_1E_2)E_2)E_2)$

(28) $^3E_1(^2E_1(^1A_1(E_1E_1)E_1)E_2)$

(27) $^3E_1(^2E_3(^1E_2(A_1E_2)E_1)E_2)$

(26) $^3E_1(^2E_1(^1E_2(A_1E_2)E_1)E_2)$

(25) $^3E_1(^2E_1(^1E_1(A_1E_1)E_2)E_2)$

(24) $^3E_1(^2A_1(^1E_1(A_1E_1)E_1)E_1)$

(23) $^3E_1(^2E_1(^1A_1(A_1A_1)E_1)E_2)$

(22) $^1E_1(^2E_1(^1E_1(E_1E_2)E_2)E_2)$

(21) $^1E_1(^2E_1(^1A_1(E_1E_1)E_1)E_2)$

(20) $^1E_1(^2E_1(^1E_2(A_1E_2)E_1)E_2)$

(19) $^1E_1(^2E_1(^1E_1(A_1E_1)E_2)E_2)$

(18) $^1E_1(^2A_1(^1E_1(A_1E_1)E_1)E_1)$

(17) $^1E_1(^2E_1(^1A_1(A_1A_1)E_1)E_2)$

(16) $^3A_2(^2E_2(^1E_1(E_1E_2)E_1)E_2)$

(15) $^3A_2(^2E_2(^1A_1(E_1E_1)E_2)E_2)$

(14) $^3A_2(^2E_2(^1E_1(A_1E_1)E_1)E_2)$

(13) $^3A_2(^2E_2(^1A_1(A_1A_1)E_2)E_2)$

(12) $^3A_2(^2E_1(^1A_1(A_1A_1)E_1)E_1)$

(11) $^1A_2(^2E_2(^1E_1(A_1E_1)E_1)E_2)$

(10) $^5A_1(^4E_2(^3A_2(E_1E_1)E_2)E_2)$

(9) $^3A_1(^2E_2(^1E_1(E_1E_2)E_1)E_2)$

(8) $^3A_1(^2E_2(^1E_1(A_1E_1)E_1)E_2)$

(7) $^1A_1(^2E_2(^1A_1(E_2E_2)E_2)E_2)$

(6) $^1A_1(^2E_2(^1E_1(E_1E_2)E_1)E_2)$

(5) $^1A_1(^2E_2(^1A_1(E_1E_1)E_2)E_2)$

(4) $^1A_1(^2E_1(^1A_1(E_1E_1)E_1)E_1)$

(3) $^1A_1(^2E_2(^1E_1(A_1E_1)E_1)E_2)$

(2) $^1A_1(^2E_2(^1A_1(A_1A_1)E_2)E_2)$

(1) $^1A_1(^2E_1(^1A_1(A_1A_1)E_1)E_1)$

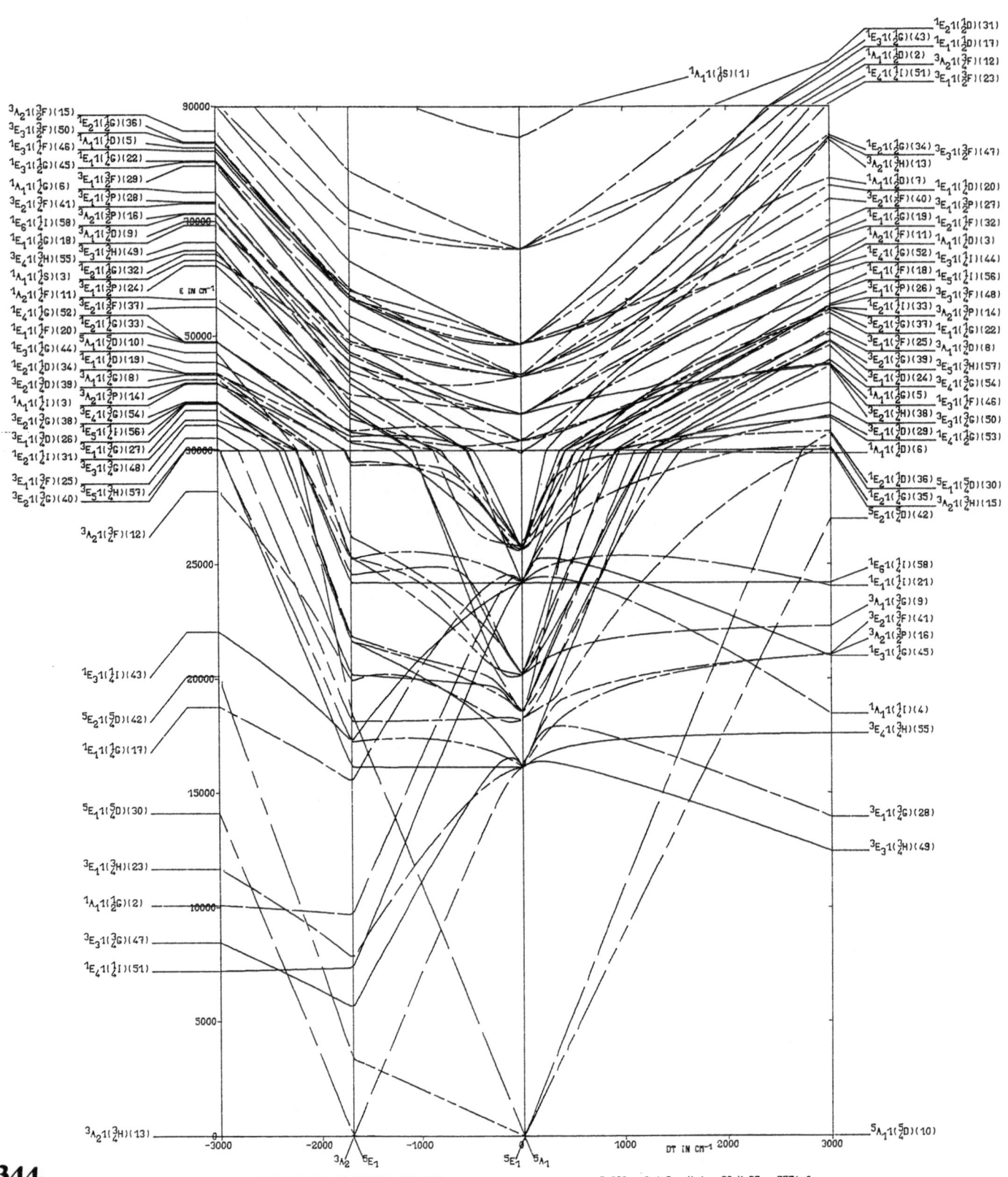

6 D ELECTRONS, CYLINDRICAL SYMMETRY   B=806   C=4×B   K=1   DS=K×DT   ZETA=0

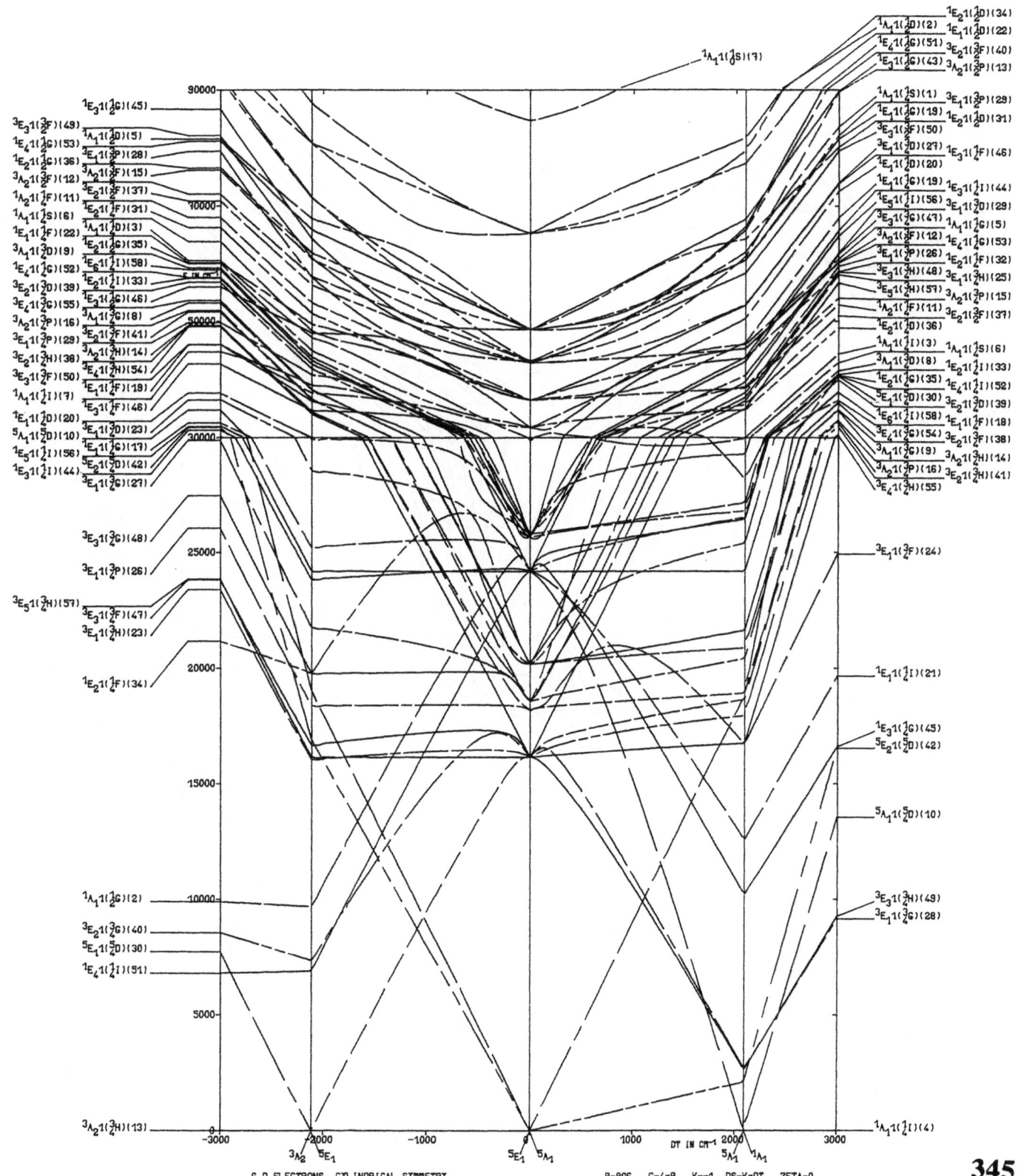

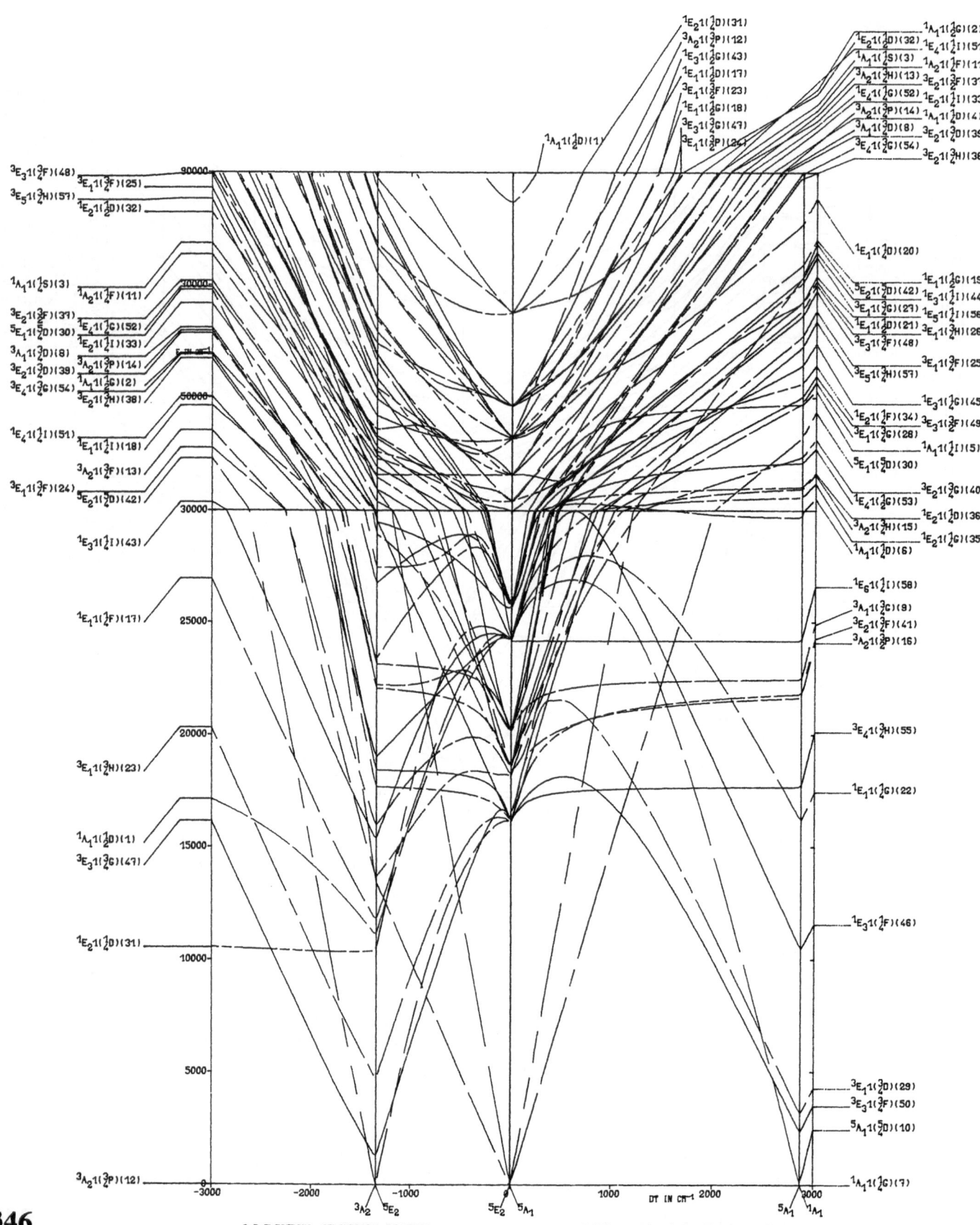

6 D ELECTRONS, CYLINDRICAL SYMMETRY   B=806   C=4×B   K=5   DS=K×DT   ZETA=0

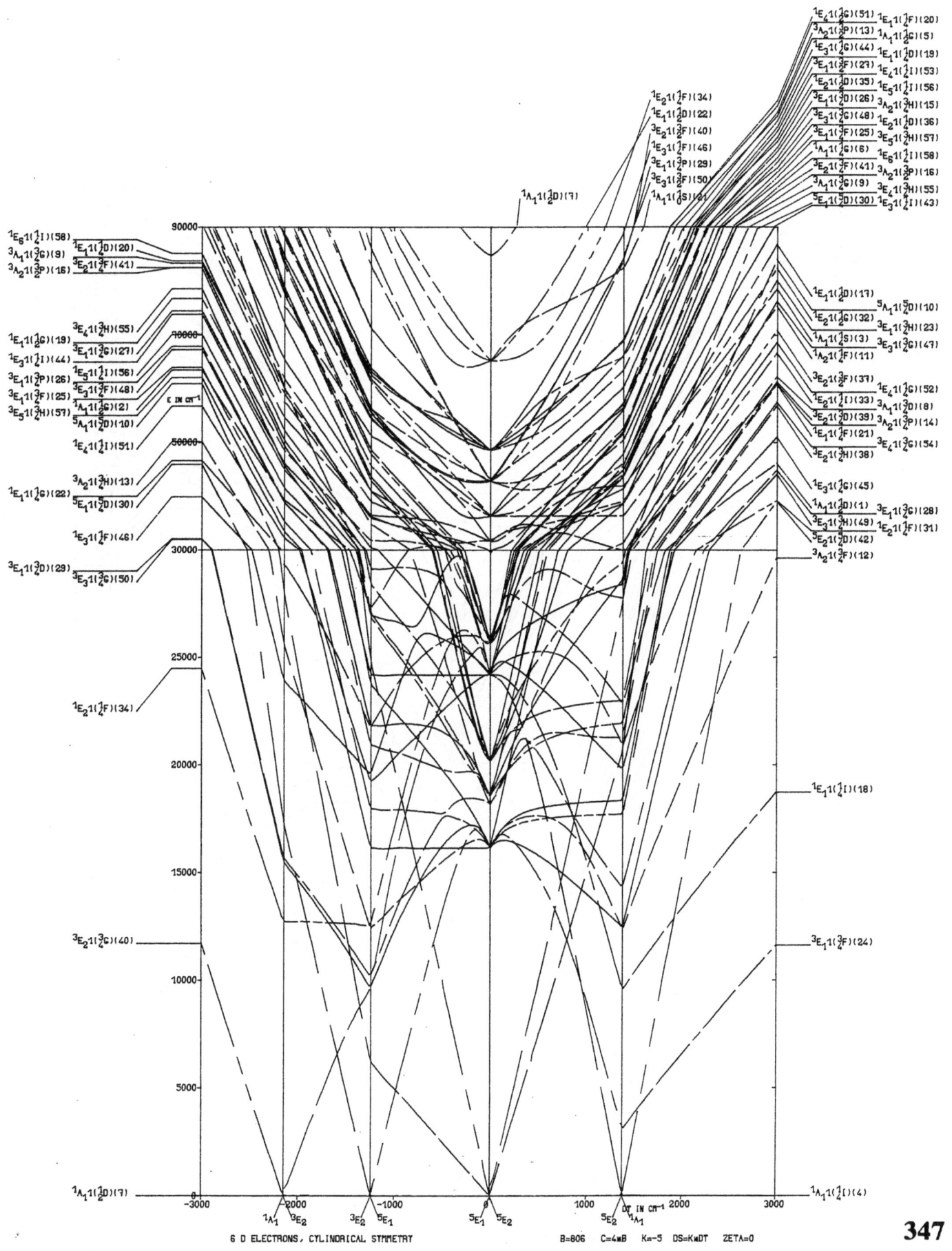

6 D ELECTRONS, CYLINDRICAL SYMMETRY   B=806   C=4×B   K=-5   DS=K×DT   ZETA=0

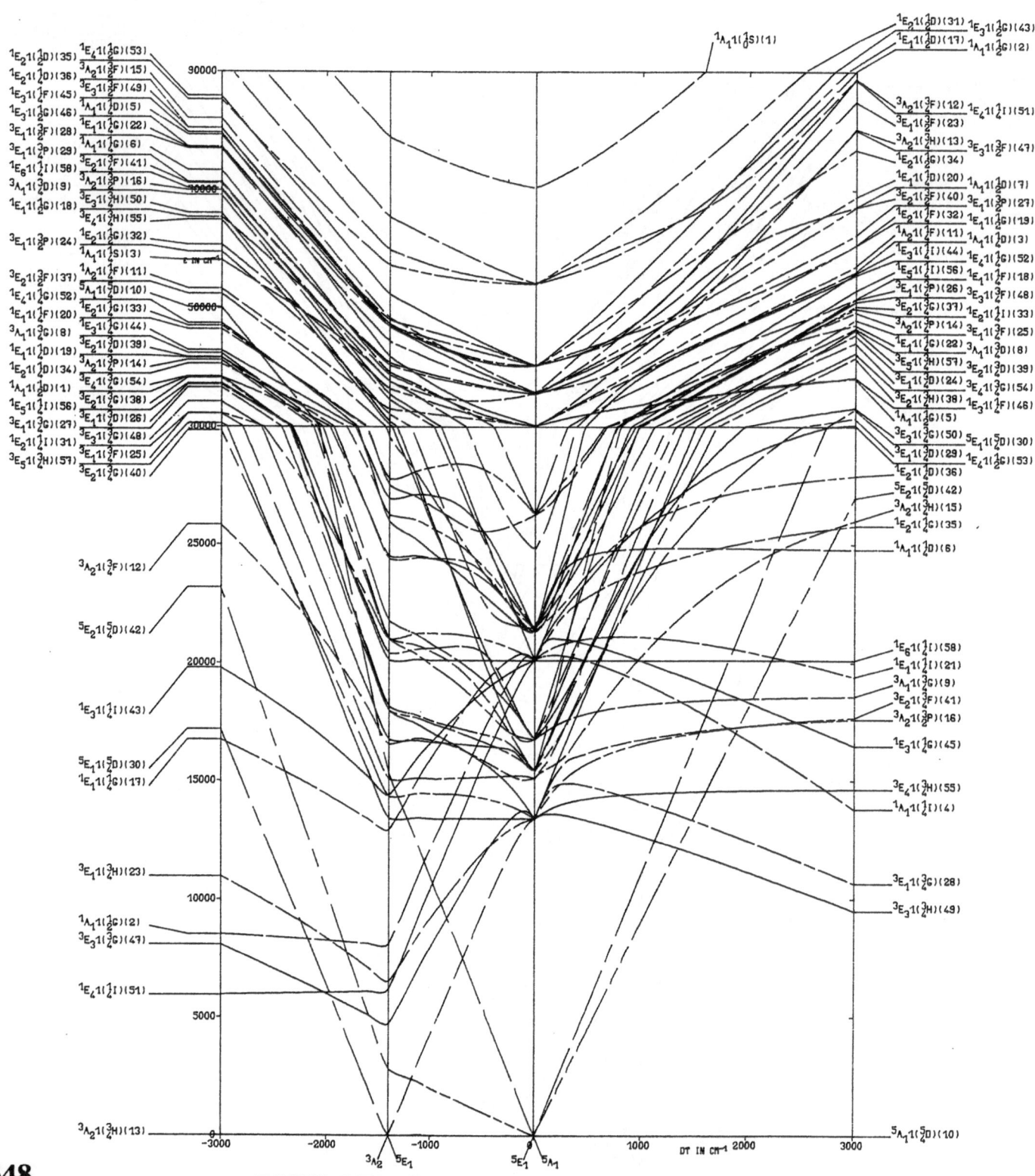

348

6 D ELECTRONS, CYLINDRICAL SYMMETRY   B=670   C=4×B   K=1   DS=K×DT   2ETA=0

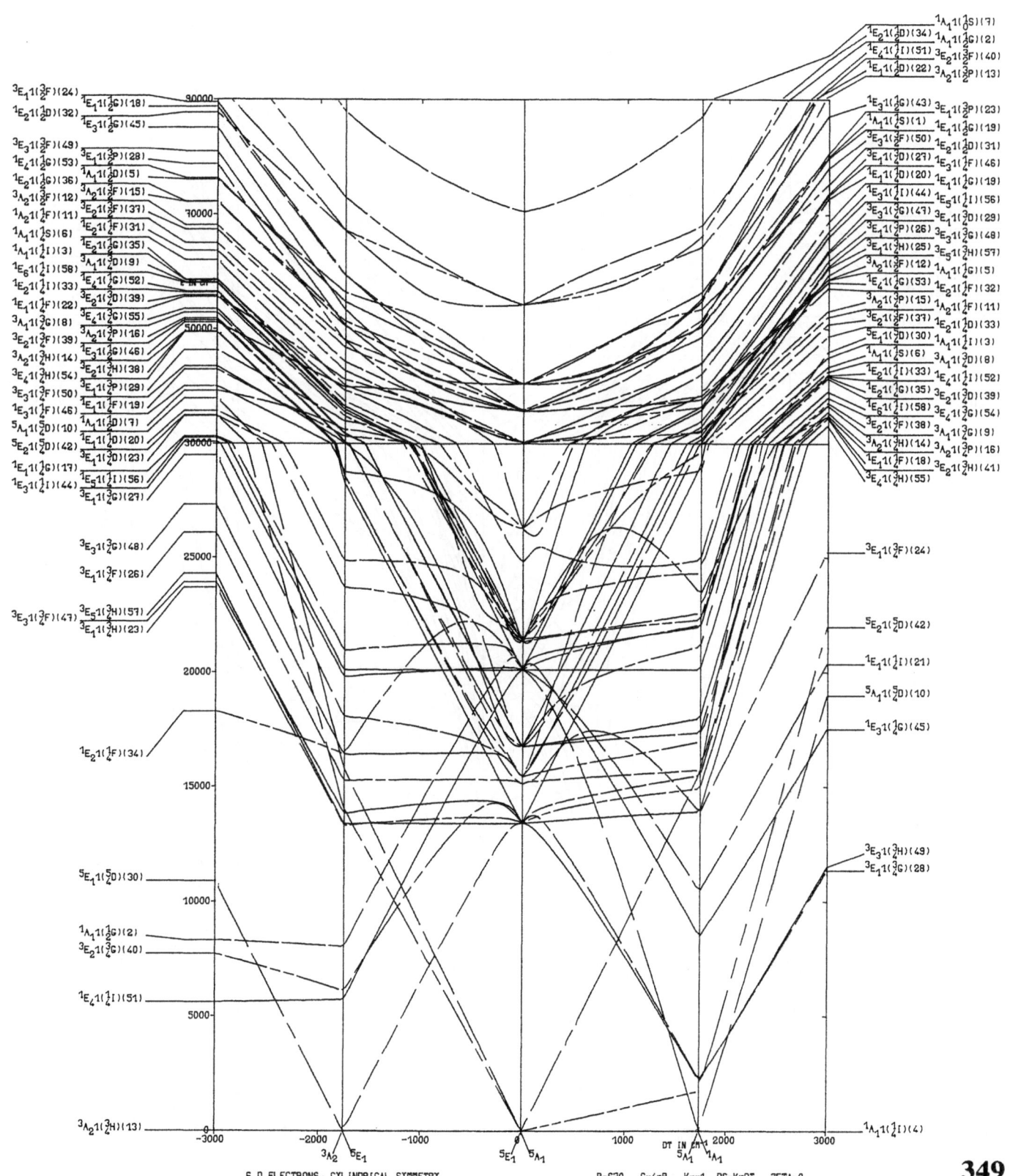

6 D ELECTRONS, CYLINDRICAL SYMMETRY   B=670   C=4*B   K=-1   DS=K*DT   ZETA=0

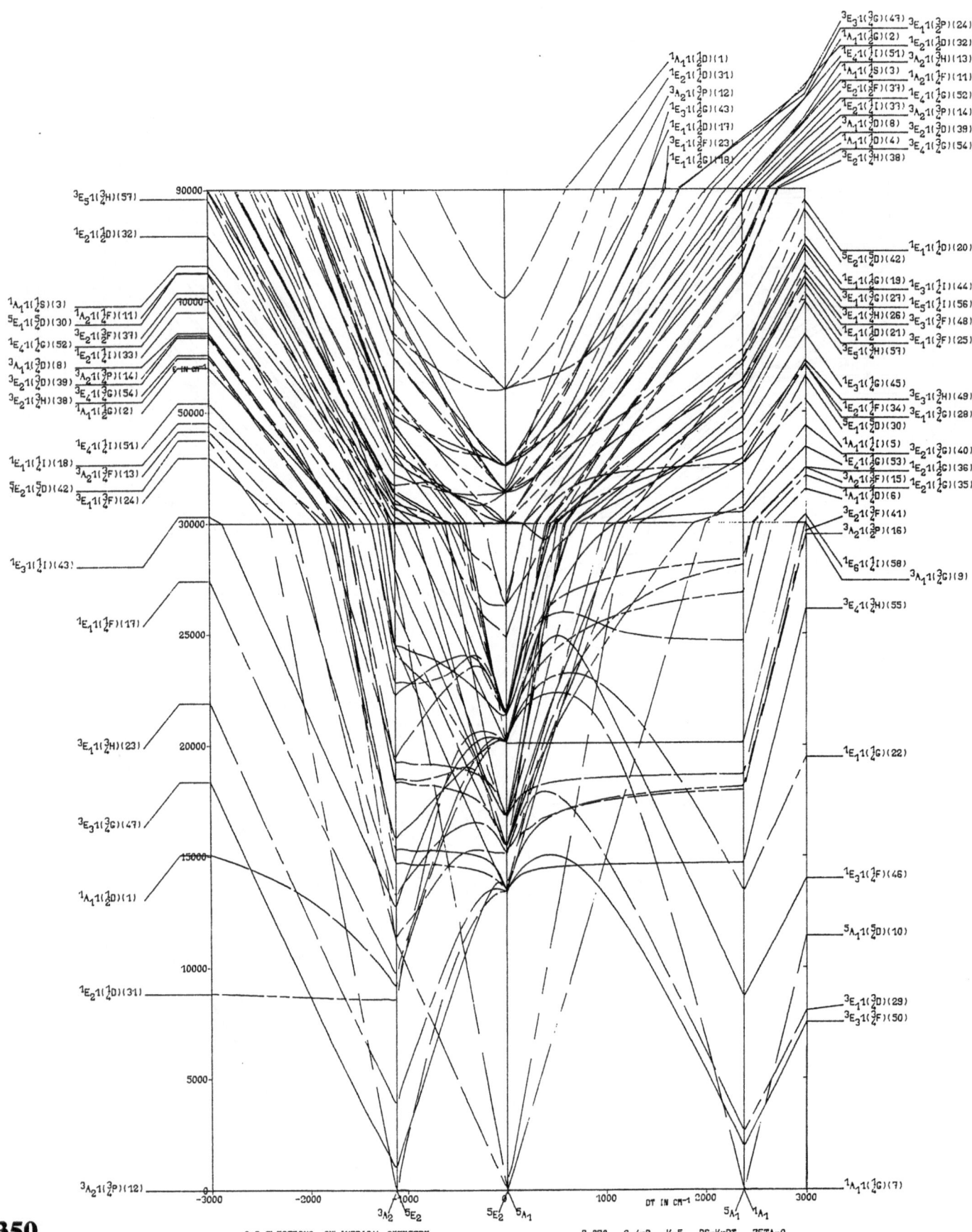

6 D ELECTRONS, CYLINDRICAL SYMMETRY   B=670   C=4xB   K=5   DS=KxDT   ZETA=0

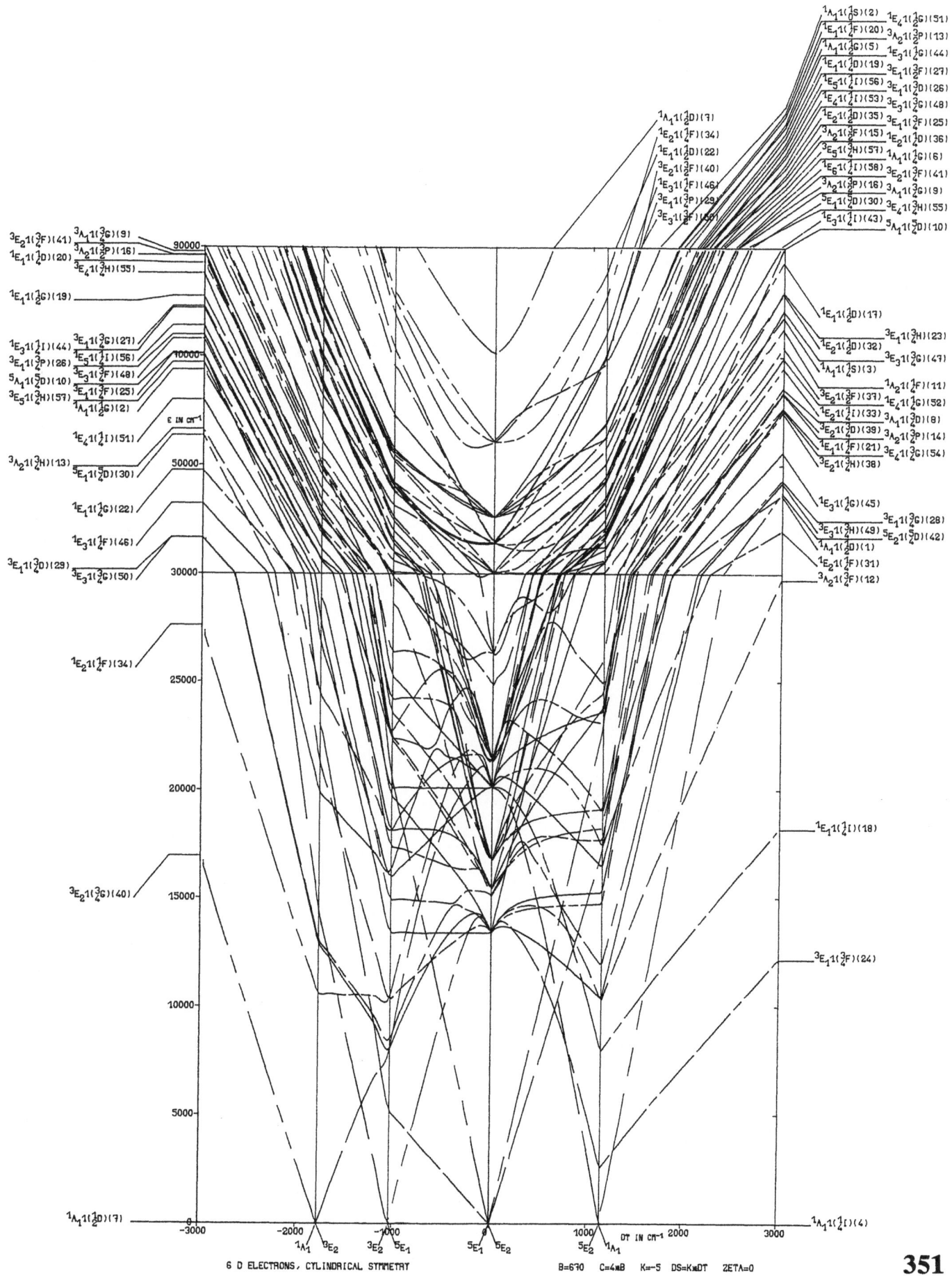

6 D ELECTRONS, CYLINDRICAL SYMMETRY   B=670   C=4×B   K=-5   DS=K×DT   ZETA=0

351

7 D ELECTRONS, TETRAGONAL SYMMETRY

B=825 C=4*B ZETA=0

ENERGY AS FUNCTION OF DQ
DT=0, 500, -500, 1000, -1000
K=1, -1, 5, -5

ENERGY AS FUNCTION OF DT
DQ=920
K=1, -1, 5, -5

(37) $^4E\ (B_1B_2(^3A_2)E\ )$
(36) $^4E\ (A_1B_2(^3B_2)E\ )$
(35) $^4E\ (A_1B_1(^3B_1)E\ )$
(34) $^4B_2(B_1E\ (^3E\ )E\ )$
(33) $^4B_1(B_2E\ (^3E\ )E\ )$
(32) $^4A_2(A_1E\ (^3E\ )E\ )$
(31) $^4A_2(A_1B_1(^3B_1)B_2)$
(30) $^2E\ (E\ E\ (^1A_1)E\ )$
(29) $^2E\ (B_2B_2(^1A_1)E\ )$
(28) $^2E\ (B_1B_2(^3A_2)E\ )$
(27) $^2E\ (B_1B_2(^1A_2)E\ )$
(26) $^2E\ (B_1B_1(^1A_1)E\ )$
(25) $^2E\ (A_1B_2(^3B_2)E\ )$
(24) $^2E\ (A_1B_2(^1B_2)E\ )$
(23) $^2E\ (A_1B_1(^3B_1)E\ )$
(22) $^2E\ (A_1B_1(^1B_1)E\ )$
(21) $^2E\ (A_1A_1(^1A_1)E\ )$
(20) $^2B_2(B_2E\ (^1E\ )E\ )$
(19) $^2B_2(B_1E\ (^1E\ )E\ )$
(18) $^2B_2(B_1B_1(^1A_1)B_2)$
(17) $^2B_2(A_1E\ (^1E\ )E\ )$
(16) $^2B_2(A_1A_1(^1A_1)B_2)$
(15) $^2B_1(B_2E\ (^1E\ )E\ )$
(14) $^2B_1(B_1E\ (^1E\ )E\ )$
(13) $^2B_1(B_1B_2(^1A_2)B_2)$
(12) $^2B_1(A_1E\ (^1E\ )E\ )$
(11) $^2B_1(A_1A_1(^1A_1)B_1)$
(10) $^2A_2(B_2E\ (^1E\ )E\ )$
(9) $^2A_2(B_1E\ (^1E\ )E\ )$
(8) $^2A_2(A_1E\ (^1E\ )E\ )$
(7) $^2A_2(A_1B_1(^3B_1)B_2)$
(6) $^2A_2(A_1B_1(^1B_1)B_2)$
(5) $^2A_1(B_2E\ (^1E\ )E\ )$
(4) $^2A_1(B_1E\ (^1E\ )E\ )$
(3) $^2A_1(A_1E\ (^1E\ )E\ )$
(2) $^2A_1(A_1B_2(^1B_2)B_2)$
(1) $^2A_1(A_1B_1(^1B_1)B_1)$

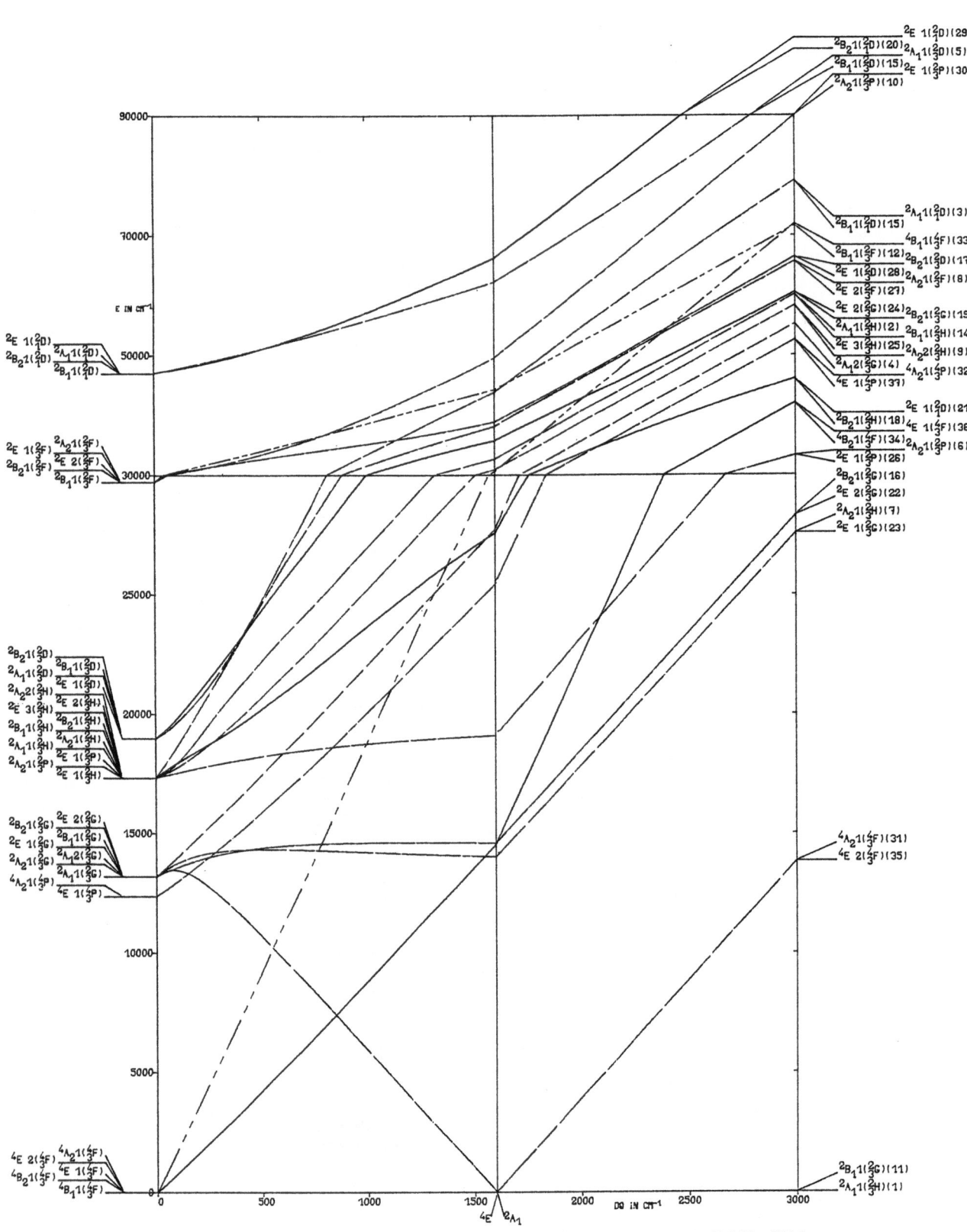

7 D ELECTRONS, TETRAGONAL SYMMETRY

B=825   C=4×B   DT=0   K=0   DS=K×DT   ZETA=0

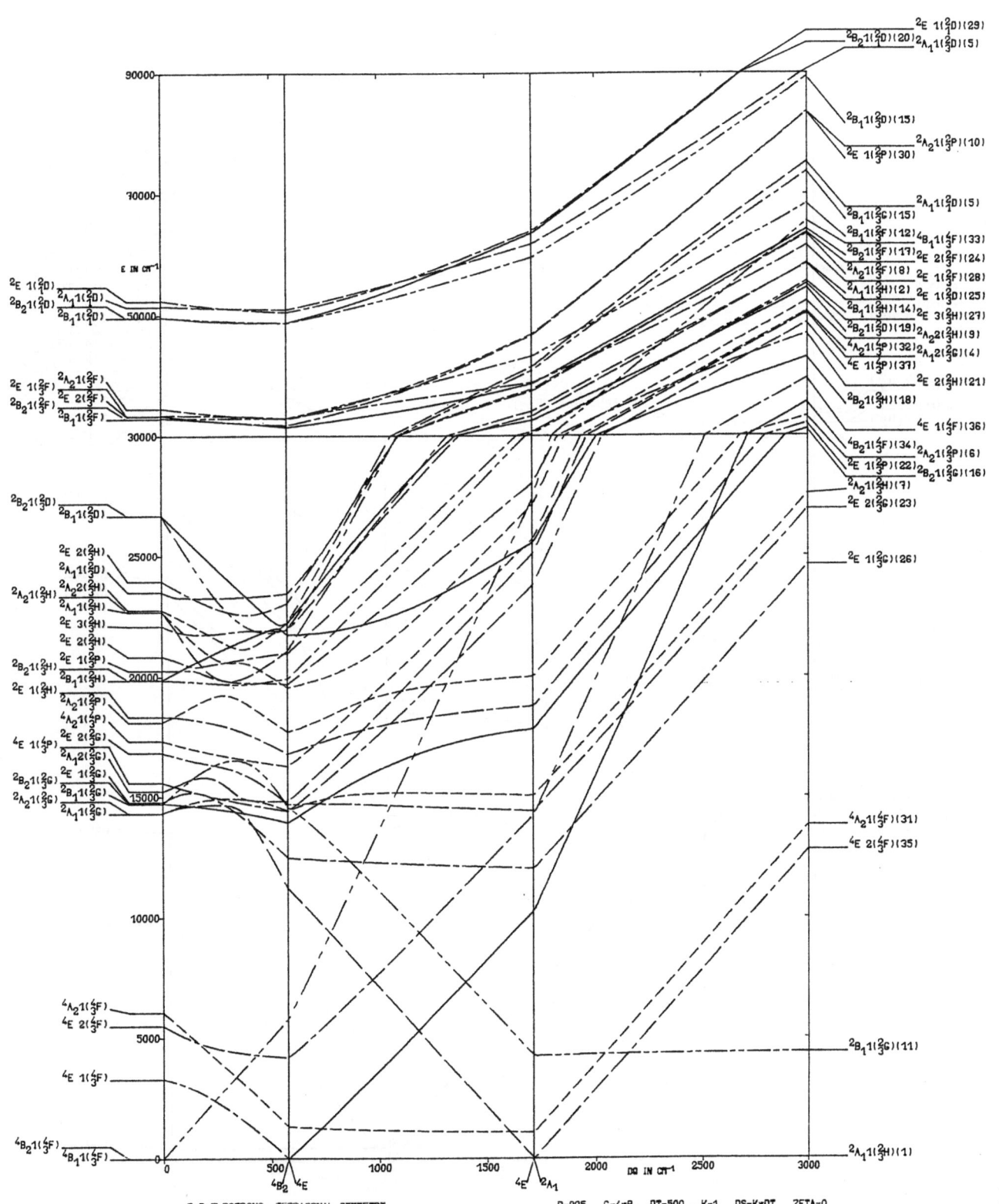

7 D ELECTRONS, TETRAGONAL SYMMETRY   B=825   C=4×B   DT=500   K=1   DS=K×DT   ZETA=0

355

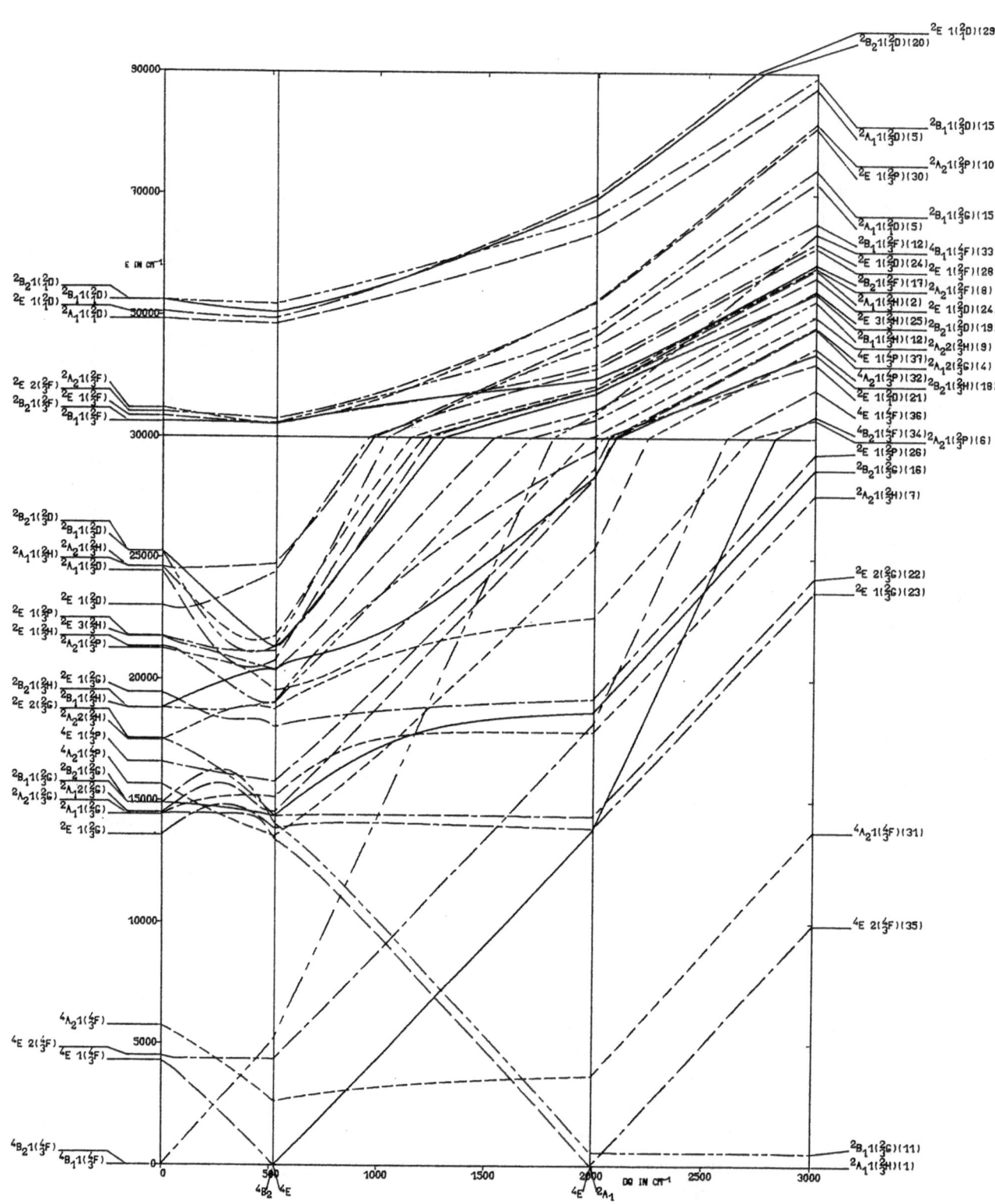

7 D ELECTRONS, TETRAGONAL SYMMETRY

B=825   C=4×B   DT=500   K=-1   DS=K×DT   2ETA=0

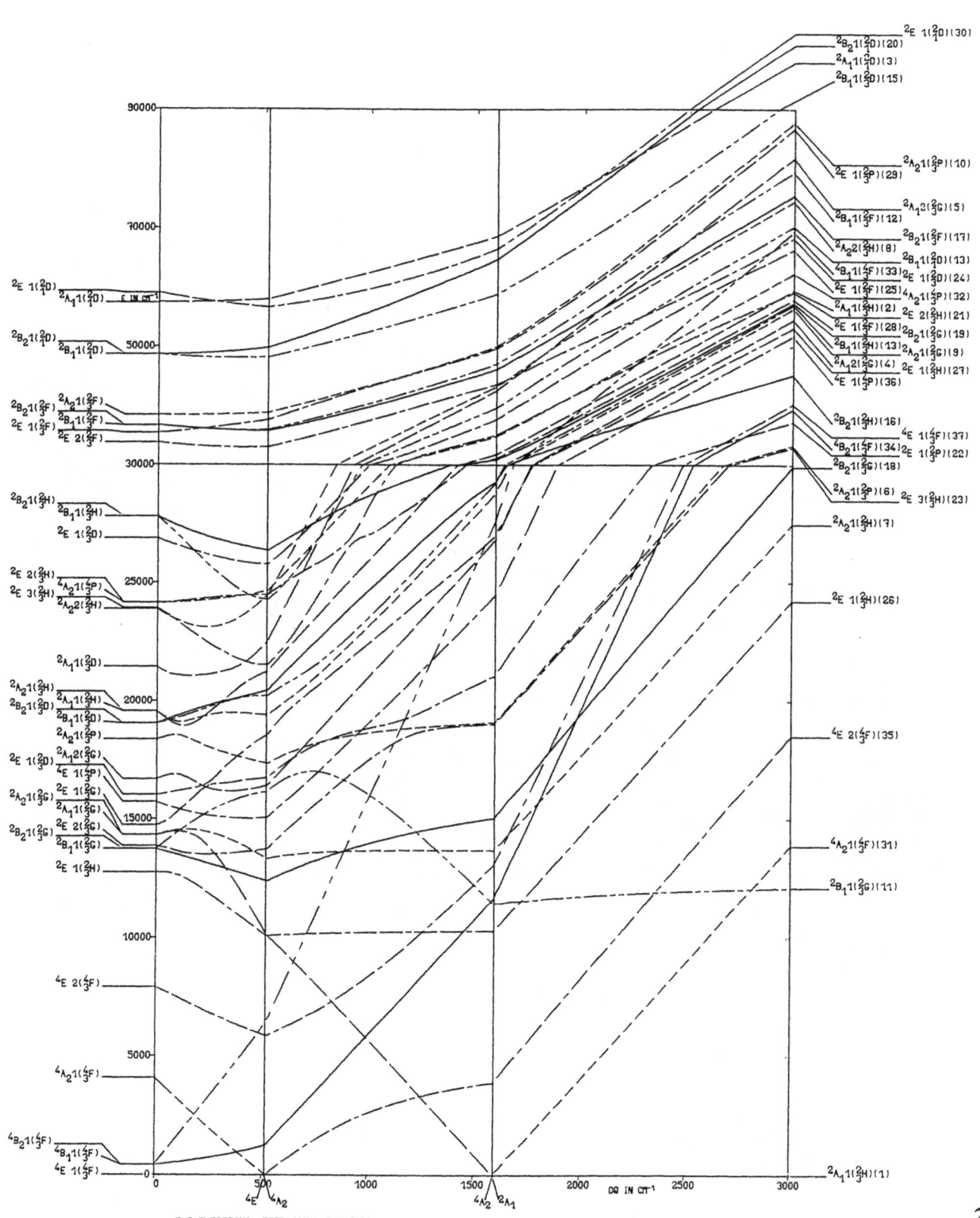

7 D ELECTRONS, TETRAGONAL SYMMETRY   B=825   C=4×B   DT=500   K=5   DS=K×DT   ZETA=0

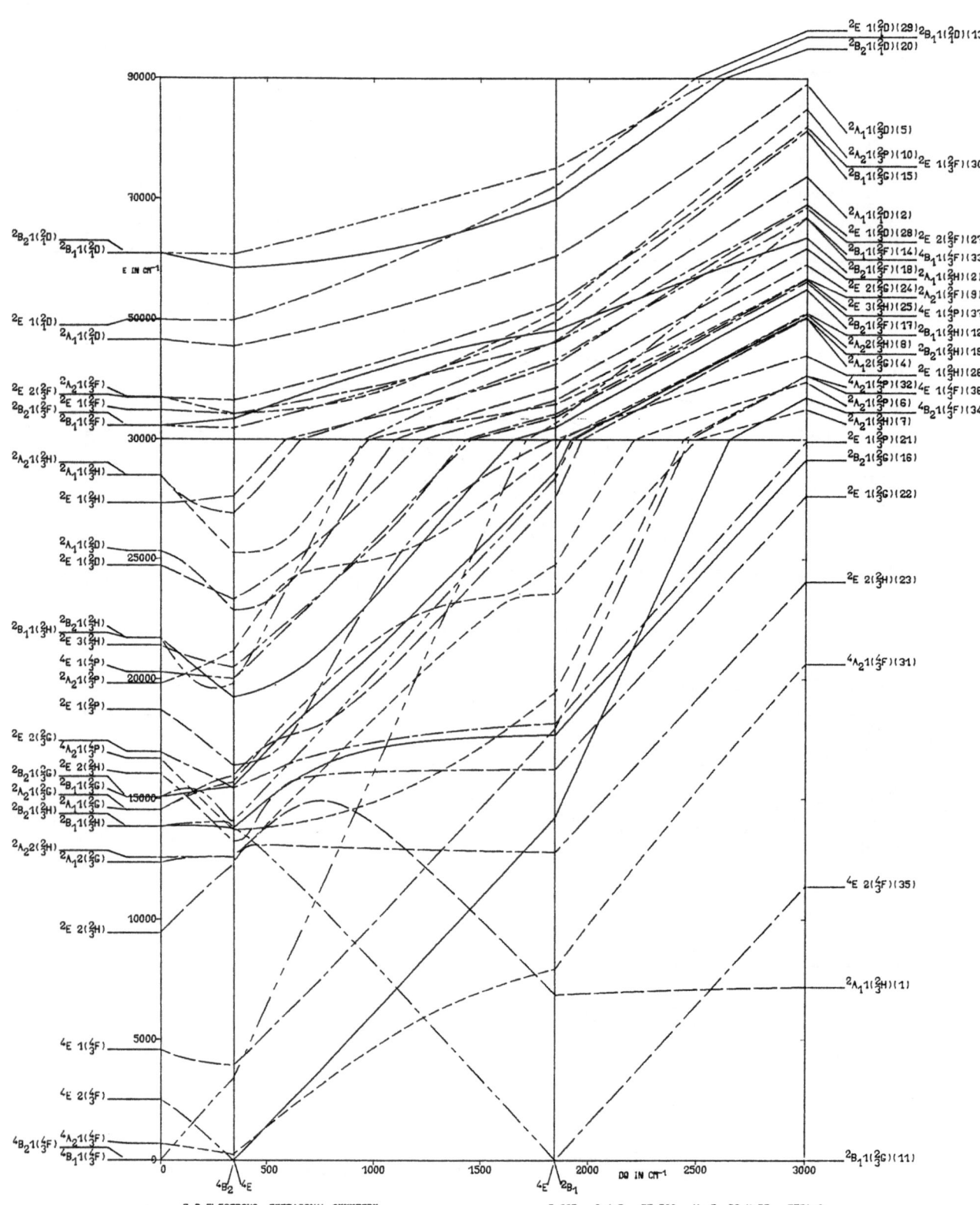

7 D ELECTRONS, TETRAGONAL SYMMETRY   B=825   C=4×B   DT=500   K=-5   DS=K×DT   2ETA=0

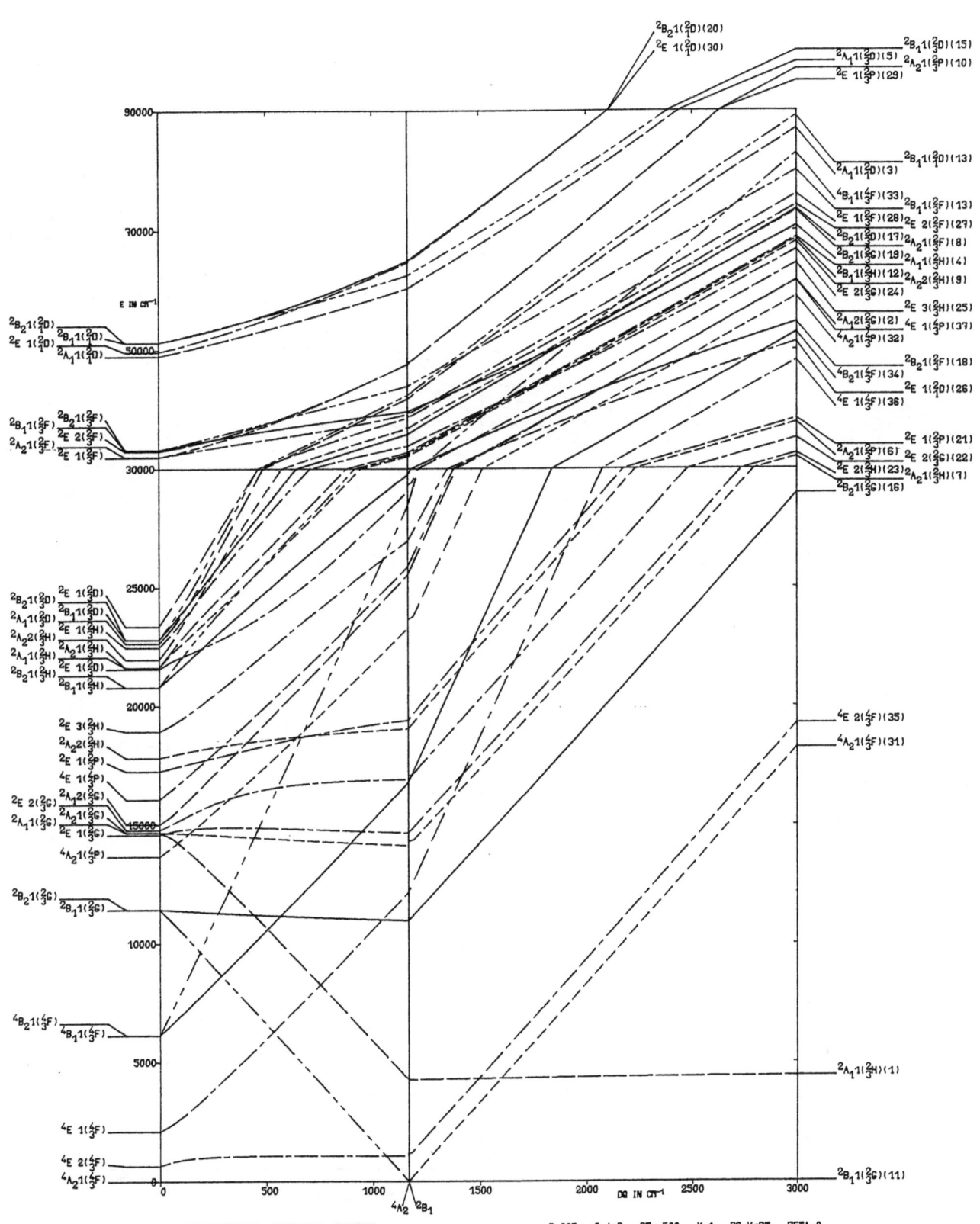

7 D ELECTRONS, TETRAGONAL SYMMETRY   B=825 C=4×B DT=-500 K=1 DS=K×DT 2ETA=0

359

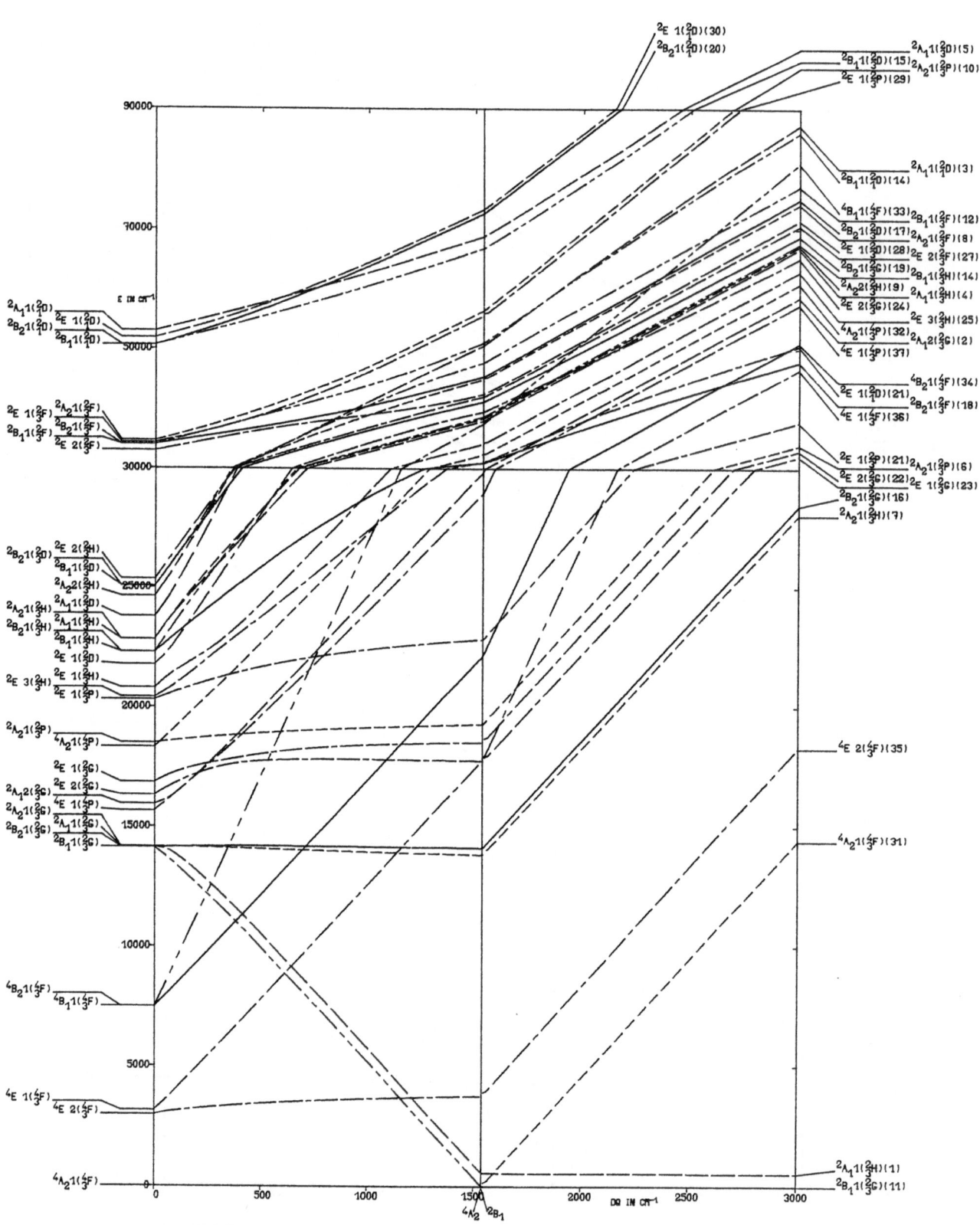

7 D ELECTRONS, TETRAGONAL SYMMETRY

B=825   C=4*B   DT=-500   K=-1   DS=K*DT   ZETA=0

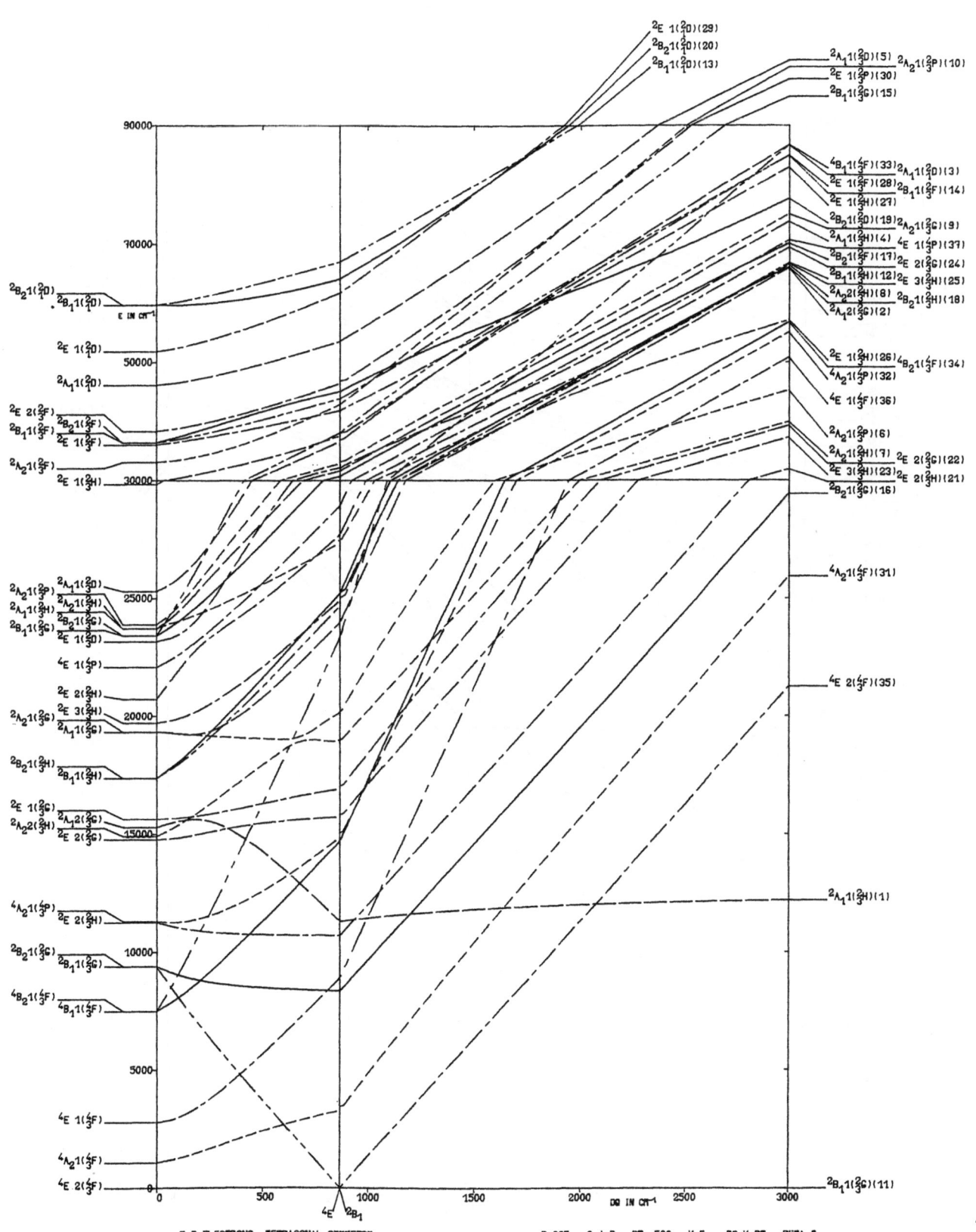

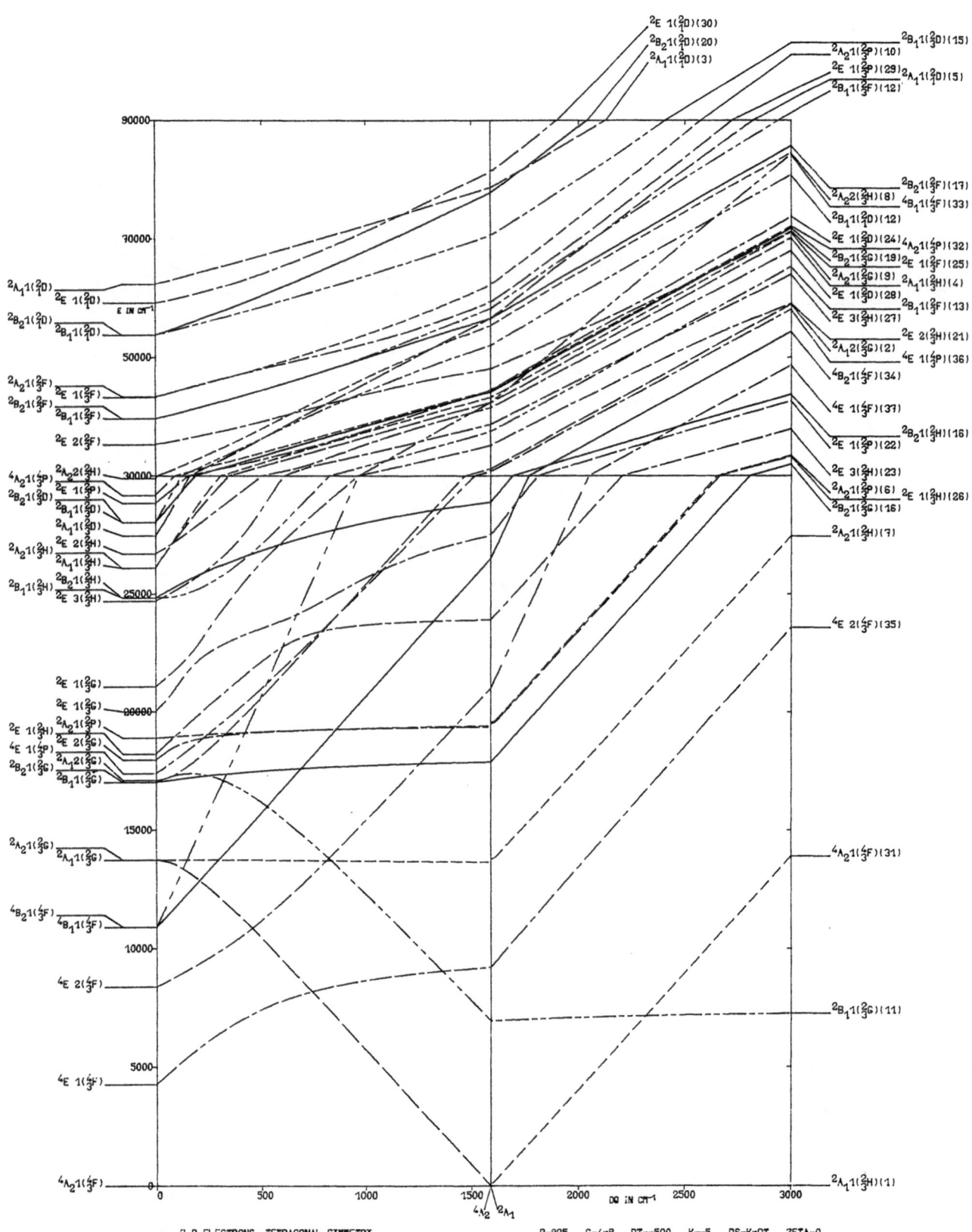

9 D ELECTRONS, TETRAGONAL SYMMETRY

B=825   C=4×B   DT=-500   K=-5   DS=K×DT   ZETA=0

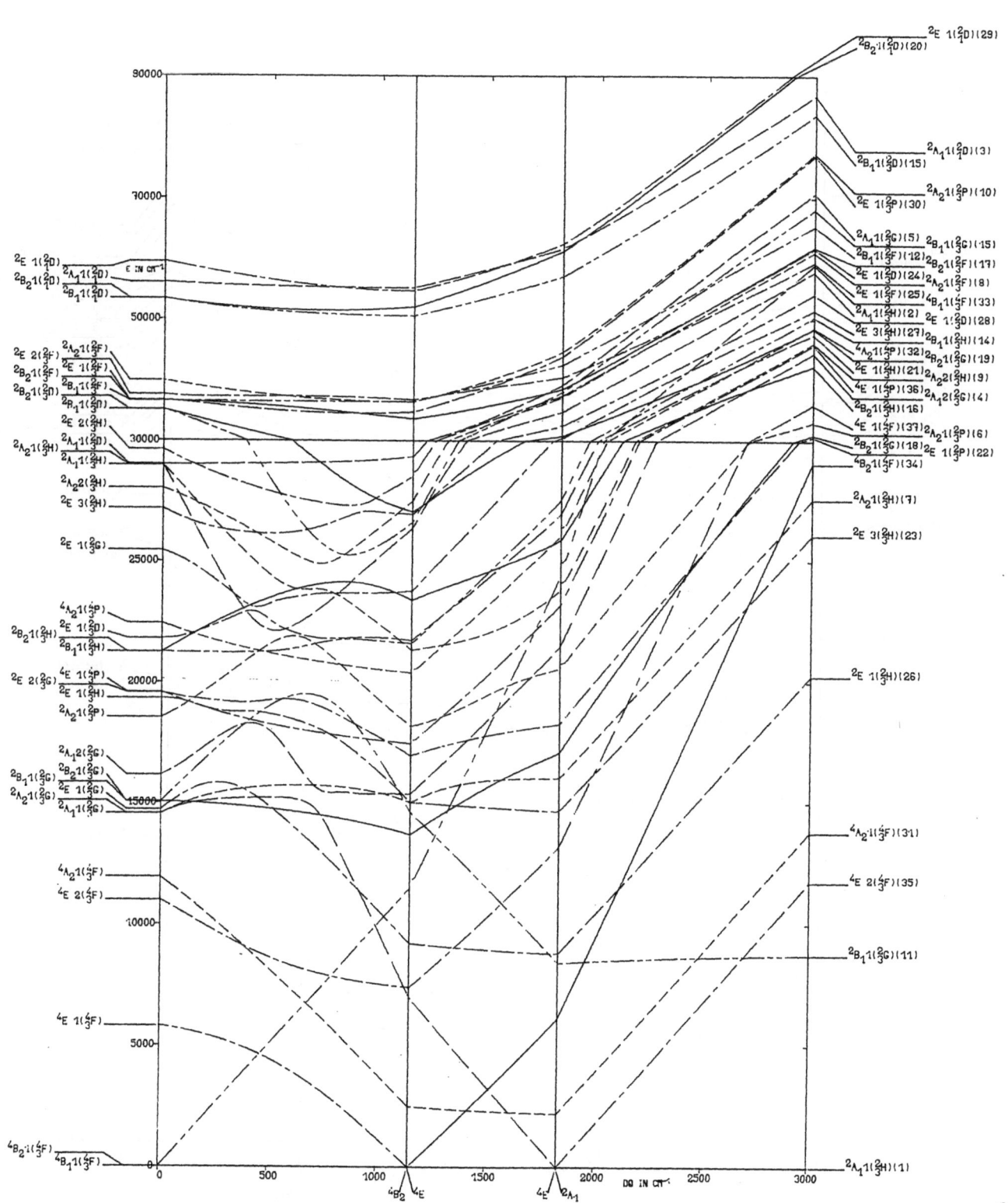

7 D ELECTRONS, TETRAGONAL SYMMETRY

B=825   C=4×B   DT=1000   K=1   DS=K×DT   2ETA=0

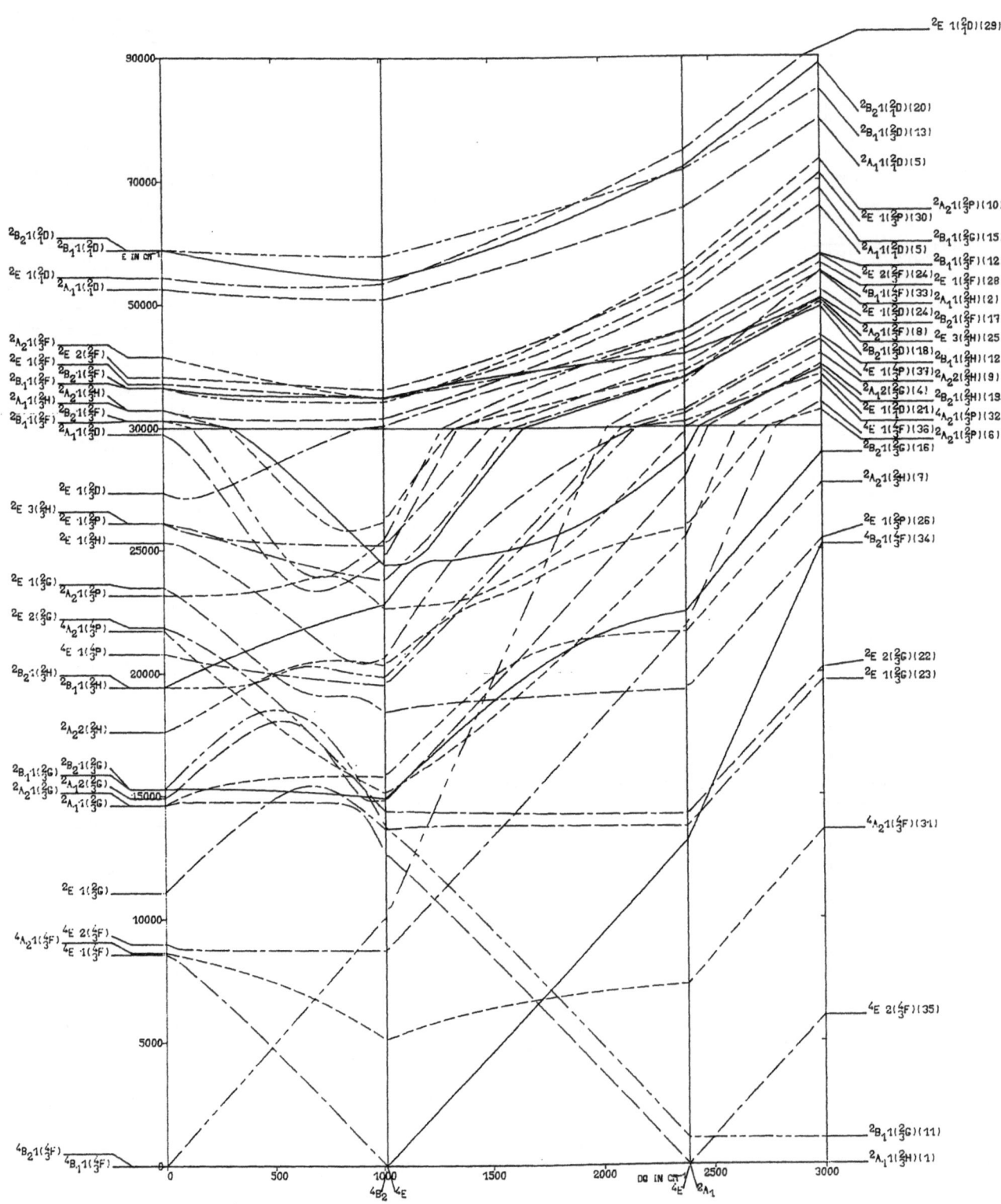

364

7 D ELECTRONS, TETRAGONAL SYMMETRY

B=825   C=4×B   DT=1000   K=-1   DS=K×DT   ZETA=0

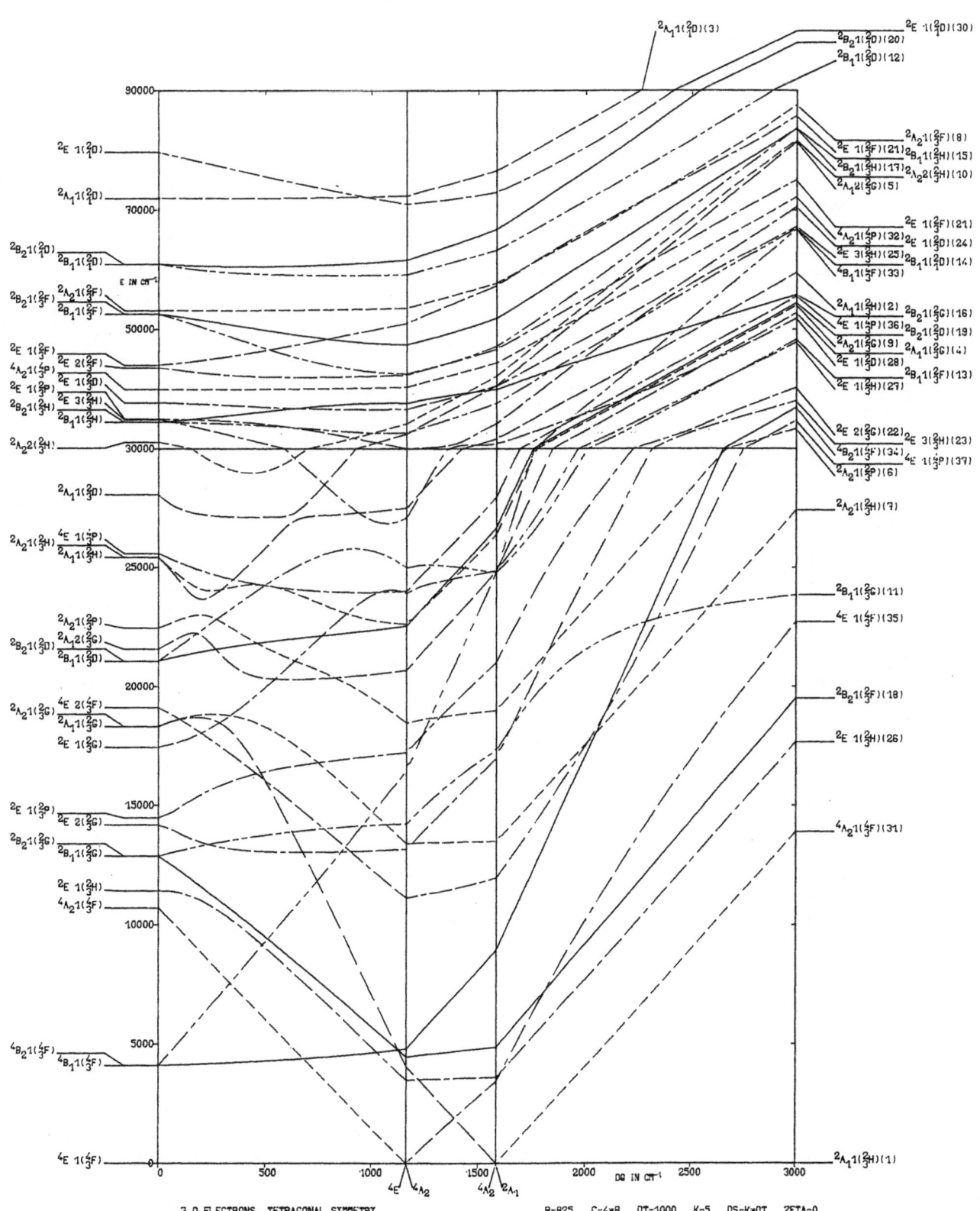

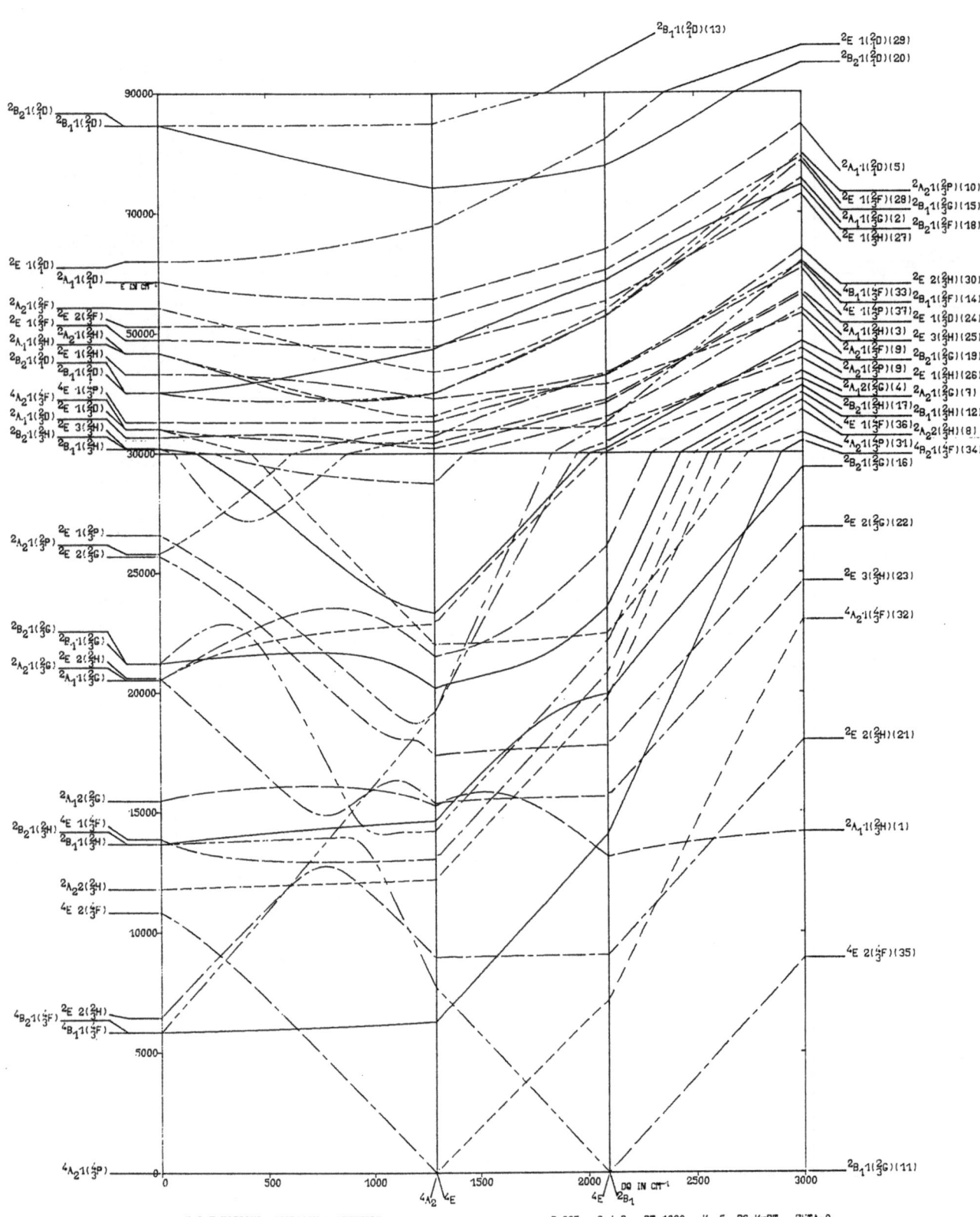

7 D ELECTRONS, TETRAGONAL SYMMETRY   B=825   C=4×B   DT=1000   K=-5   DS=K×DT   ZETA=0

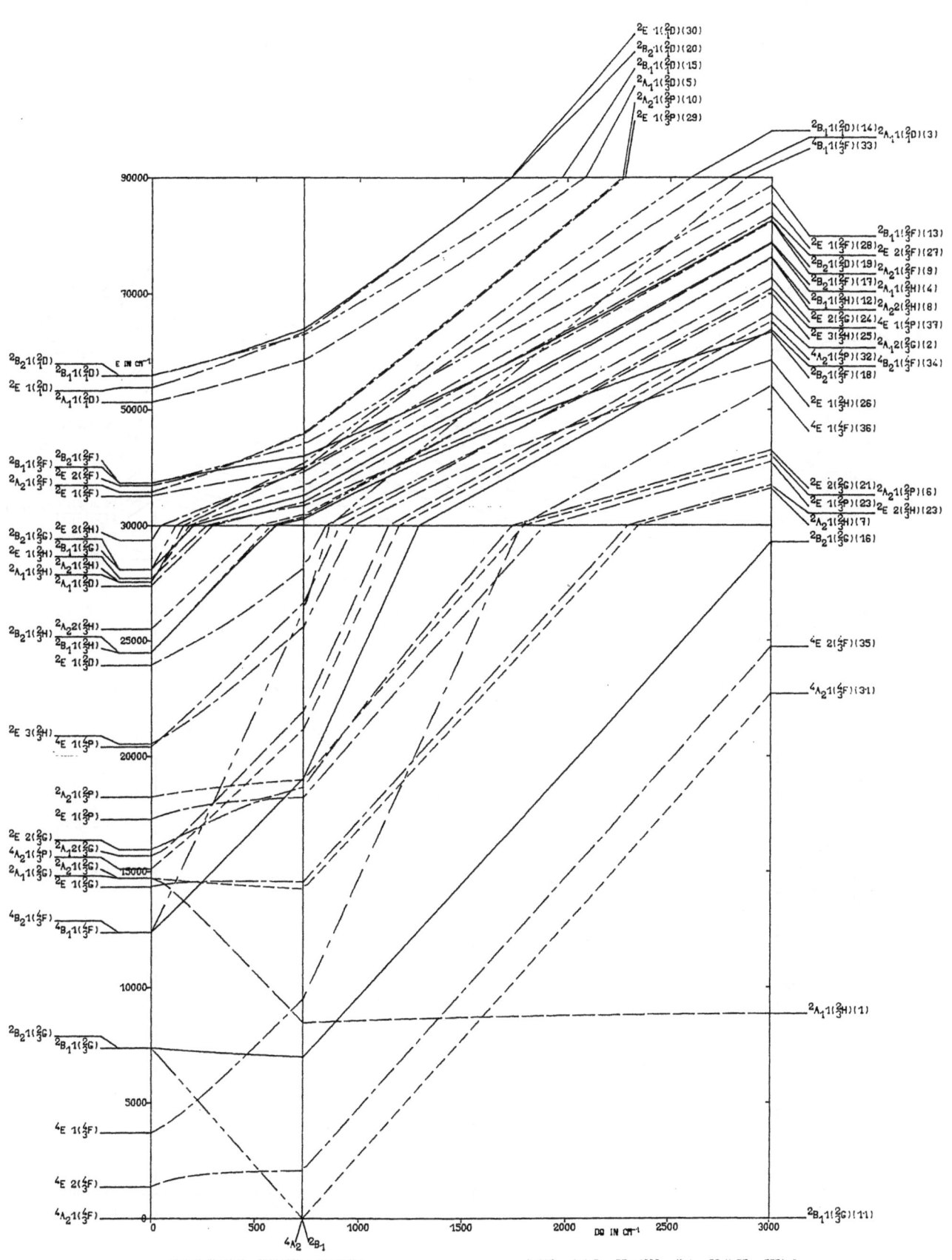

7 D ELECTRONS, TETRAGONAL SYMMETRY   B=825   C=4*B   DT=-1000   K=1   DS=K*DT   ZETA=0

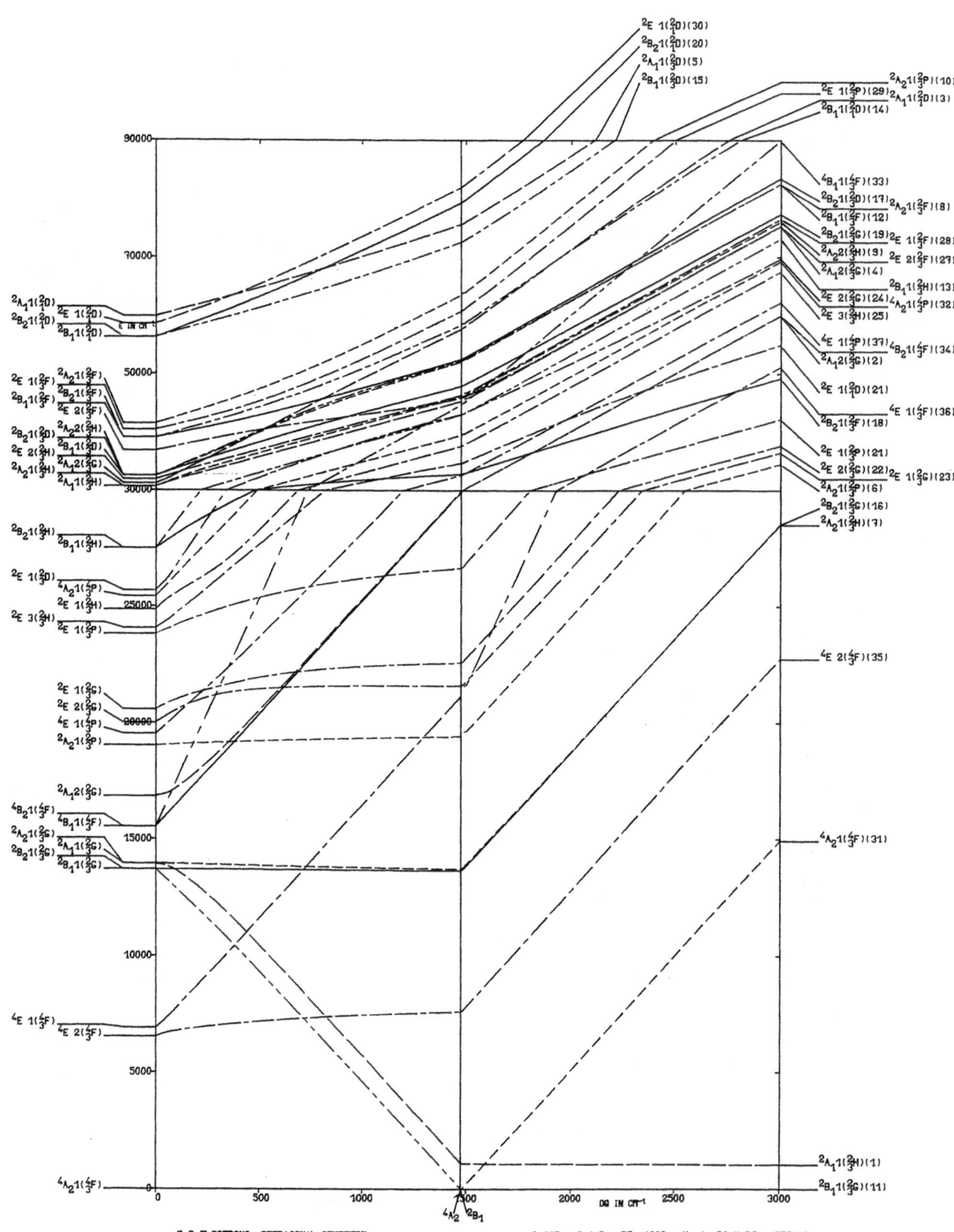

368

3 D ELECTRONS, TETRAGONAL SYMMETRY   B=825 C=4×B DT=−1000 K=−1 DS=K×DT ZETA=0

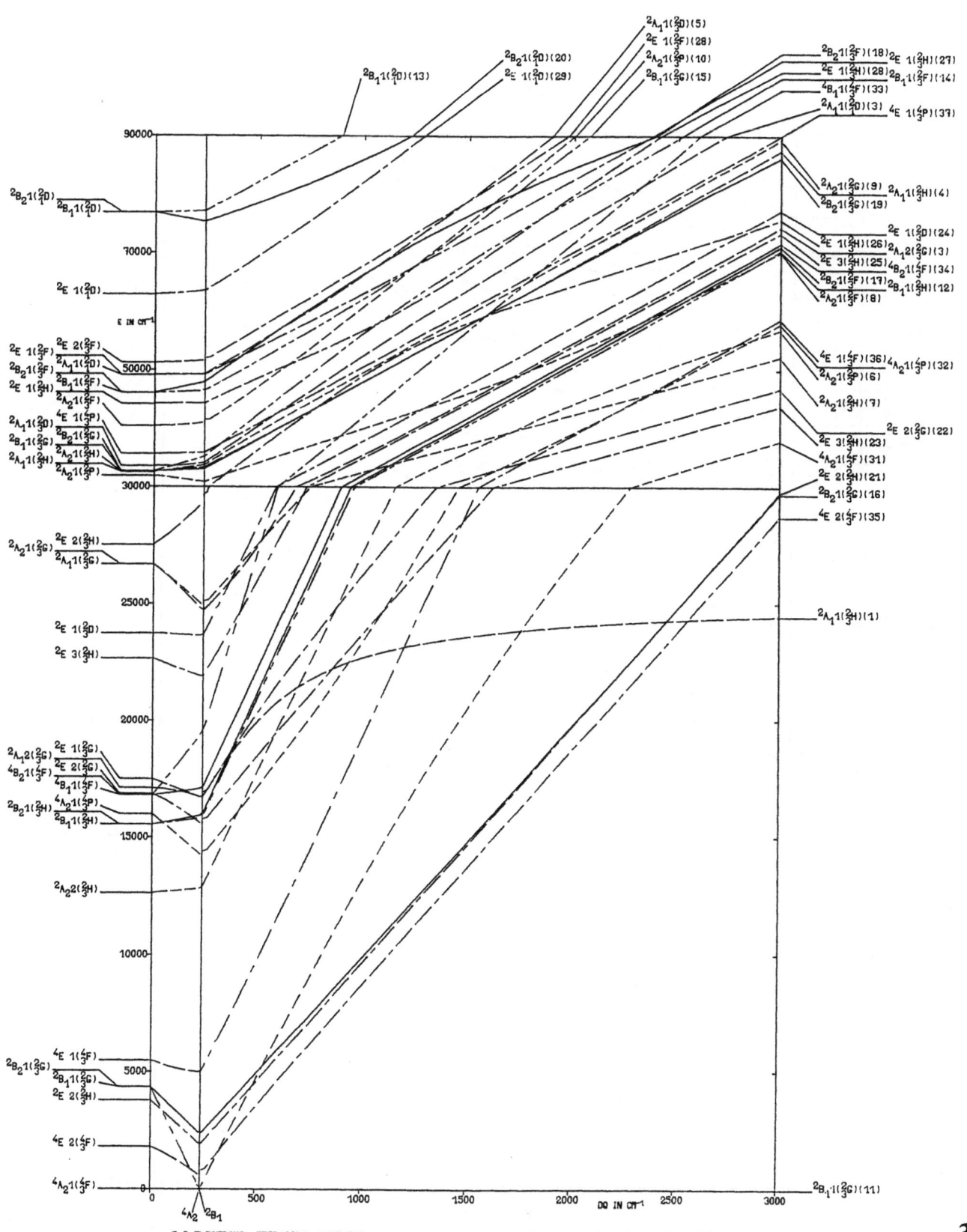

7 D ELECTRONS, TETRAGONAL SYMMETRY   B=825   C=4×B   DT=-1000   K=5   DS=K×DT   ZETA=0

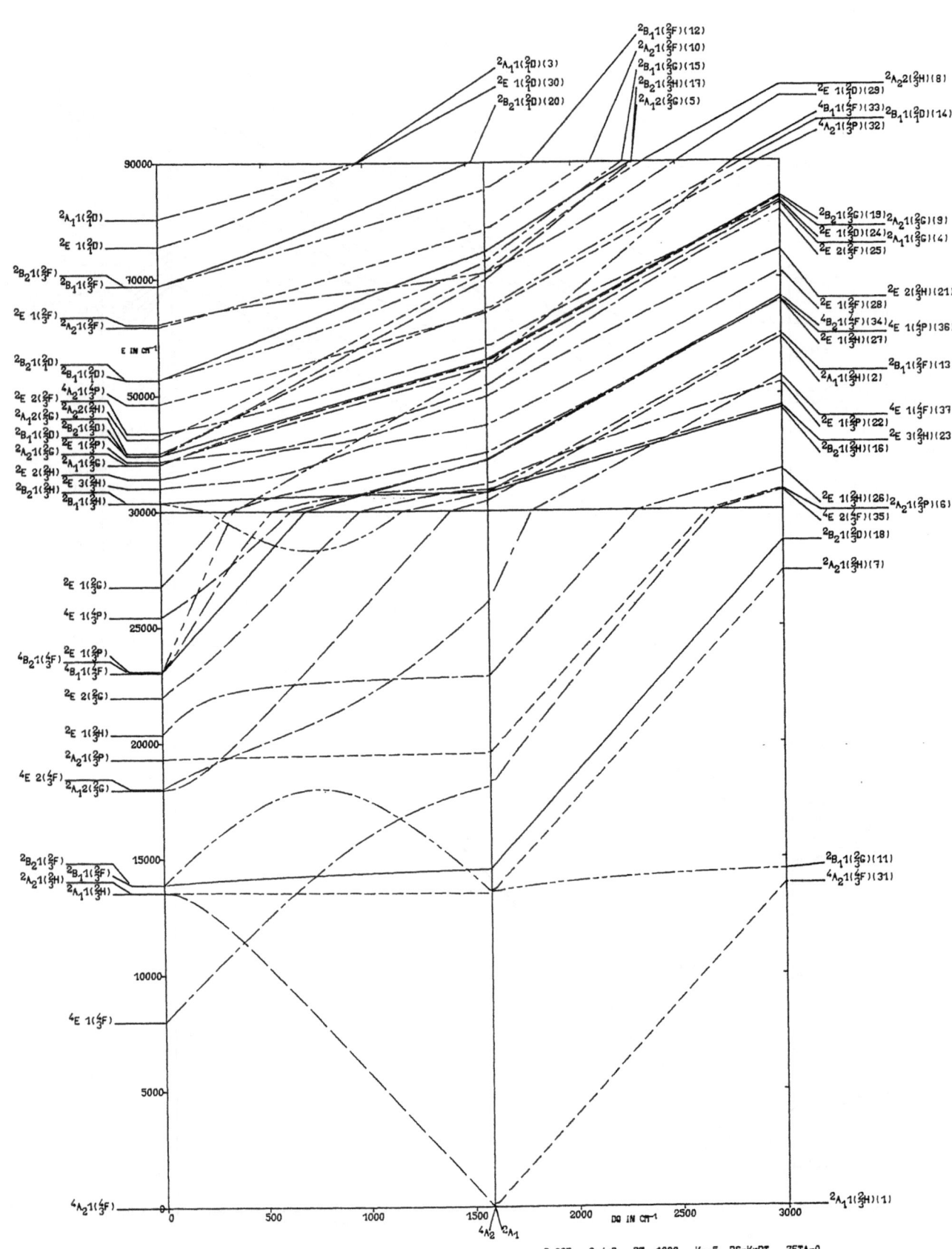

370

3 D ELECTRONS, TETRAGONAL SYMMETRY

B=825   C=4×B   DT=-1000   K=-5   DS=K×DT   ZETA=0

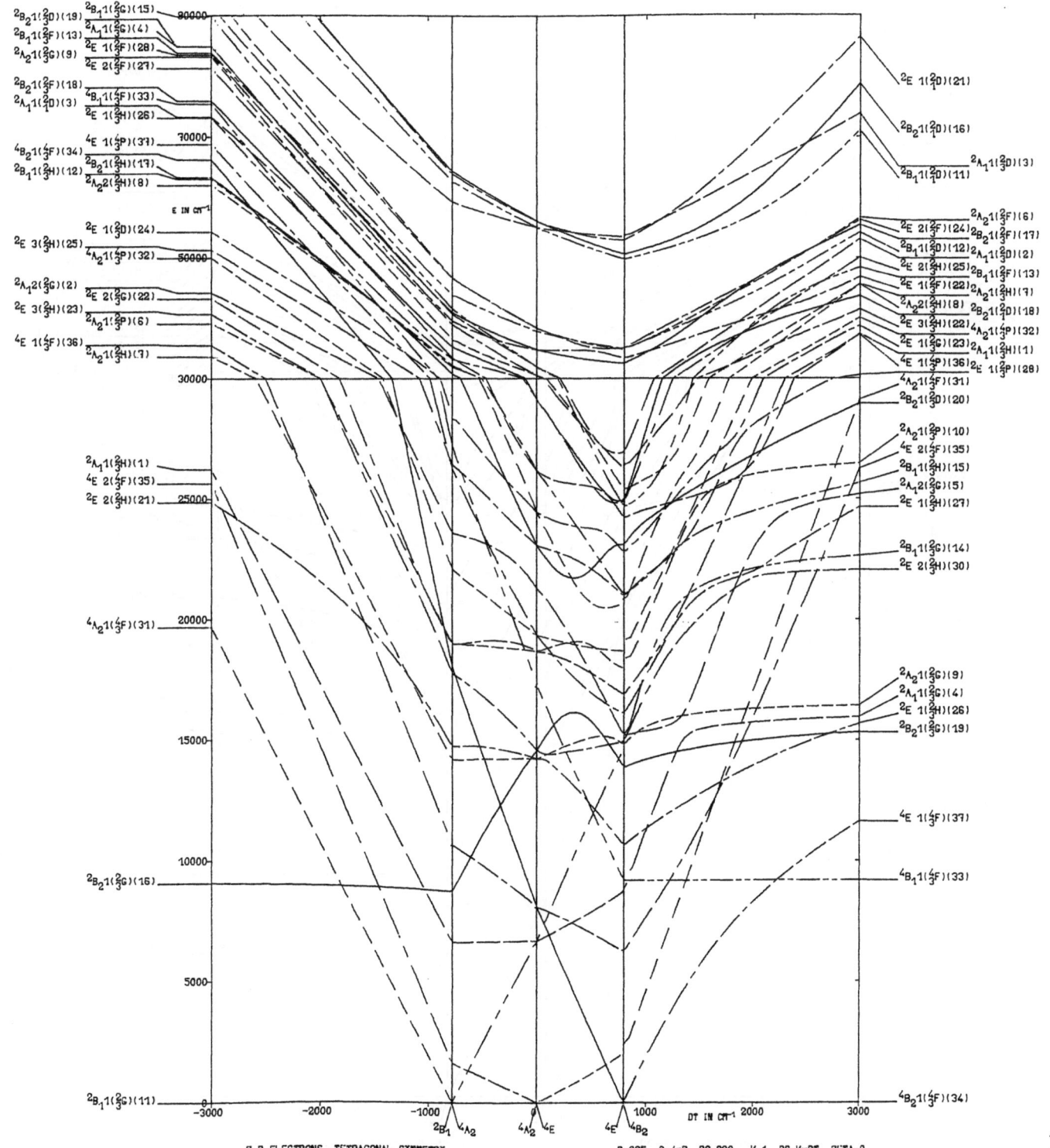

7 D ELECTRONS, TETRAGONAL SYMMETRY   B=825   C=4×B   DQ=920   K=1   DS=K×DT   ZETA=0

371

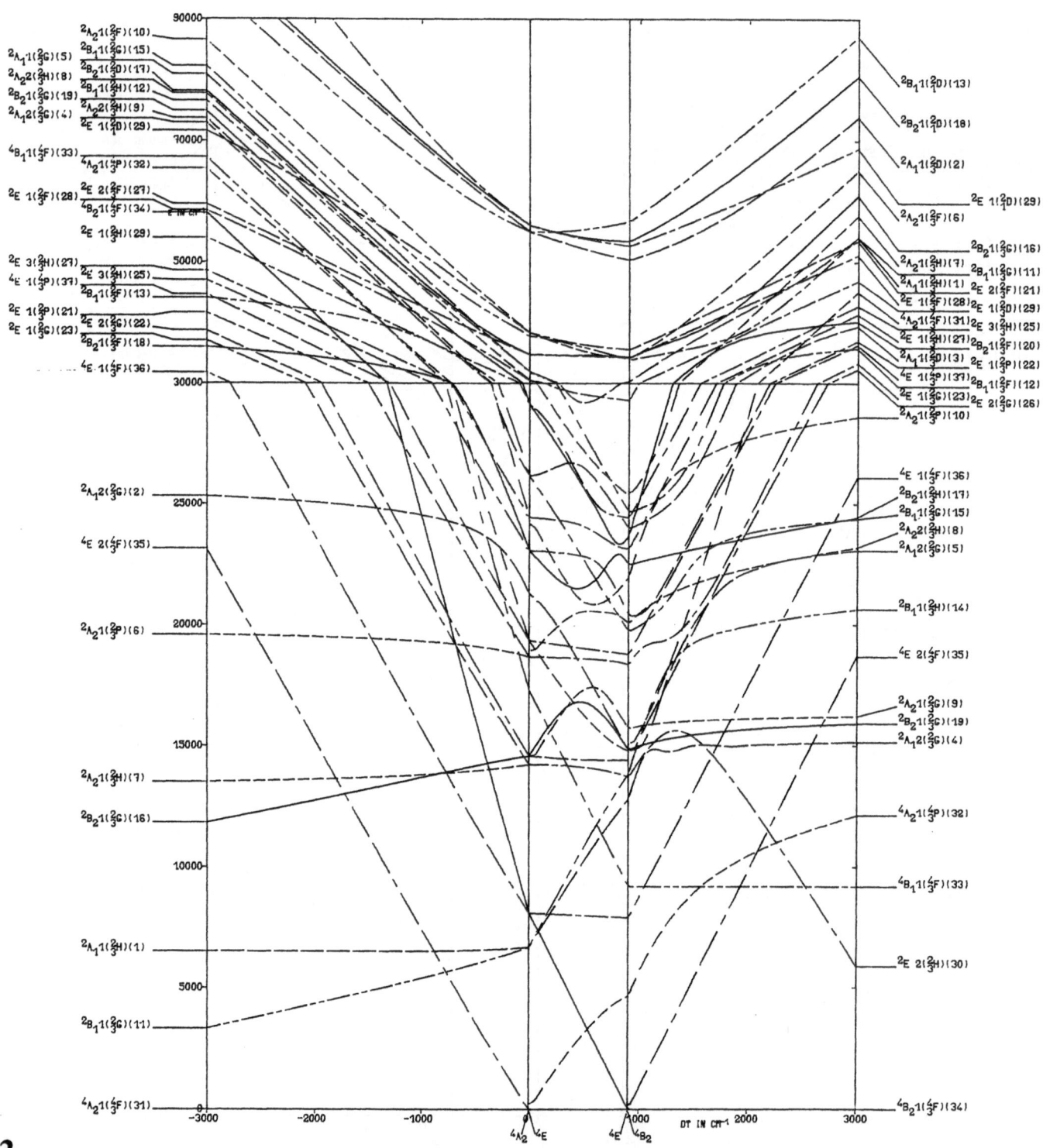

$^2A_21(^2_3F)(10)$
$^2A_11(^2_3G)(5)$
$^2A_22(^2_3H)(8)$
$^2B_21(^2_3G)(19)$
$^2A_12(^2_3G)(4)$

$^4B_11(^4_3F)(33)$

$^2E\ 1(^2_3F)(28)$

$^2E\ 3(^2_3H)(27)$
$^4E\ 1(^4_3P)(37)$

$^2E\ 1(^2_3P)(21)$
$^2E\ 1(^2_3G)(23)$

$^2A_12(^2_3G)(2)$

$^4E\ 2(^4_3F)(35)$

$^2A_21(^2_3P)(6)$

$^2A_21(^2_3H)(7)$

$^2B_21(^2_3G)(16)$

$^2A_11(^2_3H)(1)$

$^2B_11(^2_3G)(11)$

$^4A_21(^4_3F)(31)$

$^2B_11(^2_3G)(15)$
$^2B_11(^2_3D)(17)$
$^2B_11(^2_3H)(12)$
$^2A_22(^2_3H)(9)$
$^2E\ 1(^2_3D)(29)$

$^4A_21(^4_3P)(32)$

$^2E\ 2(^2_3F)(27)$
$^4B_21(^4_3F)(34)$
$^2E\ 1(^2_3H)(29)$

$^2E\ 3(^2_3H)(25)$
$^2B_11(^2_3F)(13)$
$^2E\ 2(^2_3G)(22)$
$^2B_21(^2_3F)(18)$
$^4E\ 1(^4_3F)(36)$

E IN CM^{-1}

90000

80000

70000

60000

50000

40000

30000

25000

20000

15000

10000

5000

$^2B_11(^2_3D)(13)$

$^2B_21(^2_3D)(18)$

$^2A_11(^2_3D)(2)$

$^2E\ 1(^2_3D)(29)$

$^2A_21(^2_3F)(6)$

$^2B_21(^2_3G)(16)$
$^2A_21(^2_3H)(7)$ $^2B_11(^2_3G)(11)$
$^2A_11(^2_3H)(1)$ $^2E\ 2(^2_3F)(21)$
$^2E\ 1(^2_3F)(28)$ $^2E\ 1(^2_3D)(29)$
$^4A_21(^4_3F)(31)$ $^2E\ 3(^2_3H)(25)$
$^2E\ 1(^2_3P)(27)$ $^2B_21(^2_3F)(20)$
$^2A_11(^2_3D)(3)$ $^2E\ 1(^2_3P)(22)$
$^4E\ 1(^4_3P)(37)$ $^2B_11(^2_3F)(12)$
$^2E\ 1(^2_3G)(23)$ $^2E\ 2(^2_3G)(26)$

$^2A_21(^2_3P)(10)$

$^4E\ 1(^4_3F)(36)$
$^2B_21(^2_3H)(17)$
$^2B_11(^2_3G)(15)$
$^2A_22(^2_3H)(8)$
$^2A_12(^2_3G)(5)$

$^2B_11(^2_3H)(14)$

$^4E\ 2(^4_3F)(35)$

$^2A_21(^2_3G)(9)$
$^2B_21(^2_3G)(19)$
$^2A_12(^2_3G)(4)$

$^4A_21(^4_3P)(32)$

$^4B_11(^4_3F)(33)$

$^2E\ 2(^2_3H)(30)$

$^4B_21(^4_3F)(34)$

-3000   -2000   -1000   0   1000   DT IN CM^{-1} 2000   3000

$^4A_2$ ^4E ^4E $^4B_2$

7 D ELECTRONS, TETRAGONAL SYMMETRY   B=825 C=4×B DQ=920   K=-1 DS=K×DT 2ETA=0

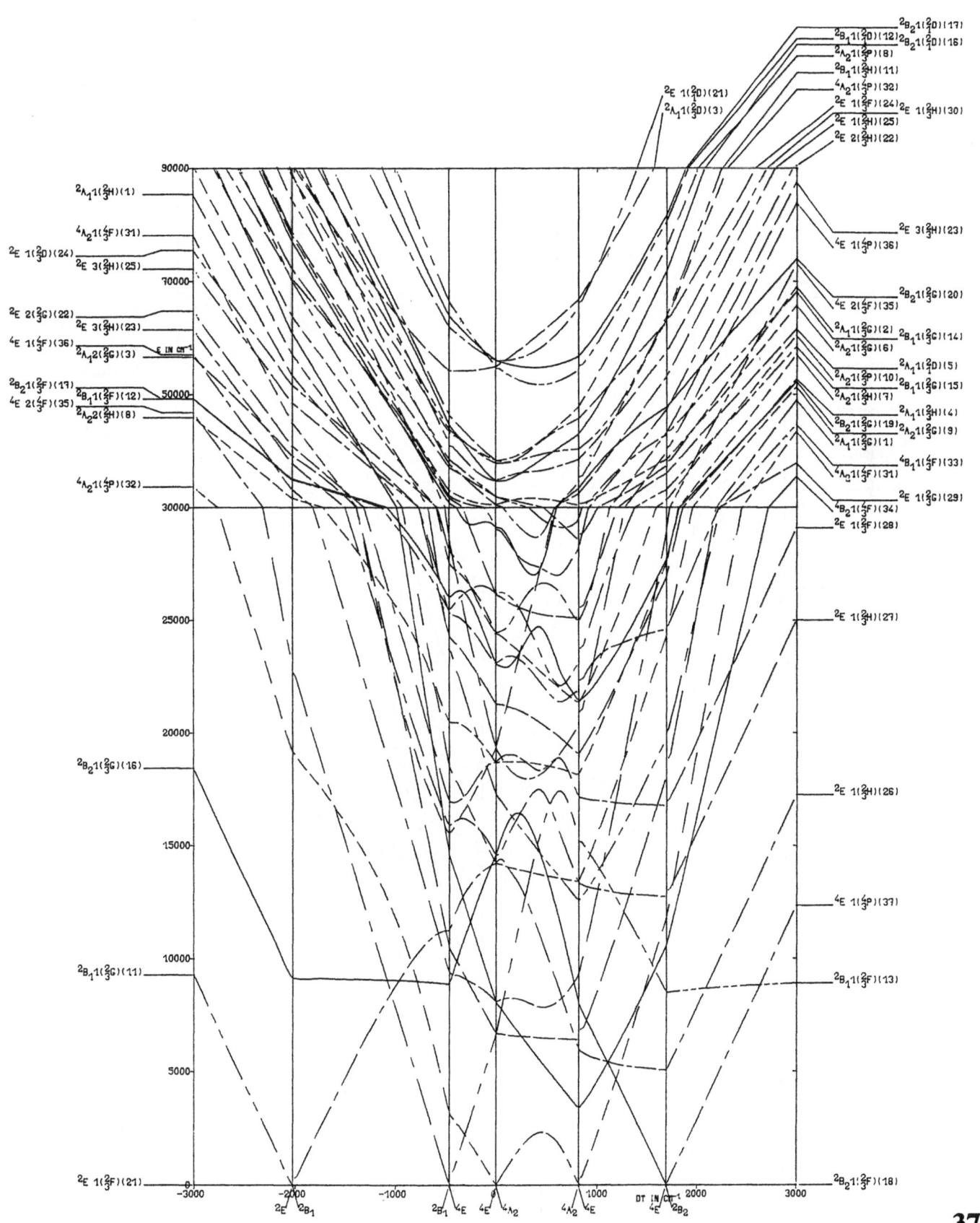

2E 1(²⁄₃D)(21)
2A₁1(²⁄₃D)(3)

2A₁1(²⁄₃H)(1)

4A₂1(⁴⁄₃F)(31)

2E 1(²⁄₃D)(24)
2E 3(²⁄₃H)(25)

2E 2(²⁄₃G)(22)
2E 3(²⁄₃H)(23)
4E 1(⁴⁄₃F)(36)
2A₁2(²⁄₃G)(3)

E IN CM⁻¹

2B₂1(²⁄₃F)(17)
2B₁1(²⁄₃F)(12)
4E 2(⁴⁄₃F)(35)
2A₂2(²⁄₃H)(8)

4A₂1(⁴⁄₃P)(32)

2B₂1(²⁄₃G)(16)

2B₁1(²⁄₃G)(11)

2E 1(²⁄₃F)(21)

2E   2B₁   2B₁ 4E 4A₂   4A₂ 4E   4E 2B₂

DT IN CM⁻¹

-3000   -2000   -1000   0   1000   2000   3000

2B₂1(²⁄₃D)(17)
2B₁1(²⁄₃D)(12) 2B₂1(²⁄₃D)(16)
2A₂1(²⁄₃P)(8)
2B₁1(²⁄₃H)(11)
4A₂1(⁴⁄₃P)(32)
2E 1(²⁄₃F)(24) 2E 1(²⁄₃H)(30)
2E 1(²⁄₃H)(25)
2E 2(²⁄₃H)(22)

2E 3(²⁄₃H)(23)

4E 1(⁴⁄₃P)(36)

2B₂1(²⁄₃G)(20)

4E 2(⁴⁄₃F)(35)
2A₁1(²⁄₃G)(2)
2A₂1(²⁄₃G)(6) 2B₁1(²⁄₃G)(14)
2A₁1(²⁄₃D)(5)
2A₂1(²⁄₃G)(10) 2B₁1(²⁄₃G)(15)
2A₂1(²⁄₃G)(7)
2A₁1(²⁄₃G)(4)
2B₂1(²⁄₃G)(19) 2A₂1(²⁄₃G)(9)
2A₁1(²⁄₃G)(1)
4B₁1(⁴⁄₃F)(33)
4A₂1(⁴⁄₃F)(31)
2E 1(²⁄₃G)(29)
4B₂1(⁴⁄₃F)(34)
2E 1(²⁄₃F)(28)

2E 1(²⁄₃H)(27)

2E 1(²⁄₃H)(26)

4E 1(⁴⁄₃P)(37)

2B₁1(²⁄₃F)(13)

2B₂1(²⁄₃F)(18)

7 D ELECTRONS, TETRAGONAL SYMMETRY   B=825 C=4×B DQ=920 K=5   DS=K×DT ZETA=0

373

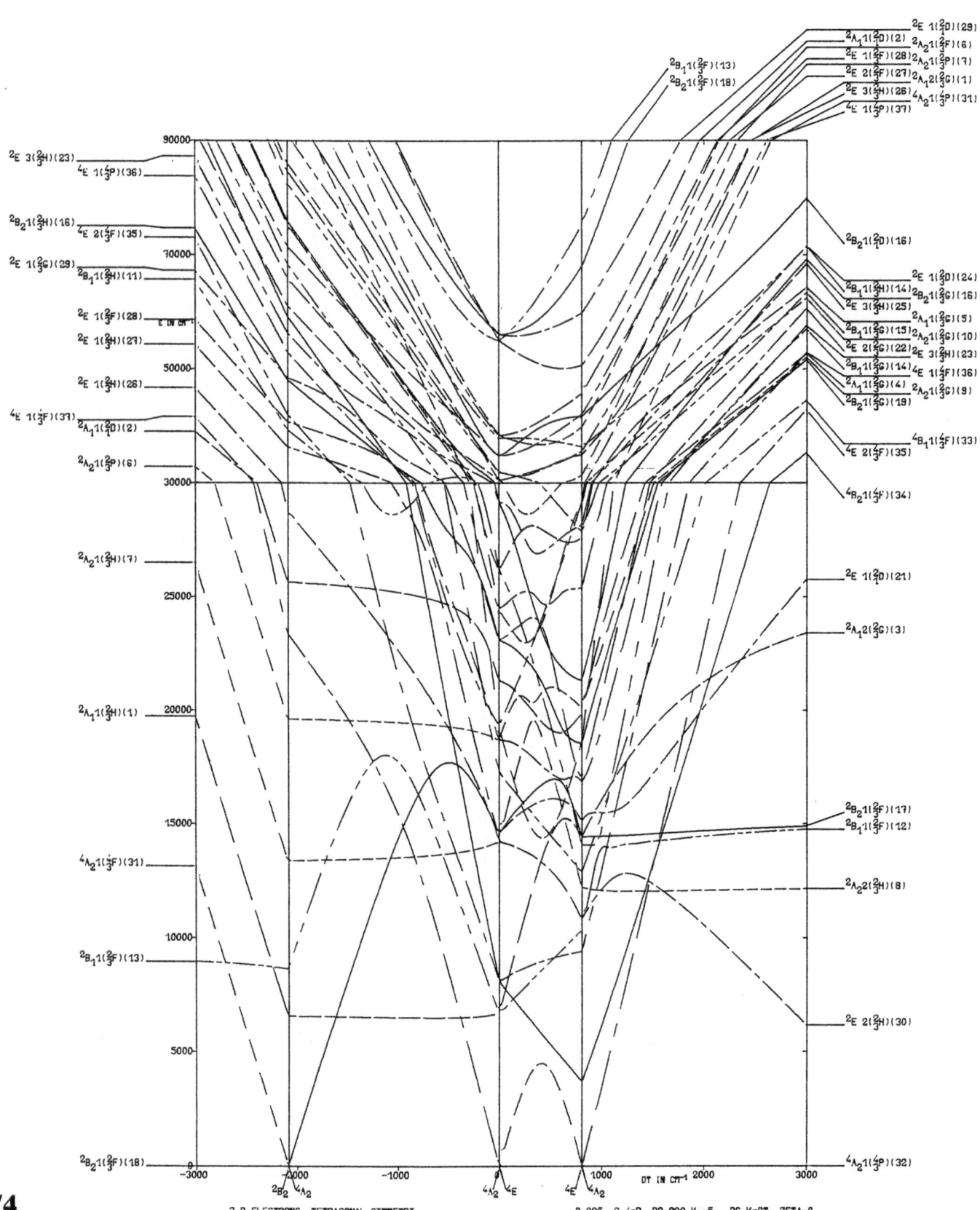

$^2E\ 3(^2_3H)(23)$
$^4E\ 1(^4_3P)(36)$

$^2B_21(^2_3H)(16)$
$^4E\ 2(^2_3F)(35)$

$^2E\ 1(^2_3G)(29)$
$^2B_11(^2_3H)(11)$

$^2E\ 1(^2_3F)(28)$

$^2E\ 1(^2_3H)(27)$

$^2E\ 1(^2_3H)(26)$

$^4E\ 1(^4_3F)(37)$
$^2A_11(^2_3D)(2)$

$^2A_21(^2_3P)(6)$

$^2A_21(^2_3H)(7)$

$^2A_11(^2_3H)(1)$

$^4A_21(^4_3F)(31)$

$^2B_11(^2_3F)(13)$

$^2B_21(^2_3F)(18)$

$^2B_11(^2_3F)(13)$
$^2B_21(^2_3F)(18)$

$^2E\ 1(^2_1D)(29)$
$^2A_11(^2_1D)(2)\ \ ^2A_21(^2_3F)(6)$
$^2E\ 1(^2_3F)(28)\ ^2A_21(^2_3P)(7)$
$^2E\ 2(^2_3F)(27)\ ^2A_12(^2_1G)(1)$
$^2E\ 3(^2_3H)(26)\ ^4A_21(^4_3P)(31)$
$^4E\ 1(^4_3P)(37)$

$^2B_21(^2_1D)(16)$

$^2B_11(^2_3F)(14)\ ^2B_21(^2_1G)(16)$
$^2E\ 3(^2_3H)(25)$
$^2A_11(^2_3G)(5)$
$^2B_11(^2_3G)(15)\ ^2A_21(^2_3G)(10)$
$^2E\ 2(^2_3G)(22)\ ^2E\ 3(^2_3H)(23)$
$^2B_11(^2_3G)(14)\ ^4E\ 1(^4_3F)(36)$
$^2A_11(^2_3G)(4)\ ^2A_21(^2_3G)(9)$
$^2B_21(^2_3G)(19)$

$^4B_11(^4_3F)(33)$
$^4E\ 2(^2_3F)(35)$

$^4B_21(^4_3F)(34)$

$^2E\ 1(^2_1D)(21)$

$^2A_12(^2_3G)(3)$

$^2B_21(^2_3F)(17)$
$^2B_11(^2_3F)(12)$

$^2A_22(^2_3H)(8)$

$^2E\ 2(^2_3H)(30)$

$^4A_21(^4_3P)(32)$

E IN CM

DT IN CM^{-1}

$^2B_2\ ^4A_2$ $^4A_2\ ^4E$ $^4E\ ^4A_2$

7 D ELECTRONS, TETRAGONAL SYMMETRY   B=825 C=4×B DQ=920 K=-5   DS=K×DT   ZETA=0

7 D ELECTRONS, TRIGONAL SYMMETRY

B=825 C=4*B ZETA=0

ENERGY AS FUNCTION OF DQ
DT=0, 500, -500, 1000, -1000
K=1, -1, 5, -5

ENERGY AS FUNCTION OF DT
DQ=920
K=1, -1, 5, -5

(33) $^{4}E\ (E_E E_E(^{3}A_2)E_T)$
(32) $^{4}E\ (E_T E_T(^{3}A_2)E_E)$
(31) $^{4}E\ (A_1 E_T(^{3}E\ )E_E)$
(30) $^{4}A_2(E_E E_E(^{3}A_2)A_1)$
(29) $^{4}A_2(A_1 E_T(^{3}E\ )E_E)$
(28) $^{4}A_2(E_T E_T(^{3}A_2)A_1)$
(27) $^{4}A_1(A_1 E_T(^{3}E\ )E_E)$
(26) $^{2}E\ (E_E E_E(^{1}A_1)E_E)$
(25) $^{2}E\ (E_E E_E(^{3}A_2)E_T)$
(24) $^{2}E\ (E_E E_E(^{1}E\ )E_T)$
(23) $^{2}E\ (E_E E_E(^{1}A_1)E_T)$
(22) $^{2}E\ (E_T E_T(^{3}A_2)E_E)$
(21) $^{2}E\ (E_T E_T(^{1}E\ )E_E)$
(20) $^{2}E\ (E_T E_T(^{1}A_1)E_E)$
(19) $^{2}E\ (E_T E_T(^{1}A_1)E_T)$
(18) $^{2}E\ (E_E E_E(^{1}E\ )A_1)$
(17) $^{2}E\ (A_1 E_T(^{3}E\ )E_E)$
(16) $^{2}E\ (A_1 E_T(^{1}E\ )E_E)$
(15) $^{2}E\ (E_T E_T(^{1}E\ )A_1)$
(14) $^{2}E\ (A_1 A_1(^{1}A_1)E_E)$
(13) $^{2}E\ (A_1 A_1(^{1}A_1)E_T)$
(12) $^{2}A_2(E_E E_E(^{1}E\ )E_T)$
(11) $^{2}A_2(E_T E_T(^{1}E\ )E_E)$
(10) $^{2}A_2(E_E E_E(^{3}A_2)A_1)$
(9) $^{2}A_2(A_1 E_T(^{3}E\ )E_E)$
(8) $^{2}A_2(A_1 E_T(^{1}E\ )E_E)$
(7) $^{2}A_2(E_T E_T(^{3}A_2)A_1)$
(6) $^{2}A_1(E_E E_E(^{1}E\ )E_T)$
(5) $^{2}A_1(E_T E_T(^{1}E\ )E_E)$
(4) $^{2}A_1(E_E E_E(^{1}A_1)A_1)$
(3) $^{2}A_1(A_1 E_T(^{3}E\ )E_E)$
(2) $^{2}A_1(A_1 E_T(^{1}E\ )E_E)$
(1) $^{2}A_1(E_T E_T(^{1}A_1)A_1)$

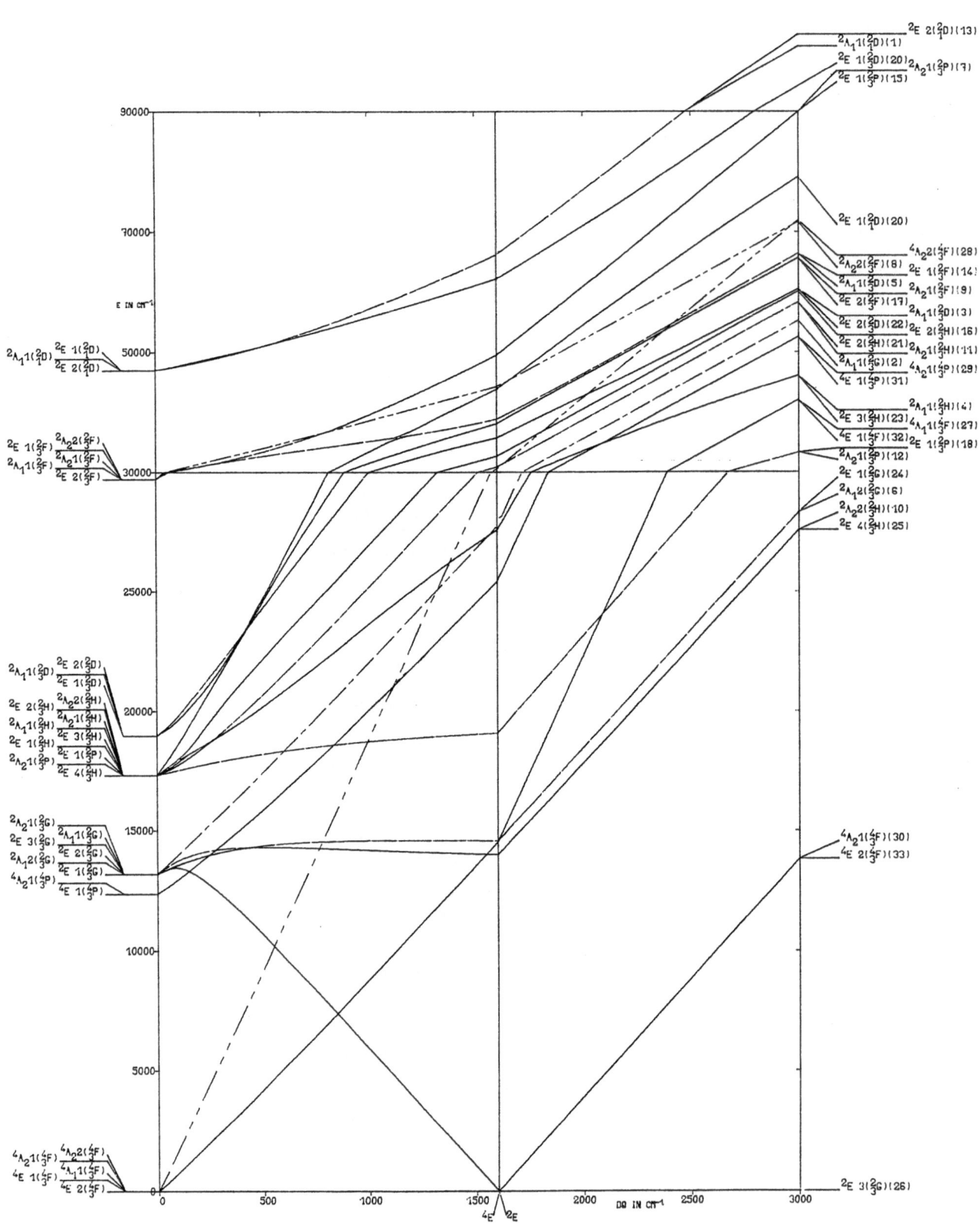

D ELECTRONS, TRIGONAL SYMMETRY

B=825 C=4*B DT=0 K=0 DS=K*DT ZETA=0

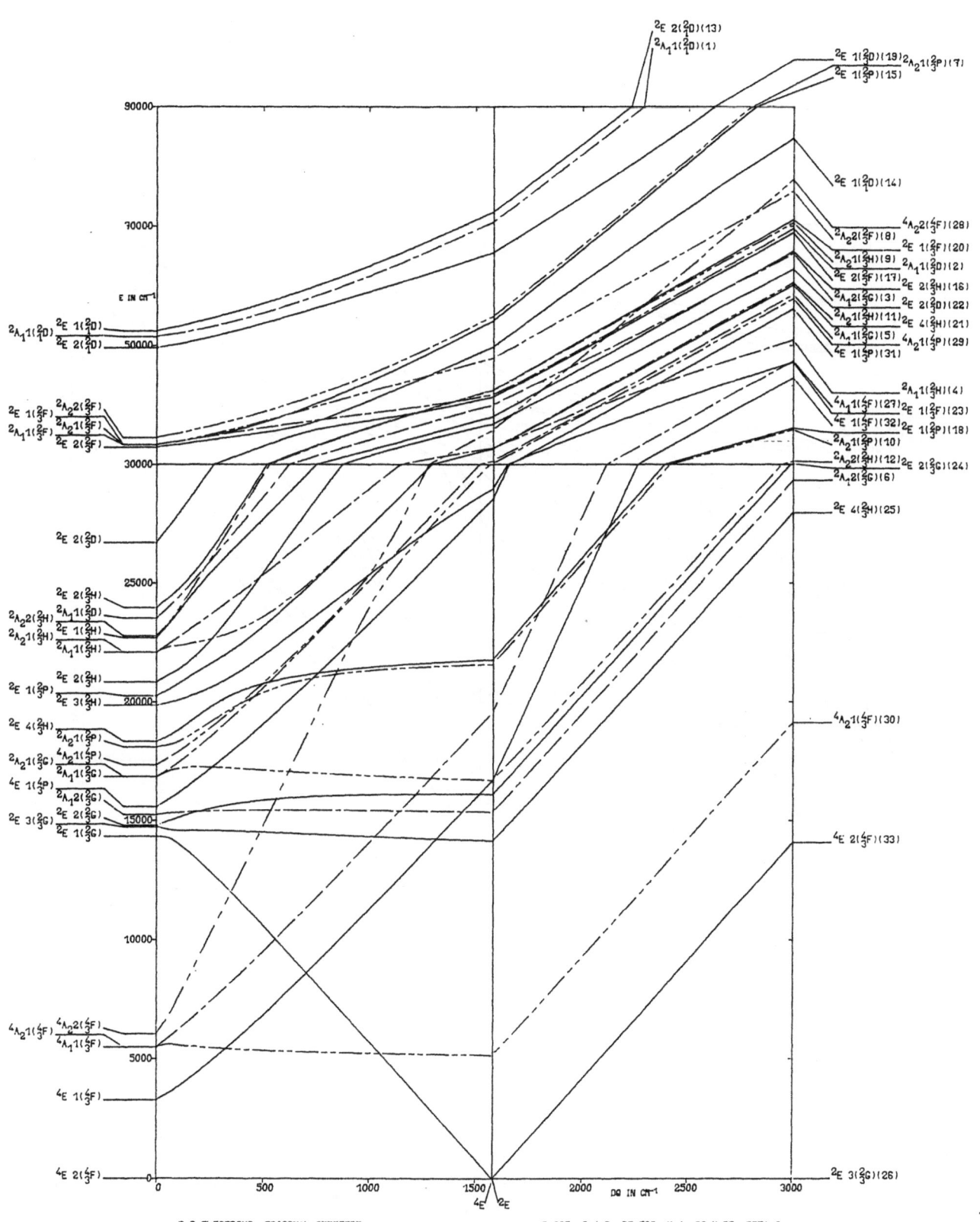

7 D ELECTRONS, TRIGONAL SYMMETRY B=825 C=4×B DT=500 K=1 DS=K×DT ZETA=0

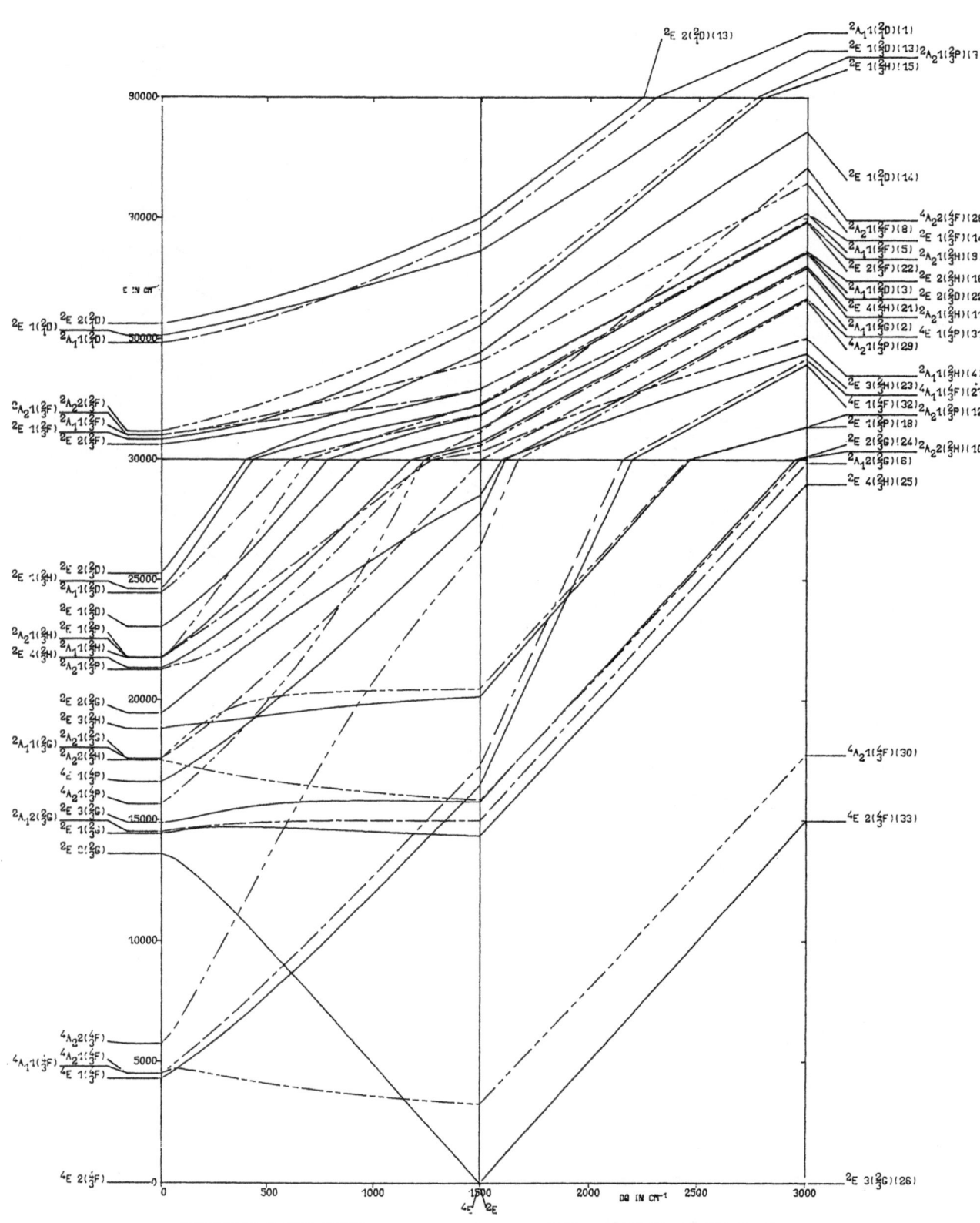

7 D ELECTRONS, TRIGONAL SYMMETRY

B=825 C=4×B DT=500 K=-1 DS=K×DT ZETA=0

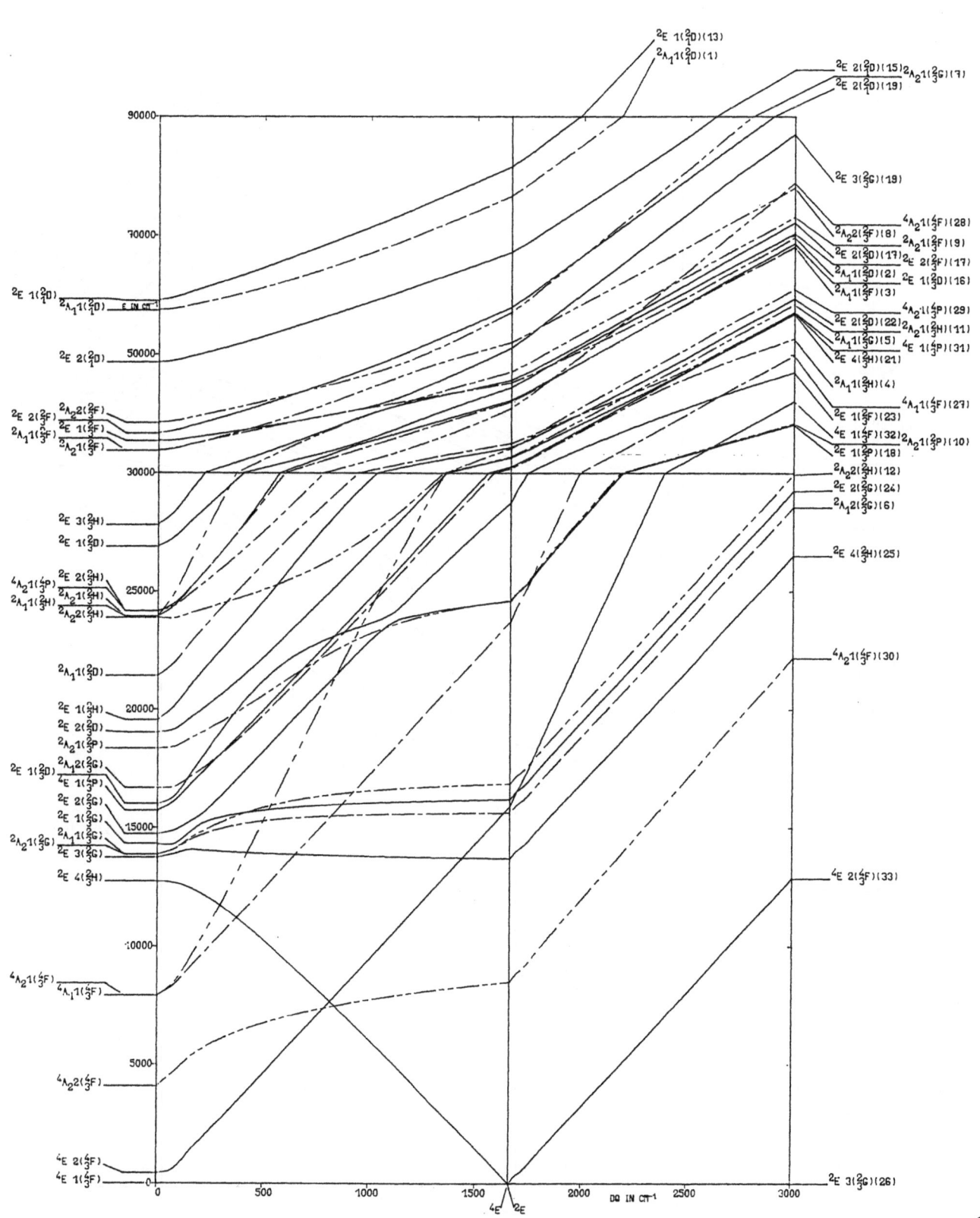

7 D ELECTRONS, TRIGONAL SYMMETRY

B=825 C=4×B DT=500 K=5 DS=K×DT ZETA=0

379

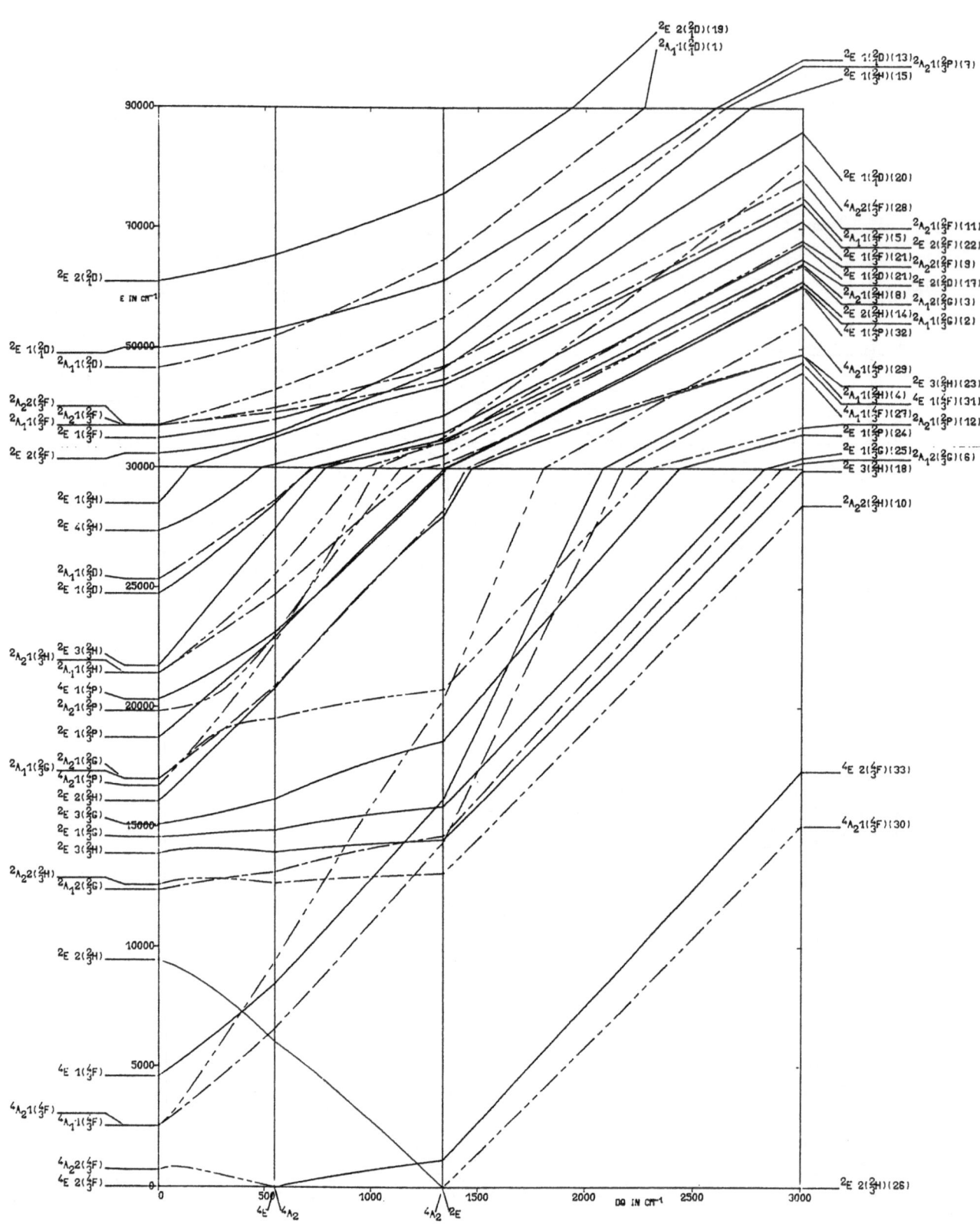

380

7 D ELECTRONS, TRIGONAL SYMMETRY

B=825 C=4×B DT=500 K=-5 DS=K×DT ZETA=0

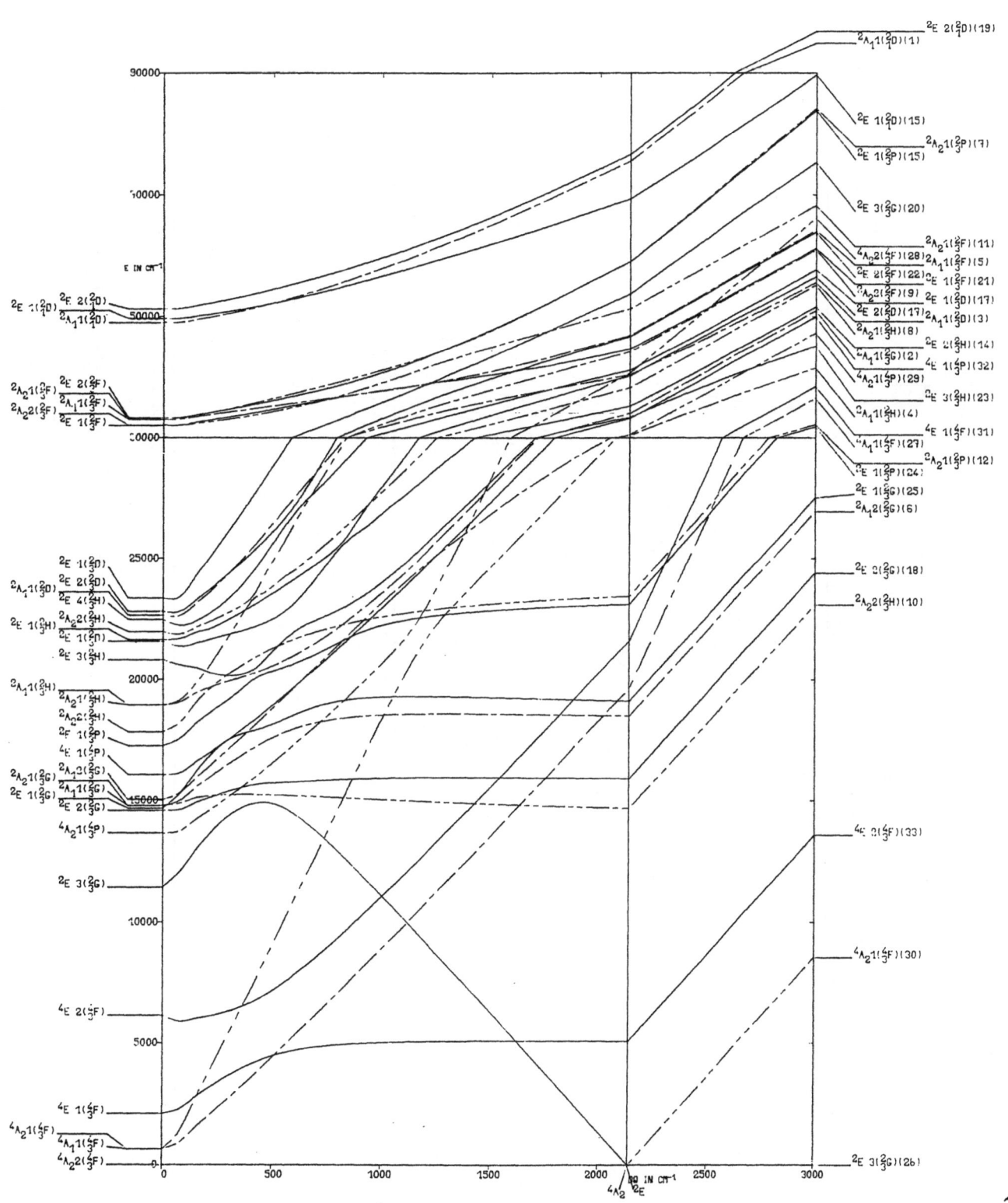

3 D ELECTRONS, TRIGONAL SYMMETRY   B=825 C=4*B DT=-500 K=1 DS=K*DT ZETA=0

381

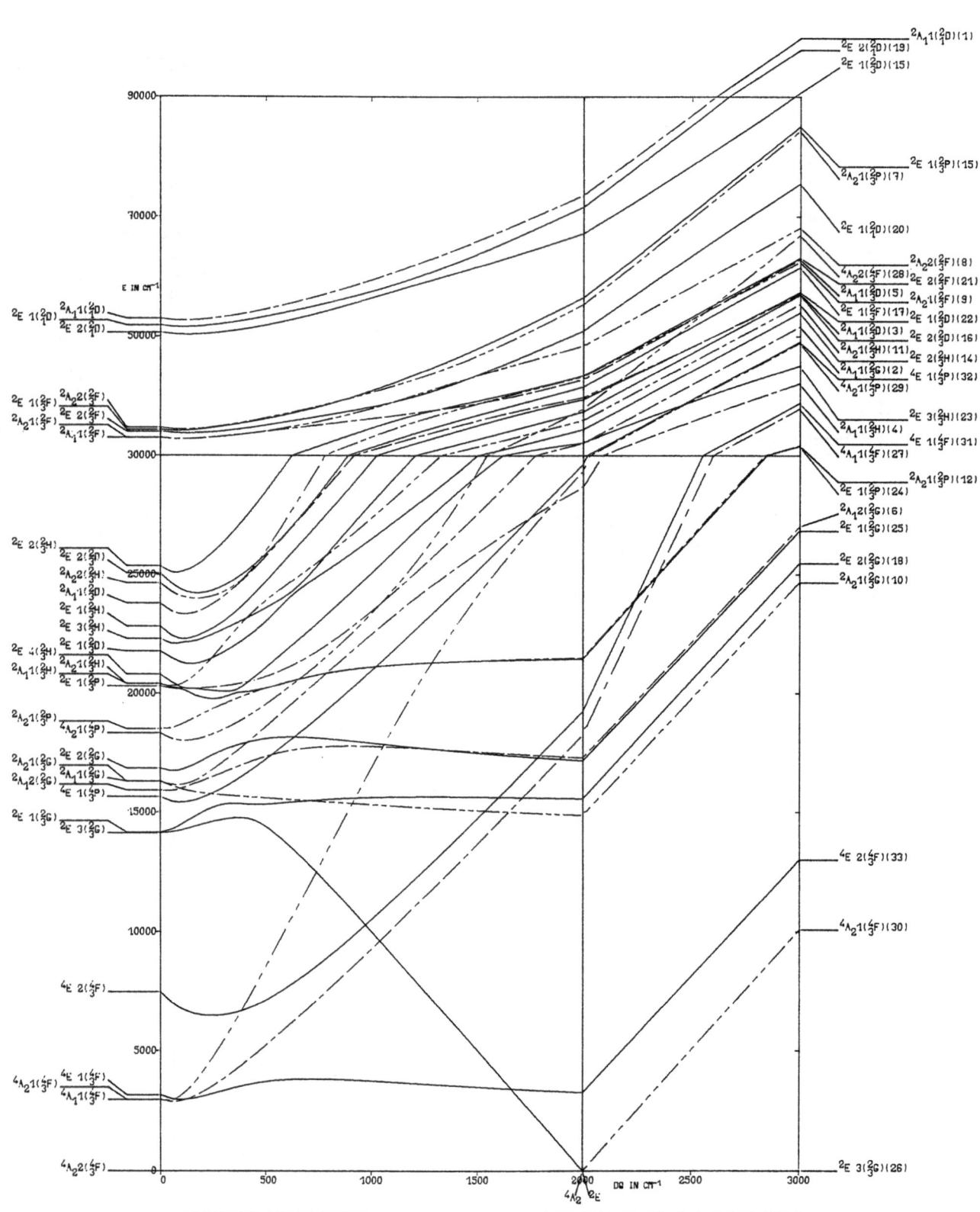

7 D ELECTRONS, TRIGONAL SYMMETRY

B=825 C=4*B DT=-500 K=-1 DS=K*DT ZETA=0

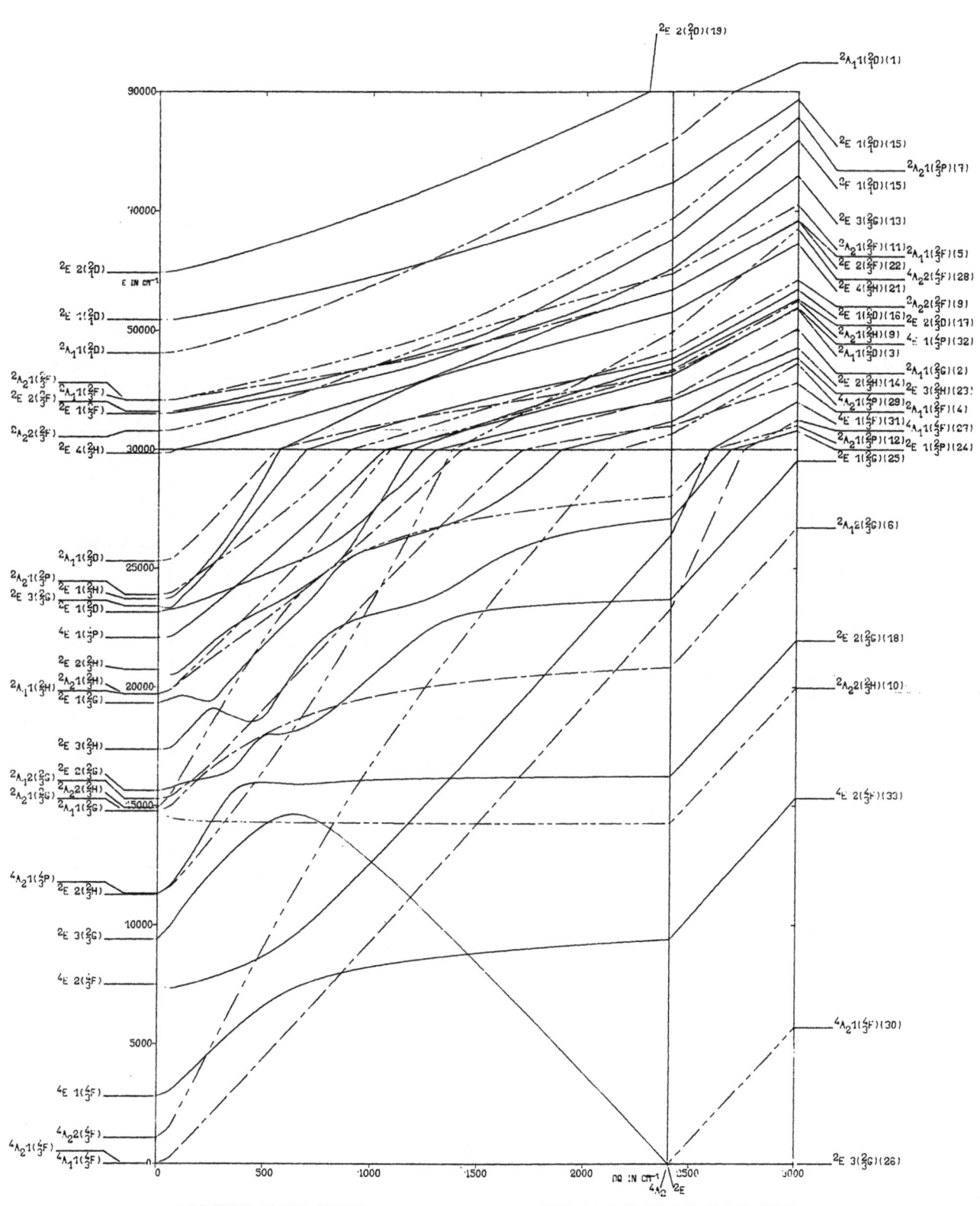

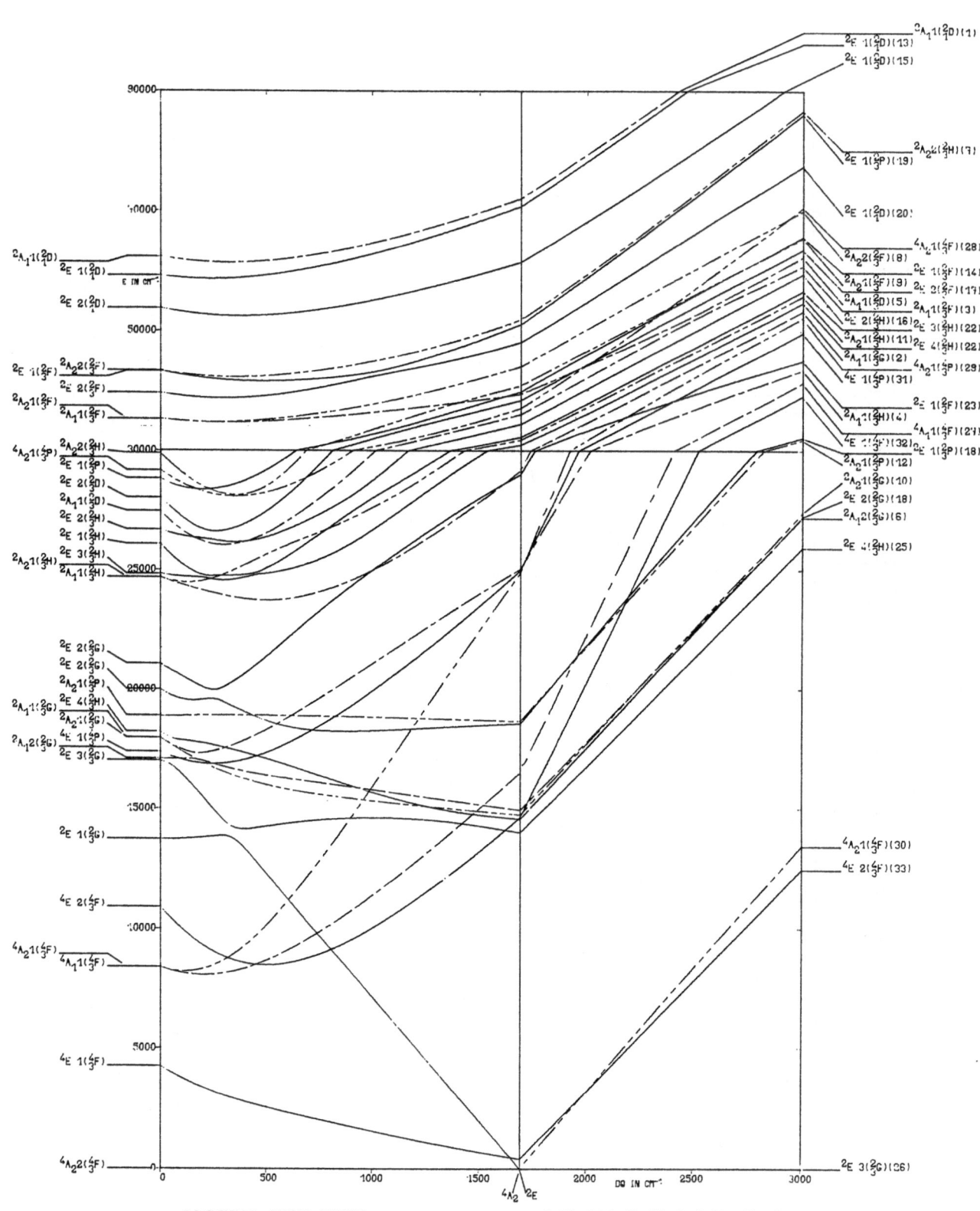

7 D ELECTRONS, TRIGONAL SYMMETRY

P=825 C=4×B DT=-500 K=-5 DS=K×DT ZETA=0

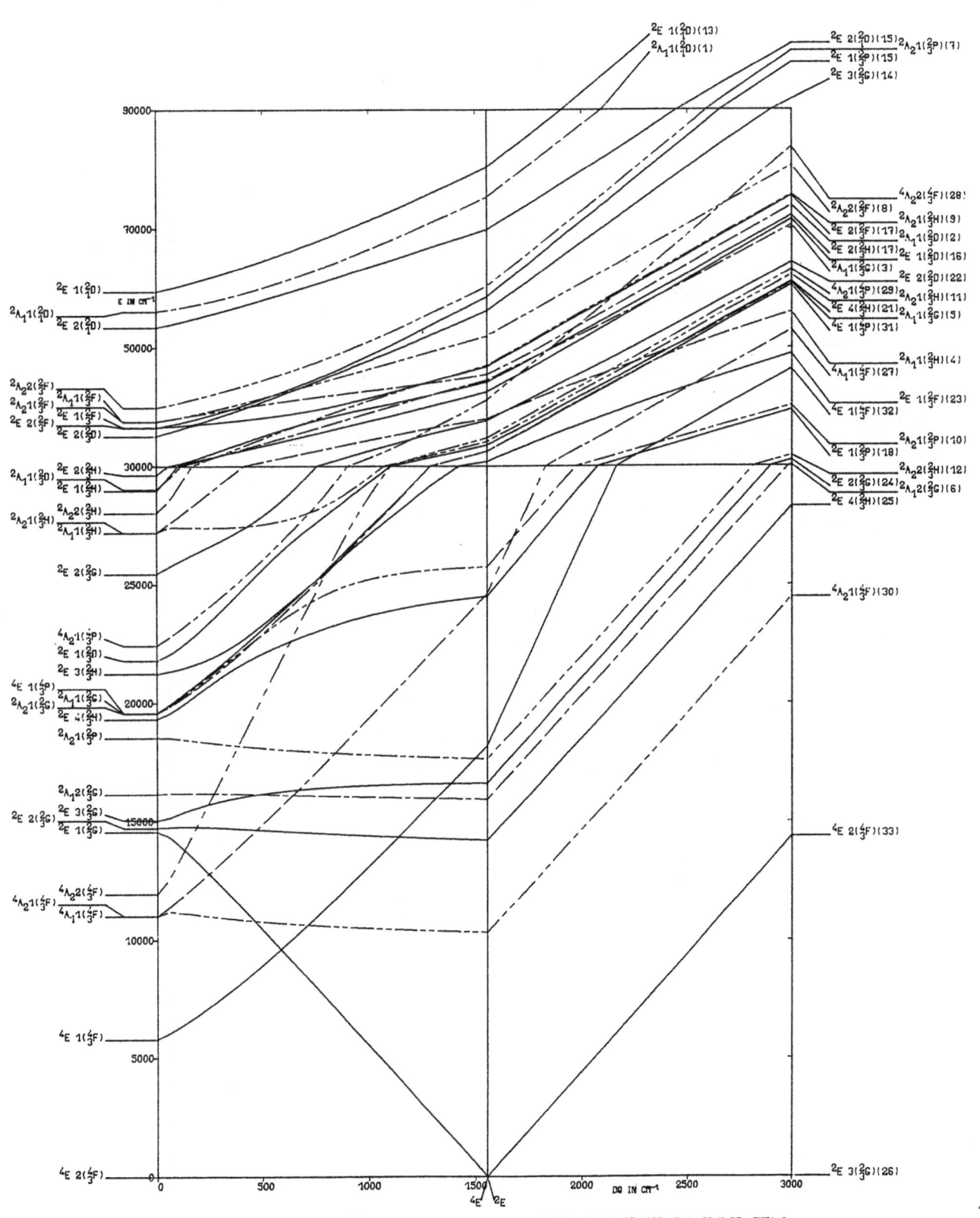

2E 1(²₂D)(13)
2A₁1(²₂D)(1)

2E 2(²₂D)(15) 2A₂1(²₃P)(7)
2E 1(²₃P)(15)
2E 3(²₂G)(14)

4A₂2(⁴₃F)(28)
2A₂1(²₃H)(9)
2E 2(²₂F)(17) 2A₁1(²₂D)(2)
2E 2(²₂H)(17) 2E 1(²₂D)(16)
2A₁1(²₂G)(3) 2E 2(²₂D)(22)
2A₂1(²₃P)(29) 2A₂1(²₃H)(11)
2E 4(²₃H)(21) 2A₁1(²₂G)(5)
4E 1(⁴₃P)(31)
2A₁1(²₃H)(4)
2E 1(²₂F)(23)
4E 1(⁴₃F)(32)
2A₂1(²₃P)(10)
2A₂2(²₃H)(12)
2E 2(²₂G)(24) 2A₁2(²₂G)(6)
4E 4(²₃H)(25)

4A₂1(⁴₃F)(30)

4E 2(⁴₃F)(33)

2E 3(²₂G)(26)

2E 1(²₂D)
2A₁1(²₂D)
2E 2(²₂D)

2A₂2(²₂F) 2A₁1(²₂F)
2A₂1(²₂F) 2E 1(²₂F)
2E 2(²₂F) 2E 2(²₂D)

2A₁1(²₂D) 2E 2(²₂H)
2E 1(²₂H)

2A₂1(²₃H) 2A₂2(²₃H)
2A₁1(²₃H)

2E 2(²₂G)

4A₂1(⁴₃P)
2E 1(²₂H)
2E 3(²₂H)
4E 1(⁴₃P) 2A₁1(²₂G)
2A₂1(²₃G) 2E 4(²₂H)
2A₂1(²₃P)

2A₁2(²₂G)
2E 2(²₂G) 2E 3(²₂G)
2E 1(²₂G)

4A₂1(⁴₃F) 4A₂2(⁴₃F)
4A₁1(⁴₃F)

4E 1(⁴₃F)

4E 2(⁴₃F)

E IN CM⁻¹

90000
70000
50000
30000
25000
20000
15000
10000
5000
0

0   500   1000   1500   2000   2500   3000
DQ IN CM⁻¹

4E 2E

⁷ D ELECTRONS, TRIGONAL SYMMETRY

B=825   C=4*B   DT=1000   K=1   DS=K*DT   ZETA=0

385

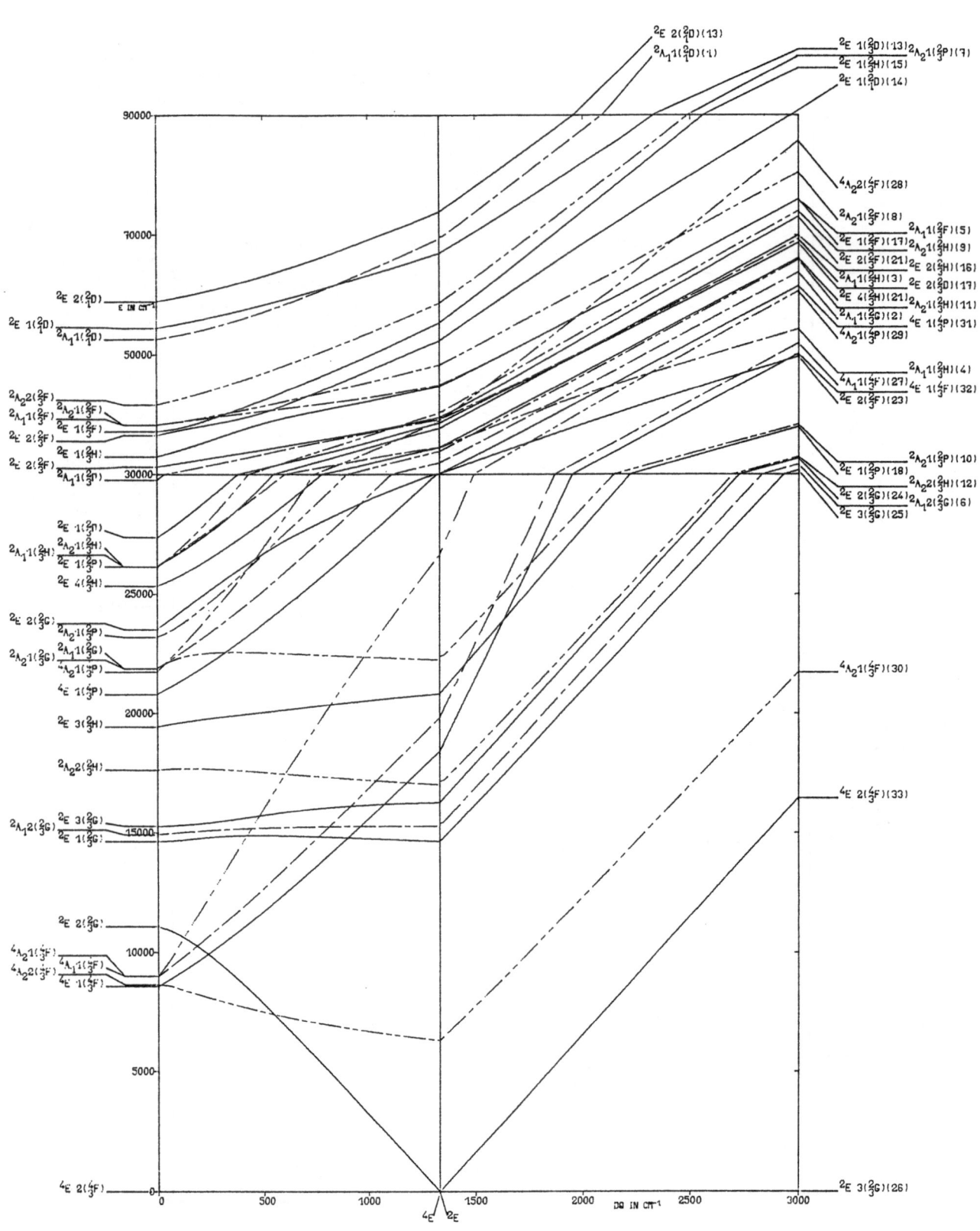

9 D ELECTRONS, TRIGONAL SYMMETRY

B=825   C=4×B   DT=1000   K=-1   DS=K×DT   ZETA=0

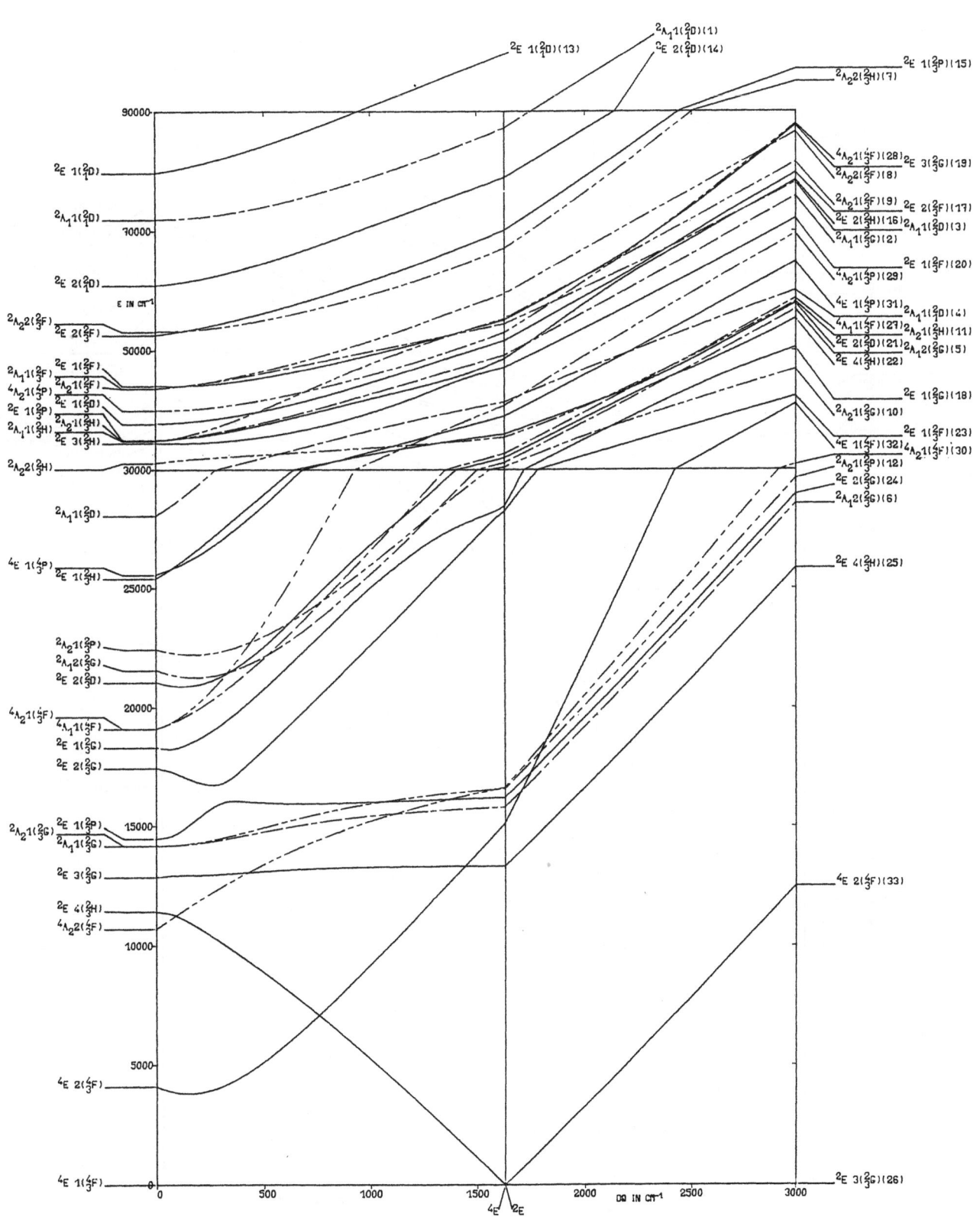

9 D ELECTRONS, TRIGONAL SYMMETRY   B=825   C=4×B   DT=1000   K=5   DS=K×DT   ZETA=0

387

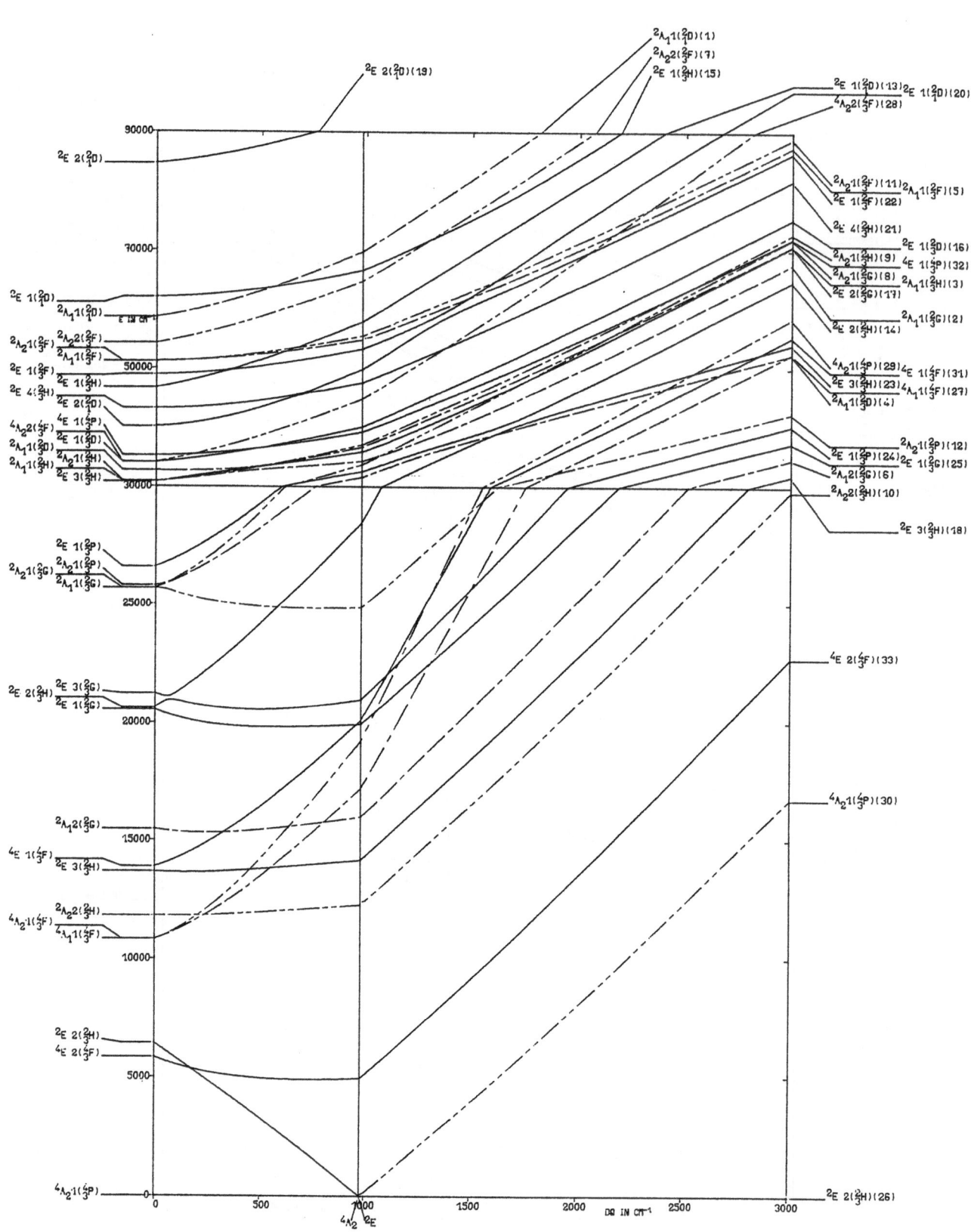

3 D ELECTRONS, TRIGONAL SYMMETRY

B=825 C=4×B DT=1000 K=-5 DS=K×DT ZETA=0

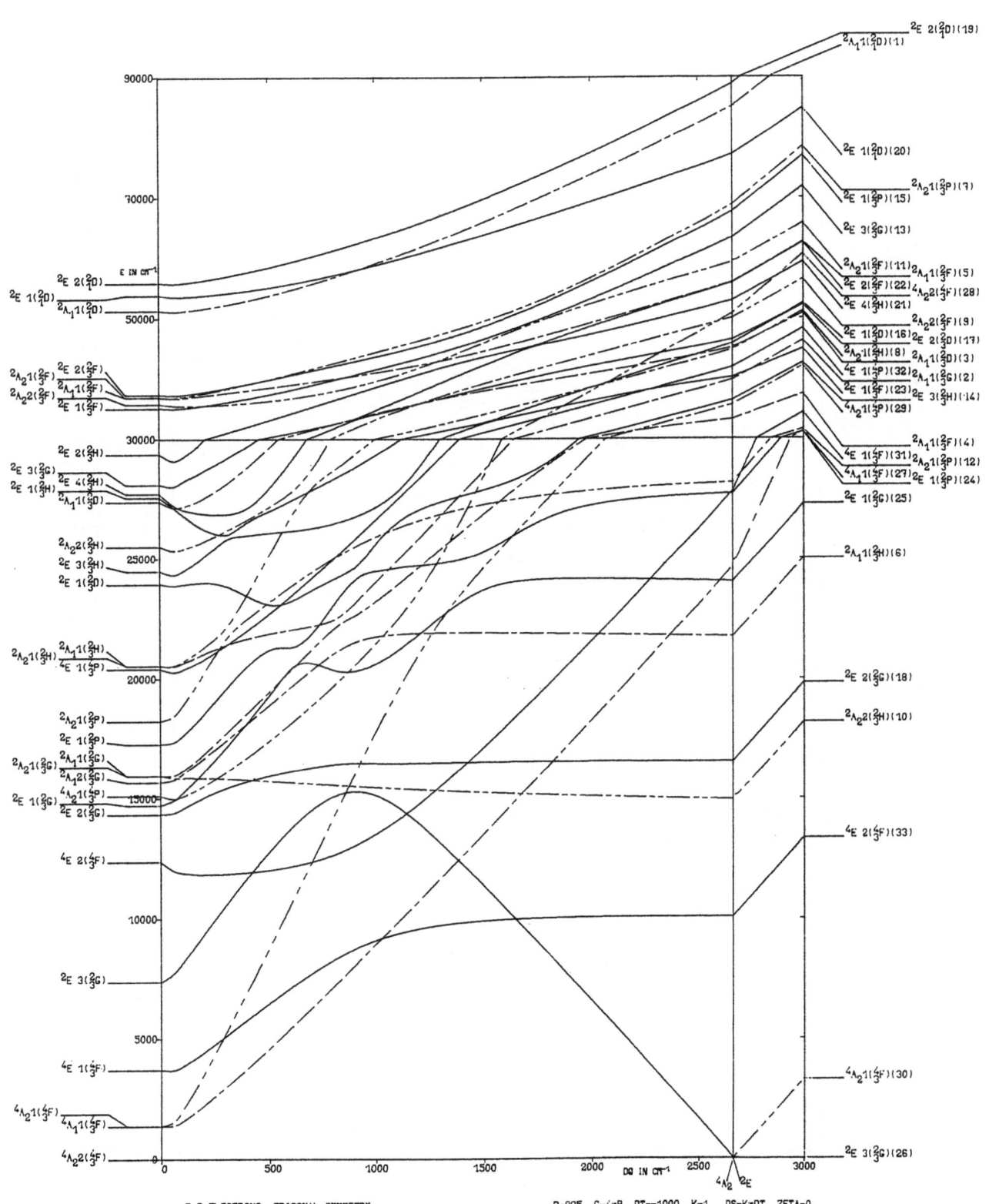

7 D ELECTRONS, TRIGONAL SYMMETRY   B=825 C=4*B DT=-1000 K=1 DS=K*DT ZETA=0

389

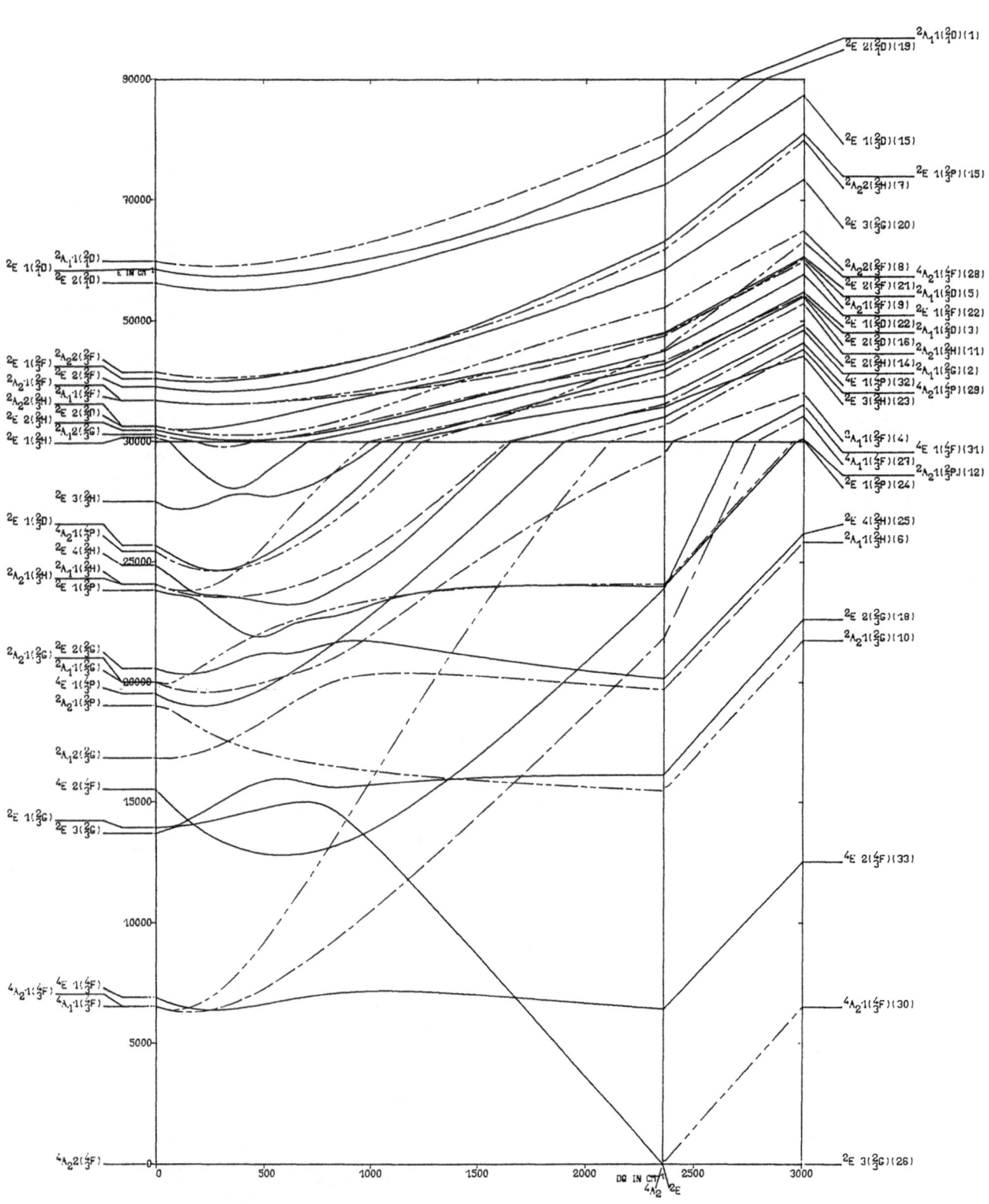

3 D ELECTRONS, TRIGONAL SYMMETRY   B=825   C=4∗B   DT=-1000   K=-1   DS=K∗DT   ZETA=0

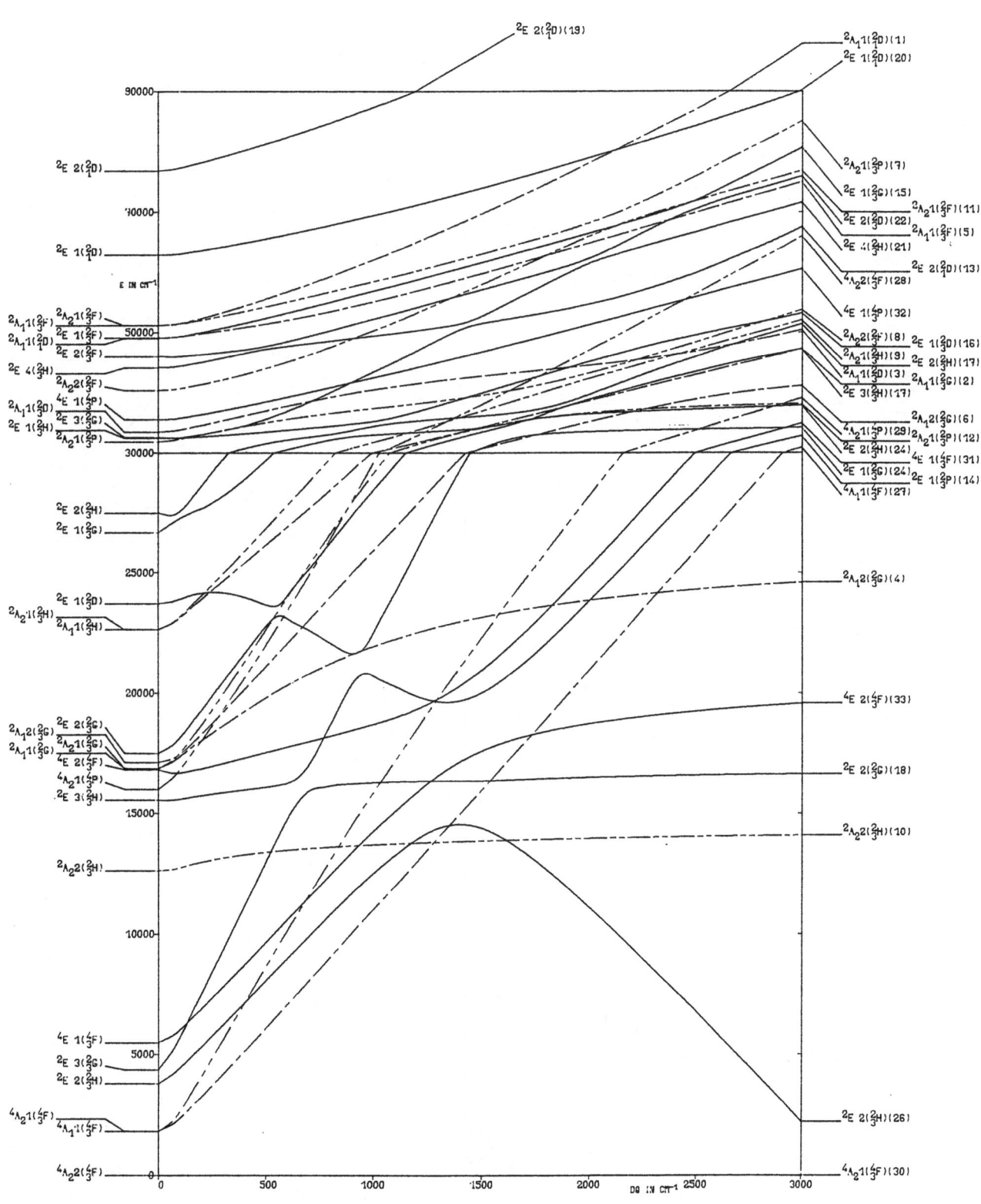

7 D ELECTRONS, TRIGONAL SYMMETRY   B=825 C=4×B DT=-1000 K=5   DS=K×DT ZETA=0

391

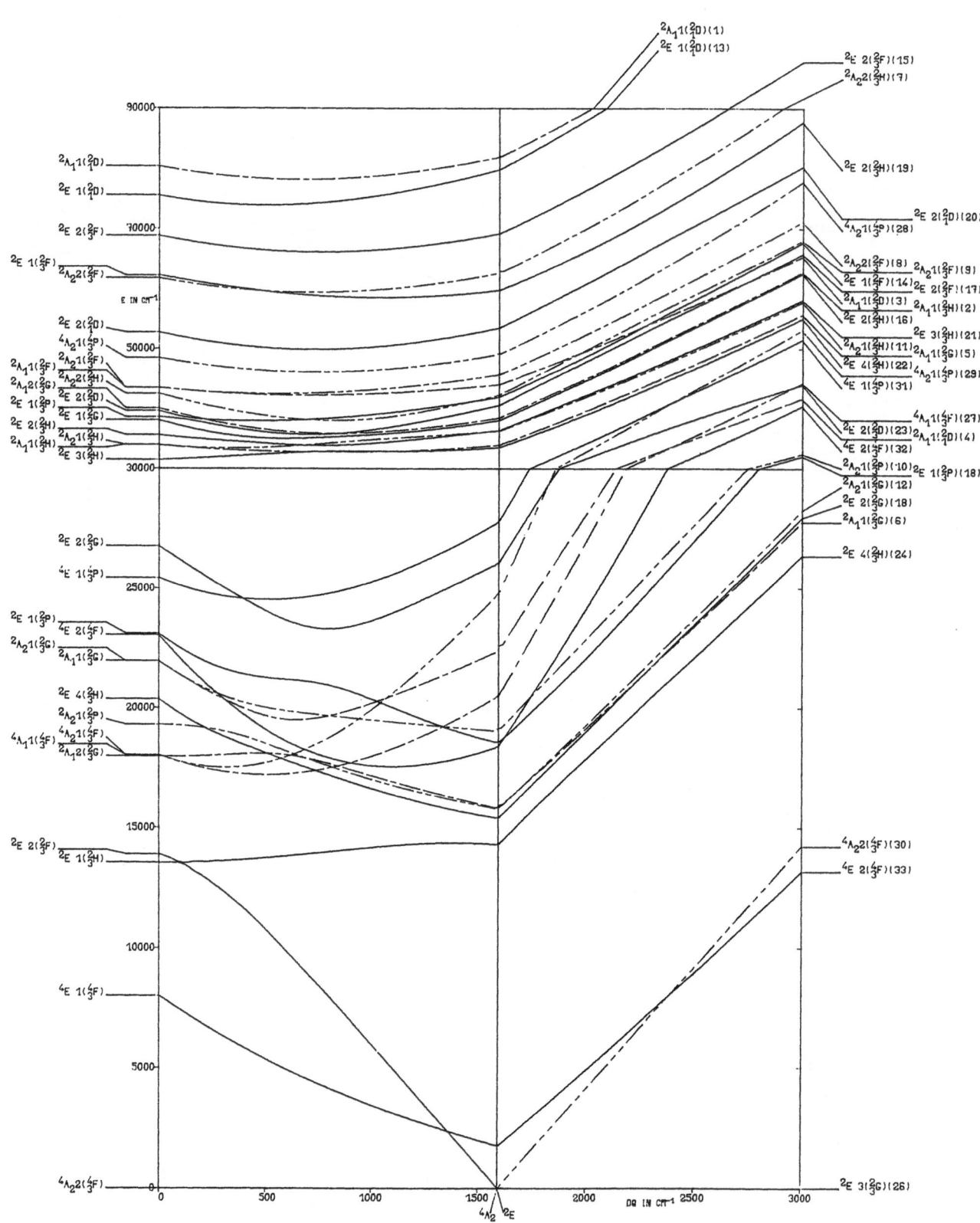

7 D ELECTRONS, TRIGONAL SYMMETRY   B=825 C=4xB DT=-1000 K=-5 DS=KxDT ZETA=0

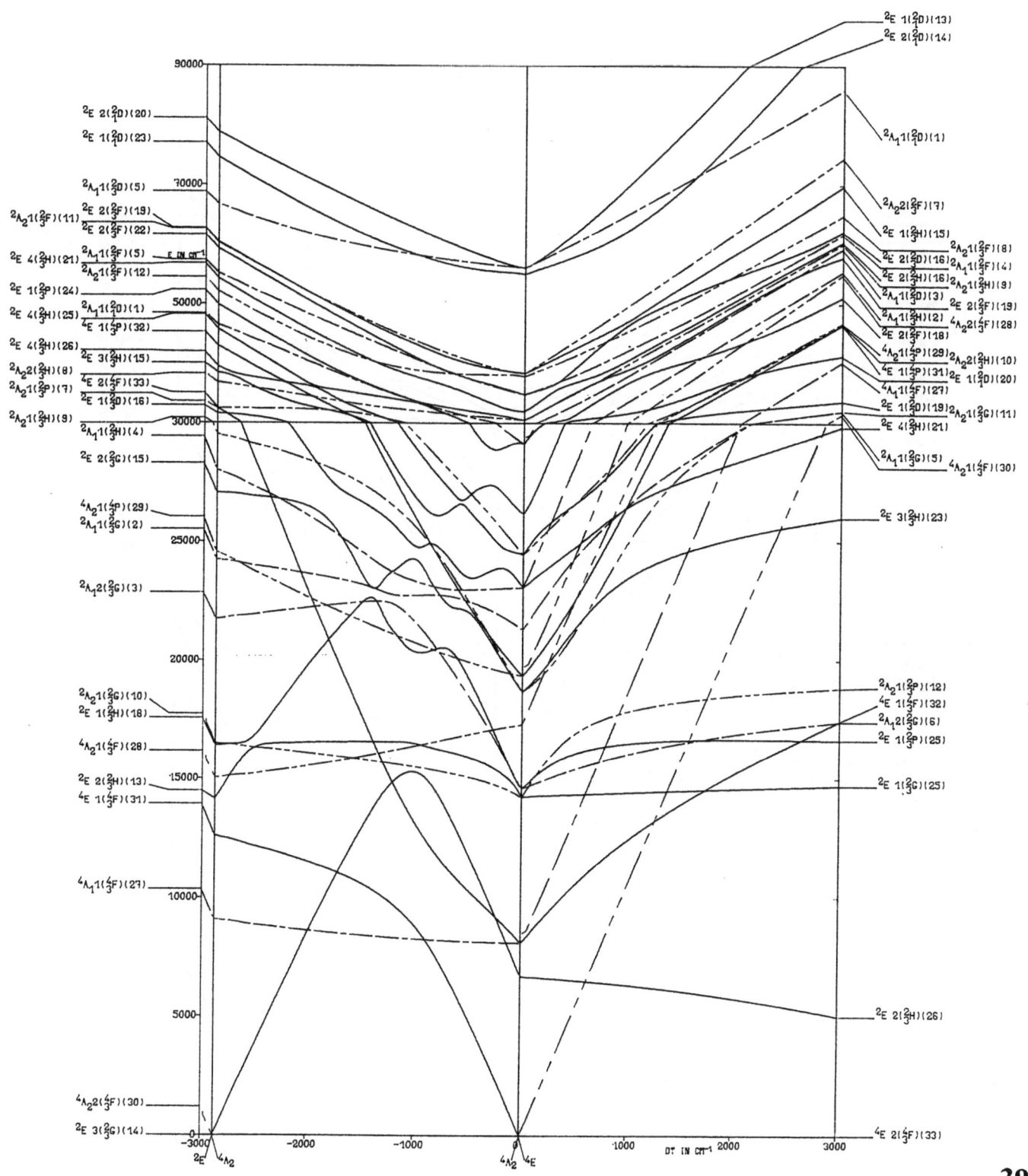

7 D ELECTRONS, TRIGONAL SYMMETRY   B=825 C=4×B DQ=920 K=1 DS=K×DT 2ETA=0

393

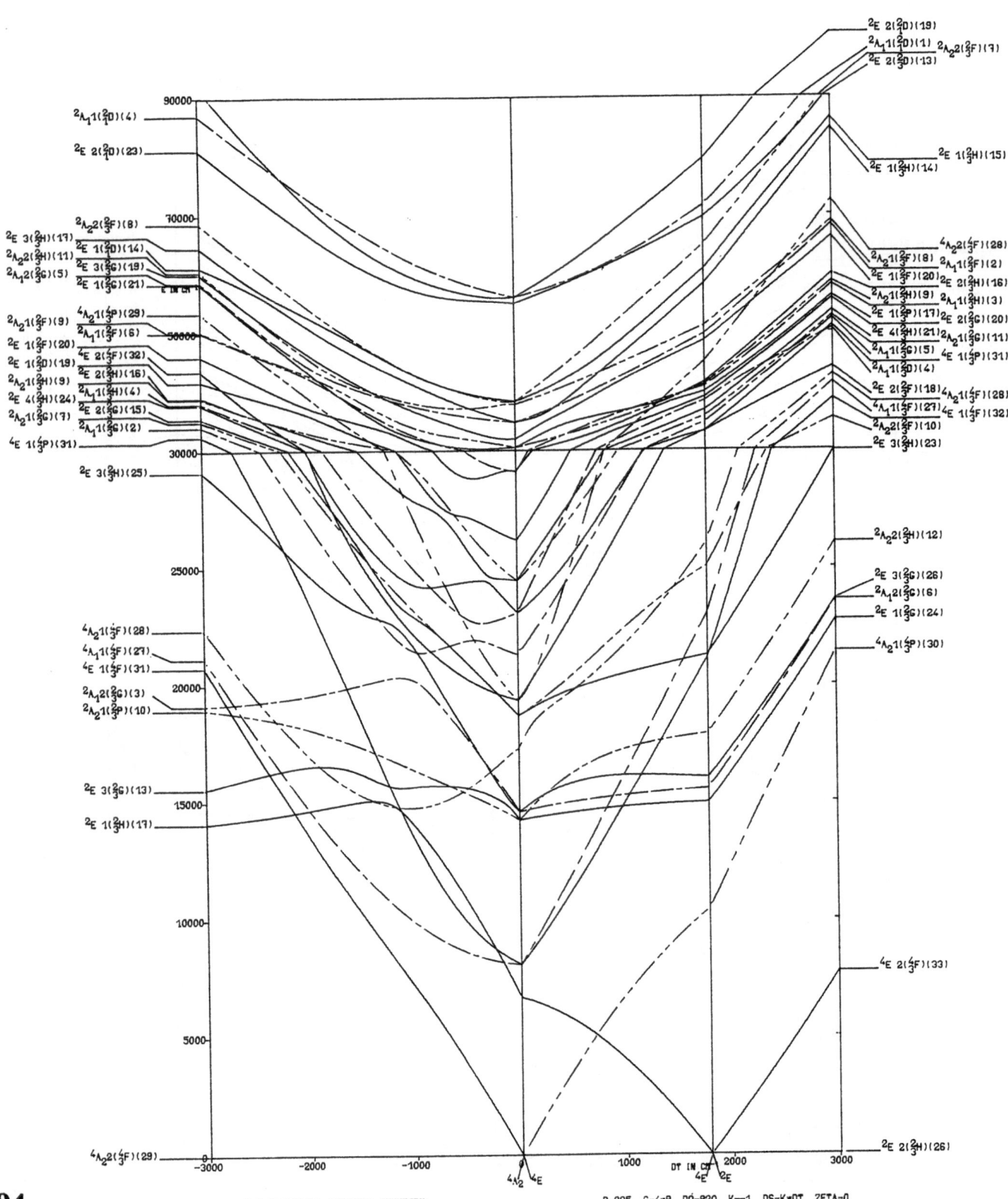

7 D ELECTRONS, TRIGONAL SYMMETRY

B=825 C=4×B DQ=920 K=-1 DS=K×DT ZETA=0

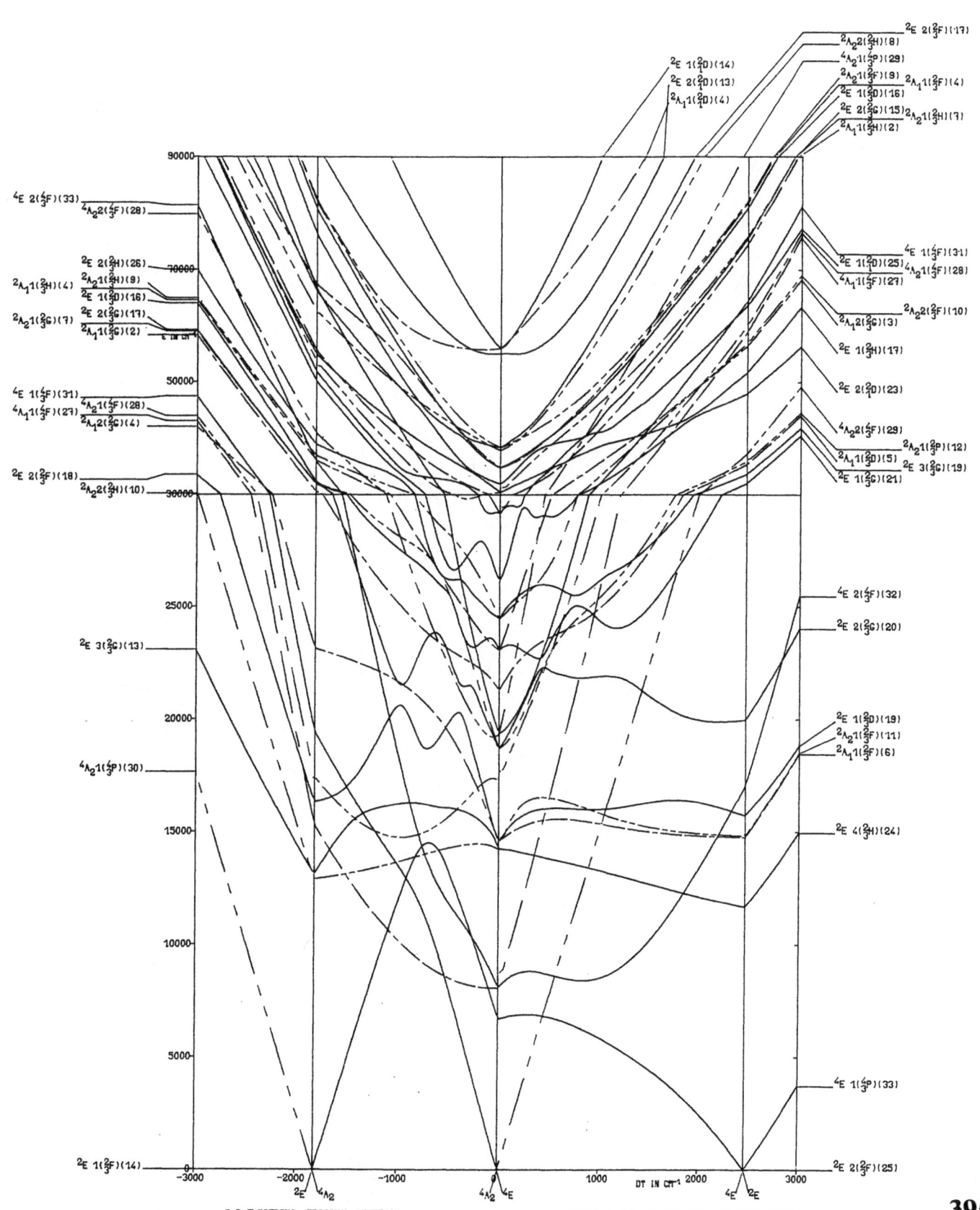

$^4E\ 2(\frac{4}{3}F)(33)$ ___
$^4A_2 2(\frac{4}{3}F)(28)$ ___

$^2E\ 2(\frac{2}{3}H)(26)$ ___
$^2A_1 1(\frac{2}{3}H)(4)$ ___ $^2A_2 1(\frac{2}{3}H)(9)$ ___
$^2E\ 1(\frac{2}{3}D)(16)$ ___
$^2A_2 1(\frac{2}{3}G)(7)$ ___ $^2E\ 2(\frac{2}{3}G)(17)$ ___
$^2A_1 1(\frac{2}{3}G)(2)$ ___

$^4E\ 1(\frac{4}{3}F)(31)$ ___ $^4A_2 1(\frac{4}{3}F)(28)$ ___
$^4A_1 1(\frac{4}{3}F)(27)$ ___ $^2A_1 2(\frac{2}{3}G)(4)$ ___

$^2E\ 2(\frac{2}{3}F)(18)$ ___
$^2A_2 2(\frac{2}{3}H)(10)$ ___

$^2E\ 3(\frac{2}{3}G)(13)$ ___

$^4A_2 1(\frac{4}{3}P)(30)$ ___

$^2E\ 1(\frac{2}{3}F)(14)$ ___

$^2E\ 1(\frac{2}{3}D)(14)$
$^2E\ 2(\frac{2}{3}D)(13)$
$^2A_1 1(\frac{2}{3}D)(4)$

$^2A_2 2(\frac{2}{3}H)(8)$
$^4A_2 1(\frac{4}{3}P)(29)$
$^2A_2 1(\frac{2}{3}F)(9)$ $^2A_1 1(\frac{2}{3}F)(4)$
$^2E\ 1(\frac{2}{3}D)(16)$
$^2E\ 2(\frac{2}{3}G)(15)$ $^2A_2 1(\frac{2}{3}H)(7)$
$^2A_1 1(\frac{2}{3}H)(2)$

$^2E\ 2(\frac{2}{3}F)(17)$

$^4E\ 1(\frac{4}{3}F)(31)$
$^2E\ 1(\frac{2}{3}D)(25)$ $^4A_2 1(\frac{4}{3}F)(28)$
$^4A_1 1(\frac{4}{3}F)(27)$
$^2A_1 2(\frac{2}{3}G)(3)$
$^2A_2 2(\frac{2}{3}F)(10)$

$^2E\ 1(\frac{2}{3}H)(17)$

$^2E\ 2(\frac{2}{3}D)(23)$

$^4A_2 2(\frac{4}{3}F)(29)$ $^2A_2 1(\frac{2}{3}P)(12)$
$^2A_1 1(\frac{2}{3}D)(5)$ $^2E\ 3(\frac{2}{3}G)(19)$
$^2E\ 1(\frac{2}{3}G)(21)$

$^4E\ 2(\frac{4}{3}F)(32)$

$^2E\ 2(\frac{2}{3}G)(20)$

$^2E\ 1(\frac{2}{3}D)(19)$
$^2A_2 1(\frac{2}{3}F)(11)$
$^2A_1 1(\frac{2}{3}F)(6)$

$^2E\ 4(\frac{2}{3}H)(24)$

$^4E\ 1(\frac{4}{3}P)(33)$

$^2E\ 2(\frac{2}{3}F)(25)$

90000
70000
50000
30000
25000
20000
15000
10000
5000
0

E IN CM

-3000   -2000   -1000   0   1000   DT IN CM⁻¹ 2000   3000

^2E $^4A_2$ $^4A_2$ ^4E ^4E ^2E

7 D ELECTRONS, TRIGONAL SYMMETRY

B=825   C=4πB   DQ=920   K=5   DS=KπDT   2ETA=0

395

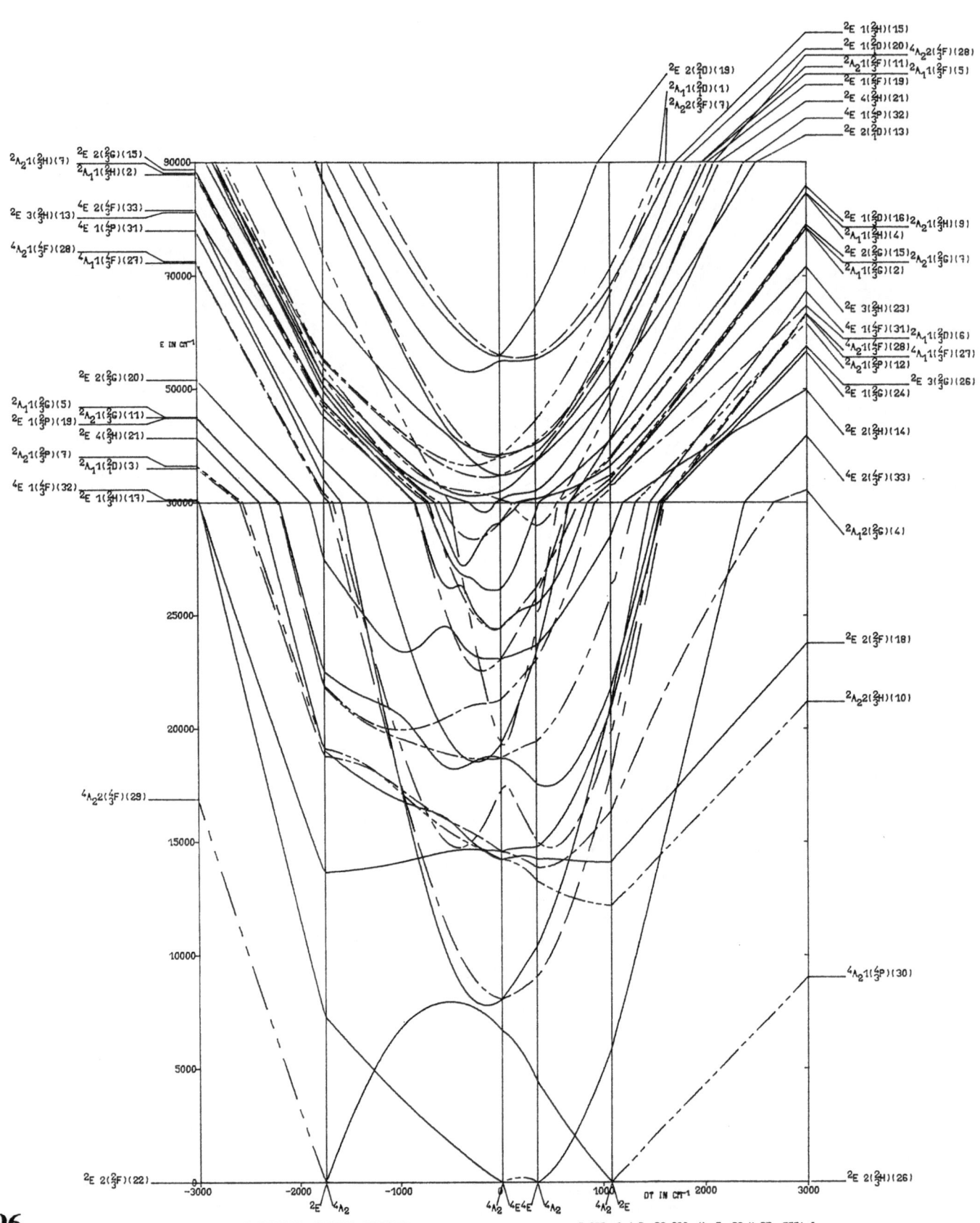

$^2A_21(^2_3H)(7)$
$^2A_11(^2_3H)(2)$

$^2E\ 2(^2_3G)(15)$

$^2E\ 3(^2_3H)(13)$
$^4E\ 1(^4_3P)(31)$

$^4E\ 2(^4_3F)(33)$

$^4A_21(^4_3F)(28)$
$^4A_11(^4_3F)(27)$

$^2E\ 2(^2_3G)(20)$

$^2A_11(^2_3G)(5)$
$^2E\ 1(^2_3P)(19)$
$^2A_21(^2_3G)(11)$
$^2E\ 4(^2_3H)(21)$

$^2A_21(^2_3P)(7)$
$^2A_11(^2_1D)(3)$
$^4E\ 1(^4_3F)(32)$
$^2E\ 1(^2_3H)(17)$

$^4A_22(^4_3F)(29)$

$^2E\ 2(^2_3F)(22)$

$^2E\ 2(^2_1D)(19)$
$^2A_11(^2_1D)(1)$
$^2A_22(^2_3F)(7)$

E IN CM^{-1}

90000
80000
70000

50000

30000

25000

20000

15000

10000

5000

0

^2E $^4A_2$ $^4A_2$ $^4E^4E$ $^4A_2$ $^4A_21$ ^2E
-3000   -2000   -1000   0   1000   DT IN CM^{-1} 2000   3000

$^2E\ 1(^2_3H)(15)$
$^2E\ 1(^2_1D)(20)$$^4A_22(^4_3F)(28)$
$^2A_21(^4_3F)(11)$$^2A_11(^2_3F)(5)$
$^2E\ 1(^2_3F)(19)$
$^2E\ 4(^2_3H)(21)$
$^4E\ 1(^4_3P)(32)$
$^2E\ 2(^2_1D)(13)$

$^2E\ 1(^2_1D)(16)$$^2A_21(^2_3H)(9)$
$^2A_11(^2_3H)(4)$
$^2E\ 2(^2_3G)(15)$$^2A_21(^2_3G)(7)$
$^2A_11(^2_3G)(2)$

$^2E\ 3(^2_3H)(23)$
$^4E\ 1(^4_3F)(31)$$^2A_11(^2_1D)(6)$
$^4A_21(^4_3F)(28)$$^4A_11(^4_3F)(27)$
$^2A_21(^2_3P)(12)$
$^2E\ 3(^2_3H)(26)$
$^2E\ 1(^2_3G)(24)$

$^2E\ 2(^2_3H)(14)$

$^4E\ 2(^4_3F)(33)$

$^2A_12(^2_3G)(4)$

$^2E\ 2(^2_3F)(18)$

$^2A_22(^2_3H)(10)$

$^4A_21(^4_3P)(30)$

$^2E\ 2(^2_3H)(26)$

396

7 D ELECTRONS, TRIGONAL SYMMETRY

B=825   C=4*B   DQ=920   K=-5   DS=K*DT   ZETA=0

$B=825$ $C=4*B$ $ZETA=0$

ENERGY AS FUNCTION OF DT
$K= 1, -1, 5, -5$

(29) $^2E_5(^1E_3(E_1E_2)E_2)$
(28) $^2E_4(^1E_2(E_1E_1)E_2)$
(27) $^2E_4(^1E_2(A_1E_2)E_2)$
(26) $^4E_3(^3E_1(A_1E_1)E_2)$
(25) $^2E_3(^1E_1(E_1E_2)E_2)$
(24) $^2E_3(^1E_2(A_1E_2)E_1)$
(23) $^2E_3(^1E_1(A_1E_1)E_2)$
(22) $^4E_2(^3A_2(E_1E_1)E_2)$
(21) $^2E_2(^1A_1(E_2E_2)E_2)$
(20) $^2E_2(^1E_1(E_1E_2)E_1)$
(19) $^2E_2(^1A_1(E_1E_1)E_2)$
(18) $^2E_2(^1E_1(A_1E_1)E_1)$
(17) $^2E_2(^1A_1(A_1A_1)E_2)$
(16) $^4E_1(^3E_1(E_1E_2)E_2)$
(15) $^4E_1(^3E_1(A_1E_1)E_2)$
(14) $^2E_1(^1E_3(E_1E_2)E_2)$
(13) $^2E_1(^1E_1(E_1E_2)E_2)$
(12) $^2E_1(^1A_1(E_1E_1)E_1)$
(11) $^2E_1(^1E_2(A_1E_2)E_1)$
(10) $^2E_1(^1E_1(A_1E_1)E_2)$
(9) $^2E_1(^1A_1(A_1A_1)E_1)$
(8) $^4A_2(^3E_2(A_1E_2)E_2)$
(7) $^4A_2(^3E_1(A_1E_1)E_1)$
(6) $^2A_2(^1E_2(E_1E_1)E_2)$
(5) $^2A_2(^1E_2(A_1E_2)E_2)$
(4) $^2A_2(^1E_1(A_1E_1)E_1)$
(3) $^2A_1(^1E_2(E_1E_1)E_2)$
(2) $^2A_1(^1E_2(A_1E_2)E_2)$
(1) $^2A_1(^1E_1(A_1E_1)E_1)$

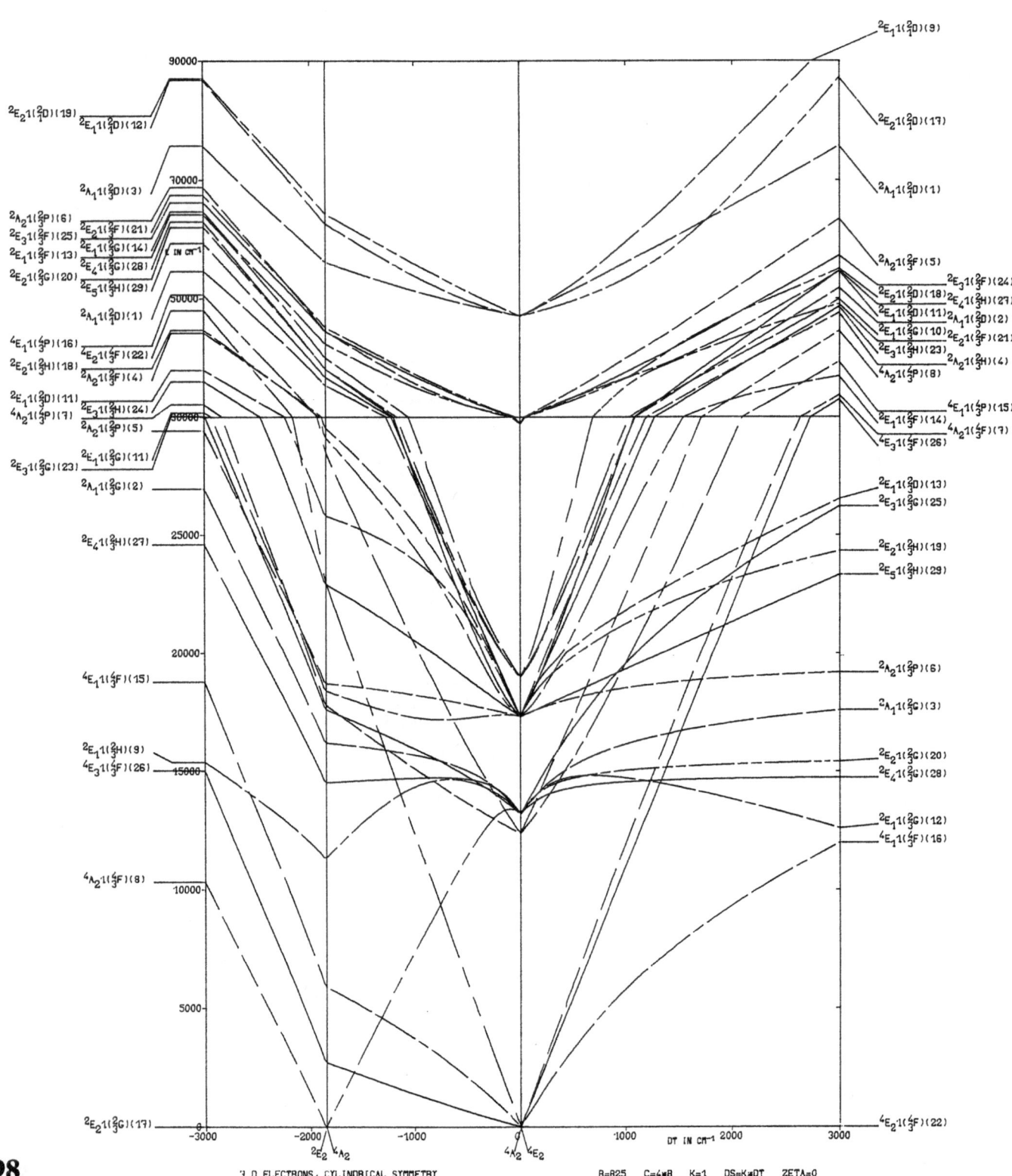

7 D ELECTRONS, CYLINDRICAL SYMMETRY   B=825   C=4×B   K=1   DS=K×DT   ZETA=0

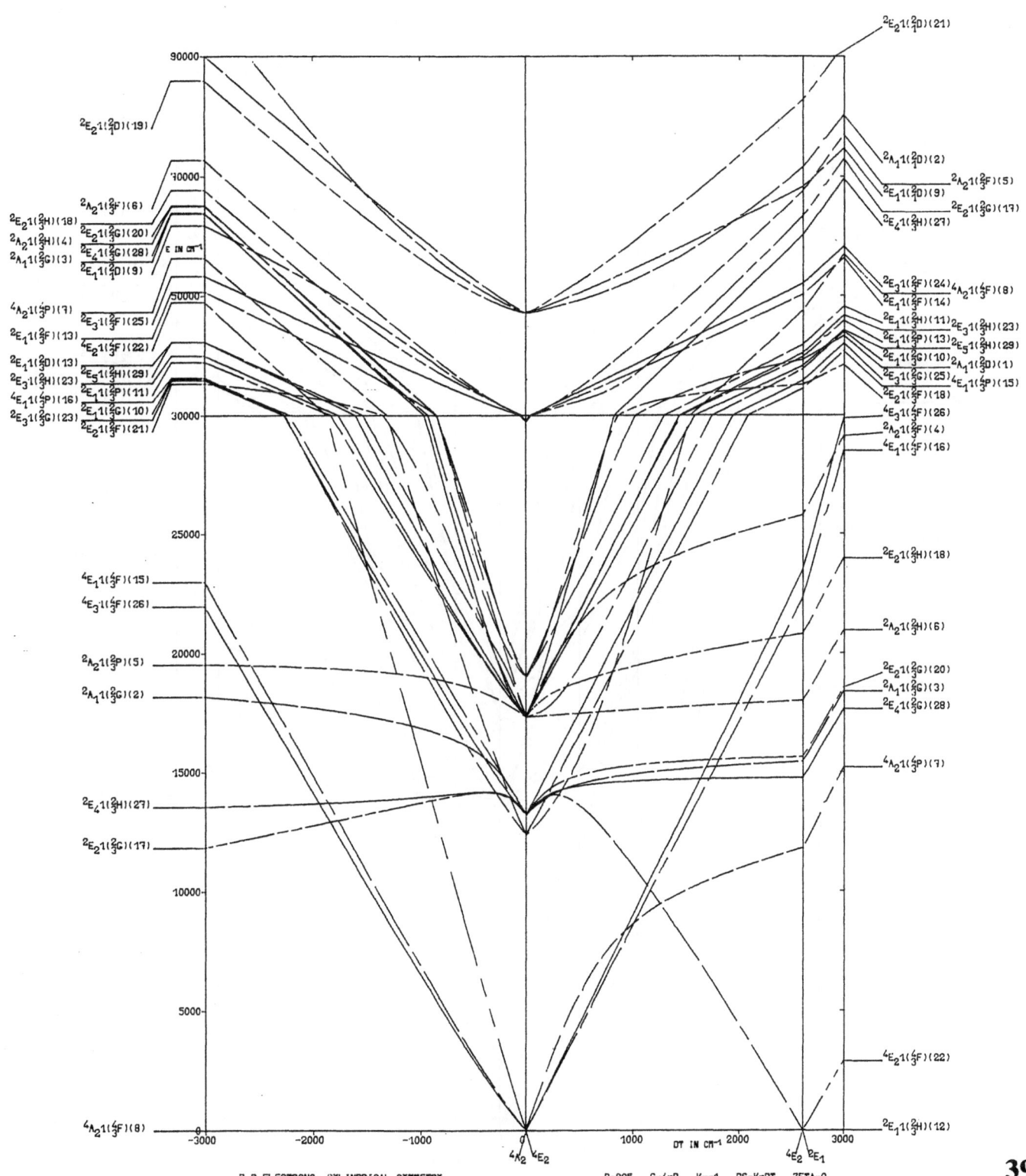

7 D ELECTRONS, CYLINDRICAL SYMMETRY   B=825   C=4*B   K=-1   DS=K*DT   ZETA=0

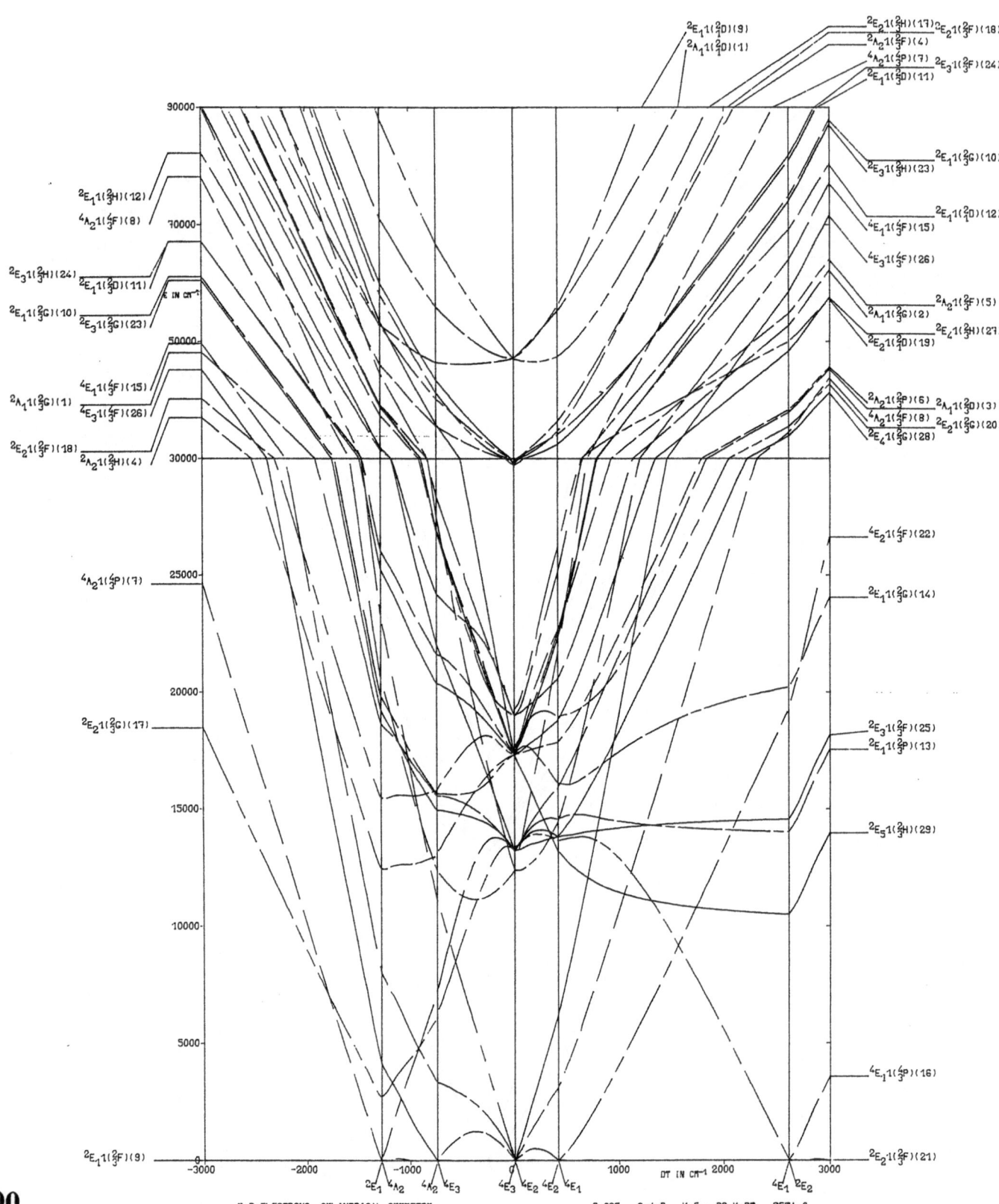

7 D ELECTRONS, CYLINDRICAL SYMMETRY   B=825   C=4×B   K=5   DS=K×DT   ZETA=0

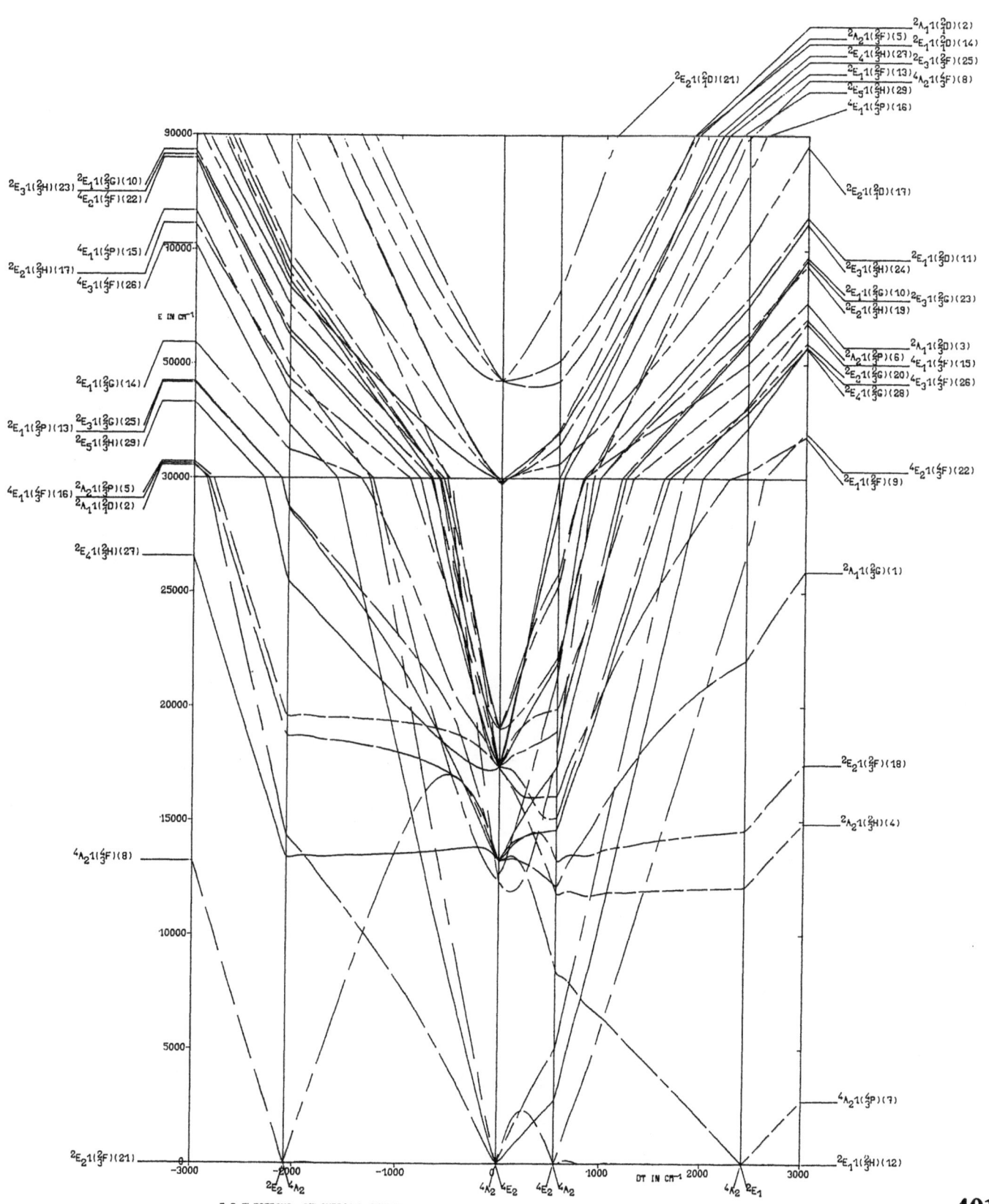

$^2E_3 1(^2_3H)(23)$ $^2E_1 1(^2_3G)(10)$
$^4E_2 1(^2_3F)(22)$

$^2E_2 1(^2_3H)(17)$ $^4E_1 1(^4_3P)(15)$
$^4E_3 1(^2_3F)(26)$

E IN CM^{-1}

$^2E_1 1(^2_3G)(14)$

$^2E_1 1(^2_3P)(13)$ $^2E_3 1(^2_3G)(25)$
$^2E_5 1(^2_3H)(29)$

$^4E_1 1(^4_3F)(16)$ $^2A_2 1(^2_3P)(5)$
$^2A_1 1(^2_3D)(2)$

$^2E_4 1(^2_3H)(27)$

$^4A_2 1(^4_3F)(8)$

$^2E_2 1(^2_3F)(21)$

$^2E_2 1(^2_3D)(21)$

$^2A_2 1(^2_3F)(5)$ $^2A_1 1(^2_3D)(2)$
$^2E_1 1(^2_3D)(14)$
$^2E_4 1(^2_3H)(27)$ $^2E_3 1(^2_3F)(25)$
$^2E_1 1(^2_3F)(13)$ $^4A_2 1(^4_3F)(8)$
$^2E_5 1(^2_3H)(29)$
$^4E_1 1(^4_3P)(16)$

$^2E_2 1(^2_3D)(17)$

$^2E_1 1(^2_3D)(11)$

$^2E_1 1(^2_3H)(24)$

$^2E_4 1(^2_3G)(10)$ $^2E_3 1(^2_3G)(23)$
$^2E_2 1(^2_3H)(19)$

$^2A_1 1(^2_3D)(3)$
$^2A_2 1(^2_3P)(6)$ $^4E_1 1(^4_3F)(15)$
$^2E_1 1(^2_3G)(20)$ $^4E_3 1(^2_3F)(26)$
$^2E_4 1(^2_3G)(28)$

$^4E_2 1(^4_3F)(22)$
$^2E_1 1(^2_3F)(9)$

$^2A_1 1(^2_3G)(1)$

$^2E_2 1(^2_3F)(18)$

$^2A_2 1(^2_3H)(4)$

$^4A_2 1(^4_3P)(7)$

$^2E_1 1(^2_3H)(12)$

$^2E_2$ $^4A_2$ $^4A_2$ $^4E_2$ $^4E_2$ $^4A_2$ $^4A_2$ $^2E_1$

8 D ELECTRONS, TETRAGONAL SYMMETRY

B=905   C=4×B   ZETA=650

ENERGY AS FUNCTION OF DQ
DT=0, 500, -500, 1000, -1000
K=1, -1, 5, -5

ENERGY AS FUNCTION OF DT
DQ=850
K=1, -1, 5, -5

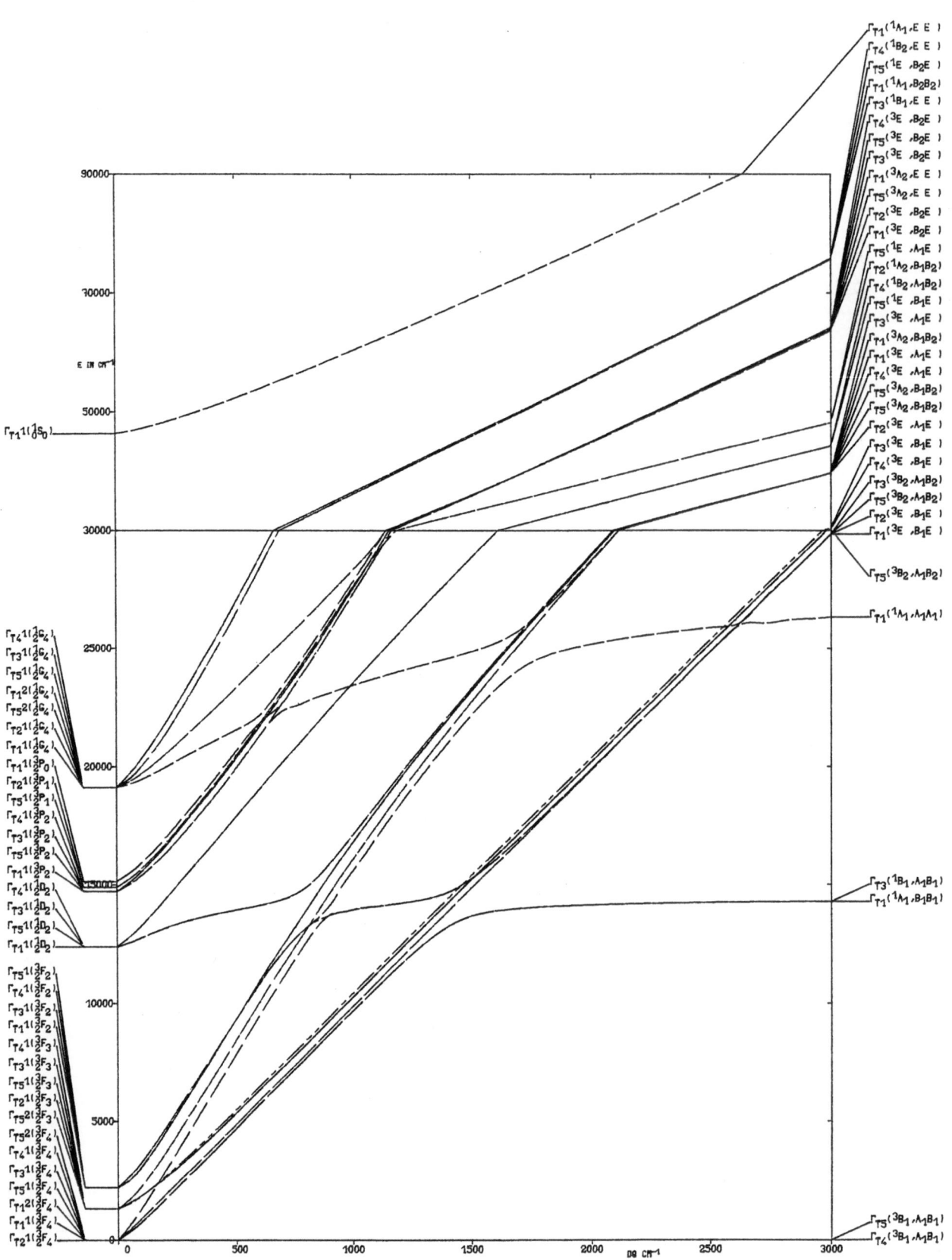

$\Gamma_{T1}1(^1_0S_0)$

$\Gamma_{T4}1(^1_2G_4)$
$\Gamma_{T3}1(^1_2G_4)$
$\Gamma_{T5}1(^1_2G_4)$
$\Gamma_{T1}2(^1_2G_4)$
$\Gamma_{T5}2(^1_2G_4)$
$\Gamma_{T2}1(^1_2G_4)$
$\Gamma_{T1}1(^1_2G_4)$
$\Gamma_{T1}1(^3_2P_0)$
$\Gamma_{T2}1(^3_2P_1)$
$\Gamma_{T5}1(^3_2P_1)$
$\Gamma_{T4}1(^3_2P_2)$
$\Gamma_{T3}1(^3_2P_2)$
$\Gamma_{T5}1(^3_2P_2)$
$\Gamma_{T1}1(^3_2P_2)$
$\Gamma_{T4}1(^1_2D_2)$
$\Gamma_{T3}1(^1_2D_2)$
$\Gamma_{T5}1(^1_2D_2)$
$\Gamma_{T1}1(^1_2D_2)$

$\Gamma_{T5}1(^3_2F_2)$
$\Gamma_{T4}1(^3_2F_2)$
$\Gamma_{T3}1(^3_2F_2)$
$\Gamma_{T1}1(^3_2F_2)$
$\Gamma_{T4}1(^3_2F_3)$
$\Gamma_{T3}1(^3_2F_3)$
$\Gamma_{T5}1(^3_2F_3)$
$\Gamma_{T2}1(^3_2F_3)$
$\Gamma_{T5}2(^3_2F_3)$
$\Gamma_{T5}2(^3_2F_4)$
$\Gamma_{T4}1(^3_2F_4)$
$\Gamma_{T3}1(^3_2F_4)$
$\Gamma_{T5}1(^3_2F_4)$
$\Gamma_{T1}2(^3_2F_4)$
$\Gamma_{T1}1(^3_2F_4)$
$\Gamma_{T2}1(^3_2F_4)$

$\Gamma_{T1}(^1A_1 , E\ E\ )$
$\Gamma_{T4}(^1B_2 , E\ E\ )$
$\Gamma_{T5}(^1E\ , B_2 E\ )$
$\Gamma_{T1}(^1A_1 , B_2 B_2)$
$\Gamma_{T3}(^1B_1 , E\ E\ )$
$\Gamma_{T4}(^3E\ , B_2 E\ )$
$\Gamma_{T5}(^3E\ , B_2 E\ )$
$\Gamma_{T3}(^3E\ , B_2 E\ )$
$\Gamma_{T1}(^3A_2 , E\ E\ )$
$\Gamma_{T5}(^3A_2 , E\ E\ )$
$\Gamma_{T2}(^3E\ , B_2 E\ )$
$\Gamma_{T1}(^3E\ , B_2 E\ )$
$\Gamma_{T5}(^1E\ , A_1 E\ )$
$\Gamma_{T2}(^1A_2 , B_1 B_2)$
$\Gamma_{T4}(^1B_2 , A_1 B_2)$
$\Gamma_{T5}(^1E\ , B_1 E\ )$
$\Gamma_{T3}(^3E\ , A_1 E\ )$
$\Gamma_{T1}(^3A_2 , B_1 B_2)$
$\Gamma_{T4}(^3E\ , A_1 E\ )$
$\Gamma_{T5}(^3A_2 , B_1 B_2)$
$\Gamma_{T5}(^3A_2 , B_1 B_2)$
$\Gamma_{T2}(^3E\ , A_1 E\ )$
$\Gamma_{T3}(^3E\ , B_1 E\ )$
$\Gamma_{T4}(^3E\ , A_1 E\ )$
$\Gamma_{T3}(^3B_2 , A_1 B_2)$
$\Gamma_{T5}(^3B_2 , A_1 B_2)$
$\Gamma_{T2}(^3E\ , B_1 E\ )$
$\Gamma_{T1}(^3E\ , B_1 E\ )$

$\Gamma_{T5}(^3B_2 , A_1 B_2)$

$\Gamma_{T1}(^1A_1 , A_1 A_1)$

$\Gamma_{T3}(^1B_1 , A_1 B_1)$
$\Gamma_{T1}(^1A_1 , B_1 B_1)$

$\Gamma_{T5}(^3B_1 , A_1 B_1)$
$\Gamma_{T4}(^3B_1 , A_1 B_1)$

E IN CM⁻¹

90000

70000

50000

30000

25000

20000

15000

10000

5000

0

0   500   1000   1500   2000   DB CM⁻¹   2500   3000

8 D ELECTRONS, TETRAGONAL SYMMETRY   B=905 C=4*B DT=0 K=0   DS=K*DT   ZETA=650

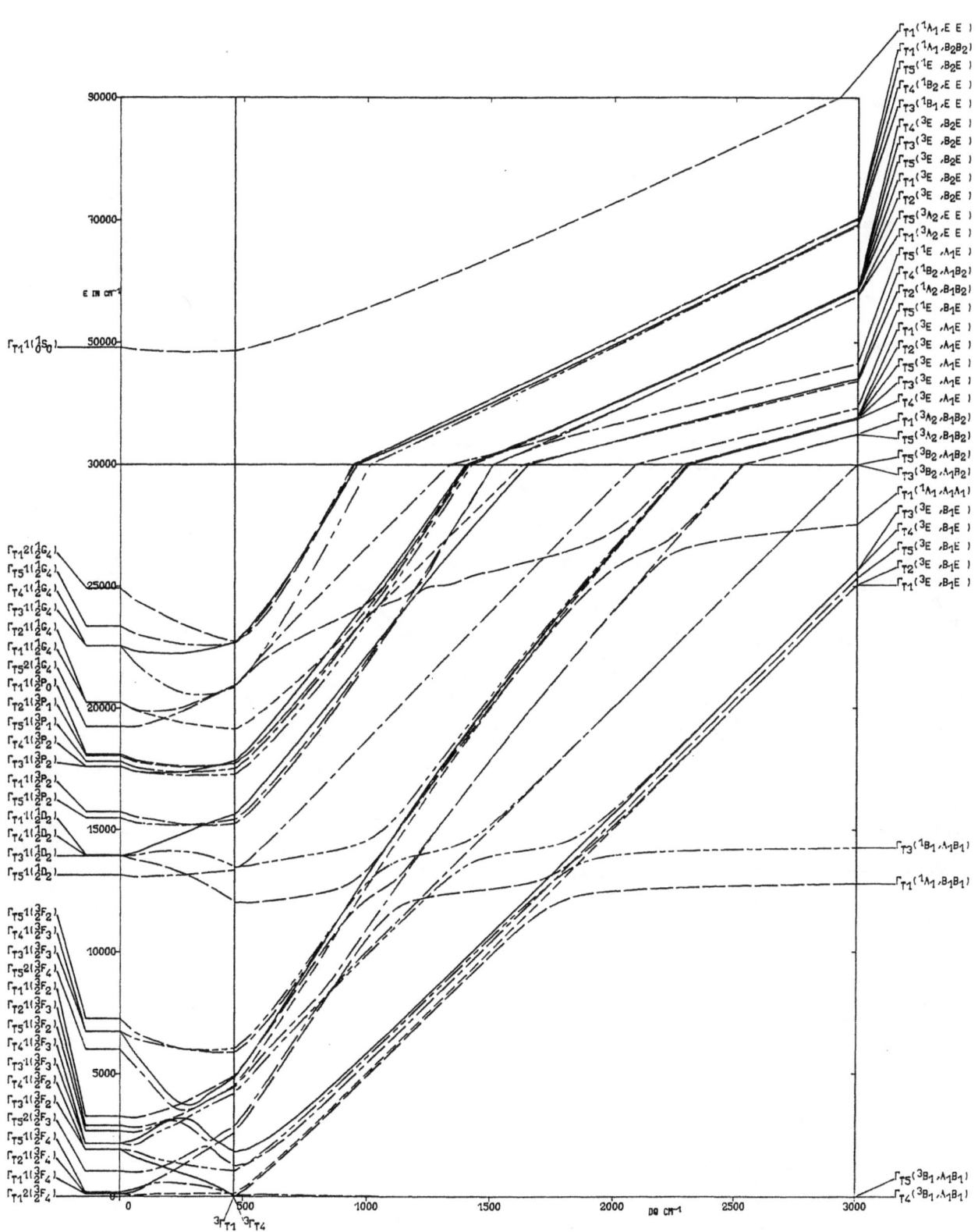

8 D ELECTRONS, TETRAGONAL SYMMETRY   B=905 C=4×B DT=500 K=1 DS=K×DT ZETA=650

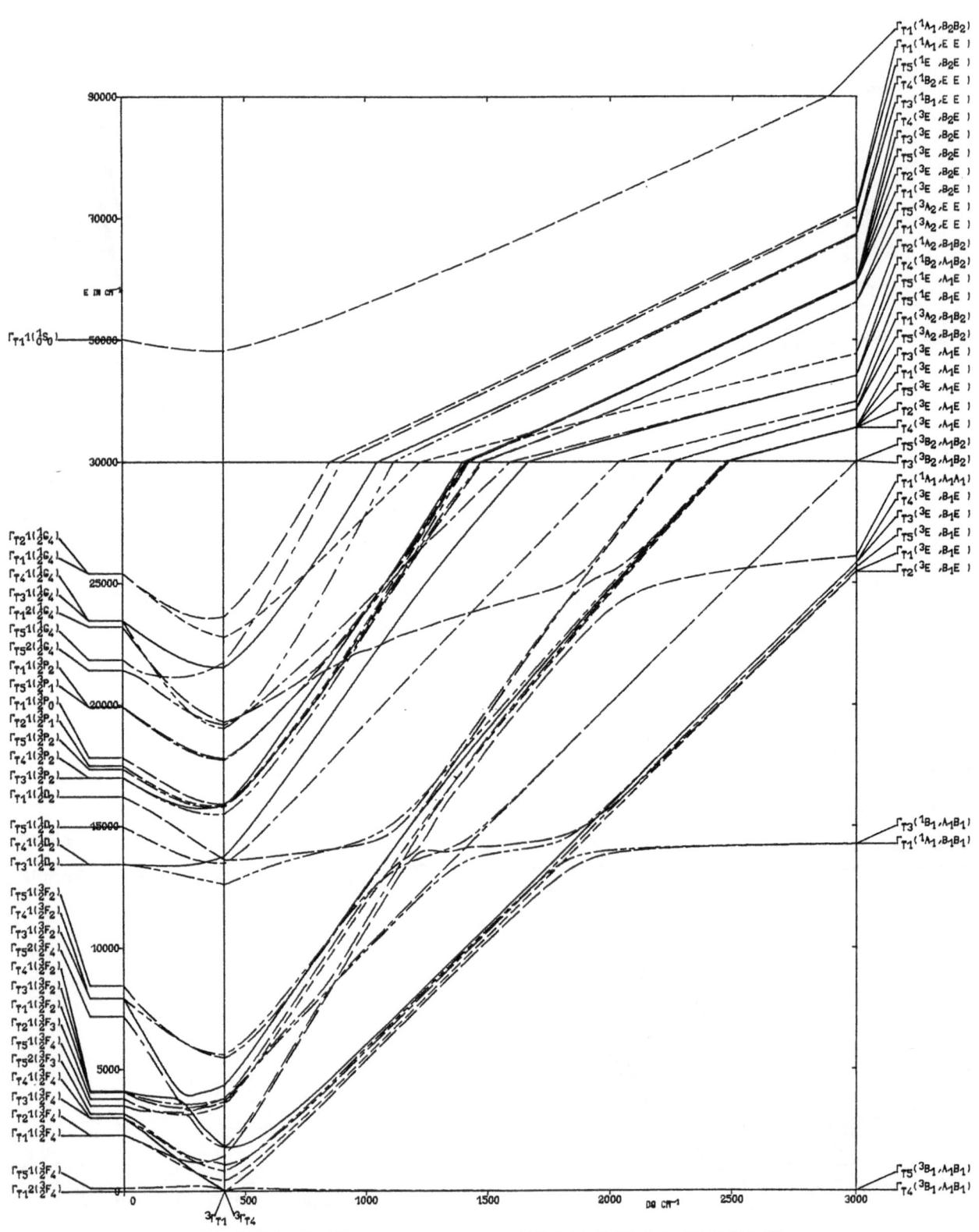

8 D ELECTRONS, TETRAGONAL SYMMETRY

B=905 C=4×B DT=500 K=-1 DS=K×DT ZETA=650

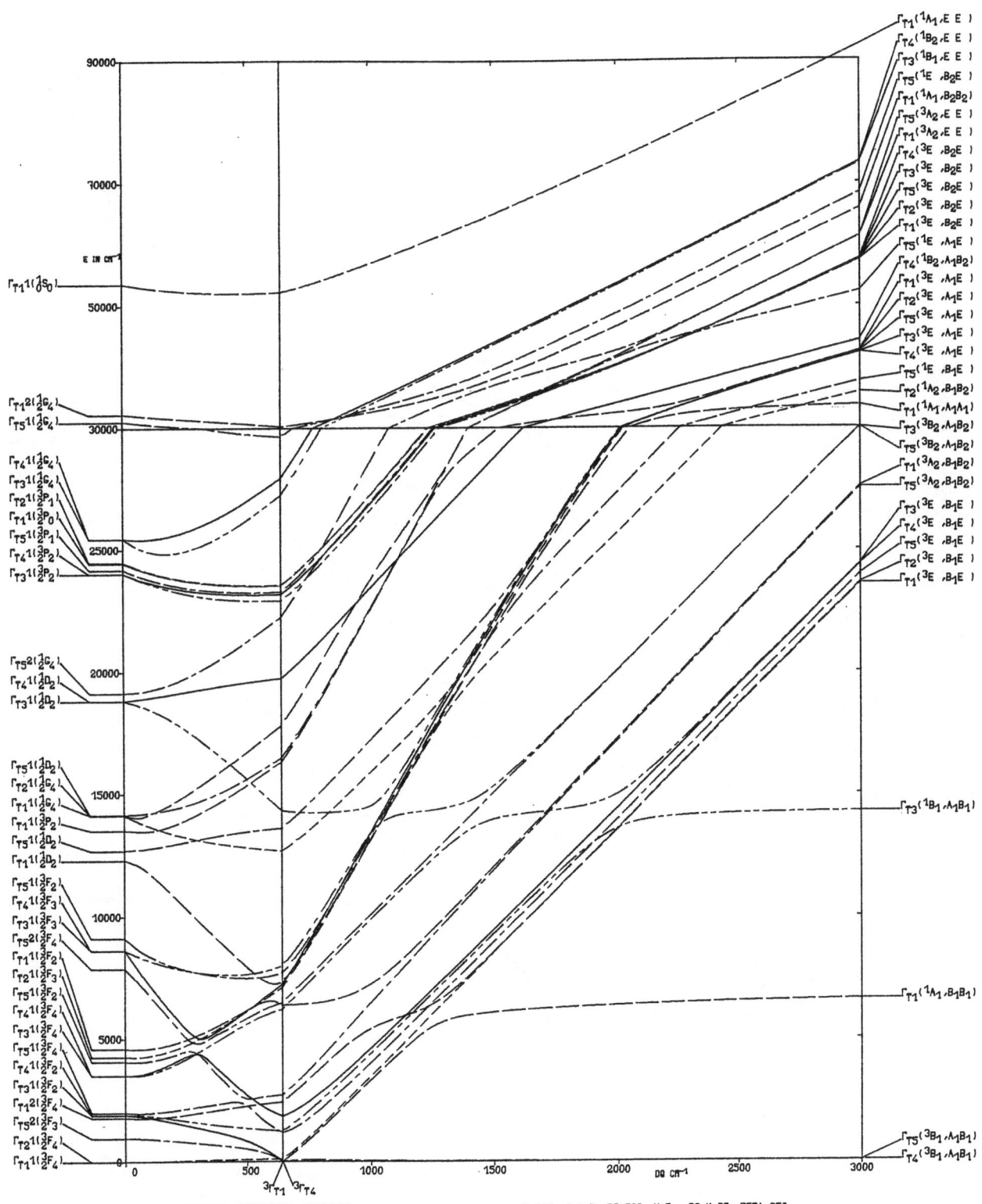

8 D ELECTRONS, TETRAGONAL SYMMETRY

B=905 C=4*B DT=500 K=5 DS=K*DT ZETA=650

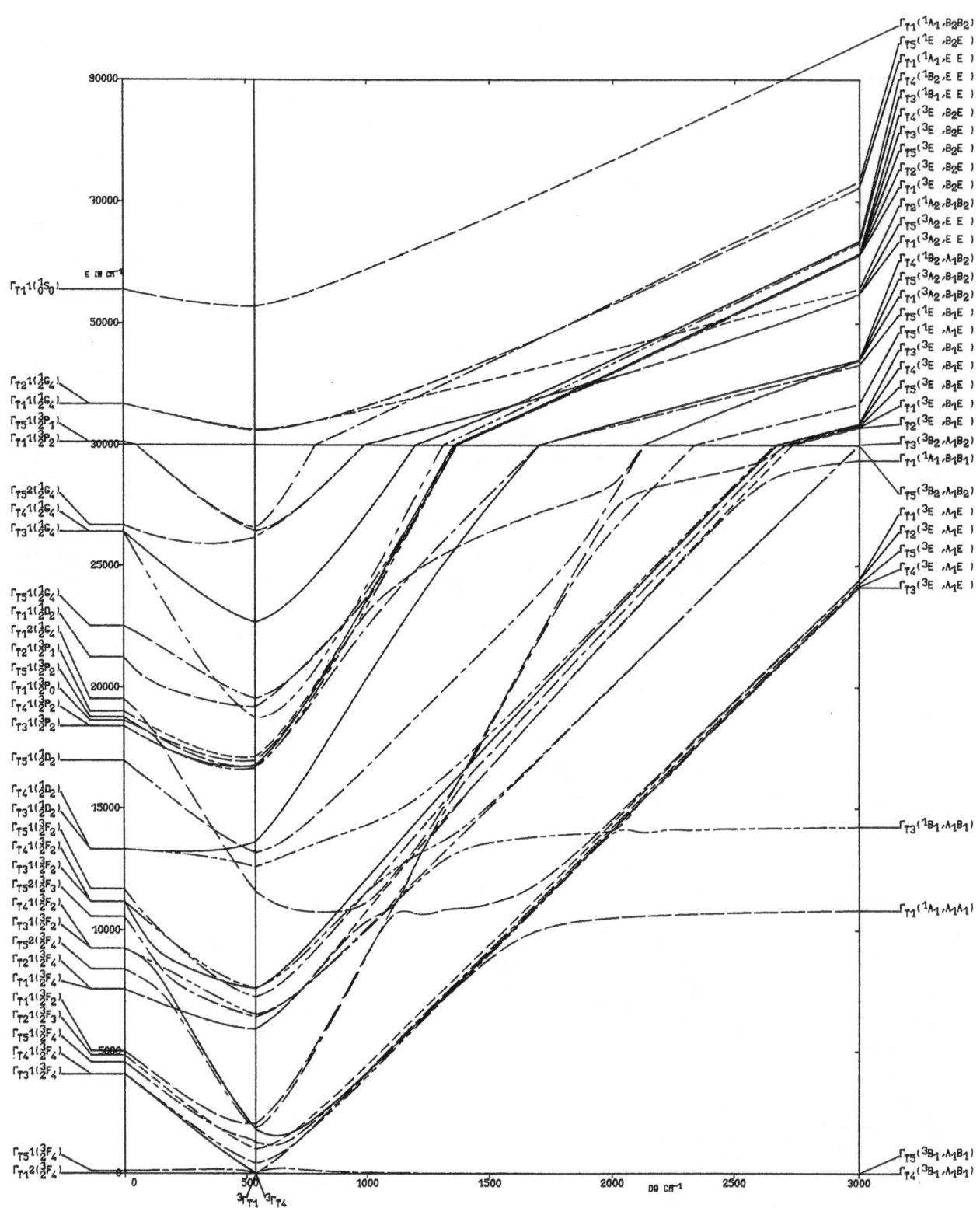

90000

70000

50000

$\Gamma_{T1}1(^1_0S_0)$

$\Gamma_{T2}1(^1_2G_4)$
$\Gamma_{T1}1(^1_2G_4)$
$\Gamma_{T5}1(^3_2P_1)$
$\Gamma_{T1}1(^3_2P_2)$ 30000

$\Gamma_{T5}2(^1_2G_4)$
$\Gamma_{T4}1(^1_2G_4)$
$\Gamma_{T3}1(^1_2G_4)$

$\Gamma_{T5}1(^1_2G_4)$ 25000
$\Gamma_{T1}1(^1_2D_2)$
$\Gamma_{T1}2(^1_2G_4)$
$\Gamma_{T2}1(^3_2P_1)$
$\Gamma_{T5}1(^3_2P_2)$
$\Gamma_{T1}1(^3_2P_0)$ 20000
$\Gamma_{T4}1(^3_2P_2)$
$\Gamma_{T3}1(^3_2P_2)$

$\Gamma_{T5}1(^1_2D_2)$

$\Gamma_{T4}1(^1_2D_2)$
$\Gamma_{T3}1(^1_2D_2)$ 15000
$\Gamma_{T5}1(^3_2F_2)$
$\Gamma_{T4}1(^3_2F_2)$
$\Gamma_{T3}1(^3_2F_2)$
$\Gamma_{T5}2(^3_2F_3)$
$\Gamma_{T4}1(^3_2F_2)$
$\Gamma_{T3}1(^3_2F_2)$ 10000
$\Gamma_{T5}2(^3_2F_3)$
$\Gamma_{T2}1(^3_2F_4)$
$\Gamma_{T1}1(^3_2F_4)$
$\Gamma_{T1}1(^3_2F_2)$
$\Gamma_{T2}1(^3_2F_3)$
$\Gamma_{T5}1(^3_2F_4)$ 5000
$\Gamma_{T4}1(^3_2F_4)$
$\Gamma_{T3}1(^3_2F_4)$

$\Gamma_{T5}1(^3_2F_4)$
$\Gamma_{T1}2(^3_2F_4)$ 0

E IN CM⁻¹

$\Gamma_{T1}(^1A_1,B_2B_2)$
$\Gamma_{T5}(^1E,B_2E)$
$\Gamma_{T1}(^1A_1,E E)$
$\Gamma_{T4}(^1B_2,E E)$
$\Gamma_{T3}(^1B_1,E E)$
$\Gamma_{T4}(^3E,B_2E)$
$\Gamma_{T3}(^3E,B_2E)$
$\Gamma_{T5}(^3E,B_2E)$
$\Gamma_{T2}(^3E,B_2E)$
$\Gamma_{T1}(^3E,B_2E)$
$\Gamma_{T2}(^1A_2,B_1B_2)$
$\Gamma_{T5}(^3A_2,E E)$
$\Gamma_{T1}(^3A_2,E E)$
$\Gamma_{T4}(^1B_2,A_1B_2)$
$\Gamma_{T5}(^3A_2,B_1B_2)$
$\Gamma_{T1}(^3A_2,B_1B_2)$
$\Gamma_{T5}(^1E,B_1E)$
$\Gamma_{T5}(^1E,A_1E)$
$\Gamma_{T3}(^3E,B_1E)$
$\Gamma_{T4}(^3E,B_1E)$
$\Gamma_{T5}(^3E,B_1E)$
$\Gamma_{T1}(^3E,B_1E)$
$\Gamma_{T2}(^3E,B_1E)$
$\Gamma_{T3}(^3B_2,A_1B_2)$
$\Gamma_{T1}(^1A_1,B_1B_1)$

$\Gamma_{T5}(^3B_2,A_1B_2)$

$\Gamma_{T4}(^3E,A_1E)$
$\Gamma_{T1}(^3E,A_1E)$
$\Gamma_{T5}(^3E,A_1E)$
$\Gamma_{T4}(^3E,A_1E)$
$\Gamma_{T3}(^3E,A_1E)$

$\Gamma_{T3}(^1B_1,A_1B_1)$

$\Gamma_{T1}(^1A_1,A_1A_1)$

$\Gamma_{T5}(^3B_1,A_1B_1)$
$\Gamma_{T4}(^3B_1,A_1B_1)$

0   500   1000   1500   2000   2500   3000
          $^3\Gamma_{T1}$ $^3\Gamma_{T4}$ DQ CM⁻¹

408

8 D ELECTRONS, TETRAGONAL SYMMETRY   B=905 C=4*B DT=500 K=-5 DS=K*DT ZETA=650

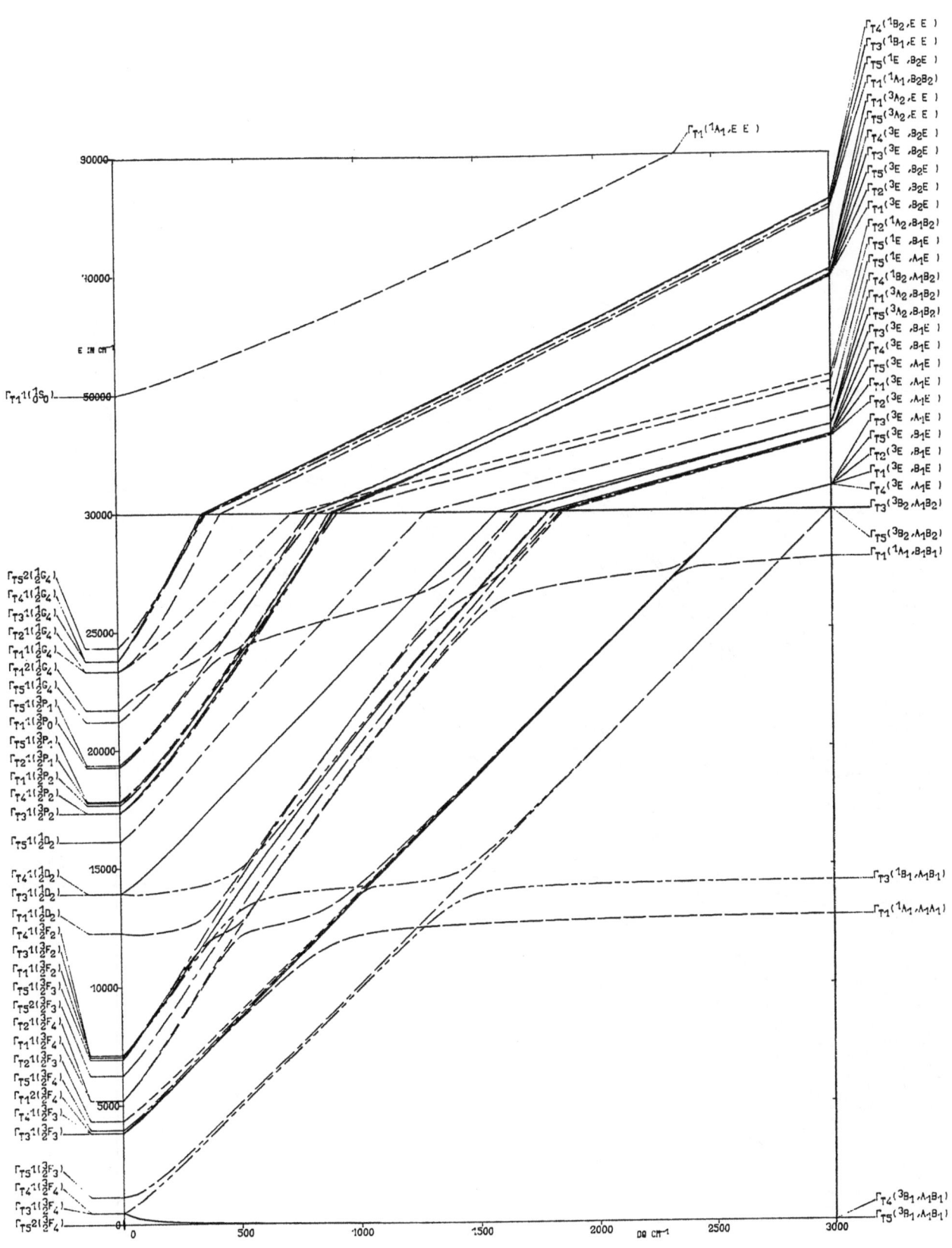

8 D ELECTRONS, TETRAGONAL SYMMETRY   B=905 C=4*B DT=-500 K=1   DS=K*DT ZETA=650

409

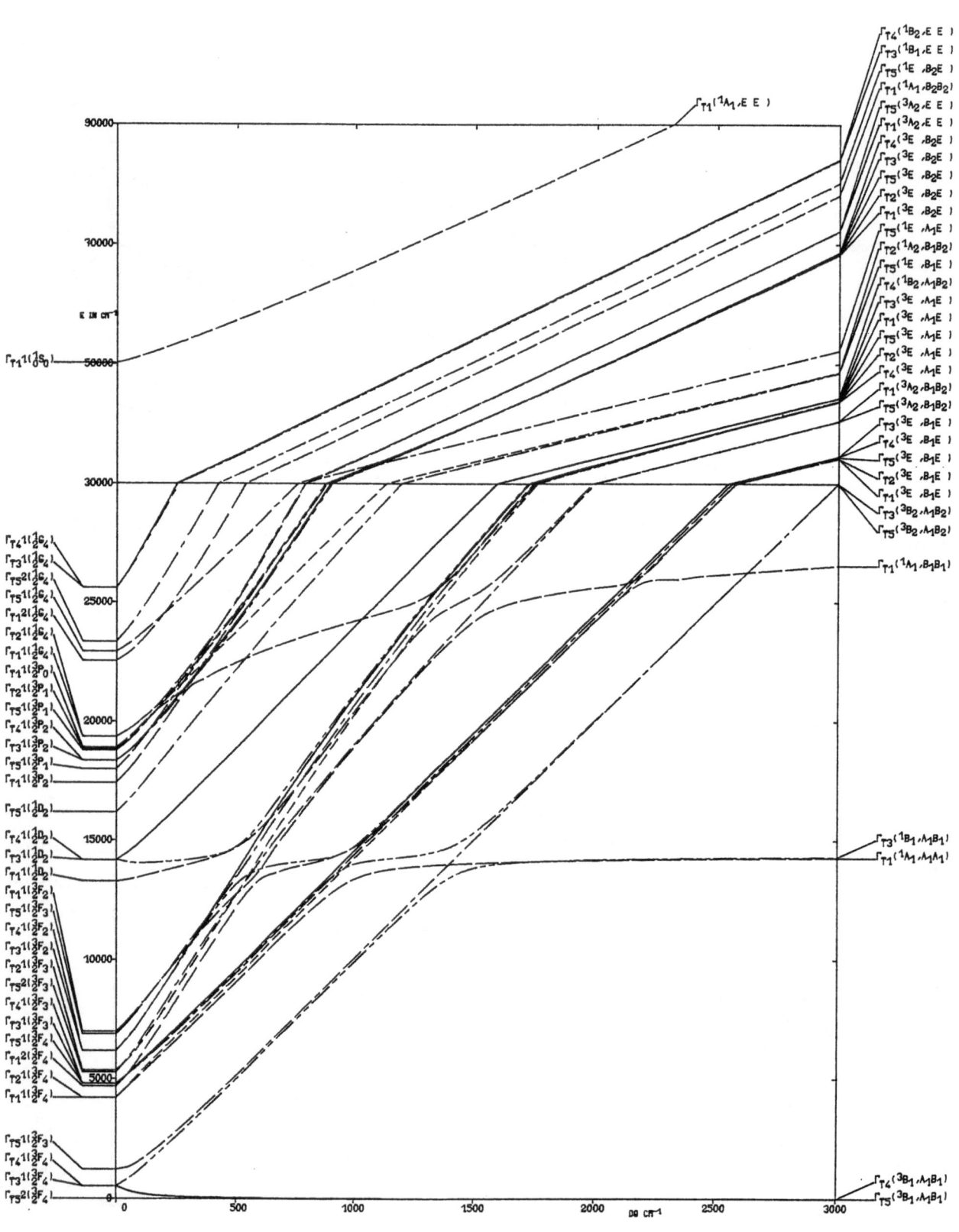

8 D ELECTRONS, TETRAGONAL SYMMETRY   B=905   C=4xB   DT=-500   K=-1   DS=KxDT   ZETA=650

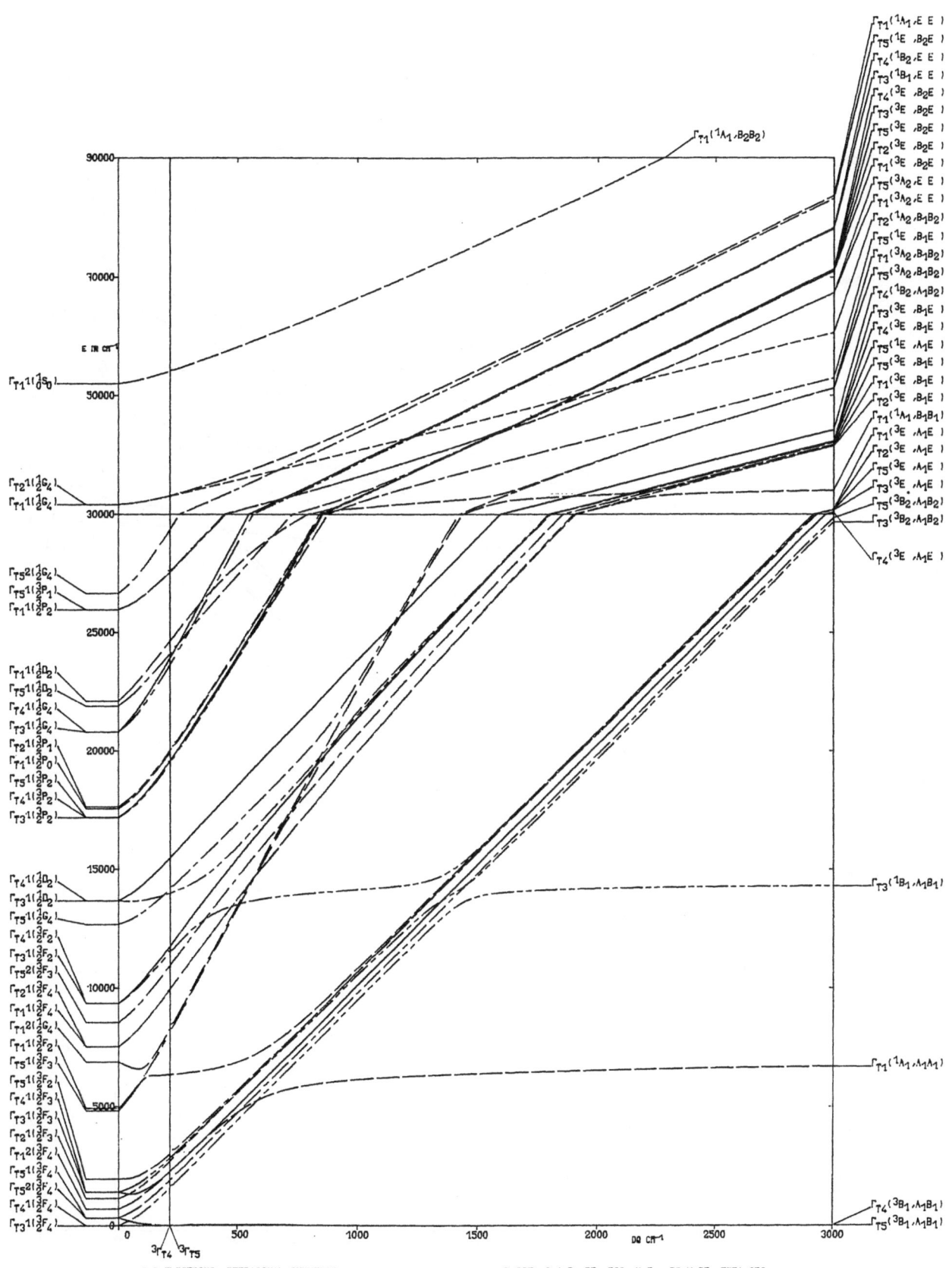

8 D ELECTRONS, TETRAGONAL SYMMETRY   B=905 C=4×B DT=-500 K=5   DS=K×DT ZETA=650

411

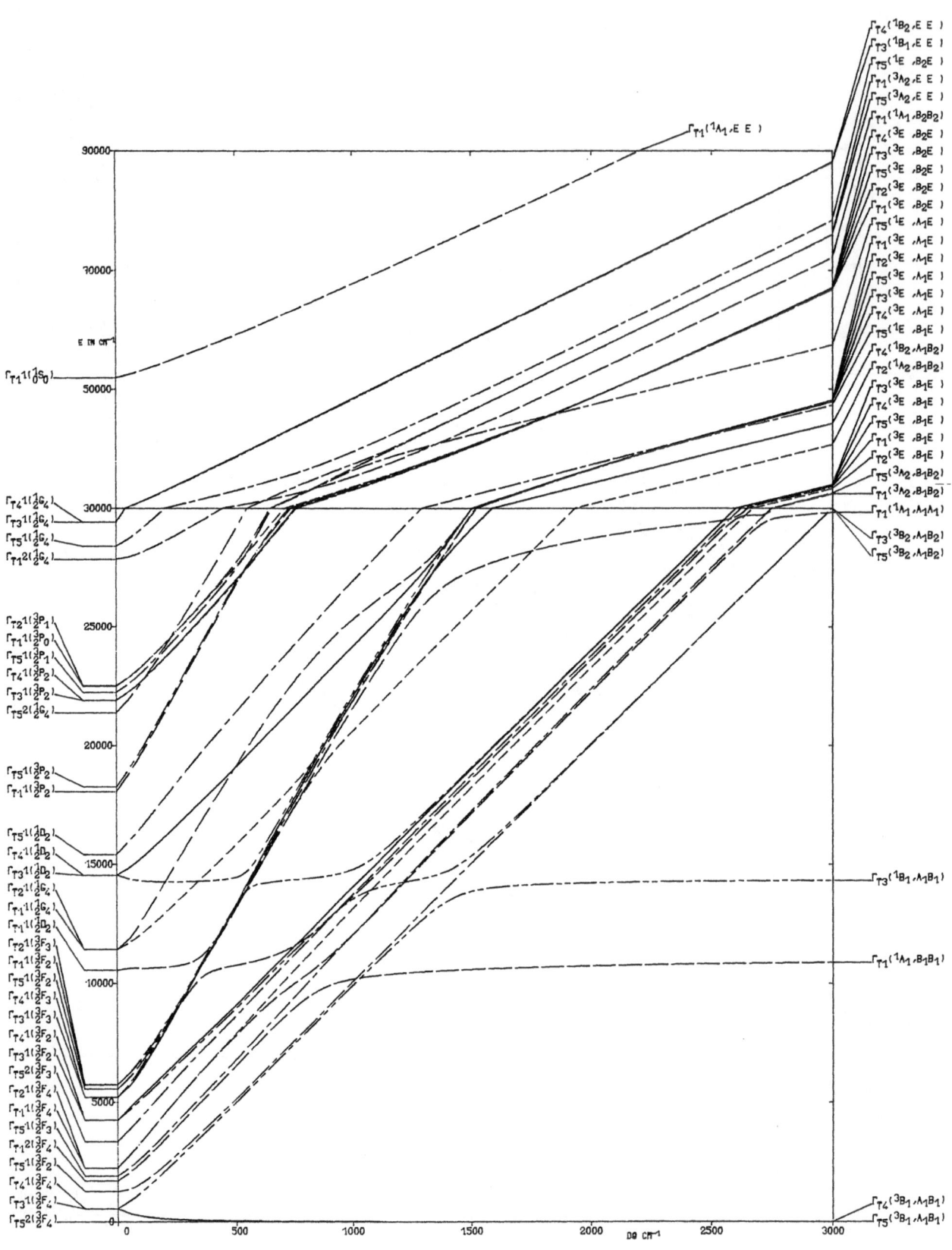

8 D ELECTRONS, TETRAGONAL SYMMETRY   B=905 C=4*B DT=-500 K=-5 DS=K*DT ZETA=650

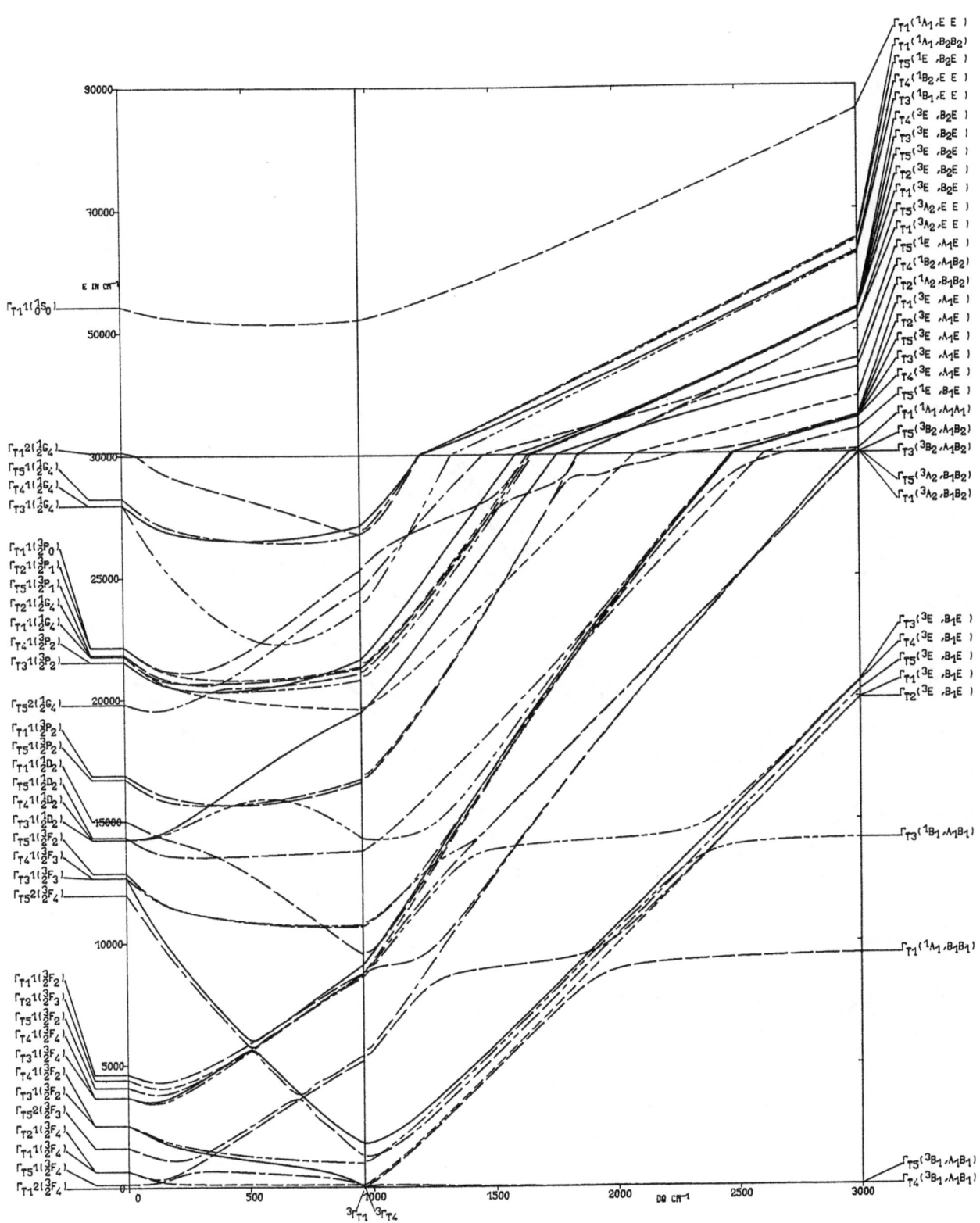

8 D ELECTRONS, TETRAGONAL SYMMETRY   B=905 C=4×B DT=1000 K=1   DS=K×DT ZETA=650

413

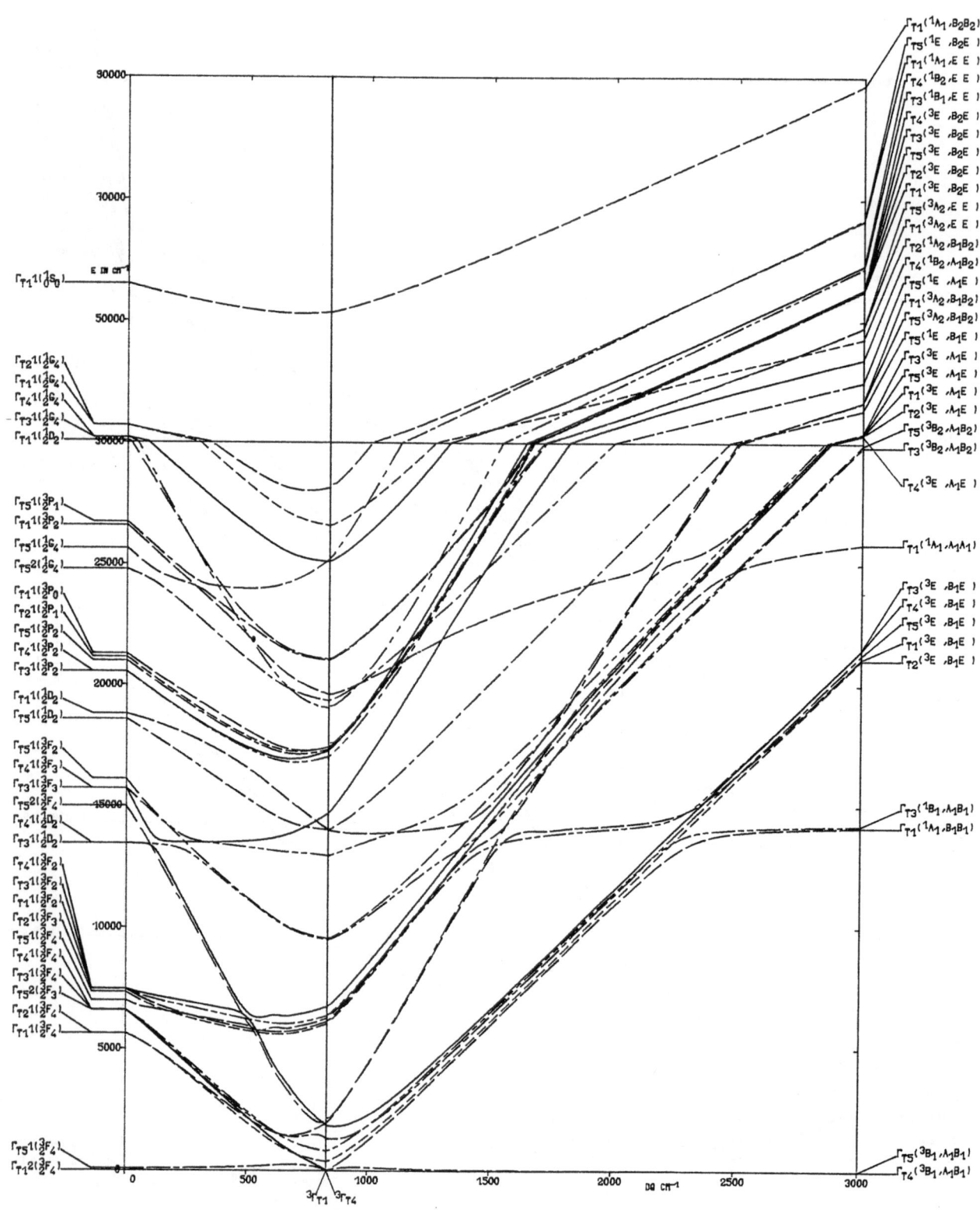

$\Gamma_{T1}{}^1(^1S_0)$

$\Gamma_{T2}{}^1(^1G_4)$
$\Gamma_{T1}{}^1(^1G_4)$
$\Gamma_{T4}{}^1(^1G_4)$
$\Gamma_{T3}{}^1(^1G_4)$
$\Gamma_{T1}{}^1(^1D_2)$

$\Gamma_{T5}{}^1(^3P_1)$
$\Gamma_{T1}{}^1(^3P_0)$
$\Gamma_{T5}{}^1(^1G_4)$
$\Gamma_{T5}{}^2(^1G_4)$

$\Gamma_{T1}{}^1(^3P_0)$
$\Gamma_{T2}{}^1(^3P_1)$
$\Gamma_{T5}{}^1(^3P_2)$
$\Gamma_{T4}{}^1(^3P_2)$
$\Gamma_{T3}{}^1(^3P_2)$

$\Gamma_{T1}{}^1(^1D_2)$
$\Gamma_{T5}{}^1(^1D_2)$

$\Gamma_{T5}{}^1(^3F_2)$
$\Gamma_{T4}{}^1(^3F_3)$
$\Gamma_{T3}{}^1(^3F_3)$
$\Gamma_{T5}{}^2(^3F_4)$
$\Gamma_{T4}{}^1(^1D_2)$
$\Gamma_{T3}{}^1(^1D_2)$

$\Gamma_{T4}{}^1(^3F_2)$
$\Gamma_{T3}{}^1(^3F_2)$
$\Gamma_{T1}{}^1(^3F_2)$
$\Gamma_{T2}{}^1(^3F_3)$
$\Gamma_{T5}{}^1(^3F_4)$
$\Gamma_{T4}{}^1(^3F_4)$
$\Gamma_{T3}{}^1(^3F_4)$
$\Gamma_{T5}{}^2(^3F_3)$
$\Gamma_{T2}{}^1(^3F_4)$
$\Gamma_{T1}{}^1(^3F_4)$

$\Gamma_{T5}{}^1(^3F_4)$
$\Gamma_{T1}{}^2(^3F_4)$

$\Gamma_{T1}(^1A_1,B_2B_2)$
$\Gamma_{T5}(^1E\ ,B_2E\ )$
$\Gamma_{T1}(^1A_1,E\ E\ )$
$\Gamma_{T4}(^1B_2,E\ E\ )$
$\Gamma_{T3}(^1B_1,E\ E\ )$
$\Gamma_{T4}(^3E\ ,B_2E\ )$
$\Gamma_{T3}(^3E\ ,B_2E\ )$
$\Gamma_{T5}(^3E\ ,B_2E\ )$
$\Gamma_{T2}(^3E\ ,B_2E\ )$
$\Gamma_{T1}(^3E\ ,B_2E\ )$
$\Gamma_{T5}(^3A_2,E\ E\ )$
$\Gamma_{T1}(^3A_2,E\ E\ )$
$\Gamma_{T2}(^1A_2,B_1B_2)$
$\Gamma_{T4}(^1B_2,A_1B_2)$
$\Gamma_{T5}(^1E\ ,A_1E\ )$
$\Gamma_{T1}(^3A_2,B_2B_2)$
$\Gamma_{T5}(^3A_2,B_2B_2)$
$\Gamma_{T5}(^1E\ ,B_1E\ )$
$\Gamma_{T3}(^3E\ ,A_1E\ )$
$\Gamma_{T5}(^3E\ ,A_1E\ )$
$\Gamma_{T1}(^3E\ ,A_1E\ )$
$\Gamma_{T2}(^3E\ ,A_1E\ )$
$\Gamma_{T5}(^3B_2,A_1B_2)$
$\Gamma_{T3}(^3B_2,A_1B_2)$

$\Gamma_{T4}(^3E\ ,A_1E\ )$

$\Gamma_{T1}(^1A_1,A_1A_1)$

$\Gamma_{T3}(^3E\ ,B_1E\ )$
$\Gamma_{T4}(^3E\ ,B_1E\ )$
$\Gamma_{T5}(^3E\ ,B_1E\ )$
$\Gamma_{T1}(^3E\ ,B_1E\ )$
$\Gamma_{T2}(^3E\ ,B_1E\ )$

$\Gamma_{T3}(^1B_1,A_1B_1)$
$\Gamma_{T1}(^1A_1,B_1B_1)$

$\Gamma_{T5}(^3B_1,A_1B_1)$
$\Gamma_{T4}(^3B_1,A_1B_1)$

30000

70000

50000

30000

25000

20000

15000

10000

5000

0   500   1000   1500   2000   2500   3000

$^3\Gamma_{T1}$ $^3\Gamma_{T4}$

DQ cm^{-1}

E IN cm^{-1}

8 D ELECTRONS, TETRAGONAL SYMMETRY

B=905 C=4×B DT=1000 K=-1 DS=K×DT ZETA=650

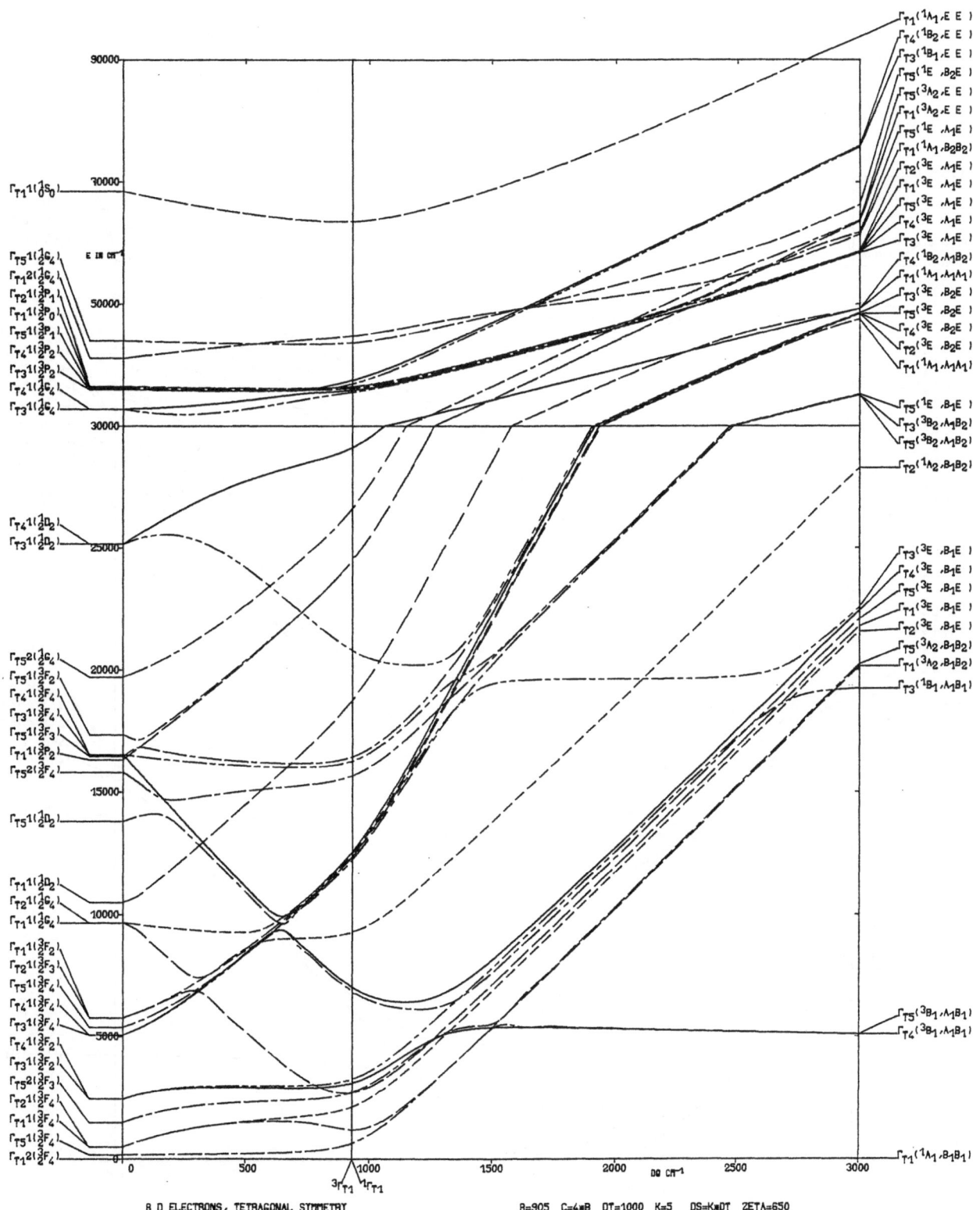

$\Gamma_{T1}({}^1_0S_0)$

$\Gamma_{T5}1({}^1_2G_4)$
$\Gamma_{T1}2({}^1_2G_4)$
$\Gamma_{T2}1({}^3_2P_1)$
$\Gamma_{T1}1({}^3_2P_0)$
$\Gamma_{T5}1({}^3_2P_1)$
$\Gamma_{T4}1({}^3_2P_2)$
$\Gamma_{T3}1({}^3_2P_2)$
$\Gamma_{T4}1({}^1_2G_4)$
$\Gamma_{T3}1({}^1_2G_4)$

$\Gamma_{T4}1({}^1_2D_2)$
$\Gamma_{T3}1({}^1_2D_2)$

$\Gamma_{T5}2({}^1_2G_4)$
$\Gamma_{T3}2({}^3_2F_2)$
$\Gamma_{T4}1({}^3_2F_4)$
$\Gamma_{T3}1({}^3_2F_4)$
$\Gamma_{T5}1({}^3_2F_3)$
$\Gamma_{T1}1({}^3_2P_2)$
$\Gamma_{T5}2({}^3_2F_4)$

$\Gamma_{T5}1({}^1_2D_2)$

$\Gamma_{T1}1({}^1_2D_2)$
$\Gamma_{T2}1({}^1_2G_4)$
$\Gamma_{T1}1({}^1_2G_4)$

$\Gamma_{T1}1({}^3_2F_2)$
$\Gamma_{T2}1({}^3_2F_3)$
$\Gamma_{T5}1({}^3_2F_4)$
$\Gamma_{T4}1({}^3_2F_4)$
$\Gamma_{T3}1({}^3_2F_4)$
$\Gamma_{T4}1({}^3_2F_2)$
$\Gamma_{T3}1({}^3_2F_2)$
$\Gamma_{T5}2({}^3_2F_3)$
$\Gamma_{T2}1({}^3_2F_4)$
$\Gamma_{T1}1({}^3_2F_4)$
$\Gamma_{T5}1({}^3_2F_4)$
$\Gamma_{T2}1({}^3_2F_4)$

$\Gamma_{T1}({}^1A_1 , E\ E\ )$
$\Gamma_{T4}({}^1B_2 , E\ E\ )$
$\Gamma_{T3}({}^1B_1 , E\ E\ )$
$\Gamma_{T5}({}^1E\ , B_2 E)$
$\Gamma_{T5}({}^3A_2 , E\ E\ )$
$\Gamma_{T1}({}^3A_2 , E\ E\ )$
$\Gamma_{T5}({}^1E\ , A_1 E)$
$\Gamma_{T1}({}^1A_1 , B_2 B_2)$
$\Gamma_{T2}({}^3E\ , A_1 E)$
$\Gamma_{T1}({}^3E\ , A_1 E)$
$\Gamma_{T5}({}^3E\ , A_1 E)$
$\Gamma_{T4}({}^3E\ , A_1 E)$
$\Gamma_{T3}({}^3E\ , A_1 E)$

$\Gamma_{T4}({}^1B_2 , A_1 B_2)$
$\Gamma_{T1}({}^1A_1 , A_1 A_1)$
$\Gamma_{T3}({}^3E\ , B_2 E)$
$\Gamma_{T5}({}^3E\ , B_2 E)$
$\Gamma_{T4}({}^3E\ , B_2 E)$
$\Gamma_{T2}({}^3E\ , B_2 E)$
$\Gamma_{T1}({}^1A_1 , A_1 A_1)$

$\Gamma_{T5}({}^1E\ , B_1 E)$
$\Gamma_{T3}({}^3B_2 , A_1 B_2)$
$\Gamma_{T5}({}^3B_2 , A_1 B_2)$

$\Gamma_{T2}({}^1A_2 , B_1 B_2)$

$\Gamma_{T3}({}^3E\ , B_1 E)$
$\Gamma_{T4}({}^3E\ , B_1 E)$
$\Gamma_{T5}({}^3E\ , B_1 E)$
$\Gamma_{T1}({}^3E\ , B_1 E)$
$\Gamma_{T2}({}^3E\ , B_1 E)$
$\Gamma_{T5}({}^3A_2 , B_1 B_2)$
$\Gamma_{T1}({}^3A_2 , B_1 B_2)$
$\Gamma_{T3}({}^1B_1 , A_1 B_1)$

$\Gamma_{T5}({}^3B_1 , A_1 B_1)$
$\Gamma_{T4}({}^3B_1 , A_1 B_1)$

$\Gamma_{T1}({}^1A_1 , B_1 B_1)$

8 D ELECTRONS, TETRAGONAL SYMMETRY   B=905   C=4*B   DT=1000   K=5   DS=K*DT   ZETA=650

$3\Gamma_{T1}$ $1\Gamma_{T1}$

DQ CM⁻¹

E IN CM⁻¹

415

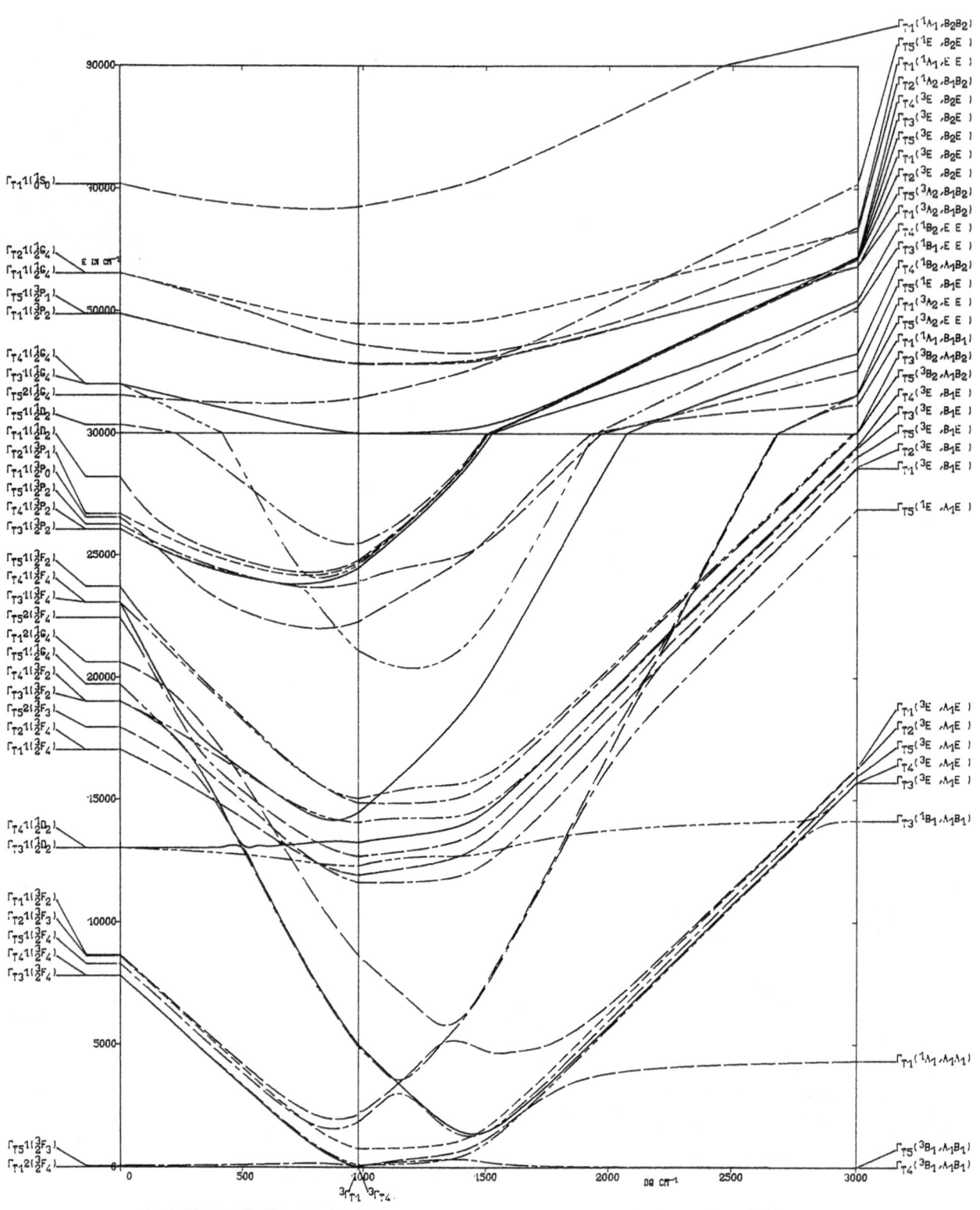

90000

$\Gamma_{T_1}1(^1S_0)$ 70000

$\Gamma_{T_2}1(^1_2G_4)$
$\Gamma_{T_1}1(^1_2G_4)$ E IN CM⁻¹

$\Gamma_{T_5}1(^3_2P_1)$
$\Gamma_{T_1}1(^3_2P_2)$ 50000

$\Gamma_{T_4}1(^1_2G_4)$
$\Gamma_{T_3}1(^1_2G_4)$
$\Gamma_{T_5}2(^1_2G_4)$
$\Gamma_{T_5}1(^1_2D_2)$
$\Gamma_{T_1}1(^1_2D_2)$ 30000
$\Gamma_{T_2}1(^3_2P_1)$
$\Gamma_{T_1}1(^3_2P_0)$
$\Gamma_{T_5}1(^3_2P_2)$
$\Gamma_{T_4}1(^3_2P_2)$
$\Gamma_{T_3}1(^3_2P_2)$

$\Gamma_{T_5}1(^3_2F_2)$ 25000
$\Gamma_{T_4}1(^3_2F_4)$
$\Gamma_{T_3}1(^3_2F_4)$
$\Gamma_{T_5}2(^3_2F_4)$
$\Gamma_{T_1}2(^1_2G_4)$
$\Gamma_{T_5}1(^1_2G_4)$
$\Gamma_{T_4}1(^3_2F_2)$ 20000
$\Gamma_{T_3}1(^3_2F_2)$
$\Gamma_{T_5}2(^3_2F_3)$
$\Gamma_{T_2}1(^3_2F_4)$
$\Gamma_{T_1}1(^3_2F_4)$

15000

$\Gamma_{T_4}1(^1_2D_2)$
$\Gamma_{T_3}1(^1_2D_2)$

$\Gamma_{T_1}1(^3_2F_2)$
$\Gamma_{T_2}1(^3_2F_3)$ 10000
$\Gamma_{T_5}1(^3_2F_4)$
$\Gamma_{T_4}1(^3_2F_4)$
$\Gamma_{T_3}1(^3_2F_4)$

5000

$\Gamma_{T_5}1(^3_2F_3)$
$\Gamma_{T_1}2(^3_2F_4)$ 0

0   500   1000   1500   2000   DQ CM⁻¹   2500   3000

$3\Gamma_{T_1}$ $3\Gamma_{T_4}$

Right side labels:
$\Gamma_{T_1}(^1A_1, B_2B_2)$
$\Gamma_{T_5}(^1E, B_2E)$
$\Gamma_{T_1}(^1A_1, E E)$
$\Gamma_{T_2}(^3A_2, B_1B_2)$
$\Gamma_{T_4}(^3E, B_2E)$
$\Gamma_{T_3}(^3E, B_2E)$
$\Gamma_{T_5}(^3E, B_2E)$
$\Gamma_{T_2}(^3E, B_2E)$
$\Gamma_{T_5}(^3A_2, B_1B_2)$
$\Gamma_{T_1}(^3A_2, B_1B_2)$
$\Gamma_{T_4}(^3B_2, E E)$
$\Gamma_{T_1}(^3B_1, E E)$
$\Gamma_{T_4}(^3B_2, A_1B_2)$
$\Gamma_{T_5}(^1E, B_1E)$
$\Gamma_{T_1}(^3A_2, E E)$
$\Gamma_{T_5}(^3A_2, E E)$
$\Gamma_{T_1}(^1A_1, B_1B_1)$
$\Gamma_{T_3}(^3B_2, A_1B_2)$
$\Gamma_{T_5}(^3B_2, A_1B_2)$
$\Gamma_{T_4}(^3E, B_1E)$
$\Gamma_{T_3}(^3E, B_1E)$
$\Gamma_{T_5}(^3E, B_1E)$
$\Gamma_{T_2}(^3E, B_1E)$
$\Gamma_{T_1}(^3E, B_1E)$

$\Gamma_{T_5}(^1E, A_1E)$

$\Gamma_{T_1}(^3E, A_1E)$
$\Gamma_{T_2}(^3E, A_1E)$
$\Gamma_{T_5}(^3E, A_1E)$
$\Gamma_{T_4}(^3E, A_1E)$
$\Gamma_{T_3}(^3E, A_1E)$

$\Gamma_{T_3}(^1B_1, A_1B_1)$

$\Gamma_{T_1}(^1A_1, A_1A_1)$

$\Gamma_{T_5}(^3B_1, A_1B_1)$
$\Gamma_{T_4}(^3B_1, A_1B_1)$

416   8 D ELECTRONS, TETRAGONAL SYMMETRY   B=905   C=4×B   DT=1000   K=-5   DS=K×DT   ZETA=650

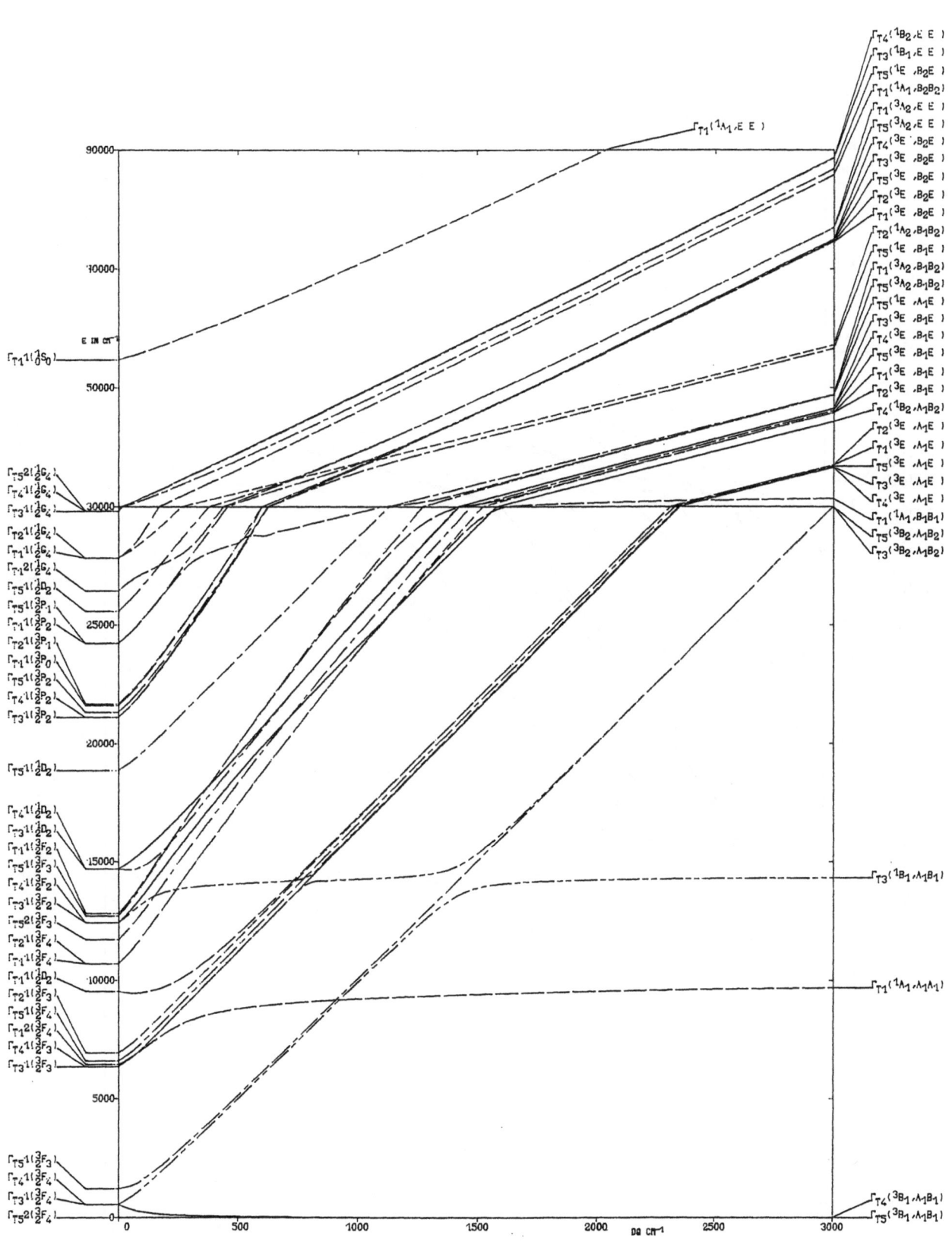

8 D ELECTRONS, TETRAGONAL SYMMETRY   B=905 C=4xB DT=-1000 K=1 DS=KxDT 2ETA=650

417

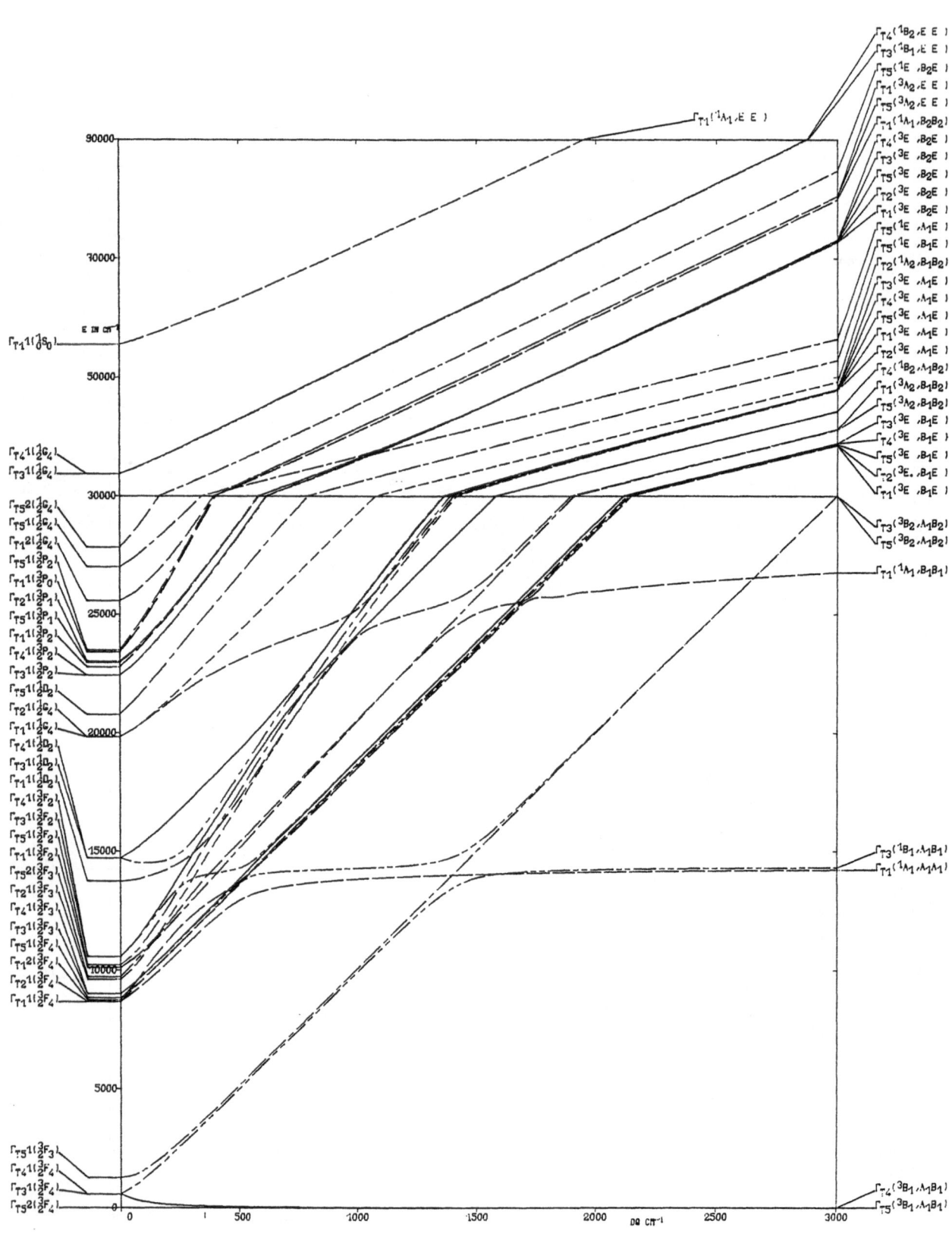

90000

70000

E IN CM⁻¹

$\Gamma_{T1}1(^1S_0)$

50000

$\Gamma_{T4}1(^1G_4)$
$\Gamma_{T3}1(^1G_4)$

30000

$\Gamma_{T5}2(^1G_4)$
$\Gamma_{T5}1(^1G_4)$
$\Gamma_{T1}2(^1G_4)$
$\Gamma_{T5}1(^3P_2)$
$\Gamma_{T1}1(^3P_0)$
$\Gamma_{T2}1(^3P_1)$
$\Gamma_{T5}1(^3P_1)$
$\Gamma_{T1}1(^3P_2)$
$\Gamma_{T4}1(^3P_2)$
$\Gamma_{T3}1(^3P_2)$

25000

$\Gamma_{T5}1(^1D_2)$
$\Gamma_{T2}1(^1C_4)$
$\Gamma_{T1}1(^1C_4)$

20000

$\Gamma_{T4}1(^1D_2)$
$\Gamma_{T3}1(^1D_2)$
$\Gamma_{T1}1(^1D_2)$
$\Gamma_{T4}1(^3F_2)$
$\Gamma_{T3}1(^3F_2)$
$\Gamma_{T5}1(^3F_2)$
$\Gamma_{T1}1(^3F_2)$

15000

$\Gamma_{T5}2(^3F_3)$
$\Gamma_{T2}1(^3F_3)$
$\Gamma_{T4}1(^3F_3)$
$\Gamma_{T3}1(^3F_3)$
$\Gamma_{T5}1(^3F_4)$
$\Gamma_{T1}2(^3F_4)$
$\Gamma_{T2}2(^3F_4)$
$\Gamma_{T1}1(^3F_4)$

10000

5000

$\Gamma_{T5}1(^3F_3)$
$\Gamma_{T4}1(^3F_4)$
$\Gamma_{T3}1(^3F_4)$
$\Gamma_{T5}2(^3F_4)$

0

0   500   1000   1500   2000   DQ CM⁻¹   2500   3000

$\Gamma_{T1}(^1A_1,EE)$

$\Gamma_{T4}(^1B_2,EE)$
$\Gamma_{T3}(^1B_1,EE)$
$\Gamma_{T5}(^1E,B_2E)$
$\Gamma_{T1}(^3A_2,EE)$
$\Gamma_{T5}(^3A_2,EE)$
$\Gamma_{T1}(^1A_1,B_2B_2)$
$\Gamma_{T4}(^3E,B_2E)$
$\Gamma_{T3}(^3E,B_2E)$
$\Gamma_{T5}(^3E,B_2E)$
$\Gamma_{T2}(^3E,B_2E)$
$\Gamma_{T1}(^3E,B_2E)$
$\Gamma_{T5}(^1E,A_1E)$
$\Gamma_{T5}(^1E,B_1E)$
$\Gamma_{T2}(^1A_2,B_1B_2)$
$\Gamma_{T3}(^3E,A_1E)$
$\Gamma_{T4}(^3E,A_1E)$
$\Gamma_{T5}(^3E,A_1E)$
$\Gamma_{T1}(^3E,A_1E)$
$\Gamma_{T2}(^3E,A_1E)$
$\Gamma_{T5}(^1B_2,A_1B_2)$
$\Gamma_{T1}(^3A_2,B_1B_2)$
$\Gamma_{T5}(^3A_2,B_1B_2)$
$\Gamma_{T3}(^3E,B_1E)$
$\Gamma_{T4}(^3E,B_1E)$
$\Gamma_{T5}(^3E,B_1E)$
$\Gamma_{T2}(^3E,B_1E)$
$\Gamma_{T1}(^3E,B_1E)$

$\Gamma_{T3}(^3B_2,A_1B_2)$
$\Gamma_{T5}(^3B_2,A_1B_2)$

$\Gamma_{T1}(^1A_1,B_1B_1)$

$\Gamma_{T3}(^1B_1,A_1B_1)$
$\Gamma_{T1}(^1A_1,A_1A_1)$

$\Gamma_{T4}(^3B_1,A_1B_1)$
$\Gamma_{T5}(^3B_1,A_1B_1)$

8 D ELECTRONS, TETRAGONAL SYMMETRY

B=905 C=4×B DT=-1000 K=-1 DS=K×DT ZETA=650

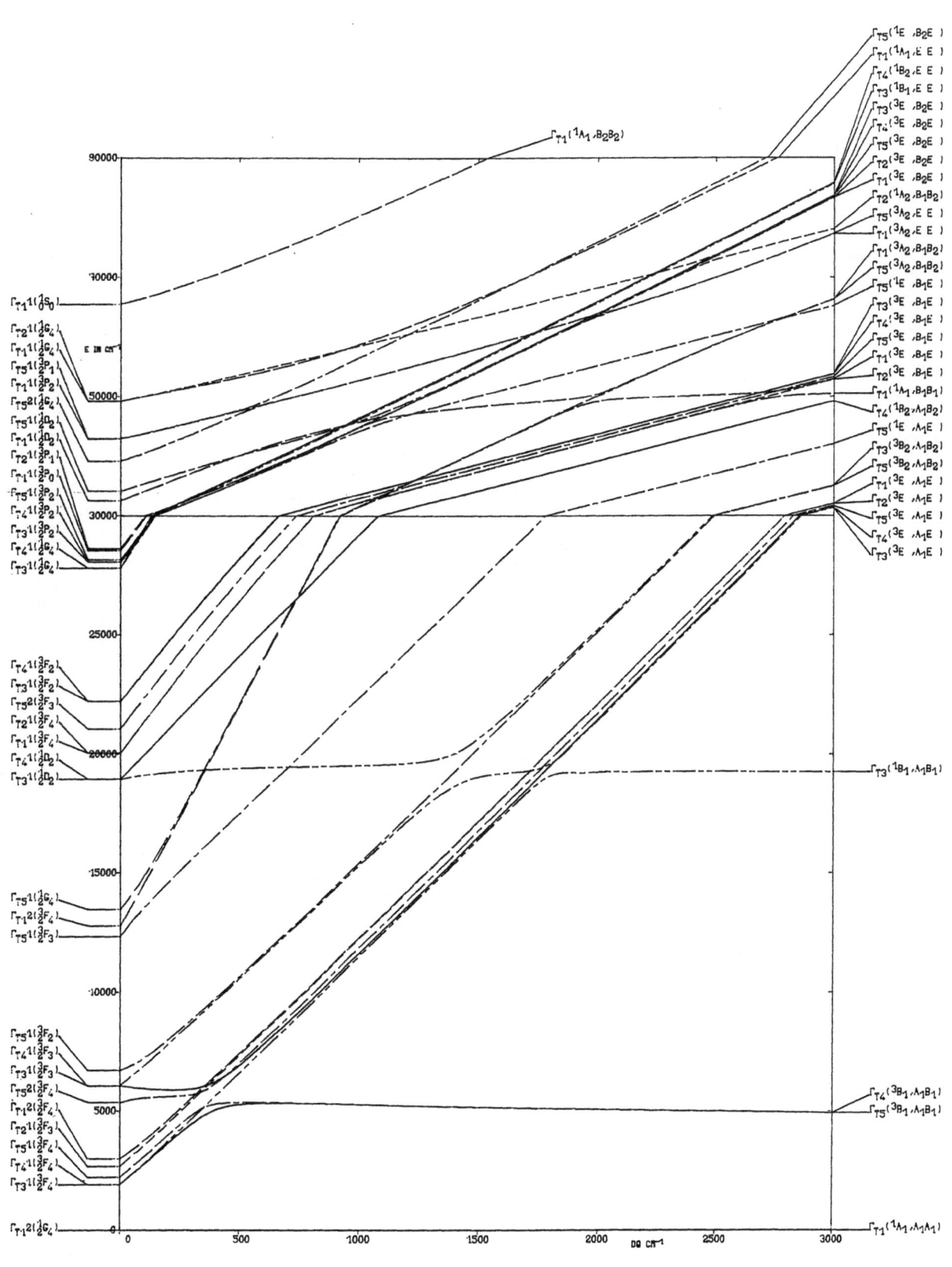

8 D ELECTRONS, TETRAGONAL SYMMETRY   B=905 C=4xB DT=-1000 K=5   DS=KxDT ZETA=650

419

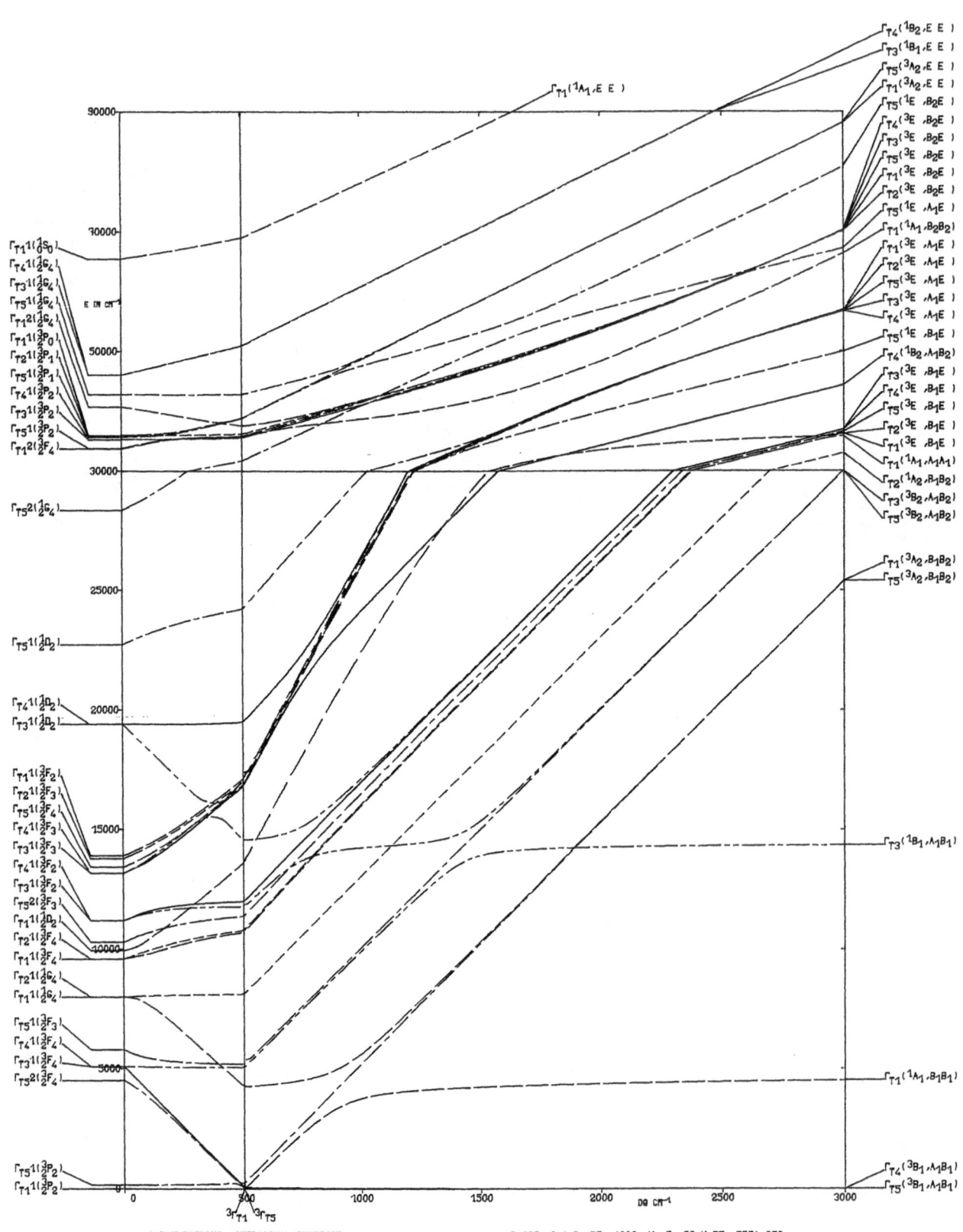

$\Gamma_{T1}1({}^1_0S_0)$
$\Gamma_{T4}1({}^1_2G_4)$
$\Gamma_{T3}1({}^1_2G_4)$
$\Gamma_{T5}1({}^1_2G_4)$
$\Gamma_{T4}2({}^1_2G_4)$
$\Gamma_{T1}1({}^3_2P_0)$
$\Gamma_{T2}1({}^3_2P_1)$
$\Gamma_{T5}1({}^3_2P_1)$
$\Gamma_{T4}1({}^3_2P_2)$
$\Gamma_{T3}1({}^3_2P_2)$
$\Gamma_{T5}1({}^3_2P_2)$
$\Gamma_{T1}2({}^1_2F_4)$

$\Gamma_{T5}2({}^1_2G_4)$

$\Gamma_{T5}1({}^1_2D_2)$

$\Gamma_{T4}1({}^1_2D_2)$
$\Gamma_{T3}1({}^1_2D_2)$

$\Gamma_{T1}1({}^3_2F_3)$
$\Gamma_{T2}1({}^3_2F_3)$
$\Gamma_{T5}1({}^3_2F_4)$
$\Gamma_{T4}1({}^3_2F_3)$
$\Gamma_{T3}1({}^3_2F_3)$
$\Gamma_{T4}1({}^3_2F_2)$
$\Gamma_{T3}1({}^3_2F_2)$
$\Gamma_{T5}2({}^3_2F_3)$
$\Gamma_{T1}1({}^1_2D_2)$
$\Gamma_{T2}1({}^1_2F_4)$
$\Gamma_{T1}1({}^3_2F_4)$
$\Gamma_{T2}1({}^1_2G_4)$
$\Gamma_{T1}1({}^1_2G_4)$
$\Gamma_{T5}1({}^3_2F_3)$
$\Gamma_{T4}1({}^3_2F_4)$
$\Gamma_{T3}1({}^3_2F_4)$
$\Gamma_{T5}2({}^3_2F_4)$

$\Gamma_{T5}1({}^3_2P_2)$
$\Gamma_{T1}1({}^3_2P_2)$

E IN CM^{-1}

$\Gamma_{T1}({}^1A_1,E\ E\ )$

$\Gamma_{T4}({}^1B_2,E\ E\ )$
$\Gamma_{T3}({}^1B_1,E\ E\ )$
$\Gamma_{T5}({}^3A_2,E\ E\ )$
$\Gamma_{T1}({}^3A_2,E\ E\ )$
$\Gamma_{T5}({}^1E\ ,B_2E\ )$
$\Gamma_{T4}({}^3E\ ,B_2E\ )$
$\Gamma_{T3}({}^3E\ ,B_2E\ )$
$\Gamma_{T5}({}^3E\ ,B_2E\ )$
$\Gamma_{T1}({}^3E\ ,B_2E\ )$
$\Gamma_{T2}({}^3E\ ,B_2E\ )$
$\Gamma_{T5}({}^1E\ ,A_1E\ )$
$\Gamma_{T1}({}^1A_1,B_2B_2)$
$\Gamma_{T1}({}^3E\ ,A_1E\ )$
$\Gamma_{T2}({}^3E\ ,A_1E\ )$
$\Gamma_{T5}({}^3E\ ,A_1E\ )$
$\Gamma_{T1}({}^3E\ ,A_1E\ )$
$\Gamma_{T4}({}^3E\ ,A_1E\ )$
$\Gamma_{T5}({}^1E\ ,B_1E\ )$
$\Gamma_{T4}({}^1B_2,A_1B_2)$
$\Gamma_{T3}({}^3E\ ,B_1E\ )$
$\Gamma_{T4}({}^3E\ ,B_1E\ )$
$\Gamma_{T5}({}^3E\ ,B_1E\ )$
$\Gamma_{T2}({}^3E\ ,B_1E\ )$
$\Gamma_{T1}({}^1A_1,A_1A_1)$
$\Gamma_{T2}({}^1A_2,B_1B_2)$
$\Gamma_{T3}({}^3B_2,A_1B_2)$
$\Gamma_{T5}({}^3B_2,A_1B_2)$

$\Gamma_{T1}({}^3A_2,B_1B_2)$
$\Gamma_{T5}({}^3A_2,B_1B_2)$

$\Gamma_{T3}({}^1B_1,A_1B_1)$

$\Gamma_{T1}({}^1A_1,B_1B_1)$

$\Gamma_{T4}({}^3B_1,A_1B_1)$
$\Gamma_{T5}({}^3B_1,A_1B_1)$

90000

70000

50000

30000

25000

20000

15000

10000

5000

0

0   500   1000   1500   2000   2500   3000
      $3\Gamma_{T1}$ $3\Gamma_{T5}$ DQ CM^{-1}

420

8 D ELECTRONS, TETRAGONAL SYMMETRY

B=905 C=4*B DT=-1000 K=-5 DS=K*DT ZETA=650

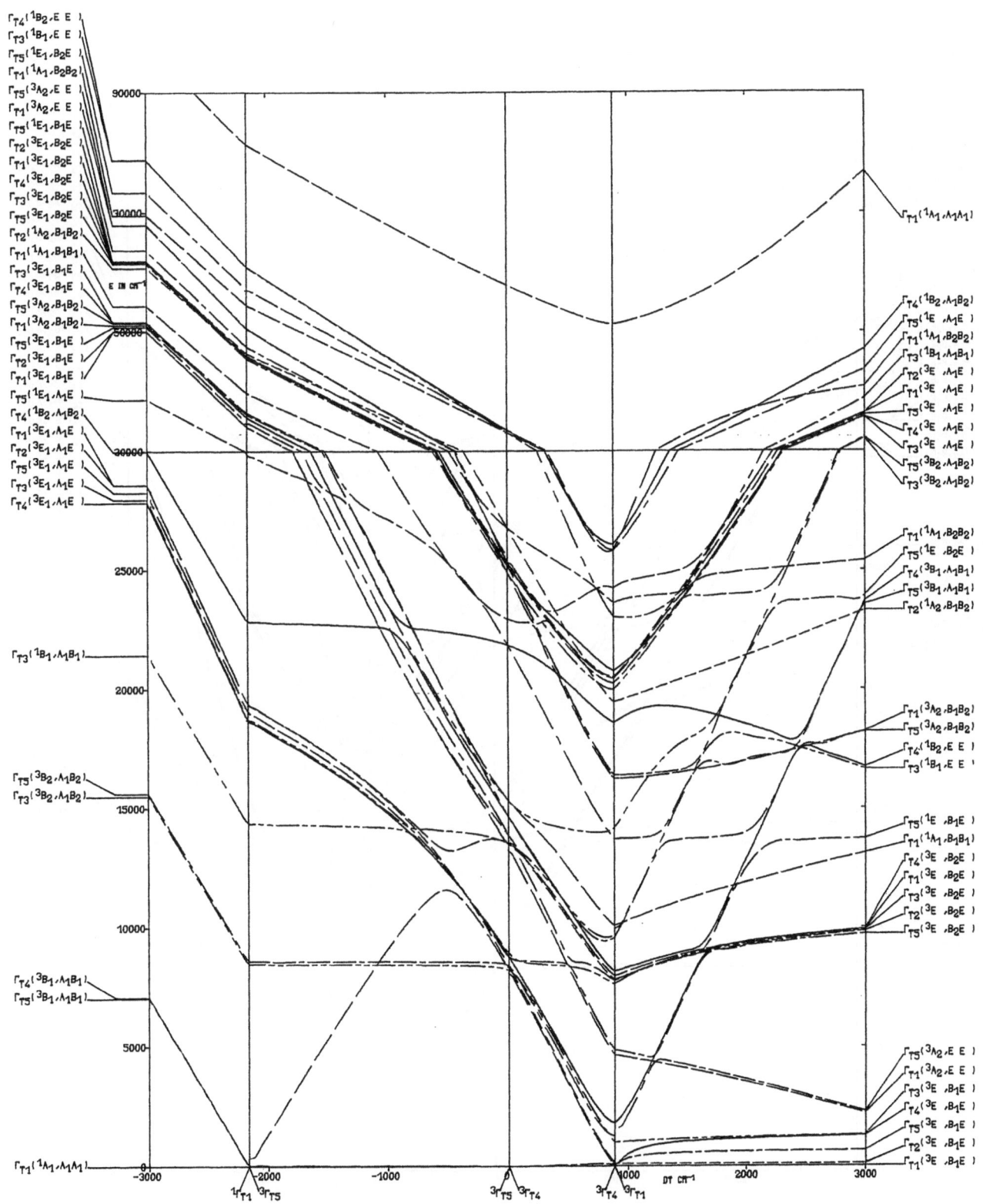

$\Gamma_{T4}(^1B_2, E\ E)$
$\Gamma_{T3}(^1B_1, E\ E)$
$\Gamma_{T5}(^1E_1, B_2E)$
$\Gamma_{T1}(^1A_1, B_2B_2)$
$\Gamma_{T5}(^3A_2, E\ E)$
$\Gamma_{T1}(^3A_2, E\ E)$
$\Gamma_{T5}(^3E_1, B_1E)$
$\Gamma_{T2}(^3E_1, B_2E)$
$\Gamma_{T1}(^3E_1, B_2E)$
$\Gamma_{T4}(^3E_1, B_2E)$
$\Gamma_{T3}(^3E_1, B_2E)$
$\Gamma_{T5}(^3E_1, B_2E)$
$\Gamma_{T2}(^3A_2, B_2B_2)$
$\Gamma_{T1}(^1A_1, B_1B_1)$
$\Gamma_{T3}(^1A_1, B_1E)$
$\Gamma_{T4}(^3E_1, B_1E)$
$\Gamma_{T5}(^3A_2, B_1B_2)$
$\Gamma_{T1}(^3A_2, B_1B_2)$
$\Gamma_{T5}(^3E_1, B_1E)$
$\Gamma_{T2}(^3E_1, B_1E)$
$\Gamma_{T1}(^3E_1, B_1E)$
$\Gamma_{T5}(^1E_1, A_1E)$
$\Gamma_{T4}(^1B_2, A_1B_2)$
$\Gamma_{T1}(^3E_1, A_1E)$
$\Gamma_{T2}(^3E_1, A_1E)$
$\Gamma_{T5}(^3E_1, A_1E)$
$\Gamma_{T3}(^3E_1, A_1E)$
$\Gamma_{T4}(^3E_1, A_1E)$

$\Gamma_{T3}(^1B_1, A_1B_1)$

$\Gamma_{T5}(^3B_2, A_1B_2)$
$\Gamma_{T3}(^3B_2, A_1B_2)$

$\Gamma_{T4}(^3B_1, A_1B_1)$
$\Gamma_{T5}(^3B_1, A_1B_1)$

$\Gamma_{T1}(^1A_1, A_1A_1)$

Right side labels:
$\Gamma_{T1}(^1A_1, A_1A_1)$
$\Gamma_{T4}(^1B_2, A_1B_2)$
$\Gamma_{T5}(^1E, A_1E)$
$\Gamma_{T1}(^1A_1, B_2B_2)$
$\Gamma_{T3}(^1B_1, A_1B_1)$
$\Gamma_{T2}(^3E, A_1E)$
$\Gamma_{T1}(^3E, A_1E)$
$\Gamma_{T5}(^3E, A_1E)$
$\Gamma_{T4}(^3E, A_1E)$
$\Gamma_{T3}(^3E, A_1E)$
$\Gamma_{T5}(^3B_2, A_1B_2)$
$\Gamma_{T3}(^3B_2, A_1B_2)$
$\Gamma_{T1}(^1A_1, B_2B_2)$
$\Gamma_{T5}(^1E, B_2E)$
$\Gamma_{T4}(^3B_1, A_1B_1)$
$\Gamma_{T5}(^3B_1, A_1B_1)$
$\Gamma_{T2}(^1A_2, B_1B_2)$
$\Gamma_{T1}(^3A_2, B_1B_2)$
$\Gamma_{T5}(^3A_2, B_1B_2)$
$\Gamma_{T4}(^1B_2, E\ E)$
$\Gamma_{T3}(^1B_1, E\ E)$
$\Gamma_{T5}(^1E, B_1E)$
$\Gamma_{T1}(^1A_1, B_1B_1)$
$\Gamma_{T4}(^3E, B_2E)$
$\Gamma_{T1}(^3E, B_2E)$
$\Gamma_{T3}(^3E, B_2E)$
$\Gamma_{T2}(^3E, B_2E)$
$\Gamma_{T5}(^3E, B_2E)$
$\Gamma_{T5}(^3A_2, E\ E)$
$\Gamma_{T1}(^3A_2, E\ E)$
$\Gamma_{T3}(^3E, B_1E)$
$\Gamma_{T4}(^3E, B_1E)$
$\Gamma_{T5}(^3E, B_1E)$
$\Gamma_{T2}(^3E, B_1E)$
$\Gamma_{T1}(^3E, B_1E)$

Bottom axis labels: $^1\Gamma_{T1}$ $^3\Gamma_{T5}$ $^3\Gamma_{T5}$ $^3\Gamma_{T4}$ $^3\Gamma_{T4}$ $^3\Gamma_{T1}$

X-axis: -3000 -2000 -1000 0 1000 DT CM⁻¹ 2000 3000

8 D ELECTRONS, TETRAGONAL SYMMETRY   B=905 C=4×B DQ=850 K=1   DS=K×DT ZETA=650

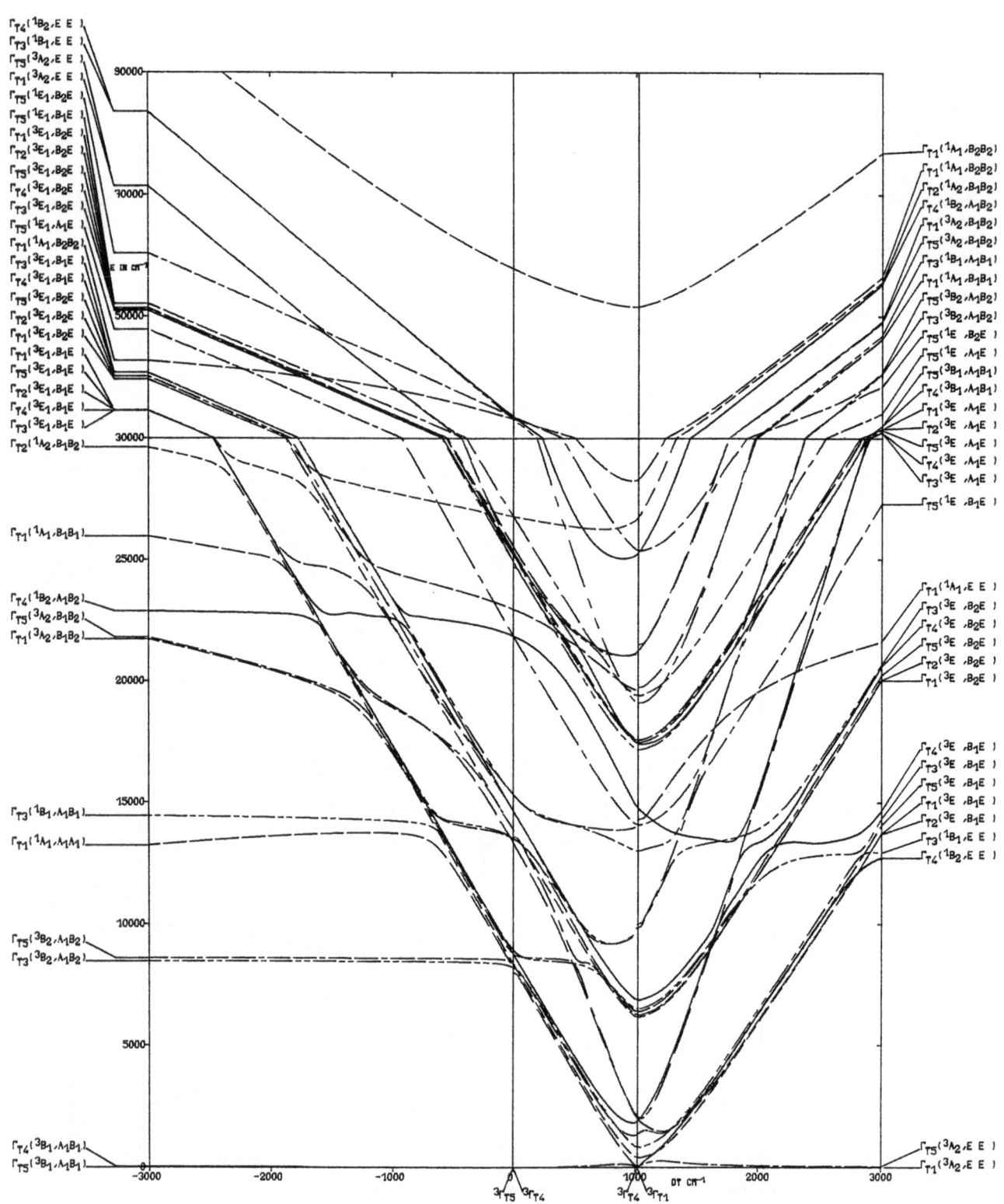

$\Gamma_{T4}(^1B_2, E\ E\ )$
$\Gamma_{T4}(^1B_1, E\ E\ )$
$\Gamma_{T5}(^3A_2, E\ E\ )$
$\Gamma_{T1}(^3A_2, E\ E\ )$
$\Gamma_{T5}(^1E_1, B_2E)$
$\Gamma_{T5}(^1E_1, B_1E)$
$\Gamma_{T1}(^3E_1, B_2E)$
$\Gamma_{T2}(^3E_1, B_2E)$
$\Gamma_{T3}(^3E_1, B_2E)$
$\Gamma_{T4}(^3E_1, B_2E)$
$\Gamma_{T3}(^3E_1, B_2E)$
$\Gamma_{T5}(^1E_1, A_1E)$
$\Gamma_{T1}(^1A_1, B_2B_2)$
$\Gamma_{T3}(^3E_1, B_1E)$
$\Gamma_{T4}(^3E_1, B_1E)$
$\Gamma_{T5}(^3E_1, B_2E)$
$\Gamma_{T2}(^3E_1, B_2E)$
$\Gamma_{T1}(^3E_1, B_2E)$
$\Gamma_{T1}(^3E_1, B_1E)$
$\Gamma_{T5}(^3E_1, B_1E)$
$\Gamma_{T2}(^3E_1, B_1E)$
$\Gamma_{T4}(^3E_1, B_1E)$
$\Gamma_{T3}(^3E_1, B_1E)$
$\Gamma_{T2}(^1A_2, B_1B_2)$

$\Gamma_{T1}(^1A_1, B_1B_1)$

$\Gamma_{T4}(^1B_2, A_1B_2)$
$\Gamma_{T5}(^3A_2, B_1B_2)$
$\Gamma_{T1}(^3A_2, B_1B_2)$

$\Gamma_{T3}(^1B_1, A_1B_1)$

$\Gamma_{T1}(^1A_1, A_1A_1)$

$\Gamma_{T5}(^3B_2, A_1B_2)$
$\Gamma_{T3}(^3B_2, A_1B_2)$

$\Gamma_{T4}(^3B_1, A_1B_1)$
$\Gamma_{T5}(^3B_1, A_1B_1)$

$\Gamma_{T1}(^1A_1, B_2B_2)$
$\Gamma_{T1}(^1A_1, B_2B_2)$
$\Gamma_{T2}(^1A_2, B_1B_2)$
$\Gamma_{T4}(^1B_2, A_1B_2)$
$\Gamma_{T1}(^3A_2, B_1B_2)$
$\Gamma_{T5}(^3A_2, B_1B_2)$
$\Gamma_{T3}(^1B_1, A_1B_1)$
$\Gamma_{T1}(^1A_1, A_1B_1)$
$\Gamma_{T5}(^3B_2, A_1B_2)$
$\Gamma_{T3}(^3B_2, A_1B_2)$
$\Gamma_{T5}(^1E, B_2E\ )$
$\Gamma_{T5}(^1E, A_1E\ )$
$\Gamma_{T5}(^3B_1, A_1B_1)$
$\Gamma_{T4}(^3B_1, A_1B_1)$
$\Gamma_{T1}(^3E, A_1E\ )$
$\Gamma_{T2}(^3E, A_1E\ )$
$\Gamma_{T4}(^3E, A_1E\ )$
$\Gamma_{T5}(^3E, A_1E\ )$
$\Gamma_{T4}(^3E, A_1E\ )$
$\Gamma_{T3}(^3E, A_1E\ )$
$\Gamma_{T5}(^1E, B_1E\ )$

$\Gamma_{T1}(^1A_1, E\ E\ )$
$\Gamma_{T3}(^3E, B_2E\ )$
$\Gamma_{T4}(^3E, B_2E\ )$
$\Gamma_{T5}(^3E, B_2E\ )$
$\Gamma_{T2}(^3E, B_2E\ )$
$\Gamma_{T1}(^3E, B_2E\ )$

$\Gamma_{T4}(^3E, B_1E\ )$
$\Gamma_{T3}(^3E, B_1E\ )$
$\Gamma_{T5}(^3E, B_1E\ )$
$\Gamma_{T1}(^3E, B_1E\ )$
$\Gamma_{T2}(^3E, B_1E\ )$
$\Gamma_{T3}(^1B_1, E\ E\ )$
$\Gamma_{T4}(^1B_2, E\ E\ )$

$\Gamma_{T5}(^3A_2, E\ E\ )$
$\Gamma_{T1}(^3A_2, E\ E\ )$

E IN CM^{-1}

90000
70000
50000
30000
25000
20000
15000
10000
5000

$^3\Gamma_{T5}$ $^3\Gamma_{T4}$ $^3\Gamma_{T4}$ $^3\Gamma_{T1}$

-3000   -2000   -1000   1000   DT CM^{-1} 2000   3000

422

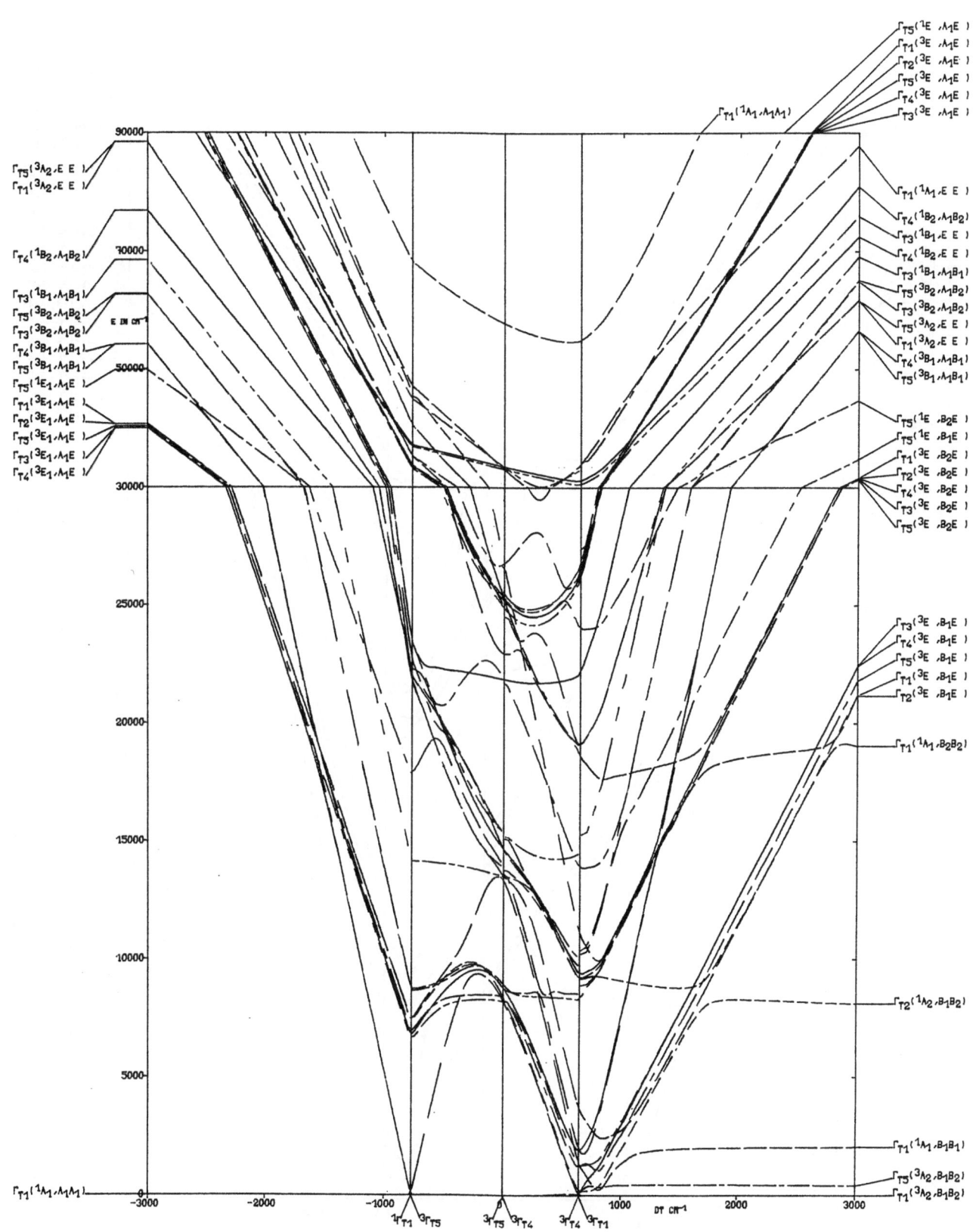

$\Gamma_{T5}(^1E , \wedge_1 E)$
$\Gamma_{T1}(^3E , \wedge_1 E)$
$\Gamma_{T2}(^3E , \wedge_1 E)$
$\Gamma_{T5}(^3E , \wedge_1 E)$
$\Gamma_{T4}(^3E , \wedge_1 E)$
$\Gamma_{T3}(^3E , \wedge_1 E)$

$\Gamma_{T1}(^1A_1 , \wedge_1 A_1)$

$\Gamma_{T5}(^3A_2 , E E)$
$\Gamma_{T1}(^3A_2 , E E)$

$\Gamma_{T1}(^1A_1 , E E)$
$\Gamma_{T4}(^1B_2 , \wedge_1 B_2)$
$\Gamma_{T3}(^1B_1 , E E)$
$\Gamma_{T4}(^1B_2 , E E)$
$\Gamma_{T3}(^1B_1 , \wedge_1 B_1)$
$\Gamma_{T5}(^3B_2 , \wedge_1 B_2)$
$\Gamma_{T5}(^3B_2 , \wedge_1 B_2)$
$\Gamma_{T3}(^3A_2 , E E)$
$\Gamma_{T1}(^3A_2 , E E)$
$\Gamma_{T4}(^3B_1 , \wedge_1 B_1)$
$\Gamma_{T5}(^3B_1 , \wedge_1 B_1)$

$\Gamma_{T4}(^1B_2 , \wedge_1 B_2)$

$\Gamma_{T3}(^1B_1 , \wedge_1 B_1)$
$\Gamma_{T5}(^3B_2 , \wedge_1 B_2)$
$\Gamma_{T3}(^3B_2 , \wedge_1 B_2)$
$\Gamma_{T4}(^3B_1 , \wedge_1 B_1)$
$\Gamma_{T5}(^3B_1 , \wedge_1 B_1)$
$\Gamma_{T5}(^1E_1 , \wedge_1 E)$

$\Gamma_{T5}(^1E , B_2 E)$
$\Gamma_{T5}(^1E , B_1 E)$
$\Gamma_{T1}(^3E , B_2 E)$
$\Gamma_{T2}(^3E , B_2 E)$
$\Gamma_{T4}(^3E , B_2 E)$
$\Gamma_{T3}(^3E , B_2 E)$
$\Gamma_{T5}(^3E , B_2 E)$

$\Gamma_{T1}(^3E_1 , \wedge_1 E)$
$\Gamma_{T2}(^3E_1 , \wedge_1 E)$
$\Gamma_{T3}(^3E_1 , \wedge_1 E)$
$\Gamma_{T4}(^3E_1 , \wedge_1 E)$

E IN CM^{-1}

$\Gamma_{T3}(^3E , B_1 E)$
$\Gamma_{T4}(^3E , B_1 E)$
$\Gamma_{T5}(^3E , B_1 E)$
$\Gamma_{T1}(^3E , B_1 E)$
$\Gamma_{T2}(^3E , B_1 E)$

$\Gamma_{T1}(^1A_1 , B_2 B_2)$

$\Gamma_{T2}(^1A_2 , B_1 B_2)$

$\Gamma_{T1}(^1A_1 , B_1 B_1)$

$\Gamma_{T1}(^1A_1 , \wedge_1 A_1)$

$\Gamma_{T5}(^3A_2 , B_1 B_2)$
$\Gamma_{T1}(^3A_2 , B_1 B_2)$

$1\Gamma_{T1}$ $3\Gamma_{T5}$ $3\Gamma_{T5}$ $3\Gamma_{T4}$ $3\Gamma_{T4}$ $3\Gamma_{T1}$

-3000   -2000   -1000   0   1000   2000   3000   DT CM^{-1}

8 D ELECTRONS, TETRAGONAL SYMMETRY   B=905 C=4*B DQ=850 K=5   DS=K*DT ZETA=650

423

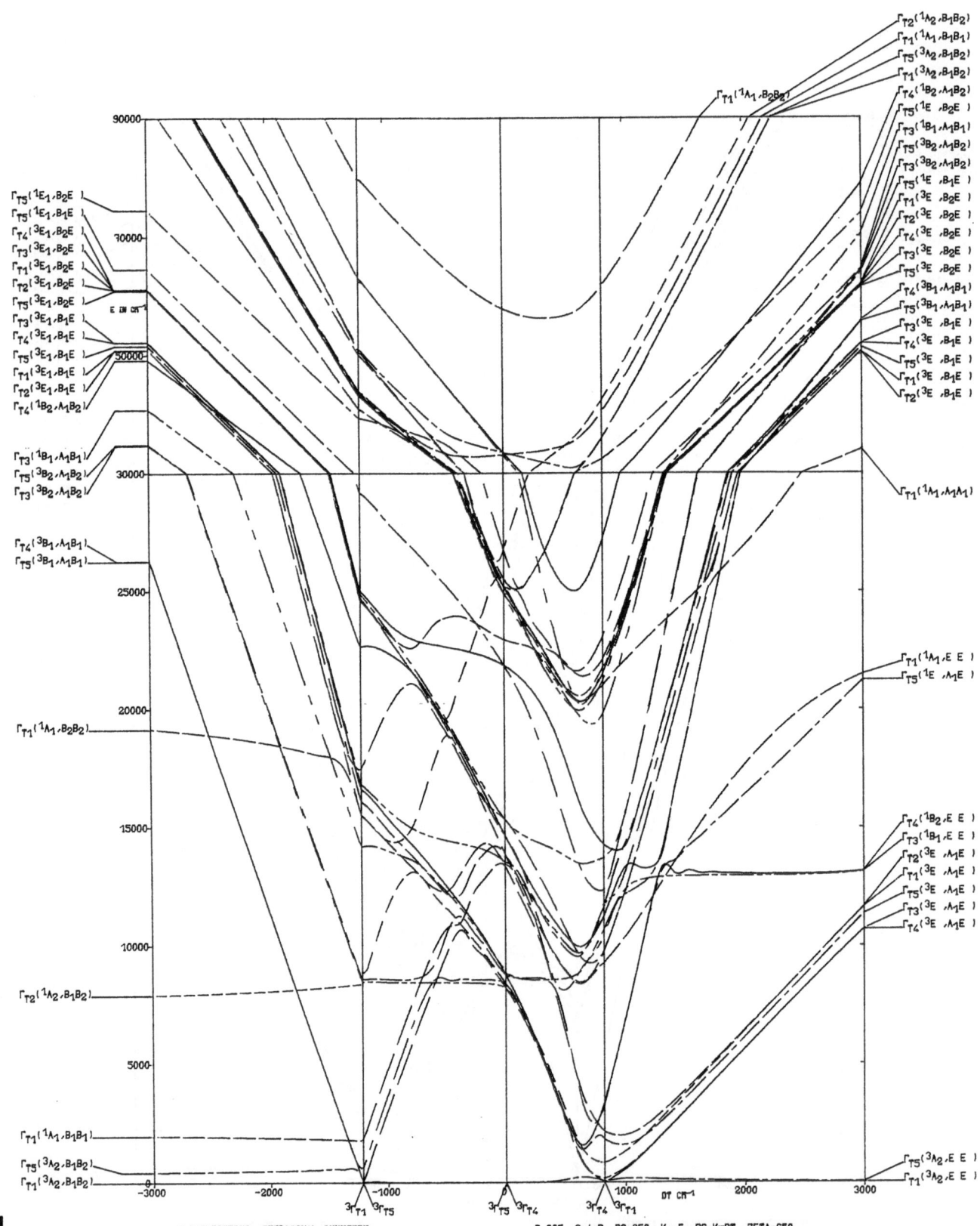

8 D ELECTRONS, TETRAGONAL SYMMETRY   B=905 C=4*B DQ=850 K=-5 DS=K*DT ZETA=650

8 D ELECTRONS, TRIGONAL SYMMETRY

B=905   C=4×B   ZETA=650

ENERGY AS FUNCTION OF DQ
DT=0, 500, -500, 1000, -1000
K=1, -1, 5, -5

ENERGY AS FUNCTION OF DT
DQ=850
K=1, -1, 5, -5

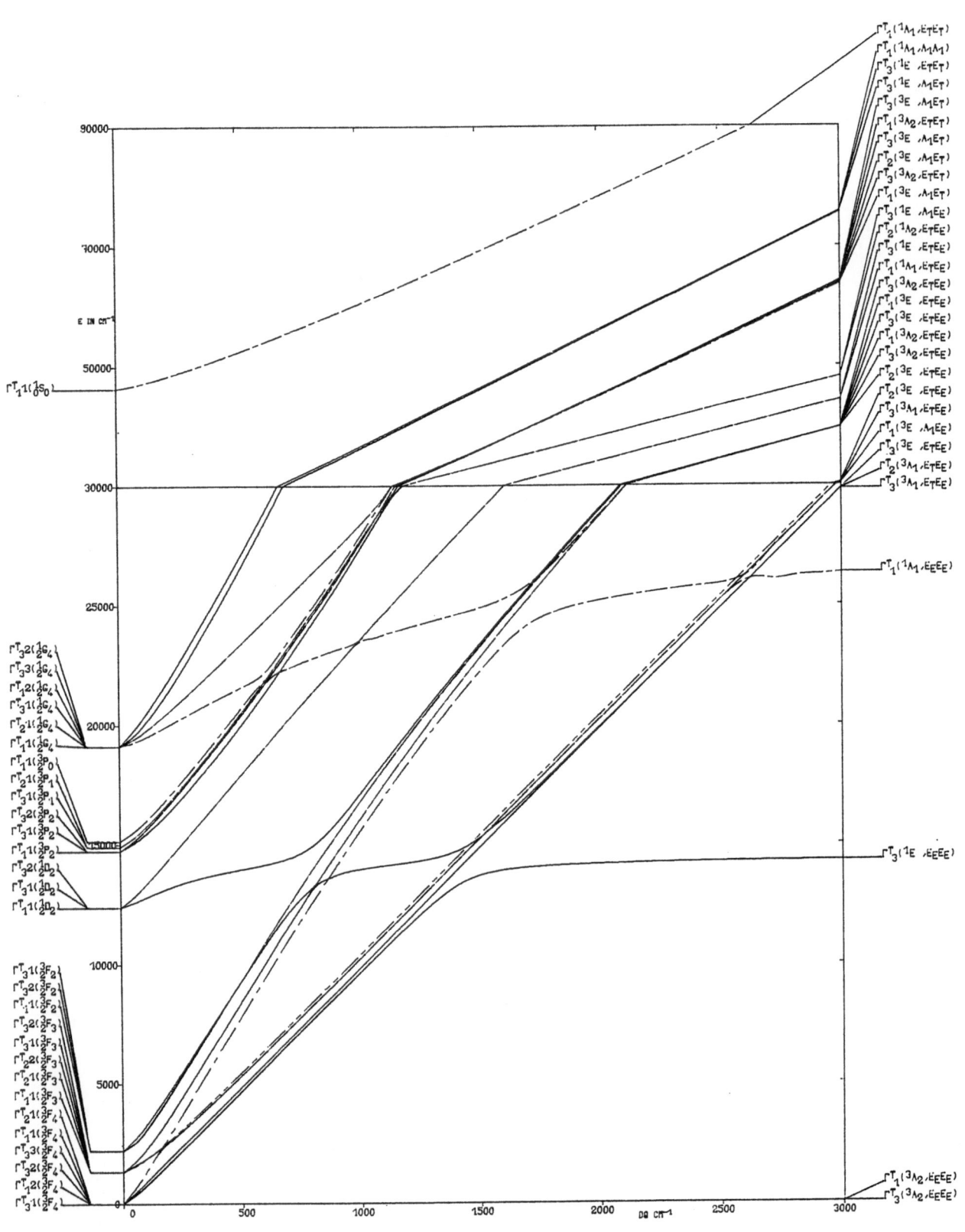

8 D ELECTRONS, TRIGONAL SYMMETRY   B=905 C=4×B DT=0 K=0 DS=K×DT ZETA=650

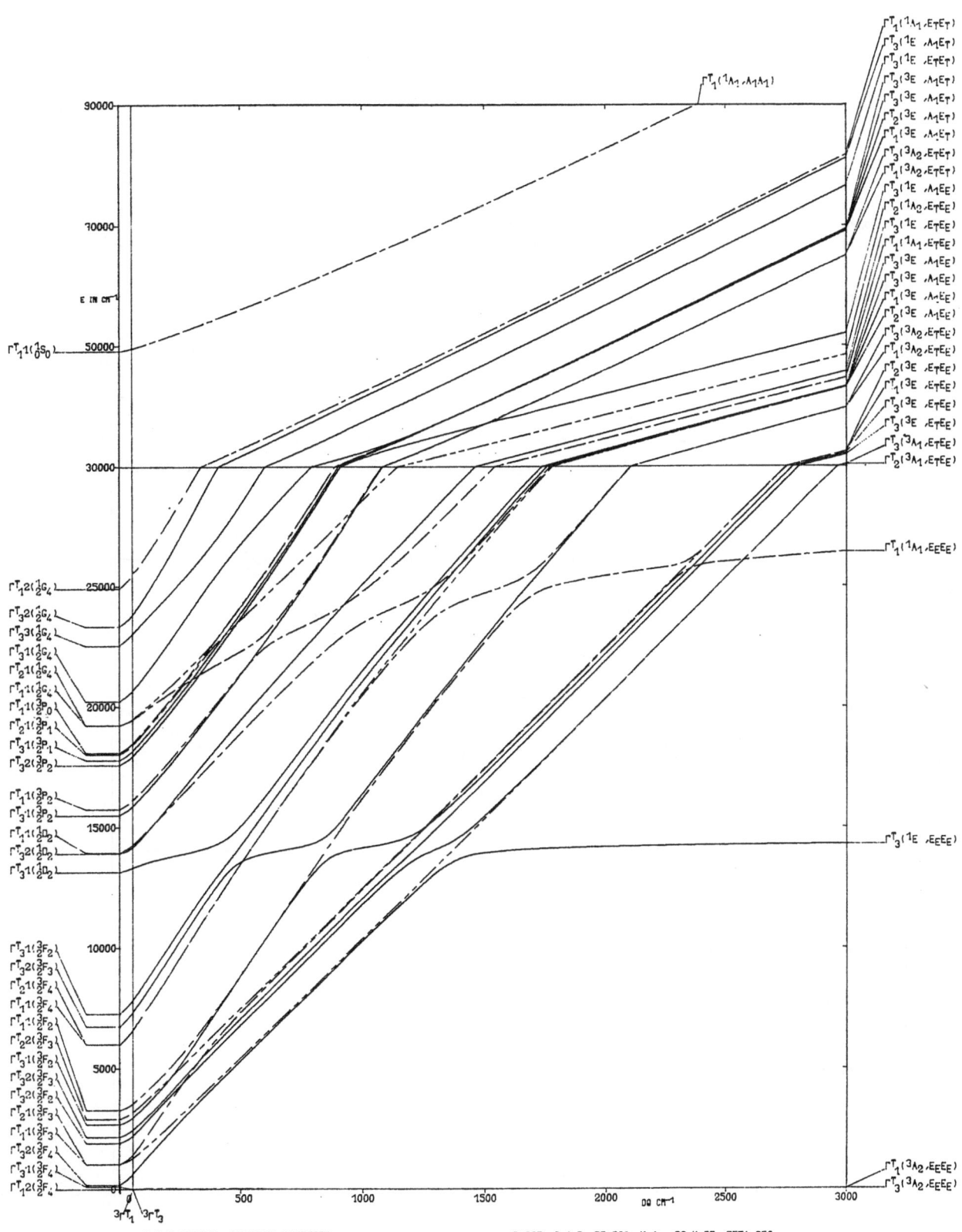

8 D ELECTRONS, TRIGONAL SYMMETRY   B=905 C=4×B DT=500 K=1 DS=K×DT ZETA=650

427

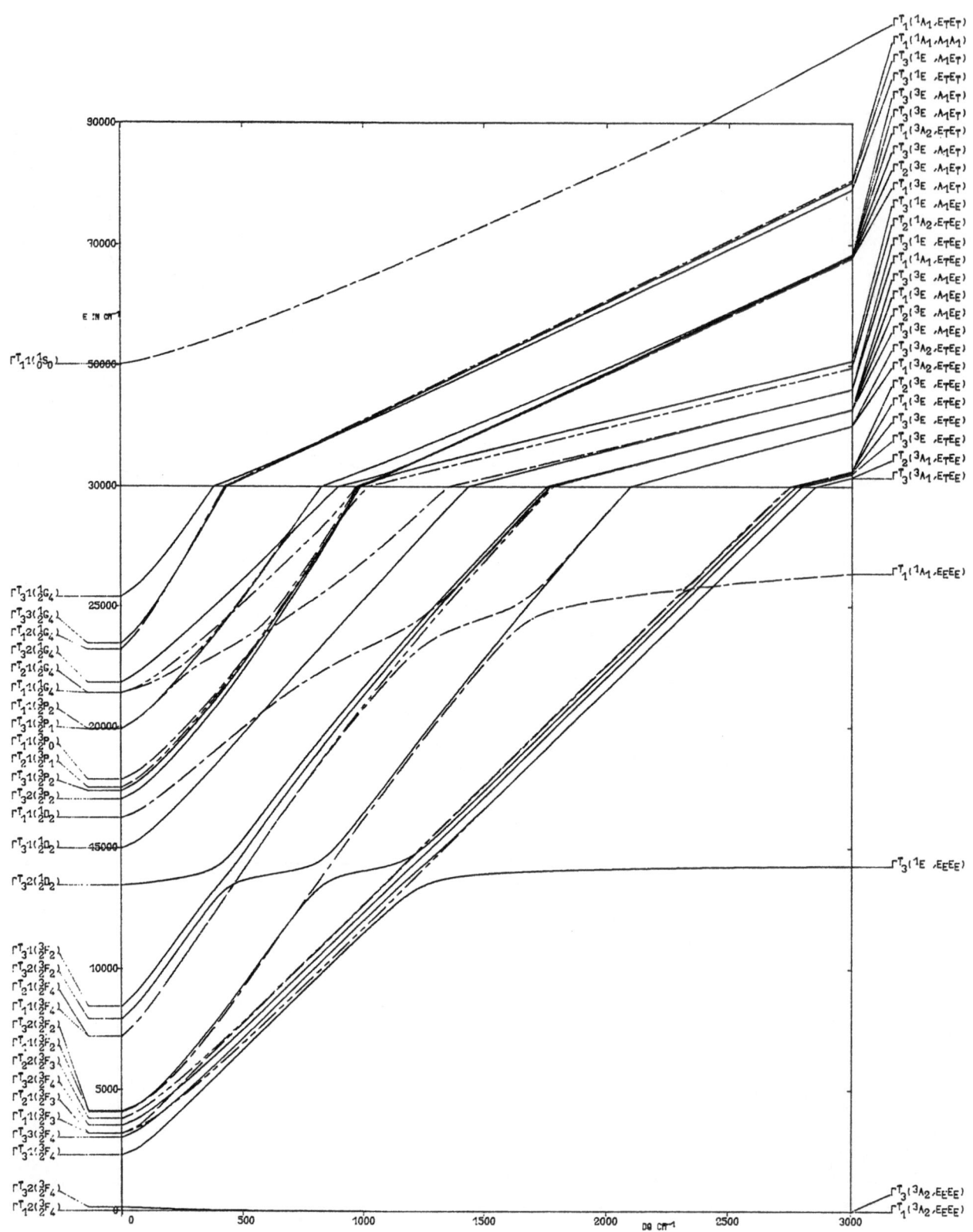

8 D ELECTRONS, TRIGONAL SYMMETRY

B=905 C=4×B DT=500 K=-1 DS=K×DT ZETA=650

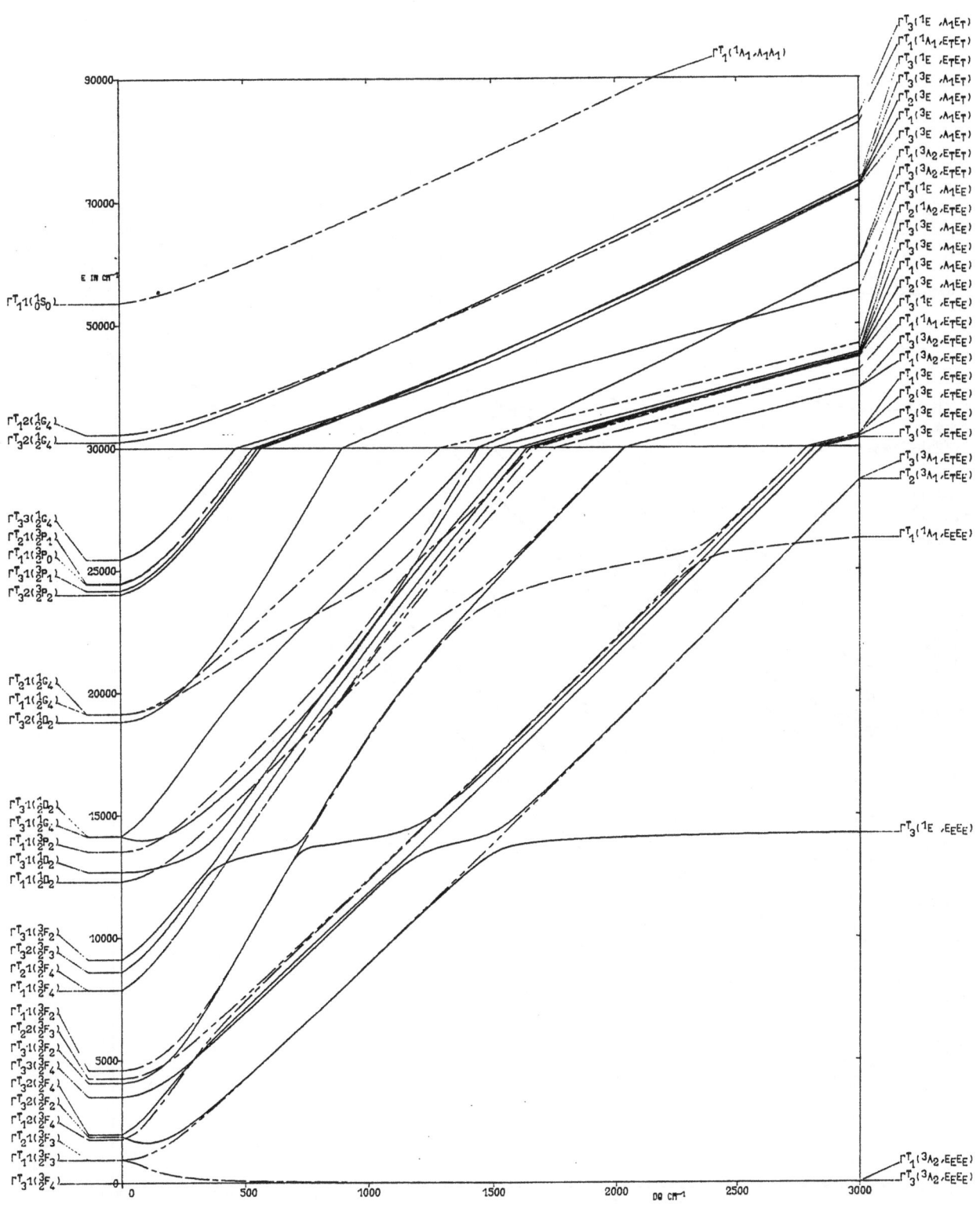

8 D ELECTRONS, TRIGONAL SYMMETRY

B=905 C=4×B DT=500 K=5 DS=K×DT ZETA=650

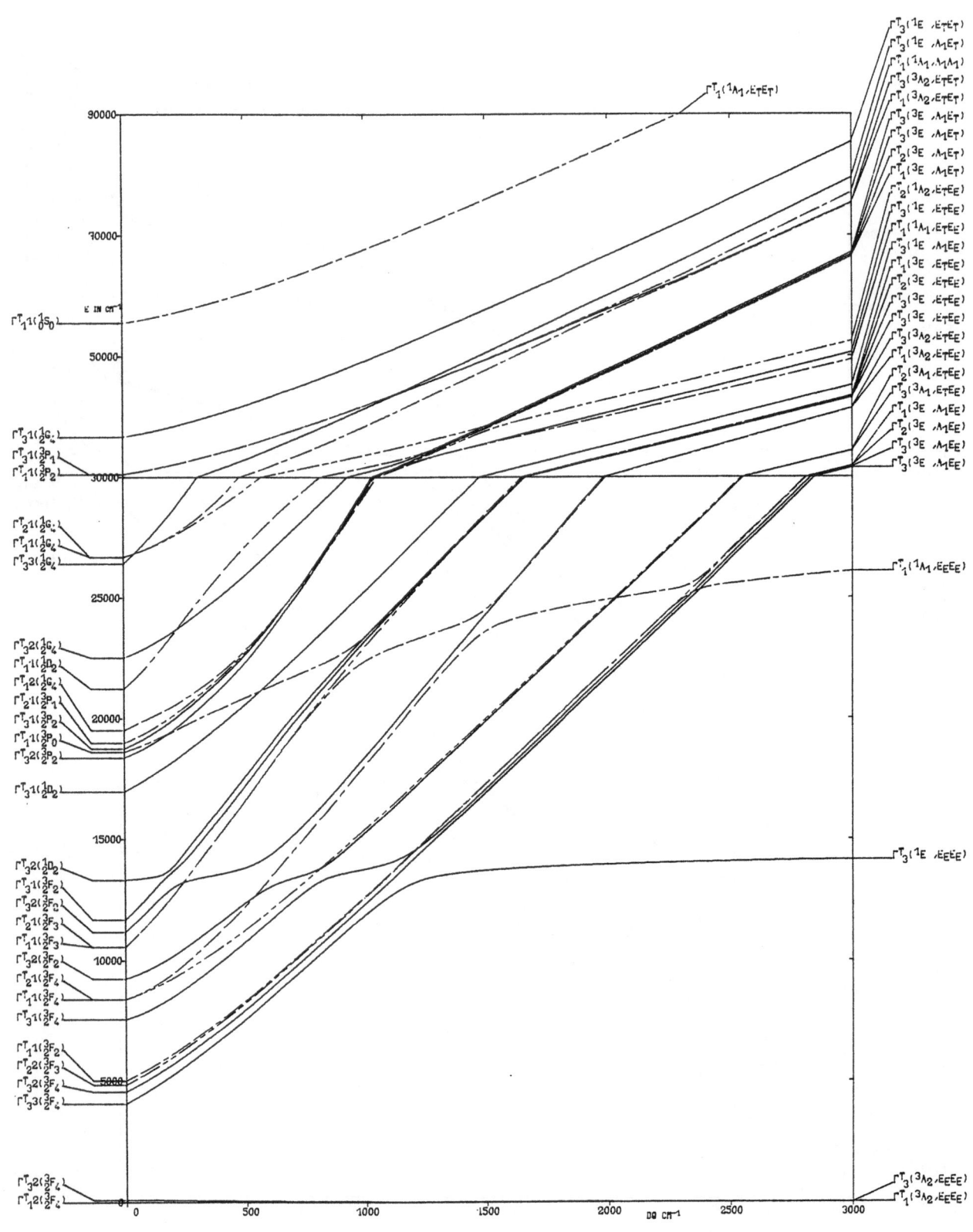

8 D ELECTRONS, TRIGONAL SYMMETRY

B=905   C=4×B   DT=500   K=-5   DS=K×DT   ZETA=650

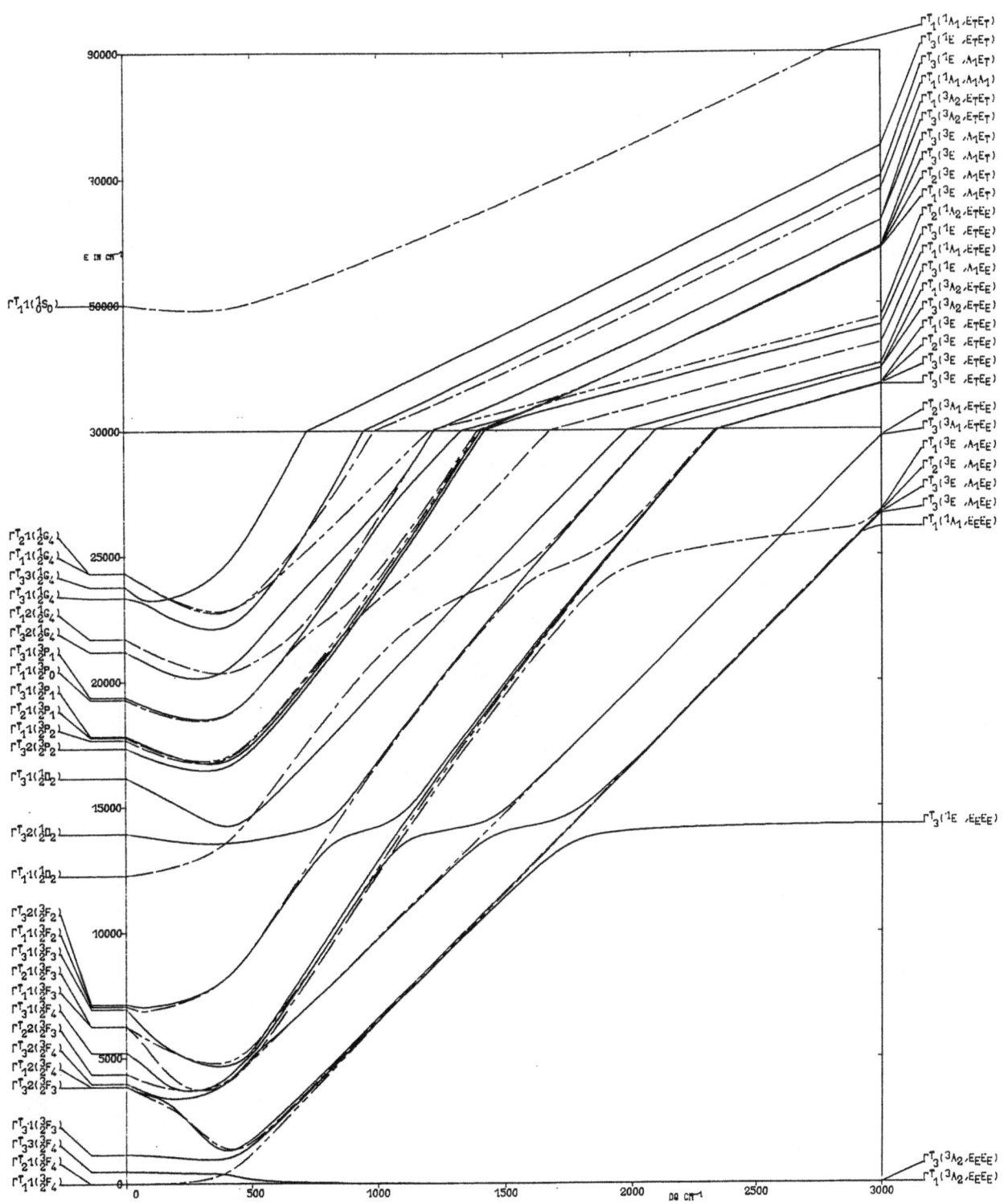

8 D ELECTRONS, TRIGONAL SYMMETRY   B=905   C=4﹡B   DT=-500   K=1   DS=K﹡DT   ZETA=650

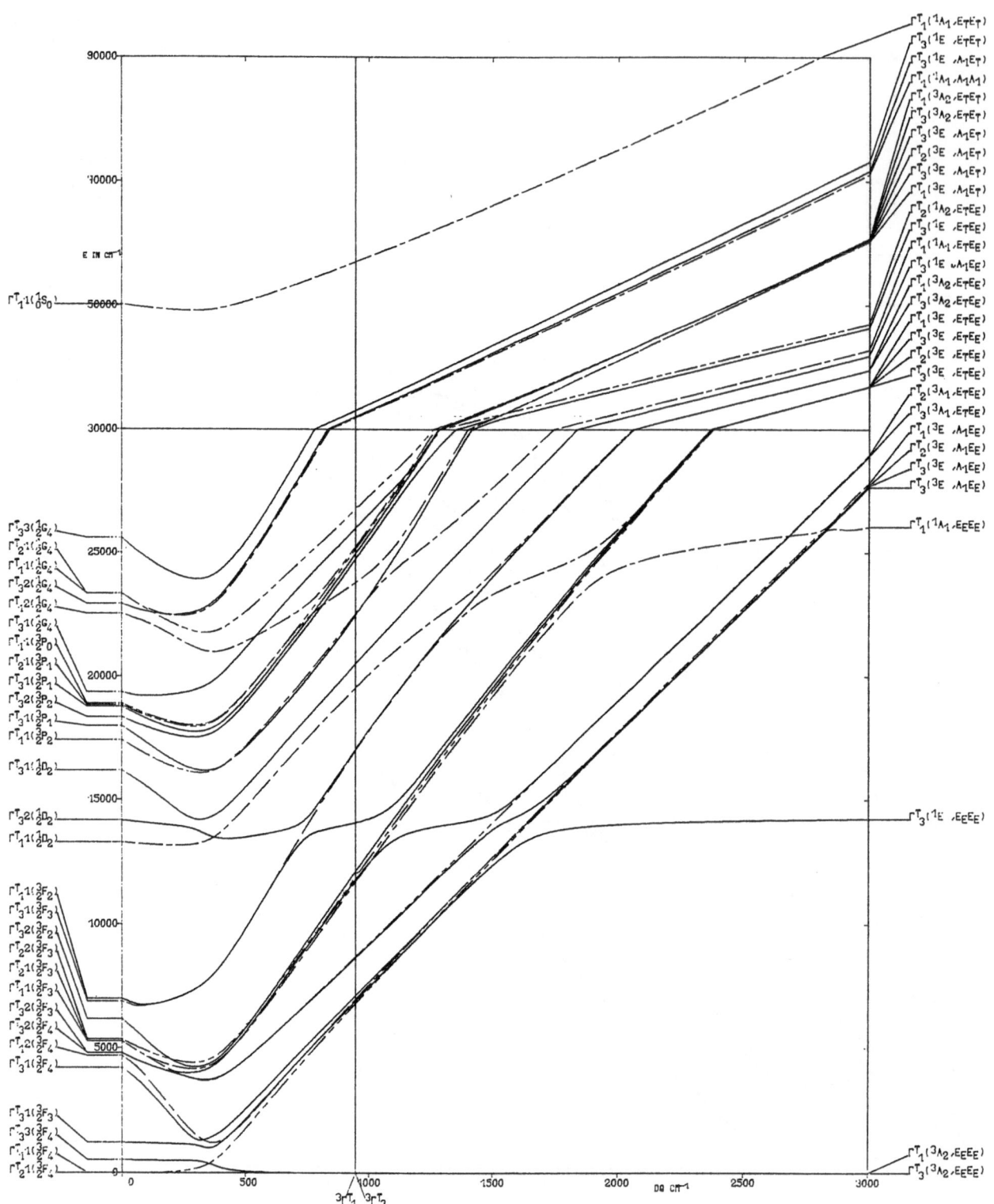

90000

70000

E IN CM⁻¹

$\Gamma_1^{\tau}1(\frac{1}{0}S_0)$ 50000

30000

$\Gamma_3^{\tau}3(\frac{1}{2}G_4)$ 25000
$\Gamma_2^{\tau}1(\frac{1}{2}G_4)$
$\Gamma_1^{\tau}1(\frac{1}{2}G_4)$
$\Gamma_3^{\tau}2(\frac{1}{2}G_4)$
$\Gamma_2^{\tau}2(\frac{1}{2}G_4)$
$\Gamma_3^{\tau}3(\frac{3}{2}P_0)$
$\Gamma_1^{\tau}1(\frac{3}{2}P_0)$ 20000
$\Gamma_2^{\tau}1(\frac{3}{2}P_1)$
$\Gamma_3^{\tau}2(\frac{3}{2}P_2)$
$\Gamma_3^{\tau}1(\frac{3}{2}P_1)$
$\Gamma_1^{\tau}1(\frac{3}{2}P_2)$
$\Gamma_3^{\tau}1(\frac{1}{2}D_2)$ 15000
$\Gamma_3^{\tau}2(\frac{1}{2}D_2)$
$\Gamma_1^{\tau}1(\frac{1}{2}D_2)$

$\Gamma_1^{\tau}1(\frac{3}{2}F_2)$
$\Gamma_3^{\tau}1(\frac{3}{2}F_2)$ 10000
$\Gamma_3^{\tau}2(\frac{3}{2}F_2)$
$\Gamma_2^{\tau}2(\frac{3}{2}F_3)$
$\Gamma_2^{\tau}1(\frac{3}{2}F_3)$
$\Gamma_1^{\tau}1(\frac{3}{2}F_3)$
$\Gamma_3^{\tau}2(\frac{3}{2}F_3)$
$\Gamma_3^{\tau}2(\frac{3}{2}F_4)$
$\Gamma_2^{\tau}2(\frac{3}{2}F_4)$ 5000
$\Gamma_3^{\tau}3(\frac{3}{2}F_4)$

$\Gamma_3^{\tau}1(\frac{3}{2}F_3)$
$\Gamma_3^{\tau}3(\frac{3}{2}F_4)$
$\Gamma_1^{\tau}1(\frac{3}{2}F_4)$
$\Gamma_2^{\tau}1(\frac{3}{2}F_4)$ 0

0   500   1000   1500   2000   2500   3000
                    $3\Gamma_1$ $3\Gamma_3$ DQ CM⁻¹

$\Gamma_1^{\tau}(^1A_1,E_TE_T)$
$\Gamma_3^{\tau}(^1E,E_TE_T)$
$\Gamma_3^{\tau}(^1E,A_1E_T)$
$\Gamma_1^{\tau}(^1A_1,A_1A_1)$
$\Gamma_3^{\tau}(^3A_2,E_TE_T)$
$\Gamma_3^{\tau}(^3A_2,E_TE_T)$
$\Gamma_3^{\tau}(^3E,A_1E_T)$
$\Gamma_2^{\tau}(^3E,A_1E_T)$
$\Gamma_3^{\tau}(^3E,A_1E_T)$
$\Gamma_1^{\tau}(^3E,A_1E_T)$
$\Gamma_2^{\tau}(^1A_2,E_TE_E)$
$\Gamma_3^{\tau}(^1E,E_TE_E)$
$\Gamma_1^{\tau}(^1A_1,E_TE_E)$
$\Gamma_3^{\tau}(^1E,A_1E_E)$
$\Gamma_3^{\tau}(^3A_2,E_TE_E)$
$\Gamma_3^{\tau}(^3A_2,E_TE_E)$
$\Gamma_1^{\tau}(^3E,E_TE_E)$
$\Gamma_3^{\tau}(^3E,E_TE_E)$
$\Gamma_2^{\tau}(^3E,E_TE_E)$
$\Gamma_3^{\tau}(^3E,E_TE_E)$
$\Gamma_2^{\tau}(^3A_1,E_TE_E)$
$\Gamma_3^{\tau}(^3A_1,E_TE_E)$
$\Gamma_1^{\tau}(^3E,A_1E_E)$
$\Gamma_2^{\tau}(^3E,A_1E_E)$
$\Gamma_3^{\tau}(^3E,A_1E_E)$

$\Gamma_1^{\tau}(^1A_1,E_EE_E)$

$\Gamma_3^{\tau}(^1E,E_EE_E)$

$\Gamma_1^{\tau}(^3A_2,E_EE_E)$
$\Gamma_3^{\tau}(^3A_2,E_EE_E)$

432

8 D ELECTRONS, TRIGONAL SYMMETRY   B=905 C=4*B DT=-500 K=-1 DS=K*DT ZETA=650

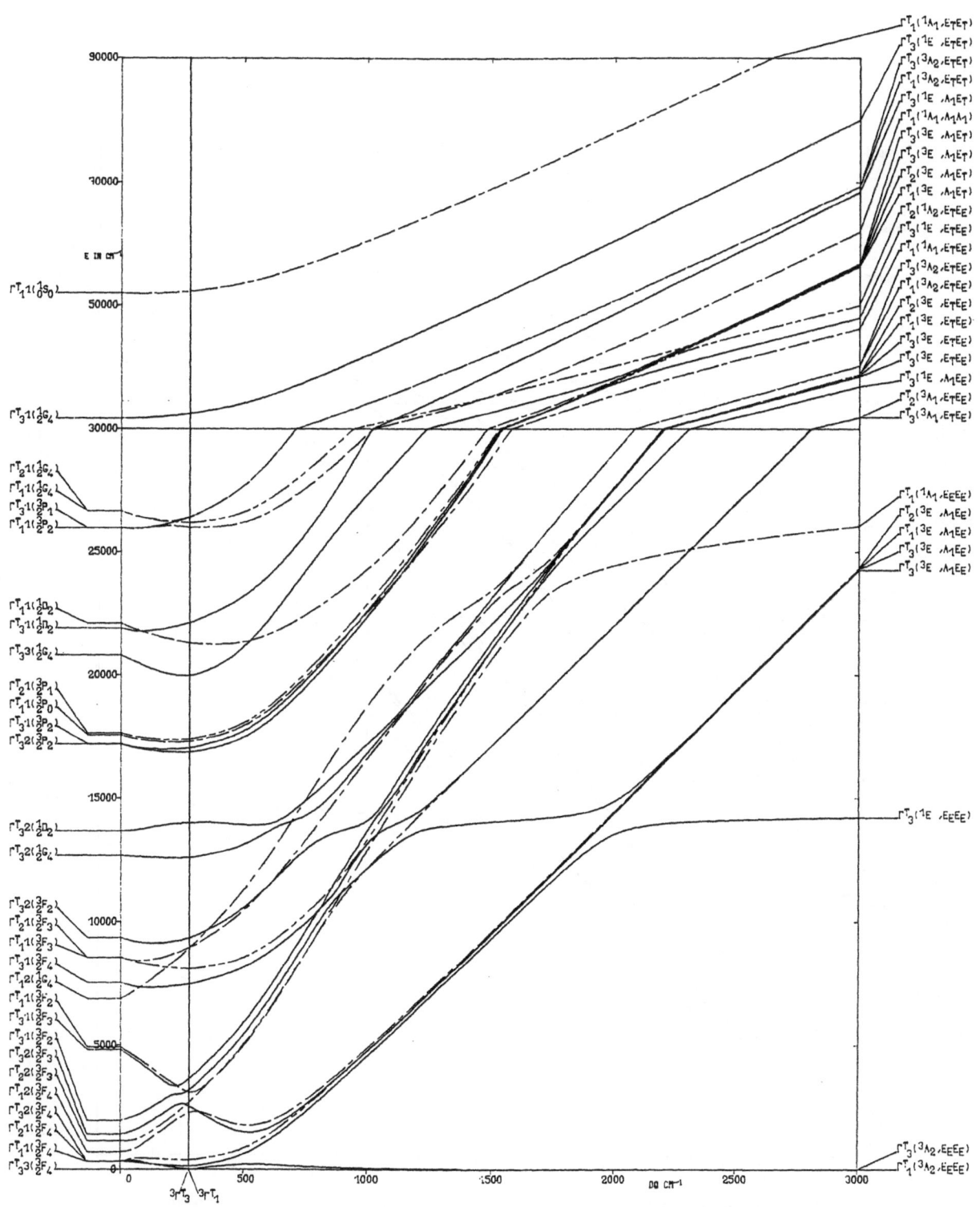

8 D ELECTRONS, TRIGONAL SYMMETRY

B=905   C=4×B   DT=-500   K=5   DS=K×DT   ZETA=650

433

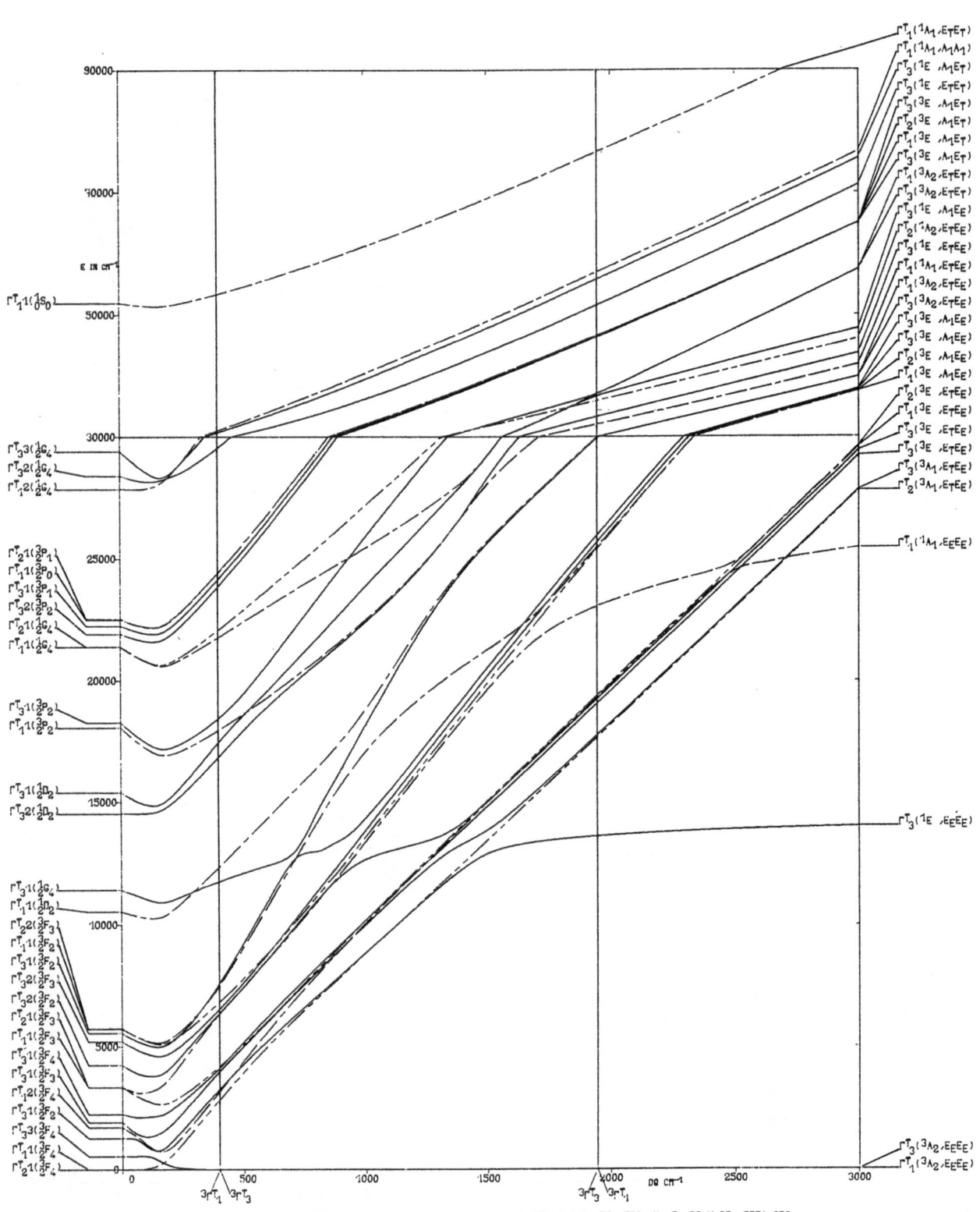

434

8 D ELECTRONS, TRIGONAL SYMMETRY

B=905 C=4×B DT=-500 K=-5 DS=K×DT ZETA=650

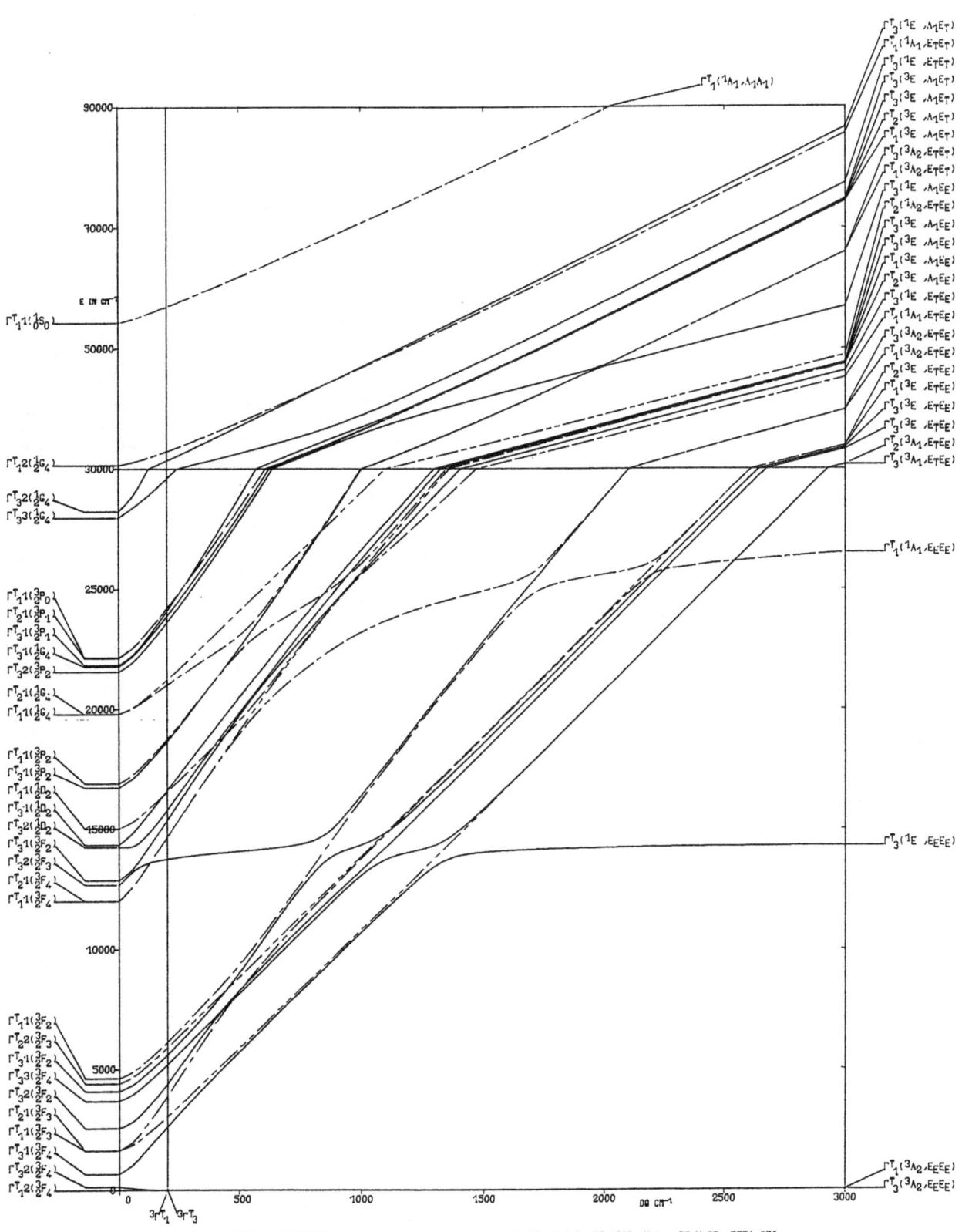

8 D ELECTRONS, TRIGONAL SYMMETRY   B=905 C=4∗B DT=1000 K=1   DS=K∗DT ZETA=650

435

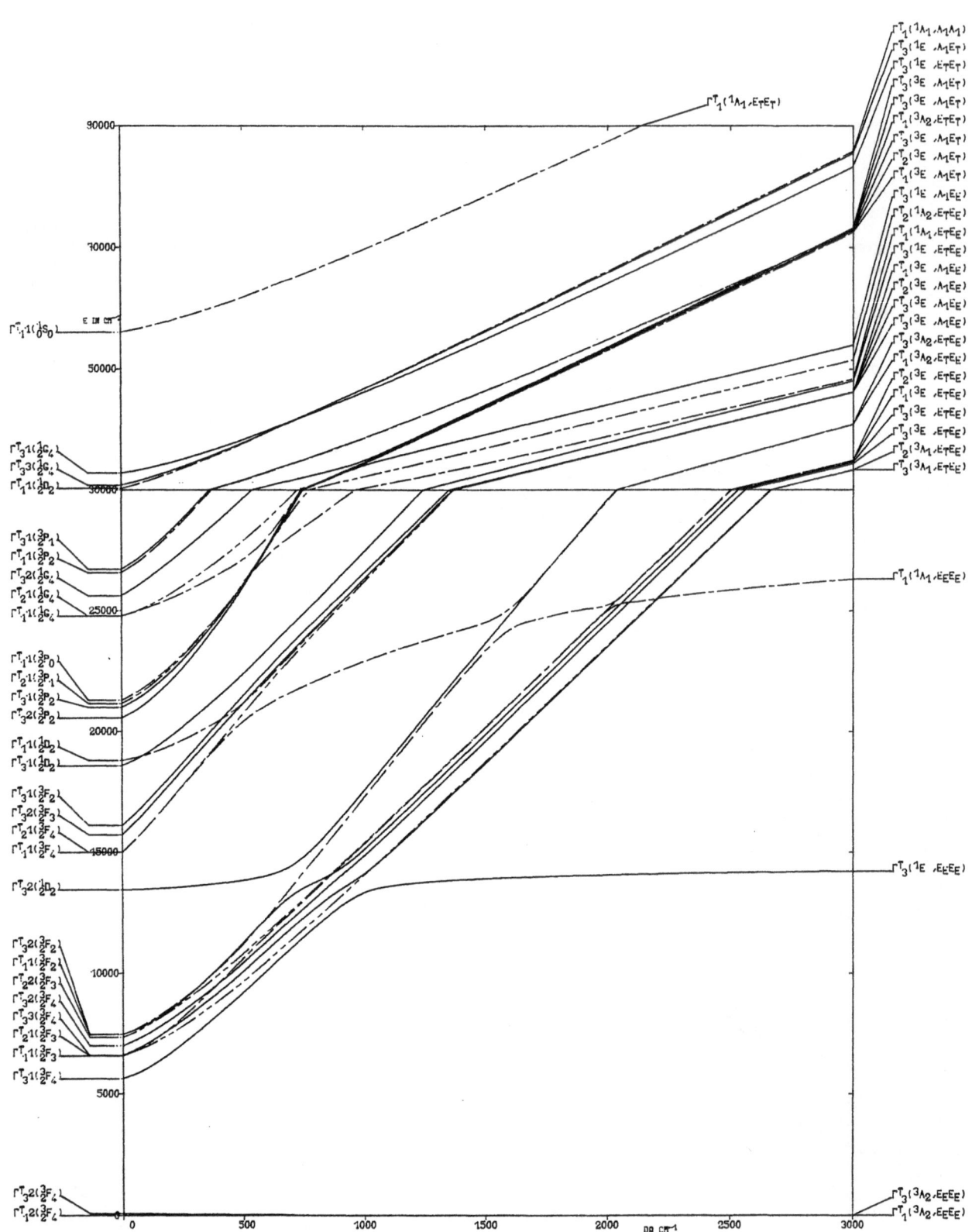

8 D ELECTRONS, TRIGONAL SYMMETRY   B=905   C=4×B   DT=1000   K=-1   DS=K×DT   ZETA=650

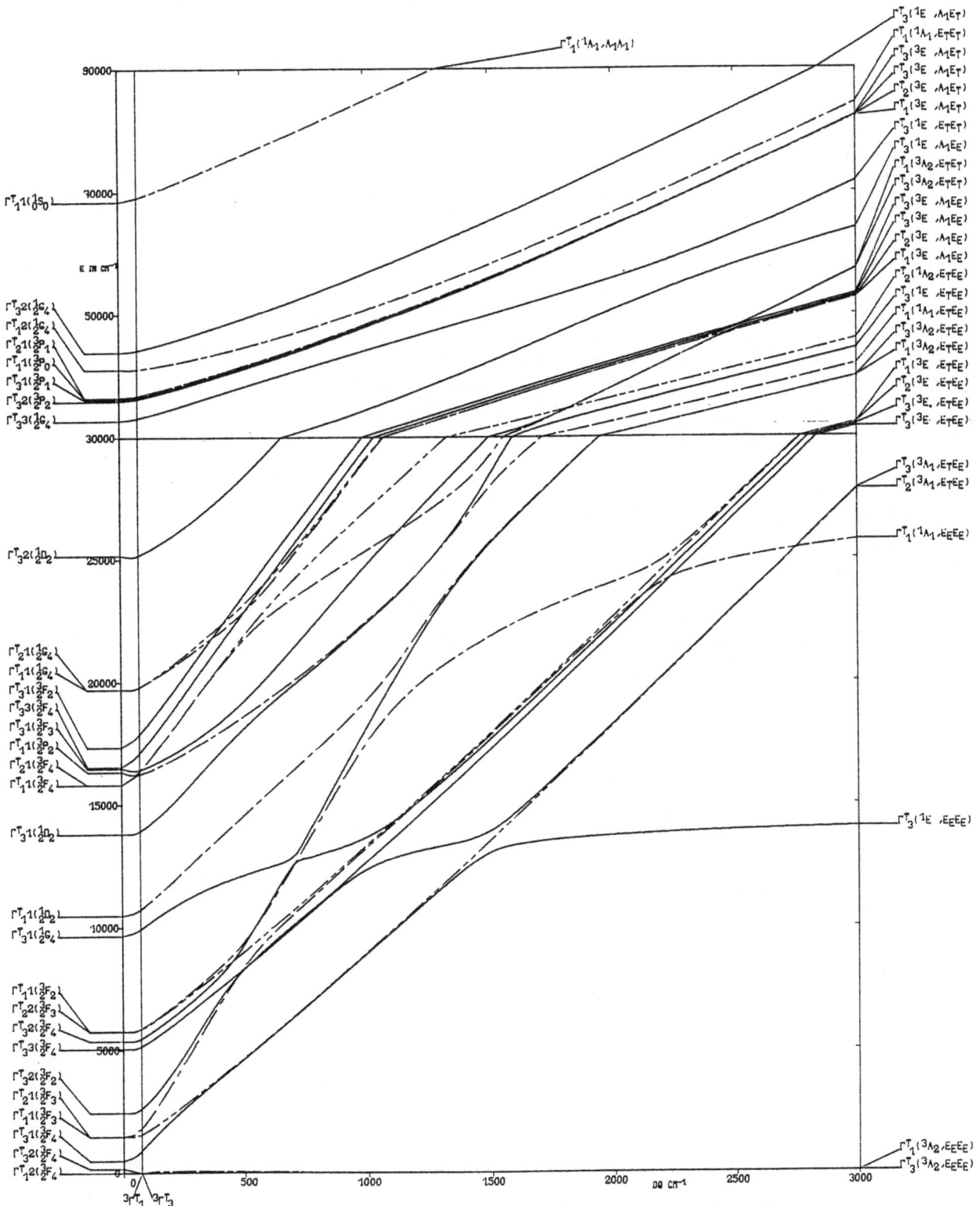

8 D ELECTRONS, TRIGONAL SYMMETRY   B=905 C=4×B DT=1000 K=5   DS=K×DT ZETA=650

437

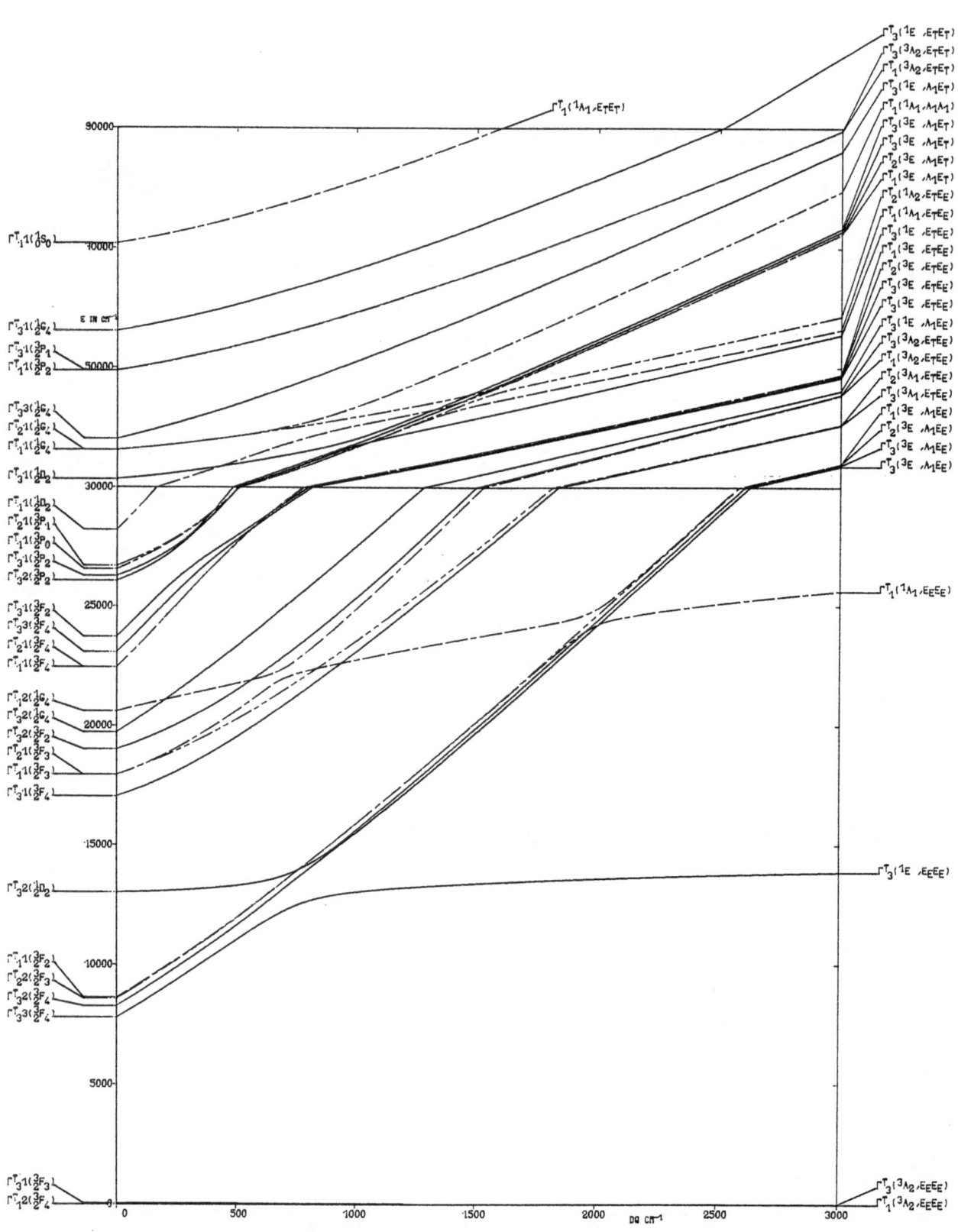

8 D ELECTRONS, TRIGONAL SYMMETRY

B=905   C=4×B   DT=1000   K=-5   DS=K×DT   ZETA=650

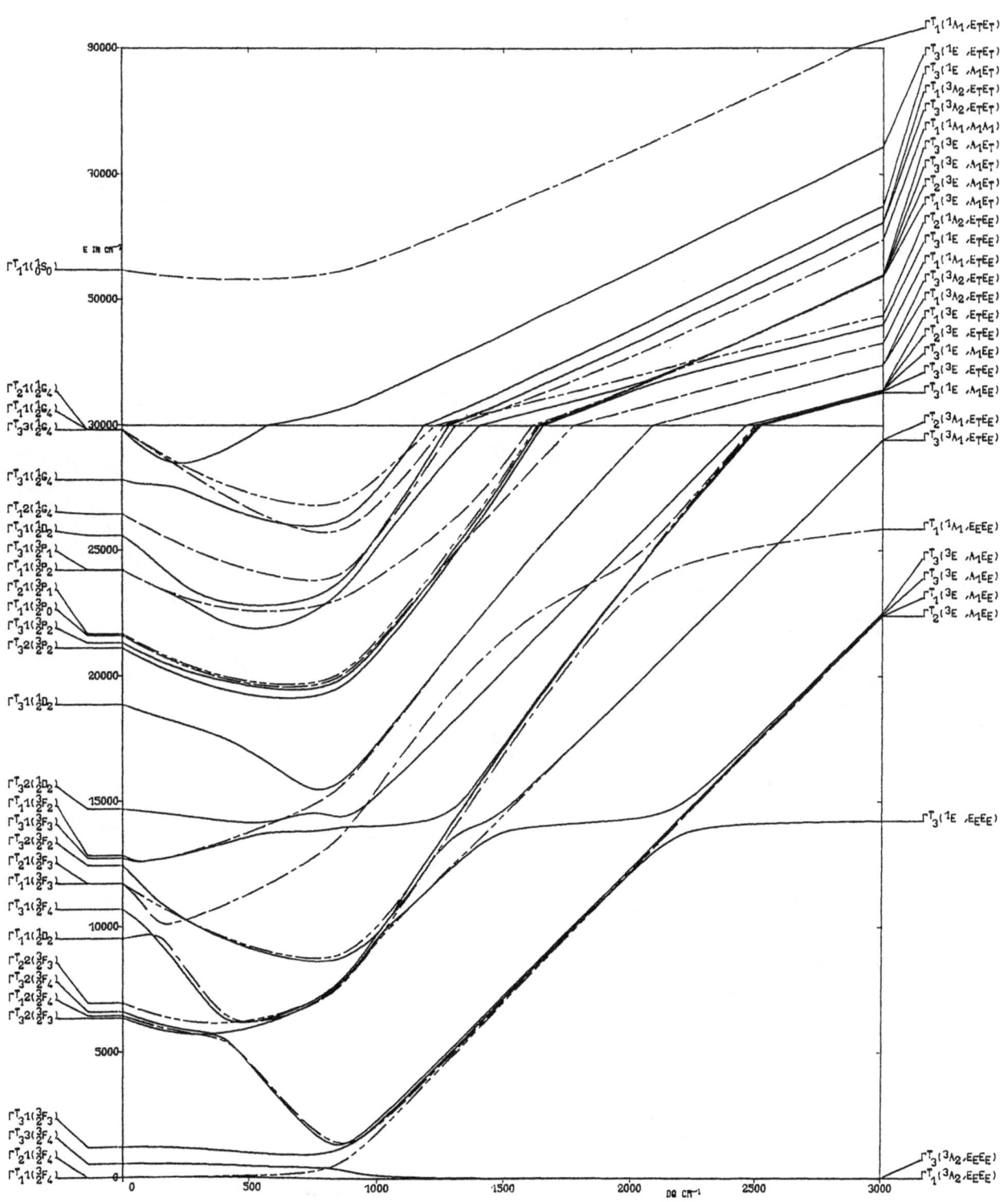

8 D ELECTRONS, TRIGONAL SYMMETRY   B=905 C=4×B DT=-1000 K=1   DS=K×DT ZETA=650

439

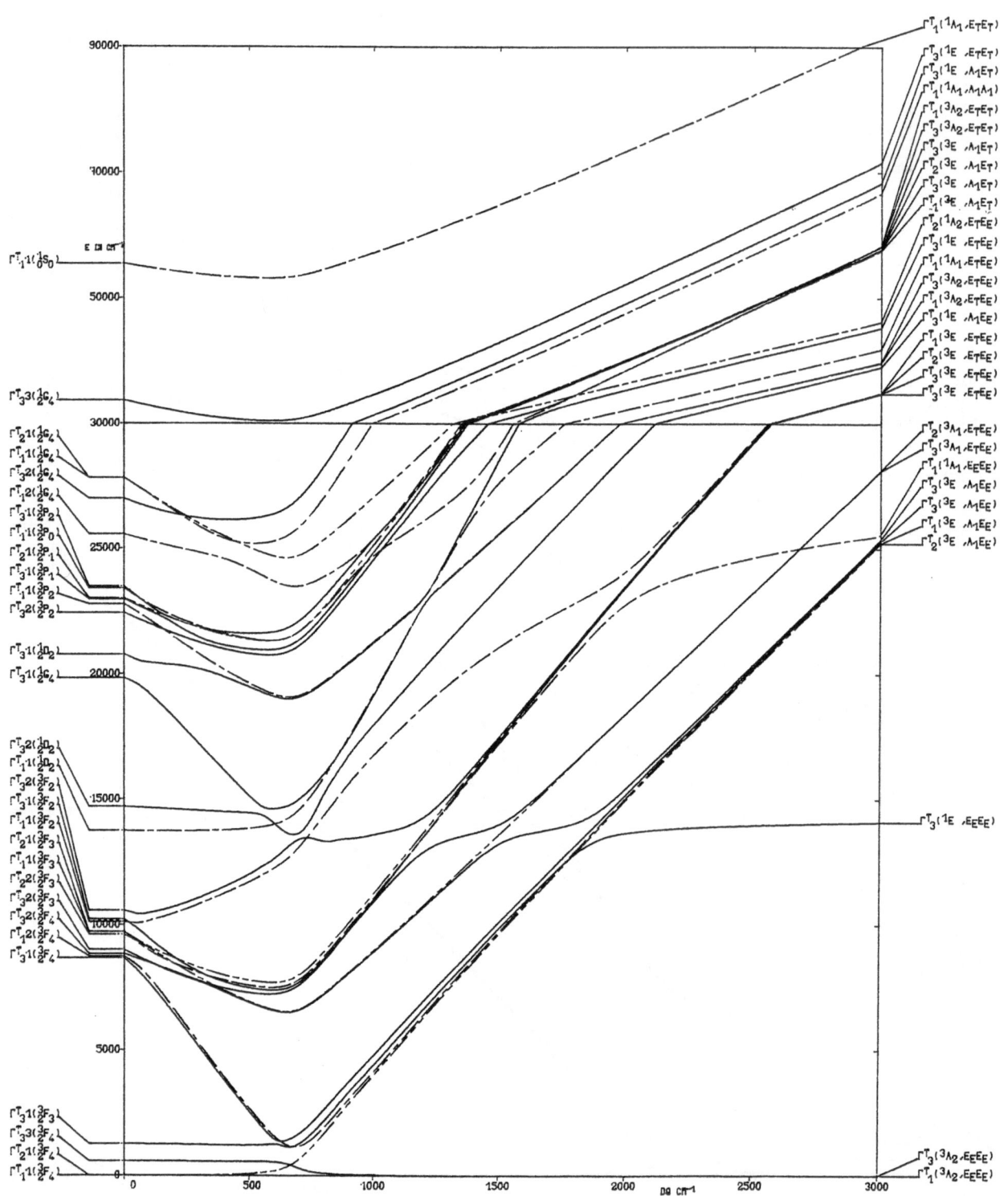

$\Gamma_1^T 1(\frac{1}{0}S_0)$

$\Gamma_3^T 3(\frac{1}{2}G_4)$

$\Gamma_2^T 1(\frac{1}{2}G_4)$
$\Gamma_1^T 1(\frac{1}{2}G_4)$
$\Gamma_3^T 2(\frac{1}{2}G_4)$
$\Gamma_1^T 2(\frac{1}{2}G_4)$
$\Gamma_3^T 1(\frac{3}{2}P_2)$
$\Gamma_1^T 1(\frac{3}{2}P_0)$
$\Gamma_2^T 1(\frac{3}{2}P_1)$
$\Gamma_3^T 1(\frac{3}{2}P_1)$
$\Gamma_1^T 1(\frac{3}{2}P_2)$
$\Gamma_3^T 2(\frac{3}{2}P_2)$

$\Gamma_3^T 1(\frac{1}{2}D_2)$

$\Gamma_3^T 1(\frac{1}{2}G_4)$

$\Gamma_3^T 2(\frac{1}{2}D_2)$
$\Gamma_1^T 1(\frac{1}{2}D_2)$
$\Gamma_3^T 2(\frac{3}{2}F_2)$
$\Gamma_1^T 1(\frac{3}{2}F_2)$
$\Gamma_2^T 1(\frac{3}{2}F_2)$
$\Gamma_1^T 1(\frac{3}{2}F_3)$
$\Gamma_2^T 2(\frac{3}{2}F_3)$
$\Gamma_2^T 1(\frac{3}{2}F_3)$
$\Gamma_3^T 2(\frac{3}{2}F_4)$
$\Gamma_1^T 2(\frac{3}{2}F_4)$
$\Gamma_3^T 1(\frac{3}{2}F_4)$

$\Gamma_3^T 1(\frac{3}{2}F_3)$
$\Gamma_3^T 3(\frac{3}{2}F_4)$
$\Gamma_2^T 1(\frac{3}{2}F_4)$
$\Gamma_1^T 1(\frac{3}{2}F_4)$

$\Gamma_1^T 1(^1A_1, E_TE_T)$
$\Gamma_3^T 1(^1E, E_TE_T)$
$\Gamma_3^T 1(^1E, A_1E_T)$
$\Gamma_1^T 1(^1A_1, A_1A_1)$
$\Gamma_3^T 1(^3A_2, E_TE_T)$
$\Gamma_3^T 2(^3A_2, E_TE_T)$
$\Gamma_3^T 3(^3E, A_1E_T)$
$\Gamma_2^T 2(^3E, A_1E_T)$
$\Gamma_3^T 3(^3E, A_1E_T)$
$\Gamma_1^T 3(^3E, A_1E_T)$
$\Gamma_2^T 2(^1A_2, E_TE_E)$
$\Gamma_3^T 3(^1E, E_TE_E)$
$\Gamma_1^T 1(^1A_1, E_TE_E)$
$\Gamma_3^T 2(^3A_2, E_TE_E)$
$\Gamma_3^T 2(^3A_2, E_TE_E)$
$\Gamma_3^T 1(^1E, A_1E_E)$
$\Gamma_3^T 2(^3E, E_TE_E)$
$\Gamma_2^T 2(^3E, E_TE_E)$
$\Gamma_3^T 3(^3E, E_TE_E)$
$\Gamma_3^T 3(^3E, E_TE_E)$

$\Gamma_2^T 1(^3A_1, E_TE_E)$
$\Gamma_3^T 1(^3A_1, E_TE_E)$
$\Gamma_1^T 1(^1A_1, E_EE_E)$
$\Gamma_3^T 3(^3E, A_1E_E)$
$\Gamma_3^T 3(^3E, A_1E_E)$
$\Gamma_1^T 1(^3E, A_1E_E)$
$\Gamma_2^T 1(^3E, A_1E_E)$

$\Gamma_3^T 1(^1E, E_EE_E)$

$\Gamma_3^T 1(^3A_2, E_EE_E)$
$\Gamma_1^T 1(^3A_2, E_EE_E)$

8 D ELECTRONS, TRIGONAL SYMMETRY

B=905 C=4×B DT=-1000 K=-1 DS=K×DT ZETA=650

440

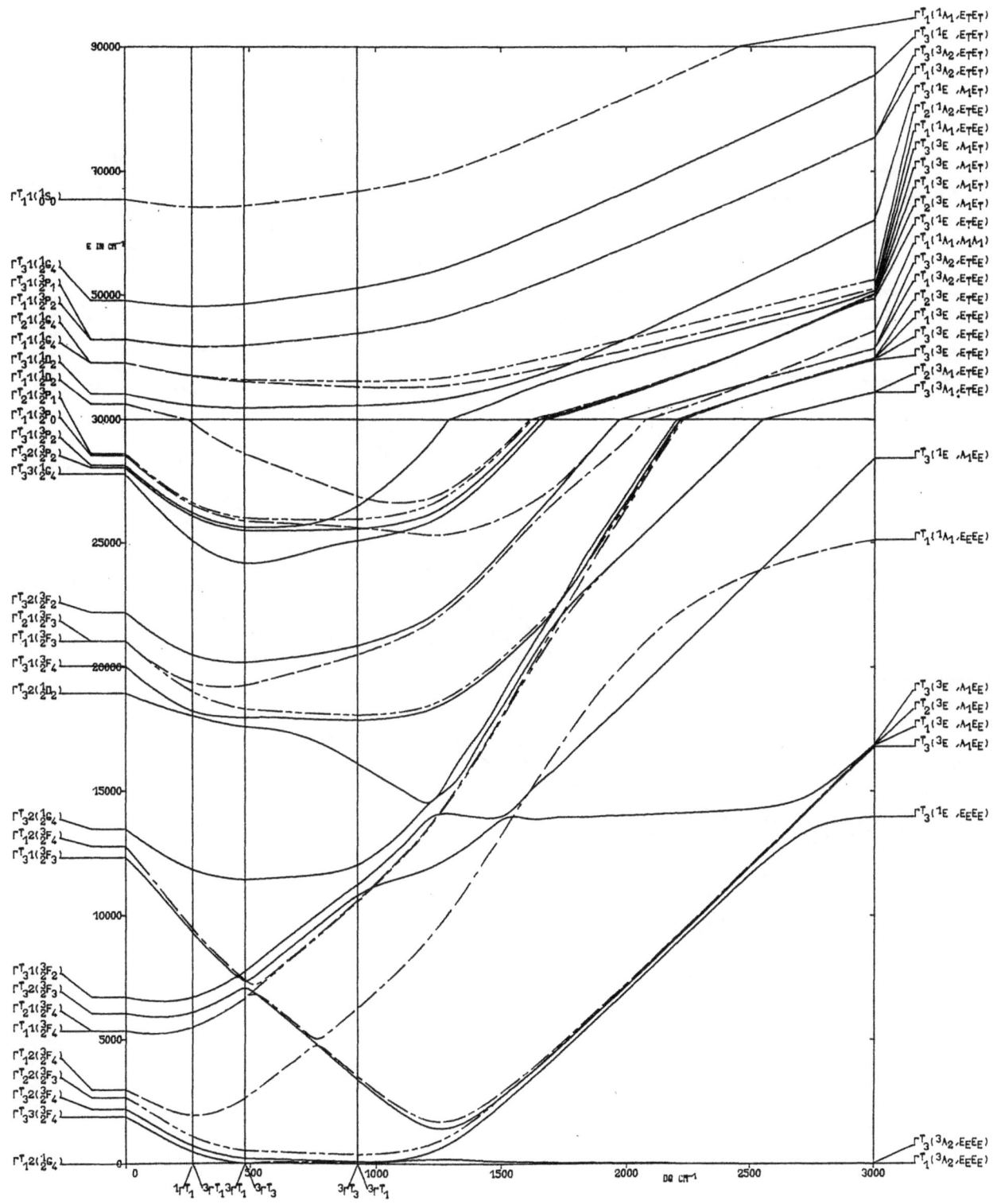

8 D ELECTRONS, TRIGONAL SYMMETRY   B=905 C=4×B DT=-1000 K=5 DS=K×DT ZETA=650

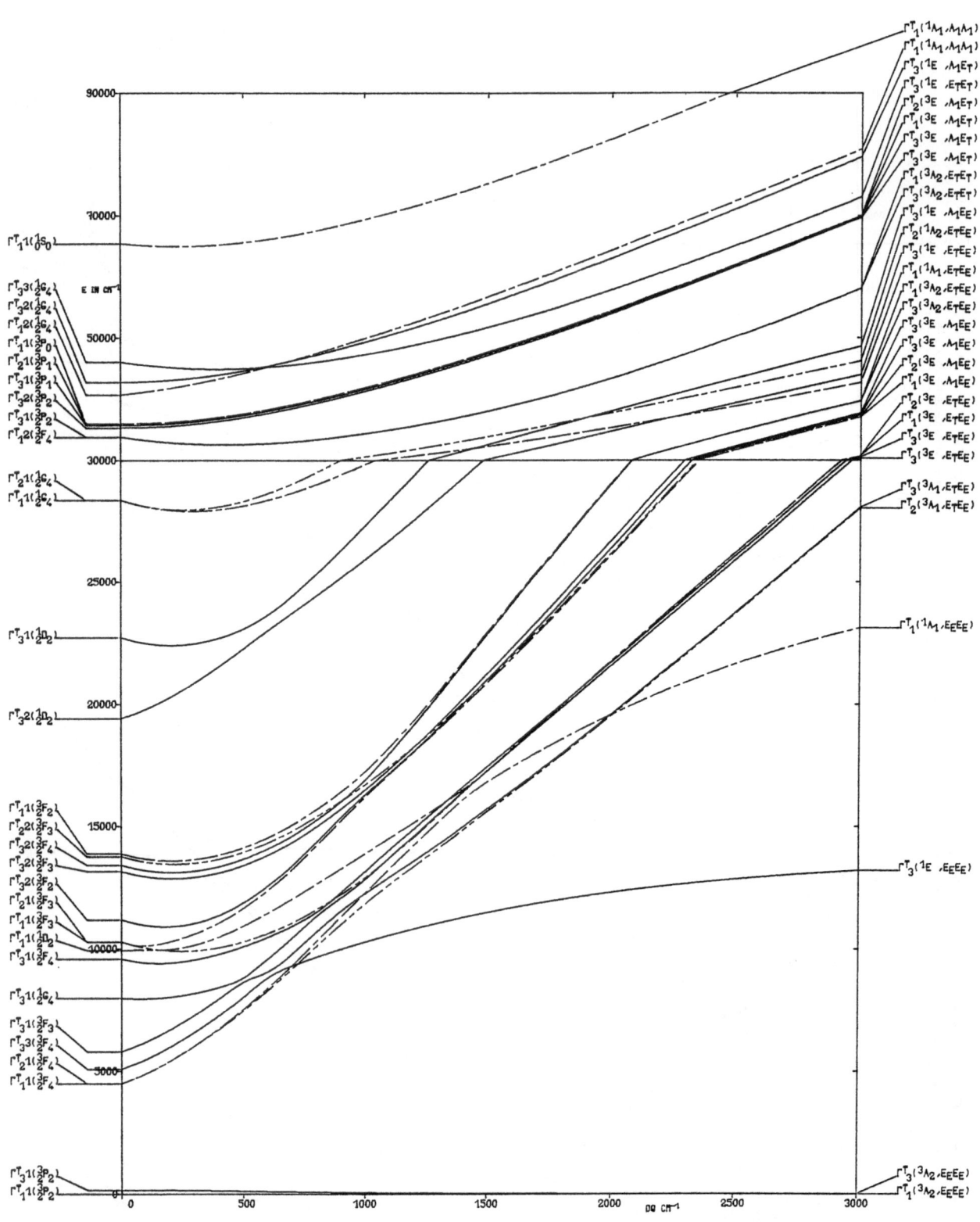

8 D ELECTRONS, TRIGONAL SYMMETRY

B=905 C=4×B DT=-1000 K=-5 DS=K×DT ZETA=650

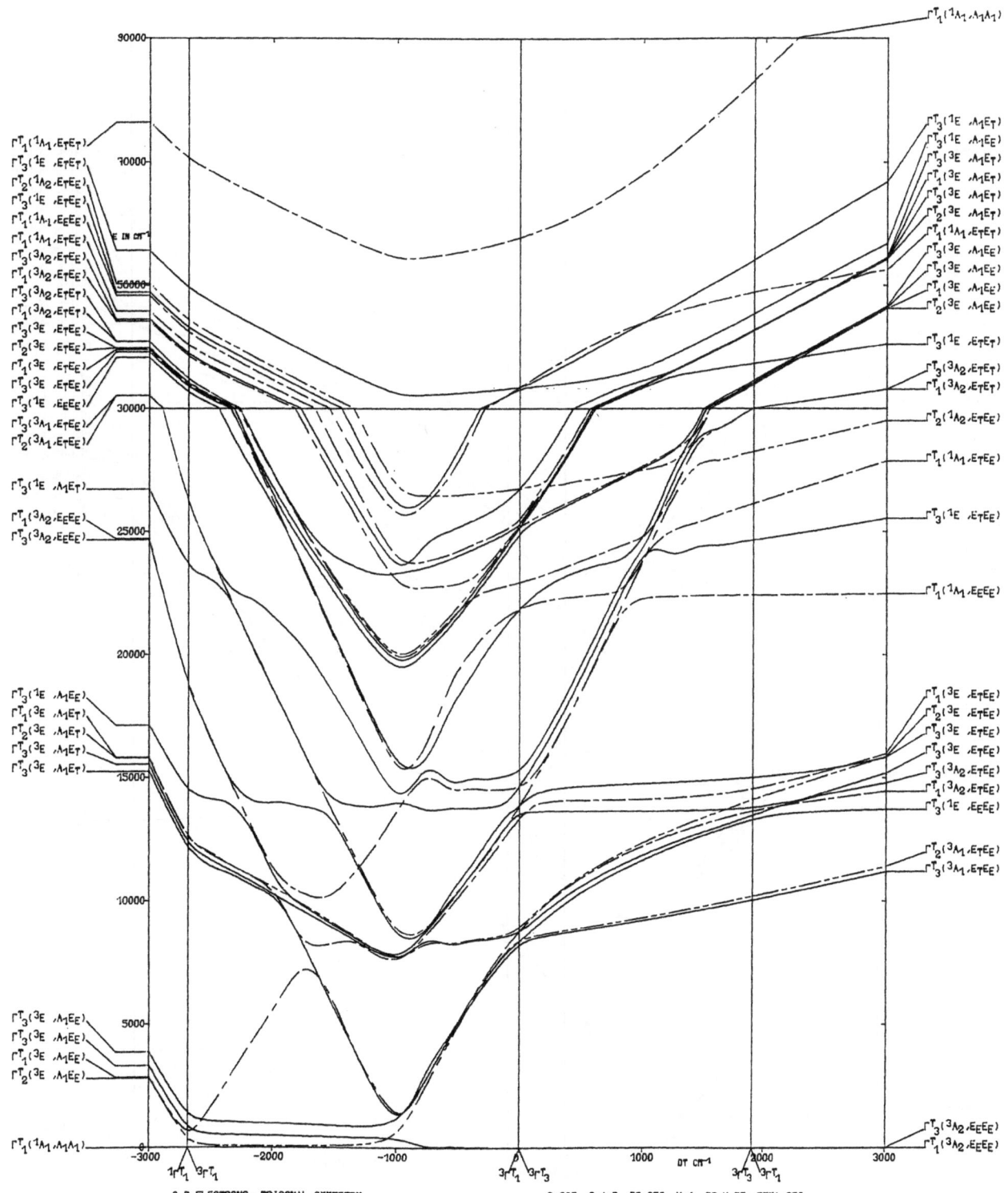

8 D ELECTRONS, TRIGONAL SYMMETRY

B=905   C=4×B   DQ=850   K=1   DS=K×DT   ZETA=650

443

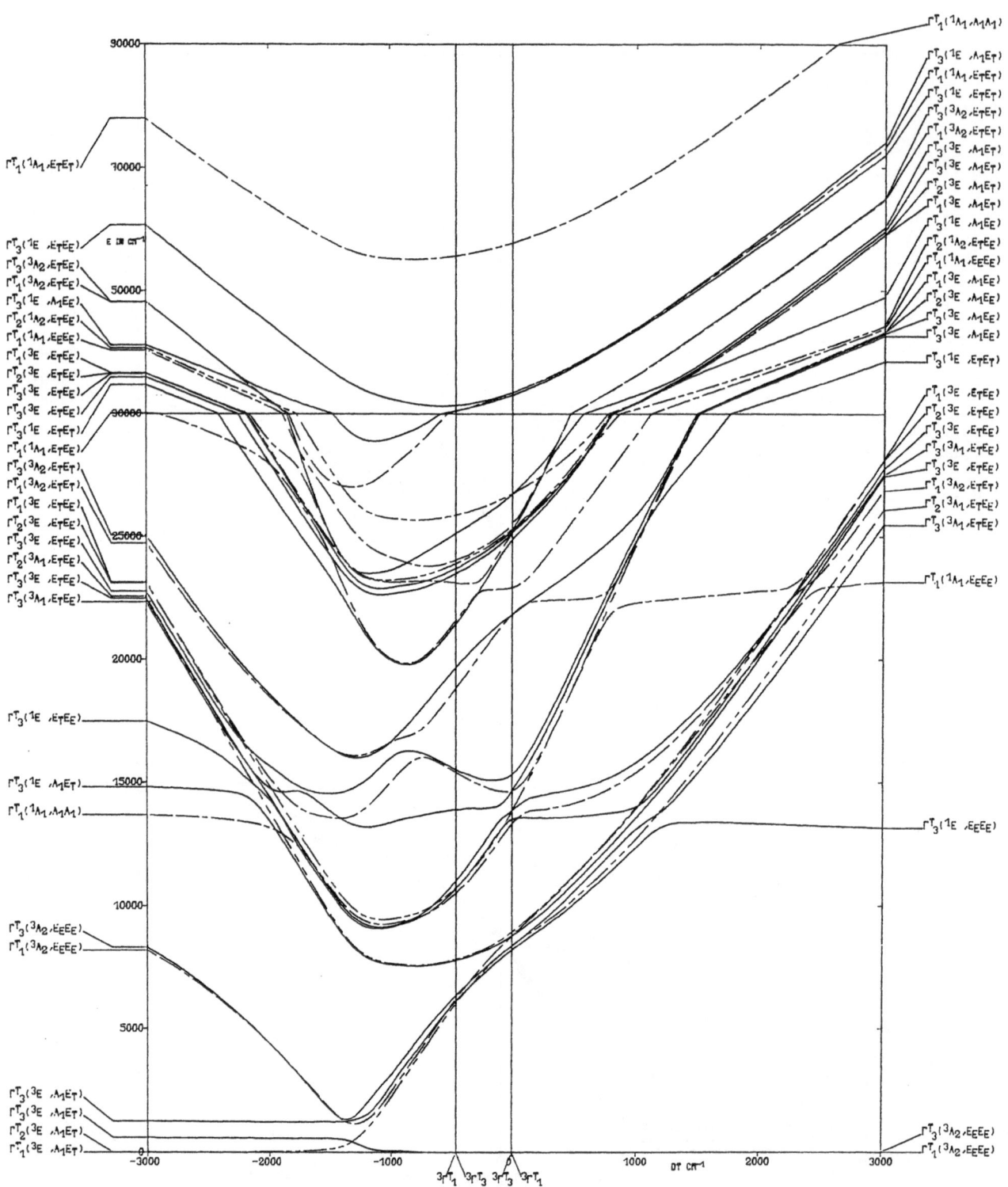

8 D ELECTRONS, TRIGONAL SYMMETRY   B=905 C=4×B DQ=850 K=-1 DS=K×DT ZETA=650

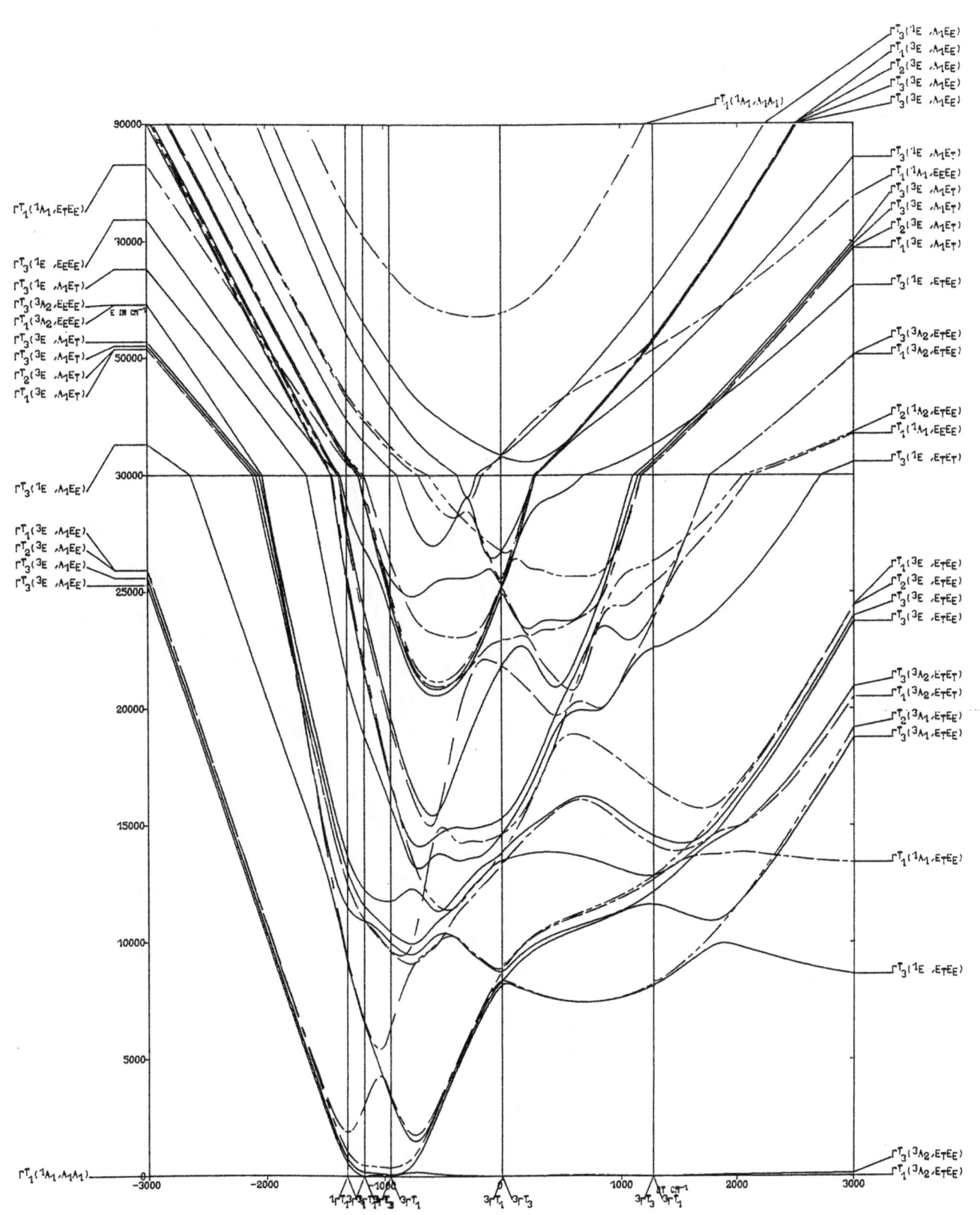

$\Gamma_1^T({}^1A_1,E_TE_E)$

$\Gamma_3^T({}^1E,E_EE_E)$

$\Gamma_3^T({}^1E,A_1E_T)$

$\Gamma_3^T({}^3A_2,E_EE_E)$
$\Gamma_1^T({}^3A_2,E_EE_E)$

$\Gamma_3^T({}^3E,A_1E_T)$
$\Gamma_3^T({}^3E,A_1E_T)$
$\Gamma_2^T({}^3E,A_1E_T)$
$\Gamma_1^T({}^3E,A_1E_T)$

$\Gamma_3^T({}^1E,A_1E_E)$

$\Gamma_1^T({}^3E,A_1E_E)$
$\Gamma_2^T({}^3E,A_1E_E)$
$\Gamma_3^T({}^3E,A_1E_E)$
$\Gamma_3^T({}^3E,A_1E_E)$

E IN CM

$\Gamma_1^T({}^1A_1,A_1A_1)$

90000

70000

50000

30000

25000

20000

15000

10000

5000

0

$\Gamma_3^T({}^1E,A_1E_E)$
$\Gamma_1^T({}^3E,A_1E_E)$
$\Gamma_2^T({}^3E,A_1E_E)$
$\Gamma_3^T({}^3E,A_1E_E)$
$\Gamma_1^T({}^1A_1,E_EE_E)$
$\Gamma_3^T({}^3E,A_1E_E)$
$\Gamma_2^T({}^3E,A_1E_E)$
$\Gamma_1^T({}^3E,A_1E_T)$

$\Gamma_3^T({}^1E,E_TE_E)$

$\Gamma_3^T({}^3A_2,E_TE_E)$
$\Gamma_1^T({}^3A_2,E_TE_E)$

$\Gamma_2^T({}^1A_2,E_TE_E)$
$\Gamma_1^T({}^1A_1,E_EE_E)$

$\Gamma_3^T({}^1E,E_TE_T)$

$\Gamma_1^T({}^3E,E_TE_E)$
$\Gamma_2^T({}^3E,E_TE_E)$
$\Gamma_3^T({}^3E,E_TE_E)$
$\Gamma_3^T({}^3E,E_TE_E)$

$\Gamma_3^T({}^3A_2,E_TE_T)$
$\Gamma_1^T({}^3A_2,E_TE_T)$

$\Gamma_2^T({}^3A_1,E_TE_E)$
$\Gamma_3^T({}^3A_1,E_TE_E)$

$\Gamma_1^T({}^1A_1,E_TE_E)$

$\Gamma_3^T({}^1E,E_TE_E)$

$\Gamma_3^T({}^3A_2,E_TE_E)$
$\Gamma_1^T({}^3A_2,E_TE_E)$

$\Gamma_1^T({}^1A_1,A_1A_1)$

-3000   -2000   -1000   1000   2000   3000

${}^1\Gamma_{T_3}{}^1\Gamma_1{}^1\Gamma_3$ ${}^3\Gamma_{T_1}$ ${}^3\Gamma_1{}^3\Gamma_3$ ${}^3\Gamma_3{}^3\Gamma_3$ ${}^3\Gamma_3{}^3\Gamma_1$

8 D ELECTRONS, TRIGONAL SYMMETRY

B=905 C=4×B DQ=850 K=5 DS=K×DT ZETA=650

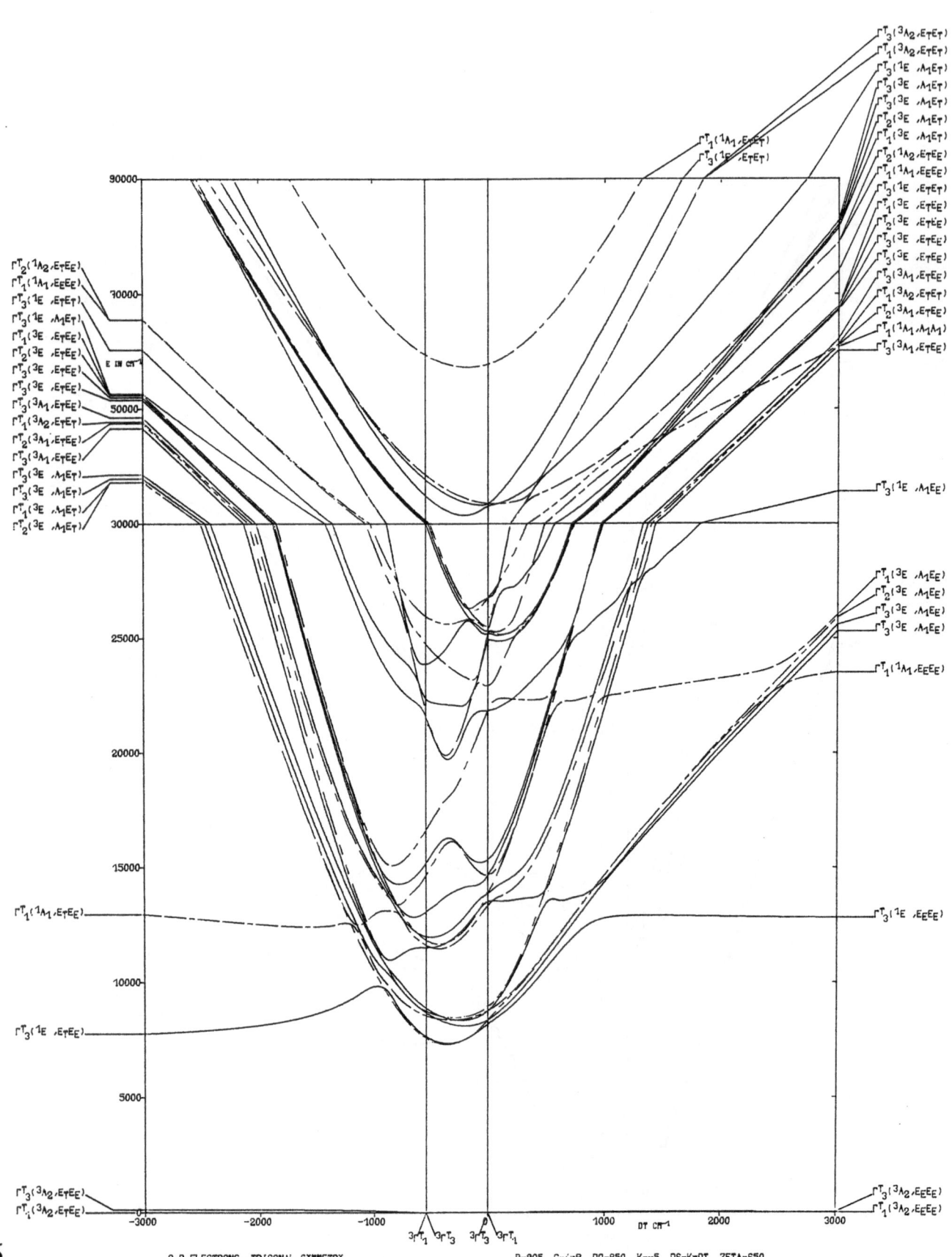

90000

$\Gamma^T_2(^1A_2,E_TE_E)$
$\Gamma^T_1(^1A_1,E_EE_E)$
$\Gamma^T_3(^1E,E_TE_T)$

70000

$\Gamma^T_3(^1E,A_1E_T)$
$\Gamma^T_1(^3E,E_TE_T)$
$\Gamma^T_2(^3E,E_TE_E)$
$\Gamma^T_3(^3E,E_TE_E)$
$\Gamma^T_3(^3E,E_TE_E)$
$\Gamma^T_3(^3A_1,E_TE_T)$
$\Gamma^T_1(^3A_2,E_TE_T)$
$\Gamma^T_2(^3A_1,E_TE_E)$
$\Gamma^T_3(^3A_1,E_TE_E)$
$\Gamma^T_3(^3E,A_1E_T)$
$\Gamma^T_3(^3E,A_1E_T)$
$\Gamma^T_1(^1E,A_1E_T)$
$\Gamma^T_2(^3E,A_1E_T)$

50000

E IN CM^{-1}

30000

25000

20000

15000

$\Gamma^T_1(^1A_1,E_TE_E)$

10000

$\Gamma^T_3(^1E,E_TE_E)$

5000

$\Gamma^T_3(^3A_2,E_TE_E)$
$\Gamma^T_1(^3A_2,E_TE_E)$

$\Gamma^T_3(^3A_2,E_TE_T)$
$\Gamma^T_1(^3A_2,E_TE_T)$
$\Gamma^T_3(^1E,A_1A_1)$
$\Gamma^T_3(^3E,A_1E_T)$
$\Gamma^T_3(^3E,A_1E_T)$
$\Gamma^T_2(^3E,A_1E_T)$
$\Gamma^T_1(^3E,A_1E_T)$
$\Gamma^T_2(^3A_2,E_TE_E)$
$\Gamma^T_1(^1A_1,E_EE_E)$
$\Gamma^T_3(^1E,E_TE_T)$
$\Gamma^T_1(^3E,E_TE_E)$
$\Gamma^T_2(^3E,E_TE_E)$
$\Gamma^T_3(^3E,E_TE_E)$
$\Gamma^T_3(^3E,E_TE_E)$
$\Gamma^T_3(^3A_1,E_TE_E)$
$\Gamma^T_2(^3A_1,E_TE_E)$
$\Gamma^T_1(^1A_1,A_1A_1)$
$\Gamma^T_3(^3A_1,E_TE_E)$

$\Gamma^T_3(^1E,A_1E_E)$

$\Gamma^T_1(^3E,A_1E_E)$
$\Gamma^T_2(^3E,A_1E_E)$
$\Gamma^T_3(^3E,A_1E_E)$
$\Gamma^T_3(^3E,A_1E_E)$

$\Gamma^T_1(^1A_1,E_EE_E)$

$\Gamma^T_3(^1E,E_EE_E)$

$\Gamma^T_3(^3A_2,E_EE_E)$
$\Gamma^T_1(^3A_2,E_EE_E)$

-3000   -2000   -1000   1000   DT CM^{-1} 2000   3000

$3\Gamma T_1$ $3\Gamma T_3$ $3\Gamma T_3$ $3\Gamma T_1$

8 D ELECTRONS, TRIGONAL SYMMETRY   B=905 C=4×B DQ=850 K=-5 DS=K×DT ZETA=650

8 D ELECTRONS, CYLINDRICAL SYMMETRY

B=905   C=4⋇B   ZETA=650

ENERGY AS FUNCTION OF DT
K=1, -1, 5, -5

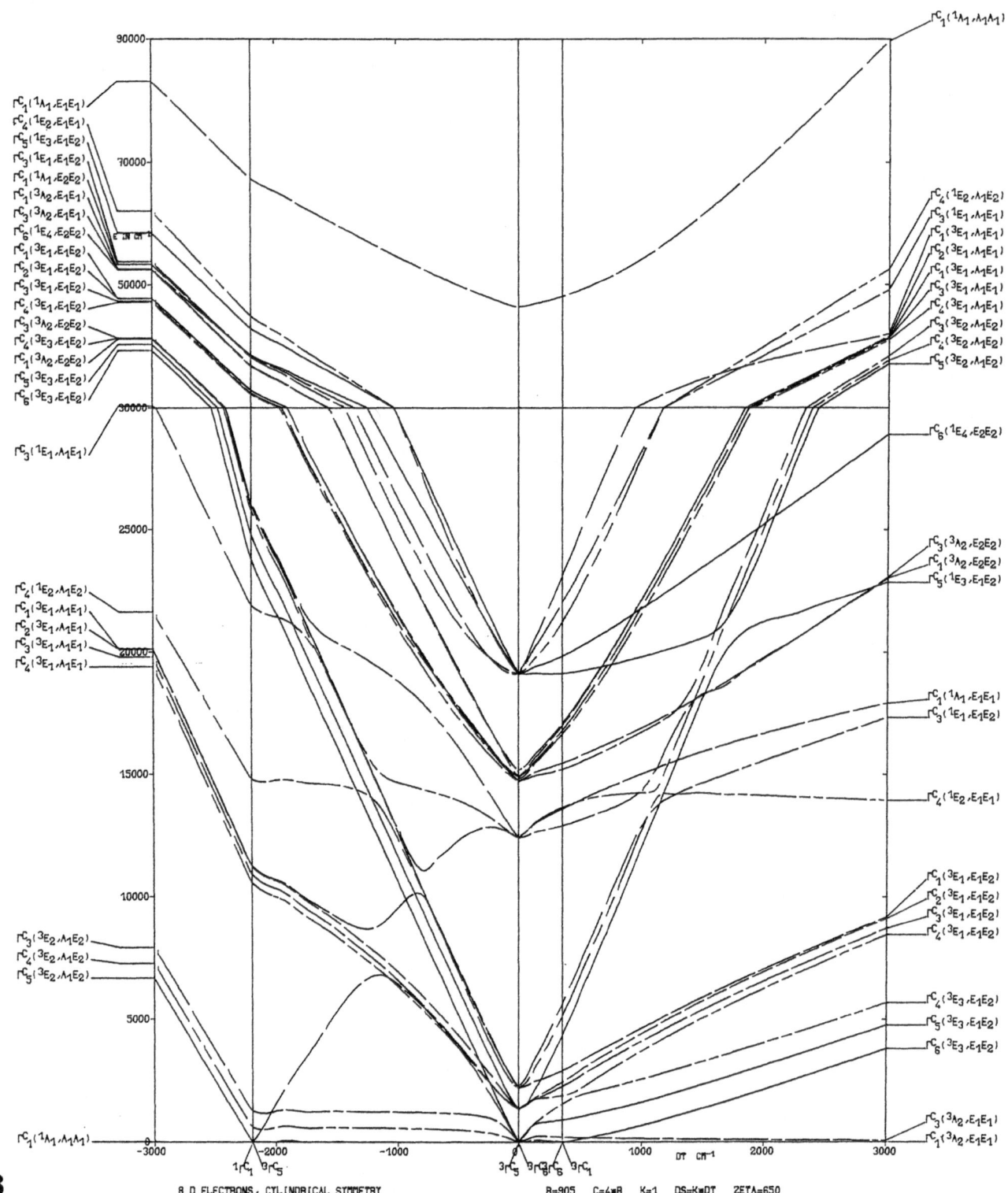

8 D ELECTRONS, CYLINDRICAL SYMMETRY   B=905   C=4×B   K=1   DS=K×DT   2ETA=650

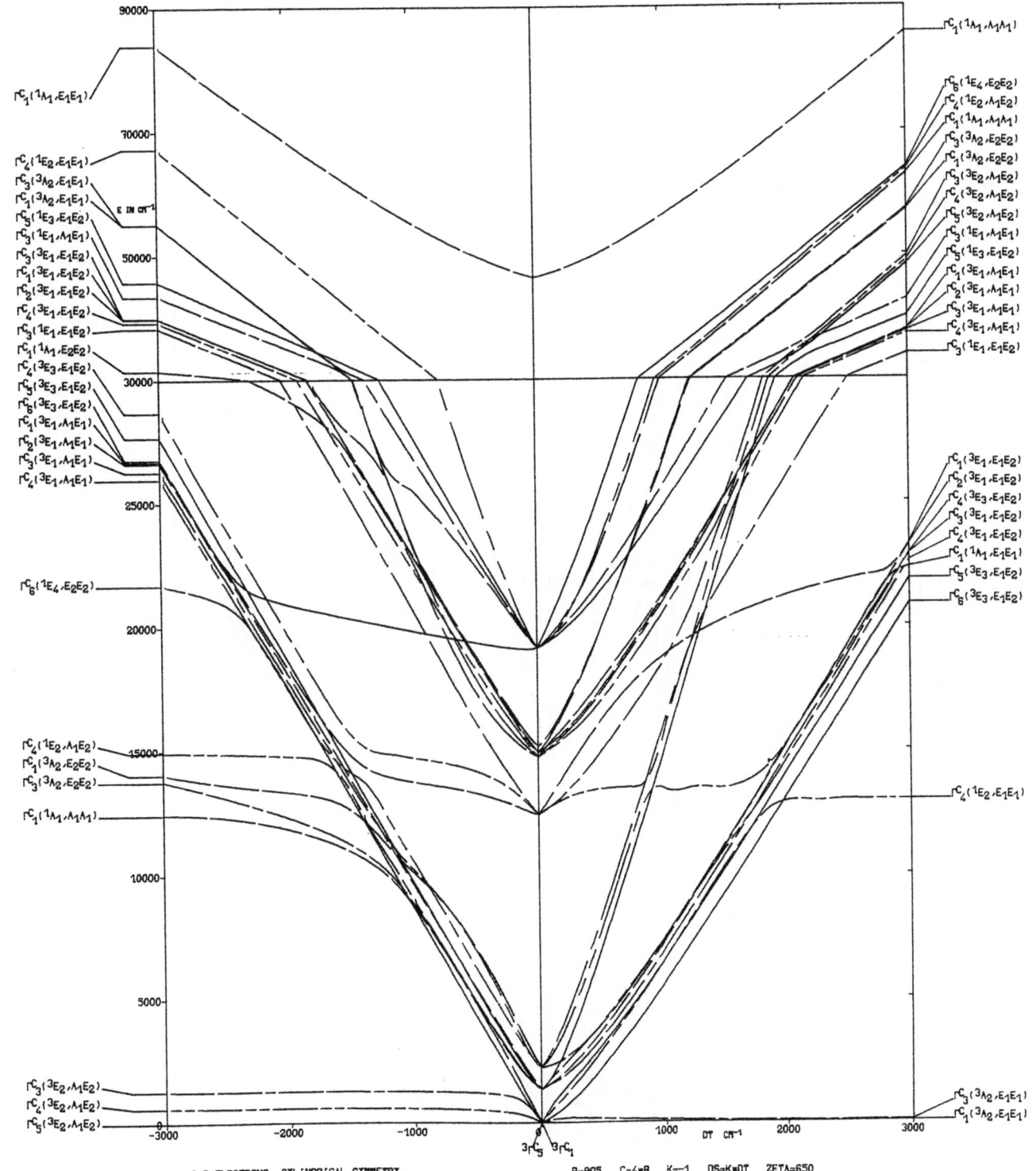

8 D ELECTRONS, CYLINDRICAL SYMMETRY   B=905   C=4×B   K=-1   DS=K×DT   ZETA=650

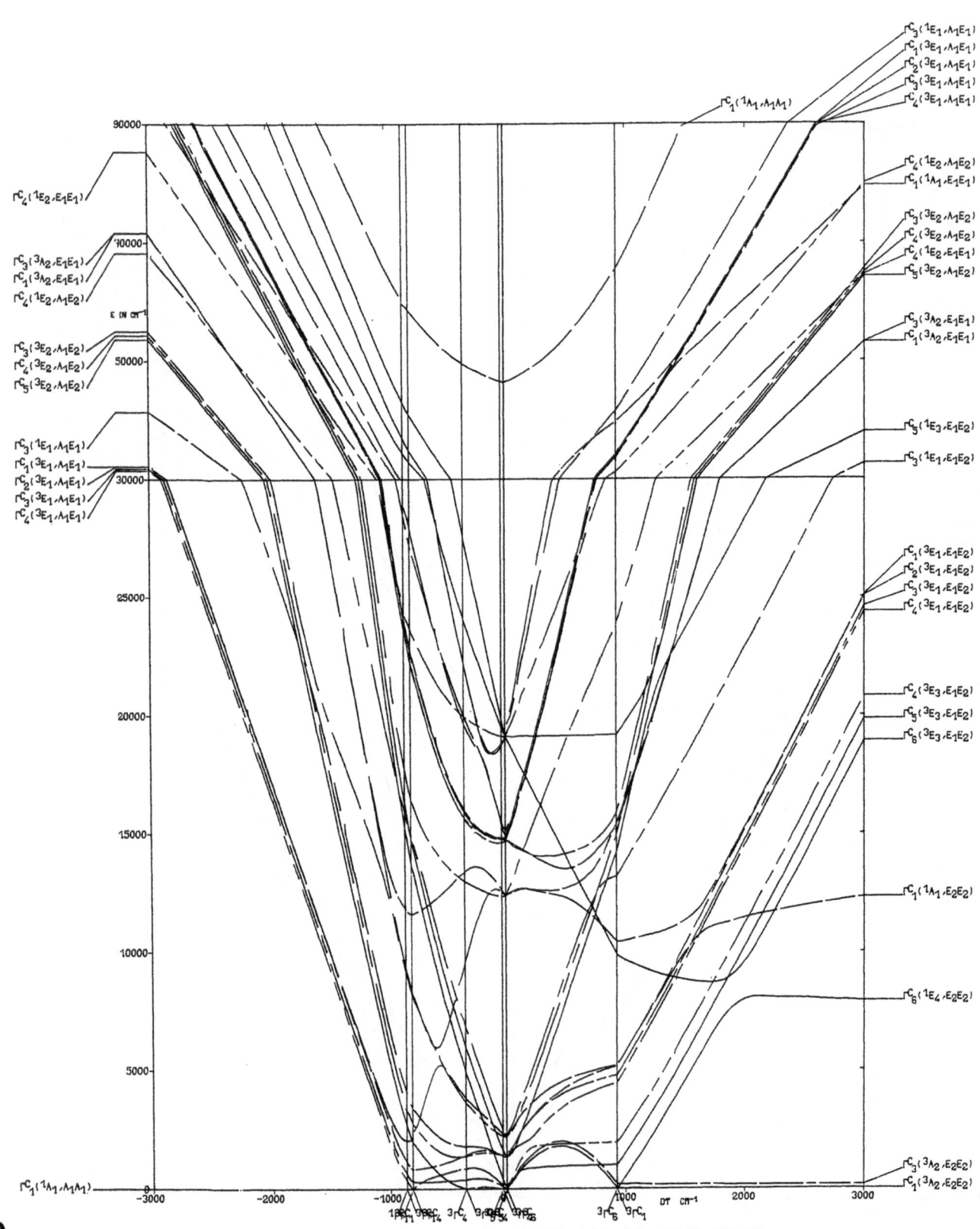

$\Gamma C_4(^1E_2, E_1E_1)$

$\Gamma C_3(^3A_2, E_1E_1)$
$\Gamma C_1(^3A_2, E_1E_1)$
$\Gamma C_1(^1E_2, A_1E_2)$

$\Gamma C_3(^3E_2, A_1E_2)$
$\Gamma C_4(^3E_2, A_1E_2)$
$\Gamma C_5(^3E_2, A_1E_2)$

$\Gamma C_3(^1E_1, A_1E_1)$
$\Gamma C_1(^3E_1, A_1E_1)$
$\Gamma C_2(^3E_1, A_1E_1)$
$\Gamma C_3(^3E_1, A_1E_1)$
$\Gamma C_4(^3E_1, A_1E_1)$

$\Gamma C_1(^1A_1, A_1A_1)$

$\Gamma C_1(^1A_1, A_1A_1)$

$\Gamma C_3(^1E_1, A_1E_1)$
$\Gamma C_1(^3E_1, A_1E_1)$
$\Gamma C_2(^3E_1, A_1E_1)$
$\Gamma C_3(^3E_1, A_1E_1)$
$\Gamma C_4(^3E_1, A_1E_1)$

$\Gamma C_4(^1E_2, A_1E_2)$
$\Gamma C_1(^1A_1, E_1E_1)$

$\Gamma C_3(^3E_2, A_1E_2)$
$\Gamma C_4(^3E_2, A_1E_2)$
$\Gamma C_4(^1E_2, E_1E_1)$
$\Gamma C_5(^3E_2, A_1E_2)$

$\Gamma C_3(^3A_2, E_1E_1)$
$\Gamma C_1(^3A_2, E_1E_1)$

$\Gamma C_5(^1E_3, E_1E_2)$

$\Gamma C_3(^1E_1, E_1E_2)$

$\Gamma C_1(^3E_1, E_1E_2)$
$\Gamma C_2(^3E_1, E_1E_2)$
$\Gamma C_3(^3E_1, E_1E_2)$
$\Gamma C_4(^3E_1, E_1E_2)$

$\Gamma C_4(^3E_3, E_1E_2)$
$\Gamma C_5(^3E_3, E_1E_2)$
$\Gamma C_6(^3E_3, E_1E_2)$

$\Gamma C_1(^1A_1, E_2E_2)$

$\Gamma C_6(^1E_4, E_2E_2)$

$\Gamma C_3(^3A_2, E_2E_2)$
$\Gamma C_1(^3A_2, E_2E_2)$

E IN CM^{-1}

DT CM^{-1}

8 D ELECTRONS, CYLINDRICAL SYMMETRY

B=905   C=4×B   K=5   DS=K×DT   ZETA=650

450

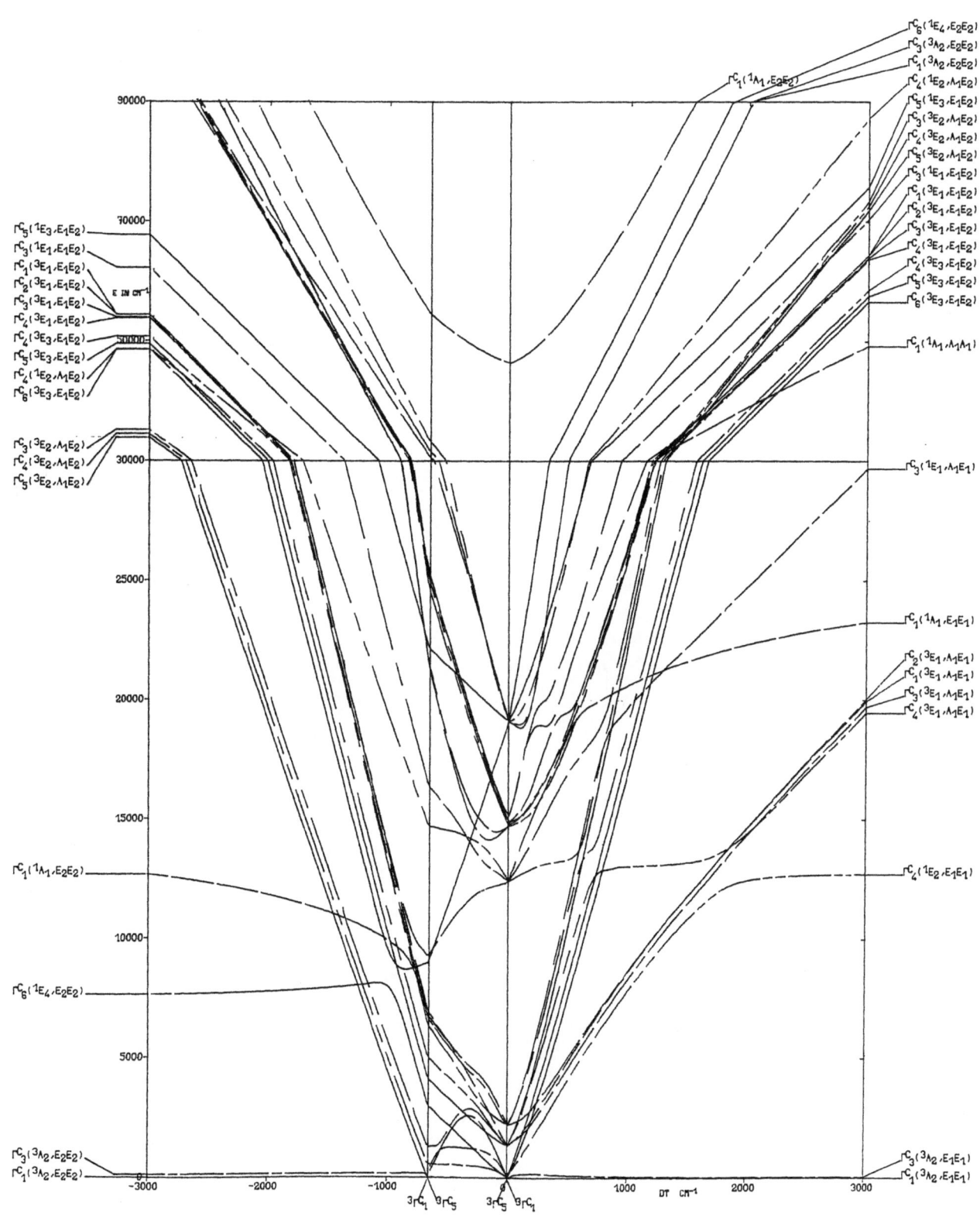

90000

70000

E IN CM⁻¹

$\Gamma^C_5(\ ^1E_3\ ,E_1E_2)$
$\Gamma^C_3(\ ^1E_1\ ,E_1E_2)$
$\Gamma^C_1(\ ^3E_1\ ,E_1E_2)$
$\Gamma^C_2(\ ^3E_1\ ,E_1E_2)$
$\Gamma^C_3(\ ^3E_1\ ,E_1E_2)$
$\Gamma^C_4(\ ^3E_1\ ,E_1E_2)$
50000
$\Gamma^C_4(\ ^3E_3\ ,E_1E_2)$
$\Gamma^C_5(\ ^3E_3\ ,E_1E_2)$
$\Gamma^C_4(\ ^1E_2\ ,A_1E_2)$
$\Gamma^C_6(\ ^3E_3\ ,E_1E_2)$

$\Gamma^C_3(\ ^3E_2\ ,A_1E_2)$
$\Gamma^C_4(\ ^3E_2\ ,A_1E_2)$ 30000
$\Gamma^C_5(\ ^3E_2\ ,A_1E_2)$

25000

20000

15000

$\Gamma^C_1(\ ^1A_1\ ,E_2E_2)$

10000

$\Gamma^C_6(\ ^1E_4\ ,E_2E_2)$

5000

$\Gamma^C_3(\ ^3A_2\ ,E_2E_2)$
$\Gamma^C_1(\ ^3A_2\ ,E_2E_2)$ 0

$\Gamma^C_1(\ ^1A_1\ ,E_2E_2)$

$\Gamma^C_6(\ ^1E_4\ ,E_2E_2)$
$\Gamma^C_3(\ ^3A_2\ ,E_2E_2)$
$\Gamma^C_1(\ ^3A_2\ ,E_2E_2)$
$\Gamma^C_4(\ ^1E_2\ ,A_1E_2)$
$\Gamma^C_5(\ ^1E_3\ ,A_1E_2)$
$\Gamma^C_3(\ ^3E_2\ ,A_1E_2)$
$\Gamma^C_4(\ ^3E_2\ ,A_1E_2)$
$\Gamma^C_5(\ ^3E_2\ ,A_1E_2)$
$\Gamma^C_3(\ ^1E_1\ ,E_1E_2)$
$\Gamma^C_1(\ ^3E_1\ ,E_1E_2)$
$\Gamma^C_2(\ ^3E_1\ ,E_1E_2)$
$\Gamma^C_3(\ ^3E_1\ ,E_1E_2)$
$\Gamma^C_4(\ ^3E_1\ ,E_1E_2)$
$\Gamma^C_4(\ ^3E_3\ ,E_1E_2)$
$\Gamma^C_5(\ ^3E_3\ ,E_1E_2)$
$\Gamma^C_6(\ ^3E_3\ ,E_1E_2)$

$\Gamma^C_1(\ ^1A_1\ ,A_1A_1)$

$\Gamma^C_3(\ ^1E_1\ ,A_1E_1)$

$\Gamma^C_1(\ ^1A_1\ ,E_1E_1)$

$\Gamma^C_2(\ ^3E_1\ ,A_1E_1)$
$\Gamma^C_1(\ ^3E_1\ ,A_1E_1)$
$\Gamma^C_3(\ ^3E_1\ ,A_1E_1)$
$\Gamma^C_4(\ ^3E_1\ ,A_1E_1)$

$\Gamma^C_4(\ ^1E_2\ ,E_1E_1)$

$\Gamma^C_3(\ ^3A_2\ ,E_1E_1)$
$\Gamma^C_1(\ ^3A_2\ ,E_1E_1)$

-3000   -2000   -1000   0   1000   2000   3000   DT CM⁻¹

3ΓC_1 3ΓC_5 3ΓC_5 3ΓC_1

8 D ELECTRONS, CYLINDRICAL SYMMETRY   B=905   C=4×B   K=-5   DS=K×DT   ZETA=650

451

INDEX